Birkhäuser

Global Tsunami Science: Past and Future. Volume III

Edited by
Alexander B. Rabinovich
Hermann M. Fritz
Yuichiro Tanioka
Eric L. Geist

Previously published in *Pure and Applied Geophysics* (PAGEOPH),
Volume 175, No. 4, 2018

Editors
Alexander B. Rabinovich
P.P. Shirshov Institute of Oceanology
Russian Academy of Sciences
Moscow, Russia

Hermann M. Fritz
School of Civil and Environmental Engineering
Georgia Institute of Technology
Atlanta, GA, USA

Yuichiro Tanioka
Institute of Seismology and Volcanology
Hokkaido University
Sapporo, Japan

Eric L. Geist
United States Geological Survey
Menlo Park, CA, USA

ISSN 2504-3625
ISBN 978-3-030-03759-8

Library of Congress Control Number: 2018967422

Cover illustration: Trace of the tsunami on land and a ship carried by the tsunami, after the 2004 Sumatra tsunami (photo: January, 2005, by Yuichiro Tanioka)

This book is published under the imprint Birkhäuser, www.birkhauser-science.com by the registered company Springer Nature Switzerland AG
The registered company address is: Gewerbestrasse 11, 6330 Cham, Switzerland

Contents

Pure Appl. Geophys.

https://doi.org/10.1007/s00024-018-1851-8

Pure and Applied Geophysics

Introduction to "Global Tsunami Science: Past and Future, Volume III"

Alexander B. Rabinovich,[1,2] Hermann M. Fritz,[3] Yuichiro Tanioka,[4] and Eric L. Geist[5]

Abstract—Twenty papers on the study of tsunamis are included in Volume III of the PAGEOPH topical issue "Global Tsunami Science: Past and Future". Volume I of this topical issue was published as PAGEOPH, vol. 173, No. 12, 2016 and Volume II as PAGEOPH, vol. 174, No. 8, 2017. Two papers in Volume III focus on specific details of the 2009 Samoa and the 1923 northern Kamchatka tsunamis; they are followed by three papers related to tsunami hazard assessment for three different regions of the world oceans: South Africa, Pacific coast of Mexico and the northwestern part of the Indian Ocean. The next six papers are on various aspects of tsunami hydrodynamics and numerical modelling, including tsunami edge waves, resonant behaviour of compressible water layer during tsunamigenic earthquakes, dispersive properties of seismic and volcanically generated tsunami waves, tsunami runup on a vertical wall and influence of earthquake rupture velocity on maximum tsunami runup. Four papers discuss problems of tsunami warning and real-time forecasting for Central America, the Mediterranean coast of France, the coast of Peru, and some general problems regarding the optimum use of the DART buoy network for effective real-time tsunami warning in the Pacific Ocean. Two papers describe historical and paleotsunami studies in the Russian Far East. The final set of three papers importantly investigates tsunamis generated by non-seismic sources: asteroid airburst and meteorological disturbances. Collectively, this volume highlights contemporary trends in global tsunami research, both fundamental and applied toward hazard assessment and mitigation.

Key words: Tsunami observations and detection, tsunami hydrodynamics and modelling, tsunami warning and hazard mitigation, asteroid tsunami, meteotsunami, tsunami probability.

[1] Department of Fisheries and Oceans, Institute of Ocean Sciences, 9860 West Saanich Road, Sidney, BC V8L 4B2, Canada. E-mail: a.b.rabinovich@gmail.com

[2] P.P. Shirshov Institute of Oceanology, Russian Academy of Sciences, 36 Nakhimovsky Pr., Moscow 117997, Russia.

[3] School of Civil and Environmental Engineering, Georgia Institute of Technology, Atlanta, GA 30332, USA. E-mail: fritz@gatech.edu

[4] Institute of Seismology and Volcanology, Hokkaido University, Sapporo, Japan. E-mail: tanioka@mail.sci.hokudai.ac.jp

[5] U.S. Geological Survey, 345 Middlefield Rd., MS 999, Menlo Park, CA 94025, USA. E-mail: egeist@usgs.gov

1. Introduction

Tsunami science has evolved significantly since two of the most destructive natural disasters that have occurred in this century: the 26 December 2004 Sumatra tsunami that killed about 230,000 people along the coasts of 14 countries in the Indian Ocean and the 11 March 2011 Tohoku (Great East Japan) tsunami that killed almost 20,000 people and destroyed the Fukushima Daiichi nuclear power plant (Satake et al. 2013a). There have also been many other devastating tsunamis over the past decade that have guided tsunami science. The 2004 tsunami killed citizens from almost 60 countries around the world and attracted extremely high international public and scientific interest to this catastrophic phenomenon. Therefore, scientists, not only from countries directly affected by this event but also from very many other countries, have come together to advance tsunami research. The global community of researchers has also expanded by discipline, adapting new developments in other sciences to study all aspects of tsunami hydrodynamics, detection, generation mechanism, hazard assessment and mitigation, and tsunami early warning.

This third and final volume of "Global Tsunami Science: Past and Future" completes a broad and encompassing overview of this critical moment in the history of tsunami science. Each of the volumes has overarching science themes that tie many of the papers together. Volume I focused on recent advances in tsunami warning and assessment, whereas a common theme in Volume II was the foundational importance of instrumental observations. In Volume III, the importance of numerical modelling on tsunami science is highlighted. Numerical modelling of tsunamis, which has its origin as early as the 1960's in a seminal paper by Aida (1969), has greatly advanced in recent decades, owing to the rapid progress of high-performance compute platforms and the development of new algorithms. Numerical modelling provides the crucial link in validating hypotheses of tsunami generation, propagation and runup, explaining unusual observations, and providing specific applications for tsunami hazard assessment and warning. The papers presented in Volume III of Global Tsunami Science are a definitive demonstration of the importance of numerical model in advancing tsunami science.

The Joint Tsunami Commission, part of the International Union of Geodesy and Geophysics (IUGG), conceived "Global Tsunami Science: Past and Future" as well as previous tsunami special issues in *Pure and Applied Geophysics*. The Joint Tsunami Commission was established following the 1960 Chile tsunami, which was generated by the largest (M_w 9.5) instrumentally recorded earthquake, propagated throughout the entire Pacific Ocean, and affected many countries in the entire ocean basin. It became obvious that tsunami investigation and effective tsunami warning is impossible without intensive international cooperation. Since 1960, the Joint Tsunami Commission has held biannual International Tsunami Symposia (ITS) and published special volumes of selected papers. Several such volumes have been published in PAGEOPH during 10 years following the 2004 Sumatra tsunami, including Satake et al. (2007, 2011a, b, 2013a, b); Cummins et al. (2008, 2009), and Rabinovich et al. (2015a, b). Two recent catastrophic tsunamis, the 2010 Chile and 2011 Tohoku, as well as other strong events which occurred around this time, attracted so much attention and brought so much new information and data that an extra, inter-session volume was collected and published (Rabinovich et al. 2014). Moreover, high interest regarding the Chilean (Illapel) earthquake and tsunami of 16 September 2015 resulted in a Topical Collection of regular PAGEOPH papers "Chile-2015" that were later published as a book (Braitenberg and Rabinovich 2017). From this point of view, the "Global Tsunami Science: Past and Future" volumes can be considered as the frontiers of tsunami science and research, as well as a record of continuous progress in tsunami warning and hazard mitigation.

Development of the topical issue "Global Tsunami Science: Past and Future" was motivated by research presented at the 26th IUGG General Assembly and sponsored by the Joint Tsunami Commission held from 22 June to 2 July 2015 in Prague, Czech Republic. At the business meeting of the Joint Tsunami Commission, it was decided to publish selected papers presented at this symposium, as well as other papers on related topics, as "Global Tsunami Science: Past and Future". Volume I (Geist et al. 2016), which was published in December 2016, comprises 25 papers. Volume II (Rabinovich et al. 2017) comprises 22 papers, which became ready for publication by August 2017. This third and final Volume III of "Global Tsunami Science" comprises 20 papers, which were prepared by March 2018. In fact, this volume includes not only papers from the 26th IUGG 2015 Assembly, but also from several other tsunami meetings of 2015–2017, in particular those presented at recent American Geophysical Union and European Geosciences Union meetings. Altogether, 67 papers published in the three volumes of "Global Tsunami Science: Past and Future" may be considered as a very significant contribution to tsunami research.

Papers in Volume III are separated into similar categories as in Volume I and II, although a new category is introduced (Paleo- and Historical Tsunami Studies) and the non-seismic tsunami studies have been expanded to include asteroid airburst tsunamis and meteotsunamis.

2. *Case Studies and Observations*

Case studies are an important part of tsunami research highlighting the hazard in specific regions and often resolving the nature of unusual source mechanisms and wave propagation and runup behaviour. In the paper by Dilmen et al. (2018), numerical modelling of the 29 September 2009 Samoa tsunami is provided for the volcanic island of Tutuila in American Samoa. The study focuses on two main issues: (1) effect of roughness variations on tsunami runup and (2) influence of shelf bathymetry variations and coral reef shields on tsunami runup and inundation.

Salaree and Okal (2018) examined the earthquake and tsunami of 13 April 1923 at Ust' Kamchatsk, northern Kamchatka, which generated a more powerful and damaging tsunami than the larger event of 03 February 1923, and qualified this event as a "tsunami earthquake". The results of numerical modelling confirmed the slow nature of the source but failed to reproduce the sharply peaked distribution of tsunami wave amplitudes along the coast.

3. *Tsunami Hazard Assessment and Uncertainty Analysis*

Numerical modelling is crucial in field of tsunami hazard assessment and uncertainty analysis. Kijko et al. (2018) develop a new empirical method to estimate tsunami probability, a topic that has been discussed extensively in the previous two volumes of "Global Tsunami Science: Past and Future". The authors combine maximum likelihood estimation of frequency–magnitude parameters with numerical models to quantify both aleatory and epistemic uncertainty. Results are demonstrated for the tsunamigenic regions of Japan, Kuril–Kamchatka and South America.

Ortiz-Huerta et al. (2018) used historical records of the 1960 Chile, 1964 Alaska and 2011 Tohoku tsunamis along the Pacific coast of Mexico to examine the validity of far-field modelling for transoceanic tsunamis generated by $M_w \sim 9.3$ thrust-fault earthquakes around the Pacific Ocean and got quite reasonable results. The computed tsunami amplitudes varied from ~ 1 to ~ 2.5 m, depending on the tsunami origin region and on the fault orientation.

Slip heterogeneity during earthquake rupture, as studied primarily through stochastic methods, has been demonstrated to have a significant effect on local tsunami wave heights. Rashidi et al. (2018) propose an alternative and novel method to incorporate slip heterogeneity into tsunami hazard calculations using observed slip distributions from recent tsunamigenic earthquakes scaled to source regions for a hazard assessment study site. The authors apply this method to the western segment of the Makran subduction zone and find that the entire Makran coast of Pakistan and Iran is most vulnerable to tsunami hazards.

4. *Tsunami Hydrodynamics and Modelling*

Although an understanding of the hydrodynamics of tsunamis originally focused on analytical and laboratory/experimental studies, recent research has been more focused on the use of numerical models. In the first paper of this section, Geist (2018) examines the effect of nonlinear resonance among tsunami edge-wave modes, incorporating more generalized initial conditions of amplitude and phase than in previous studies. This is made possible through the use of numerical differential equation solvers that are flexible enough to account for abrupt jumps in phase. Another resonance issue associated with tsunamis is related to pressure waves in the water column near the source region. This is addressed by Cecioni and Bellotti (2018) who identify under what conditions resonant behaviour occurs in relation to earthquake source parameters. Accurate modelling of such waves may have important implications for early detection and warning.

Tanioka et al. (2018) estimated the fault length, width and other parameters of the 2016 El-Salvador–Nicaragua outer-rise earthquake (M_w 6.9) from the tsunami record obtained at the ocean bottom pressure sensor DART 32411 off Central America. The dispersive character of the observed tsunami waveform enabled the authors to constrain the fault size and the seismic moment that were in good agreement with the estimates from the Global CMT catalogue. Sandanbata et al. (2018) determined the tsunami velocity field of the 2015 volcanic tsunami earthquake near Torishima in Japan using a newly developed ray tracing method for dispersive tsunamis. The validity of their results was confirmed by the comparison with observed tsunami waveforms recorded at bottom pressure gauges. They also estimated the tsunami amplitude at the source by Green's law and found that this law underestimates actual tsunami amplitudes.

Didenkulova and Pelinovsky (2018) analytically examined the nonlinear problem of tsunami wave runup on a vertical wall in a tidal environment and got the exact solution for runup height as a function of the incident wave height. The solution was used to investigate the influence of tides on runup characteristics.

Fuentes et al. (2018) presented an analytical 1 + 1D tsunami model including the earthquake rupture velocity in the slip direction and the rise time. The static case corresponds to instantaneous rise time and infinite rupture velocity. Both parameters contribute in shifting the arrival time. Runup is strictly decreasing with the rise time based on parametric analysis. However, runup is highly amplified in a certain range of slow rupture velocities. The tsunami excitation vanishes for even lower rupture velocities.

5. *Tsunami Warning and Forecasting*

One of the major advancements in recent years toward improving the accuracy of tsunami warning is forecasting the tsunami arrival time, wave height, and inundation area soon after the tsunami has been generated. This has been made possible with the development of massive pre-computed tsunami databases and rapid, real-time analysis of observed tsunami waves. A relatively new application of tsunami numerical modelling is in tsunami preparedness and awareness exercises. Chacón-Barrantes et al. (2018) describe the role of numerical modelling performed collaboratively by different research groups as part of the annual CaribeWave preparedness exercise. The focus in this paper is specifically on CaribeWave15, which examines the tsunami threat to Colombia, Costa Rica, Panamá and Puerto Rico from sources within the North Panamá Deformed Belt. Basin-wide tsunami energy plots assess the performance of the decision support tools distributed by PTWC (Pacific Tsunami Warning Center), the tsunami service provider for the Caribbean basin.

Similar questions of effective tsunami warning along the coast of the French Riviera and Corsica are examined by Gailler et al. (2018) based on results of numerical modelling of tsunami waves generated by major earthquakes ($M_w > 8.0$) at the Hellenic plate boundary. The results show that tsunami amplitudes exceeding 50 cm can be expected at particular harbours and beaches of the examined coasts.

Percival et al. (2018) suggested a quantitative parameter ("measure") to evaluate how accurately a numerical model, adjusted based on DART buoy network measurements, forecasts tsunami inundations and how much the forecasts are degraded in accuracy when one or more buoys are inoperative. The analysis uses simulated tsunami time series collected at each buoy from selected source segments in the Short-term Inundation Forecast for Tsunamis (SIFT) database and involves a set of 1000 forecasts for each buoy/segment pair at sites offshore of impacted communities. Examples are shown for buoys off the Aleutian Islands and off the west coast of South America for impact sites at Hilo Hawai'i and along the U.S. West Coast.

Pre-computed tsunami databases, such as SIFT described above, have been used successfully in Japan and the United States for over the past decade to aid in tsunami warning. Jiménez et al. (2018) describe the development of a regional pre-computed tsunami database for the Peru coast, using subduction zone earthquake sources that range in location from southern Chile to northern Mexico. The objective of this new work is to improve the accuracy of wave height forecasts for tsunami early warning.

6. *Paleo- and Historical Tsunami Studies*

The results from numerical models are often verified against information gleaned from historical observations and paleotsunami deposits. From this point of view, paleotsunami studies are especially important. Pinegina et al. (2018) examined tsunami deposits on the coast of Avachinsky Bay, Kamchatka, and reconstructed the vertical runup and horizontal inundation for 33 tsunamis recorded over the past 4200 years, including five historical events: 1737, 1792, 1841, 1923 (Feb) and 1952. The runup heights ranged from 1.9 to 5.7 m and inundation distances from 40 to 460 m; the mean recurrence period for historical events is estimated at 56 years and for all events $\sim$ 133 years.

Similarly, Razjigaeva et al. (2018) investigated tsunami signatures over the last 700 years on Russian

Island, located near Vladivostok, Russia. They could identify tsunami deposits from 1993, 1983, 1940, 1833, 1741, 1614 (or 1644) and also from some tsunamis from the fourteenth–fifteenth centuries. Major historic tsunamis were found to be more intensive than those of the twentieth century, with inland inundation distances as large as 250 m.

7. *Non-Seismically Generated Tsunamis*

Finally, because there are few instrumental records of the direct mechanism for generating non-seismic tsunamis, numerical models play an important role in explaining the coupling between such processes as asteroid airbursts and meteorological disturbances in generating ocean waves in the tsunami spectrum.

Berger and Goodman (2018) examine whether smaller asteroids bursting in air over water can generate tsunamis that could pose a threat to distant locations. Air burst-generated tsunamis differ from tsunamis generated by asteroids that strike the ocean. The numerical simulations from this scenario, based on the shallow water equations over real bathymetry, demonstrate very little tsunami threat after a short distance from the source. The importance of compressibility and dispersion is discussed along with results from a more sophisticated model problem using the linearized Euler equations. Their calculations show that the amplitude of the pressure wave response decreases much more rapidly than the gravity waves. The blast had to be very close to shore to get a sizeable tsunami response.

Meteorological tsunamis or "meteotsunamis", which in the past had been considered as an obscure phenomena, in recent years began to attract more and more attention. Several destructive meteotsunami events occurred in 2013–2017 in various regions of the world oceans. In particular, a unique chain of hazardous events impacted sites in the Mediterranean and Black Seas during the last week of June 2014. As was shown by Šepić et al. (2015), an anomalous atmospheric system ("tumultuous atmosphere") slowly propagated eastward over the region, from Spain to Ukraine and Russia. The system supported the generation of numerous intense, small-scale atmospheric disturbances, which subsequently induced meteotsunamis. The governing parameter determining the sea-level response to atmospheric disturbances is the *Froude number*, $Fr = U/c$, i.e., the ratio of the atmospheric gravity wave speed, U, to the phase speed of long ocean waves, $c = \sqrt{gh}$, where g is the gravity acceleration and h is the water depth. It was discovered that the June 2014 extreme events occurred in open-sea regions with the most favourable conditions for meteotsunami generation, $0.9 < Fr < 1.1$ ("Proudman resonance"). One such region is the Adriatic Sea; the paper describing strong meteotsunamis affecting bays and harbours on 25–26 June 2014 was published in Volume I (Šepić et al. 2016). Two papers considering other episodes in this chain of events are presented in the current volume. Šepić et al. (2018a) examine the "Odessa tsunami" of 27 June 2014 in the northwestern part of the Black Sea and use all available observational data and results of numerical modelling to explain the exceptional character of this event. The main focus of the study of Šepić et al. (2018b) is the southwestern coast of Sicily, where a major tsunami-like event, locally known as *'marrobbio'*, impacted the harbour of Mazara del Vallo on 25–26 June 2014 and produced a *meteotsunami bore*, propagating upstream in the Mazaro River. The authors determined that this bore resulted from the combined effects of external resonance (Proudman resonance on the western Sicilian shelf) and internal resonant conditions in the estuary of the Mazaro River.

Acknowledgements

We would like to thank Dr. Renata Dmowska, the Editor-in-Chief for Topical Issues of PAGEOPH, for arranging and encouraging us to organize these topical volumes. We also thank Ms. Priyanka Ganesh, Ms. Kirthana Hariharan and Mr. Sathish Srinivasan at the Journals Editorial Office of Springer for their timely editorial assistance and Dr. Kenneth Ryan for review of this Introduction. We thank the authors who contributed papers to these topical volumes. Finally, we would like to especially thank all of the reviewers who shared their time, effort, and expertise to maintain the scientific rigour of these volumes.

References

Aida, I. (1969). Numerical experiments for the tsunami propagation–The 1964 Niigata tsunami and the 1968 Tokachi–Oki tsunami. *Bulletin of the Earthquake Research Institute, 47,* 673–700.

Berger, M., & Goodman, J. (2018). Airburst-generated tsunamis. *Pure and Applied Geophysics.* https://doi.org/10.1007/s00024-018-1827-8.

Braitenberg, C., & Rabinovich, A. B. (2017). *The Chile-2015 (Illapel) Earthquake and Tsunami* (p. 335). Basel: Birkhäuser/Springer. https://doi.org/10.1007/978-3-319-57822-4.

Cecioni, C., & Bellotti, G. (2018). On the resonant behavior of a weakly compressible water layer during tsunamigenic earthquakes. *Pure and Applied Geophysics.* https://doi.org/10.1007/s00024-018-1766-4.

Chacón-Barrantes, S., López-Venegas, A., Sánchez-Ecobar, R., & Lique-Vergara, N. (2018). A collaborative effort between Caribbean states for tsunami numerical modeling: Case study Caribe Wave 15. *Pure and Applied Geophysics.* https://doi.org/10.1007/s00024-017-1687-7.

Cummins, P. R., Kong, L. S. L., & Satake, K. (2008). Tsunami science four years after the 2004 Indian Ocean Tsunami. Part I: Modelling and hazard assessment. *Pure and Applied Geophysics, 165*(11–12), Topical Issue.

Cummins, P. R., Kong, L. S. L., & Satake, K. (2009). Tsunami science four years after the 2004 Indian Ocean Tsunami. Part II: Observation and data analysis. *Pure and Applied Geophysics, 166*(1–2), Topical Issue.

Didenkulova, I., & Pelinovsky, E. (2018). Tsunami wave run-up on a vertical wall in tidal environmental. *Pure and Applied Geophysics.* https://doi.org/10.1007/s00024-017-1744-2.

Dilmen, D. I., Roe, G. H., Wei, Y., & Titov, V. V. (2018). The role of near-shore bathymetry during tsunami inundation in a reef island setting: A case study of Tutuila Island. *Pure and Applied Geophysics.* https://doi.org/10.1007/s00024-018-1769-1.

Fuentes, M., Riquelme, S., Ruiz, J., & Campos, J. (2018). Implications on 1 + 1 D Tsunami runup modeling due to time features of the earthquake source. *Pure and Applied Geophysics.* https://doi.org/10.1007/s00024-018-1804-2.

Gailler, A., Hébert, H., Schindelé, F., & Reymond, D. (2018). Coastal amplification laws for the French Tsunami Warning Center: Numerical modeling and fast estimate of tsunami wave heights along the French Riviera. *Pure and Applied Geophysics.* https://doi.org/10.1007/s00024-017-1713-9.

Geist, E. L. (2018). Effect of dynamical phase on the resonant interaction among tsunami edge wave modes. *Pure and Applied Geophysics.* https://doi.org/10.1007/s00024-018-1796-y.

Geist, E. L., Fritz, H. M., Rabinovich, A. B., & Tanioka, Y. (2016). Introduction to "Global Tsunami Science: Past and Future Volume I". *Pure and Applied Geophysics, 173*(12), 3663–3669. https://doi.org/10.1007/s00024-016-1427-4.

Jiménez, C., Carbonel, C., & Rojas, J. (2018). Numerical procedure to forecast the tsunami parameters from a database of pre-simulated seismic unit sources. *Pure and Applied Geophysics.* https://doi.org/10.1007/s00024-017-1660-5.

Kijko, A., Smit, A., Papadopoulos, G. A., & Novikova, T. (2018). Tsunami hazard assessment of coastal South Africa based on mega-earthquakes of remote subduction zones. *Pure and Applied Geophysics.* https://doi.org/10.1007/s00024-017-1727-3.

Ortiz-Huerta, L. G., Ortiz, M., & García-Gasteélum, A. (2018). Far-field tsunami hazard assessment along the Pacific coast of Mexico by historical records and numerical simulation. *Pure and Applied Geophysics.* https://doi.org/10.1007/s00024-018-1816-y.

Percival, D. B., Denbo, D. W., Gica, E., Huang, P. Y., Mofjeld, H. O., Spillane, M. C., & Titov, V.V. (2018). Evaluating the effectiveness of DART buoy networks based on forecast accuracy. *Pure and Applied Geophysics.* https://doi.org/10.1007/s00024-018-1824-y.

Pinegina, T. K., Bazanova, L. I., Zelenin, E. A., Bourgeois, J., Kozhurin, A. I., Medvedev, I. P., & Vydrin, D.S. (2018). Holocene tsunamis in Avachinsky Bay, Kamchatka, Russia. *Pure and Applied Geophysics.* https://doi.org/10.1007/s00024-018-1830-0.

Rabinovich, A. B., Borrero, J. C., & Fritz, H. M. (2014). Tsunamis in the Pacific Ocean: 2010–2011. *Pure and Applied Geophysics, 171*(12), 3175–3538.

Rabinovich, A. B., Geist, E. L., Fritz, H. M., & Borrero, J. C. (2015a). Tsunami science: Ten years after the 2004 Indian Ocean Tsunami Volume I. *Pure and Applied Geophysics, 172*(3–4), Topical Issue.

Rabinovich, A. B., Geist, E. L., Fritz, H. M., & Borrero, J. C. (2015b). Tsunami Science: Ten Years after the 2004 Indian Ocean Tsunami Volume II. *Pure and Applied Geophysics, 172*(12), 3265–3670.

Rabinovich, A. B., Fritz, H. M., Tanioka, Y., & Geist, E. L. (2017). Introduction to "Global Tsunami Science: Past and Future Volume II". *Pure and Applied Geophysics, 174*(8), 2883–2889. https://doi.org/10.1007/s00024-017-1638-3.

Rashidi, A., Shomali, Z. H., & Farajkhah, N. K. (2018). Tsunami simulation using scaled slip models in the western Makran. *Pure and Applied Geophysics.* https://doi.org/10.1007/s00024-018-1842-9.

Razjigaeva, N. G., Ganzey, L. A., Grebennikova, T. A., Arslanov, K. A., Ivanova, E. D., Ganzey, K. S., & Kharlamov, A.. A. (2018). Historical tsunami records on Russian Island, the Sea of Japan. *Pure and Applied Geophysics.* https://doi.org/10.1007/s00024-018-1840-y.

Salaree, A., & Okal, E. A. (2018). The "tsunami earthquake" of 13 April 1923b in Northern Kamchatka: Seismological and hydrodynamic investigations. *Pure and Applied Geophysics.* https://doi.org/10.1007/s00024-018-1769-1.

Sandanbata, O., Watada, S., Satake, K., Fukao, Y., Sugioka, H., Ito, A., & Shiobahara H. (2018). Ray tracing for dispersive tsunamis and source amplitude estimation on Green's Law: Application to the 2015 Volcanic tsunami earthquake near Torishima, South of Japan. *Pure and Applied Geophysics.* https://doi.org/10.1007/s00024-017-1746-0.

Satake, K., Okal, E. A., & Borrero, J. C. (2007). Tsunami and its hazards in the Indian and Pacific oceans. *Pure and Applied Geophysics, 164*(2–3), 249–631.

Satake, K., Rabinovich, A. B., Kânoğlu, U., & Tinti, S. (2011a). Tsunamis in the World Ocean: Past, Present, and Future. Volume I. *Pure and Applied Geophysics, 168*(6–7), 963–1249.

Satake, K., Rabinovich, A. B., Kânoğlu, U., & Tinti, S. (2011b). Tsunamis in the World Ocean: Past, Present, and Future Volume II. *Pure and Applied Geophysics, 168*(11), 1913–2146.

Satake, K., Rabinovich, A. B., Dominey-Howes, D., & Borrero, J. C. (2013a). Historical and recent catastrophic tsunamis in the world: Past, present, and future. Volume I: The 2011 Tohoku Tsunami. *Pure and Applied Geophysics, 170*(6–8), Topical Issue.

Satake, K., Rabinovich, A. B., Dominey-Howes, D., & Borrero, J. C. (2013b). Historical and recent catastrophic tsunamis in the world: Past, present, and future. Volume II: Tsunamis from 1755 to 2010. *Pure and Applied Geophysics, 170*(9–10), Topical Issue.

Šepić, J., Vilibić, I., Rabinovich, A. B., & Monserrat, S. (2015). Widespread tsunami-like waves of 23-27 June in the Mediterranean and Black Seas generated by high-altitude atmospheric forcing. *Scientific Reports, 5,* 11682. https://doi.org/10.1038/srep11682.

Šepić, J., Međugorac, I., Janeković, I., Dunić, N., & Vilibić, I. (2016). Multi-meteotsunami event in the Adriatic Sea generated by atmospheric disturbances of 25–26 June 2014. *Pure and Applied Geophysics, 173,* 4117–4138.

Šepić, J., Rabinovich, A. B., & Sytov, V. N. (2018a). Odessa tsunami of 27 June 2014: Observations and numerical modelling. *Pure and Applied Geophysics*. https://doi.org/10.1007/s00024-017-1729-1.

Šepić, J., Vilibić, I., Rabinovich, A., & Tinti, S. (2018b). *Meteotsunami ("marrobbio") of 25–26 June 2014 on the southwestern coast of Sicily*. Italy: Pure and Applied Geophysics. https://doi.org/10.1007/s00024-018-1827-8.

Tanioka, Y., Ramirez, A. G. C., & Yamanaka, Y. (2018). Simulation of a dispersive tsunami due to the 2016 El Salvador-Nicaragua outer-rise earthquake. *Pure and Applied Geophysics*. https://doi.org/10.1007/s00024-018-1773-5.

Pure Appl. Geophys.

https://doi.org/10.1007/s00024-018-1769-1

Pure and Applied Geophysics

The Role of Near-Shore Bathymetry During Tsunami Inundation in a Reef Island Setting: A Case Study of Tutuila Island

Derya I. Dilmen,[1,2] Gerard H. Roe,[2] Yong Wei,[1] and Vasily V. Titov[1,2]

Abstract—On September 29, 2009 at 17:48 UTC, an $M_w = 8.1$ earthquake in the Tonga Trench generated a tsunami that caused heavy damage across Samoa, American Samoa, and Tonga. One of the worst hits was the volcanic island of Tutuila in American Samoa. Tutuila has a typical tropical island bathymetry setting influenced by coral reefs, and so the event provided an opportunity to evaluate the relationship between tsunami dynamics and the bathymetry in that typical island environment. Previous work has come to differing conclusions regarding how coral reefs affect tsunami dynamics through their influence on bathymetry and dissipation. This study presents numerical simulations of this event with a focus on two main issues: first, how roughness variations affect tsunami run-up and whether different values of Manning's roughness parameter, n, improve the simulated run-up compared to observations; and second, how depth variations in the shelf bathymetry with coral reefs control run-up and inundation on the island coastlines they shield. We find that no single value of n provides a uniformly good match to all observations; and we find substantial bay-to-bay variations in the impact of varying n. The results suggest that there are aspects of tsunami wave dissipation which are not captured by a simplified drag formulation used in shallow-water waves model. The study also suggests that the primary impact of removing the near-shore bathymetry in coral reef environment is to reduce run-up, from which we conclude that, at least in this setting, the impact of the near-shore bathymetry is to increase run-up and inundation.

Key words: Tsunami, numerical modeling, manning roughness, island reef environment, American Samoa.

1. Introduction

A topic of longstanding interest in tsunami research is the role of near-shore bathymetry in tsunami dynamics (Leschka et al. 2009a; Lynett 2016). One particularly important example is the complex tsunami dynamics occurring in reef environments. The impact of reef-related bathymetry in affecting the destructiveness of tsunamis on the coast has been studied numerically and experimentally. For example, a study on Tutuila Island by Roeber et al. (2010) concluded that shallow reefs, in some instances, provided little or no protection to the coastal communities, and even transformed waves into having more destructive power. Roeber et al. (2010) also observed that fringing reefs and small embayments can amplify near-shore energy and develop local oscillation modes with 2–4-min periods adjacent to the shore of Tutuila. Tsunami observations, suggesting little protection by the coral reefs at atolls and reef-surrounded islands, were reported in the literature for the Tohoku tsunami impacts (Ford et al. 2013; Titov et al. 2016). On the other hand, Baba et al. (2008) performed numerical simulations of the 2007 Solomon Islands tsunami to explore the effect of the Great Barrier Reef on tsunami wave height and found that, in the simulations in which the reef was removed, the tsunami amplitude was larger by a factor of two or more than that produced by the model with the reef. Gelfenbaum et al. (2011) performed numerical simulations of the 2009 Samoa tsunami and concluded that a better understanding of reef roughness in particular is required to predict how coral reefs affect tsunami inundation. Experimental and theoretical studies indicated that reefs have a strong bottom drag coefficient that is about an order of magnitude larger than that for sand (see, e.g., Baptista et al. 1998; Kraines et al. 1998). Motivated by those results, Kunkel et al. (2006) performed simple one-dimensional and two-dimensional

[1] NOAA Center for Tsunami Research, Pacific Marine Environmental Laboratory, UW/Joint Institute for the Study of the Atmosphere and Ocean, 7600 Sand Point Way NE, Bldg. 3, Seattle, WA 98155, USA. E-mail: Vasily.Titov@noaa.gov

[2] Department of Earth and Space Sciences, University of Washington, Seattle, WA 98195-1310, USA.

numerical simulations with idealized topography to explore the effect of bottom friction due to a reef on tsunami run-up. With bathymetry held constant, the tsunami run-up was decreased approximately 50% when the bottom-drag coefficient was increased from 0.03 to 0.1. However, the Kunkel et al. (2006) simulations also suggest the possibility that gaps between adjacent reefs can result in flow amplification and actually increase the local wave heights. Fujima (2006) suggest that the increased damage observed at some islands in Maldives resulted from the tsunami front becoming more bore-shaped as it propagated over the fringing reef, which increased the destructive force of the wave. Fernando et al. (2005) and (2008) lend support to these numerical results: coral reefs protect coastlines behind them but local absences of reefs cause local flow amplification due to gaps. Their results are based on field observations, laboratory measurements (Fernando et al. 2008), and interviews of local people in Sri Lanka after the 2004 Indian Ocean tsunami. However, their laboratory simulations treated corals as a submerged porous barrier made of a uniform array of rods, which likely oversimplifies the complex structural distribution of coral reefs. Liu et al. (2009) found that wave amplitudes can be amplified several-fold (up to tenfold), based on the near-shore bathymetry.

In this study, we extend the analysis of reef roughness and reef bathymetry to a slightly more general study of the influence of reef-related bathymetry on tsunami impact, with a numerical-model case study of the 2009 Samoa tsunami on Tutuila island. We focus on two main goals: firstly, we want to understand how roughness variations of the shelf around the island affect run-up and inundation; secondly, we perform a model sensitivity analysis varying the bathymetry where fringing coral reefs exist, to elucidate the role of such reef–island bathymetry in controlling run-up on the coastlines it shields. For this event, localized run-up exceeded 17 m, and the event was well monitored by coastal tide gauges and off-shore DART buoys and post-tsunami run-up surveys.

The tsunami field surveys of 29 September 2009 Samoa tsunami on Tutuila Island showed that the western side of the island facing the tsunami direction had the highest wave run-up although with large variations among adjacent villages only 2–3 km apart from each other; severe destruction was also apparent along the eastern and northern coastline of the island (Okal et al. 2010; Fritz et al. 2011; Borrero et al. 2011). It has proven difficult for numerical simulations of this event to successfully reproduce these large variations in tsunami run-up and inundation at many coastal villages (e.g., Dilmen et al. 2015). One challenge for numerical tsunami models in tropical settings such as this is to properly simulate the impact of the bathymetry including the pervasive barrier and fringing coral reefs (Fig. 1). In many respects, the 2009 Samoa tsunami provides an important benchmark for numerical models to simulate the tsunami dynamics in a typical submarine environment of coral-reef islands, with the ultimate goal of better understanding the tsunami risks for island communities around the world.

Our previous work (Dilmen et al. 2015) compared a tsunami simulation of this event with several observational datasets to evaluate the high variability of the tsunami impact around the Island of Tutuila. We found that the numerical model compares very well with tide gauge data. At the same time, the modeling of the inundation process clearly demonstrated the challenges for models in handling the complex bathymetric environment of coral islands. While the numerical simulation qualitatively reproduced inundation at all bays where flooding was observed, the quantitative comparison showed a tendency for the model to underestimate field-measured run-up values. We also found that there was a high degree of variability in the simulated run-up accuracy even among adjacent bays. Such model inconsistency and the measured run-up variability could not be explained by the presence of fringing reefs, nor did it correlate well with the available coral-reef damage data. There was only a tentative indication that at villages where the model underestimates run-up, coral damage is more likely to be high or very high, suggesting that the shallow-water models may overestimate energy dissipation over coral reefs.

The results implied that a more detailed understanding of two important model controls, roughness and bathymetry, would be a very useful next step in understanding the impact of reef-driven bathymetry on tsunami dynamics. Dilmen et al. (2015) also

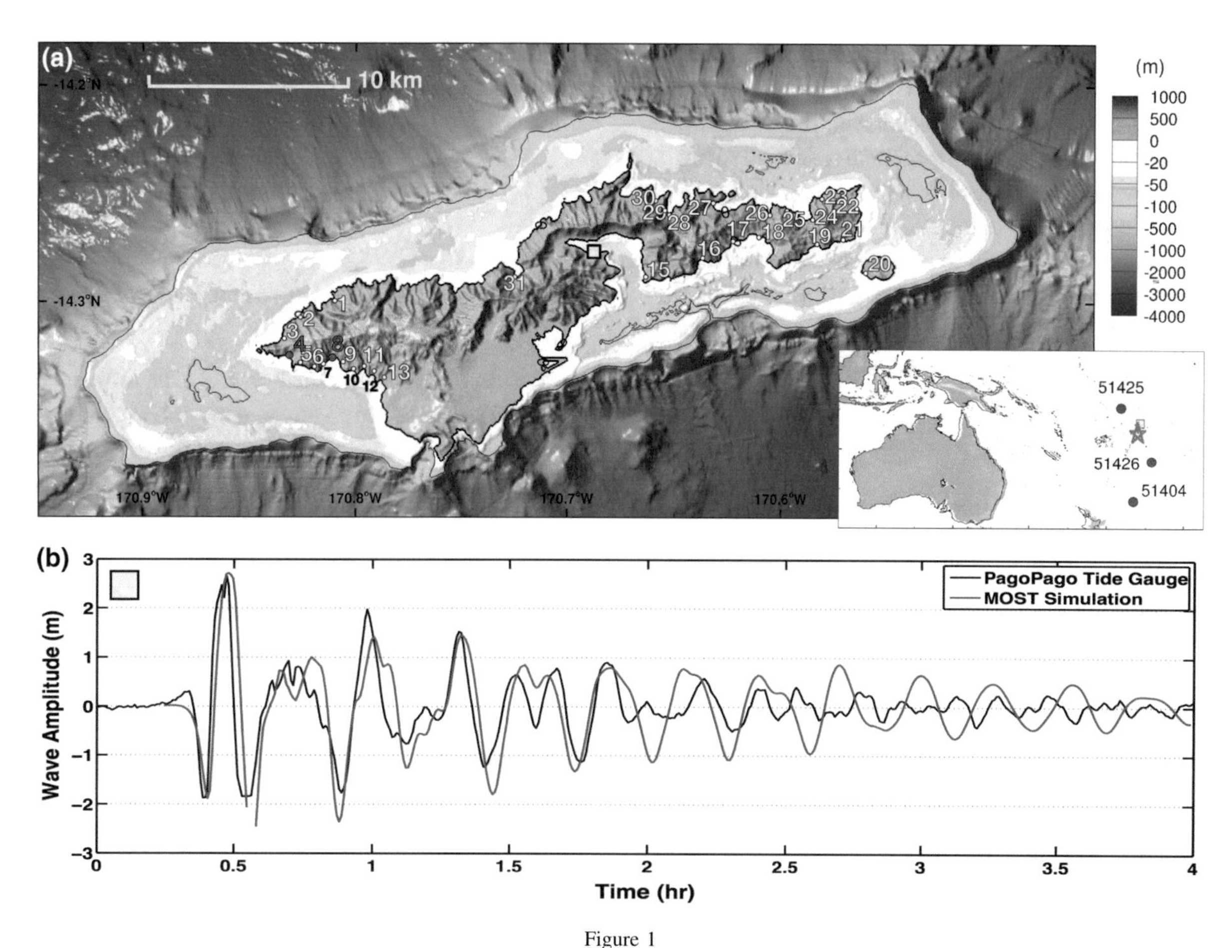

Figure 1

a Tutuila Island bathymetry. The numbers correspond to the locations of villages at which surveys were conducted, the yellow square is the location of the PagoPago tide gauge, locations of Amanave and Se'etaga are indicated in red. Black lines show the coastlines and 90 m depth contour. Inset overview map: Tutuila Island tide gauge is shown as the yellow square. Red star shows the epicenter of the 2009 Samoa Earthquake; and the red circles are the location of the DART buoys used in optimizing the earthquake fault source used in the MOST simulations. **b** A comparison of the measured and simulated tsunami time series at the Pago Pago gauge location

evaluated the relationship between the tsunami dynamics and the coral damage by numerical modeling and post-tsunami surveys. They speculated that the variability in simulated run-up differences in the high-resolution tsunami model might be due to sharp changes in bottom roughness values caused by coral reefs. One expects that in reality there are strong spatial variations in roughness values (e.g., Nunes and Pawlak 2008). It is possible then that a different roughness coefficient used in a numerical simulation might more accurately simulate the run-up. The numerical model used for these simulations employs shallow-water equations, which is the classic modeling technique used in tsunami hazard assessment and forecast. Because these equations are developed using the assumption that the velocity distribution is uniform in the vertical direction, it is unlikely those equations can adequately reproduce the complicated flows where the water depth suddenly changes near such reefs. Furthermore, the model represents dissipation using a parameterized bulk-drag formula (Eq. 1), with a single value for the Manning roughness coefficient, an empirically derived value taken from the engineering literature. Even though Dilmen et al. (2015) used a very high-resolution model for the Tutuila tsunami, there are significant differences between the observed and the modeled run-up and inundation; thus, modeling the flooding associated with this event remains a challenge. Motivated by these issues, the present study focuses on two main goals: firstly, we want to understand how roughness variations affect run-up; secondly, we perform a model sensitivity analysis by varying the bathymetry where coral reefs exist, to elucidate their role in controlling run-up on the coastlines they shield.

The remainder of the paper is organized as follows: Sect. 2 summarizes the earthquake and tsunami event; Sect. 3 presents the modeling studies and the analyses; and Sect. 4 is the summary and discussion.

2. Model Description

We simulate the 2009 Samoa tsunami using the MOST Model (Titov 1997; Titov and González 1997; Titov et al. 2016). The evaluation of the simulations, a comparison with tide gauge and offshore DART data, and the calibration of the earthquake source function were described in Dilmen et al. (2015). MOST is an established tsunami model that has been widely tested and evaluated, and it is used operationally for forecasting (e.g., Titov 2009; Titov et al. 2016) and hazard assessment (e.g., Titov et al. 2003a). There are other numerical tsunami models with alternative dynamical equations and/or numerical schemes. Recognizing the importance for intermodel evaluations (e.g., Synolakis et al. 2008), recent community efforts have focused on using models that satisfy theoretical benchmarks and case-study comparisons, such as those proposed by the National Tsunami Hazard Mitigation Program (NTHMP 2012). MOST meets the benchmarks and performs comparably to other tsunami models for the real-world case studies. The primary metrics for comparison with observations are wave run-up, inundation from post tsunami surveys at 31 villages around Tutuila, and tide-gauge data (Fig. 1).

Dilmen et al. (2015) described in detail the numerical-model setup, earthquake source, and comparison of model results with observations. A set of three, nested computational grids that zoom into the simulation area are defined around Tutuila, with successively finer spatial resolution (A-grid is outermost coarse grid, B-grid is intermediate, and C is the innermost fine-scale grid). Details of the grid resolution are listed in Table 1. Regional bathymetry and topography datasets were compiled and provided by National Center for Environmental Information (NCEI).

3. Analysis

3.1. The Impact of Changing Manning's Roughness, *n*

In the MOST numerical model, the effects of bottom friction are implemented by incorporating a basal shear stress with components (τ_{xz}, τ_{yz}) into the shallow water equations and parametrized by a drag formula:

$$(\tau_{xz}, \tau_{yz})/\rho = -C_{\mathrm{B}} \frac{\sqrt{U^2 + V^2}}{D}(U, V), \qquad (1)$$

where U, V are the components of the velocities, D is the fluid depth, and ρ is the density (Titov et al. 2003b). C_{B} is the dimensionless friction coefficient, which can in turn be related to the Manning's roughness parameter, *n,* via:

$$C_{\mathrm{B}} = \frac{gn^2}{D^{1/3}}. \qquad (2)$$

The concept of Manning's roughness was originally developed for open-channel flow. Manning's *n* values have been measured empirically for a wide variety of different materials in laboratory experiments (Chow 1959), and by large-scale field studies of river flow for fully turbulent conditions (Bricker et al. 2015). From Eq. (1), C_{B} is depth-dependent, with increasing depth implying decreased friction. For example, for $n = 0.025$ s m$^{-1/3}$, $C_{\mathrm{B}} = 0.006$ for $D = 1$ m, but $C_{\mathrm{B}} = 0.0025$ for $D = 15$ m. Bottom friction typically has the greatest impact at depths of 0–10 m, and is negligible for tsunami propagation in the deep ocean (Levin and Nosov 2016). The simple

Table 1

Computational grids for simulations

Parameters	Grid A	Grid B	Grid C
Resolution	12 arc sec (360 m)	2 arc sec (60 m)	0.3 arc sec (10 m)
Size	1110 × 781	900 × 396	2400 × 2000

form of n, its relatively straightforward implementation within shallow-water equations, and the availability of extensive datasets of experimentally verified values, have led to its widespread adoption in tsunami modeling (e.g., Imamura et al. 2008) for the representation of frictional dissipation of tsunami energy in coastal zones, for both the submerged and subaerial portions of the domain. In many studies, a value of $n = 0.025$ s m$^{-1/3}$ has been adopted as appropriate for a smooth sea bottom or land. A range of other values has also been suggested from $n \sim 0.01$ to 0.1 s m$^{-1/3}$ depending on the setting and basal conditions (Kunkel et al. 2006; Tang et al. 2009; Jaffe et al. 2010; Gelfenbaum et al. 2011; and Bricker et al. 2015).

In our previous study of the 2009 Samoa tsunami on Tutuila island (Dilmen et al. 2015), we selected $n = 0.03$ s m$^{-1/3}$. Both far-field pressure sensors (DARTs) and near-field coastal sea-level stations (tide gauges) were used to calibrate the tsunami source for this event. Simulated wave amplitudes matched well with tide-gauge observations (Fig. 1). Inundation computations from the model have been compared with point data of field run-up measurements. While the model confirmed and reproduced inundation at all 31 flooded villages, the comparison of simulated values with point observations revealed that the model underestimates the run-up at 15 village sites for this event, especially at sites behind heavily damaged coral reefs. Dilmen et al. (2015) implied that the model may have overestimated the energy dissipation over corals, possibly due to improper model roughness.

In this study, we first evaluate whether a different value of n can decrease the discrepancy between simulated and observed run-up at the 31 villages. For some villages, several separate run-up observations were made. For such villages, we took the average of all run-up observations. From the model output, we took the maximum simulated run-ups in a 3×3 grid box (30 m $\times$ 30 m) surrounding the locations of each of the observations, and averaged them. We also tried taking just the maximum simulated run-up in the 3×3 grid box, and comparing this with the maximum run-up observations at that village. Finally, we tried taking the maximum simulated run-up from the single gridpoint overlaying each run-up observation. Our results do not depend on which of these various methods we use.

We performed simulations varying n from 0.01 to 0.15 s m$^{-1/3}$, where the lowest n value represents an essentially smooth bathymetry over the reef. In Fig. 2 we present the simulations for varying n as a scatterplot of observed vs. simulated runup at 31 villages around Tutuila. The locations of the villages are shown in Fig. 1. A comprehensive summary of the data and analyses is presented in Table 2 in Appendix.

Overall, for most of the villages, the variation of runup with n is straightforward, with higher values of n having less run-up. For some villages, however, the relationship is not monotonic, and the highest simulated runup does not always occur for the lowest value of n. Thus, at individual locations, the complex patterns of refraction and reflection, and nonlinear interactions can complicate the relationship between dissipation and runup. This is also clear from the different spreads among the simulated run-ups at individual villages. At some villages, there is very little spread in simulated runup as n varies (e.g., Fagatele village, observed runup = 4.92 m, the spread is 4% of the mean simulated value), whereas

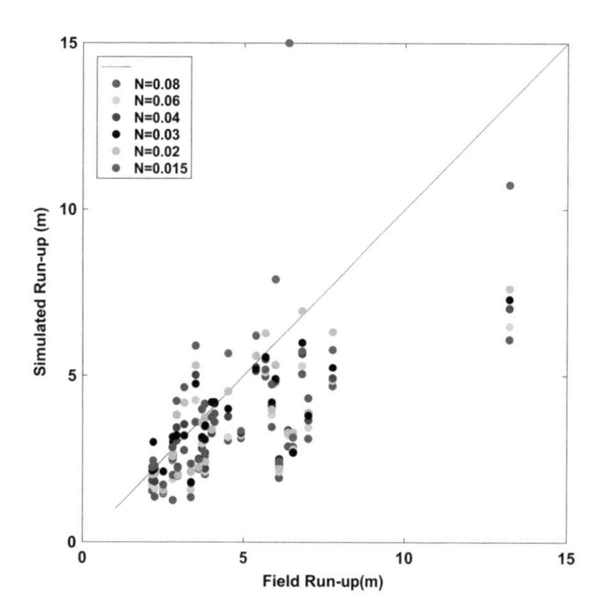

Figure 2
A comparison of model and field run-up for the simulations with varying n at 31 villages in Tutuila Island. The blue line shows the 1:1 line. Points falling on or close to the blue line show where there is good agreement between simulations and observations

at others the spread is large (e.g., Poloa village, observed runup = 13.5 m, the spread is 76% of the mean simulated value).

The results are mixed regarding whether the various values of n improve the simulated runup compared to observations. The model estimation of run-up is improved by changing n for some villages and doesn't change significantly at others. Of the total 31 villages, at 10 villages the difference between observed and estimated run-up is smallest when $n = 0.02$ s m$^{-1/3}$. At 9 villages, the closest match occurs for $n = 0.03$ s m$^{-1/3}$. At 5 villages, the closest match occurs for $n = 0.015$ s m$^{-1/3}$. For the remaining 6 villages, two villages each achieve the closest match to observations when $n = 0.04$ s m$^{-1/3}$, $n = 0.06$ s m$^{-1/3}$, and $n = 0.08$ s m$^{-1/3}$.

In our simulations, n is uniform throughout the model domain, which oversimplifies the real situation of a heterogeneous littoral and coastal environment. Equations (1) and (2) represent dissipation in these environments. An obvious next step in tsunami modeling is to evaluate whether variations in basal conditions might be represented by spatial variability in n. Introducing spatial variations in n would add tunable degrees of freedom in the model, and in practical applications it would be important not to over-constrain a model. Nonetheless, an evaluation could be performed as to whether spatial variations in n might be optimized against observations to provide agreement with detailed measurements in case studies such as ours, and it would represent a 'best-case' for the ability of equations to simulate run up in these events. Since the computational influence of friction is strongest in the very near-shore and onshore environments (the influence is inversely proportional to flow depth), even a simplified uniform-friction study like ours can provide insight for the problem. Calibrating a variable-friction map for such a variable-friction simulation represents a formidable challenge. However, general hydrodynamic studies combined with measurements of wind- and tidal-driven currents in coral-reef lagoons (where the problem appears to be the most acute) can provide needed data for variable-friction maps and hence more precise estimates of the run-up and associated currents for shallow-water wave simulations.

Next, we evaluate the spatial patterns of the maximum tsunami amplitudes for two representative villages, Se'etaga and Amanave, as a function of n. Dilmen et al. (2015) showed that, for standard parameters ($n = 0.03$ s m$^{-1/3}$), our modeling successfully simulated the run-up observations at Se'etaga, agreeing to within 2.4%. On the other hand, just 3 km southeast from Se'etaga, the simulation did a relatively poor job for the village of Amanave (the model underestimated observed run-up by 45%). This very different simulation performance illustrates the complexities of the setting, and thus made Se'etaga and Amanave suitable for our case studies.

Figure 3 shows the tsunami evolution during the propagation toward the west coast of Tutuila Island where the two villages are located, for the simulation that provided the best fit with the tide-gauge record. These snapshots of the model amplitudes illustrate the dimensions and shape of the tsunami waves at the time of the initial flooding event at the two selected village locations. The figure shows that the wavelength of the tsunami in deep water is comparable with the size of Tutuila Island. However, when the wave reaches the shallow areas of the west-coast shelf, the wavelength shortens to approximately the length of the shelf. The wave exhibits strong refraction around the island due to the dramatic bathymetry changes near the island shelf. The strong amplification of the wave amplitude and shortening of the wavelength over the shelf lead to an increased steepness of the wave front, indicating mild to strong nonlinearity during the shoaling of the tsunami near the two selected locations.

The simulated maximum wave amplitude fields are shown in Fig. 4a–d for Se'etaga, and in Fig. 4e–h) for Amanave. For Se'etaga, $n = 0.03$ s m$^{-1/3}$ provides the best agreement in run-up (5.69 m observed, 5.57 m model), whereas for Amanave, $n = 0.02$ s m$^{-1/3}$ gives closest agreement (7.74 m observed, 6.32 m model). For higher values of n, the wave-amplitude fields vary smoothly on scales of few-hundred meters. The highest values of run-up are found in the center of both bays, suggesting a refraction or focusing of wave energy there. As n is reduced from 0.04 to 0.02 s m$^{-1/3}$, some significant localized increases in wave amplitude and inundation appear, particularly in Amanave bay, where the wave amplitude approaches double that of the $n = 0.08$ s m$^{-1/3}$ simulations. For $n = 0.015$ s m$^{-1/3}$ wave amplitudes increase throughout each bay, and the increases are clearly co-located with the clusters of shallow bathymetry structures related to reefs that are dotted around the perimeter of each bay, and which can be seen in detail in Fig. 7a, c. Figure 4 also shows evidence that lower values of n have some impact offshore, with a more complex wave pattern and more small-scale structure, perhaps indicating more scattering of wave energy and a cascade to smaller scales.

Combining Eqs. 1 and 2, we find that the absolute value of the bed shear stress is proportional to the flow velocity as follows:

$$|\tau| \sim \frac{\rho n^2 \cdot |u|^2}{D^{4/3}}. \tag{3}$$

Figures 5 and 6 show the maximum values of $|u|$, the momentum flux ($\rho D|u|^2$), and $|\tau|$, as a function of the value of n, for Se'etaga and Amanave bays, respectively. In both bays, there is the same basic behavior. Lower values of n are associated with much larger velocities (top panels of Figs. 5, 6) throughout each bay. Lower values of n are also associated with higher momentum flux (middle panels of Figs. 5, 6), although the momentum-flux pattern is also affected

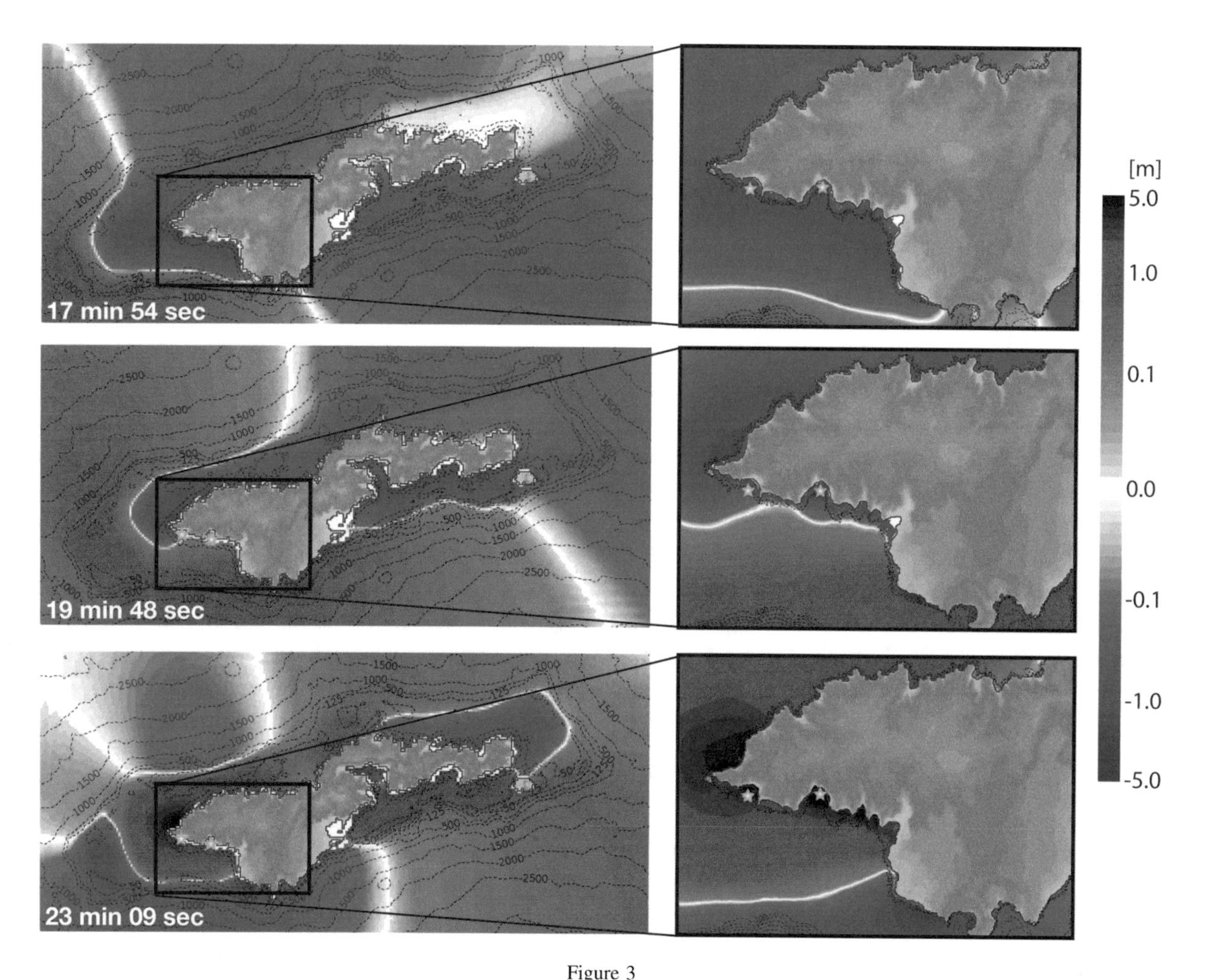

Figure 3
Model tsunami amplitudes at three different times during initial tsunami impact on the west coast of Tutuila. Bays of Se'etaga and Amanave are indicated by green stars. Positive tsunami amplitudes are colored in red, negative in blue

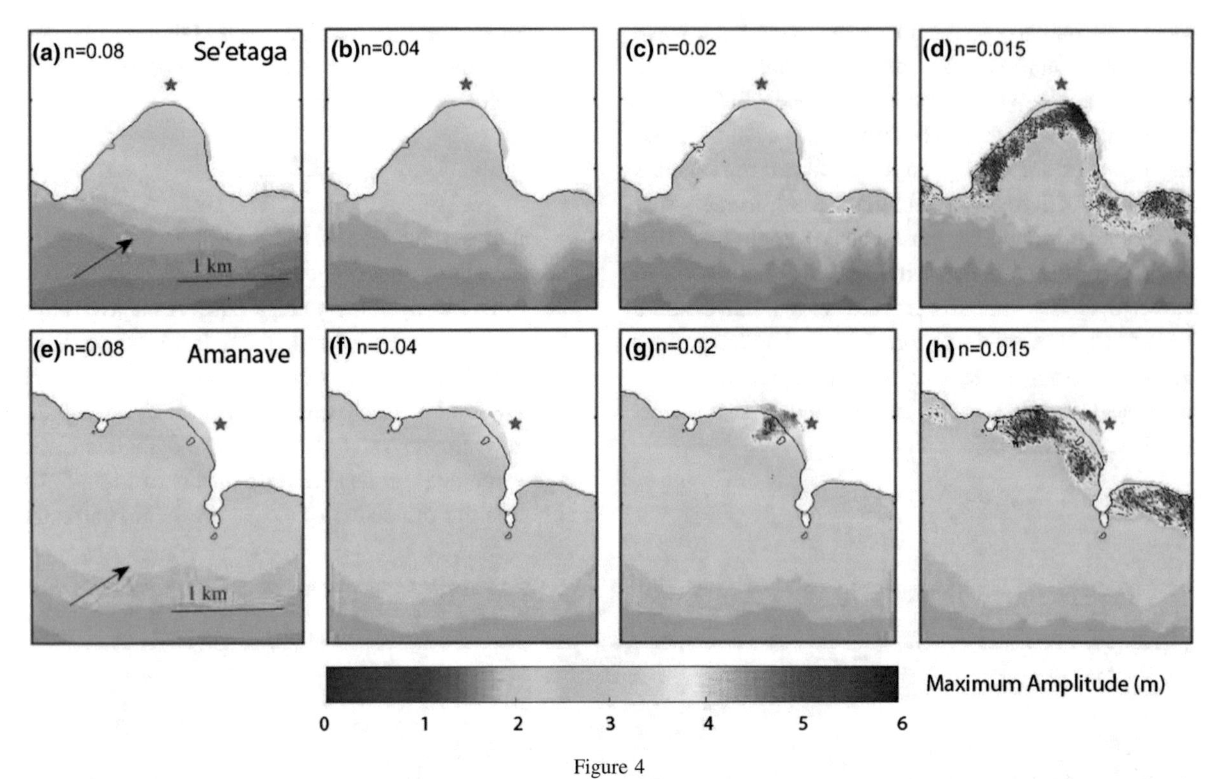

Figure 4
Simulated maximum wave amplitudes are shown for the bays near the villages of Se'etaga (top panels) and Amanave (lower panels) for $n = 0.08$, 0.04, 0.02 and 0.015 s $m^{-1/3}$. The pre-tsunami shoreline is shown as a thin black line. Colors inland of that line therefore show the degree of inundation. Village locations are shown as red stars. Black arrows indicate the tsunami direction from its source

by the shoreward-shoaling bathymetry. Note the middle panels also show that the lower values of n produce a larger momentum flux at offshore locations, showing that the greater dissipation associated with higher n is felt several km offshore. Finally, the bottom panels of Figs. 5 and 6 show the maximum computed bed shear stress from Eq. 3. It is noteworthy that the nearshore stresses are greater for lower values of n, despite the scaling in Eq. (3). The increased velocities outweigh the smaller n and larger D resulting in larger stresses. It is also clear that the largest stresses are associated with the individual reef structures seen in detail in Fig. 7b, c.

We next turn to a comparison of water-surface elevations at virtual sea-level gauges estimated from the MOST output at gridpoints near the coasts of Se'etaga and Amanave, for varying n (Fig. 7). At these villages, the maximum wave amplitude occurs during the first surge. For the Se'etaga time series, the maximum water elevations are 3.1, 3.2, and 3.3 m for $n = 0.08$, 0.04, and 0.02 m $s^{-1/3}$, respectively (Fig. 7). At Amanave, the maximum water elevations are 3.1, 3.4, and 3.9 m for $n = 0.08$, 0.04, and 0.02 m $s^{-1/3}$, respectively. The various values of n do not affect the timing of the waves, and have most impact on the amplitude of the first wave.

Both Se'etaga and Amanave time series show some evidence of resonance or constructive interference between approaching and receding waves (Fig. 7). We generally expect successive wave heights to decrease in amplitude after several waves, but at both sites wave elevations increase between the fourth and sixth waves then decrease again after the seventh wave. The time series at Se'etaga shows that the approaching and reflecting waves from the coastlines and bathymetry barriers form constructive interference after the third wave. Tsunami time series at Amanave drop off in amplitude more quickly than at Se'etaga and have a less sinusoidal form. By 1.5 h after the first waves, the tsunami has become a

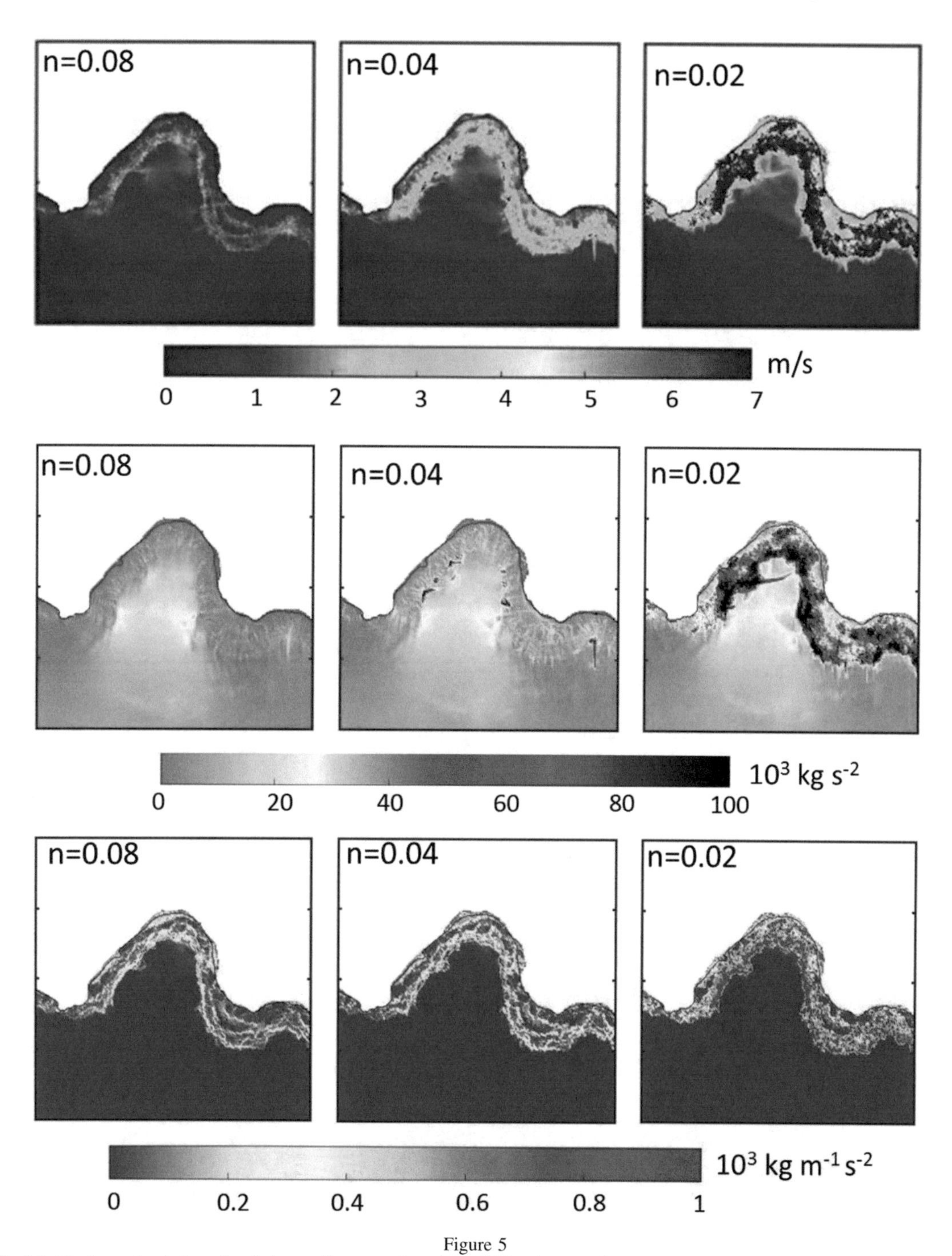

Figure 5
Dynamical fields from simulations for Se'etaga Bay as a function of Manning roughness, n. Upper panels: maximum flow velocity for $n = 0.08$, 0.04, and 0.02 s $m^{-1/3}$; middle panels: maximum momentum flux for $n = 0.08$, 0.04, and 0.02 s $m^{-1/3}$; bottom panels: maximum bed shear stress for $n = 0.08$, 0.04, and 0.02 s $m^{-1/3}$. The figure shows the maximum value of the field at every grid point, whenever it occurs in the model simulation. This accounts for the linear features seen in some of the panels

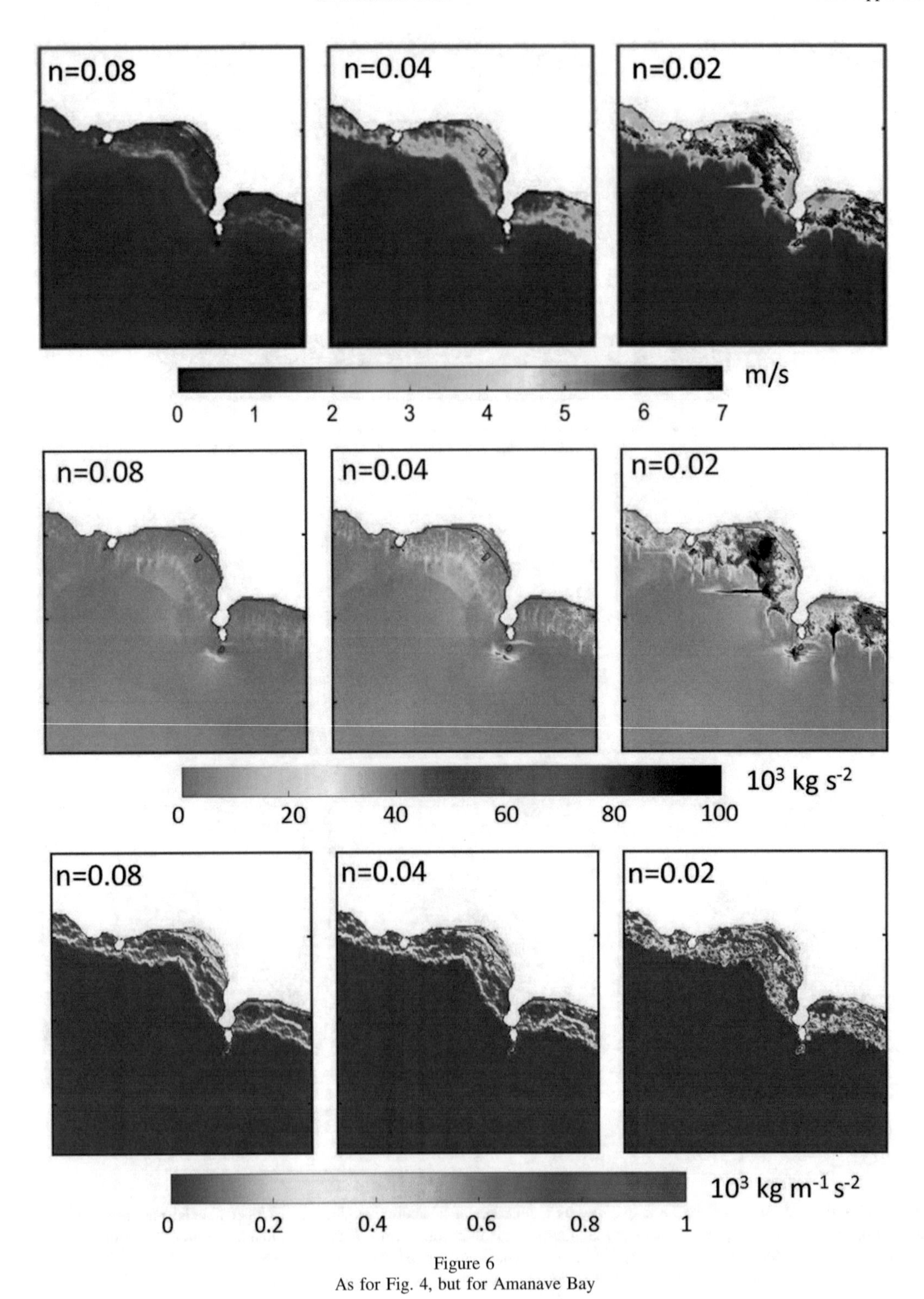

Figure 6
As for Fig. 4, but for Amanave Bay

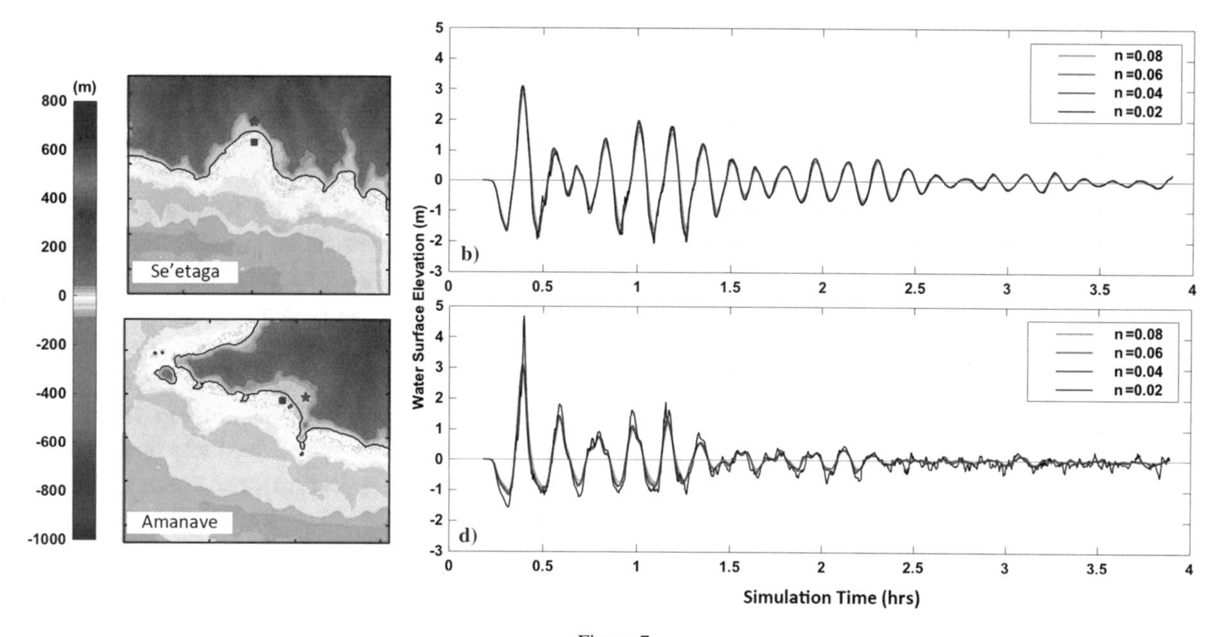

Figure 7
Comparison of water-surface elevations at selected virtual tide gauges in the near-shore regions of villages of Se'etaga (**a**, **b**) and Amanave (**c**, **d**). Maps show the location of the villages as red stars, and the locations of the virtual tide gauges are shown as blue squares. In **b** and **d** water surface elevations are given for varying n during the 4 h after the fault rupture at $t = 0$

distorted noisy wave-train. This behavior indicates more nonlinearity and more overall dissipation, compared to Se'etaga, perhaps because of the more complicated reef structures within the bay that cause the waves to be dispersed in the reef areas (Fig. 7a compared with Fig. 7c).

3.2. Changes of Island Shelf Bathymetry

In this section, we evaluate how variations in island shelf bathymetry with the associated contributions from reef structures affect the tsunami run-up and inundation inside the bays at the 31 villages around Tutuila. We generate synthetic bathymetries for the higher-resolution grids of our nested computational grid system (the B and C grids of Table 1) by progressively removing bathymetry with less than 90 m depth (i.e., bathymetry in the range 0–90 m is ultimately set equal to 90 m). As shown in Fig. 1, the 90 m contour outlines the bathymetry inside the barrier reef. Therefore, such a procedure will gradually remove all reef-related bathymetry structures, including bathymetry features related to the fringing and barrier reefs, while leaving the flat 90 m-deep shelf in front of the coastline intact. Let $z_{\mathrm{real}}(x, y)$ be the real bathymetry and let $z_{\mathrm{flat}}(x, y)$ be the extreme flattened bathymetry. We performed several experiments varying the bathymetry smoothly between these limits using the parameter, r, in the following equation:

$$z_{\mathrm{expt}}(x, y) = r z_{\mathrm{real}}(x, y) + (1 - r) z_{\mathrm{flat}}(x, y) \qquad (4)$$

Thus, $r = 1$ corresponds to the real bathymetry and $r = 0$ to the flattened bathymetry. We varied r from 0 to 1 in increments of 0.2. The resulting bathymetries are shown in Fig. 8. In all simulations, a value of Manning's roughness of $n = 0.03\ \mathrm{m\ s}^{-1/3}$ has been implemented.

We first focus on maximum wave amplitudes in simulations that implement these varying bathymetries. The wave amplitude fields near Se'etaga and Amanave are shown in Fig. 9, and the results for simulated run-up for villages around Tutuila are given in Table 3 in Appendix. In general, we find that wave amplitudes and run-up are smaller when the reef-related roughness of the original bathymetry is removed, contrary to the common belief that reefs and related bathymetric features protect reef-surrounded coasts from tsunamis (Baba et al. 2008; Kunkel et al. 2006). When there is a smooth shelf ($r = 0$), the tsunami run-up achieves its lowest value at 22 out of the 31 villages. For the remaining 9 villages, 4 villages achieve the smallest run-up when $r = 0.2$, 2 villages achieve the smallest run-up when $r = 0.4$, and 3 villages achieve the smallest run-up when $r = 0.6$ (Table 3 in Appendix).

The same basic behavior is observed in the maximum wave amplitude fields near Se'etaga and Amanave (Fig. 9). Wave amplitudes are larger for the real bathymetry than for the flattened bathymetry. The model estimates the maximum water surface elevations at Se'etaga village as 3.2, 2.2, and 2.1 m for $r = 1.0$, 0.8, and $r = 0.0$, respectively. At Amanave, the maximum water surface elevations are 3.4, 2.8, 2.1 m for $r = 1.0$, 0.8, and 0.0, respectively.

These results may seem counterintuitive at first glance, since the removal of apparently protective structures over the island shelf actually decreases the run-up and inundation for the two considered locations. While there are fairly straightforward hydrodynamic explanations for the phenomenon, the results dispute the popular misconception that any bathymetry barriers on a shelf, particularly the ones associated with barrier reef islands would provide additional protection against a tsunami. These results can only be fully explained by the nonlinear wave dynamics, since linear theory (as manifested by a Green's law, for example) would assume same amplification for waves coming from deeper water, regardless of bathymetry profile experienced along the way. The increased particle velocities of higher-amplitude waves, when amplified over reef structures, may contribute to higher run-up values. The effects of non-linear wave transformation over the reef environment appear to create more energetic flows toward the coastlines, leading to larger run-up. The inundation limits can be seen in Fig. 9 from the color shading that lies inland of the pre-tsunami shoreline. It is interesting to note that the inundation limits are not a strong function of the offshore bathymetry, despite the very large variations we have implemented. It suggests that the extent of inundation may be more sensitive to the coastal geometry and topography, and the immediate near-shore bathymetry. Indirect evidence of this finding has been

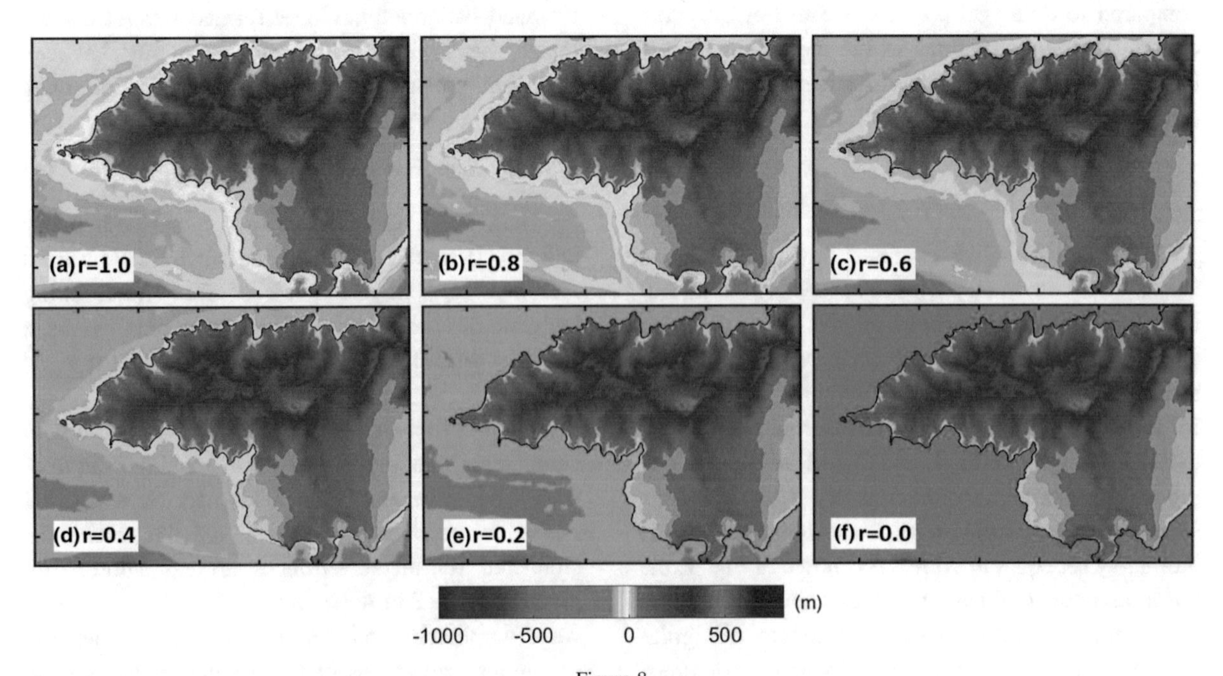

Figure 8
C grids of synthetic bathymetry generated from Eq. 3, shown for the west side of Tutuila Island. $r = 1$ gives original bathymetry; $r = 0$ gives flat shelf bathymetry from 0 m to 90 m depth

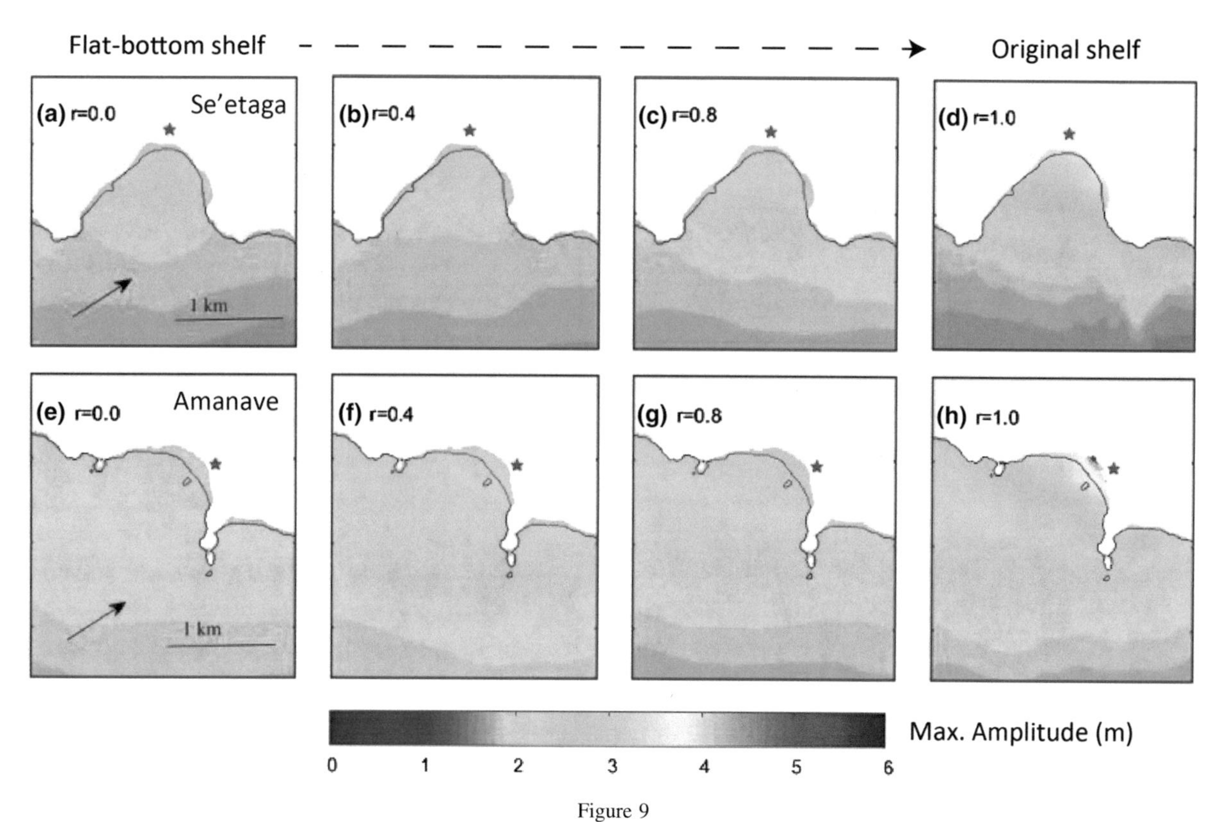

Figure 9
Simulated maximum wave amplitudes are shown for the bays near the villages of Se'etaga (top panels) and Amanave (bottom panel) for different bathymetry [$r = 0.0$ (flat-shelf), 0.4, 0.8 and 1.0 (actual)]. The pre-tsunami shoreline is shown as a thin black line. Colors inland of that line therefore show the degree of inundation. Village locations are shown as red stars. Black arrows indicate tsunami direction

documented by Wei et al. (2013), who observed very consistent inundation area predictions with fairly high variability in amplitude comparisons. Thus, hi-res topographic data (such as from LiDAR surveys) are important for accurate inundation mapping.

Simulated time series computed at the same offshore locations near Se'etaga and Amanave villages provides additional illustration and some explanation for the decreased run-up without barrier reef- and fringing reef-related bathymetric structures (Fig. 10). In contrast to the results for varying roughness, changing the bathymetry prompted significant changes in the time series. As the bathymetry is progressively removed ($r \rightarrow 0$), the first waves arrive 5–10 min earlier with generally smaller amplitudes and with shorter periods.

The time series show the significant influence of varying the bathymetry for the later tsunami waves in the two bays, due to wave resonance and shoaling. The resonance is particularly prominent in Se'etaga bay (Fig. 9a). As the bathymetry is progressively removed ($r \rightarrow 0$), the resonant amplification in the amplitude of waves 4–7 seen in Se'etaga Bay, diminishes and eventually disappears. Therefore, the resonant feature in the time series of the original bathymetry is probably related to the multiple reflections from the bathymetric features of the semi-enclosed lagoon inside the barrier reef, and not to coastal reflections. The time series also indicates that the run-up and inundation are mostly driven by the initial wave, which is the highest for all cases. Therefore, reflection and refraction are not playing significant role for the maximum run-up and inundation for the two locations. This is another indication that the nonlinear evolution of the wave over the shelf is predominantly responsible for the increased run-up for the original bathymetry.

Roeber et al. (2010) came to similar conclusion using resonant mode analysis of their numerical simulations for the 2009 Samoa tsunami around

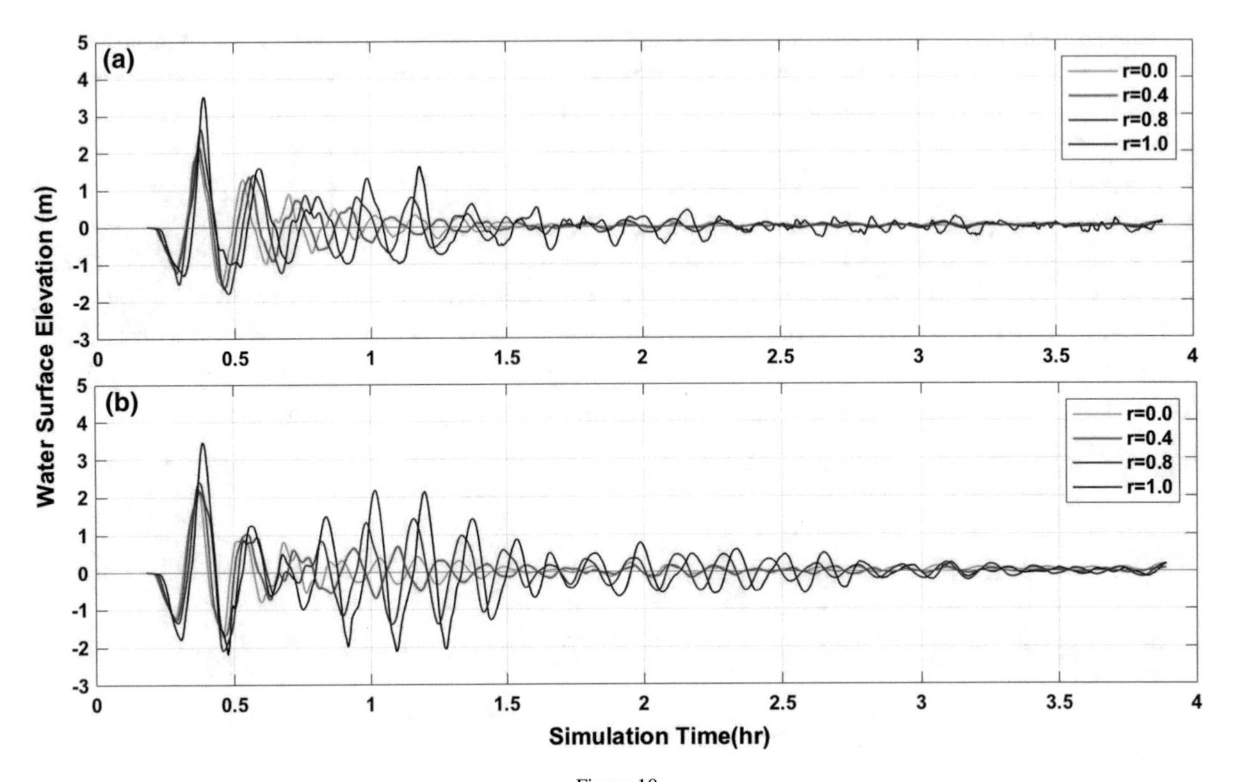

Figure 10
Comparison of water surface elevations at selected virtual tide gauges in the near-shore regions of villages of Se'etaga and Amanave. Refer to Fig. 7 for location of gauges. Water surface elevations are given for varying r in the 4 h after the fault rupture at $t = 0$

Tutuila. They found significant resonant amplifications for relatively shorter wave periods at the west coast of Tutuila, where shallow reefs provided little protection for coastal population. Our work indicates that the reef-related bathymetric structures actually increase the run-up and, therefore, tsunami impact for coastal communities there.

4. Discussion and Conclusion

Manning roughness, n, was spatially uniform in our simulations. Although almost all numerical codes can accommodate variable roughness, the use of a uniform roughness coefficient is standard practice in most current tsunami models. The main reason is the difficulty of determining the correct pattern of roughness values within the computational area. Also, the use of a lower roughness coefficient is thought to produce more conservative results (our study confirmed and quantified that notion), which is desirable for many tsunami studies. Arguably, the uniform roughness is only valid at large scales, and cannot capture the real situation of a heterogeneous reef environment. An important frontier in tsunami modeling research is whether, in pursuit of more accurate predictions, models should incorporate spatially varying n or whether, in addition, the mathematical representation of the dissipation (Eqs. 1, 2) must be reformulated.

When we remove reef-related bathymetry structures, the modelled run-up values on the coast decrease, due to non-linear wave transformation over the reef flats. A notable feature of the simulations is that varying the bathymetry has a strong impact on the wave resonance that was observed. The bathymetry plays a significant role in the constructive interference of reflected and incident waves. As the bathymetry is progressively removed, the first waves arrive earlier and the period of successive waves gets longer, consistent with faster velocities in deeper water. Interestingly, similar dynamics were noticed by Lynett (2007) for a set of one-dimensional numerical experiments with artificial bathymetry simulating the barrier and fringing reef environment. Lynett's numerical experiments showed peculiar increase of overland velocities when long obstacles were added to the one-dimensional bathymetry for weakly nonlinear waves, but not for highly nonlinear (very high amplitude) waves. While the geometry of these idealized one-dimensional experiments are different from our realistic 2-d modeling setting, the general dimensionless parameters of the 2009 Samoa tsunami wave are similar to the weakly nonlinear case of Lynett's simulations (Fig. 2). Our numerical results appear to confirm that such amplifications, wherein the reef shelf features increase the run-up, can occur in natural settings of reef islands like Tutuila. The dataset synthesized in Tables 2 and 3 in Appendix are summaries of the model results.

The role of reefs and reef–island shelf bathymetry in tsunami dynamics remains enigmatic, which echos Lynett's (2016) conclusions about the chaotic nature of flow over complex bathymetries, and the resultant challenges inherent to the deterministic modeling of such flows. Our results add to a body of literature exploring this important question. In our experiments, the reef-related bathymetry enhanced run up via non-linear tsunami transformation over reef structures. We designed our experiments to pervasively smooth and remove the reef-related small-scale bathymetry of the island shelf, rather than focusing on individual structures. To some degree, our results must depend on the proximity of the reefs to the shore. Broader reefs further from shore, such as the Great Barrier Reef example for instance, might dissipate wave energy enough to reduce onshore run up. However, given the degree of complexity we found in our simulations, it would be hazardous to conclude that reefs, and the reef–island bathymetric environment in general, provide universal protection for their inshore coastlines.

To conclude, while the impact of coral reefs on the dynamics of tsunamis impinging on tropical islands is obviously complicated and depends on many factors, our results reinforce a focus on two key issues that call for further research: first, the representation of the turbulent dissipation in terms of the governing equations and their coefficients; and second, the role of all aspects of reef–island bathymetry such as reef widths, reef types and coastal geometry on the scale of individual bays.

Acknowledgements

This publication makes use of data products provided by National Oceanic and Atmospheric Administration (NOAA), Pacific Marine Environmental Laboratory (Contribution number 4618). This publication is partially funded by the Joint Institute for the Study of the Atmosphere and Ocean (JISAO) under NOAA Cooperative Agreement NA15OAR4320063, Contribution No. 2018-0131. Hermann Fritz generously provided the post-tsunami run-up survey datasets. We are grateful to Randall Leveque, Joanne Bourgeois, Frank Gonzales, Hongqiang Zhou, Christopher Moore, Marie Eble, Diego Arcas, Lijuan Tang, Jose Borrero and the anonymous reviewer for their endless advice and help.

Appendix

These tables present MOST model output for maximum amplitude along selected villages (Fig. 1 for locations of the villages). There is a tendency that run-up increases with decreasing Manning coefficient and the model better estimates run-up (Table 2). There is also a tendency that run-up decreases with removing reefs (Table 3).

Table 2

Model, simulated run-up (m) for different Manning's roughness, n, values at 31 villages around Tutuila

Village # (Fig. 1)	Village name	Run-up field	Run-up, $n = 0.08$	Run-up, $n = 0.06$	Run-up, $n = 0.04$	Run-up, $n = 0.03$	Run-up, $n = 0.02$	Run-up, $n = 0.015$
31	LAUILI	3.36	1.35	1.58	1.78	1.80	2.12	2.35
30	AMALAU	2.50	1.46	1.49	1.52	2.12	1.55	1.72
29	AFONO1	2.18	1.92	1.98	2.06	3.00	2.20	2.45
28	AFONO2	3.80	2.04	2.11	2.21	3.50	2.41	2.68
16	AVAIO	3.70	2.81	2.95	3.12	3.15	3.59	3.98
27	MASEFAU	4.00	3.32	3.30	3.27	4.20	3.37	3.74
18	AMAUA	2.91	3.05	3.23	3.43	3.20	3.81	4.23
17	FAGAITUA	3.50	3.60	4.26	5.02	4.75	5.31	5.90
26	MASAUSI	2.79	2.46	2.48	2.50	3.15	2.56	2.84
25	SAILELE	2.95	1.99	2.01	2.01	2.25	2.00	2.23
24	AOA	2.23	1.36	1.60	1.82	2.20	2.05	2.28
19	AMOULI	3.15	2.76	3.18	3.53	3.20	4.18	4.64
23	ONENOA	3.60	2.20	2.21	2.23	2.50	2.24	2.48
21	AUASI	3.79	2.57	3.06	3.47	3.08	3.73	4.14
22	TULA	7.00	3.11	3.44	3.66	3.80	3.89	4.32
20	AUNUU	2.15	1.54	1.71	1.85	2.14	2.03	2.25
3	POLOA	13.23	6.10	6.49	7.03	7.30	7.62	10.74
4	AMANAVE	7.74	4.69	4.85	4.93	5.25	6.32	5.79
2	FAGAILII	6.00	4.77	4.78	4.84	4.90	5.33	7.90
5	FAILOLO	6.54	2.87	2.79	2.69	2.70	3.29	3.14
6	AGUGULU	6.12	1.93	2.11	2.35	2.50	2.23	2.43
7	UTUMEA	4.51	3.05	3.15	3.76	4.00	4.52	5.68
8	SE'ETAGA	5.69	4.98	5.09	5.50	5.57	6.29	5.19
1	FAGAMALO	6.39	2.88	3.22	3.37	3.30	3.30	15.00
9	NUA	4.09	3.61	3.86	4.15	4.20	3.85	3.86
10	AFAO	5.89	3.46	3.83	4.10	4.20	3.98	4.74
11	ASILII	6.81	5.05	5.30	5.66	6.00	6.94	5.74
12	AMALUIA	5.39	5.14	5.26	5.26	5.20	5.60	6.21
13	LEONE	2.80	1.26	1.89	2.58	3.00	2.66	2.01
14	FAGATELE	4.92	3.12	3.21	3.28	3.30	3.33	3.33

Table 3

Model, simulated run-up for different r values representing varying bathymetry at 31 villages around Tutuila

Village # (Fig. 1)	Villages	Run-up field	Run-up, $r = 1$	Run-up, $r = 0.8$	Run-up, $r = 0.6$	Run-up, $r = 0.4$	Run-up, $r = 0.2$	Run-up, $r = 0.0$
31	LAUILI	3.36	1.80	1.13	1.05	0.98	0.94	0.89
30	AMALAU	2.50	2.12	2.05	1.89	1.76	1.65	1.54
29	AFONO2	2.18	3.00	2.34	2.30	2.33	2.24	2.17
28	AFONO1	3.80	3.50	2.70	2.79	2.75	2.77	2.72
16	AVAIO	3.70	3.15	2.46	2.25	2.25	2.27	2.30
27	MASEFAU	4.00	4.20	2.80	2.49	2.58	2.61	2.70
18	AMAUA	2.91	3.20	1.93	1.86	1.79	1.75	1.67
17	FAGAITUA	3.50	4.75	3.14	2.78	2.86	2.88	2.95
26	MASAUSI	2.79	3.15	2.15	2.10	1.99	1.95	1.98
25	SAILELE	2.95	2.25	1.55	1.63	1.47	1.49	1.43
24	AOA	2.23	2.20	1.90	1.76	1.72	1.69	1.63
19	AMOULI	3.15	3.20	3.67	2.87	2.93	2.87	2.80
23	ONENOA	3.60	2.50	1.69	1.41	1.42	1.40	1.33
21	AUASI	3.79	3.08	2.24	2.16	2.02	2.02	1.88
22	TULA	7.00	3.80	2.38	2.22	2.07	2.07	1.98
20	AUNUU	2.15	2.14	2.07	1.90	1.78	1.78	1.67
3	POLOA	13.23	7.30	5.32	5.12	4.79	4.46	4.14
4	AMANAVE	7.74	5.25	3.29	3.06	2.85	2.73	2.59
2	FAGAILII	6.00	4.90	4.75	4.36	4.08	3.82	3.56
5	FAILOLO	6.54	2.70	2.11	2.07	2.10	2.02	1.95
6	AGUGULU	6.12	2.50	1.93	1.99	1.96	1.98	1.94
7	UTUMEA	4.51	4.00	3.12	2.86	2.86	2.88	2.92
25	SE'ETAGA	5.69	5.57	3.71	3.30	3.43	3.46	3.59
1	FAGAMALO	6.39	3.30	1.99	1.92	1.84	1.81	1.72
9	NUA	4.09	4.20	2.78	2.46	2.53	2.54	2.17
10	AFAO	5.89	4.20	2.87	2.80	2.65	2.60	2.64
11	ASILII	6.81	6.00	4.14	4.34	3.91	3.96	3.83
12	AMALUIA	5.39	5.20	4.50	4.15	4.06	4.01	3.86
13	LEONE	2.80	3.00	3.44	2.69	2.75	2.69	2.62
14	FAGATELE	4.92	3.30	2.23	1.85	1.88	1.85	1.75

References

Baba, T., Mleczko, R., Burbidge, D., Cummins, P. R., & Thio, H. K. (2008). The effect of the Great Barrier reef on the propagation of the 2007 Solomon islands tsunami recorded in Northeastern Australia. *Pure and Applied Geophysics, 165*(11), 2003–2018.

Baptista, M. A., Miranda, P. M. A., Miranda, J. M., & Victor, L. M. (1998). Constrains on the source of the 1755 Lisbon tsunami inferred from numerical modelling of historical data. *Journal of Geodynamics, 25,* 159–174.

Borrero, J.C., Greer, S.D., Lebreton, L., & Fritz, H.M. (2011). Field surveys and numerical modelling of the 2009 South Pacific and 2010 Mentawai Islands tsunamis. In *Proceedings of Coasts and Ports Conference*, Perth, Australia, September 28–30, 2011.

Bricker, J. D., Gibson, S., Takagi, H., & Imamura, F. (2015). On the need for larger Manning's roughness coefficients in depth-integrated tsunami inundation models. *Coastal Engineering Journal*. https://doi.org/10.1142/S0578563415500059.

Chow, V. T. (1959). *Open-channel hydraulics*. New York: McGraw-Hill Book Company.

Dilmen, D. I., Titov, V. V., & Roe, H. G. (2015). Evaluation of the relationship between coral damage and tsunami dynamics; case study: 2009 Samoa tsunami. *Pure and Applied Geophysics, 172*(12), 3557–3572. https://doi.org/10.1007/s00024-015-1158-y.

Fernando, H. J. S., McCulley, J. L., Mendis, S. G., & Perera, K. (2005). Coral poaching worsens tsunami destruction in Sri Lanka. *Eos Transactions American Geophysical Union, 86*(33), 301–304. https://doi.org/10.1029/2005eo330002.

Fernando, H. J. S., Samarawickrama, S. P., Balasubramanian, S., Hettiarachchib, S. S. L., & Voropayeva, S. (2008). Effects of porous barriers such as coral reefs on coastal wave propagation. *Journal of Hydro-environment Research, 1,* 187–194.

Ford, M., Becker, J. M., Merrifield, M. A., & Song, Y. T. (2013). Marshall Islands fringing reef and atoll lagoon observations of the Tohoku tsunami. *Pure and Applied Geophysics*. https://doi.org/10.1007/s00024-013-0757-8.

Fritz, H. M., Borrero, J. C., Synolakis, C. E., Okal, E. A., Weiss, R., Titov, V., et al. (2011). Insights on the 2009 South Pacific tsunami in Samoa and Tonga from field surveys and numerical simulations. *Earth-Science Reviews, 107,* 66–75.

Fujima, K. (2006). Effect of a submerged bay-mouth breakwater on tsunami behavior analyzed by 2D/3D hybrid model simulation. *Natural Hazards, 39*(2), 179–193.

Gelfenbaum, G., Apotsos, A., Stevens, A. W., & Jaffe, B. (2011). Effects of fringing reefs on tsunami inundation: American Samoa. *Earth-Science Reviews, 107,* 12–22.

Imamura, F., Goto, K., & Ohkubo, S. (2008). A numerical model for the transport of a boulder by tsunami. *Journal of Geophysical Research.* https://doi.org/10.1029/2007JC004170.

Jaffe, B. E., Gelfenbaum, G., Buckley, M. L., Steve, W., Apotsos, A., Stevens, A. W., & Richmond, B. M. (2010). The limit of inundation of the September 29, 2009, Tsunami on Tutuila, American Samoa: *U.S. Geological Survey Open File Report,* 2010–1018.

Kraines, S. B., Yanagi, T., Isobe, M., & Komiyama, H. (1998). Wind-wave driven circulation on the coral reef at Bora Bay. *Miyako Island, Coral Reefs, 17,* 133–143.

Kunkel, M., Hallberg, R. W., & Oppenheimer, M. (2006). Coral reefs reduce tsunami impact in model simulations. *Geophysical Research Letters.* https://doi.org/10.1029/2006GL027892.

Leschka, S., Kongko, W., & Larsen, O. (2009a). On the influence of nearshore bathymetry data quality on tsunami runup modeling, part I: Bathymetry. In *Proc. of the 5th International Conference on Asian and Pacific Coasts (APAC 2009)* (Vol. 1, pp. 151–156). Singapore: Soon Keat Tan, Zenhua Huang.

Levin, B. W., & Nosov, M. A. (2016). Propagation of a tsunami in the ocean and its interaction with the coast. In Levin, B. W., & Nosov M. A. (Ed.), *Physics of tsunamis* (pp. 311–358). Springer. https://doi.org/10.1007/978-3-319-24037-4.

Liu, P. L.-F., Wang, X., & Salisbury, A. J. (2009). Tsunami hazard and early warning system in South China Sea. *Journal of Asian Earth Sciences, 36*(1), 2–12.

Lynett, P. J. (2007). Effect of a shallow water obstruction on long wave runup and overland flow velocity. *Journal of Waterway, Port, Coastal, and Ocean Engineering, 133*(6), 455–462. https://doi.org/10.1061/(ASCE)0733-950X(2007)133:6(455).

Lynett, P. J. (2016). Precise prediction of coastal and overland flow dynamics: a grand challenge or a fool's errand. *Journal of Disaster Research.* https://doi.org/10.20965/jdr.2016.p0615.

NTHMP, National Tsunami Hazard Mitigation Program. (2012). *Proceedings and Results of the 2011 NTHMP Model Benchmarking Workshop.* Boulder: U.S. Department of Commerce/NOAA/NTHMP, (NOAA Special Report), 436.

Nunes, V., & Pawlak, G. (2008). Observations of bed roughness of a coral reef. *Journal of Coastal Research, 24*(2B), 39–50.

Okal, E. A., Fritz, H. M., Synolakis, C. E., Borrero, J. C., Weiss, R., Lynett, P. J., et al. (2010). Field survey of the Samoa tsunami of 29 September 2009. *Seismological Research Letters, 81*(4), 577–591. https://doi.org/10.1785/gssrl.81.4.577.

Roeber, V., Yamazaki, Y., & Cheung, K. F. (2010). Resonance and impact of the 2009 Samoa tsunami around Tutuila, American Samoa. *Geophysical Research Letters.* https://doi.org/10.1029/2010GL044419.

Synolakis, C. E., Bernard, E. N., Titov, V. V., Kânoğlu, U., González, F. I. (2008). Validation and verification of tsunami numerical models. *Pure and Applied Geophysics, 165* (11–12), 2197–2228.

Tang, L., Titov, V. V., & Chamberlin, C. D. (2009). Development, testing, and applications of site-specific tsunami inundation models for real-time forecasting. *Journal of Geophysical Research, 114,* C12025. https://doi.org/10.1029/2009jc00547.

Titov, V. V. (1997). Numerical modeling of long wave runup, *PhD Thesis*, University of Southern California, p. 141.

Titov, V. V. (2009). Tsunami forecasting. In A. N. Bernard & A. R. Robinson (Eds.), The Sea. *Tsunamis*, Chap. 12 (Vol. 15, pp. 371–400). Cambridge, MA: Harvard University Press.

Titov, V., & González, F. I. (1997). Implementation and testing of the Method of Splitting Tsunami (MOST) model. *NOAA Tech. Memo.* ERL PMEL-112 (PB98-122773), NOAA/Pacific Marine Environmental Laboratory, Seattle, WA, p. 11.

Titov, V. V., Gonzales, F. I., Bernard, E. N., Eble, M. C., Mofjeld, H. O., Newman, J. C., et al. (2003a). Real-time tsunami forecasting: Challenges and solutions. *Natural Hazards, 35*(1), 35–41.

Titov, V.V., Gonzalez, F. I., Mofjeld, H. O., & Venturato, A. J. (2003b). NOAA TIME Seattle tsunami mapping project: Procedures, data sources, and products. *NOAA Tech. Memo.* OAR PMEL-124, NTIS: PB2004-101635, p. 21.

Titov, V. V., Kânoğlu, U., & Synolakis, C. (2016). Development of MOST for real-time tsunami forecasting. *Journal of Waterway, Port, Coastal, and Ocean Engineering, 142*(6), 03116004. https://doi.org/10.1061/(asce)ww.1943-5460.0000357.

Wei, Y., Chamberlin, C., Titov, V., Tang, L., & Bernard, E. N. (2013). Modeling of the 2011 Japan tsunami—Lessons for near-field forecast. *Pure and Applied Geophysics, 170*(6–8), 1309–1331. https://doi.org/10.1007/s00024-012-0519-z.

(Received January 31, 2017, revised December 27, 2017, accepted January 4, 2018)

Pure Appl. Geophys.

https://doi.org/10.1007/s00024-017-1721-9

Pure and Applied Geophysics

The "Tsunami Earthquake" of 13 April 1923 in Northern Kamchatka: Seismological and Hydrodynamic Investigations

Amir Salaree[1] and Emile A. Okal[1]

Abstract—We present a seismological and hydrodynamic investigation of the earthquake of 13 April 1923 at Ust'-Kamchatsk, Northern Kamchatka, which generated a more powerful and damaging tsunami than the larger event of 03 February 1923, thus qualifying as a so-called "tsunami earthquake". On the basis of modern relocations, we suggest that it took place outside the fault area of the mainshock, across the oblique Pacific-North America plate boundary, a model confirmed by a limited dataset of mantle waves, which also confirms the slow nature of the source, characteristic of tsunami earthquakes. However, numerical simulations for a number of legitimate seismic models fail to reproduce the sharply peaked distribution of tsunami wave amplitudes reported in the literature. By contrast, we can reproduce the distribution of reported wave amplitudes using an underwater landslide as a source of the tsunami, itself triggered by the earthquake inside the Kamchatskiy Bight.

Key words: Tsunami, Kamchatka, landslides, simulations.

1. Introduction

In the Winter and Spring of 1923, the eastern coast of Kamchatka was the site of a series of major earthquakes, two of which generated devastating tsunamis. Their effects, investigated in detail in the years following the events by Troshin and Diagilev (1926) and later Meniaĭlov (1946) were summarized by Soloviev and Ferchev (1961). The main shock (hereafter MS) occurred on 03 February 1923 at 16:01 GMT, (03:01 local time on the 4th), in a very sparsely populated area along the Kronotskiy Bight (Fig. 1). It provoked serious damage both in Petropavlovsk-Kamchatskiy to the south, and Ust'-Kamchatsk in the north, and was followed by a powerful tsunami reaching run-up heights of 6 m, which washed away a warehouse at Ust'-Kamchatsk (Soloviev and Ferchev 1961). Incidentally, the tsunami motivated what is believed to be the first far-field warning by Jaggar (1930) at the Hawaii Volcano Observatory, which unfortunately was not heeded, leading to considerable damage and one casualty on the Island of Hawaii (Okal 2011).

Among its numerous aftershocks, the event of 13 April (O.T. 15:31 GMT) near Ust'-Kamchatsk (hereafter the UK event) stands out as featuring properties typical of a "tsunami earthquake". We recall that the name *tsunami earthquake* was introduced by Kanamori (1972) to characterize events whose tsunami is significantly larger than expected from their seismic magnitudes, especially conventional ones. Fukao (1979) and later Tanioka et al. (1997) and Polet and Kanamori (2000) have explained their properties as due to anomalously slow rupture along the fault, resulting in red-shifting of the seismic spectrum towards lower frequencies. Another, mechanically different, scenario can involve the triggering by the earthquake of a major underwater landslide, which can act as ancillary tsunami generator, a classical example being the Papua New Guinea disaster of 17 July 1998 (Synolakis et al. 2002); these are not generally considered "tsunami earthquakes". The catastrophic 1946 Aleutian tsunami is believed to have featured both an anomalously slow seismic source, and a locally devastating landslide (Okal et al. 2003; López and Okal 2006).

As hinted by the traditional magnitudes ("M_{PAS}") assigned by Gutenberg and Richter (1954) to the MS (8.4) and UK (7.2) events, the latter is clearly a smaller source from a seismological standpoint. In

[1] Department of Earth and Planetary Sciences, Northwestern University, Evanston, IL 60208, USA. E-mail: amir@earth.northwestern.edu; emile@earth.northwestern.edu

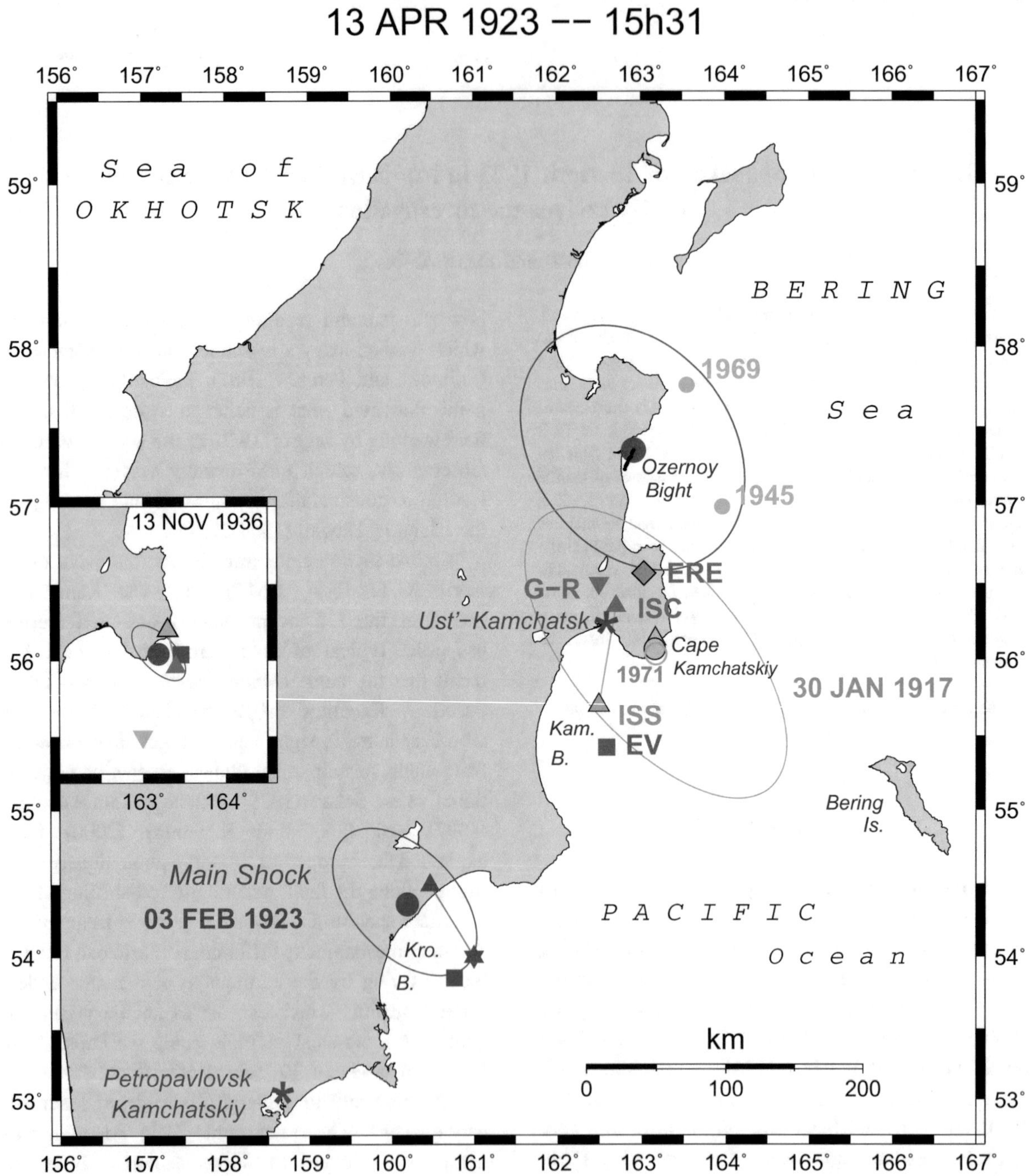

Figure 1

Relocation of the 1923 Ust'-Kamchatsk earthquake. Our preferred relocation is shown as the red dot, with associated Monte-Carlo ellipse, and small black dots identifying the negligible moveout of the epicenter as a function of constrained depth; the triangles are the initial ISS location (light gray) and the recent ISC relocation (in blue); the inverted green triangle is Gutenberg and Richter's (1954) solution, the brown square Engdahl and Villaseñor's (2002) Centennial Catalog epicenter, and the magenta diamond E.R. Engdahl's recent solution (pers. comm., 2017). For reference, we show epicentral estimates for the 1923 main shock (in dark gray with the same symbols), and for the 1917 event at Cape Kamchatskiy (in light gray). Also shown are the epicenters of the events at Ozernoy (1945 and 1969; solid green dots) and Cape Kamchatskiy (1971; open green dot). Kam. B.: Kamchatskiy Bight; Kro. B.: Kronotskiy Bight. To avoid clutter, the results for the 1936 Cape Kamchatskiy event are shown in the inset at left, on the same scale and with the same symbols as for the UK earthquake, but displaced 6° in longitude. Note the total identity of the 1917, 1936 and 1971 epicenters

Ust'-Kamchatsk, it caused only marginally more damage than the main shock, despite a clearly shorter epicentral distance. Yet, the tsunami from the UK event was significantly stronger than that of the mainshock. It ran up 11 m at Ust'-Kamchatsk, killing at least 23 people and totally destroying several canneries and other infrastructure. According to Zayakin and Luchinina (1987), the maximum run-up (20–30 m) took place in the vicinity of the First Creek ("Pervaya Rechka"), and of the cannery Nichiro which was completely destroyed, one of its cutters being deposited on a terrace 20 m high and 1 km inland. This location is estimated to be 27 km WSW of Ust'-Kamchatsk (note that their Fig. 3.10.2, p. 20, like all others, bears no scale). Inundation distances reached 7 km along the river valley at the mouths of the Kamchatka River.

It is worth noting that no mention is made of a tsunami following the aftershock on 24 February 1923 (A12 in Table 1), to which Gutenberg and Richter (1954) assigned a larger magnitude ($M_{PAS} = 7.4$) than to the UK event, thus supporting the anomalous character of the latter. In this respect, the sequence of 1923 Kamchatka events is comparable to the 1932 Mexican series (Okal and Borrero 2011): a mainshock with a significant tsunami (03 February 1923/03 June 1932), a strongest aftershock without a tsunami (24 February 1923/18 June 1932), and a very powerful tsunami during a clearly smaller aftershock (13 April 1923/22 June 1932), the latter being a typical tsunami earthquake.

These remarks motivate a detailed seismological and hydrodynamic reassessment of the UK event of 13 April 1923, which is the subject of this paper.

2. *Relocation*

We relocated the 1923 Kamchatka sequence, starting with the foreshocks of 02 February, and including all regional activity for the 6 months following the main shock. We used arrival times listed in the International Seismological Summary (ISS) and the iterative, interactive technique of Wysession

Table 1

Relocations performed in this study

No.	Date	Time (GMT)	Epicenter (°N)	Epicenter (°E)	Depth (km)	Stations Read	Stations Used	r.m.s. (s)	Magnitude
			1923 series						
F1	02 FEB (033) 1923	01:06:40.6	53.99	161.60	10	29	22	3.74	6.7 ISC
F2	02 FEB (033) 1923	05:07:41.0	54.13	160.88	10	54	45	4.71	7.2 PAS
M3	**03 FEB (034) 1923**	**16:01:47.9**	**54.36**	**160.19**	**10**	**79**	**66**	**4.51**	**8.4 PAS**
A4	03 FEB (034) 1923	17:40:59.1	53.85	162.99	10	9	7	5.06	
A5	03 FEB (034) 1923	18:42:56.8	53.26	160.76	10	27	26	4.20	
A6	03 FEB (034) 1923	18:50:46.8	54.17	163.88	10	11	10	3.60	
A7	05 FEB (036) 1923	22:23:34.8	54.07	161.70	10	8	8	3.54	
A8	11 FEB (042) 1923	22:45:54.9	55.76	161.21	103[a]	16	16	3.37	6.2 ISC
A9	12 FEB (043) 1923	01:58:41.7	54.58	162.10	10	30	25	3.65	6.4 ISC
A10	15 FEB (046) 1923	22:40:36.8	51.24	150.05	433[a]	4	4	–	
A11	18 FEB (049) 1923	23:39:54.3	56.31	161.03	21[b]	9	7	2.44	
A12	24 FEB (055) 1923	07:34:40.8	55.58	162.54	10	52	45	4.71	7.4 PAS
UK13	**13 APR (103) 1923**	**15:31:06.3**	**57.35**	**162.91**	**10**	**40**	**31**	**3.34**	**7.2 PAS**
A14	23 MAY (143) 1923	22:37:08.3	53.83	161.09	10	45	34	3.96	6.4 ISC
			Other events						
	30 JAN (031) 1917	02:45:37.5	56.07	163.16	10	47	38	3.74	8.1 PAS
	13 NOV (318) 1936	12:31:30.0	56.04	163.19	10	207	198	3.14	7.2 PAS
	15 APR (105) 1945	02:35:30.2	56.99	163.96	69[a]	112	104	3.20	7.0 PAS

F foreshock, *M* mainshock, *A* aftershock, *UK* Tsunami earthquake

[a]Floated depth: all other depths constrained

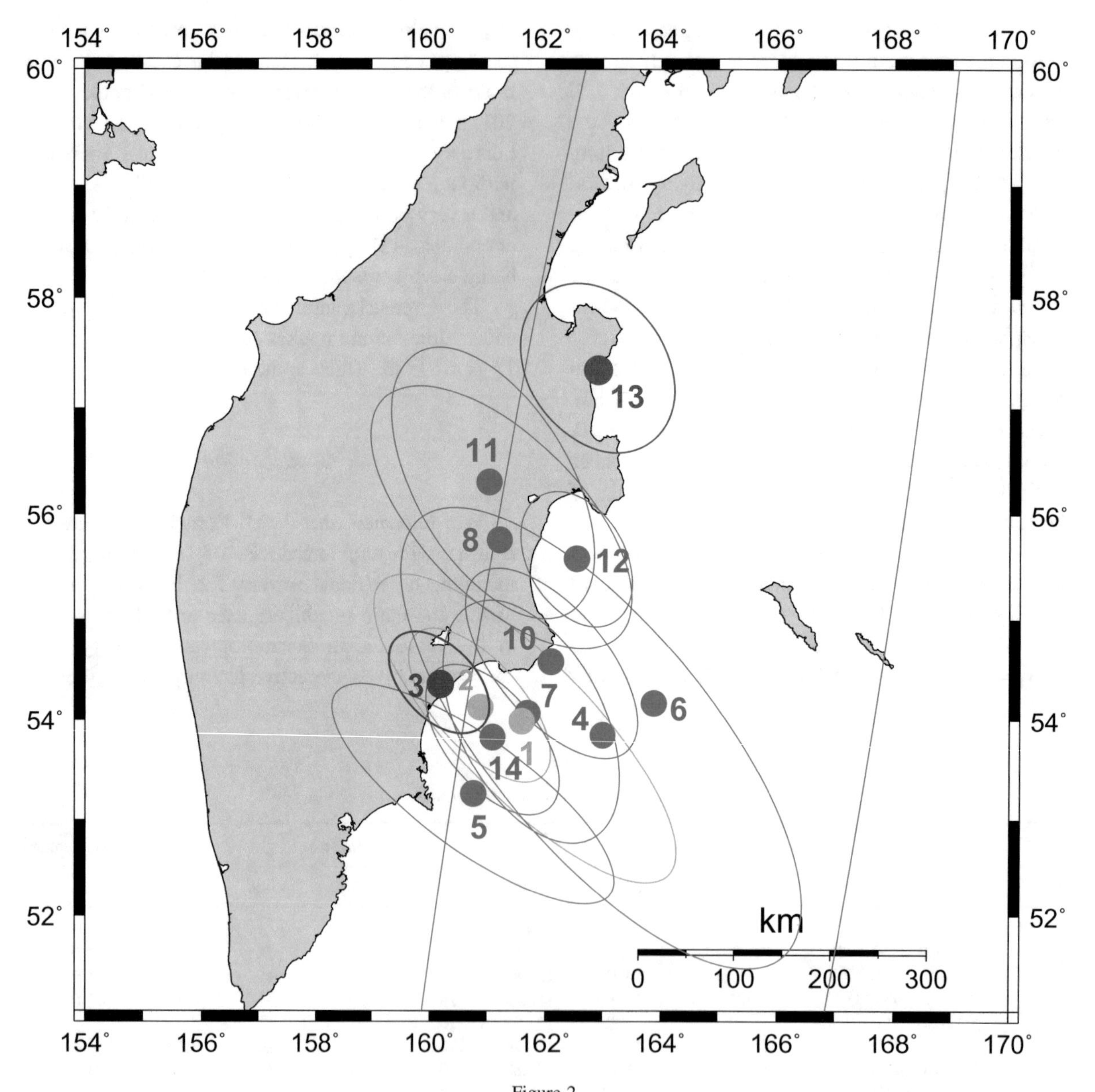

Figure 2
Relocation of the 1923 series. Individual events are shown with their numbers keyed to Table 1, and associated Monte Carlo ellipses. Foreshocks in green, Mainshock in dark gray, UK event in red. Event 10, relocating to the the Sea of Okhotsk slab, is not shown

et al. (1991), which includes a Monte Carlo algorithm injecting Gaussian noise into the dataset; for events in 1923, we use a standard deviation of the noise $\sigma_G = 8$ s. Results are listed in Table 1 and on Figs. 1 and 2. With the exception of events A8, A10 and A11, no floating depth relocations converged, and hypocentral depths were thus constrained to 10 km. We note that Event A10, initially proposed by the ISS at 55°N, 162.5°E, relocates to the Sea of Okhotsk slab, at a depth of 443 km; while the event is very poorly constrained (with only 4 arrival times), there is no reason to include it as part of the aftershock sequence, and we exclude it from Fig. 2.

As shown on Fig. 2, aftershocks A5, A14, A7, A10 and A12, reasonably well located with semi-minor axes of their Monte Carlo ellipses not exceeding 50 km in the direction parallel to the trench, would suggest a rupture zone $\sim$260 km long, to which Event A11 could also belong, and which also includes the two foreshocks (F1 and F2; shown in green on Fig. 2). We note that the largest aftershock (Event A12; $M_{PAS} = 7.4$) would, in this context, map at the edge of the fault zone, thus supporting the model proposed for smaller, intraplate earthquakes by Ebel and Chambers (2016). Event A8, relocating at a floated intermediate depth (103 km), may be triggered by stress transfer inside the slab. Events A4 and A6 are too poorly located to draw any conclusion.

Of particular interest is the relocation of the tsunami earthquake of 13 April 1923. The original ISS epicenter (55.7°N, 162.5°E) would fit at the end of the rupture area defined above, but the solution relocated as part of the GEM-ISC project (Storchak et al. 2015) moves inland, 18 km NNE from Ust'-Kamchatsk (56.36°N; 162.70°E), and only 20 km from Gutenberg and Richter's (1954) estimate ($56\frac{1}{2}$°N, $162\frac{1}{2}$°E; rounded off to the nearest 1/2°). By contrast, Engdahl and Villaseñor's (2002) location, part of their Centennial Catalogue, maps about 35 km south of the initial ISS epicenter, and more than 105 km from the revised ISC-GEM one. Our own solution, 57.35°N, 162.91°E, locates about 110 km north of the ISC-GEM epicenter, and its Monte Carlo ellipse encompasses most of Ozernoy Bight. Finally, a new solution by E.R. Engdahl (pers. comm., 2017) relocates to 56.56°N, 163.03°E, 30 km NNE of the GEM-ISC epicenter, and misses our ellipse by only 3 km. The disparity between modern relocations based a priori on the same dataset reflects its poor quality and probably results from the elimination of different sets of poor fits by various authors. Their choice remains to some extent subjective in interactive algorithms such as Wysession et al.'s (1991) where incompatible solutions are removed by the operator in an interactive process, and possibly inconsistent among fully automated ones. The situation is aggravated in the case of "tsunami earthquakes" by the slow nature of the source, which causes emergent arrivals in the near and far fields, with inconsistent picks on the part of different station operators, as described by Bell et al. (2014) in the case of the 1947 Hikurangi earthquakes in New Zealand. Finally, note that relocation at a different constrained depth affects our epicenter only marginally (a total of 16 km between depths of 5 and 100 km; solid dots on Fig. 1).

The bottom line of these relocation efforts is that the UK event most probably occurred north of the rupture zone of the MS, being triggered by Coulomb stress to the vicinity of the triple junction near Cape Kamchatskiy, or even farther north, in the Ozernoy Bight, in the direction of the epicenter of the large earthquake of 22 November 1969 (57.76°N, 163.54°E; $M_S = 7.3$). That event was given a mostly strike-slip mechanism ($\phi = 46°$; $\delta = 71°$; $\lambda = 38°$) by Fedotov et al. (1973), but a low angle thrust one ($\phi = 225°$; $\delta = 16°$; $\lambda = 97°$) by Cormier (1975), for which the auxiliary plane is close to the strike-slip plane. As detailed in Sect. 4, it generated a substantial tsunami along Ozernoy Bight (Fedotov et al. 1973; Martin et al. 2008), but recorded in Ust'-Kamchatsk with an amplitude of only 20 cm.

Note that the area of Cape Kamchatskiy was the site of two major events, on 30 January 1917 and 15 December 1971, respectively. The latter ($M_S = 7.8$) is well located at 56.04°N, 163.17°E, and was studied by Gusev et al. (1975) and Okal and Talandier (1986). Its mechanism ($\phi = 276°$; $\delta = 21°$; $\lambda = 158°$) can be interpreted as predominantly strike-slip on a shallow dipping plate, expressing the extremely oblique subduction of the Pacific plate under the Bering Sea along the westernmost part of the Aleutian-Commander island chain. Okal and Talandier (1986) obtained a moment of 6×10^{27} dyn-cm from the modeling of surface waves at mantle periods. The earthquake generated a moderate tsunami, reaching an amplitude of 47 cm on the maregraph at Ust'-Kamchatsk (Gusev et al. 1975). The 1917 event was given a magnitude $M_{\mathrm{PAS}} = 8.1$ by Gutenberg and Richter (1954). Modern relocation efforts (56.14°N, 163.17°E, ISC-GEM; 56.07°N, 163.16°E, this study) suggest a location identical to that of the 1971 event, but no tsunami was reported, despite the occurrence of the event in daylight (02:45 GMT or $\sim$13:38 solar time). In addition, a smaller earthquake took place on 13 November 1936, which we relocated at essentially the same location (56.04°N; 163.19°E; inset on Fig. 1). This event was assigned $M_{PAS} = 7.2$ by Gutenberg and Richter

(1954). An instance of anomalous wave action was described following the 1936 shock, but it took place at night during stormy weather, leading Soloviev and Ferchev (1961) to doubt its interpretation as a tsunami. Figure 1 shows a total identity of epicenters for the 1917, 1936 and 1971 events.

Finally, we also examined the earthquake of 15 April 1945 ($M_{PAS} = 7.0$) which Bourgeois et al. (2006) proposed as the origin of tsunami deposits in the southern Ozernoy Bight. Even though the ISS entry mentions “Pasadena suggests deep focus”, Gutenberg and Richter (1954) list it as shallow; a floating depth relocation does converge on (56.99°N, 163.96°E; 69 km), but the residual for constrained solutions is practically insensitive to hypocentral depth, and there is no reason to assume that the event is not shallow, especially given the absence of modern seismicity deeper than 50 km in the area.

In addition to the arrival times listed by the ISS and used in the various relocation efforts mentioned above, another constraint on the location of the UK event could be macroseismic data, i.e., the intensities at which the earthquake was felt. According to descriptions reported by Zayakin and Luchinina (1987), significant destruction took place in Ust'-Kamchatsk, suggesting local intensities of 8–9 on the MSK scale, as opposed to, e.g., 6–7 in 1971. Distances from Ust'-Kamchatsk to the various proposed epicenters are shorter for solutions in the Kamchatskiy Peninsula (19 km to the ISC-GEM epicenter, 49 km to E.R. Engdahl's new solution, 43 and 45 km to the 1917 and 1971 epicenters, 50 km to the closest point of our Monte Carlo ellipse) than to the solutions inside the rupture area of the mainshock in the Gulf of Kamchatka (58 km to the ISS solution and 90 km to the Centennial Catalog's); however, these figures are not sufficiently different to allow ruling out any of the epicenters on this basis. Further use of this approach is hampered by the fact that macroseismic data is only available from one site—Ust'-Kamchatsk—and that the probable epicentral locations are located in an area which did not support any population in 1923 (and still does not for the most part).

In conclusion, in view of the results of modern relocations (ISC-GEM, this study, E.R. Engdahl), we regard as very improbable that the UK earthquake of 13 April 1923 was a genuine aftershock of Event M3 at the northern end of its rupture zone. It could have occurred at the same location as in 1917 and 1971, but with a moment estimated at half that of the 1917 event (see Sect. 3). It would be unlikely that an earthquake of that size would have followed the 1917 event at the same location, except in an aftershock context, which would then violate Utsu's (1970) law, predicting a difference of 1.1 magnitude units (as opposed to 0.2 in the present case) between a mainshock of the size of the 1917 earthquake and its maximum aftershock.

Rather, we propose that the UK event was displaced to the north, either inside the Kamchatskiy Peninsula, as suggested by the ISC-GEM, Engdahl, and for that matter, G–R solutions, or even at the Ozernoy Bight, in the southern part of our Monte Carlo ellipse.

We also note that Lake Nerpichye, which occupies a large fraction of the Kamchatskiy Peninsula to the north of Ust'-Kamchatsk, became salinized after the April 1923 event and remained brackish for several years (Gorin and Chebanova 2011). This suggests an invasion of seawater during the tsunami, with some resilience which could have been helped by coseismic deformation of the boundaries of the lake, as would be expected from an earthquake source located under the Kamchatskiy Peninsula, and possibly under the lake itself.

3. *Estimates of Seismic Moment*

In this section, we obtain seismic moments at mantle wave periods for the principal events of the 1923 sequence, as well as for the 1969 and 1945 Ozernoy Bight, and 1917 Cape Kamchatskiy earthquakes. Table 2 lists the available waveforms. For such historical events, and given the small number of records, it is impossible to invert a moment tensor, and we simply compute a mantle magnitude $M_c = \log_{10} M_0 - 20$ (M_0 in dyn cm), corrected for a probable focal mechanism (Okal and Talandier 1989). For earthquakes MS3 (mainshock), F2 and A12, we use a geometry representative of the subduction of the Pacific plate under Kamchatka ($\phi = 210°$, $\delta = 20°$, $\lambda = 90°$); for the other events, we also use the mechanisms published for the 1971 earthquake

Table 2

Seismic records used in moment estimates

Station	Code	Instrument	Distance (°)	Azimuth (°)	Phases used
		Cape Kamchatskiy, 30 JAN 1917			
Göttingen, Germany	GTT	Wiechert	70.3	343	R_1
		Main foreshock (F2), 02 FEB 1923			
Göttingen, Germany	GTT	Wiechert	72.6	342	R_1,G_1
		Main shock (M3), 03 FEB 1923			
Cape Town, South Africa	CTO	Milne Shaw	146.4	292	G_1
De Bilt, The Netherlands	DBN	Golitsyn	71.7	344	G_1,G_2
Göttingen, Germany	GTT	Wiechert	71.4	341	G_1
La Paz, Bolivia	LPZ	Mainka	127.1	63	G_1
Strasbourg, France	STR	Wiechert	74.7	341	R_1
Uppsala, Sweden	UPP	Wiechert	62.1	340	R_1,G_1
		Main aftershock (A12), 24 FEB 1923			
Göttingen, Germany	GTT	Wiechert	70.7	342	R_1,G_1
		Tsunami Earthquake (UK13), 13 APR 1923			
De Bilt, The Netherlands	DBN	Golitsyn	69.0	346	R_1,G_1
		Ozernoy Bight 22 NOV 1969			
L'Aquila, Italy	AQU	WWSSN LP	77.0	338	R_4,G_4
Mundaring, Western Australia	MUN	WWSSN LP	97.2	219	R_1,G_1,G_2
Tucson, Arizona	TUC	WWSSN LP	61.0	75	R_3
Windhoek, Namibia	WIN	WWSSN LP	137.3	311	R_3,G_3

($\phi = 276°$, $\delta = 21°$, $\lambda = 158°$; Okal and Talandier, 1986) and for the 1969 Ozernoy shock ($\phi = 225°$, $\delta = 16°$, $\lambda = 97°$, Cormier, 1975; and $\phi = 46°$, $\delta = 71°$, $\lambda = 38°$, Fedotov et al. 1973).

- For the mainshock (MS3; Fig. 3a–c), we were able to gather records on mechanical instruments at CTO, LPZ, UPP, STR and GTT, and on the Golitsyn electromagnetic seismograph at DBN (including a second passage G_2). This dataset suggests a moment of 5×10^{28} dyn cm at the longest available periods ($T > 170$ s); Fig. 3); this value is slightly larger than proposed by Kanamori (1977) on the basis of the aftershock area (3.7×10^{28} dyn cm), but in agreement with our earlier estimate (Okal 1992) based only on the UPP records (5.5×10^{28} dyn cm). Note also that it predicts a fault length of 225 km under Geller's (1976) scaling laws, in good agreement with the proposed distribution of aftershocks (260 km; Fig. 2).
- For the foreshock (F2, 02 FEB 1923, 05:07 GMT), the Love and Rayleigh wavetrains recorded at GTT are compatible with the low-angle thrust geometry and a moment of 2×10^{27} dyn cm, at mantle frequencies (5–6 mHz).
- In the case of the main aftershock (A12, 24 FEB 1923), the only records available (the GTT Wiecherts) suggest a slightly rotated subduction mechanism ($\phi = 225°$, $\delta = 20°$, $\lambda = 90°$) with a moment of only 8×10^{26} dyn cm.
- The lone record obtained for the 1917 earthquake (R_1 at GTT; the EW record holding the Love wave being too faint to process) cannot put constraints on its mechanism, but tentatively suggests a slow source, with a very strong increase of moment with period (slope of -0.22 logarithmic units per mHz on Fig. 3d), both in the strike-slip geometry of the 1971 earthquake, and in the less probable one of a low-angle subduction. The moment at mantle periods ($\sim 6 \times 10^{27}$ dyn cm) is equivalent to that

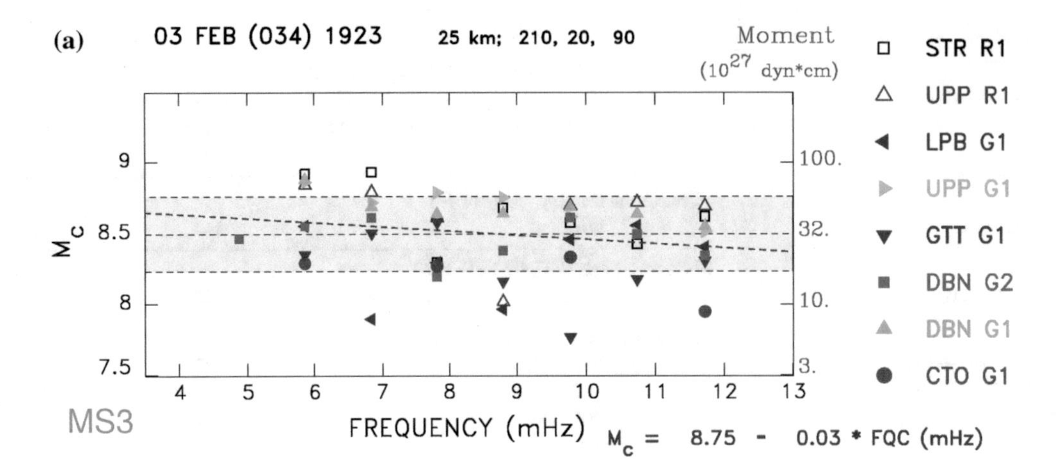

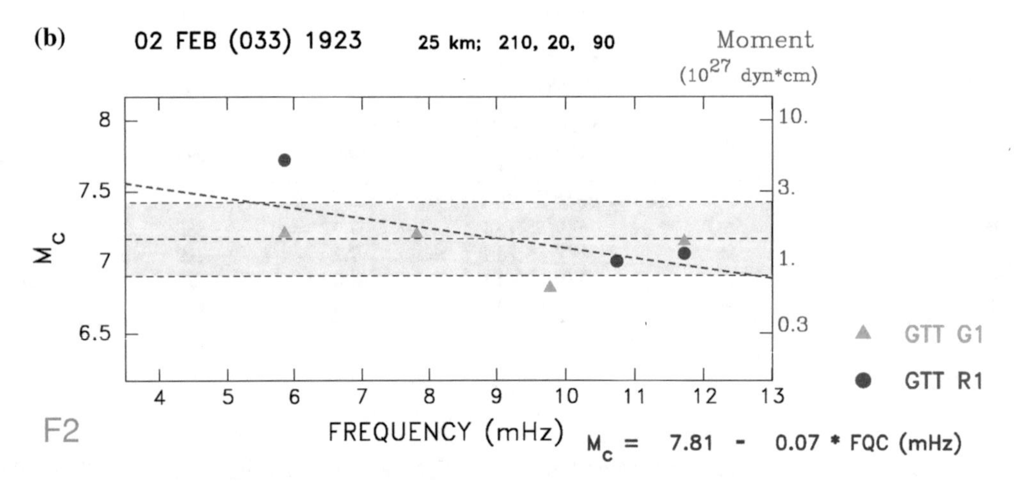

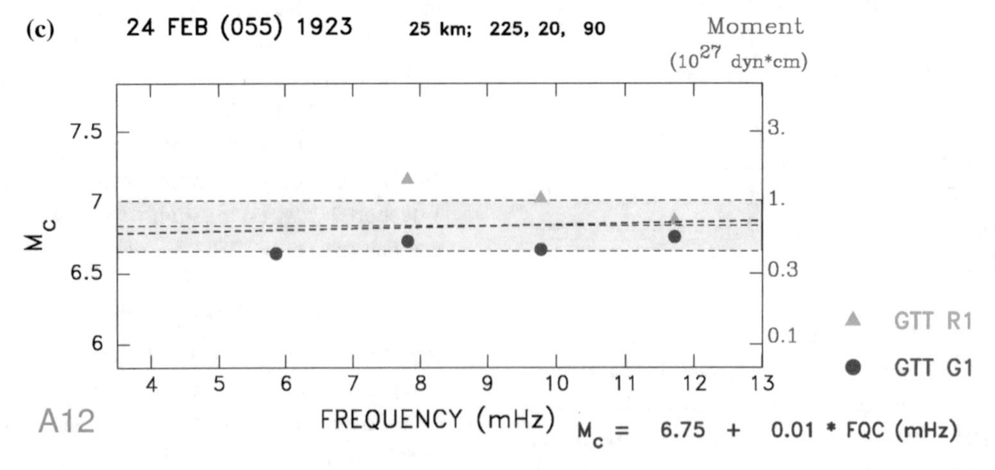

Figure 3
Mantle magnitudes corrected for focal mechanism, M_c (Okal and Talandier 1989) derived from spectral amplitudes of surface waves at mantle periods. The dashed lines and yellow band give the average and standard deviation of M_c over the entire frequency band, and the purple dashed line its linear regression. The depth and focal mechanism assumed in the correction are printed next to the date. **a** Mainshock (Event MS3); **b** Main Foreshock (Event F2); **c** Main Aftershock (Event A12); **d** 1917 Cape Kamchatskiy Event; **e** 1936 Cape Kamchatskiy Event

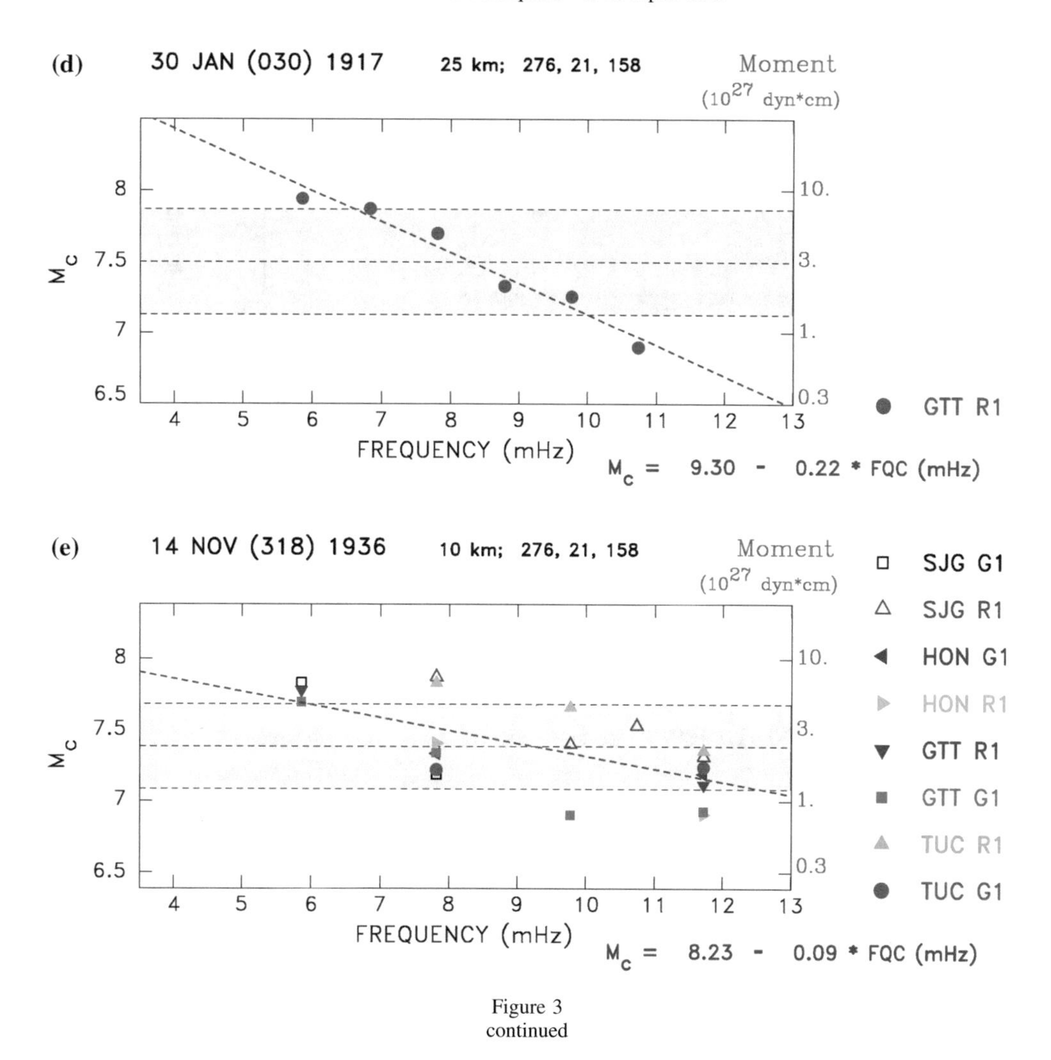

Figure 3
continued

in 1971, and the two earthquakes could then be repeats of each other, 55 years apart, with a seismic displacement of ~ 2.8 m, inferred from their common moment using Geller's (1976) scaling laws.

- In the case of the 1936 earthquake at Cape Kamchatskiy, the various spectral amplitudes are compatible with the oblique geometry of the 1971 event and a moment of about 4×10^{27} dyn cm (by contrast, a subduction geometry predicts a node of excitation of G_1 at HON which is not observed); that moment scales to 2.4 m of slip (Geller 1976) (Fig. 3e).
- In the case of the 1969 Ozernoy earthquake (Fig. 4), in addition to Fedotov et al.'s (1973) thrust mechanism, low angle thrust solutions were proposed by Cormier (1975), Stauder and Mualchin (1976) and Daughton (1990). The latter used body-wave modeling to infer a complex source with a total moment of 5×10^{27} dyn cm, but did not consider a strike-slip solution, despite the lack of constraint on the slip angle. Using a representative selection of WWSSN records, we find that the low-angle thrust mechanism proposed by Cormier (1975) leads to an unacceptable scatter of moment values; in particular, it cannot reconcile the Love-to-Rayleigh ratios at Windhoek (WIN), as it would place the station in a node of excitation of Love waves. We prefer the mostly strike-slip mechanism of Fedotov et al. (1973), which predicts a more consistent set of moment values, reaching $\sim 3 \times 10^{27}$ dyn cm at mantle wave periods.

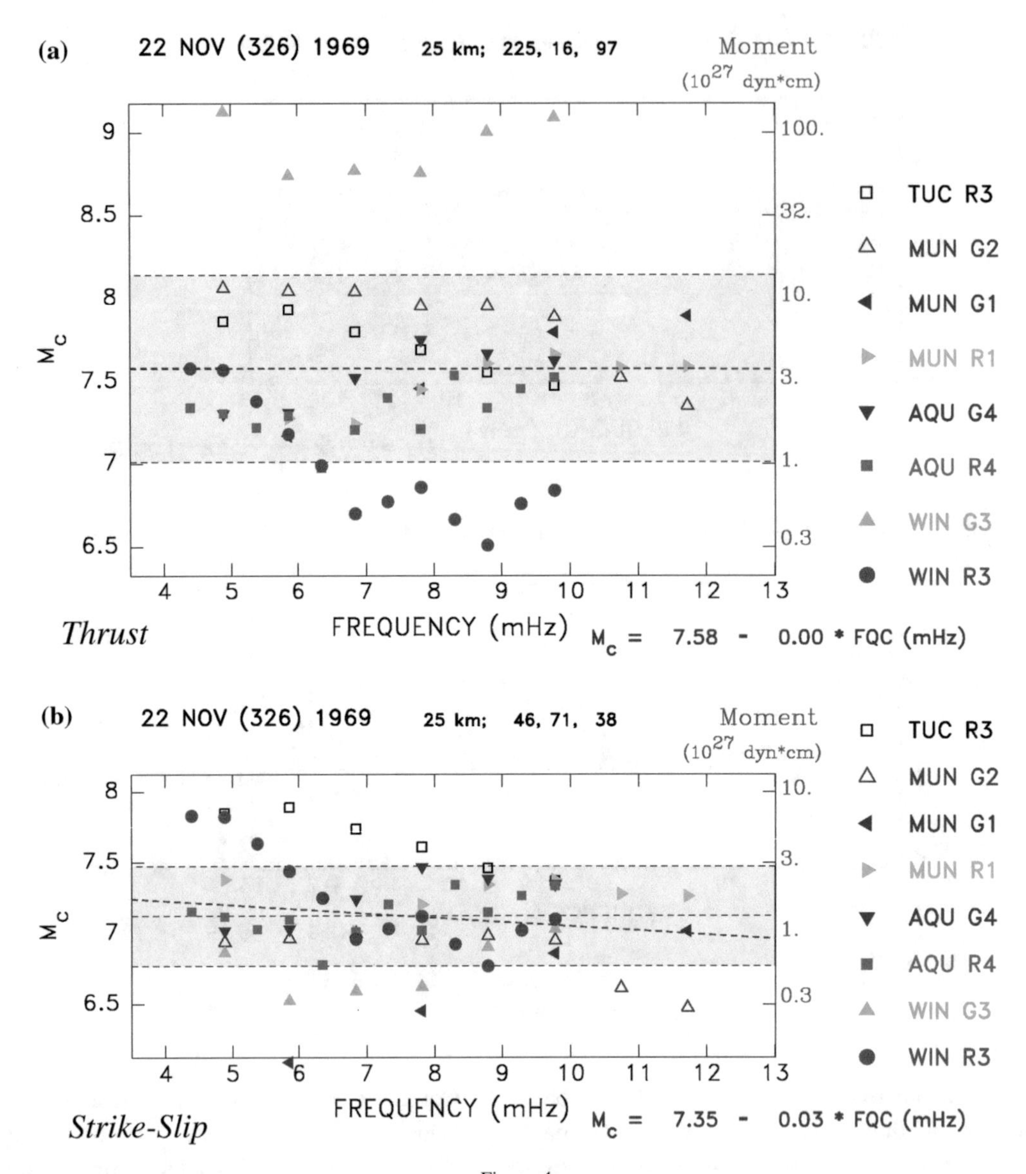

Figure 4
Same as Fig. 3 for the 1969 Ozernoy earthquake. **a** Thrust mechanism proposed by Cormier (1975); note the large scatter, notably for station WIN. **b** Preferred strike-slip mechanism (Fedotov et al. 1973); note improved fit to dataset

- By contrast, the 1945 earthquake, 90 km farther to the south, has spectral amplitudes better reconciled by Cormier's (1975) subduction geometry, than by Fedotov et al.'s strike-slip mechanism, with a suggested moment of 1×10^{27} dyn cm (Fig. 5). Note that no seismicity at a level comparable to the 1969 and 1945 events has occurred in the area since the start of the Global CMT catalog in 1976 (Dziewonski et al. 1981; Ekström et al. 2012).
- Finally, in the case of the UK tsunami earthquake of 13 April 1923, we could not obtain usable records at mantle periods from any Wiechert instruments (including at Göttingen), and could process only the two Golitsyn components at DBN. As shown on Fig. 6, the spectral amplitudes of Rayleigh and Love waves at DBN are best reconciled in the geometry of Fedotov et al.'s (1973) mechanism for the 1969 Ozernoy earthquake (with a moment of 1.0×10^{27} dyn cm

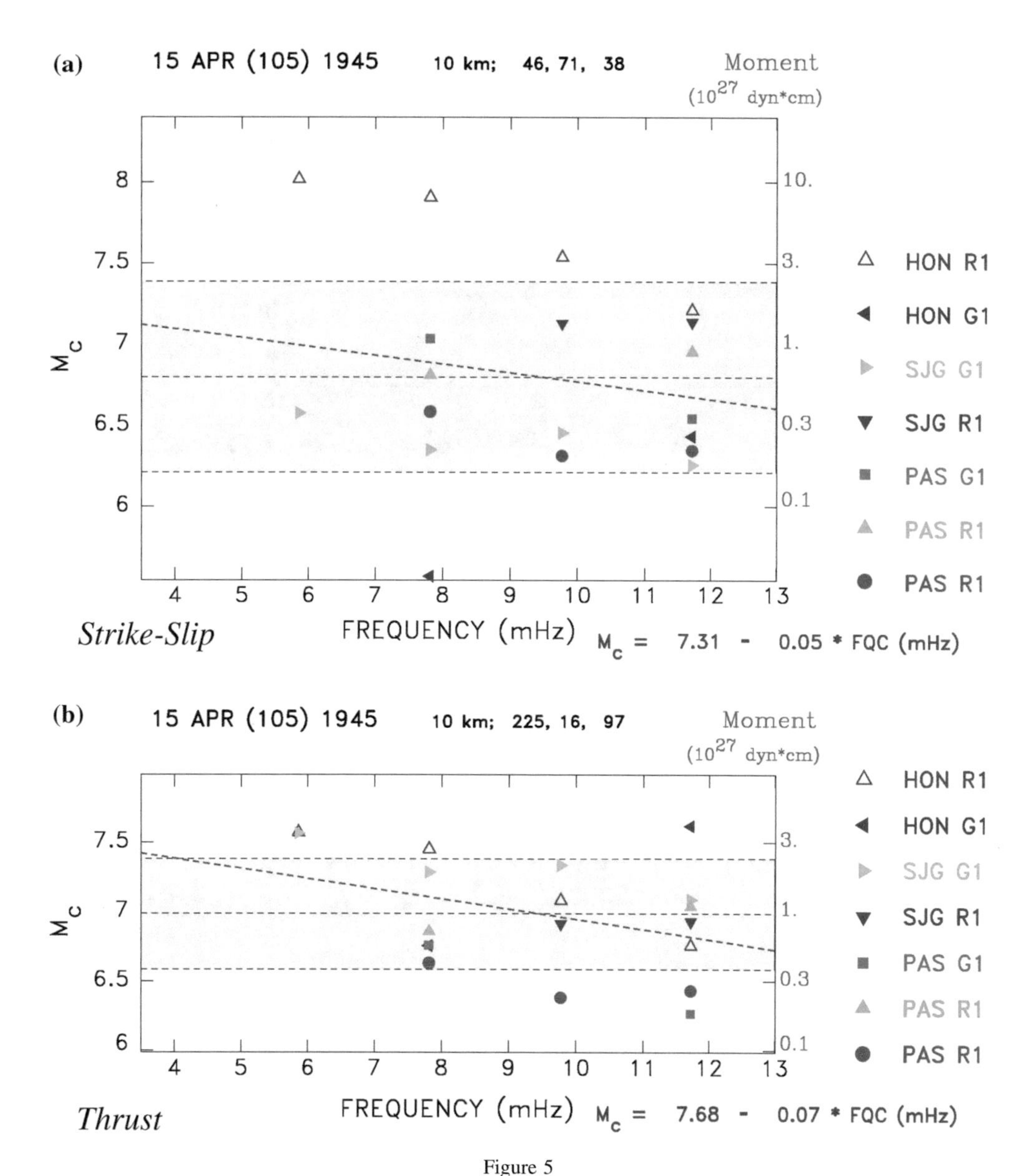

Figure 5
Same as Fig. 3 for the 1945 Ozernoy earthquake. Note that a strike-slip mechanism **a** gives a poor fit, notably to Rayleigh waves at HON. The thrust mechanism **b** is preferred

around 200 s), or of the 1971 oblique subduction (with a moment of 1.2×10^{27} dyn cm). Low-angle thrust subduction mechanisms predict much larger Love spectral amplitudes, which are not observed. This supports the model of the UK earthquake not being a genuine aftershock, but rather occurring by stress transfer outside the rupture area of the mainshock along the Kamchatka subduction zone. In the absence of adequate short-period seismograms, it was not possible to quantify the slowness of the UK event through an Energy-to-Moment parameter Θ (Newman and Okal 1998), but a remarkable aspect of Fig. 6 is the growth of moment (or magnitude M_c) with period, the slope of the regression being -0.13 logarithmic units per mHz. This number is typical if not in excess (in absolute value), of those for other tsunami earthquakes, such as the 1932 Mexican, 2010 Mentawai

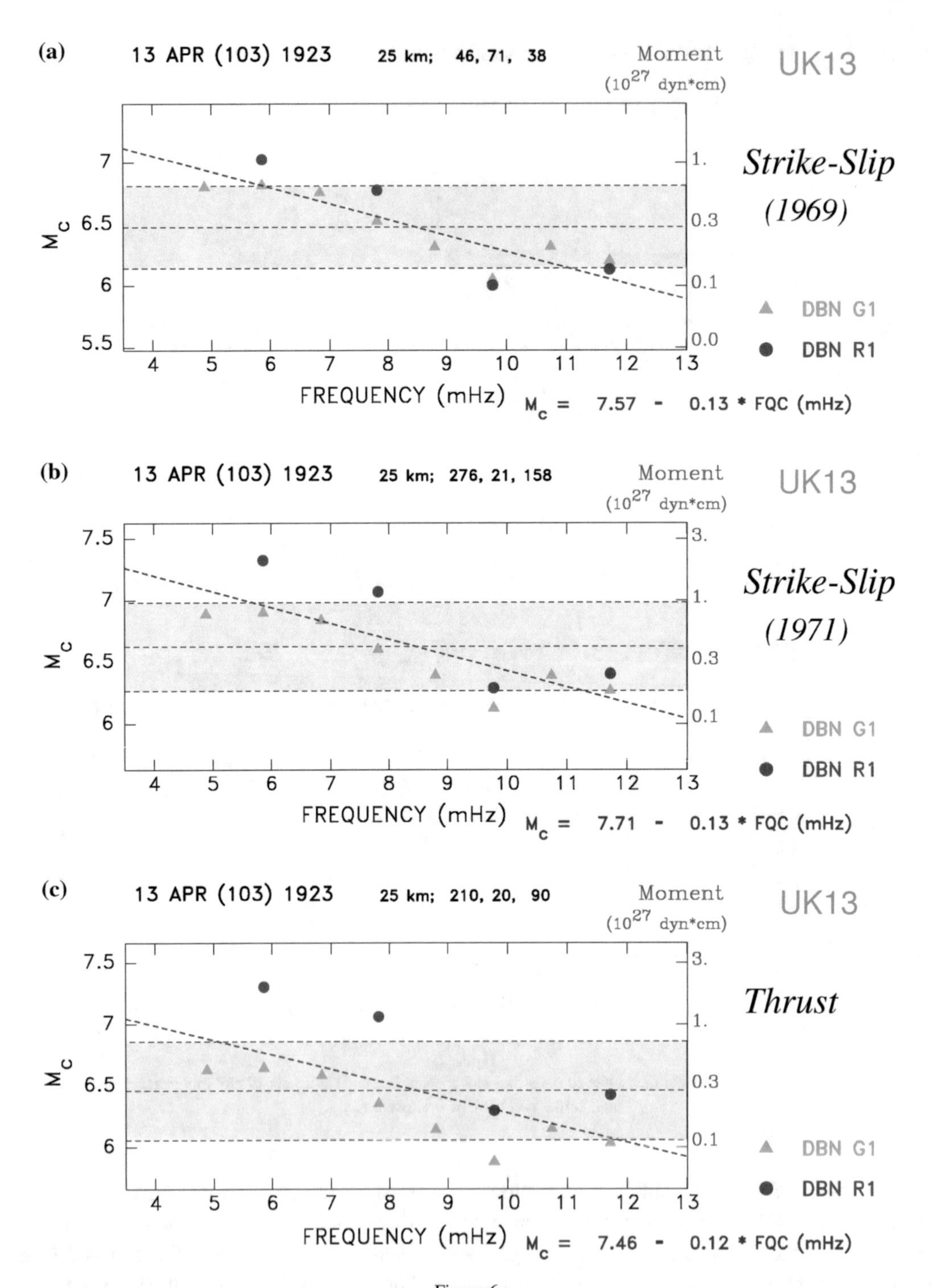

Figure 6
Same as Fig. 3 for the Ust'-Kamchatsk tsunami earthquake of 13 April 1923. Note that the spectral data fit the strike-slip geometries of the 1969 Ozernoy earthquake (**a**), and to a lesser extent of the 1971 Commander Islands event (**b**), better than the thrust fault one expected of a genuine aftershock (**c**). Note also the steep slope regressing M_c against frequency, which identifies the earthquake as slow

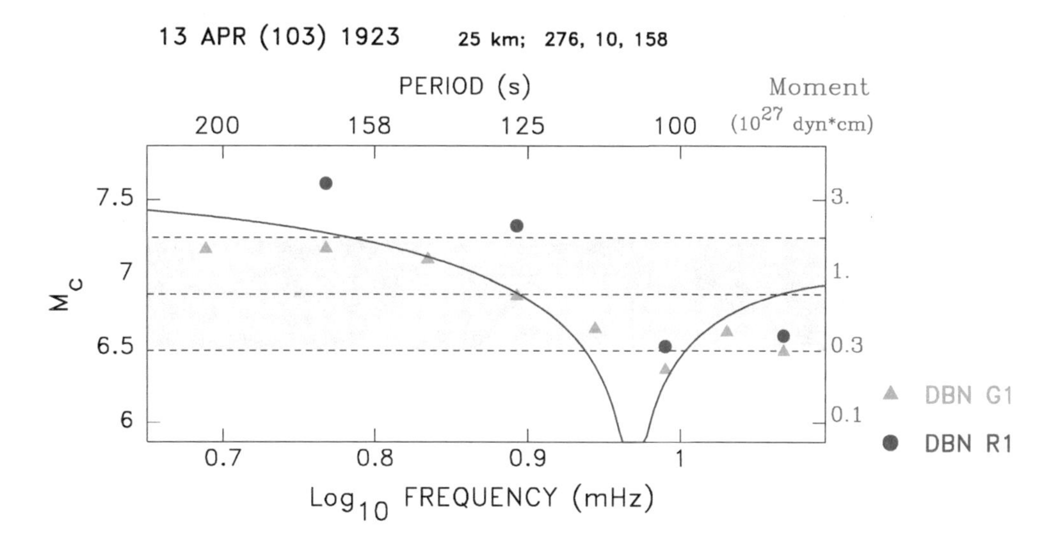

Figure 7
Same as Fig. 6b using a logarithmic scale for frequency. The solid purple line is a best fit of the spectrum of a classical sinc $\omega\tau/2$ moment release time function, suggesting a corner time $\tau \sim 100$ s

or both 1947 Hikurangi events (Okal and Borrero 2011; Okal 2011; Okal and Saloor 2017). This is in contrast to the mainshock (MS3) and the main aftershock (A12), for which the negative slopes on Fig. 3 are only – 0.03 and 0.0 logarithmic units per mHz, respectively, suggesting that those events do not feature any anomalous source slowness. Note also that the UK event has both a larger moment and a smaller magnitude M_{PAS} than the main aftershock A12. In a more classical plot using a logarithmic scale for frequency (Fig. 7), we hint at a hole in the spectrum around 10 mHz, which is confirmed by a systematic grid search for static moment and rupture time, in the form of a classic sinc $\omega\tau/2$ moment release time function, yielding $M_0 = 3 \times 10^{27}$ dyn cm and $\tau \sim 100$ s, significantly longer than expected under classical scaling laws (Okal 2003), which could be explained by a slower rupture velocity. All these remarks confirm the slow character of the UK earthquake, and its nature as a tsunami earthquake. Incidentally, we note that Gusev (2004) has quantified the 1923 UK tsunami through a "tsunami magnitude" $M_t = 8.2$, which he then assigned to the parent earthquake, in the process noting that it was larger than its conventional seismic magnitude, $M_s^{(GR)} = 7.2$. This procedure is questionable since it tacitly assumes that the tsunami was fully generated by the earthquake; it becomes very unfortunate when, in subsequent publications, this value is then equated to a moment magnitude M_w, leading for example Bourgeois and Pinegina (2017) to describe the MS and UK events as a kind of pair, when their seismological properties clearly advocate a large difference in size. In short, the UK event cannot have a moment for its dislocation as high as 2.5×10^{28} dyn cm, ($M_w = 8.2$), as used for example by Bourgeois and Pinegina (2017), since this would lead to vertical ground motion spectral amplitudes of 5 mHz Rayleigh waves at Göttingen ($\Delta = 70°$) on the order of 20 cm.s, corresponding in turn to time-domain zero-to-peak amplitudes of ~ 2 mm, or an easily detectable 1.5 cm on a horizontal mechanical instrument with a magnification of ~ 10 in that frequency band (Okal and Talandier 1989).

4. *Tsunami Investigations*

Before proceeding to simulations of the tsunami of 13 April 1923, we review tsunami data documented in the vicinity of Cape Kamchatskiy, either from actual observations during the 1923 and other

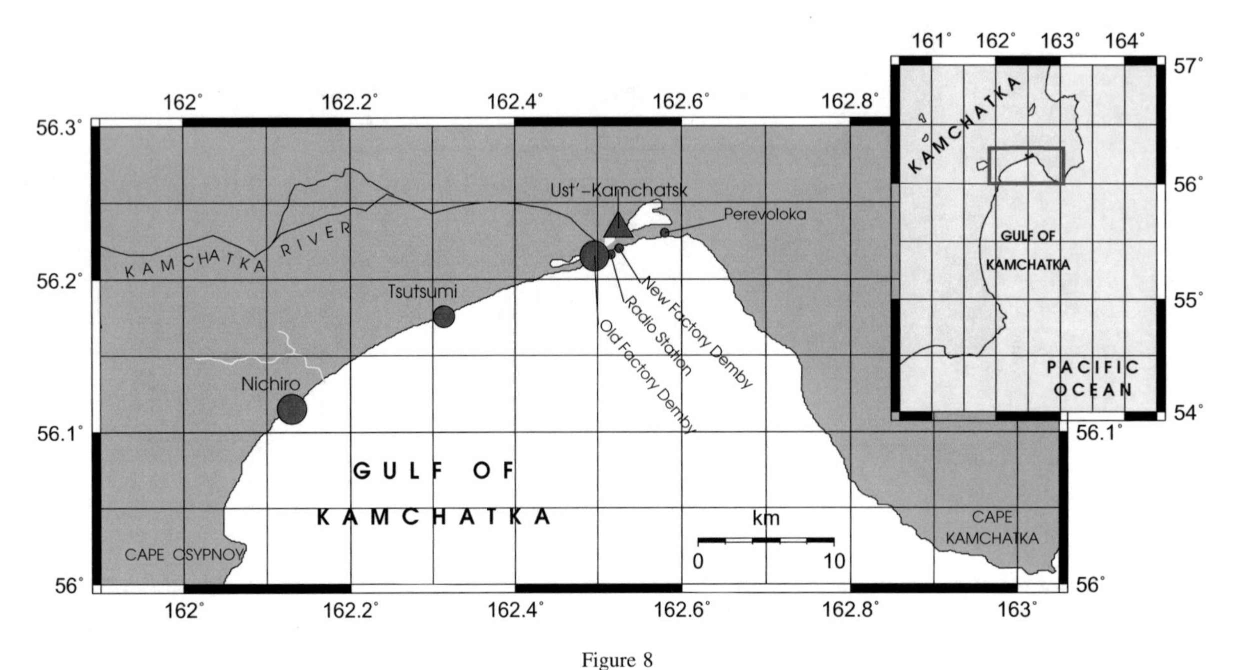

Figure 8
Area affected by the 13 April 2013 tsunami, as reported by Troshin and Diagilev (1926) and Soloviev and Ferchev (1961). The highest reported run-ups were documented on the Demby landspit (11 m) and at Nichiro (20–30 m). The yellow line is our best interpretation of Soloviev and Ferchev's First Creek

events at the very few past or present settlements along the coast, or from sedimentological work by Bourgeois et al. (2006), Martin et al. (2008) and Bourgeois and Pinegina (2017), for which age estimates can be bracketed by ash deposits from known major eruptions at the nearby volcanoes Bezymianniy and Shiveluch.

- *13 April 1923* Detailed reports of the tsunami following the UK event by Troshin and Diagilev (1926) have documented large run-up values reaching 20–30 m, with the tsunami affecting the northern coastlines of the Kamchatskiy Bight, from Cape Shubert to Cape Kamchatskiy (Fig. 8). According to Soloviev and Ferchev (1961), a first and small wave arrived at Ust'-Kamchatsk 15 min after the earthquake, "at approximately 02:15",[1] followed 15 min later by a second catastrophic wave rising up to 11 m, destroying the structures at Demby, and flowing 7 km into the mouth of the Kamchatka River; note however that Minoura et al. (1996) have argued that this exceptional penetration was facilitated by the gliding of the waves over a coastal plain covered by a thick layer of frozen snow. Higher run-up values occurred to the west of the Kamchatka River, with the maximum value of 20 m (Soloviev and Ferchev 1961) or 20–30 m (Zayakin and Luchinina 1987 reported at Nichiro, where the canneries were totally destroyed, as well as at Tsutsumi. In this context, Borisov (2002) reports (presumably based on Troshin and Diagilev (1926)) that a small cutter from the Nichiro cannery was deposited on a 30-m terrace at the location Gorbusha, which we interpret to lie 1–2 km inland, at the back of a coastal plain gently rising to an altitude of no more than 6 m. As such, the documented heights of 20–30 m do represent run-up on a sloping beach, as opposed to a splash on a steep cliff. By contrast, and according to Soloviev and Ferchev (1961), themselves quoting from Troshin and Diagilev (1926), the waves (and consequent damage) decreased eastward, notably over a distance of ~ 10 km along the spit

[1] Note here an inconsistency in absolute timing since the origin time of the event is 15:31 GMT, or 02:31 local time on the 14th. The legal time in Kamchatka was GMT+11 in 1923; it was advanced one hour in all time zones of the USSR in 1930, and hence would have been GMT+12 in 1971 (see below).

separating Lake Nerpichye from the ocean. There, the old Demby cannery at the western end was destroyed, the radio station and the new cannery at its center were only partially affected, and the water rose a mere ~ 1 m at Perevoloka, at its eastern end (Zayakin and Luchinina 1987), although this interpretation has been questioned in the context of the salinization of Lake Nerpichye in the 1920s (Gorin and Chebanova 2011). Finally, and according to Soloviev and Ferchev (1961), the tsunami reached a maximum run-up of 4 m at Nikol'skoye (Bering Island), although the exact location of this report and hence the relevant morphology of the coastline is unknown. The tsunami also produced a 30-cm wave at Hilo, Hawaii; 20 cm at Honolulu, Hawaii; 8 cm at Tofino, B.C.; 15 cm at San Francisco; and swirls in Los Angeles harbor. The latter are described as taking place between 6 and 10 a.m. on the 14th (The Los Angeles Times 1923), i.e. between 14:00 and 18:00 GMT. This can be interpreted as a delayed harbor resonance upon arrival of waves propagating outside the shallow-water approximation, as observed at distant harbors during the 2004 Sumatra tsunami Okal et al. (2006a, b). The timing would be reconciled with 100-s waves propagating at a group speed of 87 m/s in an ocean averaging 4.5 km in depth, over a distance of 7000 km (detouring around the Aleutian island chain), but the non-linear nature of the resonance precludes any quantitative interpretation. The same article (The Los Angeles Times 1923) mentions a devastating inundation of the Korean coast north of Pusan, with more than 1000 lives lost, but that disaster apparently occurred at least 24 h prior to the UK earthquake, and is, therefore, unrelated. In seeking to simulate the tsunami of 13 April 1923, the critical dataset to be modeled will consist of the local tsunami amplitudes of 11 m at Demby and more than 20 m at Nichiro, of the much weaker ones at Perevoloka, and to a lesser extent of the decimetric amplitudes in Hawaii and along the Pacific coast of North America.

- *15 December 1971* The tsunami was registered with an amplitude of 47 cm at Ust'-Kamchatsk, but was apparently not observed visually, due in part to its occurrence at low tide (Soloviev et al. 1986). Martin et al. (2008) later attributed to the 1971 event deposits identified at heights of 10 m on the Kamchatskiy Peninsula (56.19°N; 163.35°E). It is unclear how such a run-up could have remained undetected by the ~ 10 people manning the lighthouse and weather station at Cape Afrika (altitude 7–8 m), even taking into account the night-time occurrence of the event (solar time $\sim$ 19:23; legal time 20:30 GMT+12). On the other hand, and despite the seismic similarity between the events of 1917 and 1971, comparable tsunami effects in 1917 might have gone unnoticed if that location had been uninhabited, a reasonable assumption given that the lighthouse was built in 1960. In this context, it is worth noting that Bourgeois et al. (2006) describe tsunami deposits at Stolbovaya Bay (56.68°N; 162.92°E), intertwined between the tephra layers of 1854 (Shiveluch) and 1956 (Bezymianniy), for which they discount the 1923 event as a possible origin (and suggest that the deposit may result from the 1945 event), based on a comparison of the relevant run-up (5–6 m) with that reported in 1923 for Bering Island (4 m). This argument tacitly assumes a 1923 epicenter in the Pacific Ocean south of Ust'-Kamchatsk for which the distance to Stolbovaya would be greater than to Bering Island; if, as we have argued, the UK epicenter is either in the Kamchatskiy Peninsula, or north of it, the argument fails, and the deposits in question may have come from the 1923 UK event, which is also larger than the 1945 source.
- *22 November 1969* A very significant tsunami was reported north of the Kamchatskiy Peninsula, principally along the Ozernoy Bight, where run-up was described as having reached 12–15 m at a then existing meteorological station at the mouth of a river Ol'khovaya (literally, river of the alders) (Fedotov et al. 1973); the interpretation of this report is made difficult by the existence of two rivers by that name, with mouths at 57.12°N, 162.80°E and 57.62°N, 163.23°E, respectively (Anonymous 2001); the authors' Fig. 6 would support the latter location. The tsunami was also reported as far north as Lavrov Bay (60.3°N; 167.2°E), 400 km from the epicenter (Fedotov et al. 1973). By contrast, it was recorded with a

minimal amplitude (27 cm) at Ust'-Kamchatsk. Martin et al. (2008) conducted a sedimentological investigation of tsunami deposits along Ozernoy Bight, confirming run-ups on the order of 5–7 m, but casting doubt on the larger value reported by Fedotov et al. (1973) at Ust' Ol'khovaya; however, neither of the two possible locations was sampled by the authors.

- *30 January 1917* No tsunami reports are available for this earthquake, despite a daylight occurrence (origin time 02:45 GMT or 13:37 solar time), and its similarity with the 1971 event. We note however that seven out of the ten Pacific-wide maregraphs having reported the 1971 tsunami do not appear in the NOAA run-up database until 1933, with two more reporting only amplitudes larger than 10 cm until 1927, which may explain the lack of observations.

4.1. Hydrodynamic Simulations

We simulate the Kamchatka tsunami of 13 April 1923, using the MOST algorithm (Titov and Synolakis 1995; Titov and González 1997; Titov and Synolakis 1998) which solves the non-linear shallow water approximation of the Navier-Stokes equations. It has been extensively validated against actual tsunami surveys and laboratory experiments (e.g. Synolakis et al. 2008; Titov et al. 2016).

The bathymetry grid used in our simulations was digitized from Russian marine maps (Anonymous 2001), considerably more precise than available global models such as GEBCO (Fisher et al. 1982), especially in the immediate vicinity of Ust'-Kamchatsk (Fig. 9). We use several interpolations of the grid, the finest one at a resolution of 7.5 arc-seconds for smaller scale sources. The simulation time series were recorded at 41 virtual gauges located near the shoreline of the Gulf of Kamchatka.

4.1.1 Modeling an Earthquake Source

Following standard practice, we calculated static displacements from the earthquake source, to be used as initial values of the hydrodynamic simulations, using Mansinha and Smylie's (1971) algorithm. In the absence of a fully constrained focal solution for the UK event, we assumed the focal mechanism proposed by Fedotov et al. (1973) for the 22 November 1969 event, and approximated the fault dimensions from the event's size using earthquake scaling laws (Geller 1976). Because we could only use records from one station in our calculation of the seismic moment from mantle waves ($M_0 = 3 \times 10^{27}$ dyn cm), we used three magnitudes of $M_w = 7.0$, $M_w = 7.6$, and $M_w = 8.0$, with the second magnitude as our preferred result (Fig. 7), and the former and the latter as lower and higher end members.

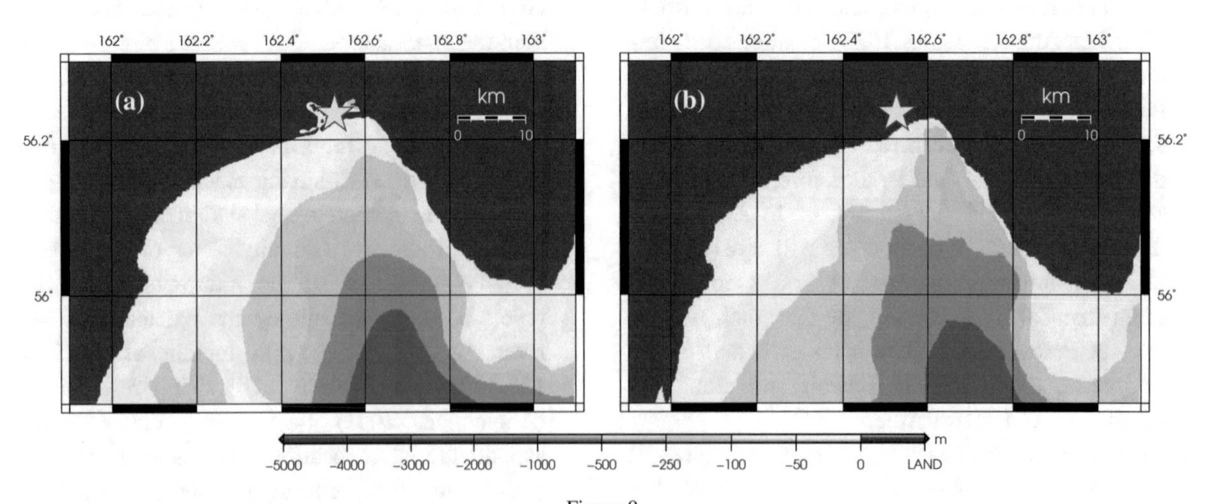

Figure 9
Bathymetry of the Gulf of Kamchatka **a** from the GEBCO global model (Fisher et al. 1982) and **b** digitized from a Russian marine map of Kamchatka (Anonymous 2001). The yellow stars in both figures depict Ust'-Kamchatsk

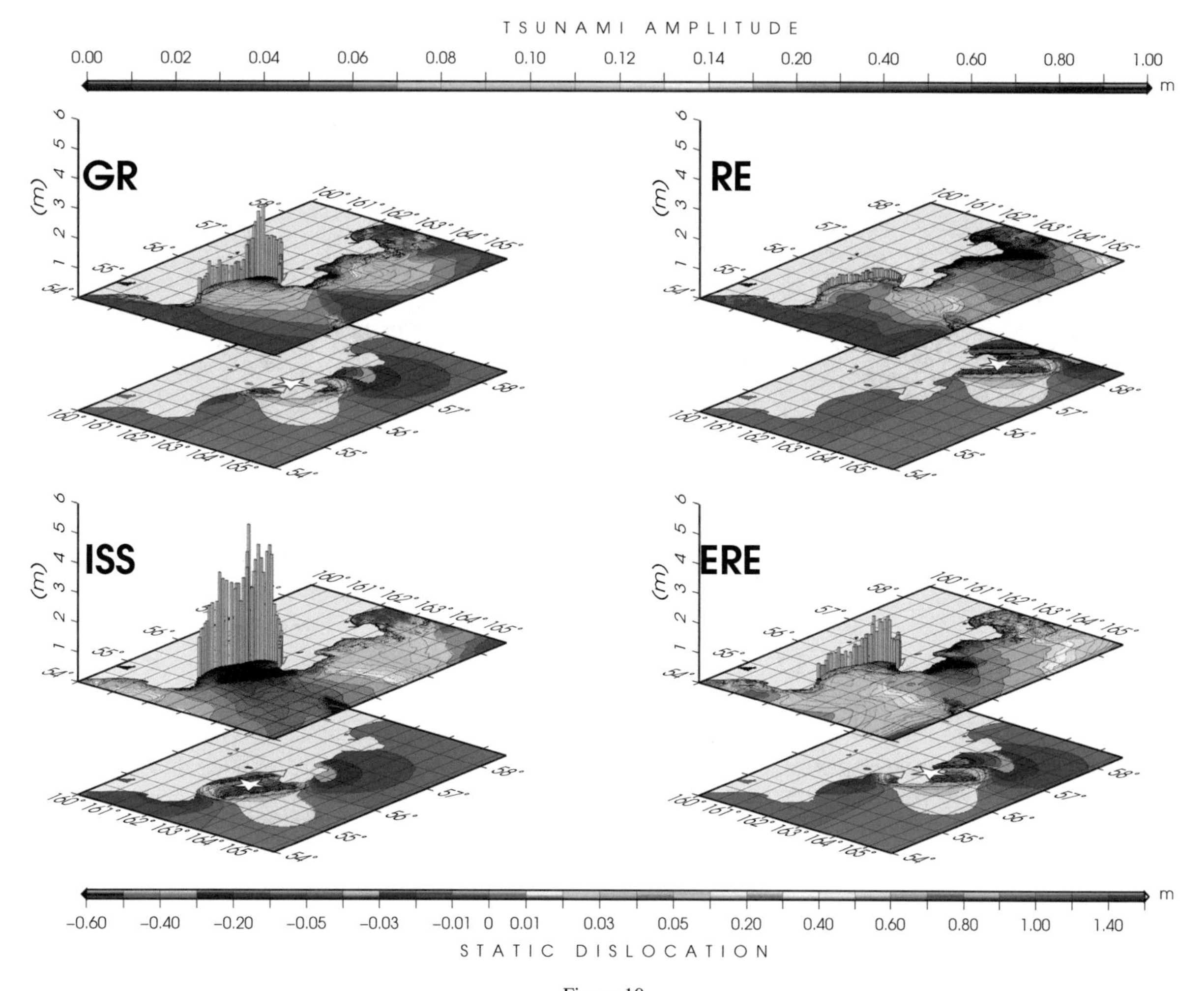

Figure 10
Static displacement maps (bottom slices) as well as simulation results (top slices) for the maximum magnitude ($M_w = 8.0$) considered in this study. Results for each scenario are labeled accordingly. Triangles and stars in all figures represent Ust’-Kamchatsk and the corresponding epicenter in each scenario. The gray vertical columns show the maxima of calculated amplitudes at the 41 gauges along the coastline of Gulf of Kamchatka

As to the event’s location, we tested four different scenarios as discussed in Sect. 2. We label the location scenarios as GR (Gutenberg–Richter), ISS, ERE (Engdahl relocation)) and RE (the relocated epicenter in this study) as shown in Figs. 1 and 10. In the case of the ISS location, we use the thrusting mechanism in Fig. 6c.

We used time steps of $\Delta t = 1$ s during 12-hr windows in a grid with a resolution of 23 arc-seconds to satisfy the CFL conditions (Courant et al. 1928). This resolution is valid for the coarse grid as prescribed e.g. by Shuto et al. (1986) and Titov and Synolakis (1995, 1997). We ended the calculations at a depth of 4 m near the coastlines and, therefore, assumed the calculated amplitudes at the virtual gauges along the coastline as proportional but not equal to the actual documented run-ups.

Figure 10 shows the static displacements (bottom slices) as well as fields of maximum calculated amplitudes (top slices) for a maximum assumed magnitude of $M_w = 8.0$, for all four possible locations.

Simulations from our relocated epicenter (RE) or the customized Engdahl relocation (ERE) produce

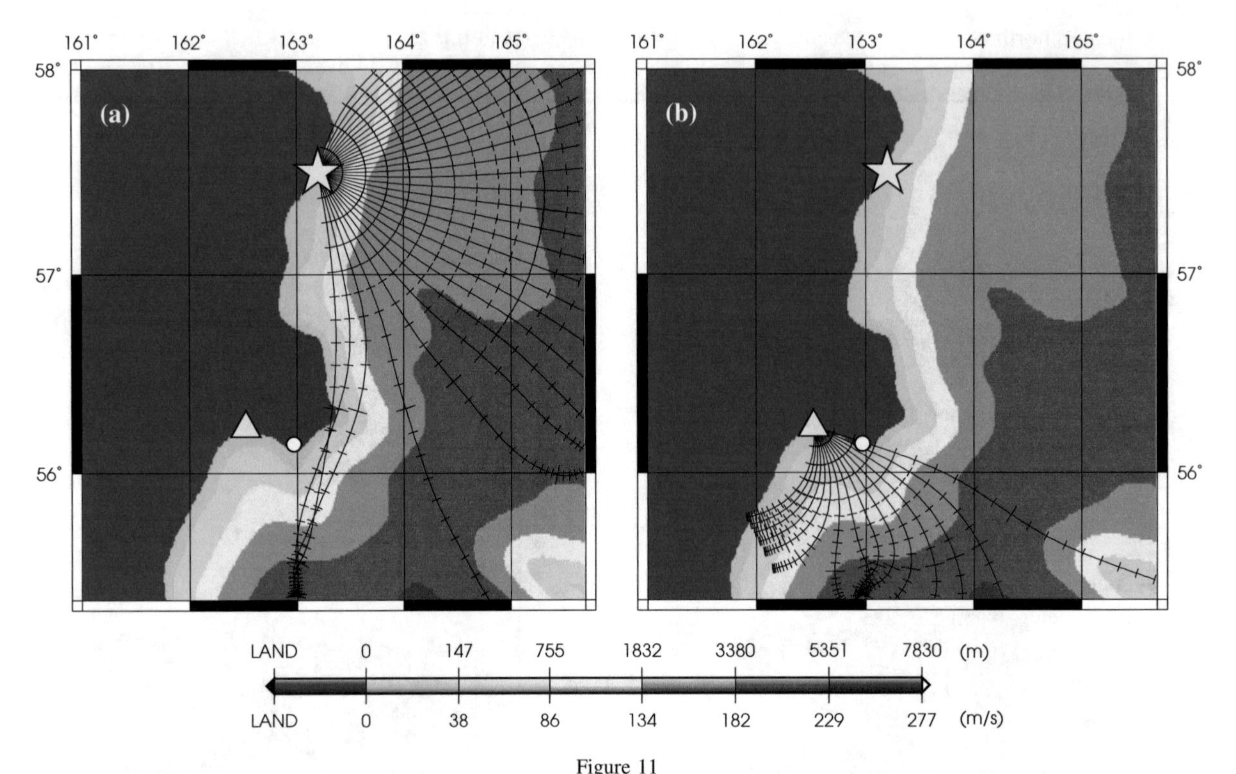

Figure 11
Tsunami ray tracing experiment from our relocated epicenter. Black curves represent tsunami rays, with the tick marks at every minute. The background undispersed velocity field is calculated from our bathymetry grid. The scale is measured for both depth (top) and velocity (bottom). In (**a**) the source is set at the relocated epicenter (RE, star), while (**b**) is a reverse diagram centered on the receiver at Ust'-Kamchatsk (triangle). The white dots represent Cape Kamchatskiy

only minimal amplitudes in the Gulf of Kamchatka (maximum ~ 1.5 m), making it improbable to reach the observed run-up amplitudes; the ISS source would generate higher amplitudes, but their distribution along the coast remains broad, illustrating the dimensions of the source (Okal and Synolakis 2004). The GR scenario (top left on Fig. 10) gives a better distribution of amplitudes along the coast, but they remain too weak (maximum ~ 2.4 m) to justify an observed run-up of 20 m on a very flat beach. Thus, none of the four possible scenarios yields satisfactory fields of amplitudes along the Kamchatskiy Bight.

4.1.1.1 Ray-Tracing Experiments The inability of a northern source, such as RE or ERE, to efficiently inundate the coasts of the Kamchatskiy Bight can be further examined by ray-tracing experiments, following the techniques of Woods and Okal (1987) and Satake (1988). We apply Jobert and Jobert's (1983) solution of the 2-D Eikonal equations on a heterogeneous sphere using a field of variable propagation velocities in the shallow-water approximation, to trace tsunami rays from our relocated epicenter using a 4th-order Runge-Kutta scheme. While some energy would be expected to diffract around the Kamchatskiy Peninsula in violation of geometrical optics (assumed by the ray-tracing methodology), this experiment provides a general illustration of the expected distribution of wave energy in the region.

As shown in Fig. 11a, a tsunami originating at our relocated epicenter (RE; yellow star) does not reach Ust'-Kamchatsk (yellow triangle), the rays approaching the Kamchatskiy Peninsula being focused by the shallow bathymetry into the deeper parts of the Kamchatskiy Bight, its northern coast thus being masked from the source by the Peninsula. In addition, using the concept of seismic reciprocity (Aki and Richards 2002), we can predict that no high amplitudes would be observed in Ust'-Kamchatsk from any source located north of Cape Kamchatskiy (Fig. 11b). Note that this result is independent of the nature of the tsunami source, and would equally apply to a landslide source, as studied below.

4.1.2 Landslide Sources

Having ruled out any of the legitimate earthquake scenarios as the source of the 13 April 1923 tsunami, we next consider the possibility of a landslide source. Generation of major tsunamis by underwater landslides triggered by large earthquakes has been documented in numerous instances, including the catastrophic events of 1998 in Papua New Guinea (Synolakis et al. 2002) and 1946 in the Aleutians (Kanamori 1985; Fryer and Watts 2001; Okal et al. 2003; von Huene et al. 2014). Even earthquakes with epicenters located significantly onland have created tsunamigenic offshore landslides, with recent examples documented in 1954 and 1980 at Orléansville/El Asnam (Soloviev et al. 1992), 1990 on the southwestern slopes of the Caspian Sea (Salaree and Okal 2015), and 2013 in Pakistan (Hoffmann et al. 2014), the record triggering distance (900 km) being for the 1910 Rukwa earthquake (Ambraseys 1991).

Modeling landslides as potential sources for this event is motivated by the local nature of the tsunami, whose maximum amplitudes are concentrated along a short (~ 35 km) stretch of relatively straight coastline between Cape Osypnoy and Ust'-Kamchatsk (Fig. 8), beyond which the amplitudes fall sharply reaching only 1 m at Perevoloka (Soloviev and Ferchev 1961). While the dataset of reported run-ups is insufficient to compute a formal aspect ratio of its distribution, this observation fits qualitatively Okal and Synolakis' (2004) model.

The triggering of an underwater landslide by an earthquake requires two conditions: an adequate slope to carry the slide and a sufficient acceleration to destabilize its material. Regarding the former, we have obtained numerically the gradient of the bathymetric grid used in our simulations and present in Fig. 12 its modulus and azimuth (direction of steepest descent), following the technique used in previous studies (Okal et al. 2014; Salaree and Okal 2015). Underwater slumps have generally been observed on slopes between $\sim 3\%$ and $\sim 6\%$, the upper bound reflecting the capacity of the slope to hold the precarious material; they can also take place on slopes as low as $\sim 1\%$ in very shallow waters (e.g. Skempton, 1953; Prior et al. 1982; Brunsden and Prior, 1984). Figure 12 shows that such slopes are amply documented in the Kamchatskiy Bight. In addition, we note an abundance of aerial slides along the eastern shore of the Kamchatskiy Bight, between Ust'-Kamchatsk and Cape Kamchatskiy (Fig. 13), suggesting the possibility of such mass failures on the sea floor, under the assumption of a morphological continuity between onland and offshore slopes, as suggested by Kawata et al. (1999) in the case of the 1998 Papua New Guinea event.

As for the peak accelerations expected from the UK earthquake at the proposed slide locations, they will obviously depend on the epicenter, focal geometry and source spectrum of the event, neither of which is fully constrained. Simulation of peak ground acceleration, using the algorithm by Campbell and Bozorgnia (2003), for the RE and ERE epicenters, predicts maximum vertical accelerations of, respectively, $0.03 \pm 0.01g$ and $0.09 \pm 0.03g$ in the northern Gulf of Kamchatka (Fig. 14) for $M_w = 7.6$. Although the former does not seem sufficient to trigger a landslide, the latter is about the triggering threshold (0.1 g) as proposed by Keefer (1984). Possibly larger seismic moments such as $M_0 = 1.0 \times 10^{28}$ or $M_w \approx 8$ may increase these values, although the slow nature of the source may limit their efficiency at the high frequencies controlling ground acceleration. Notwithstanding this reservation and irrespective of its exact source parameters, we note that the earthquake was felt at MMI VI in Ust'-Kamchatsk (Soloviev and Ferchev 1961), corresponding to peak accelerations of 0.1 to 0.2 g (Wald et al. 1999) which provide ample justification to the hypothesis of an underwater landslide in the Kamchatskiy Bight, only 30 km away, a location with ample sediment discharge by the Kamchatka River draining the volcanic province of Central Kamchatka (Kuksina and Chalov 2012).

In order to simulate the landslide tsunami, we used the block model approach for landslides as proposed by Synolakis et al. (2002). Following Salaree and Okal (2015) we model the slide as an instantaneous hydrodynamic dipole consisting of a trough and a hump of respective (positive) amplitudes

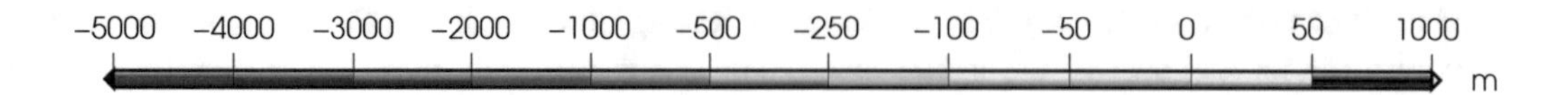
-5000 -4000 -3000 -2000 -1000 -500 -250 -100 -50 0 50 1000
m

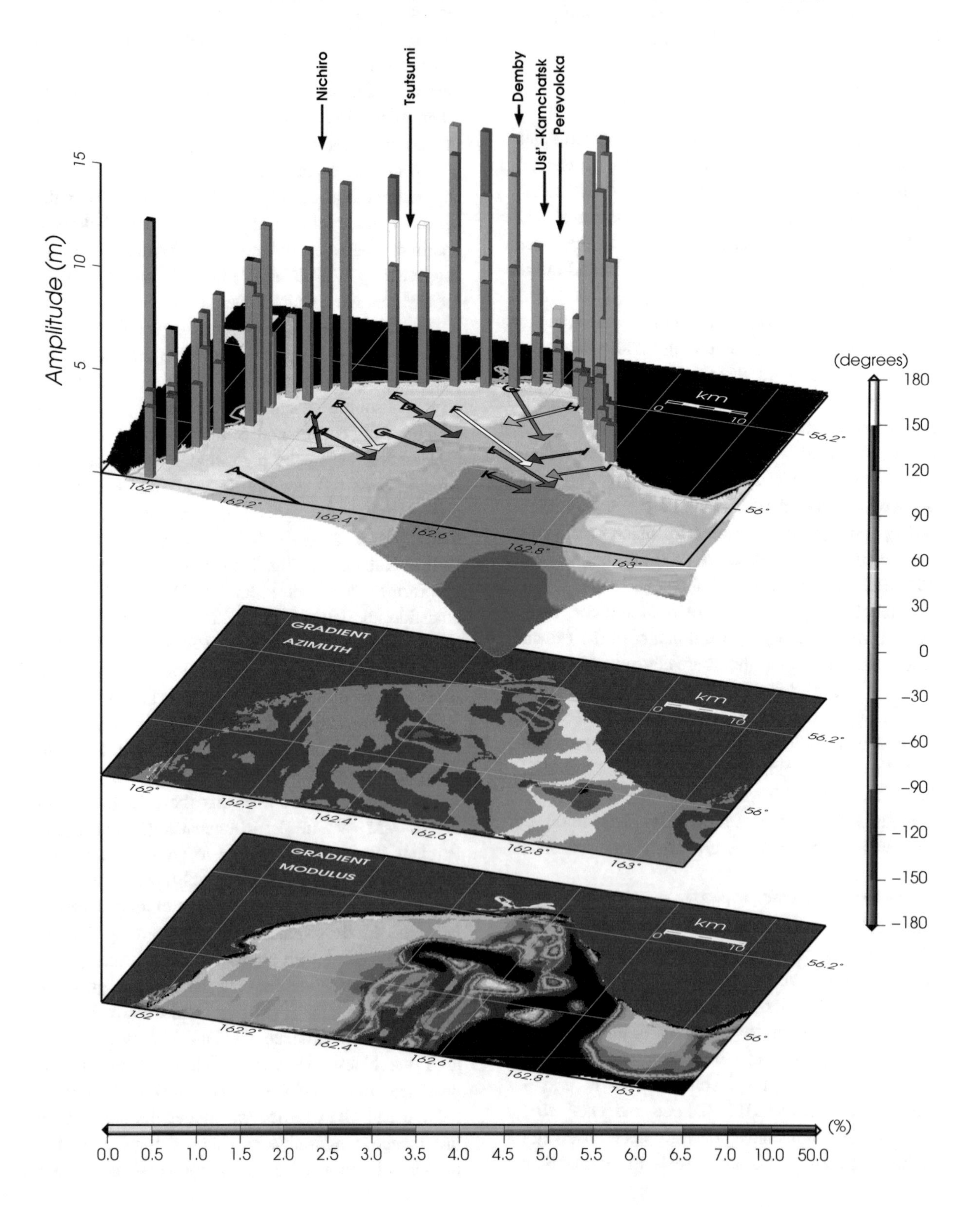
Nichiro
Tsutsumi
Demby
Ust'-Kamchatsk
Perevoloka
Amplitude (m)
15
10
5
km
0
10
56.2°
56°
162°
162.2°
162.4°
162.6°
162.8°
163°
(degrees)
180
150
120
90
60
30
0
-30
-60
-90
-120
-150
-180
GRADIENT
AZIMUTH
GRADIENT
MODULUS
(%)
0.0 0.5 1.0 1.5 2.0 2.5 3.0 3.5 4.0 4.5 5.0 5.5 6.0 6.5 7.0 10.0 50.0

◀Figure 12
(Bottom) Modulus of the gradient field calculated from bathymetry; (middle) azimuth of the bathymetry gradient field; (top) 3-D bathymetry of the northern Gulf of Kamchatka with arrows representing the designed landslide dipoles in tsunami simulations. Maximum simulated amplitudes from the landslide scenarios at each of the 41 gauges are shown as vertical columns which are color-coded to match their corresponding arrows

η_- and η_+, separated by a lever of length l in the direction x of steepest descent, with a Gaussian profile along x and a sech^2 one across it, the initial deformation thus taking the form (Okal et al. 2009)

$$\eta(x, y; t = 0_+) = -\eta_- \exp(-\alpha_- x^2) \cdot \text{sech}^2(\gamma_- y) + \eta_+ \exp(-\alpha_+ (x - l)^2) \cdot \text{sech}^2(\gamma_+ y) \quad (1)$$

We applied time steps of $\Delta t = 0.4$ s over 12 h time spans in our digitized grid interpolated down to 7.5 arc-seconds in order to accommodate a sufficient number of grid points per wavelength (e.g. Shuto et al. 1986). We designed 14 landslide scenarios (Table 3) following the slope map of the Gulf of Kamchatka calculated from our bathymetry grid as shown on the two bottom slices in Fig. 12. The columns in the upper slice in Fig. 12 indicate the maximum calculated amplitudes at 25 virtual gauges along northern margins of the Kamchatskiy Bight. These columns are color coded to match their corresponding slide scenario in the Bight. As readily seen in Fig. 15 and detailed below, only slide scenario (N) provides an acceptable wave amplitude offshore of Nichiro, and an appropriate fall-off farther northeast along the shore.

In order to analyze the stability of the model (N) solution, we varied the dipole length from 5 to 15 km and its azimuth from 150° to 160°, in unit increments and recorded the time series at Nichiro. We then selected maximum amplitudes at 25 gauges (Fig. 15). The result, in Fig. 16, shows that the suggested dipole is stable, in the sense that the highest amplitudes are concentrated around Nichiro with amplitudes falling to 1–2 m at Perevoloka, as best demonstrated with the average of all the curves (red curve in Fig. 16).

In conclusion of this section, Model (N), featuring an underwater slide initiating at 56.065°N, 162.200°E and sliding 10 km along a local slope of 1–2%, in the azimuth 155°, with a total volume of 0.4 km^3, gives a satisfactory model to the fundamental observations available for the tsunami of 13 April 1923, namely the peak run-up of $\sim$20 m at Nichiro and its rapid northeastward decay, reaching only 1–2 m at Perevoloka. It could easily have been triggered by

Table 3

Dipole model parameters

Model	Head (trough)					Lever		Hump		
	Lat. (°N)	Lon. (°N)	η_- (m)	α_- (km^{-2})	γ_- (km^{-1})	l (km)	Az (°)	η_+ (m)	α_+ (km^{-2})	γ_+ (km^{-1})
A	55.90	162.14	25	0.2	1.4	15	120	15	0.12	1.16
B	56.10	162.23	15	0.7	2.5	13	140	9	0.4	2.0
C	56.05	162.35	15	0.2	1.4	8	110	9	0.12	1.16
D	56.12	162.36	25	0.2	1.4	10	125	15	0.12	1.16
E	56.14	162.32	15	0.2	1.4	7.2	125	9	0.12	1.16
F	56.13	162.45	25	0.2	1.4	17	130	15	0.12	1.16
G	56.20	162.52	15	0.2	1.4	14	145	9	0.12	1.16
H	56.18	162.66	15	1.2	0.8	9	220	9	0.57	0.7
I	56.08	162.75	25	0.2	1.4	7	230	15	0.12	1.16
J	56.05	162.81	25	0.2	1.4	7	230	15	0.12	1.16
K	55.99	162.61	25	0.2	1.4	6.6	115	15	0.12	1.16
L	56.05	162.58	25	0.2	1.4	12	125	15	0.12	1.16
M	56.03	162.23	25	0.10	0.8	10	120	15	0.25	0.30
N	**56.06**	**162.20**	**25**	**0.08**	**0.8**	**10**	**155**	**17**	**0.25**	**0.30**

Figure 13
Examples of onland slides on the coastal slopes of the Gulf of Kamchatka (image courtesy: Google Earth)

an earthquake of moment 3×10^{27} dyn cm located in the southern part of the Ozernoy Bight or the northern Kamchatskiy Peninsula, which encompasses the ISC and ERE solutions as well as the southern part of our RE confidence ellipse.

The field of wave heights from landslide sources is known to decay with lateral distance along a beach in the near field as a result of their smaller dimension as compared to dislocations. However, this property was derived by Okal and Synolakis (2004) only in the simple case of a bathymetry with translational symmetry along the shore. We explore on Fig. 17 the possible additional influence of laterally varying bathymetry, as we note on Fig. 15 an asymmetric trend in the decay of amplitudes away from Nichiro, which is faster to the NE than to the SW. We extend our ray-tracing experiments to the case of Landslide model (N), schematized as a dipolar source consisting of a trough (T) and a hump (H), by shooting rays equally spaced at regular 4° degree intervals in azimuth from each of the poles, using time steps of 20 s. Figure 17b is an interpretation of the dataset on frame (a), obtained by color-coding the density of points in (a) per $0.05° \times 0.05°$ area. Both frames show that the actual bathymetry of the bight acts to strengthen the decay of the wave in the northeast direction, towards Ust'-Kamchatsk from its maximum around Nichiro; they predict a slower decay, and thus larger amplitudes towards the SW, where however the coast quickly becomes rugged and elevated, suggesting that it was most probably uninhabited in 1923 and that any wave action would have remained unreported. This experiment provides physical insight into the general pattern of directivity of the wavefield of the preferred Landslide source (N) for the tsunami of 13 April 1923.

Another potential datum from the historical reports of the UK tsunami is the time interval of 15 min separating the two major waves at Ust'-Kamchatsk (Soloviev and Ferchev 1961), which could be

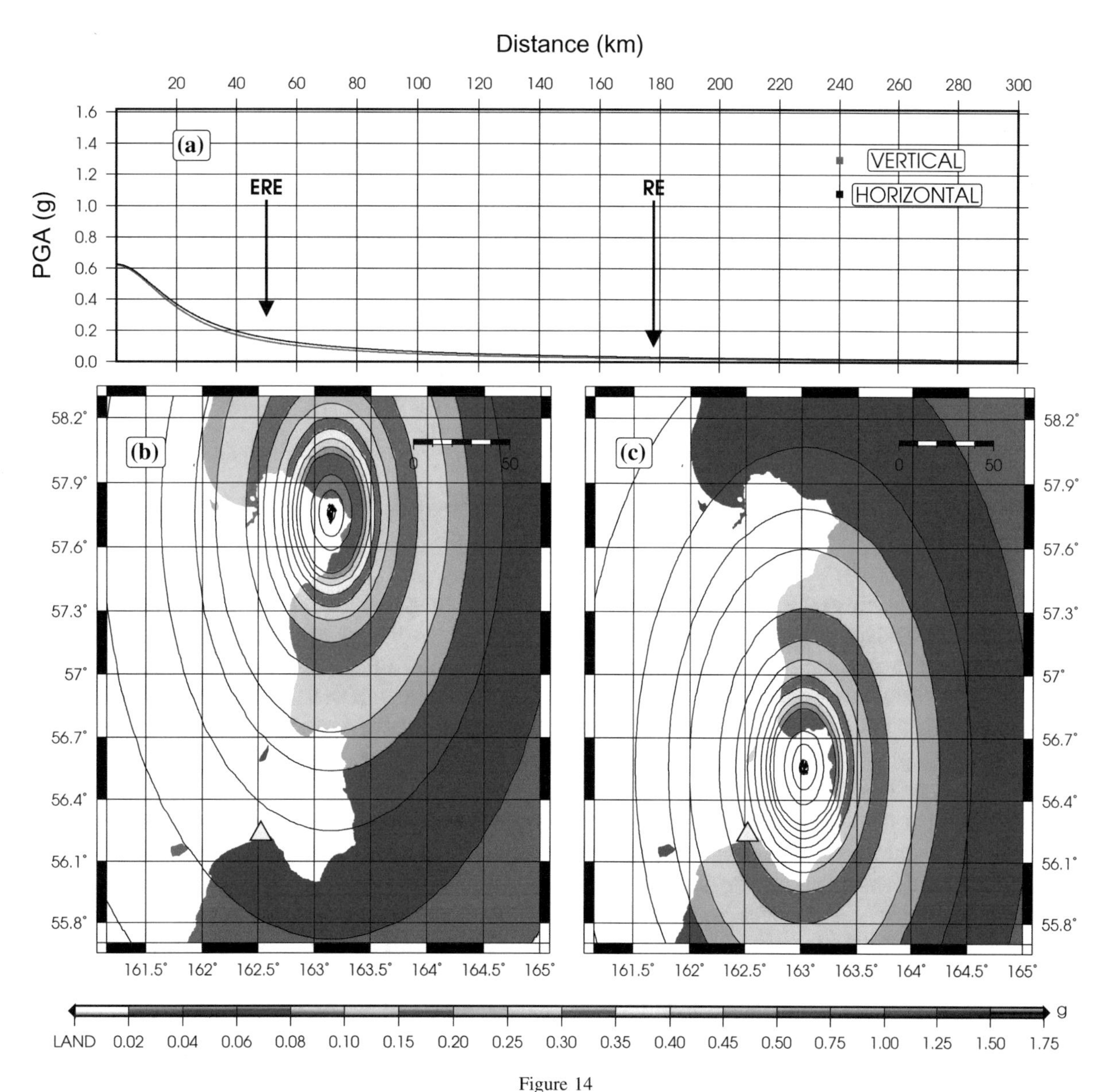

Figure 14
Calculated peak ground acceleration (PGA) at 10 Hz in units of g for the 13 April 1923 event, assuming a thrust mechanism with a magnitude of $M_w = 7.6$ and a focal depth of 10 km. **a** The attenuation curve showing acceleration decay with distance. Distances to the Engdahl relocation (ERE) and the relocated epicenter from this study (RE) to Ust'-Kamchatsk are depicted with arrows; **b** the PGA from the ERE epicenter; **c** the PGA from the RE epicenter. Ust'-Kamchatsk is shown with yellow triangles in **b** and **c**

interpreted as expressing the dominant period in the spectrum of the local tsunami wave. As such, it is tempting to use this observation as a further constraint on the source of the 1923 tsunami. In simple terms, the spectrum of an earthquake-generated tsunami should be controlled primarily by the dimensions of the source, principally its fault width W; we have verified that wavetrains simulated off Ust'-Kamchatsk for our various earthquake scenarios have spectra peaked around 1450 and 2100 s, respectively, for the average and upper bound moments. By contrast, landslide tsunamis, emanating from spatially smaller sources, should have higher frequency spectra. However, it has long been known that tsunamis recorded at shorelines have spectra strongly affected by the natural frequencies of bights

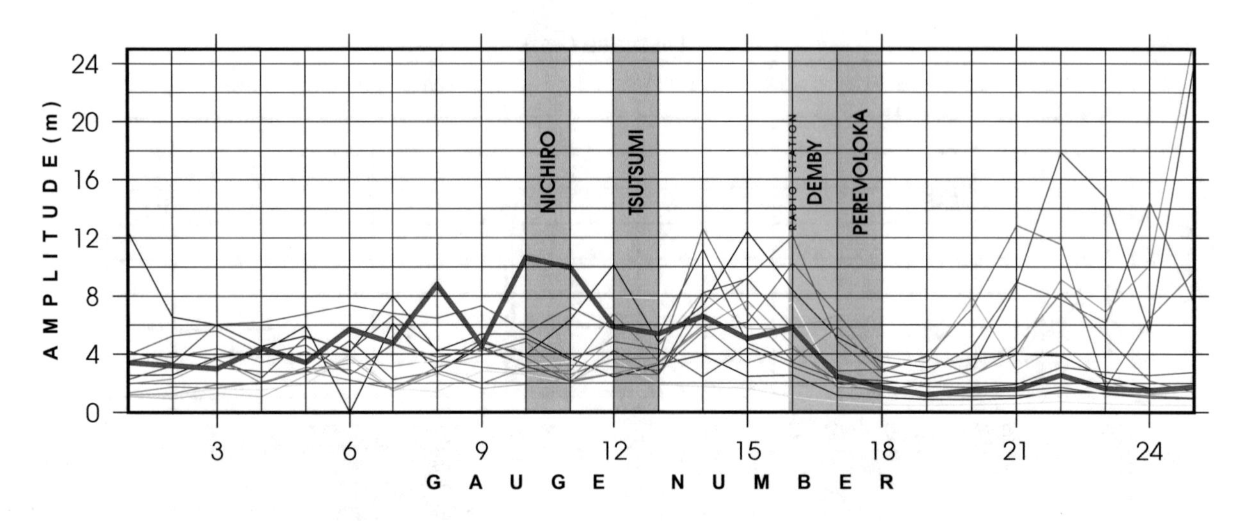

Figure 15
Tsunami heights recorded at 41 virtual gauges for the 14 landslide scenarios. The curves are color-coded to match their corresponding slides in Fig. 12

and bays, an idea already expressed by Omori (1902). Rabinovich (1997) has shown that the resulting spectral peak of a tsunami record is a complex combination of source and receiver properties, which can be unraveled only using a sufficiently rich dataset, comprising for example the spectral properties of background noise in the absence of the tsunami signal. In this context, a single datum for what may be interpreted as a dominant period cannot provide additional constraints on the source of the tsunami. We also note that the lone available report ("Fifteen minutes after the first wave, a second wave moved in"), compiled from witness memories by Troshin and Diagilev (1926), months if not years after the fact, cannot be taken as more than an order of magnitude of the time between the two waves, and clearly lacks the precision required to be used as a scientific datum.

5. Conclusion

Our principal conclusions on the Ust'-Kamchatsk earthquake and tsunami of 13 April 1923 can be summarized as follows:

- The UK earthquake of 13 April 1923 is confirmed as an anomalously slow event, featuring a weak spectrum at short periods, and a seismic moment increasing significantly at mantle frequencies (5 mHz). In this context, it is comparable to other "tsunami earthquakes" occurring in the aftermath of a major subduction event, such as during the 1963 Kuril and 1932 Mexican series.
- On the other hand, the majority of modern relocation efforts (ISC-GEM, E.R. Engdahl's, and our own solution) suggest that the UK earthquake did not constitute a genuine aftershock of the main event of 03 February, but rather took place north of the Kamchatka subduction zone, possibly at Cape Kamchatskiy, more probably inside the Kamchatskiy Peninsula, or even as far away as the Ozernoy Bight. The latter could also help interpret tsunami deposits identified by Bourgeois et al. (2006) at Stolbovaya Bay as dating back to between 1854 and 1956.
- The scarce dataset of available mantle wave records is difficult to reconcile with the subduction geometry of the mainshock, further suggesting that the UK earthquake is not part of a classic aftershock sequence. On their basis, the static seismic moment is estimated at 3×10^{27} dyn cm, suggesting a seismic slip on the order of 2.2 m.
- In this context, we note that the slip released during the 1971 earthquake (estimated at 2.8 m) is in excellent agreement with that predicted by plate tectonics models for the 35 years elapsed since the 1936 event (NUVEL: 7.8 cm/year in the azimuth

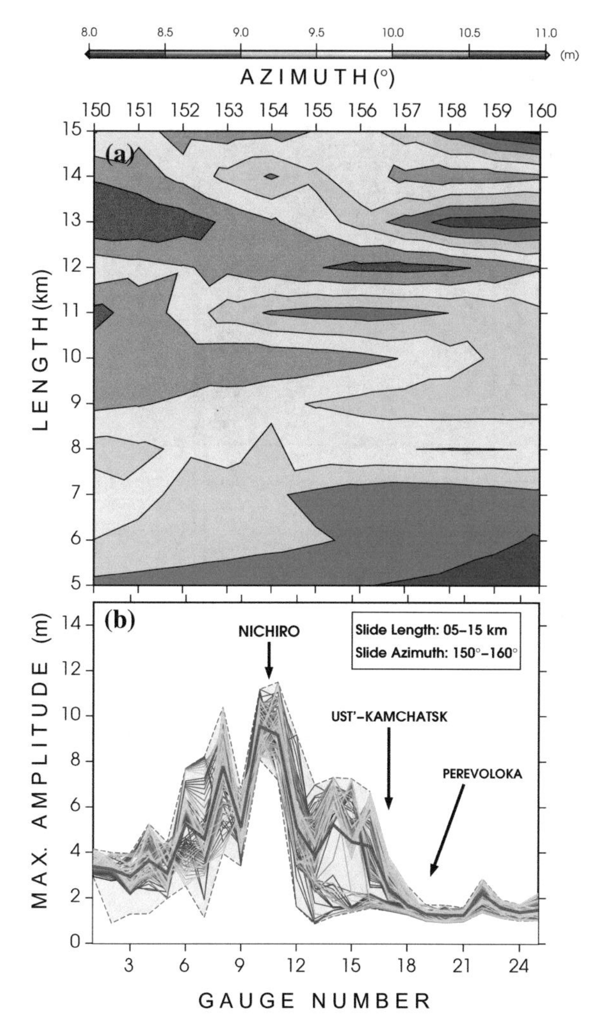

Figure 16
a Maximum calculated amplitudes at Nichiro as a function of dipole length and azimuth; **b** maximum amplitudes at the 25 virtual gauges shown in Figs. 12 and 15 while varying slide dipole length and azimuth. The red curve is connecting the average values at each gauge. The shaded area depicts the range between the maxima and minima (the red dashed curves)

311° (DeMets et al. 1990); REVEL: 7.5 cm/year in the azimuth 306° (Sella et al. 2002)), while the slip accumulated between the previous tsunami documented in the Commander Islands in 1849 (Soloviev and Ferchev 1961) and 1936, would amount to 6.5–6.8 m, a figure intermediate between the estimated slips of the 1917–1936 doublet (5.2 m) and of a possible 1917–1923–1936 trio (7.4 m). On this basis, it is not possible to exclude the common location of the 1917, 1936 and 1971 events as an epicenter for the UK event.

- While the dataset of reported run-up values from the UK tsunami remains scant, it is characterized by very large (20 m or more) values at Nichiro, falling down to 2 m or less at Perevoloka, about 40 km away. This combination of high run-ups and peaked distribution cannot be simulated using any legitimate seismic source, especially since a rapid fall-off with distance along the beach constitutes a robust discriminant between dislocative and landslide sources (Okal and Synolakis 2004).
- While the seismic dislocation must have contributed to the tsunami, we show that it is insufficient to account for the distribution of observed amplitudes. Rather, we give a model of generation of the UK tsunami by a landslide which satisfactorily reproduces this distribution, with an acceptable offshore wave amplitude at Nichiro (12 m). The source parameters for the landslide are consistent with the known gradient field of bathymetry, with the macroseismic effects of the earthquake reported at Ust'-Kamchatsk, and with the ubiquitous presence of aerial landslides along the nearby elevated shorelines.

The present study exacerbates the strong diversity of environments and mechanisms for "tsunami earthquakes". As previously reviewed (Okal 2008; Okal and Saloor 2017), these events have generally been described as falling into two categories: "Aftershock tsunami earthquakes" (ATEs) occurring in the aftermath of mega events, and "Primary tsunami earthquakes" (PTEs) taking place as mainshocks. Typical examples of ATEs are the 20 October 1963 Kuril or 22 June 1932 Mexican earthquakes, and of PTEs the 1992 Nicaragua or 1994 and 2006 Java events.

Previous work had already pointed out diverse patterns for ATEs, with some events (e.g., Mentawai 2010) simply rupturing an extension of the plate interface upwards from the mainshock (Hill et al. 2012) while others (e.g., Kuril 1963 (20 Oct.) and 1975) would take place by stress transfer along splay faults in accretionary prisms, as originally proposed by Fukao (1979). The 1923 Kamchatka tsunami earthquake presents a variation to the latter, where the stress transfer can occur across a plate boundary, into an a priori different tectonic environment. In

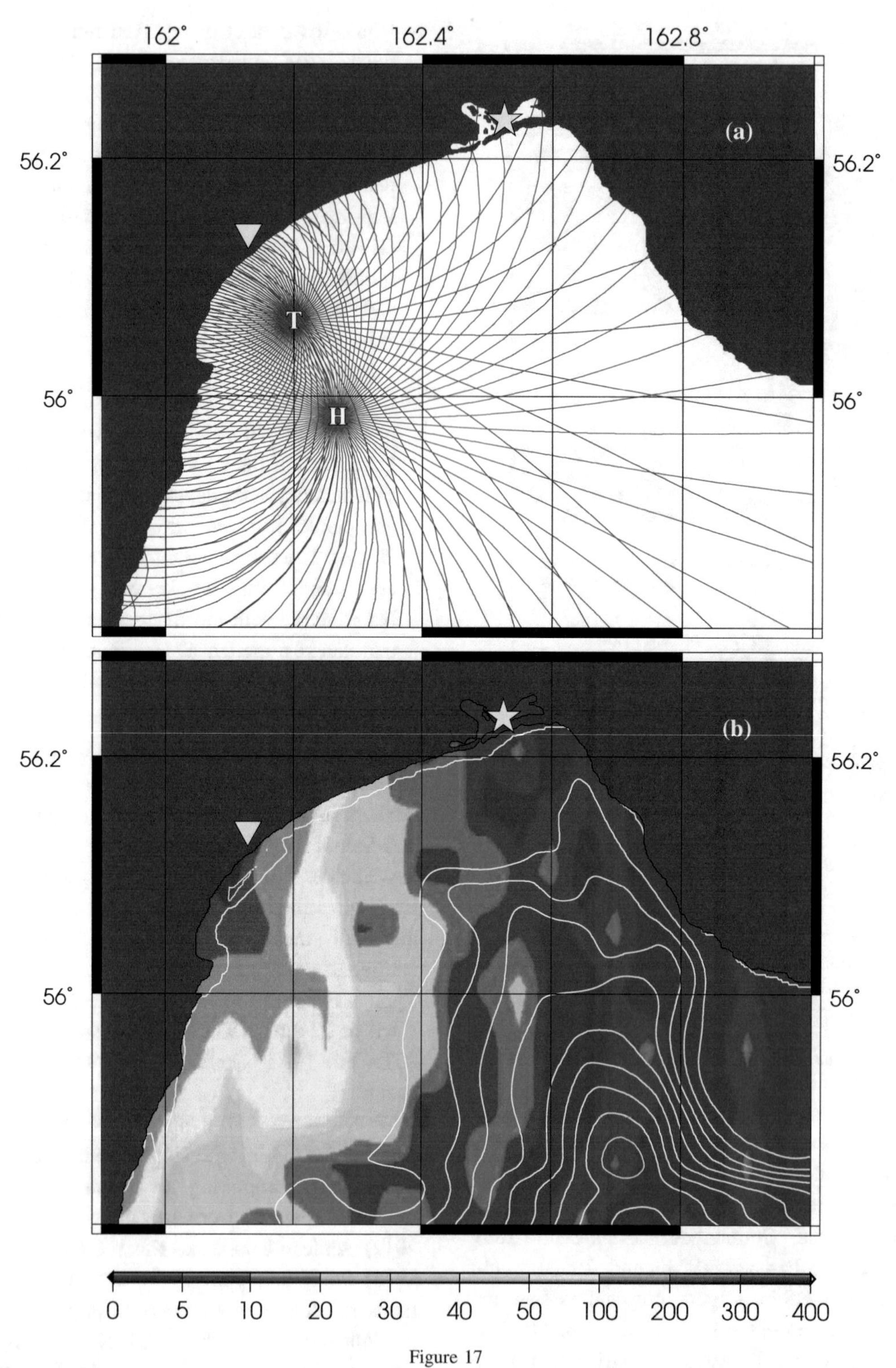

Figure 17

a Ray-tracing of the slide dipole. Blue and red lines show rays propagating, respectively, from the trough (T) and the hump (H) of the dipole; **b** number of ray points per $0.05^\circ \times 0.05^\circ$ area calculated from (**a**). The yellow stars and inverted triangles represent Ust'-Kamchatsk and Nichiro, respectively

addition, our study provides new evidence for the role of underwater landslides, triggered by earthquakes, in the generation of tsunamis with catastrophic amplitudes on nearby shorelines. In this respect, the UK event is comparable to the great 1946 Aleutian one, which similarly combined a very slow seismic source and an underwater landslide, even though that earthquake would qualify as a PTE, and featured a much larger seismic moment.

Finally, we note that the reported maximum run-up (at least 20 m) and inundation (7 km into the mouth of the Kamchatka River) were larger than measured along a similarly flat shoreline during the 1998 Papua New Guinea tsunami (15 m and at most 1 km (Synolakis et al. 2002)); thus a comparable societal disaster was avoided only thanks to the sparse population of the affected area. Considering the fundamentally non-linear nature of the sliding process, and our obviously limited knowledge of the relevant population of precarious masses with potential for destabilization on the sea floor, the 1923 Ust'-Kamchatsk earthquake casts an ominous augury in terms of seismic hazard for coastal communities in seismic areas worldwide.

Acknowledgements

We thank Bob Engdahl for a customized relocation of the Ust'-Kamchatsk event. We are grateful to Editor A. Rabinovich and two anonymous reviewers for their constructive comments. Olga Yakovenko helped optimize transliteration of Russian names. This research was partly supported by the National Science Foundation, under Grant Number OCE-13-31463 to the University of Pittsburgh; we thank Louise Comfort for her leadership in that joint venture. Some figures were drafted using the GMT software (Wessel and Smith 1991).

References

Aki, K., & Richards, P. G. (2002). *Quantitative seismology*. University Science Books.

Ambraseys, N. (1991). The Rukwa earthquake of 13 December 1910 in East Africa. *Terra Nova*, *3*(2), 202–211.

Anonymous. (2001). *Province of Kamchatka map*. Moscow: Federal Service of Geodesy and Geography of Russia.

Bell, R., Holden, C., Power, W., Wang, X., & Downes, G. (2014). Hikurangi margin tsunami earthquake generated by slow seismic rupture over a subducted seamount. *Earth and Planetary Science Letters*, *397*, 1–9.

Borisov, V. I. (2002). Forgotten Tragedy, www.kamchatsky-krai.ru/geography/article/borisov.htm, Accessed on October 23, 2017 [in Russian].

Bourgeois, J., & Pinegina, T. K. (2017). 1997 Kronotsky earthquake and tsunami and their predecessors, Kamchatka, Russia. *Natural Hazards and Earth System Sciences Discussions* (submitted).

Bourgeois, J., Pinegina, T. K., Ponomareva, V., & Zaretskaia, N. (2006). Holocene tsunamis in the southwestern Bering Sea, Russian Far East, and their tectonic implications. *Geological Society of America Bulletin*, *118*(3–4), 449–463.

Brunsden, D., & Prior, D. B. (1984). *Slope instability*. New York: Wiley.

Campbell, K. W., & Bozorgnia, Y. (2003). Updated near-source ground-motion (attenuation) relations for the horizontal and vertical components of peak ground acceleration and acceleration response spectra. *Bulletin of the Seismological Society of America*, *93*(1), 314–331.

Cormier, V. F. (1975). Tectonics near the junction of the Aleutian and Kuril-Kamchatka arcs and a mechanism for middle Tertiary magmatism in the Kamchatka basin. *Geological Society of America Bulletin*, *86*(4), 443–453.

Courant, R., Friedrichs, K., & Lewy, H. (1928). Über die partiellen Differenzengleichungen der mathematischen Physik. *Mathematische Annalen*, *100*(1), 32–74.

Daughton, T. M. (1990). *Focal mechanism of the 22 November 1969 Kamchatka earthquake from teleseismic waveform analysis*. Honors Thesis, Colorado College, Colorado Springs. 70 pp.

DeMets, C., Gordon, R. G., Argus, D., & Stein, S. (1990). Current plate motions. *Geophysical Journal International*, *101*(2), 425–478.

Dziewonski, A., Chou, T.-A., & Woodhouse, J. (1981). Determination of earthquake source parameters from waveform data for studies of global and regional seismicity. *Journal of Geophysical Research: Solid Earth*, *86*(B4), 2825–2852.

Ebel, J. E., & Chambers, D. W. (2016). Using the locations of $M \geq 4$ earthquakes to delineate the extents of the ruptures of past major earthquakes. *Geophysical Journal International*, *207*(2), 862–875.

Ekström, G., Nettles, M., & Dziewoński, A. (2012). The global CMT project 2004–2010: Centroid-moment tensors for 13,017 earthquakes. *Physics of the Earth and Planetary Interiors*, *200*, 1–9.

Engdahl, E. R., & Villaseñor, A. (2002). Global seismicity: 1900–1999.

Fedotov, C., Gusev, A., Zobin, B., Kondratienko, A., & Chepkynas, K. (1973). The Ozernovskiy earthquake and tsunami of 22 (23) November 1969. *Earthquakes in the USSR in the year 1969*, 195–208. [in Russian].

Fisher, R., Jantsch, M., & Comer, R. (1982). General Bathymetric Chart of the Oceans (GEBCO).

Fryer, G. J., & Watts, P. (2001). Motion of the Ugamak slide, probable source of the tsunami of 1 April 1946. In *Proceedings of the International Tsunami Symposium* (pp. 683–694).

Fukao, Y. (1979). Tsunami earthquakes and subduction processes near deep-sea trenches. *Journal of Geophysical Research: Solid Earth*, *84*(B5), 2303–2314.

Geller, R. J. (1976). Scaling relations for earthquake source parameters and magnitudes. *Bulletin of the Seismological Society of America*, *66*(5), 1501–1523.

Gorin, S., & Chebanova, V. (2011). Salinization-related transformation of hydrological regime and benthos in the Nerpichye and Kultuchnoe Lakes, at the Kamchatka River estuary. In *V.V. Levandinov Memorial Lectures* (pp. 119–128). Vladivostok: Russian Academy of Sciences [in Russian].

Gusev, A. A. (2004). Schematic map of the source zones of large Kamchatka earthquakes of the instrumental epoch. In *Complex seismological and geophysical investigations of Kamchatka* (pp. 75–80). Petropavlovsk–Kamchatsky. [in Russian].

Gusev, A. A., Zobin, V. M., Kondratenko, A. M., & Shumilina, L. S. (1975). The earthquake of Ust'-Kamchatsk, 15 XII. *Earthquakes in the USSR in the year 1971* (pp. 172–184) [in Russian].

Gutenberg, B., & Richter, C. F. (1954). *Seismicity of the Earth and associated phenomena*. Princeton, NJ: Princeton Univ. Press.

Hill, E. M., Borrero, J. C., Huang, Z., Qiu, Q., Banerjee, P., Natawidjaja, D. H., et al. (2012). The 2010 $M_w = 7.8$ Mentawai earthquake: Very shallow source of a rare tsunami earthquake determined from tsunami field survey and near-field GPS data. *Journal of Geophysical Research: Solid Earth*, *117*, B06402.

Hoffmann, G., Al-Yahyai, S., Naeem, G., Kociok, M., & Grützner, C. (2014). An Indian Ocean tsunami triggered remotely by an onshore earthquake in Balochistan, Pakistan. *Geology*, *42*(10), 883–886.

Jaggar, T. (1930). *Volcano letter*, *274*, 1–4.

Jobert, N., & Jobert, G. (1983). An application of ray theory to the propagation of waves along a laterally heterogeneous spherical surface. *Geophysical Research Letters*, *10*(12), 1148–1151.

Kanamori, H. (1972). Mechanism of tsunami earthquakes. *Physics of the Earth and Planetary Interiors*, *6*(5), 346–359.

Kanamori, H. (1977). The energy release in great earthquakes. *Journal of Geophysical Research*, *82*(20), 2981–2987.

Kanamori, H. (1985). Non-double-couple seismic source. In *Proc: XXIIIrd Gen. Assemb. Intl. Assoc. Seismol. Phys. Earth Inter* (p. 425)

Kawata, Y., Benson, B. C., Borrero, J. C., Borrero, J. L., Davies, H. L., Lange, W. P., et al. (1999). Tsunami in Papua New Guinea was as intense as first thought. *Eos, Transactions American Geophysical Union*, *80*(9), 101–105.

Keefer, D. K. (1984). Landslides caused by earthquakes. *Geological Society of America Bulletin*, *95*(4), 406–421.

Kuksina, L., & Chalov, S. (2012). The suspended sediment discharge of the rivers running along territories of contemporary volcanism in Kamchatka. *Geography and Natural Resources*, *33*(1), 67–73.

López, A. M., & Okal, E. A. (2006). A seismological reassessment of the source of the 1946 Aleutian 'tsunami' earthquake. *Geophysical Journal International*, *165*(3), 835–849.

Mansinha, L., & Smylie, D. (1971). The displacement fields of inclined faults. *Bulletin of the Seismological Society of America*, *61*(5), 1433–1440.

Martin, M. E., Weiss, R., Bourgeois, J., Pinegina, T. K., Houston, H., & Titov, V. V. (2008). Combining constraints from tsunami modeling and sedimentology to untangle the 1969 Ozernoi and 1971 Kamchatskii tsunamis. *Geophysical Research Letters*, *35*, L01610.

Menialov, A. A. (1946). Tsunamis in the Ust'-Kamchatsk region. *Bull. Kamchatka Volcanol. Stn.* (Vol. 12) [in Russian].

Minoura, K., Gusiakov, V., Kurbatov, A., Takeuti, S., Svendsen, J., Bondevik, S., et al. (1996). Tsunami sedimentation associated with the 1923 Kamchatka earthquake. *Sedimentary Geology*, *106*(1–2), 145–154.

Newman, A. V., & Okal, E. A. (1998). Teleseismic estimates of radiated seismic energy: The E/M_0 discriminant for tsunami earthquakes. *Journal of Geophysical Research: Solid Earth*, *103*(B11), 26885–26898.

Okal, E. (2008). The excitation of tsunamis by earthquakes. In E. Bernard & A. Robinson (Eds.) *The Sea: Ideas and observations on progress in the study of the seas* (pp. 137–177). Cambridge: Harvard Univ. Press.

Okal, E. A. (1992). Use of the mantle magnitude M_m for the reassessment of the moment of historical earthquakes. *Pure and Applied Geophysics*, *139*(1), 17–57.

Okal, E. A. (2003). Normal mode energetics for far-field tsunamis generated by dislocations and landslides. *Pure and Applied Geophysics*, *160*(10), 2189–2221.

Okal, E. A. (2011). Tsunamigenic earthquakes: past and present milestones. *Pure and Applied Geophysics*, *168*(6–7), 969–995.

Okal, E. A., & Borrero, J. C. (2011). The 'tsunami earthquake' of 1932 June 22 in Manzanillo, Mexico: seismological study and tsunami simulations. *Geophysical Journal International*, *187*(3), 1443–1459.

Okal, E. A., & Saloor, N. (2017). Historical tsunami earthquakes in the Southwest Pacific: An extension to $\Delta > 80°$ of the energy-to-moment parameter Θ. *Geophysical Journal International*, *210*(2), 852–873.

Okal, E. A., & Synolakis, C. E. (2004). Source discriminants for near-field tsunamis. *Geophysical Journal International*, *158*(3), 899–912.

Okal, E. A., & Talandier, J. (1986). *T*-wave duration, magnitudes and seismic moment of an earthquake—Application to tsunami warning. *Journal of Physics of the Earth*, *34*(1), 19–42.

Okal, E. A., & Talandier, J. (1989). M_m: A variable-period mantle magnitude. *Journal of Geophysical Research: Solid Earth*, *94*(B4), 4169–4193.

Okal, E. A., Plafker, G., Synolakis, C. E., & Borrero, J. C. (2003). Near-field survey of the 1946 Aleutian tsunami on Unimak and Sanak Islands. *Bulletin of the Seismological Society of America*, *93*(3), 1226–1234.

Okal, E. A., Fritz, H. M., Raad, P. E., Synolakis, C., Al-Shijbi, Y., & Al-Saifi, M. (2006a). Oman field survey after the December 2004 Indian Ocean tsunami. *Earthquake Spectra*, *22*(S3), 203–218.

Okal, E. A., Fritz, H. M., Raveloson, R., Joelson, G., Pančošková, P., & Rambolamanana, G. (2006b). Madagascar field survey after the December 2004 Indian Ocean tsunami. *Earthquake Spectra*, *22*(S3), 263–283.

Okal, E. A., Synolakis, C. E., Uslu, B., Kalligeris, N., & Voukouvalas, E. (2009). The 1956 earthquake and tsunami in Amorgos, Greece. *Geophysical Journal International*, *178*(3), 1533–1554.

Okal, E. A., Visser, J. N., & de Beer, C. H. (2014). The Dwarskersbos, South Africa local tsunami of August 27, 1969: Field survey and simulation as a meteorological event. *Natural Hazards*, *74*(1), 251–268.

Omori, F. (1902). On tsunamis around Japan. *Rep. Imper. Earthq. Comm.*, *34*, 5–79 [in Japanese].

Polet, J., & Kanamori, H. (2000). Shallow subduction zone earthquakes and their tsunamigenic potential. *Geophysical Journal International*, *142*(3), 684–702.

Prior, D. B., Bornhold, B. D., Coleman, J. M., & Bryant, W. R. (1982). Morphology of a submarine slide, Kitimat Arm, British Columbia. *Geology*, *10*(11), 588–592.

Rabinovich, A. B. (1997). Spectral analysis of tsunami waves: Separation of source and topography effects. *Journal of Geophysical Research: Oceans*, *102*(C6), 12663–12676.

Salaree, A., & Okal, E. A. (2015). Field survey and modelling of the Caspian Sea tsunami of 1990 June 20. *Geophysical Journal International*, *201*(2), 621–639.

Satake, K. (1988). Effects of bathymetry on tsunami propagation: Application of ray tracing to tsunamis. *Pure and Applied Geophysics*, *126*(1), 27–36.

Sella, G. F., Dixon, T. H., & Mao, A. (2002). REVEL: A model for recent plate velocities from space geodesy. *Journal of Geophysical Research: Solid Earth*, *107*(2081).

Shuto, N., Suzuki, T., & Hasegawa, K. (1986). A study of numerical techniques on the tsunami propagation and run-up. *Science of Tsunami Hazard*, *4*, 111–124.

Skempton, A. (1953). Soil mechanics in relation to geology. *Proceedings of the Yorkshire Geological Society*, *29*(1), 33–62.

Soloviev, S., Go, C. N., & Kim, K. S. (1986). Catalog of tsunamis in the Pacific, 1969–1982, *Moscow: USSR Academy of Sciences, Soviet Geophysical Committee*, [in English; translation: Amerind Publishing Co., New Delhi, 1998].

Soloviev, S., Campos-Romero, M., & Plink, N. (1992). Orléansville tsunami of 1954 and El Asnam tsunami of 1980 in the Alboran Sea (Southwestern Mediterranean Sea). *Izvestiya Earth Phys*, *28*(9), 739–760.

Soloviev, S. L., & Ferchev, M. D. (1961). Summary of data on tsunamis in the USSR, Bull. Council Seism. Acad. *USSR*, *9*, 23–55.

Stauder, W., & Mualchin, L. (1976). Fault motion in the larger earthquakes of the Kurile-Kamchatka Arc and of the Kurile-Hokkaido corner. *Journal of Geophysical Research*, *81*(2), 297–308.

Storchak, D., Di Giacomo, D., Engdahl, E., Harris, J., Bondár, I., Lee, W., et al. (2015). The ISC-GEM global instrumental earthquake catalogue (1900–2009): Introduction. *Physics of the Earth and Planetary Interiors*, *239*, 48–63.

Synolakis, C., Bernard, E., Titov, V., Kânoğlu, U., & González, F. (2008). Validation and verification of tsunami numerical models. *Pure and Applied Geophysics*, *165*(11–12), 2197–2228.

Synolakis, C. E., Bardet, J.-P., Borrero, J. C., Davies, H. L., Okal, E. A., Silver, E. A., Sweet, S., & Tappin, D. R. (2002). The slump origin of the 1998 Papua New Guinea tsunami. *Proceedings of the Royal Society of London, Series A*, *458*, 763–789.

Tanioka, Y., Ruff, L., & Satake, K. (1997). What controls the lateral variation of large earthquake occurrence along the Japan Trench? *Island Arc*, *6*(3), 261–266.

The Los Angeles Times. (1923). *15 April 1923*.

Titov, V. & González, F. (1997). *Implementation and Testing of the Method of Splitting Tsunami (MOST) Model*, US Department of Commerce, National Oceanic and Atmospheric Administration, Environmental Research Laboratories, Pacific Marine Environmental Laboratory.

Titov, V., Kânoğlu, U., & Synolakis, C. (2016). Development of MOST for real-time tsunami forecasting. *Journal of Waterway, Port, Coast and Oceanic Engineering*, *142*, 03116004-1–03116004-16.

Titov, V. V., & Synolakis, C. E. (1995). Modeling of breaking and nonbreaking long-wave evolution and runup using VTCS-2. *Journal of Waterway, Port, Coastal, and Ocean Engineering*, *121*(6), 308–316.

Titov, V. V., & Synolakis, C. E. (1997). Extreme inundation flows during the Hokkaido-Nansei-Oki tsunami. *Geophysical Research Letters*, *24*(11), 1315–1318.

Titov, V. V., & Synolakis, C. E. (1998). Numerical modeling of tidal wave run-up. *Journal of Waterway, Port, Coastal, and Ocean Engineering*, *124*(4), 157–171.

Troshin, A. N., & Diagilev, G. A. (1926). *The Ust' Kamchatsk earthquake of April 13, 1923*. Library Institute Physics Earth, USSR Academy of Sciences, Moscow [in Russian].

Utsu, T. (1970). Aftershocks and earthquake statistics (1): Some parameters which characterize an aftershock sequence and their interrelations. *Journal of the Faculty of Science, Hokkaido University. Series 7, Geophysics*, *3*(3), 129–195.

von Huene, R., Kirby, S., Miller, J., & Dartnell, P. (2014). The destructive 1946 Unimak near-field tsunami: New evidence for a submarine slide source from reprocessed marine geophysical data. *Geophysical Research Letters*, *41*(19), 6811–6818.

Wald, D. J., Quitoriano, V., Heaton, T. H., & Kanamori, H. (1999). Relationships between peak ground acceleration, peak ground velocity, and modified Mercalli intensity in California. *Earthquake Spectra*, *15*(3), 557–564.

Wessel, P., & Smith, W. H. (1991). Free software helps map and display data, *Eos, Transactions American Geophysical Union*, *72*(41), 441 and 445–446.

Woods, M. T., & Okal, E. A. (1987). Effect of variable bathymetry on the amplitude of teleseismic tsunamis: A ray-tracing experiment. *Geophysical Research Letters*, *14*(7), 765–768.

Wysession, M. E., Okal, E. A., & Miller, K. L. (1991). Intraplate seismicity of the Pacific Basin, 1913–1988. *Pure and Applied Geophysics*, *135*(2), 261–359.

Zayakin, Y., & Luchinina, A. (1987). *Catalogue of Tsunamis on Kamchatka*. Obninsk: VNIIGMI-MTSD. [in Russian].

(Received September 17, 2017, revised October 31, 2017, accepted November 11, 2017)

Pure Appl. Geophys.

https://doi.org/10.1007/s00024-017-1727-3

Pure and Applied Geophysics

Tsunami Hazard Assessment of Coastal South Africa Based on Mega-Earthquakes of Remote Subduction Zones

Andrzej Kijko,[1] Ansie Smit,[1] Gerassimos A. Papadopoulos,[2] and Tatyana Novikova[2]

Abstract—After the mega-earthquakes and concomitant devastating tsunamis in Sumatra (2004) and Japan (2011), we launched an investigation into the potential risk of tsunami hazard to the coastal cities of South Africa. This paper presents the analysis of the seismic hazard of seismogenic sources that could potentially generate tsunamis, as well as the analysis of the tsunami hazard to coastal areas of South Africa. The subduction zones of Makran, South Sandwich Island, Sumatra, and the Andaman Islands were identified as possible sources of mega-earthquakes and tsunamis that could affect the African coast. Numerical tsunami simulations were used to investigate the realistic and worst-case scenarios that could be generated by these subduction zones. The simulated tsunami amplitudes and run-up heights calculated for the coastal cities of Cape Town, Durban, and Port Elizabeth are relatively small and therefore pose no real risk to the South African coast. However, only distant tsunamigenic sources were considered and the results should therefore be viewed as preliminary.

Key words: seismic hazard assessment, tsunami hazard, tsunamigenic source, South Africa.

1. Introduction

The death and destruction in the wake of the mega-earthquakes and tsunamis of 26 December 2004 in Sumatra and 11 March 2011 in Tōhoku, Japan, serve as a reminder of the necessity for timely assessment of the potential tsunami hazard faced by susceptible coastlines. It is accepted generally that the extremely high death toll could have been avoided if an effective Indian Ocean tsunami warning system had been in place, similar to the Pacific Tsunami Warning Center (PTWC) in Hawaii and the National Tsunami Warning Center (NTWC) in Alaska. In response to the tragedy in Sumatra, the Indian Ocean Tsunami Early Warning System (IOTWS) was created.

The Intergovernmental Oceanographic Commission of UNESCO (IOC-UNESCO) was tasked with coordinating the establishment of the early warning system. In June 2005, during the 23rd session of the IOC, the Intergovernmental Coordination Group for the Indian Ocean Tsunami Warning and Mitigation System (ICG/IOTWS) was formally established. Today, the group has 28 member states, including South Africa. Many countries participate through national tsunami warning centres. However, three Regional Tsunami Service Providers (Australia, India and Indonesia) are the primary source of tsunami advisories for the Indian Ocean (http://iotic.ioc-unesco.org/indian-ocean-tsunami-warning-system/tsunami-early-warning-centres/57/regional-tsunami-service-providers; last access 16 Oct. 2017; see also Thomalla and Larsen 2010).

The 2004 event served to renew research interest into the mega-transoceanic seismic and tsunami hazard the Indian Ocean is facing. Most relevant research focuses on individual countries or regions, with two notable exceptions. A report by Berryman et al. (2013) was published by the global earthquake model (GEM) initiative and provides a detailed description of subduction zones worldwide, as well as estimates of maximum possible magnitudes. A study by Burbidge et al. (2009) focussed specifically on the Indian Ocean, with a panel of experts compiling a report on the potential tsunami effect of particular subduction zones in the coastal cities of the Indian

[1] University of Pretoria Natural Hazard Centre, Africa, Department of Geology, University of Pretoria, Private Bag X20, Hatfield, Pretoria 0028, South Africa. E-mail: andrzej.kijko@up.ac.za; ansie.smit@up.ac.za

[2] Institute of Geodynamics, National Observatory of Athens, P.O. Box 20048, Thissio, 11810 Athens, Greece. E-mail: papadop@noa.gr; tatyana@noa.gr

Ocean. It was recommended that all the susceptible countries identified by this study invest in comprehensive and thorough tsunami hazard analyses, which should include regional inundation models.

In South Africa, there is a significant lack of recorded information on tsunamis that have affected the country and, currently, only six events have been identified as tsunamis (Table 1). The most recent event, attributed to the 2004 mega-transoceanic tsunami, affected parts of the eastern coast of Africa. In South Africa, maximum wave heights between 0.75 m and 0.9 m were observed at Simon's Bay, Cape Town, and Saldanha, with two drownings being reported. Of the tsunami events listed in Table 1, this event is the only one for which there is conclusive evidence. The event of 1809 is usually associated with the magnitude 6.3 Milnerton earthquake that is commonly believed to have triggered a submarine landslide (Hartnady and Botha 2007). As regards the rest of the events, the sources of the abnormal wave heights are unknown, as these waves cannot be associated with any recorded local or distant earthquakes at this time. Possible local tsunami sources, which could have contributed to such events, include the steeply sloped sediments on the Western Cape continental margin (Dingle 1980; Dingle et al. 1987; Dingle and Robson 1992; Wefer et al. 1998) and the possible existence of mud volcanoes south of the Walvis Ridge (Ben-Avraham et al. 2002). Although a recent study by Salzmann and Green (2012) dismissed the possibility of local tsunamis being attributable to landslide events on the northern KwaZulu-Natal coastline, questions remain over the likelihood of submarine landslides in the southern and south-eastern coastline of South Africa. However, owing to the inferior quality of the available data, these events were excluded from consideration in our analyses.

The aim of the current study was to assess the potential tsunami hazard to the South African coastline. For this purpose, we assessed earthquake recurrence parameters for each of the identified tsunamigenic source zones. Provision was made in all the calculations for the incompleteness of the earthquake catalogues, the uncertainty of the earthquake magnitude determination, and the uncertainty of the applied earthquake occurrence model.

Estimated parameters were used to calculate the maximum expected wave height through the numerical simulation of tsunamis for "worst-case" and "realistic" scenarios. The worst-case scenarios depict hypothetical, yet possible scenarios, while realistic scenarios concern particular tsunami events that affected the coastline in the past. In our case, worst-case scenarios include mega-earthquakes associated with various subduction zones around the Indian Ocean, while a realistic scenario is the large tsunami produced by the mega-earthquake of 1833 generated in the Sumatra subduction zone. The aim was to evaluate the maximum possible impact, based on the

Table 1

Observed tsunami and tsunami-like events in South Africa

Date	Location	Source	Comments	References
04/12/1809	Milnerton	Earthquake/landslide	Fish observed in the streets of Cape Town	Hartnady and Botha (2007)
24/08/1883	Port Elizabeth	Krakatau volcanic eruption	Maximum wave height approximately 1.4 m	Choi et al. (2003)
26/08/1969	Dwarskersbos, West Coast	Unknown	Abnormal wave run-up, approx. 5 m	Council for Geoscience
11/05/1981	Agulhas Bank	Slumping (?)	–	Council for Geoscience
(20-21)/08/2008	Cape West Coast	Unknown	Abnormal wave	Council for Geoscience
26/12/2004	Eastern coast of South Africa (e.g. Port Elizabeth, Port Nolloth) and as far as Saldanha Bay, Cape Town, Simon's Bay)	26 December 2004, Sumatra–Andaman earthquake	Maximum wave heights of 0.75–0.9 m at Simon's Bay, Cape Town, and Saldanha; two people drowned	Joseph et al. (2006); Okal and Hartnady (2009)

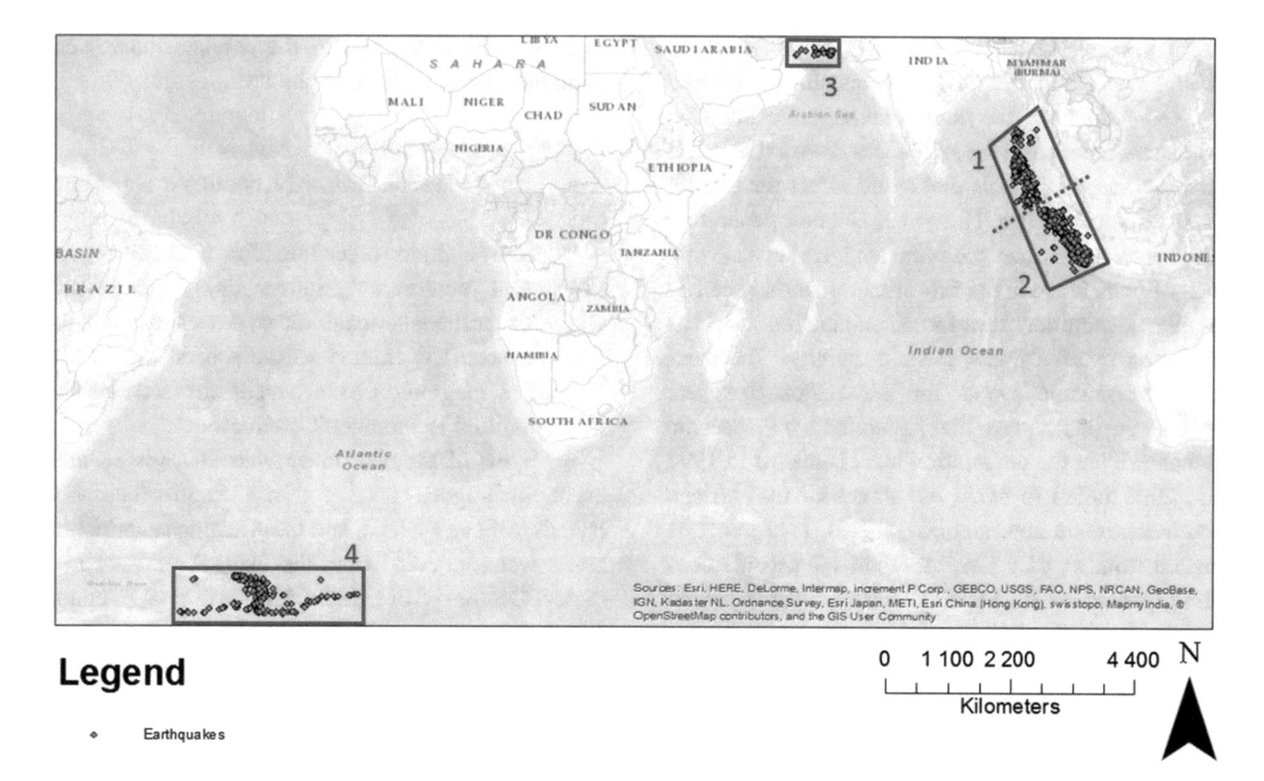

Figure 1
Four tsunamigenic subduction zones considered to have the potential to generate mega-tsunamis that could affect the South African coast: (1) Andaman Subduction Zone, (2) Sumatra Island, (3) Makran, and (4) South Sandwich Islands

available information and the results of several numerical tests. Other authors have followed a similar approach in their research, including Burbidge et al. (2009), González et al. (2009), Power et al. (2007), Sørensen et al. (2012), and Brizuela et al., (2014).

After the brief introduction in Sect. 1, a description of the identified tsunamigenic sources and the applied methodologies is presented in Sect. 2. The results of the earthquake and tsunami hazard assessment are provided in Sects. 3 and 4, with the discussion and conclusions provided in Sect. 5.

2. *Identification of Source Zones*

According to Burbidge et al. (2009), the subduction zones of the South Sandwich Islands, Sumatra, and, to a lesser extent, the Andaman Islands are considered the main contributors to the tsunami hazard to the coastal cities of South Africa. In this paper, we discuss zone-characteristic seismic hazard analysis of the subduction zones of Andaman, Sumatra Island, a section of the Sunda Arc in Indian Ocean, the South Sandwich Islands in the southern Atlantic Ocean, and the Makran subduction zones in the Persian Gulf and the Gulf of Oman (Fig. 1). Okal and Synolakis (2008) conducted similar analysis of the far-field tsunami hazard from mega-thrust earthquakes in the Indian Ocean related to the seismic sources located at the boundaries of the 2004 Sumatra–Andaman rupture. Such sources run along the southern coast of Sumatra and in the Andaman-Myanmar region, along the Makran coast of Pakistan and Iran, and farther along the southern coast of Java. These authors investigated how the variation in the earthquake focal parameters and the mode of seismic slip could affect the influence of far-field tsunamis. They concluded that if the seismic moment of the earthquake remained the same, the tsunami characteristics in the far-field domain appeared remarkably robust with respect to the variations in the properties of the parent earthquakes.

In the current study, following Burbidge et al. (2009), we assumed that the distant earthquake sources of Makran, the South Sandwich Islands, and Sumatra–Andaman were subduction zones that could generate mega-tsunamis that could affect the coastal areas of South Africa. Near-field tsunami-generating sources were excluded from the analysis because of a lack of confidence in the information. During the last couple of centuries, Java trench subduction zone has exhibited relatively low seismic activity. The catalogue of seismic events for this region does not provide any earthquake that generated a transoceanic tsunami. Even the devastating local tsunamis in 1994 and 2006 failed to reach the shores of the African continent. (Okal and Synolakis 2008). Java trench is known from its very low, less than 1% seismic coupling, the ratio of seismic energy release by an earthquake to the total energy accumulated. It suggests that most of the accumulated tectonic energy in the Java trench is released aseismically, by aseismic deformation, as, for example, fault creep (Ruff and Kanamori 1980; Javier et al. 1993; Ruff and Tichelaar 1996; Okal and Synolakis 2008 and references therein). The subduction zone of Java was therefore not included in our analysis of the Sumatra–Andaman tsunamogenic zone. However, the potential to generate mega-tsunami by Java trench is worth to study as a separate issue, due to the potential socio-economic effect of tsunami in the area.

3. *Seismic Hazard Assessment*

As the applied procedure for tsunami hazard assessment requires only information on the distribution of earthquake magnitudes, our analysis of the seismic hazard is limited to the assessment of the respective magnitude exceedances for each of the four identified tsunamigenic sources.

In terms of moment magnitude M_W, the seismic event catalogues used for the South Sandwich Islands, Sumatra Island, and the Andaman subduction zones were collected from the International Seismological Centre—GEM (ISC-GEM) historical (Version 1.0) and complete (Version 2.01) catalogues (Storchak et al. 2013). The catalogue from Karimiparidari et al. (2013) was used for the Makran subduction zone. Table 2 presents a summary of each of the applied seismic catalogues, which include geographical coordinates, time span, number of events, level of completeness, as well as the maximum observed magnitudes for each of the identified regions. The geographical coordinates were taken from Berryman et al. (2013), with the exception of

Table 2

Summary of earthquake event catalogue used for the Makran, South Sandwich Islands (SSI), Sumatra Island, and the Andaman subduction zone

Coordinates	Makran [24.38°N; 57.06°E] to [26.05°N; 65.03°E]	SSI [40.00°S; 63.00°W] to [10.00°S; 55.00°W]	Sumatra Island [8.17°S; 96.20°E] to [1.35°N; 104.58°E]	Andaman [1.35°N; 92.07°E] to [13.72°N; 96.20°E]
Time span	1438–2002	1921–2011	1681–2011	1837–2011
Historic catalogues				
Date	1438–1969	1921–1969	1681–1969	1837–1969
Number of events	4	2	10	5
Level of completeness	6.0	7.5	7.5	7.5
Standard error in magnitude	0.2	0.3	0.3	0.3
Maximum magnitude observed	8.0	8.1	9.0	7.9
Complete catalogues				
Date	1970–2002	1970–2011	1970–2011	1970–2011
Number of events	23	269	281	213
Level of completeness	4.4	5.5	5.5	5.5
Standard error in magnitude	0.1	0.1	0.1	0.1
Maximum magnitude observed	5.5	7.4	8.5	9.0
Maximum observed magnitude	8.1	8.1	9.0	9.0

Table 3

Estimated zone-characteristic seismic hazard parameters for the four identified tsunamigenic zones of Makran, South Sandwich Islands (SSI), Sumatra Island, and the Andaman subduction zones

	Makran	SSI	Sumatra Island	Andaman
Method of maximum magnitude estimation	N-P-G[a]	K-S-B[b]	K-S-B[b]	N-P-G[a]
Maximum magnitude ($\hat{m}_{\max}$)	8.4 ± 0.5	8.2 ± 0.3	9.2 ± 0.3	9.4 ± 0.4
Minimum magnitude ($\hat{m}_{\min}$)	4.40	5.50	5.50	5.50
Gutenberg–Richter *b*-value	1.01 ± 0.08	1.00 ± 0.06	1.00 ± 0.05	1.07 ± 0.06
Mean seismic activity rate (λ) [earthquakes per year]	0.6 ± 0.2	6.5 ± 1.6	5.2 ± 1.0	4.7 ± 1.0

[a] Nonparametric Gaussian maximum magnitude estimation (Kijko and Singh 2011)

[b]Kijko–Sellevoll–Bayes maximum magnitude estimation (Kijko and Singh 2011)

Table 4

Estimated zone-specified seismic hazard parameters for the four identified tsunamigenic zones of Makran, South Sandwich Islands (SSI), Sumatra Island, and the Andaman subduction zones

Magnitude	Return period (years)			
Tsunamigenic zone	Makran	SSI	Sumatra	Andaman
7.0	318	4	5	6
7.5	862	13	11	16
8.0	3170	90	30	43

the South Sandwich Islands, for which the investigated area was increased to include more events.

The seismic recurrence parameters, the mean annual seismic activity rate λ, and the Gutenberg–Richter *b*-value ($b = \beta \ln(10)$) were estimated by using the maximum likelihood procedure (MLE), as defined in Kijko et al. (2016). We made provision for the incompleteness of the data and the uncertainty associated with the magnitude estimation, as well as the uncertainty of the applied earthquake occurrence models. Aleatory uncertainty in the earthquake occurrence models is introduced by assuming that both λ and the *b*-value are random variables, each described by the gamma distribution. The approach results in the replacement of the classic frequency-magnitude Gutenberg–Richter relation and the Poisson distribution, describing the temporal earthquake occurrence of events, by the mixture-gamma distributions (Benjamin 1968; Campbell 1982, 1983).

An unbiased estimator of the zone-characteristic (seismogenic source) maximum possible earthquake magnitude $m_{\max}$ was determined by using the parametric procedure, hereafter referred to as Kijko–Sellevoll–Bayes (K–S–B), (Kijko and Singh 2011) for the subduction zones of South Sandwich Island and Sumatra. However, the nonparametric Gaussian (N-P-G) procedure provides more reliable $m_{\max}$ estimates for the Makran and Andaman subduction zones. Nonparametric methods make the least number of assumptions about the underlying model distribution (Kijko 2004) and prove useful in instances where standard parametric procedures provide incoherent results. The underlying complexity in the earthquake magnitude distribution, as well as the number of events in the catalogue (Makran), could explain probably why Andaman and Makran perform better with the nonparametric method.

The zone-characteristic seismic hazard is expressed in terms of mean return periods and the probabilities of being exceeded at least once in a specified time interval for specified earthquake magnitudes. The results, expressed in terms of zone-characteristic hazard curves, showing the probability for a wide range of magnitudes (M_W) to be exceeded in 1, 5, 10, and 25 years, are presented in Table 3.

The range of estimated earthquake maximum magnitudes (M_w) is [7.9, 8.9] for Makran, [7.9, 8.5] for the South Sandwich Islands, [8.9, 9.5] for Sumatra Island, and [9.0, 9.8] for the Andaman subduction zone. Table 4 provides the respective estimates of the

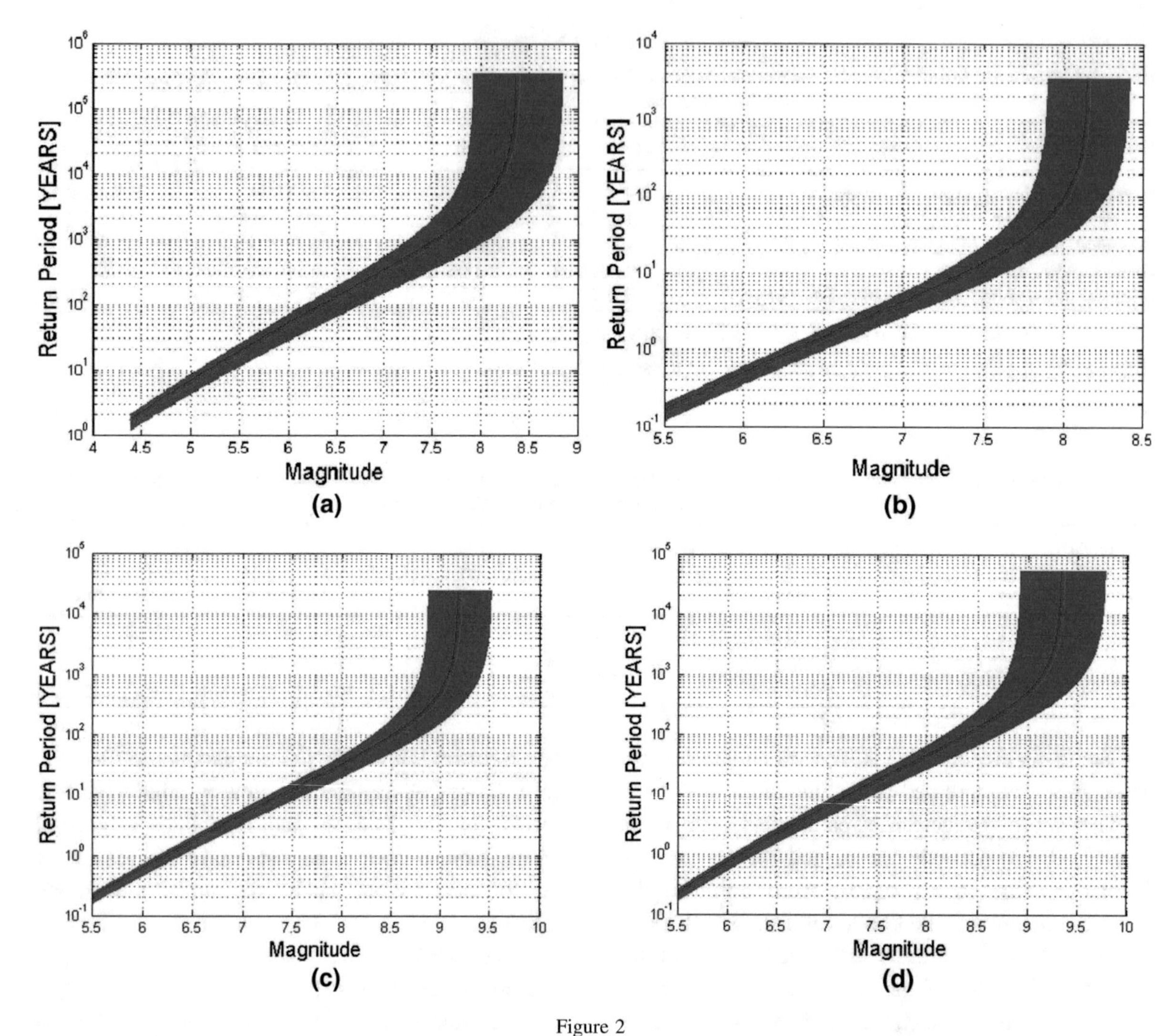

Figure 2
Mean return periods for earthquake magnitude for the subduction zones of a Makran, b the South Sandwich Islands, c Sumatra Island and d the Andaman subduction zone. The red curve shows the mean return period, whereas the two blue curves indicate the mean return period ± one standard deviation

return periods for earthquakes with magnitudes 7.0, 7.5, and 8.0. The most frequent (lowest return period) threat is posed by Sumatra Island, the South Sandwich Islands, and the Andaman subduction zones.

Figure 2 and Table 4 illustrate the return period for a given moment magnitude (M_w) for Makran, South Sandwich Islands, Sumatra Island, and the Andaman subduction zone, respectively. As an example, the return period of earthquake occurrence with magnitude equal to or greater than 7.0 to occur in the Makran subduction zone is approximately every 318 years, every 4 years for South Sandwich Island and every 5 and 6 years, respectively, for Sumatra Island, and the Andaman subduction zone. The confidence intervals for the probability of exceedance and return periods (± one standard deviation) are shown in Fig. 2.

Comparing the above results with those of similar investigations confirms the results of the other researchers to a considerable extent. The report by Berryman et al. (2013), under the auspices of the Faulted Earth Global Component for the Global Earthquake Model (GEM) project, provides extensive and in-depth analysis of parameters impacting the

Table 5

Focal parameters of the earthquakes that were applied for tsunami hazard assessment based on the realistic and worst-case scenarios (RC realistic-case and WC worst-case) for the subduction zones of Makran, South Sandwich Islands and Sumatra–Andaman

Fault parameters	Makran		SSI			Sumatra–Andaman	
Coordinates	[24.38°N; 57.06°E] to [26.05°N; 65.03°E]		[40°S; 63°W] to [10°S; 55°W]			[8.17°S; 96.20°E] to [13.72°N; 104.58°E]	
	RC	WC	RC	WC1	WC2	RC	WC
Fault length (km)	326	326	191	191	191	943	550
Fault width (km)	105	105	78	78	78	191	175
Mean displacement (m)	4.92	4.92	2.62	2.62	2.62	16.97	13
Strike (degree)	236	280	38	272	130	338	280
Hypocenter depth (km)	27	27	35	15	35	25	25
Coordinates of hypocenter (deg.)	24.61°N; 60.1°E	25.0°N; 62.3°E	27.3°S; 55.5°W	27.3°S; 55.5°W	26.9°S; 58.1°W	3.3°N; 95.8°E	8.3°N; 98.0°E
Dip angle (degree)	7 (2–27)	7	50	57	50	8 (8–15)	7
Slip rake (degree)	90	90	95	71	95	90	90
Magnitude (M_W)	7.5	8.2	7.5	8.13	8.3	9.1	8.9

tsunami potential of an area. This report identifies ranges (minimum, maximum, and average) for the Gutenberg–Richter *b*-value and the maximum magnitude for worldwide subduction zones. Determining the maximum earthquake magnitude is based on the rupture length of the different fault segments, as described by McCaffrey (2008). The maximum magnitude estimated for Makran, South Sandwich Islands, Sumatra, and Andaman does not differ significantly from the results obtained by Berryman et al. (2013). As regards the Makran subduction zone, the estimated m_{max} was lower than the M_w 8.7 suggested by Berryman et al. (2013) and Zaman et al. (2012). Moreover, these two studies provide a *b*-value that is much lower and appears to deviate from what is normally observed in areas with tectonic origin seismicity. Therefore, it is prudent to assume that the four identified subduction zones could generate mega-transoceanic tsunamis that could potentially affect the South African coastline and, by implication, the eastern part of the African coastline.

4. Tsunami Hazard Assessment

Maximum tsunami wave amplitudes for realistic and worst-case scenarios were estimated by employing numerical techniques (described below) for the three selected sites along the South African coast, namely the metropolitan cities of Durban, Port Elizabeth, and Cape Town. During the numerical tests, the Sumatra and Andaman zones were combined and analysed as one zone.

The parameters used in the regional modelling of tsunami amplitudes are summarized in Table 5. The values were obtained from Ioualalen et al. (2006), Okal and Synolakis (2008), and Okal and Hartnady (2009), as well as from a number of hypothetical tests for the worst-case scenarios. The geographic longitudes and latitudes of the three hypothetical tide gauges are 31.044°E and 29.858°S for Durban, 25.604°E and 33.948°S for Port Elizabeth, and 18.411°E and 33.895°S for Cape Town.

Numerical simulations for all the scenarios were performed using the GEOWAVE application (Watts et al. 2003), which is a combination of TOPICS (Tsunami Open and Progressive Initial Conditions System) and FUNWAVE. TOPICS uses a variety of curve fitting techniques and was designed (Grilli and Watts 1999) as a simulation tool to provide approximate surface water elevations and velocities as initial conditions for the tsunami propagation model. The FUNWAVE (Wei and Kirby 1995; Wei et al. 1995) numerical model performs wave propagation simulation based on the fully nonlinear Boussinesq theory, allowing the user to obtain accurate run-up and inundation data at the same time. The use of GEOWAVE for tsunami simulations has been validated

Table 6

Calculated offshore tsunami amplitudes (m) from the realistic and worst-case scenario models with homogeneous and heterogeneous slip for the considered seismic zones (RC = realistic-case and WC = worst-case) for the subduction zones of Makran, South Sandwich Islands, and Sumatra–Andaman

Site	Amplitudes (m)						
	Makran		SSI			Sumatra–Andaman	
	RC	WC	RC	WC1	WC2	RC	WC
Cape Town	0.003	0.06	0.011	0.02	0.29	0.03	0.037
Port Elizabeth	0.006	0.07	0.037	0.05	0.06	0.04	0.078
Durban	0.009	0.13	0.001	0.008	0.11	0.06	0.21

well by previous case studies of tsunamis generated by earthquakes (Day et al. 2005; Grilli et al. 2007; Ioualalen et al. 2006, 2007), pyroclastic flows (Watts and Waythomas 2003; Novikova et al. 2011), underwater landslides (Watts et al. 2003; Day et al. 2005; Greene et al. 2005), and debris flow (Walder et al. 2003).

The earthquake tsunami source is described by the standard half-plane solution for an elastic dislocation with maximum slip Δ (Okada 1985). Accordingly, a planar fault is defined through horizontal length L and width W, with the centroid being located at the earthquake epicentre (x_0, y_0), and depth d. The Okada (1985) solution is implemented in the TOPICS software application. Both homogeneous and heterogeneous (certain number of subfaults with different dimensions and values of slips along the respective faults) were considered in our study. However, it was found that the resulting wave amplitudes at the three sites practically do not differ if applied to the homogeneous or heterogeneous slip models.

The definition of the grid is based mainly on 2 (minute) resolution bathymetry, and topography (ETOPO 2′ database, National Geophysical Data Center, NOAA, USA). This grid covers the region from 60°S to 25°N and from 28°W to 120°E. The simulation time step was established based on the size of the constructed grid. The maximum tsunami wave amplitudes for realistic (RC) and hypothetical worst-case (WC) scenarios were calculated numerically for each of the three identified sites on the South African coast. The focal parameters used during the simulations are provided in Table 5.

4.1. Realistic Scenarios

The predicted tsunami amplitudes for the realistic scenarios at the three South African coastal sites are extremely small (Table 6). In fact, the largest amplitudes, being in the order of a few centimetres (Fig. 6, Appendix), were expected to be generated by an earthquake that occurred at the Sumatra–Andaman tsunamigenic zone. Even lower amplitudes were obtained from the Makran (Fig. 8, Appendix) and South Sandwich Islands (Fig. 9, Appendix) sources. The parameters of these seismic sources are given in Table 5. The results indicate clearly that the tsunami hazard related to the three distant seismic sources is minimal.

To investigate more complex and, perhaps, more realistic scenarios, we also employed the model for heterogeneous slip distribution to the seismic faults. The model for heterogeneous slip did not change the resulting far-field tsunami wave amplitudes. In fact, the calculated amplitudes for all three locations had the same values as those obtained from a model with homogeneous seismic slip. However, in the case of the Sumatra–Andaman zone, the energy directivity effect, as observed during the 2004 mega-tsunami, was reproduced better by employing the heterogeneous slip model (Fig. 7, Appendix).

4.2. Worst-Case Scenarios

Running number of scenarios with various source parameters for every seismic zone, we selected the cases when the tsunami influence at the considered coastal zone will be maximum. We considered "worst-case" conditions in terms of source orientation and the direction of wave energy radiation, i.e. in cases where wave energy is directed towards the investigated site, and we selected the hypocenter location that would maximize the phenomenon. Final values of those tests are presented in Table 5.

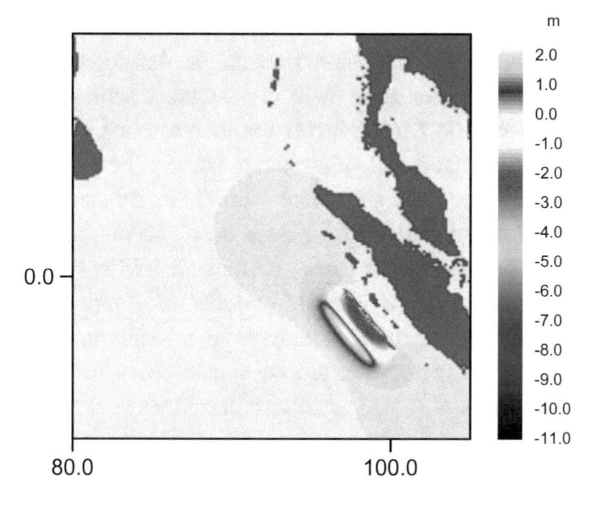

Figure 3
Static fault displacement generated by the 1833 Sumatra earthquake, which is used for modelling the worst-case scenario

4.2.1 Sumatra–Andaman Tsunamigenic Zone

The worst-case hazard scenario for the coasts of Kenya, Tanzania, and South Africa would be a repeat of the 25 November 1833 great earthquake (Fig. 3), with the expected impact likely more severe than that of the 26 December 2004 event. Okal and Hartnady (2009) attributed this to the more southerly azimuth of the energy directivity pattern. The parameters for a scenario such as the 1833 earthquake are given in Table 5. Accordingly, in this study, calculations were performed by using the parameters of the 1833 earthquake with homogeneous seismic slip. This scenario is considered a worst-case scenario for a tsunami generated by the Sumatra–Andaman zone.

Based on the 1833 scenario, offshore tsunami amplitudes range between 0.037 and 0.21 m (Table 6). The distribution of the amplitudes is shown in Fig. 4, with the initial surface elevation presented in Fig. 3. The wave amplitude values were shown to be 1.2–3.5 times larger than those produced by the Sumatra 2004 seismic model. The expected impact of such a tsunami would be considerably stronger than that from the 2004 tsunami. This is ascribed principally to the more southerly azimuth of the energy directivity pattern, controlled by the geometry of the earthquake source.

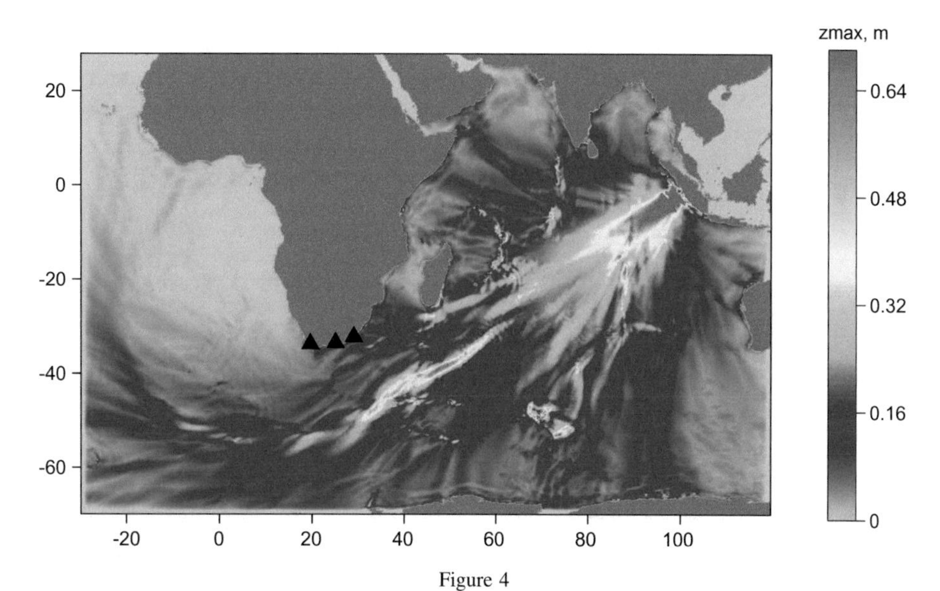

Figure 4
Estimated maximum surface elevation owing to the 1833 Sumatra earthquake with homogeneous seismic slip. The earthquake parameters are given in Table 5. The triangles from east to west represent Durban, Port Elizabeth, and Cape Town

4.2.2 Makran Tsunamigenic Zone

We conducted a simulation of the worst-case tsunami scenario generated by the Makran tsunamigenic zone, assuming the occurrence of an earthquake with a homogeneous slip and focal parameters, as shown in Table 5. The configuration of this subduction zone can be found in Byrne et al. (1992). The corresponding static co-seismic fault displacement is illustrated in Fig. 10 (Appendix). However, in the selection process for the worst-case scenario, the different directions (strike) of the fault were tested, with the aim to direct (focus) tsunami energy towards the South African coast (Fig. 11, Appendix). The final fault strike selected was equal to 28$\mathring{0}$. The estimated tsunami amplitudes are listed in Table 6. Although still insignificant, they are slightly larger than those obtained by the realistic scenarios. The tsunami impact to the selected sites in this scenario remains minimal.

4.2.3 South Sandwich Islands Tsunamigenic Zone

To assess the worst-case scenario for a tsunami of which the source is off the South Sandwich Islands (Fig. 12, Appendix), we employed the earthquake parameters for homogeneous seismic slip, as provided in Table 5. In addition, a number of tests were performed by varying the hypocentre depth and the strike of the earthquake fault, with the aim to direct maximum tsunami energy towards the South African coast (Fig. 13, Appendix). This scenario is referred to as worst-case scenario 1. It was found that by keeping the seismic moment unchanged, the variation of the earthquake fault strike did not affect the tsunami amplitudes obtained at the three selected sites. The amplitudes obtained for an earthquake fault strike of 130° are listed in Table 6, and the amplitude distribution is shown in Fig. 13 and Appendix.

A tsunami was simulated for this region in accordance with the study of Okal and Hartnady (2009). The simulation considered a tsunami-generating earthquake with the parameters as shown in Table 3 of their work. This is an alternative scenario, referred to as worst-case scenario 2, and assumed that an interplate thrust could generate an earthquake at the southern corner of the arc. According to Okal and Hartnady (2009), this is the most dangerous tsunami scenario for the African and South African coast. The strong effect of tsunami wave generated during such event would be the result of source directivity and wave focusing during its propagation. The results of our numerical experiment are similar to those obtained by Okal and Hartnady (2009). Their study indicated wave amplitude equal to 0.3 m offshore from Cape Town. Our calculations resulted amplitudes of 0.29 m in Cape Town, 0.06 m in Port Elizabeth, and 0.11 m in Durban (Fig. 5).

The Sumatra–Andaman zone is considered the most dangerous zone, as it can generate earthquakes of magnitude 7.0 and larger for shorter return periods compared with the other three zones. These earthquakes can generate a high amplitude tsunami that could affect the South African coast. In the case of the Makran seismic zone, earthquakes with similar magnitudes have return periods of more than 300 years. For this tsunamigenic region, only the worst-case scenario could be considered a threat to South Africa. A 90-year return period was calculated for the South Sandwich Island subduction zone for an earthquake of magnitude 7.0 or higher. Earthquakes with lower magnitudes that occur more frequently cannot generate a tsunami with a significant amplitude to affect the South African coast negatively.

Estimates that are more accurate of tsunami run-up values for the South African coast (Table 7) can be obtained only after the implementation of the local model, with the incorporation of the high-resolution sources of near-shore bathymetry and coastal area topography. Such application could increase the offshore wave amplitude and, consequently, could improve the run-ups values.

5. Conclusions

Failing to quantify hazards accurately could have devastating effects on the infrastructure of a country, its economic stability, and the quality of life of its people (World Bank 2005). The general perception is that South Africa is safe from the impact of a mega-transoceanic tsunamis. The intention of this work was to discuss this assumption by combining seismic and tsunami hazard procedures into a unified study and to quantify the tsunami threat to the South Africa

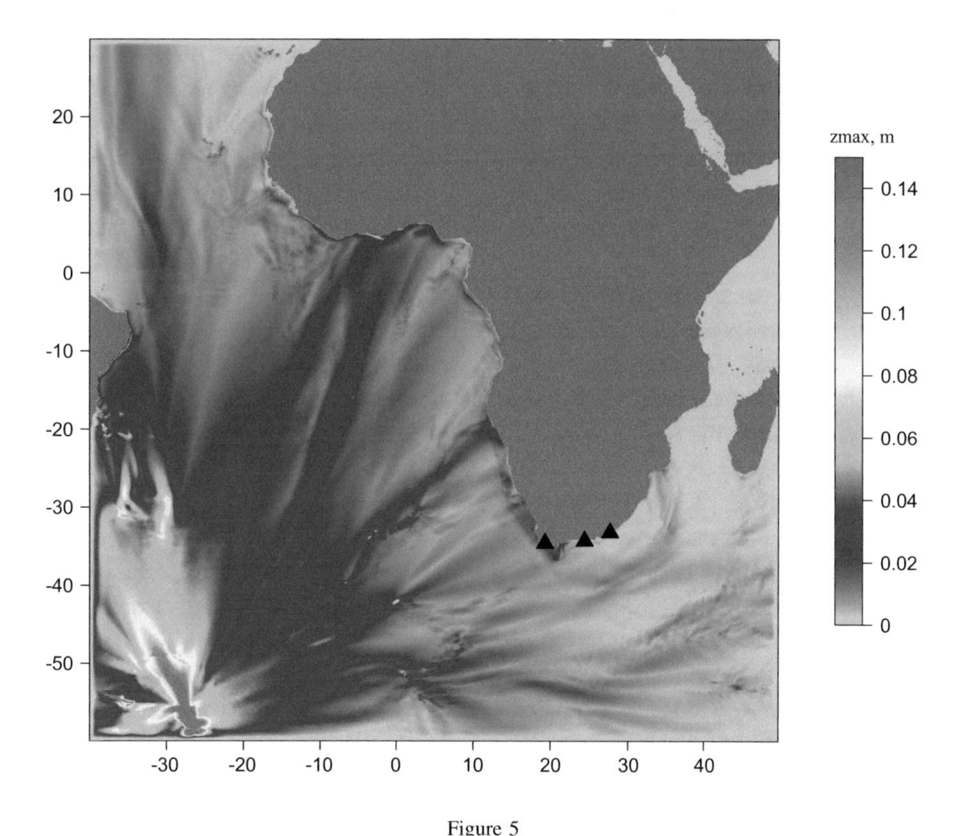

Figure 5
Estimated maximum surface elevation (m) owing to worst-case scenario 2 for the South Sandwich Islands earthquake, with homogeneous seismic slip and fault strike equal to 130°. The earthquake parameters are as given in Table 5. Triangles are as in Fig. 4

Table 7

Simulated tsunami run-up (RU) in metres calculated (right columns) for the worst-case scenarios (WC)

Site	Makran	SSI		Sumatra–Andaman
	WC RU	WC1 RU	WC2 RU	WC RU
Cape Town	0.83	0.4	2.5	0.6
Port Elizabeth	0.93	0.15	0.8	1.0
Durban	1.43	0.19	1.3	2.0

Values in parentheses for the South Sandwich Islands source represent the run-up values obtained for worst-case scenario 2

coastline. As shown, the Makran, South Sandwich Island, Sumatra, and the Andaman subduction zones could generate mega-earthquakes that could result in transoceanic tsunamis of amplitudes that could have a negative impact on the coastline.

Three factors were identified as major contributors to the uncertainty of the tsunami hazard estimates. First, uncertainty is introduced by the application of relatively short earthquake catalogues. The seismic event dataset of the South Sandwich Islands does not span 100 years, and the catalogues for the Makran, Sumatra, and the Andaman subduction zones span 564, 330, and 174 years, respectively. It is quite likely that the full seismic cycle of a mega-earthquake magnitude would be larger than the time spans of each of these catalogues. Since the assessment of zone-characteristic seismic hazard parameters did not include prehistoric-earthquake information, the results are uncertain and potentially could underestimate the zone-characteristic, maximum possible earthquake magnitudes and, at the same time, overestimate the return periods. Future research is required, where the available information should be utilized by applying the Bayesian formalism.

The second form of data uncertainty is that only a small fraction of large earthquakes in subduction zones generate mega-tsunamis. The probability of a significant earthquake occurrence induced by the Makran subduction zone during the next 25 years is extremely small. Additional investigation is required to determine a realistic value of the fraction of earthquakes capable of generating mega-tsunamis, and in what way this would affect the relevant seismic hazard recurrence parameters.

The third factor of uncertainty is the inclusion of local tsunamigenic sources in the calculations. Only distant tsunamigenic sources were considered owing to the size and inferior quality of historical records on local tsunamis. As pointed out in a report by ESKOM (2009), a tsunami generated by an offshore landslide is considered the largest unknown risk factor for the South African coast. There is increasing recognition of the role that submarine and coastal landslides could play in the generation of powerful tsunamis. Accordingly, future research should incorporate such tsunami sources, using, for example, the information provided by Roberts (2008) and Luger (2010) on the evidence for South African submarine slumps.

Several conclusions can be drawn from the results of the numerical tsunami simulation for the regional model. The simulated realistic scenarios of tsunami hazard for Durban, Port Elizabeth, and Cape Town by the three analysed subduction zones showed relatively small tsunami amplitudes, which do not represent a significant risk to the South African coast. This is also true for the worst-case scenarios, with the estimated wave amplitudes not being changed significantly; however, the worst-case scenarios better represent the observed effect of energy directivity.

Our results are consistent to a large extent with the outcomes of similar investigations performed by other researchers, who applied different simulation techniques. By incorporating additional sources of information into the local model, such as high-resolution near-shore bathymetry and the coastal topography of each selected site, assessments that are more accurate could be obtained. It is expected that after incorporating the local models, the run-up in the three analysed coastal sites could reach 2 m, which is a considerable height from a tsunami risk perspective.

The results of the tsunami hazard assessment for the coast of South Africa indicate that the tsunami threat appears to be important subject for further detailed study including also aseismic sources and employing finer bathymetric sources for propagation model. These results and the factor of data uncertainty suggest that additional research and superior quality data are required to increase the accuracy of the tsunami hazard assessment. Therefore, the results provided in this article should be treated as preliminary only.

Acknowledgements

The project was financially supported by both the Nuclear Structural Engineering (Pty) and the National Research Foundation through the Technology and Human Resources for Industry Programme project (THRIP) TP2011061400009. The authors would like to extend their gratitude to Professor H. Gupta for valuable insights into the possible seismogenic sources that could have an effect on South Africa.

Appendix

See Figs. 6, 7, 8, 9, 10, 11, 12 and 13.

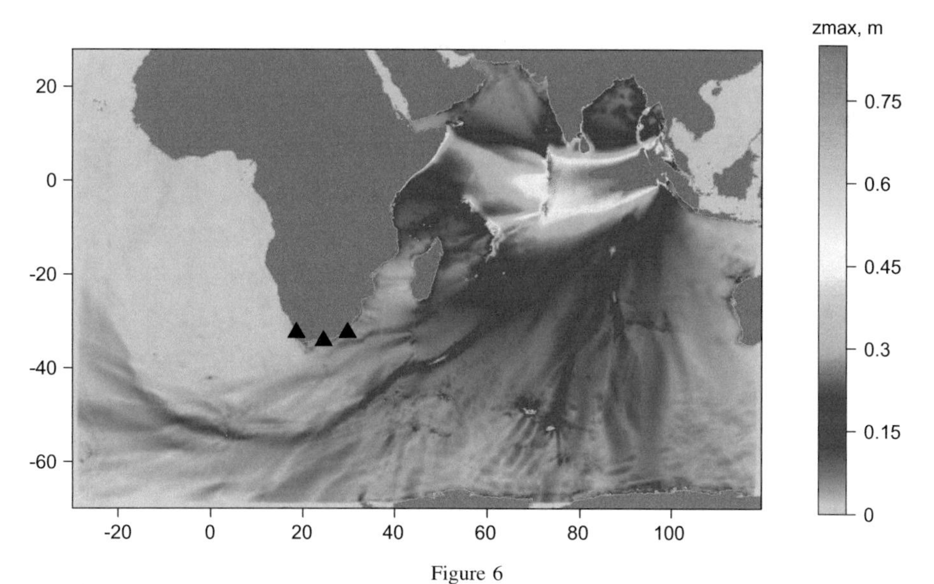

Figure 6
Maximum surface elevation (m) at all times after an earthquake with homogeneous seismic slip generated in the Sumatra–Andaman zone. The earthquake parameters are listed in Table 5. Triangles from east to west represent Durban, Port Elizabeth, and Cape Town

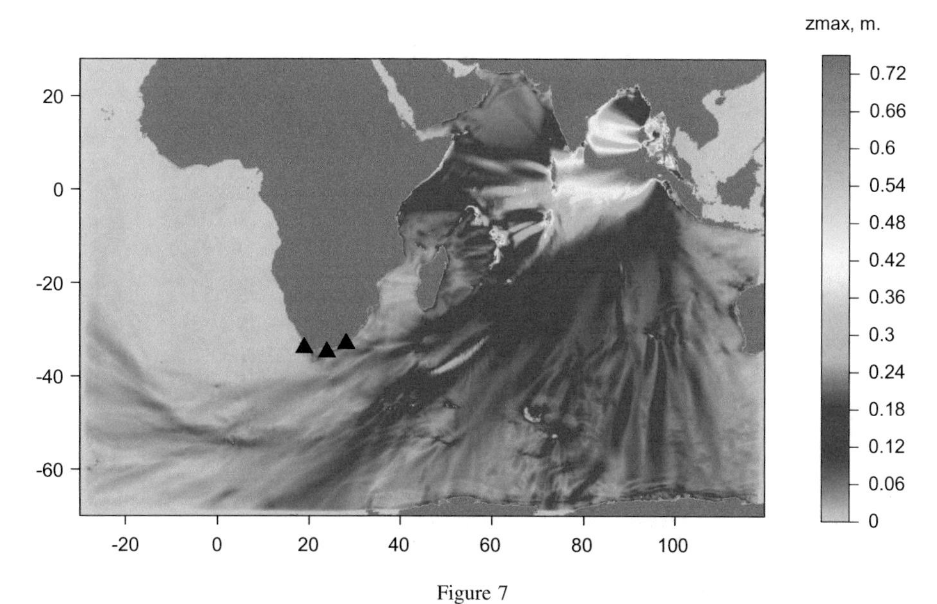

Figure 7
Maximum surface elevation (m) at all times after an earthquake with heterogeneous seismic slip at the Sumatra–Andaman tsunamigenic zone. Earthquake parameters are listed in Table 5. Triangles are as in Fig. 6

Figure 8
Maximum surface elevation (m) at all times after an earthquake with homogeneous slip generated in the Makran tsunamigenic zone. The earthquake parameters are listed in Table 5. Triangles are as in Fig. 6

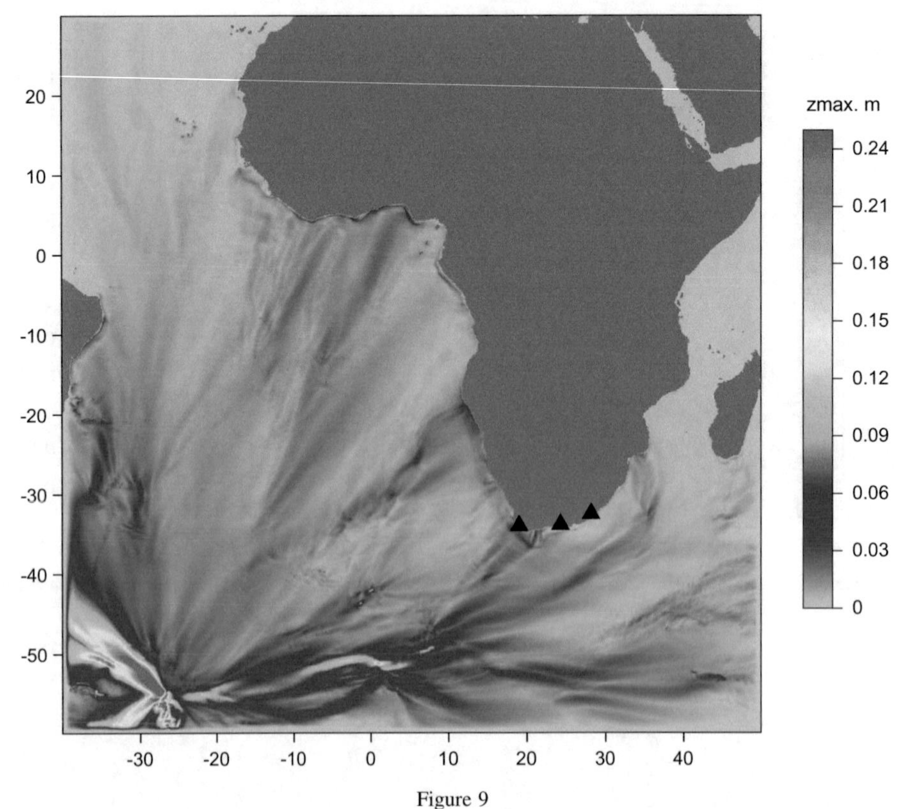

Figure 9
Maximum surface elevation (m) at all times owing to homogeneous seismic slip produced by an earthquake at the South Sandwich Islands zone. The earthquake parameters are listed in Table 5. Triangles are as in Fig. 6

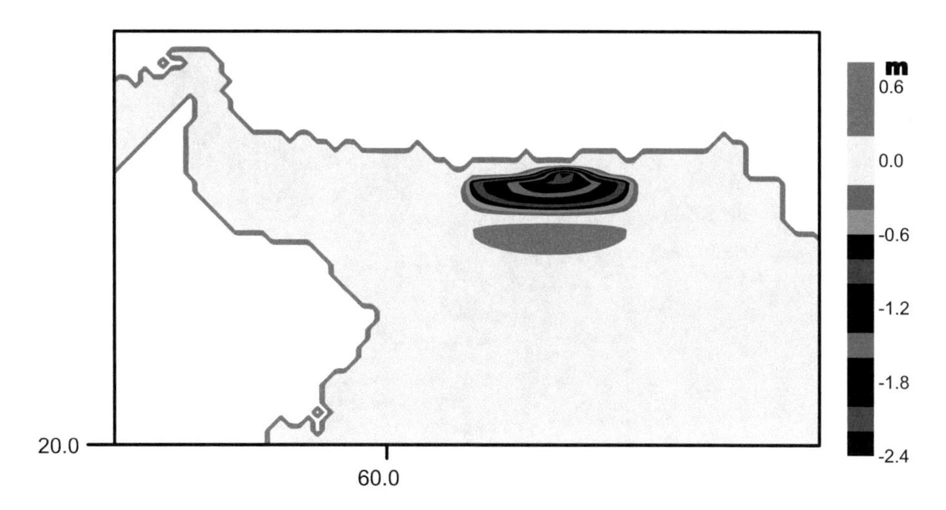

Figure 10
Static fault displacement generated by an earthquake occurring at the Makran tsunamigenic zone, which is used for modelling the worst-case scenario, with parameters listed in Table 5, but for a strike equal to 280°

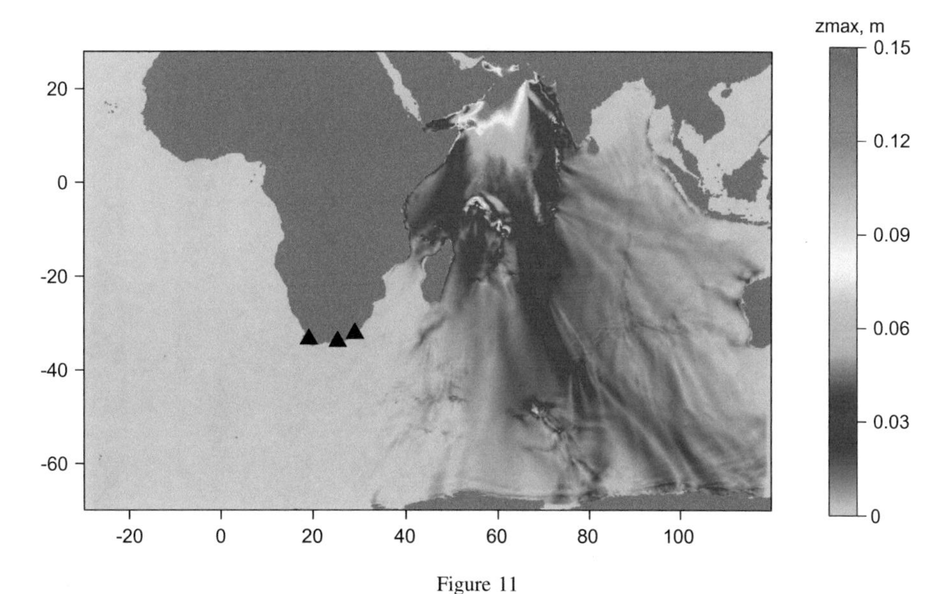

Figure 11
Estimated maximum surface elevation owing to a worst-case scenario earthquake occurring in the Makran tsunamigenic zone, with homogeneous seismic slip, and fault strike of 280°. The earthquake parameters are listed in Table 5. Triangles are as in Fig. 6

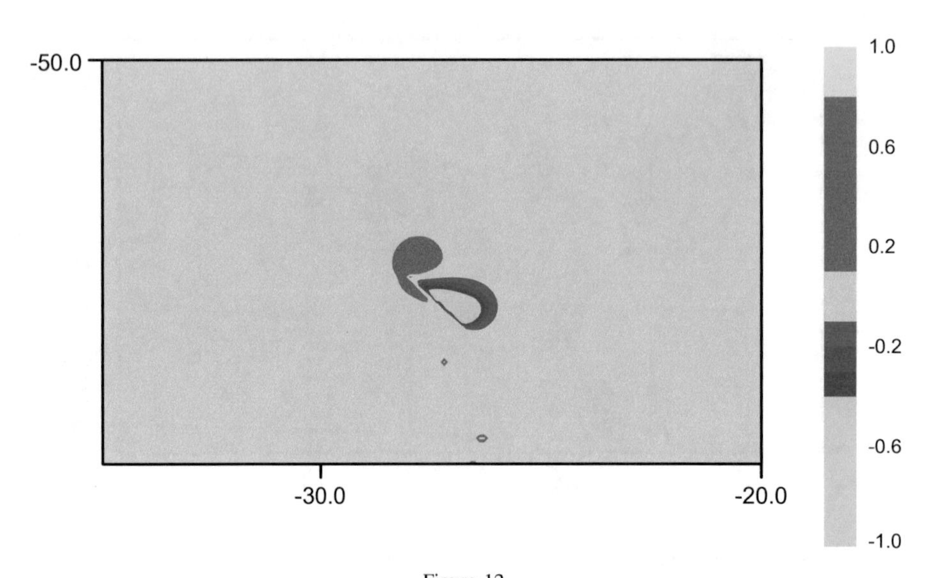

Figure 12
Static fault displacement of the South Sandwich Islands earthquake used as an initial condition for worst scenario 1. The earthquake parameters are listed in Table 5, with earthquake fault strike of 130°

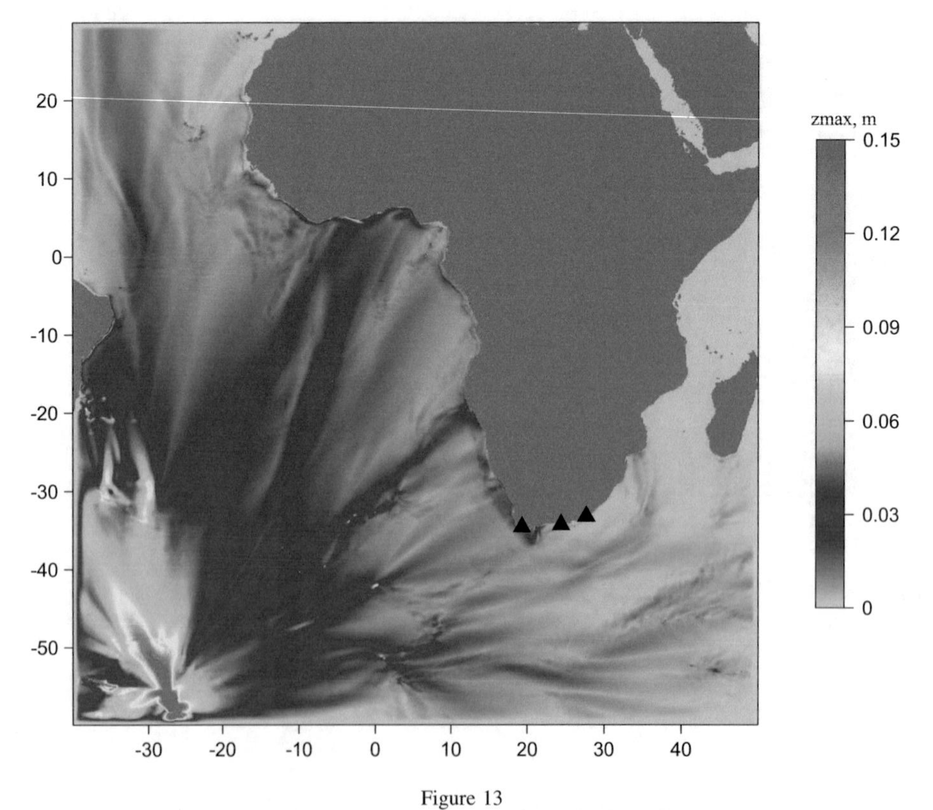

Figure 13
Estimated maximum surface elevation (m) owing to worst scenario 1 for the South Sandwich Islands earthquake, with homogeneous seismic slip and fault strike equal to 130°. The earthquake parameters are listed in Table 5. Triangles are as in Fig. 6

References

Ben-Avraham, Z., Smith, G., Reshef, M., & Jungslager, E. (2002). Gas hydrate and mud volcanoes on the southwest African continental margin off South Africa. *Geology, 30*(10), 927–930.

Benjamin, J. R. (1968). Probabilistic models for seismic forces design. *Journal of the Structural Division, 94*(ST5), 1175–1196.

Berryman, K., Wallace, L., Hayes, G., Bird, P., Wang, K., Basili, R., Lay, T., Stein, R., Sagiya, T., Rubin, C., Barreintos, S., Kreemer, C., Litchfield, N., Pagani, M., Gledhill, K., Haller, K., and Costa, C. (2013). The GEM faulted earth subduction characterisation project. Report produced in the context of the GEM faulted earth global components.

Brizuela, B., Armigliato, A., & Tinti, S. (2014). Assessment of tsunami hazards for the Central American Pacific coast from southern Mexico to northern Peru. *Natural Hazards and Earth Systems Sciences, 14,* 1889–1903.

Burbidge, D. R., Cummins, P. R., Latief, H., Mleczko, R., Mokhtari, M., Natawidjaja, D., Rajendran, C. P., & Thomas, C. (2009). A probabilistic tsunami hazard assessment of the Indian Ocean Nations. Australian Government, Geoscience Australia, Professional Opinion, No. 2009/11.

Byrne, D. E., Sykes, L., & Davis, D. (1992). Great thrust earthquakes and aseismic slip along the plate boundary of the Makran subduction zone. *Journal of Geophysical Research, 97,* 449–478.

Campbell, K. W. (1982). Bayesian analysis of extreme earthquake occurrences. Part I. Probabilistic hazard model. *Bulletin of the Seismological Society of America, 72,* 1689–1705.

Campbell, K. W. (1983). Bayesian analysis of extreme earthquake occurrences. Part II. Application to the San Jacinto fault zone of southern California. *Bulletin of the Seismological Society of America, 73,* 1099–1115.

Choi, B. H., Pelinovsky, E., Kim, K. O., & Lee, J. S. (2003). Simulation of the trans-oceanic tsunami propagation due to the 1883 Krakatau volcanic eruption. *Natural Hazards and Earth System Sciences, 3*(5), 321–332.

Day, S. J., Watts, P., Grilli, S. T., & Kirby, J. T. (2005). Mechanical models of the 1975 Kalapana, Hawaii earthquake and tsunami. *Marine Geology, 215,* 59–92.

Dingle, R. V. (1980). Large allochthonous sediment masses and their role in the construction of the continental slope and rise off south western Africa. *Marine Geology, 37,* 333–354.

Dingle, R. V., Birch, G. F., Bremner, J. M., De Decker, R. H., Du Plessis, A., Engelbrecht, J. C., et al. (1987). Deep-sea sedimentary environments around southern Africa, South-East Atlantic and South-West Indian Oceans. *Annals of the South African Museum, 98,* 1–27.

Dingle, R. V., & Robson, S. H. (1992). South western Africa continental rise: Structural and sedimentary evolution. In C. W. Poag & P. C. Graciansky (Eds.), *Geologic evolution of atlantic continental rises* (pp. 62–76). New York: Van Nostrand.

ESKOM. (2009). ESKOM nuclear sites site safety reports. Numerical modelling of coastal processes. Duynefontein (p. 17). Report No. 1010/4/101. September 2009.

González, F. I., Geist, E. L., Jaffe, B., Kânoğlu, U., Mofjeld, H., Synolakis, C. E., et al. (2009). Probabilistic tsunami hazard assessment at seaside, Oregon, for near-and far-field seismic sources. *Journal of Geophysical Research: Oceans, 114,* C11.

Greene, H. G., Murai, L. Y., Watts, P., Maher, N. A., Fisher, M. A., Paull, C. E., et al. (2005). Submarine landslides in the Santa Barbara Channel as potential tsunami sources. *Natural Hazards and Earth System Science, 6,* 63–88.

Grilli, S. T., Ioualalen, M., Asavanant, J., Shi, F., Kirby, J. T., & Watts, P. (2007). Source constraints and model simulation of the December 26, 2004 Indian Ocean tsunami. *Journal of Waterway, Port, Coastal, and Ocean Engineering, 133*(6), 414–428.

Grilli, S. T., & Watts, P. (1999). Modelling of waves generated by a moving submerged body: Applications to underwater landslides. *Engineering Analysis with Boundary Elements, 23*(8), 645–656.

Hartnady, C. J. H., & Botha, J. (2007). Earthquake risk assessment for the Western Cape. Department of Local Government and Housing, Western Cape Provincial Government.

Ioualalen, M., Asavanant, J., Shi, F., Kirby, J. T., & Watts, P. (2007). Modeling the 26 December 2004 Indian Ocean tsunami: Case study of impact in Thailand. *Journal of Geophysical Research, 112,* C07024. https://doi.org/10.1029/2006JC003850.

Ioualalen, M., Pelletier, B., Regnier, M., & Watts, P. (2006). Numerical modelling of the 26th November 1999 Vanuatu tsunami. *Journal of Geophysical Research, 1111,* C06030. https://doi.org/10.1029/2005JC003249.

Joseph, A., Odametey, J. T., Nkebi, E. K., Pereira, A., Prabhudesai, R. G., Mehra, P., et al. (2006). The 26 December 2004 Sumatra tsunami recorded on the coast of West Africa. *African Journal of Marine Science, 28*(3–4), 705–712.

Karimiparidari, S., Zaré, M., Memarian, H., & Kijko, A. (2013). Iranian earthquakes, a uniform catalog with moment magnitudes. *Journal of Seismology, 17*(3), 897–911.

Kijko, A. (2004). Estimation of the maximum earthquake magnitude mmax. *Pure and Applied Geophysics, 161,* 1–27.

Kijko, A., & Singh, M. (2011). Statistical tools for maximum possible earthquake magnitude estimation. *Acta Geophysica, 59,* 674–700.

Kijko, A., Smit, A., & Sellevoll, M. A. (2016). Estimation of earthquake hazard parameters from incomplete data files, Part III, incorporation of uncertainty of earthquake-occurrence model. *Bulletin of the Seismological Society of America.* https://doi.org/10.1785/0120150252.

Luger, S. A. (2010). Eskom nuclear sites site safety reports: Numerical modelling of coastal processes Thyspunt. Prestedge Retief Dresner Wijnberg (Pty) Ltd Report 1010/2/102.

McCaffrey, R. (2008). Global frequency of magnitude 9 earthquakes. *Geology, 36*(3), 263–266.

Novikova, T., Papadopoulos, G. A., & McCoy, F. W. (2011). Modelling of Tsunami generated by the Giant Late Bronze age eruption of Thera, South Aegean Sea, Greece. *Geophysical Journal International, 186,* 665–680. https://doi.org/10.1111/j.1365-246X.2011.05062.

Okada, Y. (1985). Surface deformation due to shear and tensile faults in a half-space. *Bulletin of the Seismological Society of America, 75,* 1135–1154.

Okal, E. A., & Hartnady, C. J. (2009). The South Sandwich Islands earthquake of 27 June 1929: Seismological study and inference on tsunami risk for the South Atlantic. *South African Journal of Geology, 112*(3–4), 359–370.

Okal, E. A., & Synolakis, C. E. (2008). Far-field tsunami hazard from mega-thrust earthquakes in the Indian Ocean. *Geophysical Journal International, 172*(3), 995–1015.

Power, W., Downes, G., & Stirling, M. (2007). Estimation of tsunami hazard in New Zealand due to South American earthquakes. *Pure and Applied Geophysics, 164,* 547–564. https://doi.org/10.1007/s00024-006-0166-3.

Roberts, D. L. (2008). Nuclear siting investigation programme: Potential sources of tsunami along the South African coast. Council for Geoscience 2008–0220.

Ruff, L. J., & Kanamori, H. (1980). Seismicity and the subduction process. *Physics of the Earth and Planetary Interiors, 23,* 240–252.

Salzmann, L., & Green, A. (2012). Boulder emplacement on a tectonically stable, wave-dominated coastline, Mission Rocks, northern KwaZulu-Natal, South Africa. *Marine Geology, 323,* 95–106.

Sørensen, M. B., Spada, M., Babeyko, A., Wiemer, S., & Grünthal, G. (2012). Probabilistic tsunami hazard in the Mediterranean Sea. *Journal of Geophysical Research.* https://doi.org/10.1029/2010JB008169.

Storchak, D. A., Di Giacomo, D., Bondár, I., Engdahl, E. R., Harris, J., Lee, W. H. K., et al. (2013). Public release of the ISC-GEM global instrumental earthquake catalogue (1900–2009). *Seismological Research Letters, 84*(5), 810–815. https://doi.org/10.1785/0220130034.

Thomalla, F., & Larsen, R. K. (2010). Resilience in the context of tsunami early warning systems and community disaster preparedness in the Indian Ocean region. *Environmental Hazards, 9*(3), 249–265.

Walder, J., Watts, P., Sørensen, O., & Janssen, K. (2003). Tsunamis generated by subaerial mass flows. *Journal of Geophysical Research, 108*(B5), 1–19.

Watts, P., Grilli, S. T., Kirby, J. T., Fryer, G. J., & Tappin, D. R. (2003). Landslide tsunami case studies using a Boussinesq model and a fully nonlinear tsunami generation model. *Natural Hazards and Earth System Science, 3*(5), 391–402.

Watts, P., & Waythomas, C. F. (2003). Theoretical analysis of tsunami generation by pyroclastic flows. *Journal of Geophysical Research, 108*(B12), 1–19.

Wefer, G., Berger, W. H., Richter, C., et al. (1998). Shipboard scientific party, 1998, Site 1086 In G. Wefer, W. H. Berger, C. Richter et al., Proceedings of the ocean drilling program, initial reports (Vol. 175, p. 429). College Station, TX (Ocean Drilling Program). https://doi.org/10.2973/odp.proc.ir.175.1998.

Wei, G., & Kirby, J. T. (1995). Time-depended numerical code for extended Boussinesq equations. *Journal of Waterway, Port, Coastal, and Ocean Engineering, 121*(5), 251–261.

Wei, G., Kirby, J. T., Grilli, S. T., & Subramanya, R. (1995). A fully nonlinear Boussinesq model for free surface waves. Part 1: Highly nonlinear unsteady waves. *Journal of Fluid Mechanics, 294,* 71–92.

Word Bank (2005). World Bank East Asia and Pacific economic update 2011, Vol. 1; World Bank response to the Tsunami disaster, February 2, 2005.

Zaman S., Ornthammarath, T., Warnitchai, P. (2012). Probabilistic seismic hazard maps for Pakistan. 15 WCEE Lisbon 2012.

(Received August 4, 2017, revised November 14, 2017, accepted November 16, 2017)

Pure Appl. Geophys.

https://doi.org/10.1007/s00024-018-1816-y

Pure and Applied Geophysics

Far-Field Tsunami Hazard Assessment Along the Pacific Coast of Mexico by Historical Records and Numerical Simulation

Laura G. Ortiz-Huerta,[1] Modesto Ortiz,[2] and Alejandro García-Gastélum[3]

Abstract—Historical records of the Chile (22 May 1960), Alaska (27 March 1964), and Tohoku (11 March 2011) tsunamis recorded along the Pacific Coast of Mexico are used to investigate the goodness of far-field tsunami modeling using a focal mechanism consisting in a uniform slip distribution on large thrust faults around the Pacific Ocean. The Tohoku 2011 tsunami records recorded by Deep ocean Assessment and Reporting of Tsunami (DART) stations, and at coastal tide stations, were used to validate transoceanic tsunami models applicable to the harbors of Ensenada, Manzanillo, and Acapulco on the coast of Mexico. The amplitude resulting from synthetic tsunamis originated by $M_w \sim 9.3$ earthquakes around the Pacific varies from $\sim$ 1 to $\sim$ 2.5 m, depending on the tsunami origin region and on the directivity due to fault orientation and waveform modification by prominent features of sea bottom relief.

Key words: Pacific Coast of Mexico, far-field tsunami hazard, numerical modeling, historical records.

1. Introduction

According to instrumental records of historical and recent far-field tsunamis along the Pacific Coast of Mexico, only far-field tsunamis produced by the large earthquakes of Chile (22 May 1960, M_w 9.5), Alaska (27 March 1964, M_w 9.3), and Tohoku, Japan (11 March 2011, M_w 9.0) are known to have reached water levels up to 1.5 m above the still water level (Figs. 1, 2). Although damages in Mexico caused by these tsunamis have consisted in the interruption of harbor operations and the cessation of civil activities, we cannot discard the possibility of a major tsunami hazard by large far-field tsunamis originated in other regions. Therefore, we investigate the tsunami hazard via the numerical simulation of tsunamis produced by large synthetic earthquakes ($M_w \sim 9.3$) in the tsunami source regions around the Pacific. Since rupture areas ($L \times W$) of the last three large earthquakes in the Pacific were estimated to be 300–500 × 150–200 km^2 (Tohoku 2011: Ammon et al. 2011; Gusman et al. 2012; Yoshida et al. 2011); 700–800 × 150–250 km^2 (Alaska 1964; Plafker 1965); 800 × 200 km^2 (Chile 1960 Kanamori and Cipar 1974), we evaluated the goodness of far-field tsunami modeling by prescribing a focal mechanism consisting of a uniform slip distribution on thrust faults of a fixed area (600 × 180 km^2) around the Pacific. We are aware that heterogeneities in the slip distribution of large infrequent rupture areas are the rule rather than the exception (Lay and Kanamori 2011); therefore, the goodness of the far-field tsunami modeling is validated here with the Chile 1960, Alaska 1964, and Tohoku 2011 tsunami records along the coast of Mexico.

2. Numerical Modeling of the Tohoku 2011 Tsunami

A large number of coastal and open ocean records of the Tohoku 2011 tsunami have allowed the validation of the numerical modeling of the tsunami, as well as the inversion of tsunami data, to estimate the tsunami source heterogeneities (e.g., Fujii et al. 2011; Gusman et al. 2012; Hayashi et al. 2011; Lay et al. 2011a; Maeda et al. 2011), which are in agreement with seismic and GPS estimates of the focal

[1] Instituto de Investigaciones Oceanológicas, Universidad, Autónoma de Baja California (UABC), Ensenada, Baja California, Mexico.
[2] Departamento de Oceanografía Física, Centro de Investigación Científica y de Educación Superior de Ensenada (CICESE), Ensenada, Baja California, Mexico. E-mail: ortizf@cicese.mx
[3] Facultad de Ciencias Marinas, Universidad, Autónoma de Baja California (UABC), Ensenada, Baja California, Mexico.

Reprinted from the journal

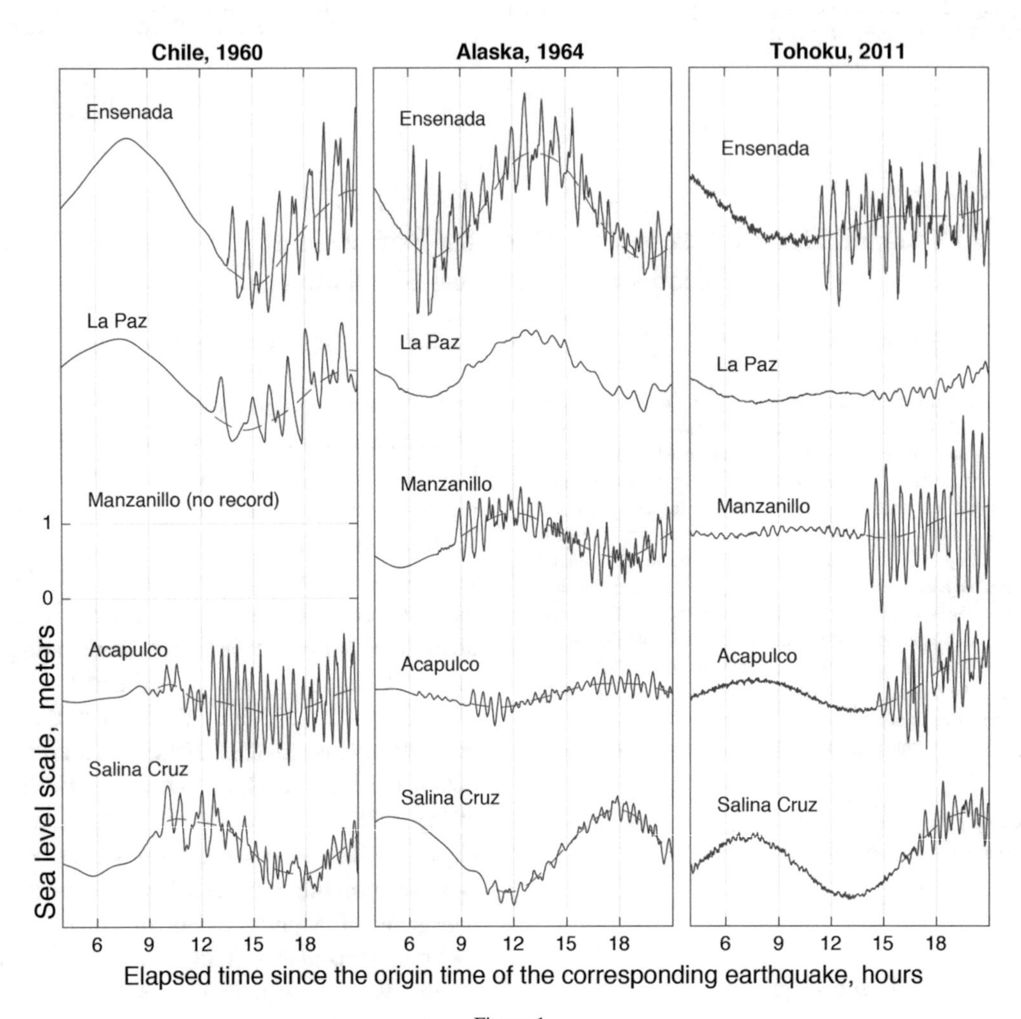

Figure 1
Historical records of the three large far-field tsunamis recorded at coastal tide stations along the Pacific Coast of Mexico (dashed lines indicate the corresponding predicted tide)

mechanism of the earthquake (Ammon et al. 2011; Hayes 2011; Ide et al. 2011; Koper et al. 2011; Lay et al. 2011b; Shao et al. 2011; Simons et al. 2011; Yoshida et al. 2011; Iinuma et al. 2011; Ito et al. 2011; Ozawa et al. 2011; Yue and Lay 2011). Here, the numerical modeling of the Tohoku 2011 tsunami is used to validate the modeling in the deep ocean as well as in the harbors of Ensenada, Manzanillo, and Acapulco, where tide gauges are in operation along the coast of Mexico.

The transoceanic modeling of the tsunami is computed with the numerical tsunami propagation model of Goto et al. (1997), which solves the vertically integrated (depth-averaged) linearized shallow water momentum and continuity equations, or linear long wave equations in a rotating Earth (e.g., Dronkers 1964; Pedlosky 1979):

$$\frac{\partial \eta}{\partial t} + \nabla \cdot \mathrm{M} = 0, \tag{1}$$

$$\frac{\partial \mathrm{M}}{\partial t} + 2\Omega \times \mathrm{M} + gh\nabla\eta = 0. \tag{2}$$

In Eqs. (1–2), t is time, η is the vertical displacement of the water surface above the still water level (the equipotential surface), h is the ocean depth, g is the gravitational acceleration, and Ω is the angular velocity of the Earth. $M = [U, V]$ is the horizontal depth-averaged volume flux vector, where $U = u(\eta + h)$ and $V = v(\eta + h)$ are the horizontal depth-averaged volume flux vectors in longitudinal

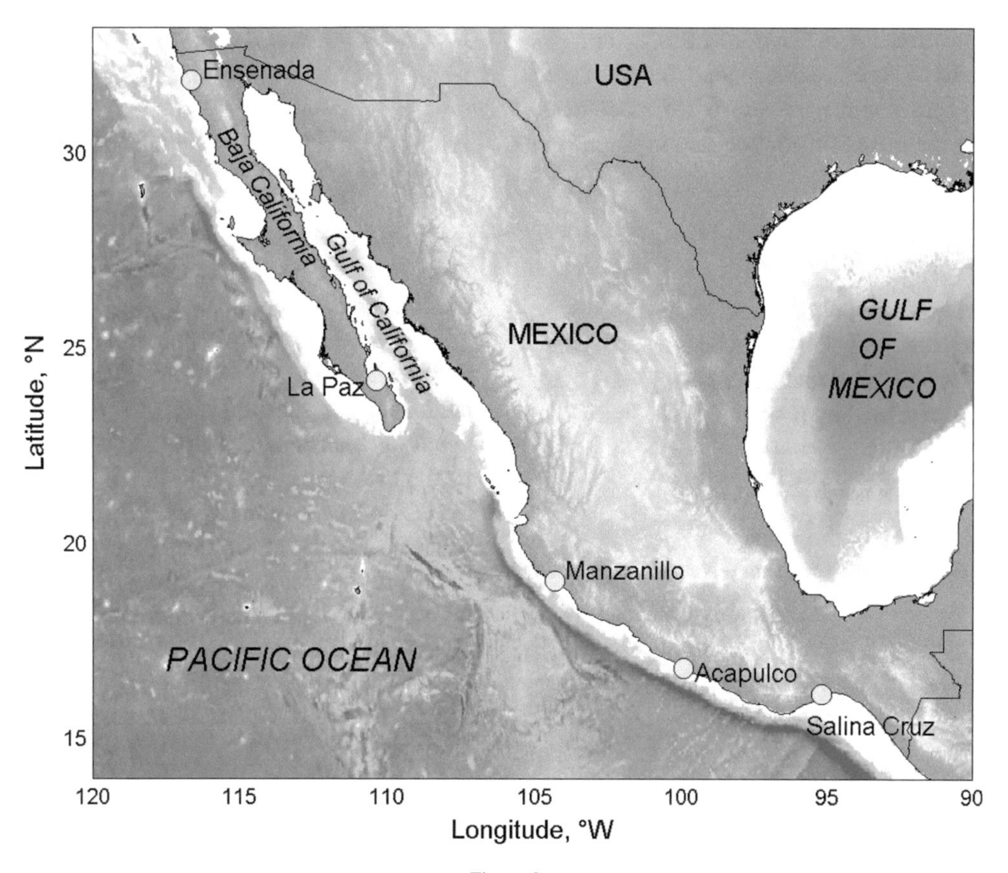

Figure 2
Coastal tide stations (yellow dots) along the Pacific Coast of Mexico that correspond to the tide records shown in Fig. 1

and latitudinal spherical coordinates, and u and v are the corresponding velocity of water particles.

The above set of Eqs. (1–2) are solved by finite differences in an explicit staggered leap-frog scheme at time steps of 2 s in a grid resolution of 2 min. The current bathymetry or domain of integration of the numerical model (Fig. 4) was extracted from the ETOPO2v2 dataset (National Geophysical Data Center, 2006) in a matrix of 5761 $\times$ 3891 nodes. In the model computations, the maximum depth along the trenches was set to 7000 m to prevent numerical instabilities. Numerical experiments that set the maximum depth to 7000 m have shown that tsunami wave deformation across the Japan Trench, as well as across several trenches, is less than 0.3% in amplitude (phase delay less than 5 s), which would be imperceptible in the far-field.

The modeling of the tsunami in the target regions is computed at time steps of 2 s in a set of interconnected nested grids within the domain of integration shown in Fig. 4. High-resolution bathymetric data (Figs. 11, 12 and 13, in the Appendix) are used around the harbors of Ensenada, Manzanillo, and Acapulco (grid resolution of $\sim$ 40 m; Division of Coasts and Harbors of the Mexican Institute of Transportation 2008), where full non-linear terms (advective and bottom friction) are included in the expressions (3–4) of Eqs. (1–2) in a rectangular coordinate system $(x,\ y)$ disregarding Coriolis terms:

$$\frac{\partial \eta}{\partial t} + \frac{\partial U}{\partial x} + \frac{\partial V}{\partial y} = 0 \tag{3}$$

$$\begin{aligned} &\frac{\partial U}{\partial t} + \frac{\partial}{\partial x}\left(\frac{U^2}{D}\right) + \frac{\partial}{\partial y}\left(\frac{UV}{D}\right) + gD\frac{\partial \eta}{\partial x} \\ &\quad + \frac{gn^2}{D^{7/3}} U\sqrt{U^2+V^2} = 0 \\ &\frac{\partial V}{\partial t} + \frac{\partial}{\partial x}\left(\frac{UV}{D}\right) + \frac{\partial}{\partial y}\left(\frac{V^2}{D}\right) \\ &\quad + gD\frac{\partial \eta}{\partial y} + \frac{gn^2}{D^{7/3}} V\sqrt{U^2+V^2} = 0. \end{aligned} \tag{4}$$

The expression for the bottom friction terms in Eq. (4) merits a brief explanation. Before depth averaging, friction terms are expressed as follows, according to the quadratic friction law in a uniform flow (e.g., Dronkers 1964):

$$\frac{\tau_x}{\rho} = \frac{C_D}{2D} u\sqrt{u^2+v^2}, \frac{\tau_y}{\rho} = \frac{C_D}{2D} v\sqrt{u^2+v^2}, \qquad (5)$$

where τ_x and τ_y are the bottom stress in the directions x and y, respectively, and ρ is the fluid density. The drag coefficient C_D and the Gauckler–Manning roughness coefficient n are related by (Goto et al. 1997):

$$n = \sqrt{\frac{C_D D^{1/3}}{2g}}. \qquad (6)$$

Consequently, bottom friction terms are expressed as follows:

$$\frac{\tau_x}{\rho} = \frac{gn^2}{D^{4/3}} u\sqrt{u^2+v^2}, \frac{\tau_y}{\rho} = \frac{gn^2}{D^{4/3}} v\sqrt{u^2+v^2}. \qquad (7)$$

By integrating (7) from the sea bottom to the water surface, friction terms appear as those presented in Eq. (4), where the Gauckler–Manning roughness coefficient is set to 0.025 (cf., Titov et al. 2003). Water depth is an important factor controlling friction terms magnitude.

The tsunami initial condition or sea surface deformation (Fig. 3) is taken as the vertical deformation of the seafloor computed by the expression given by Mansinha and Smylie (1971), as produced by the heterogeneous dislocation model, defined here as FOCAL-I, consisting of the three thrust fault segments described in Table 1.

Depth was set to 20 km in the dislocation model in order to avoid sharp forms in the sea surface

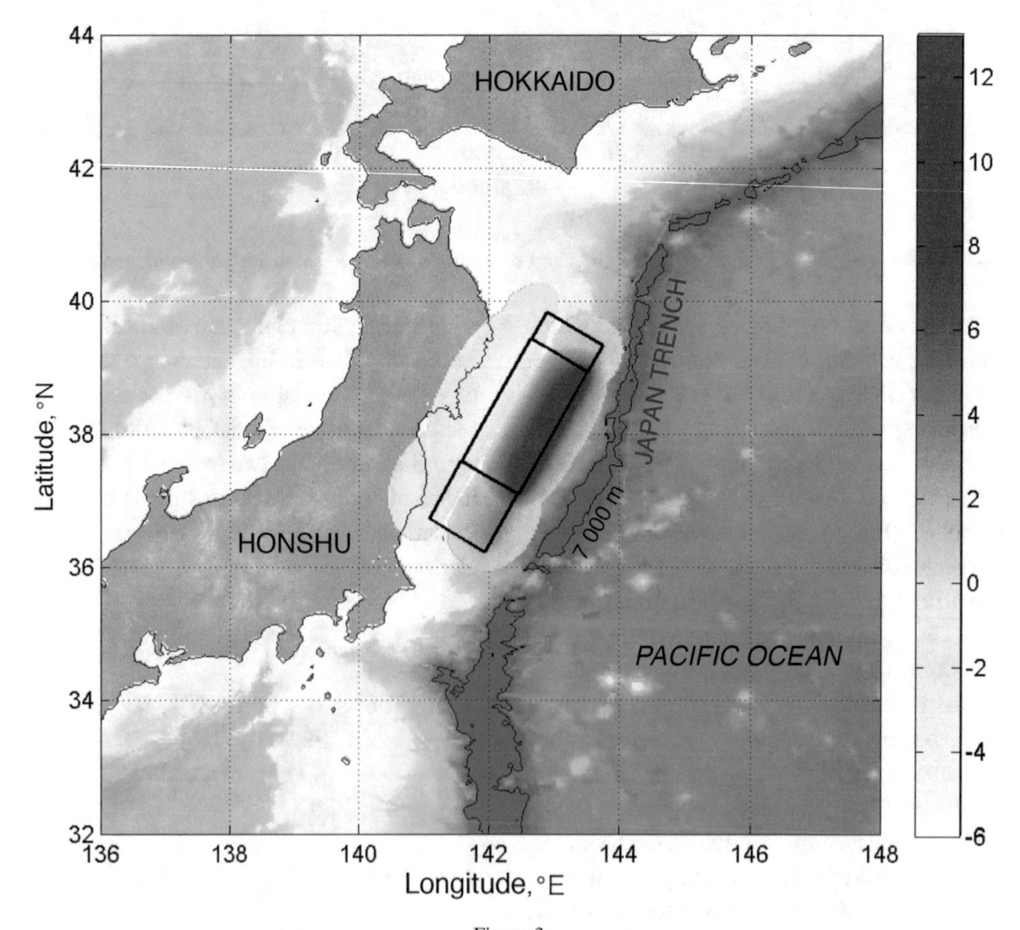

Figure 3
Vertical deformation of the seafloor (color scale, meters). Rectangles indicate the surface projection of the fault segments. The Japan Trench is indicated by the isobath of 7000 m

Table 1

Simplified focal mechanism of the Tohoku 2011 earthquake (FOCAL-I), approximately following the heterogeneous slip distribution model illustrated in Gusman et al. (2012)

Segment	Length (km)	Width (km)	Slip (m)	Depth[a] (km)	Dip (deg.)	Strike (deg.)	Rake (deg.)
Northern	50	160	5	20	15	210	90
Central	250	160	30	20	15	210	90
Southern	150	160	5	20	15	210	90

[a]Depth refers to the shallow edge of the buried fault plane

produced by a realistic shallow-depth dislocation model. Sharp shapes (short wavelengths) in the vertical deformation of the sea surface may produce numerical instabilities in the tsunami propagation model (cf., Fine et al. 2012; Gusman et al. 2012).

The results of the Tohoku 2011 tsunami propagation model using the FOCAL-I dislocation model are illustrated in Fig. 4. The results consist of the maximum peak-wave amplitude or MOM (Maximum of Maximum) values above the still water level computed in every grid point during the transoceanic tsunami propagation. In general, MOM values correspond to the height of the first wave in the open ocean, although values near the coast may correspond to the height of one of the later waves.

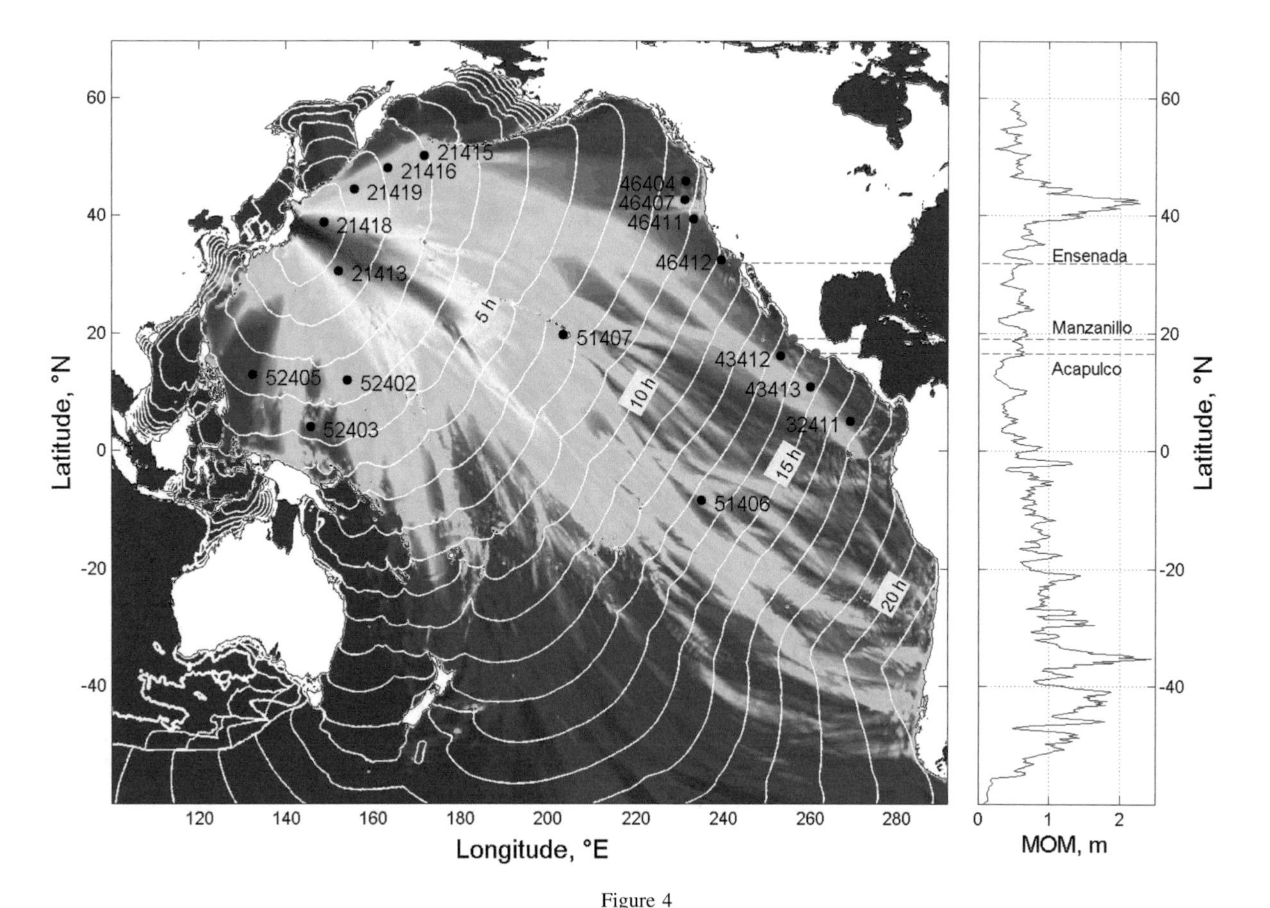

Figure 4
Tohoku 2011 tsunami propagation model. The color scale shows larger MOM values in warmer colors. The white contours indicate tsunami travel time at time steps of 1 h. Black dots indicate DART stations with their corresponding ID numbers. The right panel shows the corresponding MOM values along the Eastern Pacific Coast

Since ocean depth and grid resolution of the whole region (grid resolution 2 min) do not adequately describe nearshore bathymetry, the corresponding MOM values along the Eastern Pacific Coast (right panel in Fig. 4) are only a rough estimate. However, those values reveal potential hot spots or sites where the tsunami is amplified because of the directivity due to fault orientation and waveform modification by prominent features of sea bottom relief, which focus the tsunami and amplify the local response or normal resonance modes (e.g., Ben-Menahem and Rosenman 1972; Miyoshi 1968; Hammack 1973; Horrillo et al. 2008; Kowalik et al. 2008; Satake 1988; the directivity of tsunamis due to fault orientation and the Earth's sphericity effects are illustrated in the Appendix; Figs. 14, 15). In particular, MOM values in Fig. 4 indicate that the tsunami is focused towards the coast of Chile due to directivity and towards several hot spots along the coast of Mexico.

The synthetic tsunami records resulting from the numerical modeling of the tsunami are illustrated in Fig. 5 for comparison with the corresponding residual tide provided within the DART records. In the near field (DART stations 21413, 21418, and 21419), the height of the synthetic tsunami is within $\pm$ 75 cm of the observed one, whereas the wave height of the synthetic tsunami in the far-field is adequately reproducing the corresponding observed one within $\pm$ 4 cm ($\pm$ 20%).

It is worth mentioning that the elapsed arrival time is 6–9 min earlier in the synthetic tsunami records than in the observed tsunami data after a propagation time of 8–17 h, respectively (cf., Watada et al. 2014).

The resulting synthetic tsunami records at the coastal tide stations are illustrated in Fig. 6 for comparison with the observed tsunami in Ensenada, Manzanillo, and Acapulco. The synthetic tsunamis are adequately reproducing the wave height and period of the first 9–12 waves. However, the synthetic tsunami in Manzanillo is not adequately reproducing the second set of larger observed waves (the 11th–15th waves; elapsed time 19:00–21:00 h), where the maximum peak-wave amplitude of 1.30 m occurred at the elapsed time of 19:30 h.

Quantitative comparisons of the synthetic and observed tsunami records illustrated in Fig. 6 are given in Table 2 by computing H_s and T_s, where H_s is the significant wave height or the average height of the highest one-third waves in the corresponding record, and T_s is the significant wave period or the average zero-crossing wave period of the highest one-third waves. Individual wave heights are taken as the height, crest to trough, in a zero-crossing wave period, disregarding smaller superimposed peaks and troughs within the selected wave period by slightly smoothing the tide record. Wave heights were taken trough to crest in case the leading wave was negative in order to include the onset of the leading wave.

We applied the concept of H_s and T_s from wind-wave theory and observations (e.g., Bretschneider 1964), which even in the presence of uncorrelated phase shifts takes into account individual wave heights and their corresponding periods in the one-third average. Although the tsunami waveforms are not a stationary process, H_s and T_s were calculated based on the waves present in the first 12–15 h after the tsunami arrival time. In this case, significant wave heights and periods (Table 2) indicate that synthetic tsunami waveforms are adequately reproducing the observed ones. Synthetic H_s and T_s reproduce, within $\pm$ 25 cm ($\pm$ 19%) and $\pm$ 11 min, the observed ones in Ensenada, Manzanillo, and Acapulco and the corresponding MOM values are within $\pm$ 33 cm of the ones observed.

3. Goodness of Far-Field Tsunami Modeling

Synthetic tsunamis were produced at the location of the rupture areas of the Chile 1960, Alaska 1964, and Tohoku 2011 earthquakes, using the FOCAL-II dislocation model, defined here by a uniform slip distribution of 20 m on a rectangular large thrust fault (600×180 km^2) dipping 15° at the depth of 20 km. Although the dip angles and slip magnitudes of the earthquakes of Chile 1960 and Alaska 1964 are estimated to be 10° and 20 m, and 6–12° and 14.9–17.4 m, respectively (Kanamori and Cipar 1974; Ichinose et al. 2007), the purpose of the dislocation model FOCAL-II is to estimate the goodness of far-field tsunami modeling for large tsunamis around the Pacific by prescribing a rough dislocation model. The FOCAL-II dislocation model can

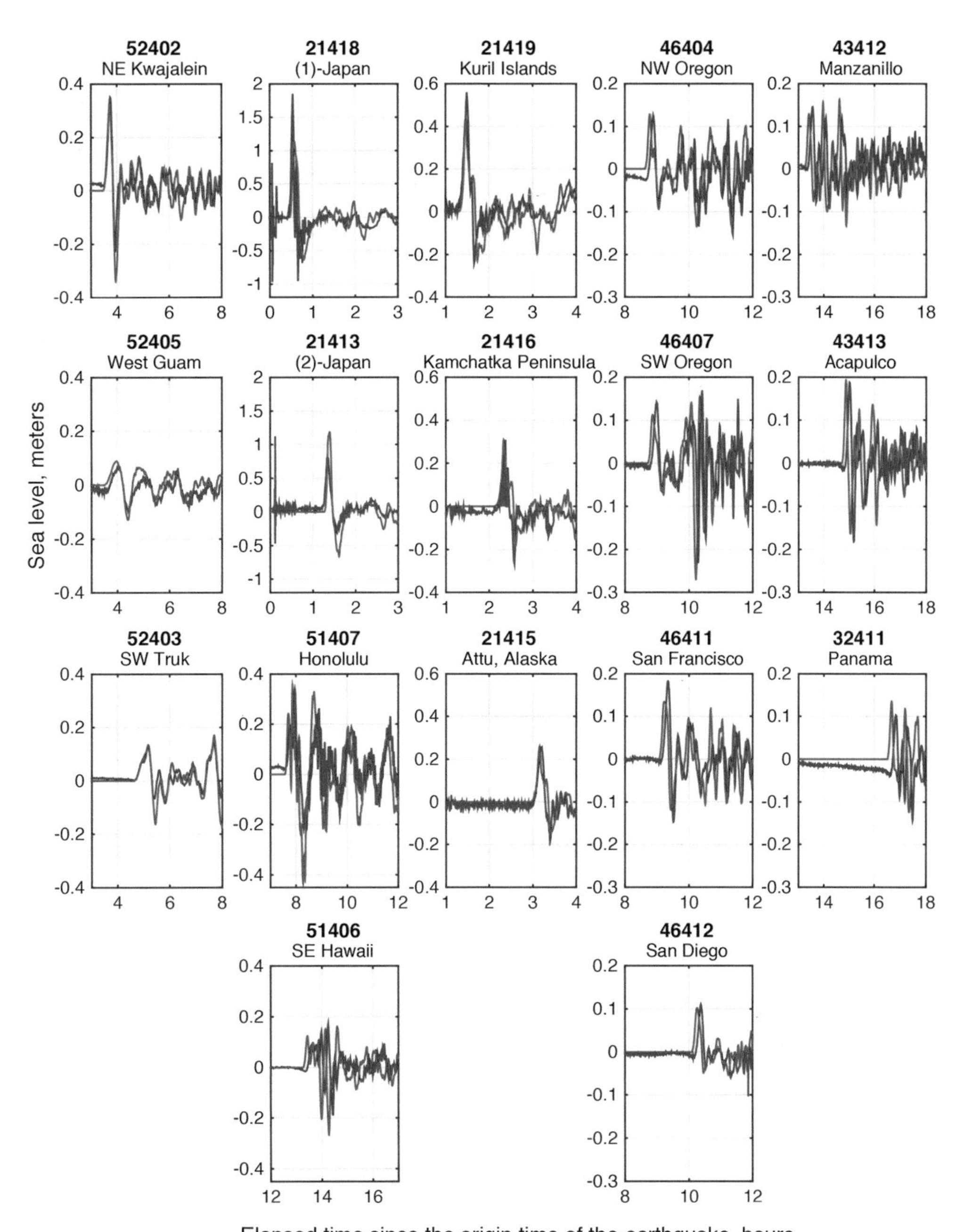

Figure 5
The Tohoku 2011 tsunami records recorded at DART stations (blue line) in comparison with the corresponding results (red line) of the tsunami propagation model. The five-digit numbers correspond to the DART stations shown in Fig. 4

represent earthquake magnitudes $M_w = [9.0, 9.2, 9.3, 9.4]$ by correspondingly varying the slip magnitude $d = [9,\ 15,\ 20,\ 30\ \text{m}]$ according to the seismic moment (M_o) and moment magnitude (M_w) expressions $M_o = \mu A d$ (μ = rigidity modulus, A = rupture area; Aki, 1966) and $M_w = (\log_{10} M_o - 9.05)/1.5$ (Hanks and Kanamori, 1979; M_o expressed in Newton meters), assuming a rigidity modulus $\mu = 4 \times 10^{10}\ \text{Nm}^{-2}$.

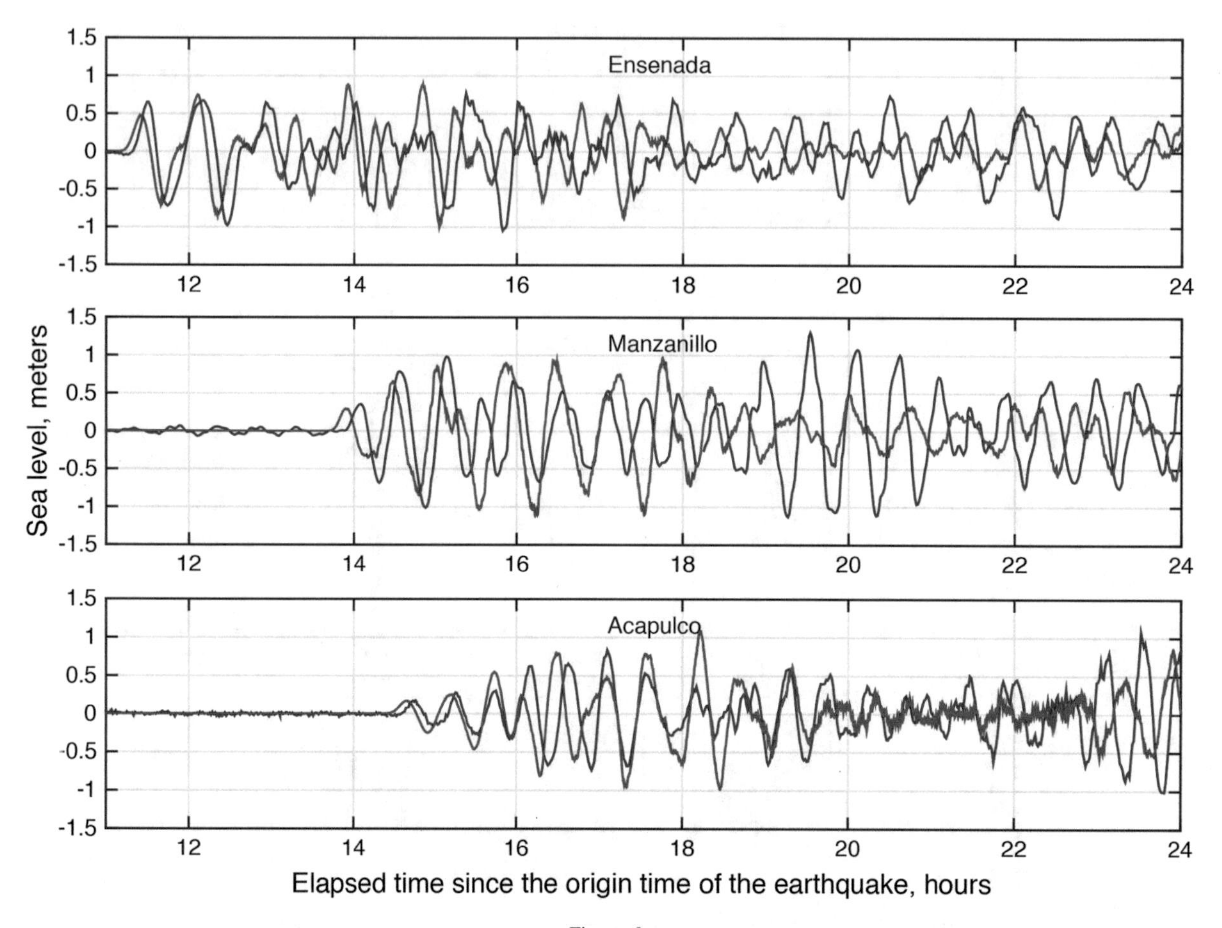

Figure 6
Tohoku 2011 tsunami records (residual tide) at the coastal tide gauges along the Pacific Coast of Mexico (blue line) in comparison with the corresponding results (red line) of the numerical modeling of the tsunami. The residual tide is obtained by subtracting the predicted tide shown in Fig. 1 from the corresponding tide records

Table 2

Comparison of the synthetic and observed tsunami records in Ensenada, Manzanillo, and Acapulco

Tide station	Number of waves (one-third)	Observed tsunami			Synthetic tsunami			Ratio, $\frac{H_{s\ synthetic}}{H_{s\ observed}}$
		MOM (m)	H_s (m)	T_s (min)	MOM (m)	H_s (m)	T_s (min)	
Ensenada (31.8483°N; 116.6180°W)	9	0.78	1.26	43	0.89	1.27	40	1.01
Manzanillo (19.0555°N; 104.3177°W)	9	1.30	1.71	34	0.97	1.46	45	0.85
Acapulco (16.8407°N; 99.8513°W)	9	0.83	0.99	31	1.08	1.18	35	1.19

Tsunami origin region: Tohoku 2011; dislocation model FOCAL-I

Statistical parameters of the Tohoku 2011 tsunami waves along the Pacific Coast of Mexico including arrival times, maximum wave heights, and periods are given in Zaytsev et al. (2017)

The resulting synthetic tsunamis produced by prescribing the FOCAL-II dislocation model on the rupture areas of the Chile 1960, Alaska 1964, and Tohoku 2011 earthquakes are illustrated in Fig. 7 and are compared with the corresponding historic tsunamis recorded in Ensenada, Manzanillo, and Acapulco.

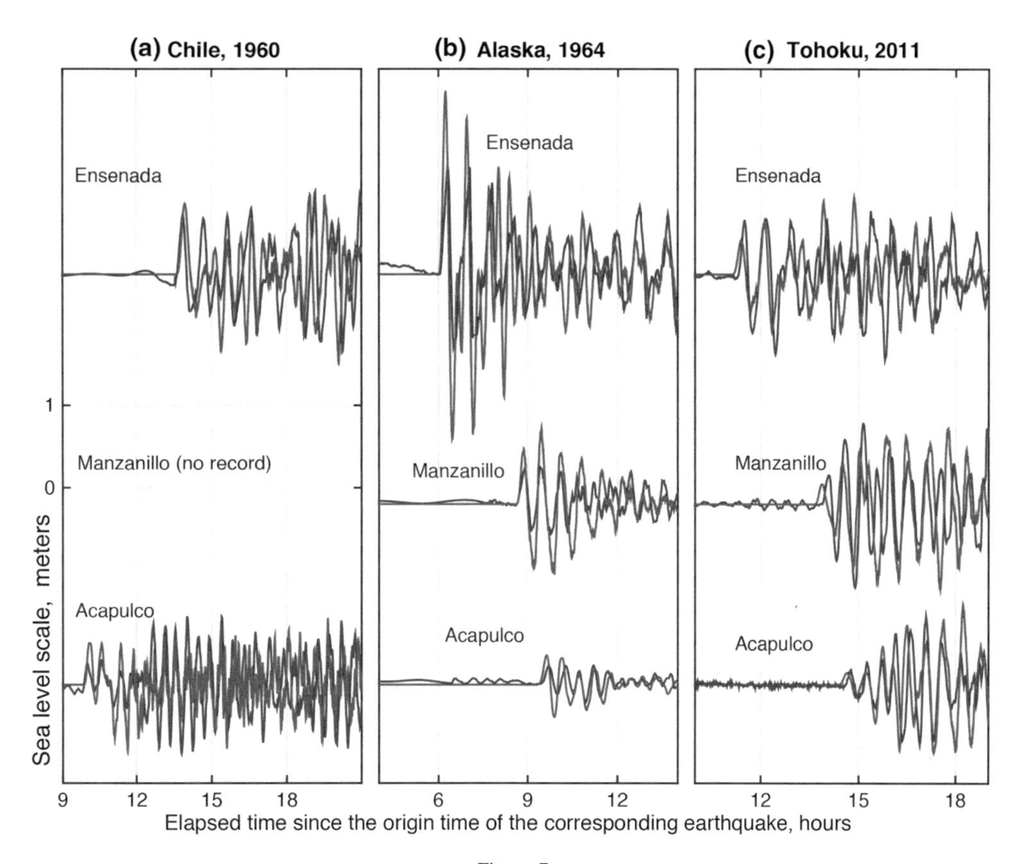

Figure 7
Historical records of the three large far-field tsunamis recorded along the Pacific Coast of Mexico (blue line) are compared with the corresponding numerical results (red line) produced by prescribing the FOCAL-II dislocation model

Quantitative comparisons of the synthetic and observed tsunami records are summarized in Tables 3, 4, and 5:

(a) The synthetic waveforms of the Chile 1960 tsunami (Fig. 7a) reproduce, within ± 9 cm (± 8%) and ± 7 min, the significant wave height and period of the corresponding tsunami records in both tide stations Ensenada and Acapulco (Table 3). Synthetic MOM values are within ± 22 cm (± 22%) of the observed ones.

(b) In the Case of the Alaska 1964 tsunami (Fig. 7b), significant wave height of the synthetic tsunami records (Table 4) is overestimated by 83 cm (64%), 25 cm (37%), and 5 cm (18%), compared to the wave heights observed in Ensenada, Manzanillo, and Acapulco, respectively. In addition, MOM values of the synthetic tsunami are overestimated by 89 cm (67%), 38 cm (79%), and 12 cm (50%), respectively, compared to MOM values observed in Ensenada, Manzanillo, and Acapulco.

Table 3

Comparison of the synthetic and observed tsunami records in Ensenada, Manzanillo, and Acapulco

Tide station	Number of waves	Observed tsunami			Synthetic tsunami			Ratio, $\frac{H_{s\,synthetic}}{H_{s\,observed}}$
		MOM (m)	H_s (m)	T_s (min)	MOM (m)	H_s (m)	T_s (min)	
Ensenada	9	1.01	1.57	57	1.23	1.58	50	1.01
Manzanillo	9	–	–	–	0.88	1.32	42	–
Acapulco	9	0.82	1.12	35	0.77	1.03	36	0.92

Tsunami origin region: Chile 1960 (region 19 in Fig. 8)

Table 4

Comparison of the synthetic and observed tsunami records in Ensenada, Manzanillo, and Acapulco

Tide station	Number of waves	Observed tsunami			Synthetic tsunami			Ratio, $\frac{H_{s\ synthetic}}{H_{s\ observed}}$
		MOM (m)	H_s (m)	T_s (min)	MOM (m)	H_s (m)	T_s (min)	
Ensenada	9	1.33	1.28	45	2.22	2.11	31	1.64
Manzanillo	9	0.48	0.66	40	0.86	0.91	44	1.37
Acapulco	9	0.24	0.30	46	0.36	0.35	32	1.18

Tsunami origin region: Alaska 1964 (region 2 in Fig. 8)

Table 5

Comparison of the synthetic and observed tsunami records in Ensenada, Manzanillo, and Acapulco

Tide station	Number of waves	Observed tsunami			Synthetic tsunami			Ratio, $\frac{H_{s\ synthetic}}{H_{s\ observed}}$
		MOM (m)	H_s (m)	T_s (min)	MOM (m)	H_s (m)	T_s (min)	
Ensenada	9	0.78	1.26	43	0.93	1.08	40	0.86
Manzanillo	9	1.30	1.71	34	0.92	1.31	43	0.77
Acapulco	9	0.83	0.99	31	0.95	1.14	36	1.15

Tsunami origin region: Tohoku 2011 (region 11 in Fig. 8)

(c) The synthetic waveforms of the Tohoku 2011 tsunami (Fig. 7c) reproduce, within ± 40 cm (± 23%) and ± 9 min, the significant wave heights and periods observed in Ensenada, Manzanillo, and Acapulco (Table 5). Synthetic MOM values are within ± 38 cm of the observed MOM values.

The goodness of the far-field Tsunami Modeling employing the FOCAL-II dislocation model is adequate for large tsunamis, such as the ones produced by the Chile 1960 and Tohoku 2011 earthquakes. In contrast, in the case of tsunamis such as the one produced by the Alaska 1964 earthquake, the model predicts a maximum amplitude of 2.22 m in Ensenada, whereas the observed MOM is 1.33 m (Table 4). This difference is likely a consequence of the 20 m uniform slip distribution prescribed in FOCAL-II that overestimates the slip magnitude of the Alaska 1964 megathrust fault (peak slip of 14.9 m; Ichinose et al. 2007). Indeed, a uniform slip magnitude of 12 meters in FOCAL-II adequately reproduced both the observed H_s and MOM values of the Alaska 1964 tsunami in Ensenada within ± 2 cm (± 1%). However, even though a slip dislocation of 12 m can be used to reproduce the observed tsunami, the uniform slip distribution of 20 m, prescribed in FOCAL-II, will continue to be used here as a threshold to estimate the far-field tsunami hazard along the coast of Mexico from the tsunami source regions around the Pacific Ocean illustrated in Fig. 8.

4. *Far-Field Tsunami Hazard Assessment*

The far-field tsunami hazard assessment along the Pacific Coast of Mexico is estimated by the numerical modeling of tsunamis produced by the dislocation model FOCAL-II along the tsunami source regions around the Pacific Ocean illustrated in Fig. 8.

The maximum synthetic tsunami amplitudes (MOM) in Ensenada, Manzanillo, and Acapulco, resulting from each one of the 23 tsunami source regions around the Pacific Ocean identified in Fig. 8, are illustrated in Fig. 9, whereas statistics and arrival times of the corresponding synthetic tsunamis are given in Tables 6, 7 and 8. Observed MOM values in Ensenada, Manzanillo, and Acapulco (Tables 3, 4 and 5) are also illustrated in Fig. 9.

In general, MOM patterns in Fig. 9 indicate the effect of directivity due to fault orientation and waveform modification by prominent features of sea bottom relief. In particular, tsunami source regions

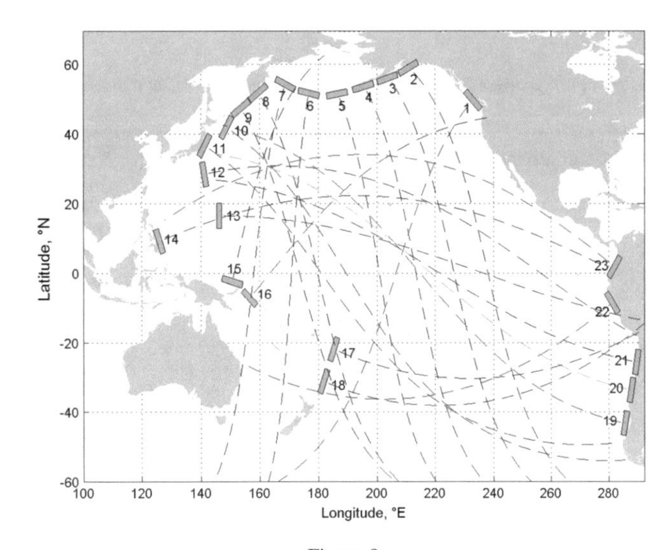

Figure 8
Numerals indicate hypothetical rupture areas (gray rectangles) along some of the tsunami source regions around the Pacific. Great Circles (dashed lines) perpendicular to the rupture areas roughly indicate the tsunami directivity due to fault orientation

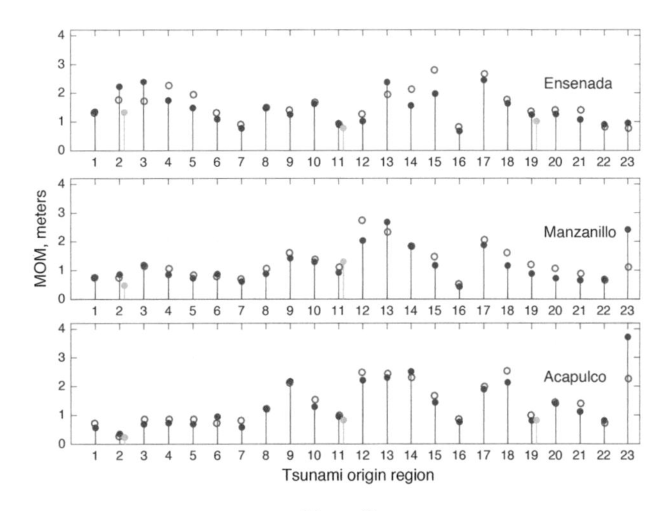

Figure 9
Synthetic MOM values (blue dots) resulting from each one of the tsunamis produced at the 23 tsunami source regions indicated in Fig. 8. Red circles represent the corresponding MOM values estimated as 1.47 times $H_s/2$. Green dots represent the observed MOM values in Ensenada, Manzanillo, and Acapulco in Tables 3, 4 and 5

2–5 in Fig. 8 produce high MOM values ($\sim$ 2 m) in Ensenada due to fault orientation and the proximity of the source regions. In contrast, MOM values produced by the same source regions 2–5 are $\sim$ 1 m in Manzanillo and Acapulco where the tsunami spreads by diffraction after passing the peninsula of Baja California (e.g., Figure 15b in the Appendix). Regional tsunamis originated in the Ecuador–Colombia subduction zone (region 23 in Fig. 8) may produce extreme MOM values on the southeast coast of Mexico due to fault orientation and proximity of the tsunami source region. Further research focused in this region is needed for particular earthquake scenarios and for regional tsunami hazard assessment (e.g., Arreaga-Vargas et al. 2005; Kelleher 1972).

Estimated MOM values (red circles in Fig. 9) were computed as 1.47 times $H_s/2$, according to the best linear fit (Fig. 10) among the set of H_s versus the corresponding MOM values in Tables 6, 7 and 8, excluding the values from region 23. Assuming a normal distribution of the residuals obtained by subtracting $1.47H_s/2$ from the corresponding MOM

Table 6

ENSENADA: Statistics of synthetic tsunamis produced around the Pacific by the FOCAL-II dislocation model. Tsunami origin regions are indicated in Fig. 8

Tsunami origin region	Elapsed tsunami arrival time (hh:mm)	MOM (m)	MOM (elapsed time) (hh:mm)	Number of waves, (one-third)	H_s (m)	T_s (min)
(1) Cascadia	03:30	1.35	13:00	10	1.56	67
(2) Kodiak, AK	06:00	2.22	06:30	9	2.11	31
(3) Alitak, AK	06:00	2.39	06:30	10	2.07	54
(4) Sand Point, AK	06:30	1.74	12:00	9	2.70	60
(5) Adak, AK	07:00	1.48	13:00	9	2.35	58
(6) Kiska, AK	07:30	1.09	11:30	9	1.59	57
(7) Commander, RU	08:30	0.77	24:00	8	1.10	55
(8) Kamchatka	09:00	1.47	12:30	9	1.78	40
(9) Kuril Islands	09:30	1.25	13:30	10	1.65	49
(10) Hokkaido	10:00	1.62	15:00	9	2.01	42
(11) Japan	11:30	0.94	15:00	9	1.08	40
(12) Ogasawara	11:30	1.02	14:00	9	1.51	42
(13) Mariana Islands	12:00	2.37	14:30	8	2.35	48
(14) Philippines	15:00	1.56	17:30	5	2.53	53
(15) Manus	13:30	1.97	17:00	5	3.34	58
(16) Solomon	13:00	0.67	14:30	6	0.95	56
(17) Tonga	11:00	2.44	14:00	7	3.19	62
(18) Kermadec	12:00	1.63	15:30	7	2.13	51
(19) South Chile	13:30	1.24	23:30	9	1.62	50
(20) Central Chile	12:30	1.26	15:00	9	1.70	45
(21) North Chile	12:00	1.07	22:30	9	1.70	59
(22) Peru	09:30	0.90	20:30	9	0.95	55
(23) Ecuador–Colombia	08:30	0.95	09:00	10	0.91	48

Table 7

MANZANILLO: Statistics of synthetic tsunamis produced around the Pacific by the FOCAL-II dislocation model

Tsunami origin region	Elapsed tsunami arrival time (hh:mm)	MOM (m)	MOM (elapsed time) (hh:mm)	Number of waves, (one-third)	H_s (m)	T_s (min)
(1) Cascadia	06:30	0.76	22:30	9	0.88	53
(2) Kodiak, AK	08:30	0.86	09:30	9	0.91	44
(3) Alitak, AK	08:30	1.19	09:30	10	1.36	42
(4) Sand Point, AK	07:00	0.85	12:30	8	1.29	47
(5) Adak, AK	09:30	0.73	12:30	9	1.00	45
(6) Kiska, AK	10:00	0.87	20:00	9	0.95	45
(7) Commander, RU	11:00	0.61	21:30	8	0.82	41
(8) Kamchatka	11:00	0.88	13:00	9	1.28	39
(9) Kuril Islands	12:00	1.42	15:30	7	1.92	43
(10) Hokkaido	12:00	1.29	18:00	7	1.67	41
(11) Japan	14:00	0.92	18:00	9	1.31	43
(12) Ogasawara	14:00	2.03	15:30	5	3.26	52
(13) Mariana Islands	12:30	2.67	16:00	6	2.80	42
(14) Philippines	17:30	1.84	19:00	5	2.21	47
(15) Manus	15:30	1.17	17:30	5	1.76	48
(16) Solomon	14:30	0.43	21:30	7	0.61	39
(17) Tonga	11:30	1.87	13:30	7	2.46	54
(18) Kermadec	12:30	1.16	15:30	9	1.91	57
(19) South Chile	10:30	0.88	15:00	8	1.41	43
(20) Central Chile	09:30	0.72	20:30	8	1.25	51
(21) North Chile	08:30	0.65	21:30	9	1.08	49
(22) Peru	04:00	0.69	06:30	10	0.80	48
(23) Ecuador–Colombia	05:00	2.41	05:30	10	1.35	52

Tsunami origin regions are indicated in Fig. 8

Table 8

ACAPULCO: statistics of synthetic tsunamis produced around the Pacific by the FOCAL-II dislocation model

Tsunami origin region	Elapsed tsunami arrival time (hh:mm)	MOM (m)	MOM (elapsed time) (hh:mm)	Number of waves (one-third)	H_s (m)	T_s (min)
(1) Cascadia	07:00	0.57	15:30	10	0.88	52
(2) Kodiak, AK	09:30	0.36	09:30	9	0.35	32
(3) Alitak, AK	09:30	0.69	10:00	11	1.01	36
(4) Sand Point, AK	09:30	0.73	20:00	11	1.04	37
(5) Adak, AK	10:00	0.69	13:30	11	1.04	34
(6) Kiska, AK	11:00	0.95	24:00	9	0.88	37
(7) Commander, RU	11:30	0.58	22:00	8	0.97	37
(8) Kamchatka	12:30	1.23	24:00	10	1.46	32
(9) Kuril Islands	13:00	2.17	16:00	9	2.55	36
(10) Hokkaido	13:30	1.29	18:00	8	1.83	35
(11) Japan	14:30	0.95	18:30	9	1.18	35
(12) Ogasawara	15:00	2.21	16:00	6	2.97	41
(13) Mariana Islands	15:00	2.30	16:30	6	2.94	37
(14) Philippines	18:00	2.51	19:30	5	2.77	36
(15) Manus	16:00	1.44	18:00	5	1.99	43
(16) Solomon	15:00	0.76	18:00	6	1.03	37
(17) Tonga	12:00	1.89	13:30	12	2.38	33
(18) Kermadec	13:00	2.13	18:00	8	3.04	45
(19) South Chile	10:00	0.81	19:30	10	1.21	38
(20) Central Chile	08:30	1.40	22:00	10	1.74	41
(21) North Chile	08:00	1.12	11:00	10	1.70	32
(22) Peru	05:30	0.81	05:30	17	0.88	31
(23) Ecuador–Colombia	04:00	3.72	05:00	10	2.69	31

Tsunami origin regions are indicated in Fig. 8

values (Fig. 10b, c), the correspondence between MOM and estimated MOM values is within a 95% confidence interval of ± 0.46 m. Note that synthetic MOM values in Ensenada, Manzanillo, and Acapulco, corresponding to tsunami origin regions 1–22, are within the 95% confidence interval of estimated MOM values. In contrast, MOM values in Manzanillo and Acapulco, corresponding to tsunami origin region 23, are significantly different from estimated MOM values. This difference may indicate spurious model results, or in this case, a strong non-stationary process due to the proximity of the tsunami source region.

5. Discussion and Conclusions

The numerical modeling presented here derives MOM values at three specific study sites (Ensenada, Manzanillo and Acapulco) under a set of $M_w \sim 9.3$ uniform slip earthquakes. Zaytsev et al. (2016, 2017) demonstrate that these three locations are hot spots with relatively high tsunami hazards from far-field tsunamis according to analysis of tide gauge data (although there are still long stretches of coast that were not instrumentally measured). The largest MOM from our modeling of far-field tsunamis (excluding the Ecuador-Columbia regional source) is ~ 2.5 m. As a precautionary threshold recommendation for the Pacific Coast of Mexico, we propose increasing this by 50% to 3.75 m above Mean Higher High Water (MHHW) in order to cover the possibility of larger magnitude earthquakes (in the simple but imprecise linear approximation this amounts to an increase in fault slip to 30 m and $M_w \sim 9.4$), the effects of nonuniform slip, the possibility of source earthquakes with different locations and geometries, and to include some allowance for currently unidentified hot spots. We stress that this is only an interim recommendation and that further work is required to increase the set of study sites and to explicitly model a wider set of source scenarios. This MOM is also an offshore value and in some specific circumstances, onshore run-up heights may be higher still.

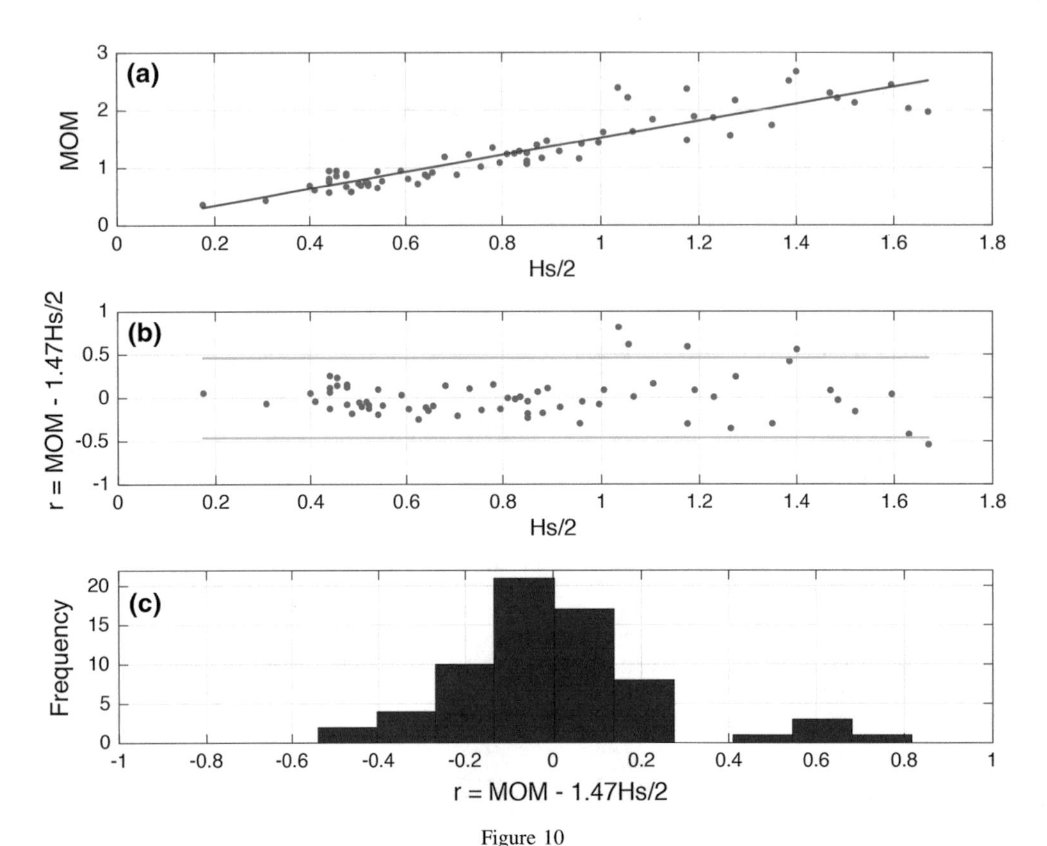

Figure 10

a Best linear fit (red line) among the set of all of the H_s versus the corresponding MOM values in Tables 6, 7 and 8. **b** Residual values (r) obtained by subtracting $1.47H_s/2$ from the corresponding MOM values. Green lines indicate the 95% confidence interval. **c** Histogram of the residual values in (**b**)

Statistics of synthetic tsunamis in Tables 6, 7 and 8 are applicable for far-field tsunami preparedness in Ensenada, Manzanillo, and Acapulco to investigate harbor response by prescribing H_s and T_s as boundary conditions for high-resolution physical as well as numerical hydrodynamic models. Regardless of the origin region, historical and synthetic tsunamis have shown that large earthquakes ($M_w \sim 9$) in the Pacific would represent a tsunami hazard for harbor operations.

Acknowledgements

We thank both anonymous reviewers for their helpful review and fructiferous discussions. L.G. Ortiz-Huerta acknowledges support from CONACyT (México) through the Ph.D. program scholarship. DART sea level data are courtesy of the National Data Buoy Center of the National Oceanic and Atmospheric Administration (NOAA). ETOPO2v2 are courtesy of the National Geophysical Data Center, 2006. M. Ortiz acknowledges J.M. Montoya of the Division of Coasts and Harbors of the Mexican Institute of Transportation (IMT) for providing high-resolution bathymetric data around the harbors of Ensenada, Manzanillo, and Acapulco. The tsunami records along the coast of Mexico are courtesy of the Joint Sea Level Network Operations of Mexico, operated by CICESE, IMT, SEMAR, and UNAM.

Appendix

Bathymetry

High resolution near shore bathymetry used in the tsunami model around the harbors of Ensenada, Manzanillo, and Acapulco (see Figs. 11, 12, 13).

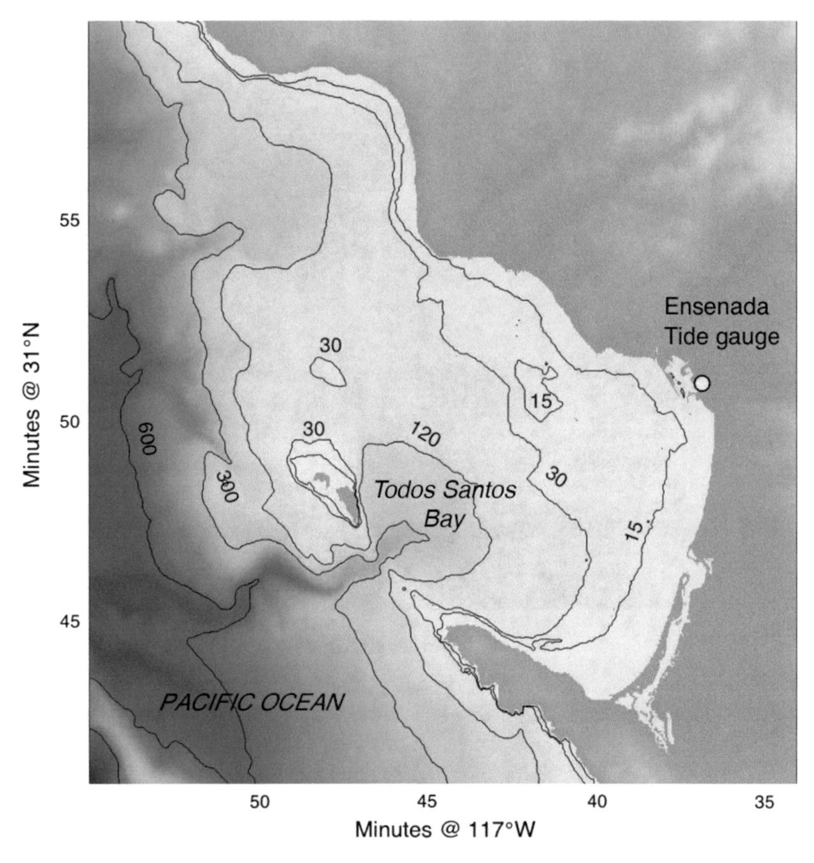

Figure 11
Ensenada bathymetric map. Isobaths are indicated in meters referred to the mean sea level

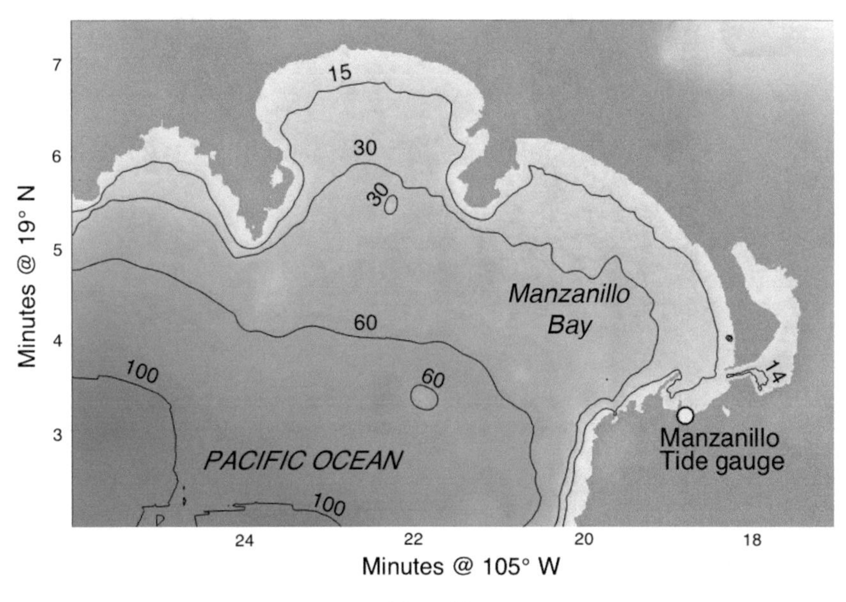

Figure 12
Manzanillo bathymetric map. Isobaths are indicated in meters referred to the mean sea level

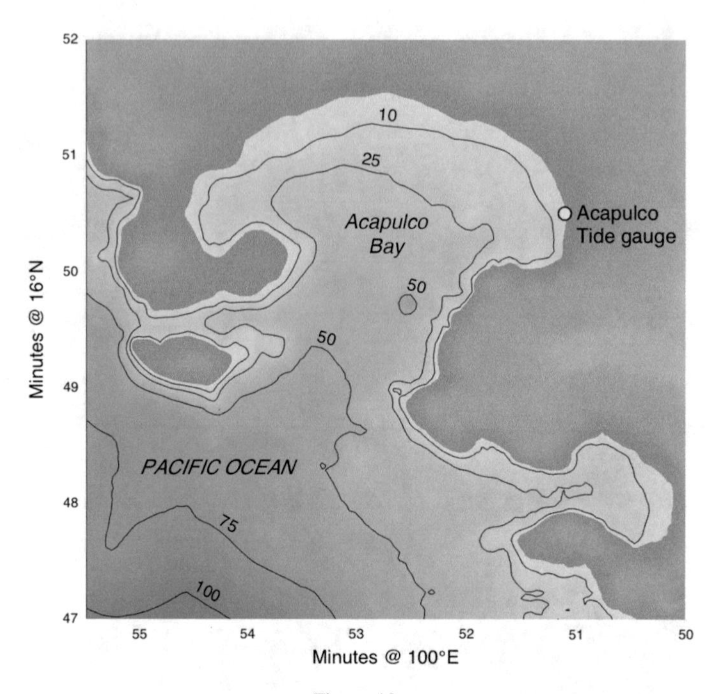

Figure 13
Acapulco bathymetric map. Isobaths are indicated in meters referred to the mean sea level

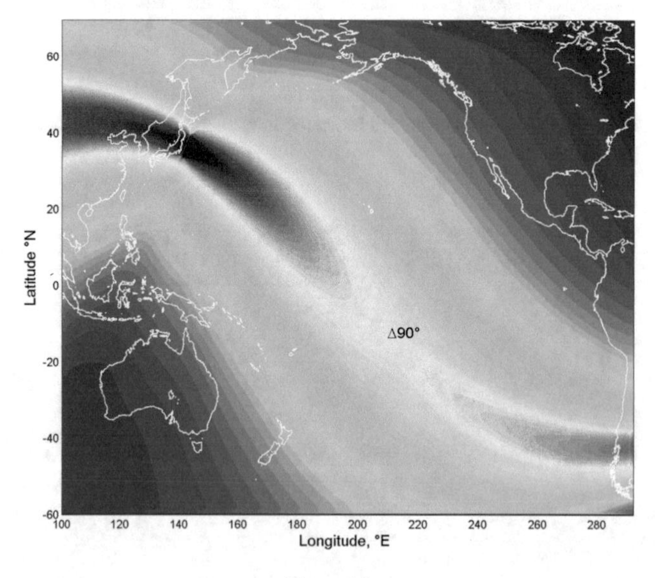

Figure 14
Directivity and Earth's sphericity effects on the Tohoku 2011 tsunami propagated in a spherical ocean of uniform depth. Land contours are traced as a geographical reference

Tsunami Directivity

The Tohoku 2011 tsunami is propagated in an ocean of uniform depth to illustrate the directivity due to fault orientation and the Earth's sphericity effects (Fig. 14). In this case, the tsunami is focused on the direction perpendicular to the rupture length along the Great Circle towards the coast of Chile, where the amplitude of the wavefront decreases

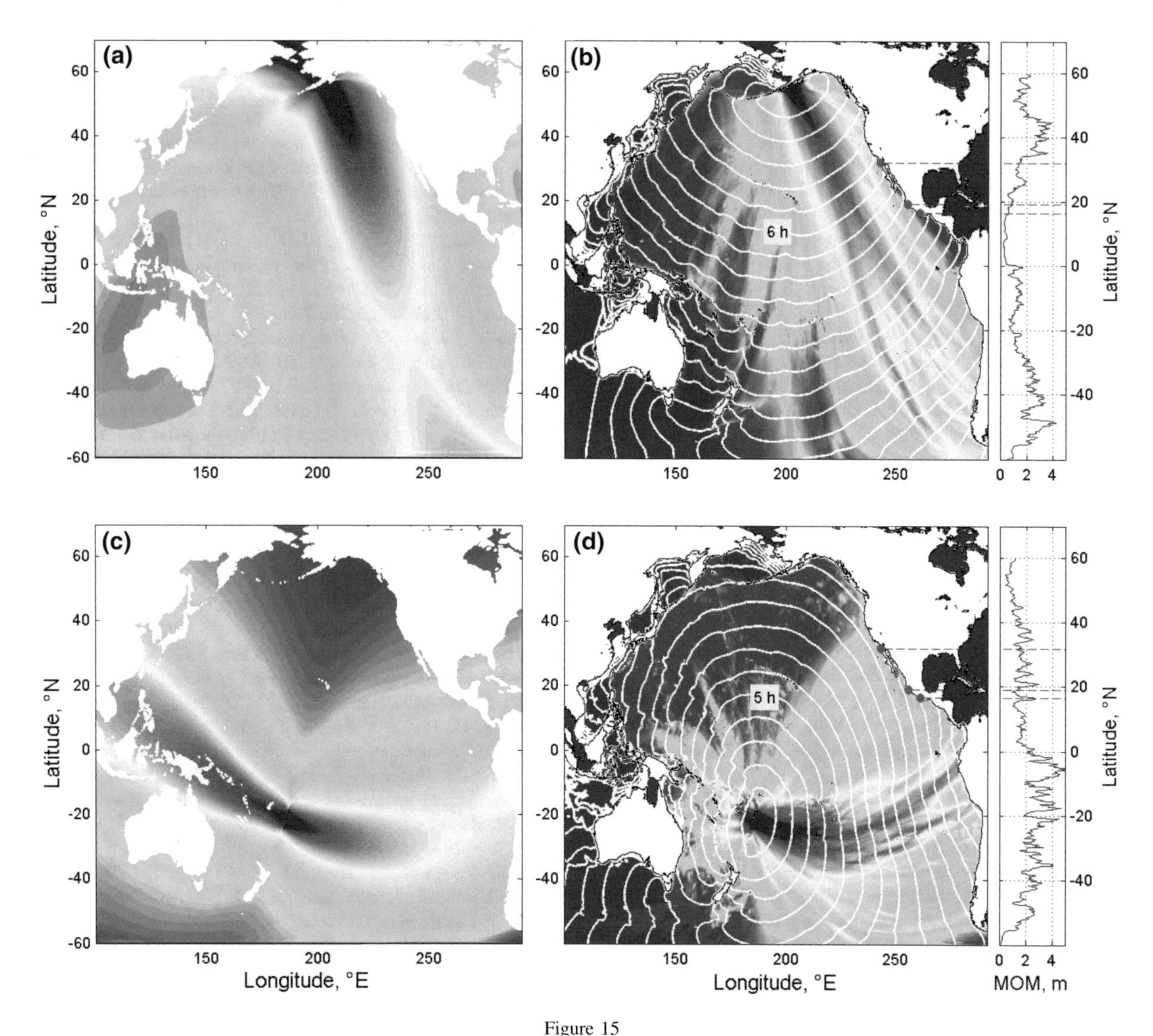

Figure 15
Tsunamis originated in the Aleutian and Tonga Trenches propagated in a spherical ocean of uniform depth (**a**, **c**) and in a spherical ocean with realistic bathymetry (**b**, **d**)

continuously up to a propagation distance of Δ90° due to the geometrical spreading of the wavefront. The amplitude recovers beyond this point towards the antipodal location (Δ180°) near the coast of Chile. In correspondence, tsunamis originated in southern Chile are focused towards Japan (cf., Okal et al. 2014) (see Fig. 14).

The following examples illustrate the effect of waveform modification by prominent features of sea bottom relief, which focus the tsunami towards potential hot spots or sites where the tsunami is amplified:

1. A synthetic tsunami originated in the Aleutian Trench (Fig. 15a, b) would be focused towards the South Pole due to directivity and Earth's sphericity effects; however, in a realistic bathymetry, the tsunami is also focused towards several hot spots along the Easter Pacific Coast.
2. A synthetic tsunami originated in the Tonga Trench (Fig. 15c, d) is also revealing hot spots along the coasts of Mexico and Chile, despite being mainly focused towards the coast of Peru.

References

Aki, K. (1966). Generation and Propagation of G Waves from the Niigata Earthquake of June 16, 1964: Part 2. Estimation of earthquake moment, released energy, and stress-strain drop from the G wave spectrum. *Bulletin of the Earthquake Research Institute, 44,* 73–88.

Ammon, C. J., Lay, T., Kanamori, H., & Cleveland, M. (2011). A rupture model of the 2011 off the Pacific Coast of Tohoku Earthquake. *Earth, Planets and Space, 63*(7), 33. https://doi.org/10.5047/eps.2011.05.015.

Arreaga-Vargas, P., Ortiz, P., & Farreras, S. F. (2005). Mapping the possible tsunami hazard as the first step towards a tsunami resistant community in Esmeraldas, Ecuador. In K. Satake (Ed.), *Tsunamis. Advances in natural and technological hazards research* (Vol. 23). Dordrecht: Springer.

Ben-Menahem, A., & Rosenman, M. (1972). Amplitude patterns of tsunami waves from submarine earthquakes. *Journal of Geophysical Research, 77*(17), 3097–3128. https://doi.org/10.1029/JB077i017p03097.

Bretschneider, C. L. (1964). Generation of waves by wind, state of the art. National Engineering Science Co. Report SN-134-6, p. 96.

Dronkers, J. J. (1964). *Tidal computations in rivers and coastal waters.* Amsterdam: North-Holland Publishing Company.

Fine, I. V., Kulikov, E. A., & Cherniawsky, J. Y. (2012). Japan's 2011 tsunami: characteristics of wave propagation from observations and numerical modelling. *Pure and Applied Geophysics, 170*(6–8), 1295–1307. https://doi.org/10.1007/s00024-012-0555-8.

Fujii, Y., Satake, K., Sakai, S. I., Shinohara, M., & Kanazawa, T. (2011). Tsunami source of the 2011 off the Pacific Coast of Tohoku Earthquake. *Earth, Planets and Space, 63*(7), 55. https://doi.org/10.5047/eps.2011.06.010.

Goto, C., Ogawa, Y., Shuto, N., and Imamura, F. (1997). IUGG/IOC Time Project, in IOC Manuals and Guides No. 35, UNESCO, Paris.

Gusman, A. R., Tanioka, Y., Sakai, S., & Tsushima, H. (2012). Source model of the great 2011 Tohoku earthquake estimated from tsunami waveforms and crustal deformation data. *Earth and Planetary Science Letters, 341,* 234–242. https://doi.org/10.1016/j.epsl.2012.06.006.

Hammack, J. L. (1973). A note on tsunamis: their generation and propagation in an ocean of uniform depth. *Journal of Fluid Mechanics, 60*(4), 769–799. https://doi.org/10.1017/S0022112073000479.

Hanks, T. C., & Kanamori, H. (1979). A moment magnitude scale. *Journal of Geophysical Research, 84*(5), 2348–2350. https://doi.org/10.1029/JB084iB05p02348.

Hayashi, Y., Tsushima, H., Hirata, K., Kimura, K., & Maeda, K. (2011). Tsunami source area of the 2011 off the Pacific Coast of Tohoku Earthquake determined from tsunami arrival times at offshore observation stations. *Earth, Planets and Space, 63*(7), 54. https://doi.org/10.5047/eps.2011.06.042.

Hayes, G. P. (2011). Rapid source characterization of the 2011 M_w 9.0 off the Pacific Coast of Tohoku earthquake. *Earth, Planets and Space, 63*(7), 4. https://doi.org/10.5047/eps.2011.05.012.

Horrillo, J., Knight, W., and Kowalik, Z. (2008). Kuril Islands tsunami of November 2006: 2. Impact at Crescent City by local enhancement. *Journal of Geophysical Research: Oceans, 113*(C01021). https://doi.org/10.1029/2007jc004404

Ichinose, G., Somerville, P., Thio, H. K., Graves, R., and O'Connell, D. (2007). Rupture process of the 1964 Prince William Sound, Alaska, earthquake from the combined inversion of seismic, tsunami, and geodetic data. *Journal of Geophysical Research: Solid Earth, 112*(B07306). https://doi.org/10.1029/2006jb004728.

Ide, S., Baltay, A., & Beroza, G. C. (2011). Shallow dynamic overshoot and energetic deep rupture in the 2011 M_w 9.0 Tohoku-Oki earthquake. *Science, 332*(6039), 1426–1429. https://doi.org/10.1126/science.1207020.

Iinuma, T., Ohzono, M., Ohta, Y., & Miura, S. (2011). Coseismic slip distribution of the 2011 off the Pacific Coast of Tohoku Earthquake (M 9.0) estimated based on GPS data—was the asperity in Miyagi-oki ruptured? *Earth, Planets and Space, 63*(7), 643–648. https://doi.org/10.5047/eps.2011.06.013.

Ito, T., Ozawa, K., Watanabe, T., & Sagiya, T. (2011). Slip distribution of the 2011 off the Pacific Coast of Tohoku Earthquake inferred from geodetic data. *Earth, Planets and Space, 63*(7), 627–630. https://doi.org/10.5047/eps.2011.06.023.

Kanamori, H., & Cipar, J. J. (1974). Focal process of the great Chilean earthquake May 22, 1960. *Physics of the Earth and Planetary Interiors, 9*(2), 128–136. https://doi.org/10.1016/0031-9201(74)90029-6.

Kelleher, J. A. (1972). Rupture zones of large South American earthquakes and some predictions. *Journal of Geophysical Research, 77*(11), 2087–2103. https://doi.org/10.1029/JB077i011p02087.

Koper, K. D., Hutko, A. R., Lay, T., Ammon, C. J., & Kanamori, H. (2011). Frequency-dependent rupture process of the 2011 M_w 9.0 Tohoku Earthquake: comparison of short-period P wave backprojection images and broadband seismic rupture models. *Earth, Planets and Space, 63*(7), 599–602. https://doi.org/10.5047/eps.2011.05.026.

Kowalik, Z., Horrillo, J., Knight, W., and Logan, T. (2008). Kuril Islands tsunami of November 2006: 1. Impact at Crescent City by distant scattering. *Journal of Geophysical Research: Oceans, 113*(C01020). https://doi.org/10.1029/2007jc004402

Lay, T., Ammon, C. J., Kanamori, H., Xue, L., Kim, M. J., et al. (2011a). Possible large near-trench slip during the 2011 M_w 9.0 off the Pacific Coast of Tohoku Earthquake. *Earth, Planets and Space, 63*(7), 687–692. https://doi.org/10.5047/eps.2011.05.033.

Lay, T., Kanamori, H., et al. (2011b). Insights from the great 2011 Japan earthquake. *Physics Today, 64*(12), 33–39. **(ISSN 0031-9228)**.

Lay, T., Yamazaki, Y., Ammon, C. J., Cheung, K. F., Kanamori, H., et al. (2011c). The 2011 M_w 9.0 off the Pacific Coast of Tohoku Earthquake: comparison of deep-water tsunami signals with finite-fault rupture model predictions. *Earth, Planets and Space, 63*(7), 52. https://doi.org/10.5047/eps.2011.05.030.

Maeda, T., Furumura, T., Sakai, S. I., & Shinohara, M. (2011). Significant tsunami observed at ocean-bottom pressure gauges during the 2011 off the Pacific Coast of Tohoku Earthquake. *Earth, Planets and Space, 63*(7), 53. https://doi.org/10.5047/eps.2011.06.005.

Mansinha, L., & Smylie, D. E. (1971). The displacement fields of inclined faults. *Bulletin of the Seismological Society of America, 61*(5), 1433–1440.

Miyoshi, H. (1968). Re-consideration on directivity of the tsunami (1). *Zisin, ii, 21*(2), 121–138. **(in Japanese)**.

National Geophysical Data Center. (2006). 2-minute Gridded Global Relief Data (ETOPO2) v2. National Geophysical Data Center, NOAA. https://doi.org/10.7289/v5j1012q

Okal, E. A., Reymond, D., & Hébert, H. (2014). From earthquake size to far-field tsunami amplitude: development of a simple formula and application to DART buoy data. *Geophysical Journal International, 196*(1), 340–356. https://doi.org/10.1093/gji/ggt328.

Ozawa, S., Nishimura, T., Suito, H., Kobayashi, T., Tobita, M., & Imakiire, T. (2011). Coseismic and postseismic slip of the 2011 magnitude-9 Tohoku-Oki earthquake. *Nature, 475*(7356), 373–376. https://doi.org/10.1038/nature10227.

Pedlosky, J. (1979). *Geophysical fluid dynamics*. New York, NY: Springer. https://doi.org/10.1007/978-1-4684-0071-7.

Plafker, G. (1965). Tectonic deformation associated with the 1964 Alaska earthquake. *Science, 148*(3678), 1675–1687. https://doi.org/10.1126/science.148.3678.1675.

Satake, K. (1988). Effects of bathymetry on tsunami propagation: application of ray tracing to tsunamis. *Pure and Applied Geophysics, 126*(1), 27–36. https://doi.org/10.1007/BF00876912.

Shao, G., Li, X., Ji, C., & Maeda, T. (2011). Focal mechanism and slip history of the 2011 Mw 9.1 off the Pacific Coast of Tohoku Earthquake, constrained with teleseismic body and surface waves. *Earth, Planets and Space, 63*(7), 553–557. https://doi.org/10.5047/eps.2011.06.028.

Simons, M., Minson, S. E., Sladen, A., Ortega, F., Jiang, J., Owen, S. E., et al. (2011). The The 2011 Magnitude 9.0 Tohoku-Oki Earthquake: mosaicking the megathrust from seconds to centuries. *Science, 332*(6), 1421–1425. https://doi.org/10.1126/science.1206731.

Titov, V.V., González, F.I., Mofjeld, H.O., & Venturato, A.J. (2003). NOAA time seattle tsunami mapping project: procedures, data sources, and products. US Department of Commerce, National Oceanic and Atmospheric Administration, Oceanic and Atmospheric Research Laboratories, Pacific Marine Environmental Laboratory.

Watada, S., Kusumoto, S., & Satake, K. (2014). Traveltime delay and initial phase reversal of distant tsunamis coupled with the self-gravitating elastic Earth. *Journal of Geophysical Research: Solid Earth, 119*(5), 4287–4310. https://doi.org/10.1002/2013JB010841.

Yoshida, Y., Ueno, H., Muto, D., & Aoki, S. (2011). Source process of the 2011 off the Pacific Coast of Tohoku Earthquake with the combination of teleseismic and strong motion data. *Earth, Planets and Space, 63*(7), 565–569. https://doi.org/10.5047/eps.2011.05.011.

Yue, H., & Lay, T. (2011). Inversion of high-rate (1 sps) GPS data for rupture process of the 11 March 2011 Tohoku earthquake (M_w 9.1). *Earthquake, 38*(March), 1–6. https://doi.org/10.1029/2011GL048700.

Zaytsev, O., Rabinovich, A. B., & Thomson, R. E. (2016). A comparative analysis of coastal and open-ocean records of the Great Chilean Tsunamis of 2010, 2014 and 2015 off the Coast of Mexico. *Pure and Applied Geophysics, 173*(12), 4139–4178. https://doi.org/10.1007/s00024-016-1407-8.

Zaytsev, O., Rabinovich, A. B., & Thomson, R. E. (2017). The 2011 Tohoku tsunami on the coast of Mexico: a case study. *Pure and Applied Geophysics, 174*(8), 2961–2986. https://doi.org/10.1007/s00024-017-1593-z.

(Received October 12, 2017, revised February 20, 2018, accepted February 22, 2018)

Pure Appl. Geophys.

https://doi.org/10.1007/s00024-018-1842-9

Pure and Applied Geophysics

Tsunami Simulations in the Western Makran Using Hypothetical Heterogeneous Source Models from World's Great Earthquakes

AMIN RASHIDI,[1] ZAHER HOSSEIN SHOMALI,[1,2] and NASSER KESHAVARZ FARAJKHAH[3]

Abstract—The western segment of Makran subduction zone is characterized with almost no major seismicity and no large earthquake for several centuries. A possible episode for this behavior is that this segment is currently locked accumulating energy to generate possible great future earthquakes. Taking into account this assumption, a hypothetical rupture area is considered in the western Makran to set different tsunamigenic scenarios. Slip distribution models of four recent tsunamigenic earthquakes, i.e. 2015 Chile M_w 8.3, 2011 Tohoku-Oki M_w 9.0 (using two different scenarios) and 2006 Kuril Islands M_w 8.3, are scaled into the rupture area in the western Makran zone. The numerical modeling is performed to evaluate near-field and far-field tsunami hazards. Heterogeneity in slip distribution results in higher tsunami amplitudes. However, its effect reduces from local tsunamis to regional and distant tsunamis. Among all considered scenarios for the western Makran, only a similar tsunamigenic earthquake to the 2011 Tohoku-Oki event can re-produce a significant far-field tsunami and is considered as the worst case scenario. The potential of a tsunamigenic source is dominated by the degree of slip heterogeneity and the location of greatest slip on the rupture area. For the scenarios with similar slip patterns, the mean slip controls their relative power. Our conclusions also indicate that along the entire Makran coasts, the southeastern coast of Iran is the most vulnerable area subjected to tsunami hazard.

Key words: Tsunami hazard, western Makran subduction zone, scaled slip distributions, heterogeneity, numerical modeling.

1. Introduction

The Makran region, with 1000-km in length, is located at the border of the Eurasian–Arabian plates, along Pakistan–Iran offshore in the northwest of Indian Ocean (Fig. 1). Since the Early Cretaceous period, the oceanic crust of Arabian plate has been subducted beneath the Eurasian plate in northward direction with $\sim$ 2 cm/year (Byrne et al. 1992; Vernant et al. 2004). The subduction zone is bounded on the eastern side by the strike-slip system of the Ornach-Nal and Chaman Faults and on the western side by the Minab–Zendan Fault system.

Makran subduction zone (MSZ) is peculiar from different perspectives, e.g. it is a typical two-dimensional (2D) subduction zone in which the rate of the subducting plate varies slightly along strike (White and Klitgord 1976; Minshull et al. 1992; Kopp et al. 2000). The Makran region is one of the most extensive subduction zones with a very wide accretionary prism ($\sim$ 350–400 km) and an extremely high sediment thickness ($\sim$ 7 km) (Kopp et al. 2000; Kukowski et al. 2001; Grando and McClay 2007; Hoffmann et al. 2013). The seismicity of Makran subduction zone is relatively low compared to other subduction zones around the world (Musson 2009). The pattern of seismicity toward the sea is significantly different in the western and eastern parts of Makran. The eastern segment of Makran has potential generating moderate to large magnitude earthquakes. In November 27, 1945, this segment produced an M_w 8.1 earthquake in Pasni-Ormara, which is one of the only two instrumentally recorded tsunamis in the Makran subduction zone; the other is the 2013 event (Heidarzadeh and Satake 2014; Hoffmann et al. 2014). The earthquake triggered a regional tsunami with waves up to 10 m. It was responsible for 4000 deaths and caused a remarkable damage along the coasts of Iran, Oman, Pakistan and India (Heck 1947; Heidarzadeh et al. 2008; Okal and Synolakis 2008; Rajendran et al. 2013). Unlike the eastern part, the western Makran is marked with almost no major

[1] Institute of Geophysics, University of Tehran, Tehran, Iran. E-mail: amin.rashidi@ut.ac.ir
[2] Department of Earth Sciences, Uppsala University, Uppsala, Sweden.
[3] Research Institute of Petroleum Industry (RIPI), Tehran, Iran.

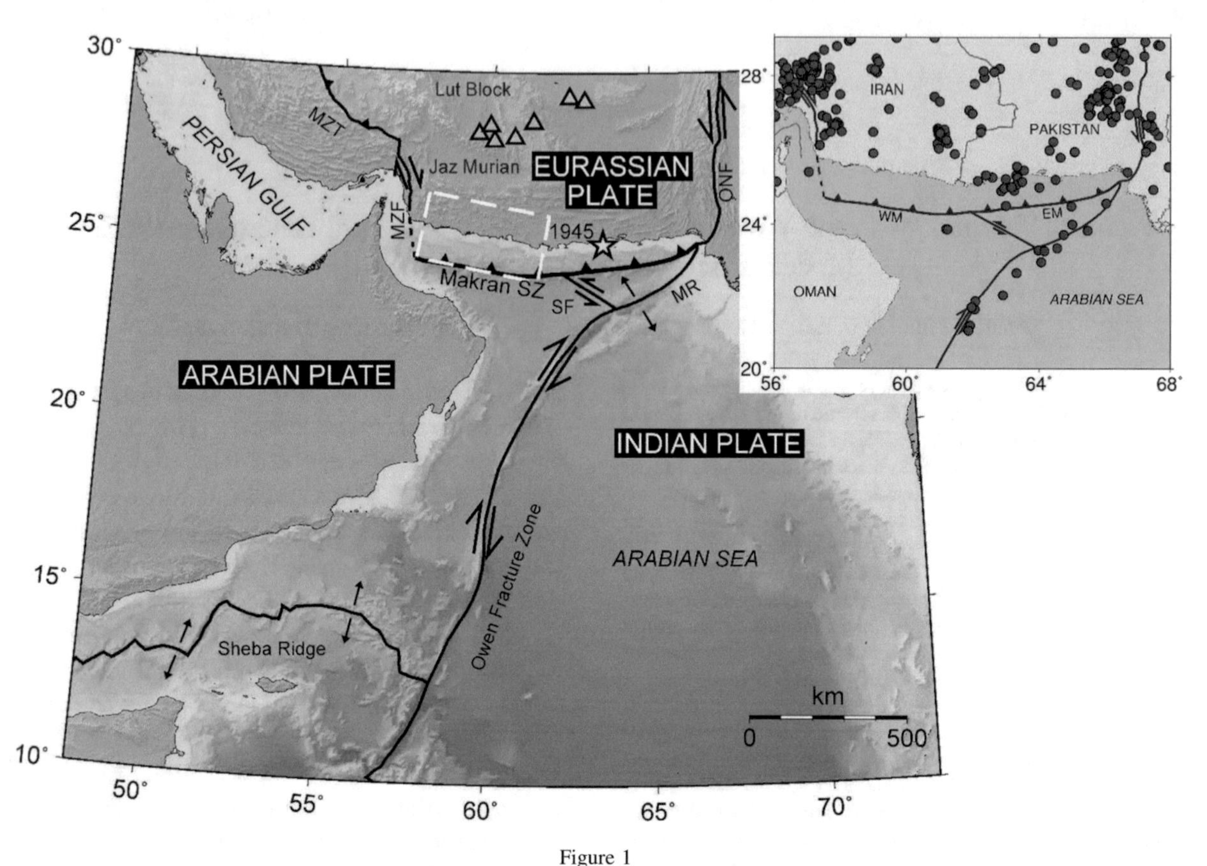

Figure 1
Tectonic features and seismicity of the Makran subduction zone (*MZT* Main Zagros Thrust, *MZF* Minab–Zendan fault, *SZ* subduction zone, *SF* Sonne fault, *MR* Murray ridge). Yellow star represents the epicenter of 1945 M_w 8.1 Makran earthquake. Orange triangles are volcanos. Dashed and solid black lines indicate plate boundaries. Dashed white lines show the western Makran fault zone adopted in this study. Inset map shows the earthquakes distribution ($M > 4.0$) in the Makran region from 1900 to 2016 from the ISC catalog (*WM* western Makran, *EM* eastern Makran)

seismicity in the offshore areas. Even though, in 1483, it might have experienced a strong earthquake which affected the Strait of Hormuz and northeastern Oman (Ambraseys and Melville 1982; Byrne et al. 1992).

Earthquake and consequent tsunami hazard in the Makran subduction zone has been subject of interest, especially after the 2004 Indian Ocean tsunami (e.g., Okal and Synolakis 2008; Heidarzadeh and Kijko 2011; Rajendran et al. 2013; Smith et al. 2013; Heidarzadeh and Satake 2014; Payande et al. 2015; Hoechner et al. 2016). Beside, several researchers have studied the 1945 Makran tsunami (e.g., Heidarzadeh et al. 2008; Neetu et al. 2011; Hoffmann et al. 2013; Heidarzadeh and Satake 2015). In some studies, possible megathrust tsunamigenic events in the Makran subduction zone were suggested. Okal and Synolakis (2008) evaluated far-field tsunami hazard for two worst case scenarios of the eastern and the entire MSZ being capable of generating megathrust earthquakes. Heidarzadeh et al. (2009) proposed two worst case scenarios including a M_w 8.6 source model of the eastern Makran and a M_w 9.0 source model of the full rupture of the Makran plate boundary and performed tsunami numerical modeling. Payande et al. (2015) defined three tsunami scenarios along the Makran subduction zone and performed numerical modeling for each scenario to assess tsunami hazard for Chabahar bay. The third scenario presents a moment magnitude of M_w 9.1. All scenarios have identical epicenter located in the Gulf of Oman and right in front of Chabahar bay.

The lack of major earthquakes for many years supports the hypothesis indicating the possibility of locking the western Makran segment in contrast to non-seismic deformation as proposed. This could indicate that this segment is capable of producing future great tsunamigenic earthquakes. If the western Makran segment is assumed to be locked, then it has important implications in the hazard assessment for the near coastal areas (Rajendran et al. 2013). Future high-magnitude earthquakes in the western Makran will provoke significant regional tsunamis that will undoubtedly pose real threats to the Makran coasts, particularly for Iran and Oman. Therefore, it is indispensable to evaluate tsunami hazard for the western Makran. Numerical modeling of tsunamis is a powerful tool for investigating the impacts of possible future tsunamis on the coastal areas. The vague seismogenic behavior of the western Makran has given no insight into the mechanism and slip distribution of possible future events. In this study, different tsunamis are numerically modeled for a range of hypothetical large earthquake scenarios in the western Makran using the scaled slip distribution models of some recent tsunamigenic earthquakes.

2. Methodology

2.1. Setting a Rupture Fault Geometry for the Western Makran

In this study, a fault zone scaled by certain slip models in numerical simulation of tsunamis is considered. It is presumed that the western Makran segment is locked and thus is capable of generating great plate boundary earthquakes. A planar fault is constructed to define the rupture geometry based on Heidarzadeh et al. (2009) and Smith et al. (2013) with a length of 400 km, a width of 210 km (Fig. 1), a focal depth of 25 km, a strike angle of 280°, and a dip angle of 7° towards north.

Smith et al. (2013) presented three rupture scenarios in the Makran region. The first scenario includes a full length of the Makran subduction zone, while the second scenario is limited to the eastern Makran segment. Each scenario has a length, a minimum (210 km) and a maximum width (355 km), a magnitude and a co-seismic slip of 10 m (Smith et al. 2013). To set the fault geometry, we take a length of 400 km and minimum width of 210 km similar to the second rupture scenario used by Smith et al. (2013). To define the strike, dip and the central depth of the fault plane, the parameters presented in Heidarzadeh et al. (2009) for the western segment of Makran are used, which are modified from Okal and Synolakis (2008).

2.2. Selecting Slip Models for the Western Makran Zone (WMZ)

In tsunami simulation process, accuracy of source model plays an important role in generation and propagation modeling steps. Earthquakes, especially large ones, have a complex rupture process that needs accurate information about space–time evolution of rupture to be modeled (Ampuero et al. 2006; Ruiz et al. 2015). This information on earthquake rupture including details of the slip distribution is critical for especially near-field tsunami hazard evaluating and forecasting (Geist 2002). Heterogeneity of slip distribution on the fault controls local tsunami wave amplitude variations (Geist and Dmowska 1999).

In the current study, tsunami simulation modeling is done using slip distribution models of four significant great earthquakes, as listed in Table 1 and shown in Fig. 2, for the western Makran subduction zone. These slip models are all scaled in the western Makran rupture area assuming the equivalent seismic moments for each hypothetical earthquake. The seismic moment is defined by the equation $M_0 = \mu AD$ (Hanks and Kanamori 1979), where A is the rupture area, D is the slip and μ is the shear modulus. We assume same value of μ, and by keeping the seismic moment (M_0) the same between a hypothetical western Makran event and the original event, we will have

$$\sum(D_iA_i) = \sum(D_{\mathrm{wm}}A_{\mathrm{wm}}) \quad (1)$$

where subscripts i and wm denote the actual tsunami events as listed in Table 1 and hypothetical western Makran event. The dimensions of subfaults along the strike and downdip are calculated by dividing WMZ length and width by the number of subfaults along the strike and dip, respectively. It is noteworthy that the geographical distribution of the events in Table 1

Table 1

List of events used for scaling the slip distribution models into the western Makran zone (WMZ)

No.	Location	Date	Epicenter location		Focal depth (km)	M_w	References	Scenario name for WMZ
			Latitude (°)	Longitude (°)				
1	Kuril Islands (Russia)	Nov 15, 2006	42.62	153.27	26	8.3	Ji (2006)	SC1
2	Tohoku-Oki (Japan)	Mar 11, 2011	38.35	142.46	29	9.0	Hayes (2011)	SC2
3	Tohoku-Oki (Japan)	Mar 11, 2011	38.19	142.68	21	9.1	Wei et al. (2012)	SC3
4	Illapel (Chile)	Sep 16, 2015	− 31.58	− 71.65	25	8.3	USGS (2015)	SC4

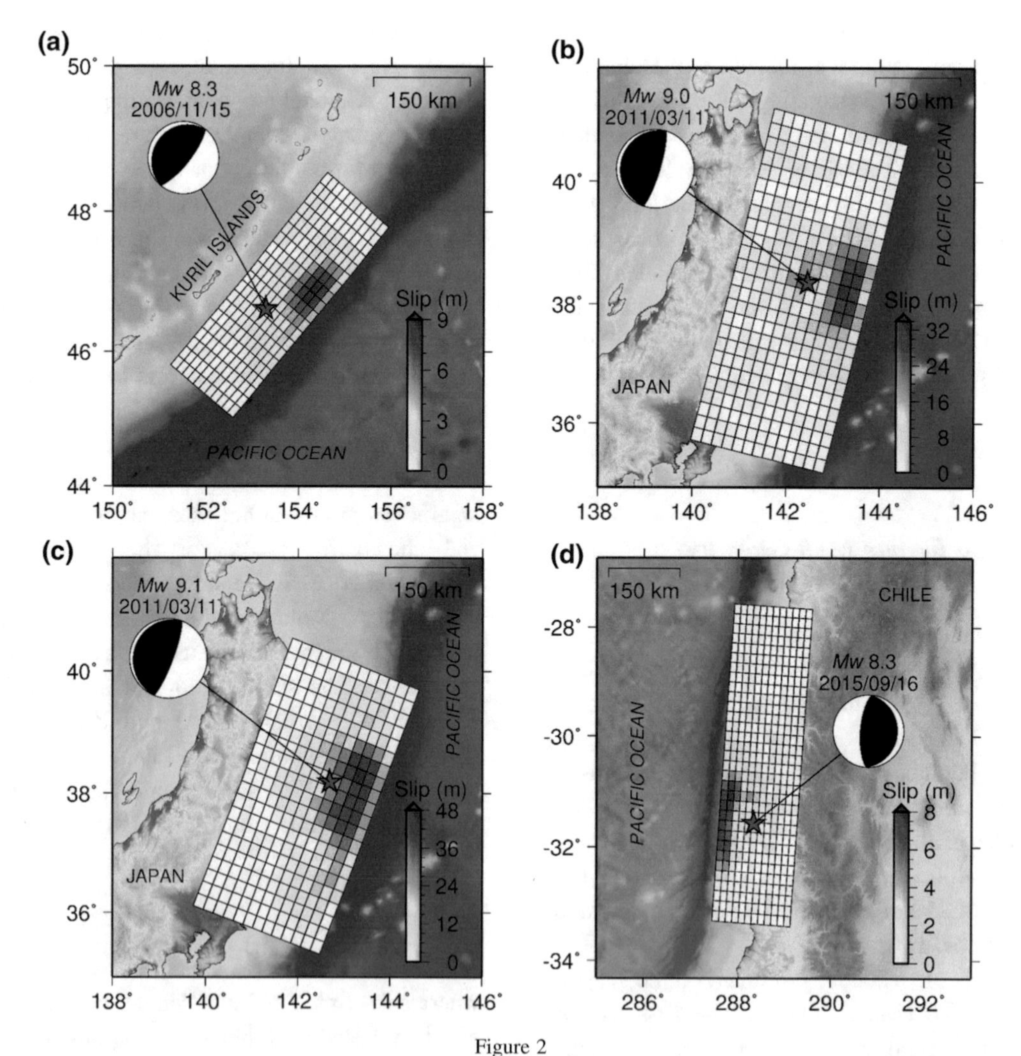

Figure 2

Source area and slip distribution for **a** the 2006 Kuril Islands earthquake based on Ji (2006), **b** the 2011 Tohoku-Oki earthquake based on Hayes (2011), **c** the 2011 Tohoku-Oki earthquake based on Wei et al. (2012), and **d** the 2015 Illapel, Chile earthquake based on USGS (2015). The star marks the hypocenter. The base image is the regional topography map

plays a key role with selection. Two different slip models of the 2011 Tohoku-Oki event are used (Table 1). Source model and other characteristics of the 2015 Illapel, Chile earthquake have been studied in detail by many researchers (e.g., Omira et al. 2016; Ye et al. 2016; Satake and Heidarzadeh 2017). However, in this study the slip model of the 2015 Illapel, Chile earthquake was taken from USGS (2015). A detail study for the 2006 Kuril Islands tsunami can also be found in Rabinovich et al. (2008) where near-source observations and modeling of the Kuril Islands tsunamis of 15 November 2006 and 13 January 2007 were studied in detail.

2.3. Setting Scenarios in the Western Makran for a Set of Scaled Slip Models

On the western Makran fault, we scale the slip pattern of four significant earthquakes as listed in Table 1 and define two different scenarios for each earthquake in the western Makran. First scenario, a scaled non-uniform slip distribution assuming the same seismic moment (M_0) and moment magnitude (M_w) proportional to the original event (type I). Second scenario presents a uniform slip (type II) by keeping M_0 and M_w same as the original event and choosing a mean value for the slip. Therefore, a total of 10 scenarios are produced and are simulated (see Table 2).

2.4. Numerical Simulation

We employ a hydrodynamic model, COMCOT (Liu et al. 1998), to simulate western Makran tsunamis. COMCOT (Liu et al. 1998) uses explicit staggered leapfrog finite difference schemes to solve both linear and non-linear shallow water equations in spherical and Cartesian coordinates (Wang 2009). COMCOT (Liu et al. 1998) is a well-known numerical modeling tool for studying and performing simulations of near-field and far-field tsunamis (e.g., Liu et al. 1994, 1995; Wang and Liu 2005, 2006; Barkan and ten Brink 2010; Heidarzadeh and Satake 2014). Tsunami modeling requires a bathymetric grid covering the region of interest and computed initial water surface displacement. The results of the numerical simulations are fields of maximum amplitudes and distributions of coastal wave amplitudes at the last water point along shorelines. Note that run-up is not calculated in this study.

The modeling area covers the northwestern Indian Ocean (from 25.40°S to 31.42°N and from 40.27°E to 84.32°E) including the Gulf of Oman, the Persian Gulf and the Arabian Sea Basin. The GEBCO 1-min bathymetry data (available at http://www.gebco.net/) are used for our simulations. Due to the lack of site-specific high-resolution maps of local bathymetry and topography, this study does not involve computing tsunami inundation on dry land. All simulations are

Table 2

List of tsunami scenarios used for tsunami numerical simulation

Scenario name	Associated event/modeler[a]	Latitude (°)	Longitude (°)	Depth (km)	Average rake (°)	Average slip (m)	M_w	M_0 (N m)	Slip distribution
SC1	Kuril Islands 2006/Ji (2006)	25.51	59.45	27	91	1.1	8.3	3.2×10^{21}	Non-uniform
SC1_2		25.51	59.45	27	91	1.1	8.3	3.2×10^{21}	Uniform
SC2	Tohoku-Oki 2011/Hayes (2011)	25.25	59.86	25	88	10.9	9.0	4.5×10^{22}	Non-uniform
SC2_2		25.25	59.86	25	88	10.9	9.0	4.5×10^{22}	Uniform
SC3	Tohoku-Oki 2011/Wei et al. (2012)	25.11	60.03	23	90	15.5	9.1	5.5×10^{22}	Non-uniform
SC3_2		25.11	60.03	23	90	15.5	9.1	5.5×10^{22}	Uniform
SC4	Illapel 2015/USGS (2015)	25.22	60.47	24	90	1.1	8.3	3.2×10^{21}	Non-uniform
SC4_2		25.22	60.47	24	90	1.1	8.3	3.2×10^{21}	Uniform

The corresponding slip models are all scaled to represent slip distribution patterns for the western Makran zone

[a]The events and the corresponding slip models are listed in Table 1

performed for a total run time of 20 h (36,000 steps) with a time step of 2 s.

The algorithm of Mansinha and Smylie (1971) is used to compute the seabed static deformation as the initial condition is required for the numerical simulations. The duration of an earthquake rupture process is very short compared to the tsunami wave period; therefore, the initial water surface can be assumed to be equal to the seafloor deformation (Baba 2003), subsequently the rupture duration is ignored. Fault parameters including the epicenter location, focal depth, fault dimensions, strike, dip angle, rake, and the amount of slip on the fault are required to compute the deformation field caused by earthquakes. For the scenarios with uniform slip distribution, the algorithm of computing the static deformation field is simply used for a rectangular finite fault. For the scenarios with non-uniform slip, each finite fault is discretized into subfaults with variable slip values, and the total vertical deformation field at the seafloor is computed by the sum of vertical displacements from each subfault (Geist 2002).

3. Results

Simulation results are presented in Figs. 3, 4, 5 and 6 for the scenarios listed in Table 2. Each figure shows the mapping area of the source models, the computed vertical seafloor deformations and the distribution of maximum absolute amplitudes of each scenario.

The computed vertical deformations of the seabed from the uniform slip distributions show a similar pattern for all scenarios. They all produce uplift near the upper edge of the source and subsidence above the deeper part of the fault zone. The subsidence fields occur on the continent since the deeper zone of the western Makran fault is located in the continental crust. Therefore, the initial sea level is estimated to be positive. The coastlines of Iran and Oman both are found in the uplifted zone.

The static deformation fields from non-uniform slip distributions exhibit more complex and heterogeneous pattern compared to uniform slip models. As a result of modeling, in most cases, the largest subsidence and uplift occur seaward. The maximum crustal uplift lies near the location of the peak-slip on the fault. The maximum vertical displacements from SC1, SC2, SC3 and SC4 scenarios are about 1.5, 15.5, 20.0 and 3.0 m, respectively. The shoreline of Iran in the north of rupture area is located in both uplift and subsidence zones while the coastline of Oman in south of the western Makran experiences only uplift. The maximum values of vertical displacement generated by complex slip distributions are about three times larger than peak-static uplift from the scenarios with uniform slip.

The maximum wave amplitude maps shown in Figs. 3, 4, 5 and 6 pose a regional risk to the coasts of Iran, Oman, Pakistan and western India. Presumably, trapping of tsunami waves inside the Gulf of Oman causes a significantly local risk mainly between the shorelines of Iran and Oman (see also Okal and Synolakis 2008). The simulation results reveal a major difference between maximum amplitude fields resulting from uniform and non-uniform slip distributions. Considering the modeling results, the produced tsunami amplitudes from uniform slip distributions are remarkably smaller than those from non-uniform slip distributions.

The Persian Gulf connected to the Gulf of Oman through the Strait of Hormuz is located in the western part of the study area. Numerical modeling shows that the Persian Gulf has less impact compared to the Gulf of Oman. Tsunamis affect Madagascar, Maldives, Seychelles and western Sri Lanka in far-field. The patterns of tsunami in far-field for all scenarios are relatively similar in character. However, the fields of maximum amplitude in near-field and far-field are noticeably different. The tsunami waves in far-field are less powerful than near-field as expected.

The ratios of the rupture area of the actual source models (Table 1) to the rupture area of the hypothetical western Makran are 0.65, 1.93, 1.63 and 1.28, respectively. Consequently, the most likely scenario for the western Makran can be SC4 (Fig. 6). Among all 10 scenarios, SC1_2 (Fig. 3) and SC4_2 (Fig. 6) lead to weakest tsunamis and SC3 (Fig. 5) causes the highest tsunami amplitudes considered as the worst case scenario in this study. Representing uniform slip distributions with identical mean slip values (1.1 m, see also Table 2), SC1_2 and SC4_2 produce a weak wavefield. Impacts from these scenarios can only be

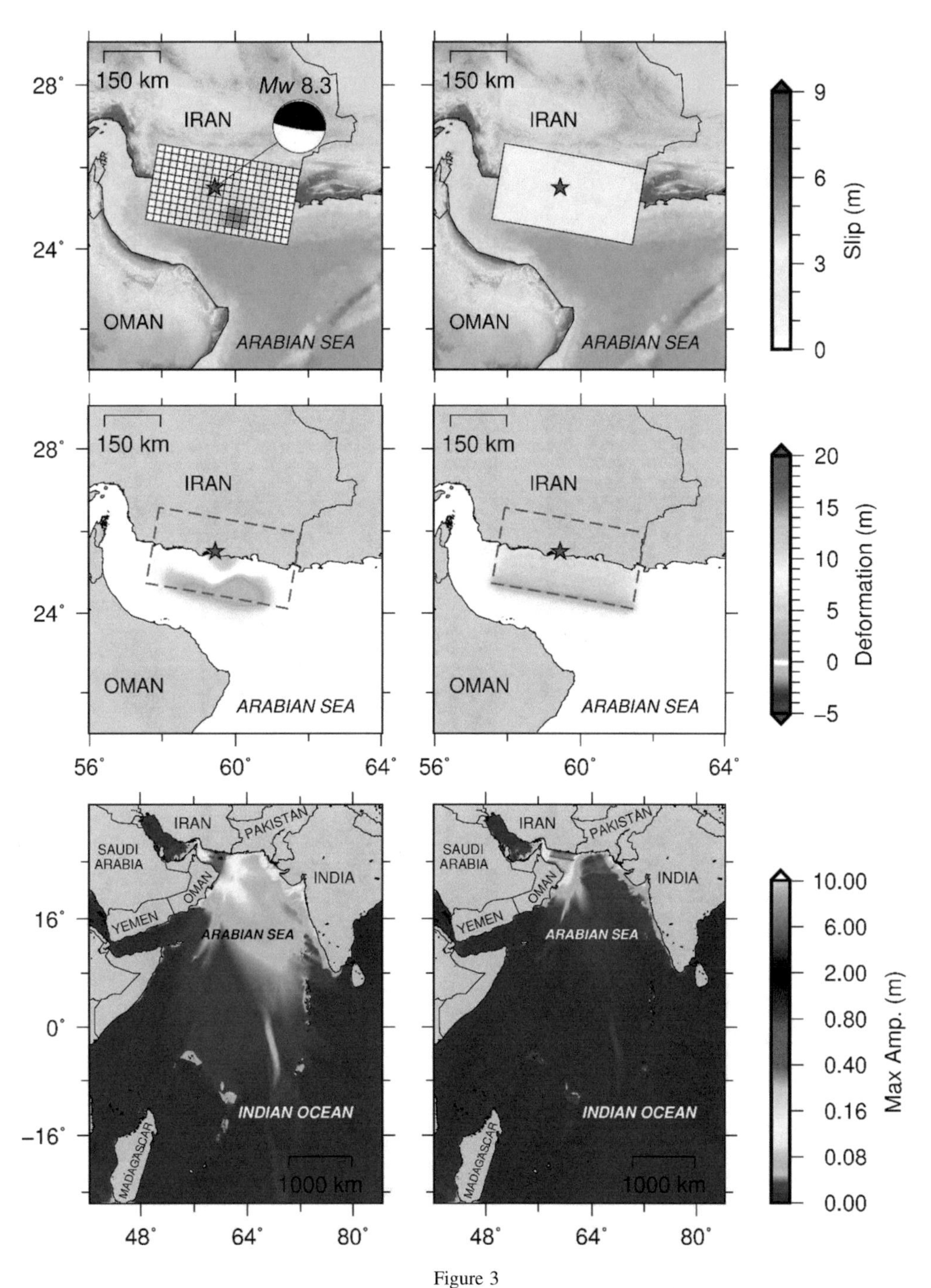

Figure 3

Simulations for the scenarios SC1 (left panel) and SC1_2 (right panel). The top frames show the source models. Purple star is the epicenter and solid contours represent uplift and subsidence. The middle frames show the computed vertical seafloor deformations of the earthquake scenarios used as initial conditions in tsunami simulations. Dashed lines represent the fault plane. The bottom frames show the maximum wave amplitude from the scenarios

observed in the Gulf of Oman. Unlike the other scenarios, the tsunami from SC3 impacts the Persian Gulf, Red Sea and eastern India more strongly. The 2006 Kuril and 2015 Chile earthquakes (Table 1) have an equal Magnitude (M_w 8.3), but SC4 generates a stronger tsunami than SC1 which is due to higher

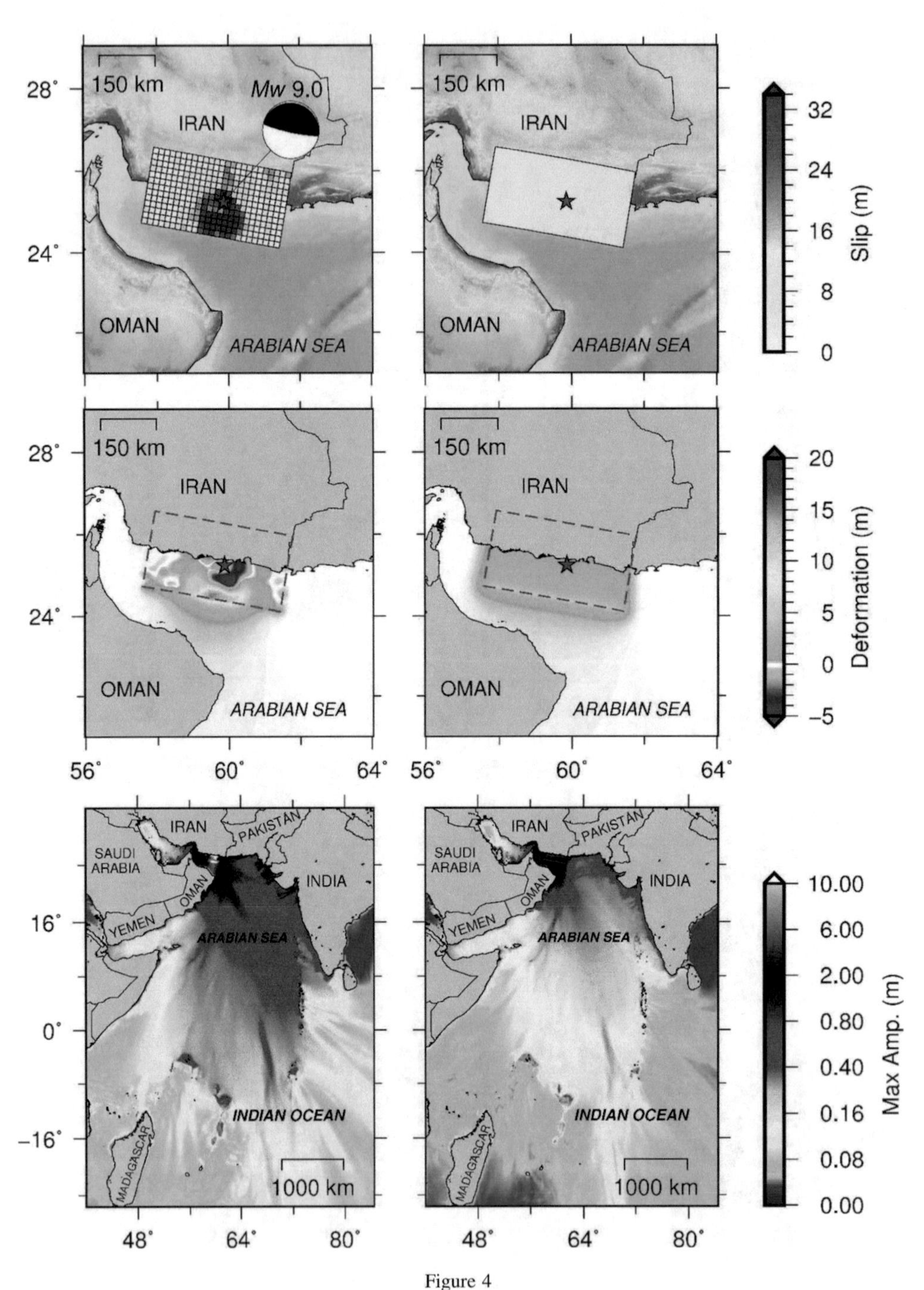

Figure 4
The same as Fig. 3 for the scenarios SC2 (left panel) and SC2_2 (right panel)

peak-slip in SC4. SC4 causes a partly considerable regional risk. Except the continental shelf of western India, the distant areas remain untouched by SC4.

The distribution of coastal amplitudes along the coastlines of Iran, Pakistan, Oman and India for SC1, SC2, SC3 and SC4 scenarios is shown in Fig. 7. The

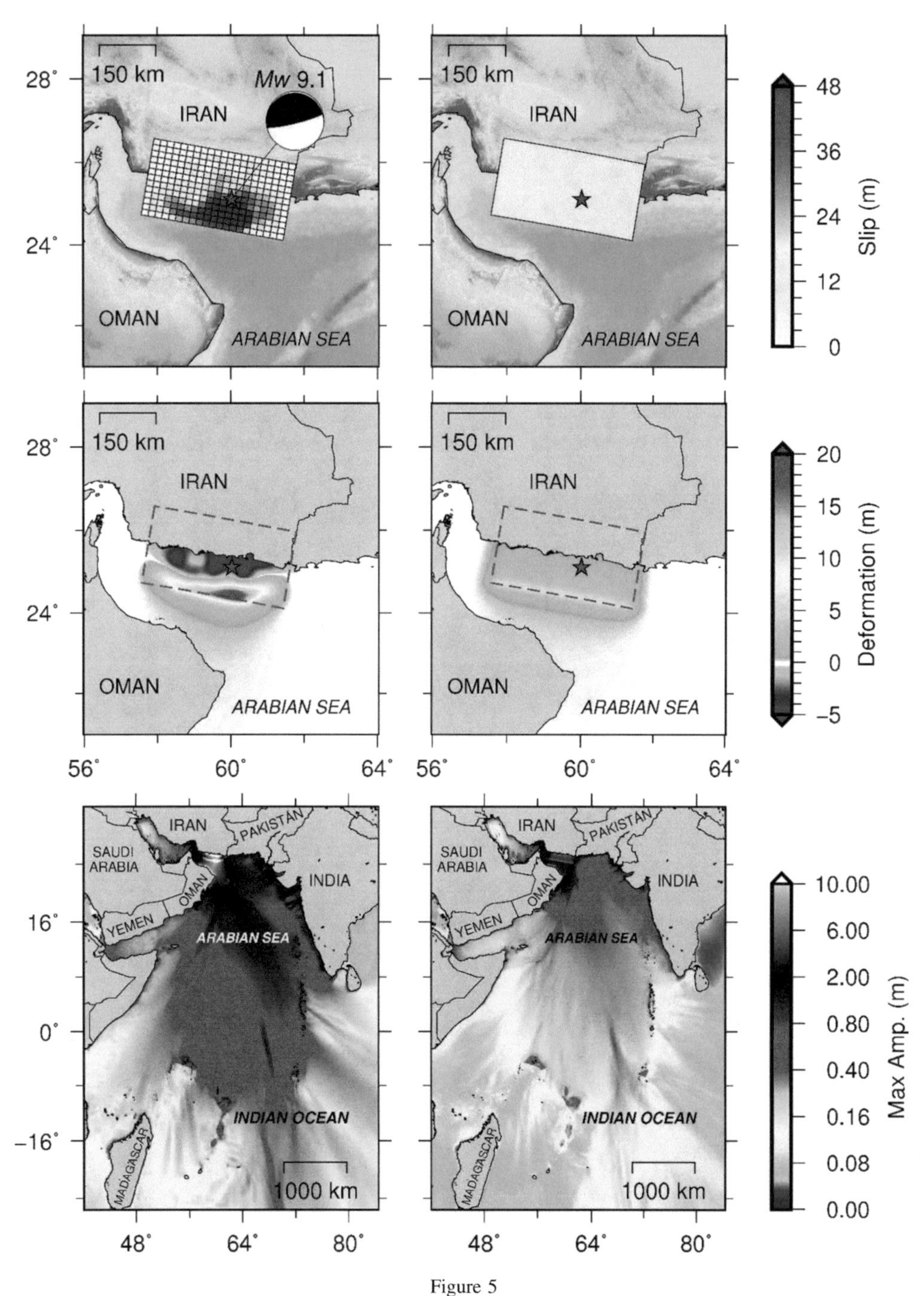

Figure 5
The same as Fig. 3 for the scenarios SC3 (left panel) and SC3_2 (right panel)

results reveal that the entire coasts of southeastern Iran, Oman, Pakistan and western India are affected by all scenarios. SC3 causes the largest coastal wave amplitudes. Among the shores, Iran experiences the highest tsunami waves. Coastal wave amplitudes along the coastline of Iran increase from west toward east. The most affected area by tsunamis along the Iranian shoreline is located between Chabahar and

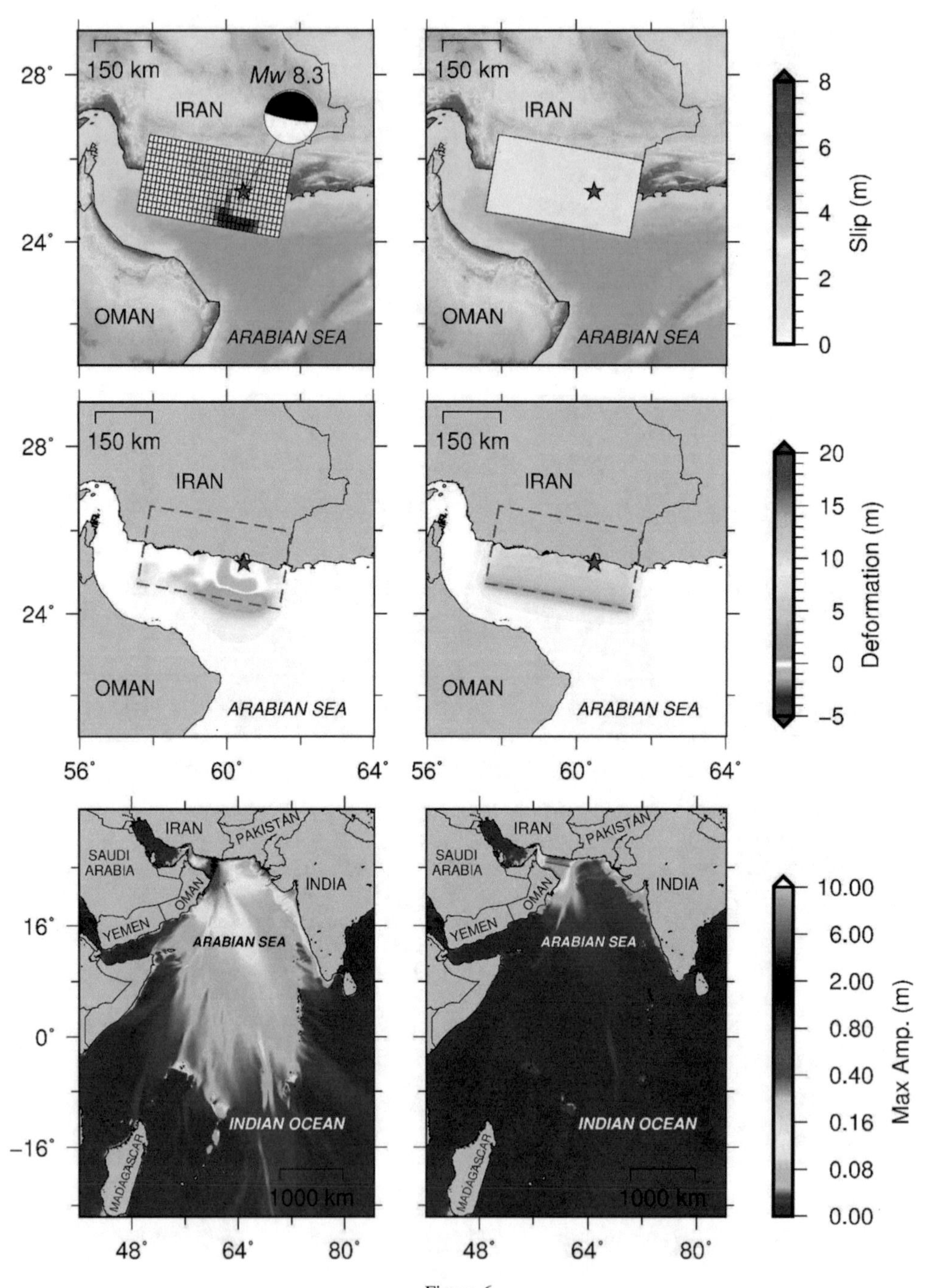

Figure 6
The same as Fig. 3 for the scenarios SC4 (left panel) and SC4_2 (right panel)

Kereti. The maximum coastal amplitudes reached from SC1, SC2, SC3 and SC4 are 4, 41, 52 and 5 m, respectively, along the Iranian shoreline. The shoreline of Oman, located in the south of western Makran, is also severely impacted by tsunamis. An area between Muscat and Sur in northeastern Oman sees the highest tsunami waves (see Fig. 7). The maximum coastal amplitudes resulting from SC1, SC2,

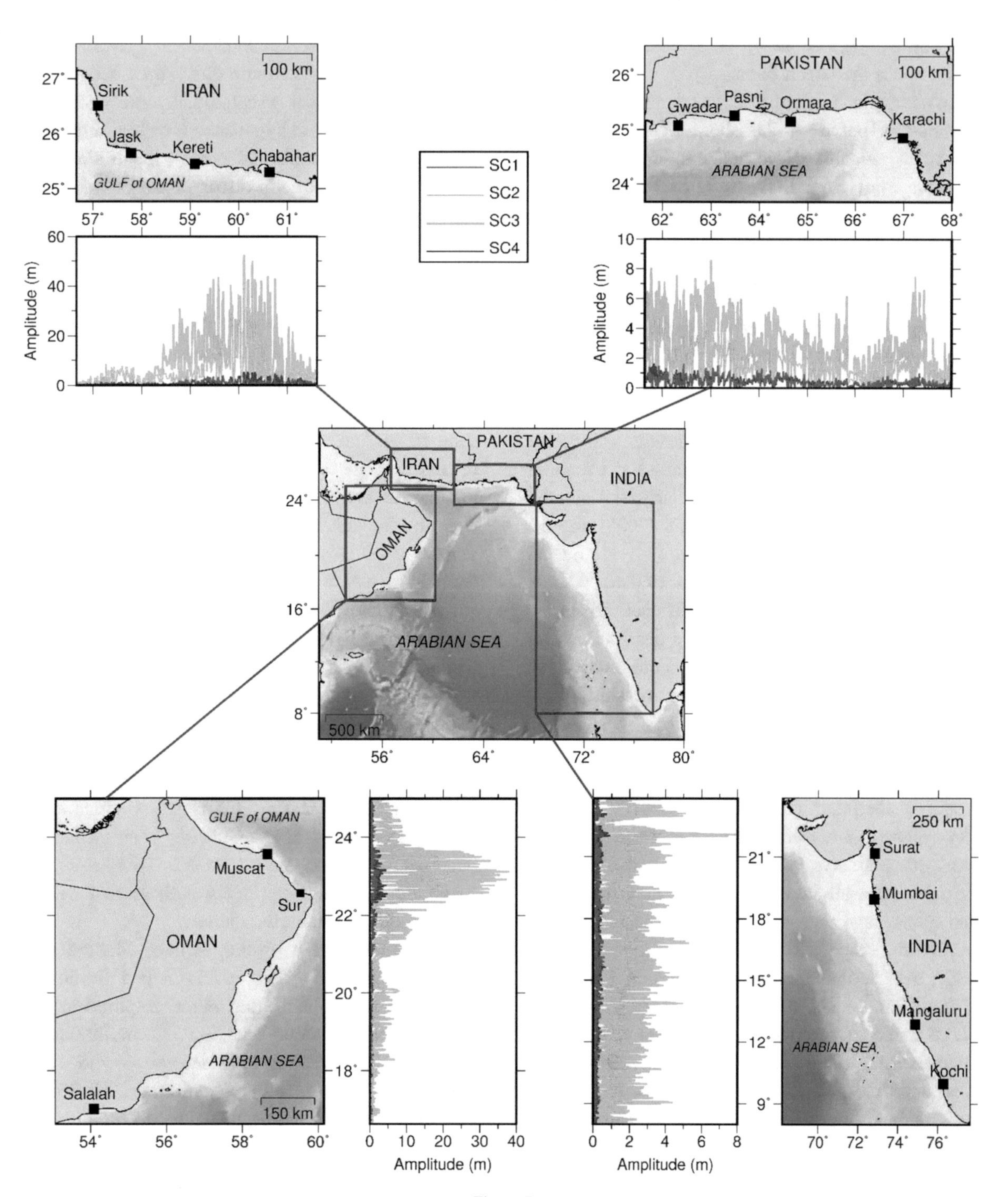

Figure 7
Distribution of coastal wave amplitudes along the shores of Iran, Oman, Pakistan and India for the four earthquake models in this study (SC1, SC2, SC3 and SC4)

SC3 and SC4 are about 4, 23, 38 and 5 m, respectively, along the Oman coast.

In comparison with the coasts of Iran and Oman, a less risk is posed to the shore of Pakistan. The maximum amounts of coastal amplitude along the Pakistan coast from SC1, SC2, SC3 and SC4 are about 2, 5, 9 and 2 m, respectively. The effect of tsunamis on relatively distant shore of India is considerable. The entire western coast of India is uniformly exposed to tsunami risk from different scenarios, unless for an area in the north of Surat, observing the greatest coastal amplitude. The largest computed coastal amplitude from SC1, SC2, SC3 and SC4 is about 1, 4, 8 and 1 m, respectively.

Below we also compare our results with other numerical results done for magnitude 9.0 events in the Makran subduction zone. Similarly, the simulation results by Okal and Synolakis (2008) for the eastern and the entire MSZ show a significant regional risk posed to the shores of Iran, Oman, Pakistan and western India. Note that Okal and Synolakis (2008) used uniform slip distributions for tsunami scenarios, thus, their results are comparable to uniform slip models of the present study, e.g., SC2_2 (M_w 9.0) and SC3_2 (M_w 9.1), as shown in Fig. 4.

The first scenario in Heidarzadeh et al. (2009) produces maximum coastal amplitudes of 3, 3 and 10 m along the coastlines of Iran, Oman and Pakistan, respectively. For their second scenario, the maximum coastal amplitudes along the coastlines of Iran, Oman and Pakistan are in the order of 24, 18 and 20 m, respectively. However, in the present study, the maximum amplitudes along the coastlines of Iran, Oman and Pakistan under SC2 (M_w 9.0) are 41, 23 and 5 m, whereas SC3 (M_w 9.1) causes maximum coastal amplitudes of 52, 38 and 9 m along those coastlines. Note that our results shown in Fig. 7 are obtained based on non-uniform slip for the western part of Makran, which is larger than those reported by Heidarzadeh et al. (2009) that is based on uniform slip distribution and different source area.

Payande et al. (2015) report the maximum coastal amplitude of 18 m for Chabahar port. While in the present study, the maximum coastal amplitude in Chabahar port is 8 m under SC2 and 5 m under SC2_2. The maximum coastal amplitudes in Chabahar port caused by SC3 and SC3_2 are 38 and 8 m, respectively. The main differences between the results of the present research and those from other studies are due to the hypocenter location, dimension of source area (length and width), mean slip value and heterogeneity of slip distribution.

4. Sensitivity Analysis

The source parameters of the scenarios are based on different studies, thus are associated with some uncertainties. Changing any parameter can affect tsunami amplitudes, especially for near-field. Therefore, to evaluate the uncertainties associated to the results, a range of uncertainty is considered for each parameter. The sensitivity analysis is performed to examine the effects of source parameters on tsunami amplitudes. Alternatively, we change the parameters from SC1 to SC4 scenarios through sensitivity analysis. Results of the sensitivity analysis of SC3 being the worst case scenario are presented in Fig. 8. The results are displayed as diagrams showing the maximum tsunami amplitudes in the selected stations (Chabahar, Muscat, Karachi, Mumbai, Dubai and Kish) as a function of variation in the fault parameters. The analysis is performed in near-field. Chabahar and Muscat are the closest stations to the fault zone, located in the Gulf of Oman basin, perpendicular to the fault's trend. The computed maximum tsunami amplitudes at these two stations are significantly more than others.

The strike angle of the western Makran fault changed from 270° to 295° in this context to study the sensitivity of tsunami maximum amplitudes. The results that illustrate maximum amplitudes in Chabahar and Muscat are increasing with strike contrary to Karachi, Mumbai and Dubai. No significant effect is observed on the tsunamis amplitude at Kish by changing the strike angle. In this sensitivity analysis study, the dip angle of the fault ranged from 2° to 89°. The results indicate that changing the dip angle highly affects computed tsunami amplitudes. For most stations, the maximum amplitudes are changed smoothly from an uptrend to a downtrend. The maximum tsunami amplitudes reach their peak values at dip angles lower than 45°. It means a low-

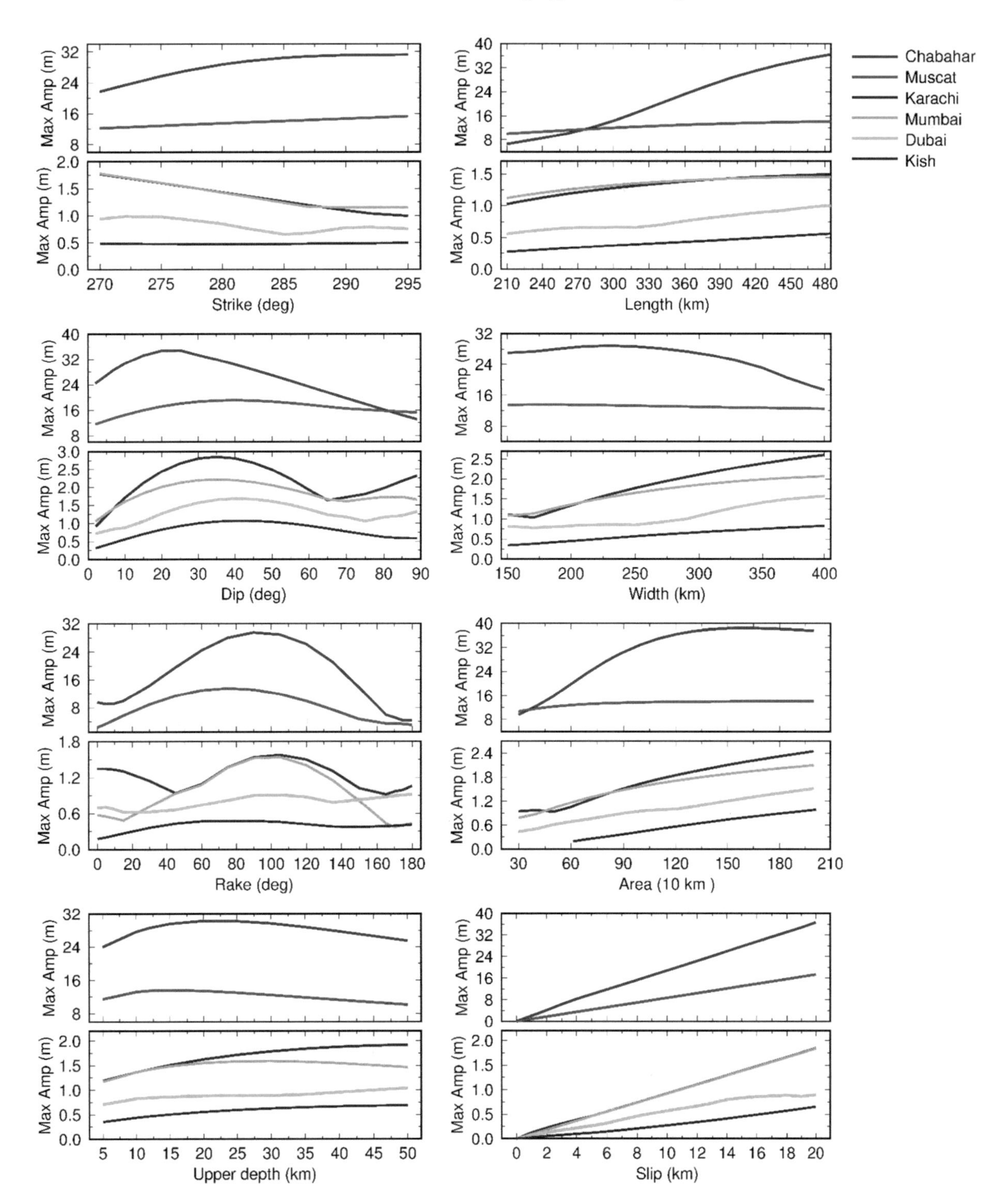

Figure 8
Comparison between the maximum tsunami amplitudes in the selected stations as a function of variation in the fault parameters for SC3 scenario

angle fault geometry for the western Makran may cause more powerful tsunamis.

To study the influence of mean rake angle through the sensitivity analysis, it is changed from 0° to 180°. By changing the slip angle, it remarkably has an effect on computed maximum amplitudes at Chabahar, Muscat, Karachi and Mumbai. The graphs for the stations have an almost symmetric shape with highest maximum amplitudes in slip angles between 75° and 105°. No remarkable effect is obvious at Dubai and Kish. The sensitivity of the tsunami amplitudes to top depth of the source for a range of values from 5 to 50 km is examined. Increasing the upper depth of the fault has no considerable impact on the tsunami amplitudes. It can be seen that the tsunami amplitudes tend to increase uniformly over the fault length. Increasing the fault width does not give similar results to the fault length. While a narrower rupture produces higher amplitudes at Chabahar, a wider fault zone generates higher amplitudes at other stations. The impact of fault area on tsunami amplitudes is also studied by increasing the length and width simultaneously. Extending the fault area mainly affects the maximum tsunami amplitudes in Chabahar. Increasing the mean slip value from 0 to 20 m linearly increases the computed tsunami maximum amplitude at all stations.

5. Conclusions

Heterogeneous slip models from four known subduction zones around the world are used and adapted in the western Makran zone. The maximum uplift produced by SC1, SC2, SC3 and SC4 scenarios are about 1.5, 15.5, 20.0 and 3.0 m, respectively. The maximum uplift from heterogeneous slip models is about three times greater than homogeneous slip models. The maximum tsunami amplitudes from heterogeneous slip distributions are considerably larger than uniform slip distributions. Our numerical modeling results reveal that SC3 is the strongest tsunami scenario. Tsunami waves generated by SC3 can impressively affect the Persian Gulf, the Red Sea and eastern India. Among the different shores, the southeastern Iran is exposed to highest risk. The simulation results show the peak coastal amplitude from non-uniform source models varies from 4 to 52 m along the Iranian shoreline. The impact of tsunamis on the coast of Oman is also very strong. The maximum coastal amplitude resulting from heterogeneous slip models ranges between 4 and 38 m along the Oman coast. The maximum values of coastal amplitudes change between 2 and 9 m along the shoreline of Pakistan and range from 1 to 8 m along the western Indian coast. The results of our sensitivity analysis indicate that changing the dip angle and mean rake has the most impact on tsunami wave amplitudes.

Acknowledgements

The authors would like to thank the developers of COMCOT program (Liu et al. 1998) for modeling tsunamis and also to thank the developers of GMT (Wessel and Smith 1991). They would also like to express their gratitude to the authors of the finite-fault models (Ji 2006; Hayes 2011; Wei et al. 2012; USGS 2015) used in this research for all their valuable efforts and making the models available. The authors would like to thank the editor, Alexander Rabinovich and two anonymous reviewers for their constructive comments and useful suggestions. The second author, ZHS, would like to acknowledge the financial support of University of Tehran for this research under Grant number 27875/01/10.

References

Ambraseys, N. N., & Melville, C. P. (1982). *A history of Persian earthquakes*. Britain: Cambridge University Press.

Ampuero, J.-P., Ripperger, J., & Mai, P. M. (2006). Properties of dynamic earthquake ruptures with heterogeneous stress drop. In R. Abercrombie, A. McGarr, H. Kanamori, & G. di Toro (Eds.), *Radiated energy and the physics of earthquakes, geophysical monograph, 170* (pp. 255–261). Washington, DC: American Geophysical Union.

Baba, T. (2003). Slip distributions of the 1944 Tonankai and 1946 Nankai earthquakes including the horizontal movement effect on tsunami generation. *Frontier Research on Earth Evolution, 1,* 213–218.

Barkan, R., & ten Brink, U. (2010). Tsunami simulations of the 1867 Virgin Islands earthquake: Constraints on epicenter location and fault parameters. *Bulletin of the Seismological Society of America, 100*(3), 995–1009.

Byrne, D. E., Sykes, L. R., & Davis, D. M. (1992). Great thrust earthquakes and aseismic slip along the plate boundary of the

Makran subduction zone. *Journal of Geophysical Research, 97*(B1), 449–478.

Geist, E. L. (2002). Complex earthquake rupture and local tsunamis. *Journal of Geophysical Research*. https://doi.org/10.1029/2000JB000139.

Geist, E. L., & Dmowska, R. (1999). Local tsunamis and distributed slip at the source. *Pure and Applied Geophysics, 154,* 485–512.

Grando, G., & McClay, K. (2007). Morphotectonics domains and structural styles in the Makran accretionary prism, offshore Iran. *Sedimentary Geology, 196,* 157–179.

Hanks, T. C., & Kanamori, H. (1979). A moment magnitude scale. *Journal of Geophysical Research, 84*(B5), 2348–2350. https://doi.org/10.1029/JB084iB05p02348.

Hayes, G. P. (2011). Rapid source characterization of the 2011 M_w 9.0 off the Pacific coast of Tohoku Earthquake. *Earth Planets Space, 63,* 529–534.

Heck, N. H. (1947). List of seismic sea waves. *Bulletin of the Seismological Society of America, 37*(4), 269–286.

Heidarzadeh, M., & Kijko, A. (2011). A probabilistic tsunami hazard assessment for the Makran subduction zone at the northwestern Indian Ocean. *Natural Hazards, 56*(3), 577–593.

Heidarzadeh, M., Pirooz, M. D., & Zaker, N. H. (2009). Modeling the nearfield effects of the worst-case tsunami in the Makran subduction zone. *Ocean Engineering, 36*(5), 368–376.

Heidarzadeh, M., Pirooz, M. D., Zaker, N. H., Yalciner, A. C., Mokhtari, M., & Esmaeily, A. (2008). Historical tsunami in the Makran subduction zone off the southern coasts of Iran and Pakistan and results of numerical modeling. *Ocean Engineering, 35*(8&9), 774–786.

Heidarzadeh, M., & Satake, K. (2014). Possible sources of the tsunami observed in the northwestern Indian Ocean following the 2013 September 24 M_w 7.7 Pakistan inland earthquake. *Geophysical Journal International, 199*(2), 752–766.

Heidarzadeh, M., & Satake, K. (2015). New insights into the source of the Makran tsunami of 27 November 1945 from tsunami waveforms a coastal deformation data. *Pure and Applied Geophysics, 172,* 621–640.

Hoechner, A., Babeyko, A. Y., & Zamora, N. (2016). Probabilistic tsunami hazard assessment for the Makran region with focus on maximum magnitude assumption. *Natural Hazards and Earth System Sciences, 16,* 1339–1350.

Hoffmann, G., Al-Yahyai, S., Naeem, G., Kociok, M., & Grützner, C. (2014). An Indian Ocean tsunami triggered remotely by an onshore earthquake in Balochistan, Pakistan. *Geology, 42*(10), 883–886.

Hoffmann, G., Rupprechter, M., Al Balushi, N., Grützner, C., & Reicherter, K. (2013). The impact of the 1945 Makran tsunami along the coastlines of the Arabian Sea (Northern Indian Ocean)—a review. *Zeitschrift für Geomorphologie, 57,* 257–277.

Ji, C. (2006). Rupture process of the 2006 Nov 15 magnitude 8.3—KURIL Island earthquake (revised). http://earthquake.usgs.gov/eqcenter/eqinthenews/2006/usvcam/finite_fault.php.

Kopp, C., Fruehn, J., Flueh, E. R., Reichert, C., Kukowski, N., Bialas, J., et al. (2000). Structure of the Makran subduction zone from wide-angle and reflection seismic data. *Tectonophysics, 329,* 171–191.

Kukowski, N., Schillhorn, T., Huhn, K., von Rad, U., Husen, S., & Flueh, E. R. (2001). Morphotectonics and mechanics of the central Makran accretionary wedge off Pakistan. *Marine Geology, 173,* 1–19.

Liu, P. L.-F., Cho, Y. S., Briggs, M. J., Kanoglu, U., & Synolakis, C. E. (1995). Runup of solitary waves on a circular island. *Journal of Fluid Mechanics, 302,* 259–285.

Liu, P. L.-F., Cho, Y. S., Yoon, S. B., & Seo, S. N. (1994). Numerical simulations of the 1960 Chilean tsunami propagation and inundation at Hilo, Hawaii. In M. I. El-Sabh (Ed.), *Recent Development in Tsunami Research* (pp. 99–115). Dordrecht: Kluwer Academic.

Liu, P. L. -F., Woo, S. B., & Cho, Y. S. (1998). Computer programs for tsunami propagation and inundation. Technical report, Cornell University.

Mansinha, L., & Smylie, D. E. (1971). The displacement fields of inclined faults. *Bulletin of the Seismological Society of America, 61,* 1433–1440.

Minshull, T. A., White, R. S., Barton, P. J., & Collier, J. S. (1992). Deformation at plate boundaries around the Gulf of Oman. *Marine Geology, 104,* 265–277.

Musson, R. M. W. (2009). Subduction in the western Makran: The historian's contribution. *Journal of the Geological Society, 166,* 387–391.

Neetu, S., Suresh, I., Shankar, R., Nagarajan, B., Sharma, R., Shenoi, S. S. C., et al. (2011). Trapped waves of the 27 November 1945 Makran tsunami: Observations and numerical modeling. *Natural Hazards, 59,* 1609–1618.

Okal, E. A., & Synolakis, C. E. (2008). Far-field tsunami hazard from mega-thrust earthquakes in the Indian Ocean. *Geophysical Journal International, 172,* 995–1015.

Omira, R., Baptista, M. A., & Lisboa, F. (2016). Tsunami characteristics along the Peru-Chile Trench: analysis of the 2015 M_w 8.3 Illapel, the 2014 M_w 8.2 Iquique and the 2010 M_w 8.8 Maule tsunamis in the near-field. *Pure and Applied Geophysics, 173,* 1063–1077.

Payande, A. R., Niksokhan, M. H., & Naserian, H. (2015). Tsunami hazard assessment of Chabahar bay related to megathrust seismogenic potential of the Makran subduction zone. *Natural Hazards, 76,* 161–176.

Rabinovich, A. B., Lobkovsky, L. I., Fine, I. V., Thomson, R. E., Ivelskaya, T. N., & Kulikov, E. A. (2008). Near-source observations and modeling of the Kuril Islands tsunamis of 15 November 2006 and 13 January 2007. *Advances in Geosciences, 14*(1), 105–116.

Rajendran, C. P., Rajendran, K., Shah-hosseini, M., Beni, A. N., Nautiyal, C. M., & Andrews, R. (2013). The hazard potential of the western segment of the Makran subduction zone, northern Arabian Sea. *Natural Hazards, 65*(1), 219–239.

Ruiz, J. A., Fuentes, M., Riquelme, S., Campos, J., & Cisternas, A. (2015). Numerical simulation of tsunami runup in northern Chile based on non-uniform k − 2 slip distributions. *Natural Hazards, 79*(2), 1177–1198.

Satake, K., & Heidarzadeh, M. (2017). A review of source models of the 2015 Illapel, Chile earthquake and insights from tsunami data. *Pure and Applied Geophysics, 174,* 1–9.

Smith, G. L., McNeill, L. C., Wang, K., He, J., & Henstock, T. J. (2013). Thermal structure and megathrust seismogenic potential of the Makran subduction zone. *Geophysical Research Letters, 40*(8), 1528–1533.

USGS. (2015). Preliminary Finite Fault Results for the Sep 16, 2015 M_w 8.3 46 km W of Illapel, Chile Earthquake (Version 1). http://earthquake.usgs.gov/earthquakes/eventpage/us20003k7a#finite-fault.

Vernant, Ph., Nilforoushan, F., Hatzfeld, D., Abbassi, M. R., Vigny, C., Masson, F., et al. (2004). Present-day crustal deformation and plate kinematics in the Middle East constrained by GPS measurements in Iran and northern Oman. *Geophysical Journal International, 157,* 381–398.

Wang, X. (2009). User manual for COMCOT version 1.7, first draft. Cornell University.

Wang, X., & Liu, P. L.-F. (2005). A numerical investigation of Boumerdes-Zemmouri (Algeria) earthquake and tsunami. *Computer Modeling in Engineering Science, 10*(2), 171–184.

Wang, X., & Liu, P. L.-F. (2006). Analysis of 2004 Sumatra earthquake fault plane mechanisms and Indian Ocean tsunami. *Journal of Hydraulic Research, 44*(2), 147–154.

Wei, S., Graves, R., Helmberger, D., Avouac, J. P., & Jiang, J. (2012). Sources of shaking and flooding during the Tohoku-Oki earthquake: A mixture of rupture styles. *Earth and Planetary Science Letters, 333–334,* 91–100.

Wessel, P., & Smith, W. H. F. (1991). Free software helps map and display data. *Eos, Transactions American Geophysical Union, 72*(41), 441–446.

White, R. S., & Klitgord, K. (1976). Sediment deformation and plate tectonics in the Gulf of Oman. *Earth and Planetary Science Letters, 32*(2), 199–209.

Ye, L., Lay, T., Kanamori, H., & Koper, K. D. (2016). Rapidly estimated seismic source parameters for the 16 September 2015 Illapel, Chile M_w 8.3 earthquake. *Pure and Applied Geophysics, 173,* 321–332.

(Received December 26, 2017, revised March 11, 2018, accepted March 13, 2018)

Pure Appl. Geophys.

https://doi.org/10.1007/s00024-018-1796-y

Pure and Applied Geophysics

Effect of Dynamical Phase on the Resonant Interaction Among Tsunami Edge Wave Modes

ERIC L. GEIST[1]

Abstract—Different modes of tsunami edge waves can interact through nonlinear resonance. During this process, edge waves that have very small initial amplitude can grow to be as large or larger than the initially dominant edge wave modes. In this study, the effects of dynamical phase are established for a single triad of edge waves that participate in resonant interactions. In previous studies, Jacobi elliptic functions were used to describe the slow variation in amplitude associated with the interaction. This analytical approach assumes that one of the edge waves in the triad has zero initial amplitude and that the combined phase of the three waves $\varphi = \theta_1 + \theta_2 - \theta_3$ is constant at the value for maximum energy exchange ($\varphi = 0$). To obtain a more general solution, dynamical phase effects and non-zero initial amplitudes for all three waves are incorporated using numerical methods for the governing differential equations. Results were obtained using initial conditions calculated from a subduction zone, inter-plate thrust fault geometry and a stochastic earthquake slip model. The effect of dynamical phase is most apparent when the initial amplitudes and frequencies of the three waves are within an order of magnitude. In this case, non-zero initial phase results in a marked decrease in energy exchange and a slight decrease in the period of the interaction. When there are large differences in frequency and/or initial amplitude, dynamical phase has less of an effect and typically one wave of the triad has very little energy exchange with the other two waves. Results from this study help elucidate under what conditions edge waves might be implicated in late, large-amplitude arrivals.

Key words: Tsunamis, edge waves, nonlinear, resonance, earthquake, stochastic slip.

1. Introduction

It is commonly known that the largest waves of a tsunami arriving at the coast often occur well after the first arrival. For near-field sites oblique to the source region and for far-field sites, trapped edge waves are primarily implicated in the occurrence of the largest waves (Geist 2009, 2012). The study by Munk et al. (1956) was one of the first to demonstrate that tsunamis can excite edge waves. Subsequent studies (Snodgrass et al. 1962; Munk et al. 1964) compared observed dominant frequencies in southern California with the theory of trapped waves. Miller et al. (1962) specifically looked at the energy decay of the 1960 Chile tsunami as a function of frequency in southern California, with most of this energy presumably occurring as edge waves. Edge waves are also significant resonant mechanisms for meteotsunamis (Greenspan 1956; Monserrat et al. 2006 and Vennell 2010) have been associated with hurricane landfall (Munk et al. 1956; Yankovsky 2009).

Although understanding edge waves is important from a hazard perspective, tracking different edge wave phases throughout the coda of a tsunami, as well as distinguishing these phases from other late arrivals such as basin-wide reflections and scattering (e.g., Saito et al. 2013), is observationally difficult. In fact, the time series of the coda, with *e*-folding times of 13–45 h (Rabinovich et al. 2011), has been described as a random wavefield by Geist (2009). Simplified systems of equations describing edge waves, however, focus on important interactions occurring within the coda, much the same way as for Rossby waves (Lynch 2003). These interactions may result in late, large-amplitude tsunami arrivals. Furthermore, given the wide spectrum of edge waves and associated group velocities indicated in previous studies, it is possible to have a "rogue-wave effect" through, for example, dispersive focusing (Kurkin and Pelinovsky 2002; Pelinovsky et al. 2010; Slunyaev et al. 2011).

Recent observations combined with numerical modeling have helped elucidate the phenomenology of tsunami edge waves. Many of the new observations involve analysis of current meter records. Using

[1] US Geological Survey, 345 Middlefield Rd., MS 999, Menlo Park, CA 94025, USA. E-mail: egeist@usgs.gov

acoustic Doppler current profile (ADCP) data, Sobarzo et al. (2012) is able to detect several dominant periods of energy consistent with the shore-parallel propagation of edge waves for the 2010 M_w 8.8 Maule (Chile) earthquake tsunami (see also Yamazaki and Cheung 2011). Using wavelet-based multifractal analysis, Toledo et al. (2013) suggest that a complex dynamical structure, such as the one discussed in this paper, is associated with 2010 Chile tsunami. Benjamin et al. (2016) use high-frequency Doppler radio measurements to detect edge waves offshore Oahu, Hawai'i, from the 2011 M_w 9.0 Tohoku-Oki earthquake tsunami, complementing ADCP observations of edge waves off Oahu from the 2006 Kuril tsunami by Bricker et al. (2007). In addition, Catalán et al. (2015) are able to distinguish edge wave modes from other resonant modes for the 2014 M_w 8.2 Pisgua earthquake tsunami using numerical analysis. The shelf width, defined by the 200 m isobaths in Cortés et al. (2017), appears to be a significant contributing factor to the site response for the late-arriving waves (Catalán et al. 2015; Cortés et al. 2017). Similarly, Vela et al. (2014) is able to associate dominant periods of tsunami energy with edge waves through numerical modeling for the 2003 M_w 6.9 Algeria earthquake tsunami at Majorca Island in the Mediterranean Sea.

Previous studies of tsunami edge waves include early analytical investigations, supported by observations (Kajiura 1972; Ishii and Abe 1980; Abe and Ishii 1987; Golovachev et al. 1992). Carrier (1995) and Fujima et al. (2000) calculate edge waves generated by near-field tsunamis and demonstrate that orientation and location of the source away from shore dictate which modes are excited. Kirby et al. (1998) and Dubinina et al. (2006) examined the nonlinear resonant interaction among edge wave modes in general, and Geist (2016) specifically applies their results to near-field tsunami edge waves.

The objective of this study is to determine the effect that dynamical phase has on the interactions among tsunami edge wave modes. A dynamical system in physics describes a point in geometric space subject to differential equations in time. Solution to the differential equations, either through analytical functions or computer simulations, provides information on the time history in a Hamiltonian framework. Previous studies on edge waves assumed a constant phase equal to that for maximum energy exchange. However, dynamical phase effects have been demonstrated for generalized waves (Armstrong et al. 1962; Wilhelmsson et al. 1970), with more recent specific applications to planetary waves (e.g., Kartashova and L'vov 2007; Lynch 2009). The dynamical phase effects range from shortening the period of interaction to greatly reducing the energy exchange among modes. This study is aimed at determining the specific effects of dynamical phase on tsunami edge wave interactions.

2. Background: Tsunami Edge Waves and Interaction Among Modes

Edge waves occur in discrete modes and are characterized by along-shore ray paths with an exponential decay in amplitude. The general equation for spatiotemporal amplitude $\zeta(x, y, t)$ is separated between a cross-shore $\eta(x)$ component and an oscillatory long-shore component:

$$\zeta(x,y,t) = \eta(x)e^{i(ky-\omega t)}, \quad (1)$$

where k is wavenumber and ω is angular frequency. The cross-shore component for a uniform slope is given in terms of Laguerre polynomials of order n:

$$\eta(x) = e^{-|k|x}L_n(2|k|x). \quad (2)$$

The dispersion relation for edge waves under the shallow-water assumption is given by (Leblond and Mysak 1978):

$$\omega_n^2 = gk(2n+1)\tan\beta, \quad n = 0, 1, 2, \ldots, \quad (3)$$

where β is bathymetric slope (Fig. 1). Previous observations of tsunami edge waves indicate periods that range from 20 min. (Catalán et al. 2015) to 72 min. and perhaps longer (Sobarzo et al. 2012). A few measurements of group velocities for tsunami edge waves, identified from tide gauge records, range between 17 m/s (Abe and Ishii 1987) and 65 m/s (Ishii and Abe 1980).

Rather than using simple analytic functions to describe sea floor displacement caused by earthquakes, stochastic models provide a more realistic representation of rupture complexity. Starting with

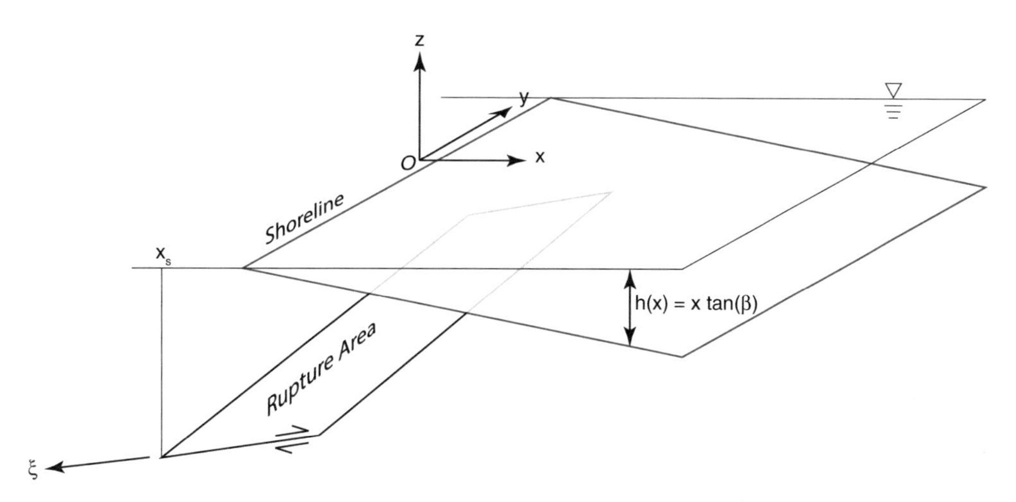

Figure 1
Geometry of uniform sloping bathymetry and buried fault rupture whose surface projection straddles the shoreline

Andrews (1980), stochastic models of static slip $D(x)$ during earthquake rupture have been designed to match and explain slip variability from seismic observations. The particular stochastic model used in this and previous studies (Geist 2012, 2016) is that of Lavallée et al. (2006, 2011) that is composed of two components: a radial wavenumber spectrum $D(k)$ and a probability model $R(k)$ that describes the random component of slip:

$$D(x) = D_0 F^{-1}[R(k)D(k)], \tag{4}$$

where D_0 is a constant and F^{-1} is the inverse Fourier transform. The wavenumber spectrum is characterized by a power-law decay (with exponent v) beyond a corner wavenumber, given in various forms (e.g., Herrero and Bernard 1994; Mai and Beroza 2002), for example:

$$D(k) \propto \frac{\Delta\sigma}{G}\frac{L}{k^{\upsilon}}, \tag{5}$$

where $\Delta\sigma$ is the static stress drop of the earthquake, L is a characteristic rupture dimension, G is the shear modulus. The probability model that is able to encompass observed high variability in slip, as indicated by Lavallée et al. (2006), is the Lévy α-stable probability family, whose characteristic function is given by:

$$\varphi(k) = \begin{cases} \exp\{i\lambda k - c|k|[1 + 2i\theta \ln|k|\mathrm{sgn}(k)/\pi]\}\alpha = 1 \\ \exp\{i\lambda k - c^{\alpha}|k|^{\alpha}[1 - i\theta\tan(\pi\alpha/2)\mathrm{sgn}(k)]\}\alpha \neq 1 \end{cases}, \tag{6}$$

where α is a stability parameter ($0 < \alpha \leq 2$), θ a skewness parameter ($-1 \leq \theta \leq 1$), λ a location parameter (real number), and c a scale parameter ($c > 0$). In Geist (2012, 2016), several different choices were used for α, including $\alpha = 2$ (Gaussian distribution) and $\alpha = 1$ (Cauchy distribution for $\theta = 0$). Sea floor displacement is calculated from the slip distribution by discretizing the rupture area and superimposing displacements from each fault cell in the rupture area using Okada's (1985) analytic expressions for a homogeneous elastic half space (Fig. 1).

The stochastic slip model is incorporated into the calculation of the evolution of edge wave amplitude by the following formula (Carrier 1995; Fujima et al. 2000):

$$\zeta(x, y, t) = \int_{-\infty}^{\infty} \sum_{n=0}^{\infty} a_n(|k|)e^{-|k|x}L_n(2|k|x)e^{-iky}(e^{-i\omega t} + e^{i\omega t})\mathrm{d}k. \tag{7}$$

In Eq. 7, a_n is calculated from the along-shore Fourier coefficients $c_k(x)$ of sea-floor displacement:

$$a_n(|k|) = |k| \int_0^\infty c_k(x) e^{-|k|x} L_n(2|k|x) \mathrm{d}x. \tag{8}$$

Nonlinear resonance among different modes in geophysical fluids has primarily been discussed in the context of Rossby waves. An early study of the interaction of planetary-scale Rossby waves in the atmosphere and ocean was made by Longuet-Higgins and Gill (1967) as well as more recent studies such as Cree and Swaters (1991), Reznik et al. (1993) and Lynch (2003, 2009). The ingredients necessary for nonlinear resonance to occur is that coupling or interaction coefficients (described below) are non-zero and a dispersion relation that allows for addition and subtraction relationships in wavenumber and frequency:

$$\begin{aligned} \pm k_l \pm k_m &= k_n \\ \pm \omega_l^p \pm \omega_m^q &= \omega_n^r, \end{aligned} \tag{9}$$

where the superscripts indicate mode number and subscripts indicates the number of the wave (1, 2, and 3).

Kenyon (1970) was one of the first studies to detail the interaction coefficients for ocean edge waves using the Hasselmann (1966) equations (see also Bretherton 1964). Kirby et al. (1998) and Dubinina et al. (2006) further develop the analytical theory of ocean edge waves, in terms of both collinear and counter-propagating waves. Dubinina et al. (2008) provide the interaction equations for edge waves propagating along a step shelf. The method of multiple scales or perturbation theory (e.g., Mei et al. 2005) is used to establish the interaction equations where

$$\zeta = \zeta^{(1)} + \varepsilon\zeta^{(2)} + \varepsilon^2\zeta^{(3)} \ldots \tag{10}$$

The slow time evolution ($T = \varepsilon t$) of edge-wave amplitude is given by (Kirby et al. 1998):

$$\begin{aligned} \frac{\partial A_n^r}{\partial T} = i \sum_l \sum_m \sum_p \sum_q \{ & {}_+V_{lmn}^{pqr} A_l^p A_m^q \delta(l+m-n) \\ & \delta(\omega_l^p + \omega_m^q - \omega_n^r) + {}_-V_{lmn}^{pqr} A_l^p A_m^{q*} \delta(l-m-n) \\ & \delta(\omega_l^p - \omega_m^q - \omega_n^r) + {}_-V_{mln}^{qpr} A_l^{p*} A_m^q \delta(m-l-n) \\ & \delta(\omega_m^q - \omega_l^p - \omega_n^r) \}, \end{aligned} \tag{11}$$

where δ is the Kronecker delta, ${}_+V$ and ${}_-V$ are the interaction coefficients for the sum and difference interactions, respectively, and A_n^r is the slowly varying amplitude for wave n and mode r (Eqs. 1 and 2). Interaction coefficients in Eq. 11 are given by (Kirby et al. 1998):

$$\begin{aligned} {}_\pm V_{lmn}^{pqr} = {} & \omega_l^p (\pm\omega_m^q) \left[8\omega_n^r \int_0^\infty (F_n^r)^2 \mathrm{d}x \right]^{-1} \\ & \int_0^\infty \Big\{ 2(\omega_l^p \pm \omega_m^q) F_l^{p'} F_m^{q'} F_n^r + \omega_l^p F_l^p F_m^{q''} F_n^r \\ & \pm \omega_m^q F_l^{p''} F_m^q F_n^r + \big[2(\omega_l^p \pm \omega_m^q) k_l^p (\mp k_m^q) \\ & - \omega_l^p (k_m^q)^2 \mp \omega_m^q (k_l^p)^2 \big] F_l^p F_m^q F_n^r \Big\} \mathrm{d}x, \end{aligned} \tag{12}$$

where $F_n^r(x)$ are the eigenmodes given by Eq. 2. The linear system $\zeta^{(1)}$ then is given by:

$$\zeta^{(1)} = \sum_n \sum_r \frac{1}{2} A_n^r(T) F_n^r(x) e^{i(k_n y - \omega_n^r t)} + \text{c.c.}, \tag{13}$$

where c.c. stands for complex conjugate. The spatiotemporal structure of the wave field summed over just several modes is quite complicated as demonstrated by Dubinina et al. (2006, 2008) and Pelinovsky et al. (2010). Geist (2016) applies results from these studies to specifically calculate edge waves generated by tsunamis in the near field using different stochastic slip models. All of the aforementioned edge-wave studies assume constant dynamical phase.

3. Dynamical Phase

The effect of dynamical phase on nonlinear resonance has been investigated as early as the 1960s for a number of different wave phenomena. The influence of phase in the nonlinear resonance of light waves was discussed by Armstrong et al. (1962) who originally provided solutions to the resonance problems in terms of Jacobi elliptic functions. Wilhelmsson et al. (1970) describe the role of phase in differentiating between interactions that are oscillating and stable and those that exhibit explosive instability (see also Alber et al. 1998). Kaup et al. (1979) provides a thorough review of three-wave interactions, including how phases adjust during evolution, and provides a general solution to the initial-value problem using the inverse scattering transform.

More recently, analytical solutions have been developed that included dynamical phase in nonlinear resonance using a Hamiltonian formulation (see Kartashova 2011). Lynch and Houghton (2004) first provide analytical solutions for both amplitudes and phases in the context of an elastic pendulum. These solutions are applied to Rossby waves, both in the atmosphere and in the ocean (Kartashova et al. 2008; Lynch 2009; Kartashova 2011). The effect of dynamical phases has also been numerically determined for various clusters of triads (Kartashova and L'vov 2008; Bustamante and Kartashova 2009a, b; Lynch 2009).

In general, the dynamical equations for the slow evolution of edge-wave amplitudes (Eq. 11) can be written in complex-value form as (Kirby et al. 1998; Dubinina et al. 2006):

$$\begin{aligned}\dot{A}_1 &= i\mu_1 A_2^* A_3\\ \dot{A}_2 &= i\mu_2 A_1^* A_3\\ \dot{A}_3 &= i\mu_3 A_1 A_2,\end{aligned} \tag{14}$$

where $A_j = a_j \exp(i\theta_j)$ (asterisk indicates complex conjugate) and μ_j are simplified expressions for the coupling coefficients. The transformation indicated by Lynch (2009)

$$B_1 = \sqrt{\mu_2\mu_3}A_1, B_2 = \sqrt{\mu_1\mu_3}A_2, B_3 = \sqrt{\mu_1\mu_2}A_3 \tag{15}$$

results in a simplified form of the triad equations:

$$\begin{aligned}\dot{B}_1 &= iB_2^* B_3\\ \dot{B}_2 &= iB_1^* B_3\\ \dot{B}_3 &= iB_1 B_2.\end{aligned} \tag{16}$$

Rather than depending on individual phases θ_j, it is shown that the dynamical phase effect depends specifically on the combination $\varphi = \theta_1 + \theta_2 - \theta_3$ (Kartashova 2011). Including dynamical phase introduces a fourth equation involving φ into the system given by Eq. 16 (Lynch and Houghton 2004):

$$\dot{\varphi} = -H_T\left(B_1^{-2} + B_2^{-2} - B_3^{-2}\right), \tag{17}$$

where H_T is the Hamiltonian given by:

$$H_T = B_1 B_1 B_1 \sin\varphi. \tag{18}$$

The Hamiltonian is a conserved quantity representing the total energy of the system (Kartashova 2011). Other independent constants of motion are given by:

$$\begin{aligned}I_{13} &= |B_1|^2 + |B_3|^2\\ I_{23} &= |B_2|^2 + |B_3|^2,\end{aligned} \tag{19}$$

which are known as the Manley–Rowe relations (Lynch and Houghton 2004; Kartashova 2011).

Previous studies of edge wave resonance (Kirby et al. 1998; Dubinina et al. 2006) provide analytical solutions at maximum energy exchange ($\varphi = 0$) and when one of the wave amplitudes is initially zero. Under these conditions, the solutions can be written in terms of Jacobi elliptic functions. Figure 2 shows an example of how amplitude slowly varies for a triad involving fundamental mode edges for waves 1 and 2 and a mode 1 edge wave for wave 3 (termed a 001 triad). In these and subsequent figures, amplitude (a) is normalized with respect to the largest initial amplitude of the three waves. Slow time (T) is normalized with respect to $\sqrt{\beta g K}$ (Dubinina et al. 2006), where K is determined from the discretization of the problem shown in Fig. 1: $K = 2\pi/N_y\Delta_y$ (see Geist 2016). Whereas wave 1 is little affected by the nonlinear interaction in this case, waves 2 and 3 dramatically exchange energy. As noted by Kirby et al. (1998), although wave 1 shows little amplitude variation, it needs to be present for the interaction to occur. It should be noted that other wave systems

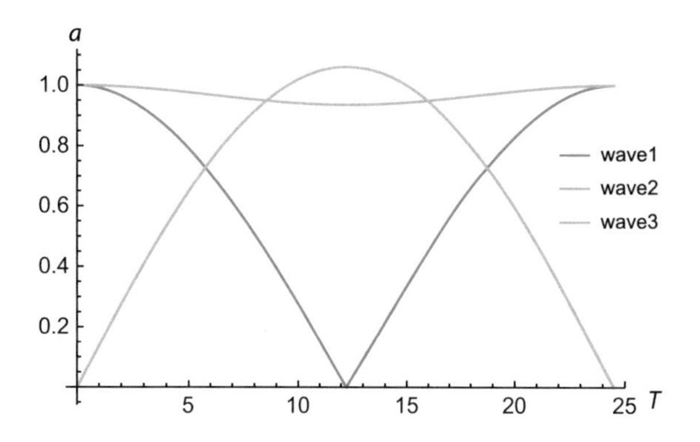

Figure 2
Slow amplitude variation associated with resonant interaction of three edge waves (001 triad) in nondimensional units (see text). Solution determined using Jacobi elliptic functions as described by Kirby et al. (1998)

have a slightly different form of the resonance equations, such as the shelf-wave study of Hsieh and Mysak (1980), where maximum energy exchange occurs at $\varphi = \pi/2$.

Rather than assuming constant phase, Bustamante and Kartashova (2009b) demonstrate the effects of using dynamical phase. For example, if $\varphi = 0$ is set as an initial condition, the resulting phase remains constant and the amplitudes can change sign. A change in the system, however, is seen for a non-zero but very small initial value of φ. In this case, the amplitudes are all positive and abrupt changes in φ are seen. The abrupt changes in φ and the double-value of inverse trigonometric functions make the dynamical system difficult to solve (Kartashova 2011). As the initial value of φ increases to $\pi/2$, Bustamante and Kartashova (2009b) indicate that the variation in amplitude and period decreases. Changes in the system are also observed for different relative values of the initial amplitudes. In the "Tsunami Applications" section, the effect of using different initial conditions is examined for tsunami edge wave interactions.

4. *Tsunami Applications*

4.1. *Description of Test Case*

To demonstrate the effects of using dynamical phase on tsunami edge wave resonance, the same source geometry beneath a uniform bathymetric slope as in Geist (2012, 2016) is used. In these studies, near-field edge waves are examined for large-magnitude continental subduction zone type earthquakes (Fig. 1), like the 2010 M_w 8.8 Chile earthquake. In Geist (2012), several probability models of stochastic slip for earthquake rupture provide heterogeneous initial conditions from which a broad range of edge-wave modes are excited. A Gaussian probability model results in fundamental-mode edge waves (Fig. 3) primarily being generated for a near-shore earthquake. In contrast, heavier-tail probability models such as a Cauchy distribution results in edge-wave energy being distributed among a number of different modes (Fig. 4). In these spectral plots, k is normalized with respect to $K = 2\pi/N_y\Delta_y$ (see Geist 2016).

In Geist (2016), a triad-search method was developed in which dominant edge-wave triad modes were found using a stochastic earthquake source. The first mode is the maximum amplitude among all modes and wavenumbers (arrow 1 in Figs. 3, 4). The second mode is the maximum amplitude that satisfies resonance conditions among all modes. The third mode of the resonant triad is assumed to have zero amplitude in Geist (2016) so that the analytical solutions could be used to investigate the period of energy exchange. In this study, the actual amplitude of the third wave (e.g., arrow 3 in Figs. 3, and 4) is used in the solution method described below, rather than assuming it is zero.

4.2. *Method*

Because dynamical phase is considered in this study, along with non-zero amplitudes for all triad modes, analytical solutions are not readily available. Equations 16 through 18 were solved numerically using LSODA as part of the NDSolve Framework in Mathematica. LSODA is a variant of LSODE (Livermore Solver for Ordinary Differential Equations), part of the public domain ODEPACK (Hindmarsh 1983), that is able to accommodate "stiff" solutions where the time step for numerical iteration becomes exceedingly small. LSODA switches between stiff and non-stiff solution methods as the solution warrants. As a test case, the results from numerical model are compared to the analytical solution shown in Fig. 2. The mean absolute error between the NDSolve solution for constant phase and the analytical solution using Jacobi elliptic functions is $O(10^{-7})$.

4.3. *Results*

To demonstrate the effect that dynamical phase has on tsunami edge wave interactions, the "characteristic" triads of Geist (2016) are first examined. These are the most commonly occurring triads using the Gaussian stochastic slip model. The characteristic triads are defined by a dominant, high-amplitude fundamental mode wave and two subharmonic waves, one of mode 0 and the other of either mode 2 or 3 (i.e., a 002 or 003 triad) that have

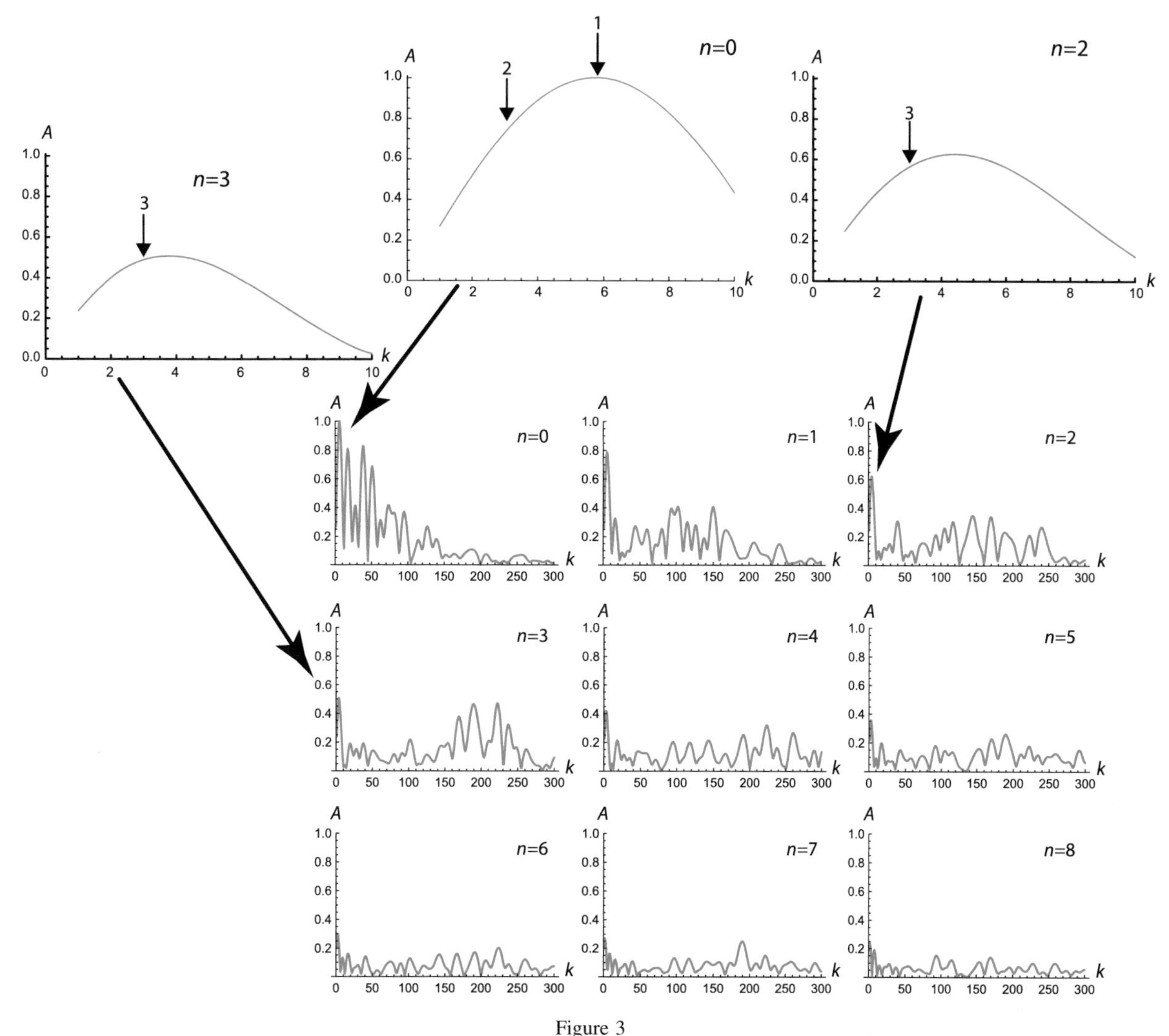

Figure 3
Spectral amplitudes for the first eight edge wave modes (below) using a Gaussian stochastic slip model. The dominant wavenumber (arrow 1) is part of a 002 or 003 resonant triad (arrows 2 and 3) shown in the close-ups (above)

approximately the same wavenumber (Fig. 3). For an initial value of φ exactly 0, the amplitude of wave 3 attains negative values (cf. Bustamante and Kartashova 2009b). However, as noted by Bustamante and Kartashova (2009b) if a very small initial value of φ is used (termed here as 0^+), similar behavior as observed for the analytical case (Fig. 2) is observed in Fig. 5a (002 triad shown by solid line and 003 triad shown by dashed line), although in this case there is energy exchange among all three waves. As explored below, this is a result of the ratios among initial amplitudes. Furthermore, because the amplitude of the third wave is not set to zero, the triad interaction is phase shifted compared to the analytical solution where the third wave is assumed to have zero amplitude initially. Plotted in Fig. 5b are the interactions of the three waves using an initial phase of $\varphi(0) = \pi/2$. The effect is to greatly decrease the variation in amplitude for all three waves and slightly decrease the interaction period. The corresponding evolution of phase for these two cases is shown in Fig. 6. In the case of 0^+ initial phase (Fig. 6a), the phase changes abruptly at the cusps in the interaction shown in Fig. 5a. Conversely, the phase changes smoothly and to a lesser degree in the case of $\varphi(0) = \pi/2$ (Fig. 6b).

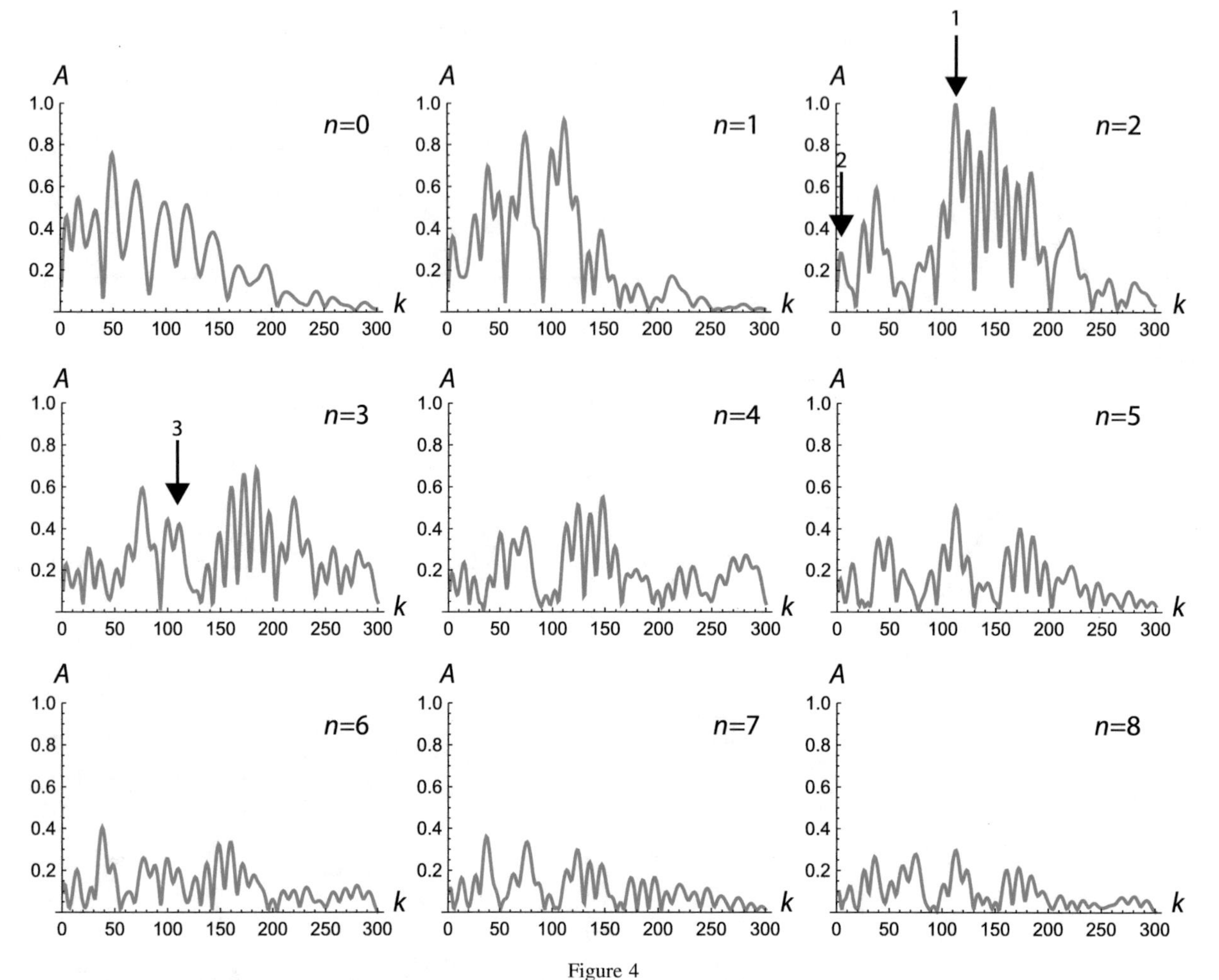

Figure 4
Spectral amplitudes for the first eight edge wave modes using a Cauchy stochastic source model that results in higher slip variability than the Gaussian source model (Fig. 3). Accordingly, tsunami energy is spread into higher modes and wavenumbers. For this case, a 223 resonant triad is dominant

Bustamante and Kartashova (2009b) demonstrate that the differences in frequencies (or wavenumbers) among edge waves in the triad control the level of interaction. The wave with the highest frequency is termed the active (*A*) mode and the other two waves are termed passive (*P*) modes. In the case of the characteristic 002 and 003 triads, the frequencies of the three wave are close together, with the *A* mode also being dominant amplitude wave (wave 1 in Fig. 3). The result is an interaction that affects all three waves.

Different behavior is seen for other possible triads, mainly those that result from using the Cauchy slip model associated with greater slip variation. Tsunami excitation using this slip model can be characterized as occurring from spatially distinct subevents, often creating edge waves with higher frequencies and higher modes (e.g., Fig. 4). A case in which the *A* mode is wave 3 of a 223 triad is shown in Fig. 7. In this case, the *A* mode primarily interacts with wave 1, which has a similar frequency, leaving wave 2 with a frequency an order of magnitude less relatively unaffected, even though both waves 1 and 2 have similar small initial amplitude. Changing the initial value for φ results in only a small effect of the interaction, with waves 1 and 3 not dropping to zero amplitude as they do for $\varphi(0) = 0^+$. The effect on φ is shown in Fig. 7c, d.

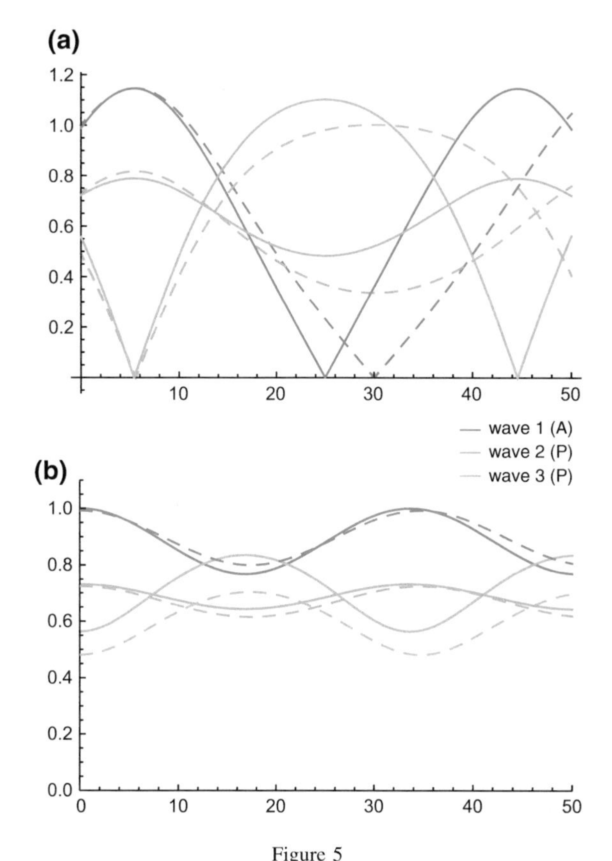

Figure 5
Amplitude variation for the characteristic triads (002—solid; 003—dashed) using dynamical phase and nonzero amplitude for wave 3. *A* active mode. *P* passive mode. **a** $\varphi(0) = 0^+$. **b** $\varphi(0) = \pi/2$

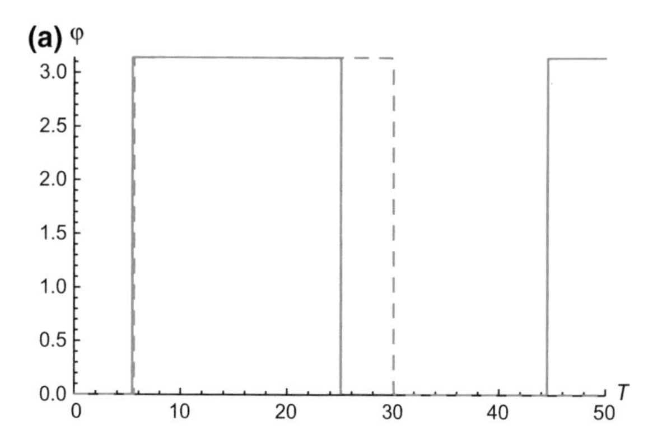

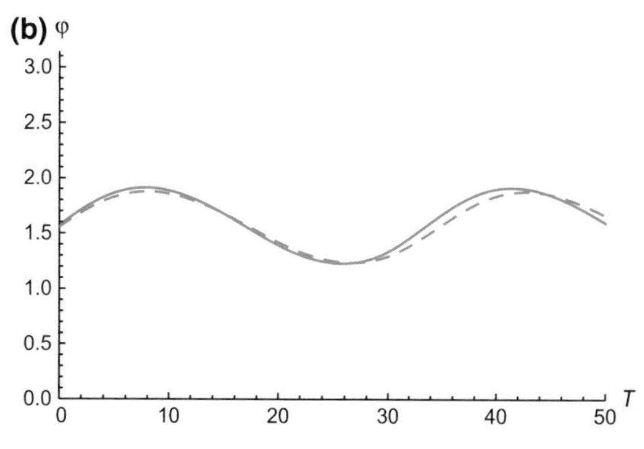

Figure 6
Variation of composite phase φ for the characteristic triads (002—solid; 003—dashed) corresponding to the amplitude variation shown in Fig. 5. **a** $\varphi(0) = 0^+$. **b** $\varphi(0) = \pi/2$

A case using the Cauchy source where wave 2 is the *A* mode and a smaller difference in frequencies are shown in Fig. 8. In this case, for an initial value of $\varphi(0) = 0^+$, the interaction is primarily between waves 2 and 3 with a small effect on wave 1, the dominant amplitude wave. Interestingly, however, when an initial value of φ is $\pi/2$, essentially no interaction occurs for all three waves (Fig. 8b). Correspondingly, there is also little change in the phase with time (Fig. 8d). For a larger difference in the initial amplitudes of waves 2 and 3, the interaction between these two waves remains significant for both starting values of φ (Fig. 9).

5. *Discussion*

The results shown above indicate that the level and characteristics of edge wave interaction depend on several factors: initial value of φ, which wave is the *A* mode, the difference in frequencies (wavenumbers) among the waves, and the amplitude ratios among the different waves. In most of the cases using a Gaussian slip model, the *A* mode is the wave with the highest amplitude (wave 1). If the initial amplitudes and frequencies of the waves are similar, then all three waves will interact (Fig. 5) and the initial value of φ will have a significant effect on the scale of interaction. This observation also holds when one of the other waves is the *A* mode (Fig. 8). There is little interaction with a *P* wave that has a much smaller frequency than the *A* wave and the other *P* wave (Fig. 7). Finally, when there is a large contrast in amplitudes between the two *P* modes, the *A* mode will primarily interact with one of the *P* modes, leaving the other *P* mode little changed (Fig. 9). For the extreme case shown in Fig. 2, where one of the waves has zero initial amplitude, dynamical phase has no effect—the numerical solution

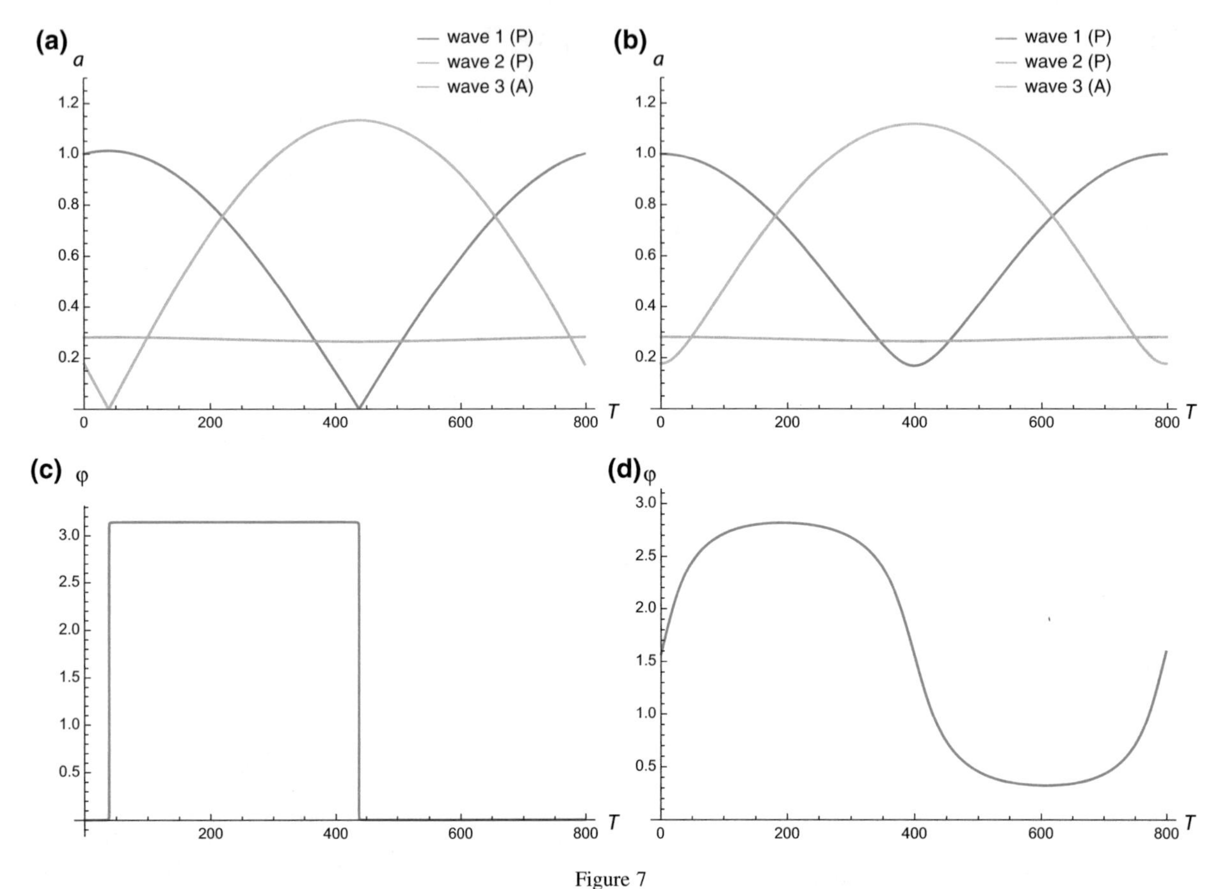

Figure 7
Variation of amplitude (**a**, **b**) and composite phase φ (**c**, **d**) for a 223 triad using a Cauchy source model. **a**, **c** $\varphi(0) = 0^{+}$. **b**, **d** $\varphi(0) = \pi/2$

indicates that φ immediately changes to a constant value $\varphi = 0,\ \pi$ no matter what the initial value is. Again, although there may be little amplitude variation in one of the waves as in Figs. 2, 7 and 9, all waves need to be present for the triad interaction to occur.

The initial values of amplitude and phase described above are linked to the sea-floor displacement field and, therefore, the slip distribution of an earthquake. The characteristic triads associated with a Gaussian slip model tend to have a large amplitude difference between the dominant amplitude mode, which also happens to be the *A* mode, and the two *P* modes. There is also less chance of having spatially distinct subevents that might give rise to nonzero values of φ, compared to heavier tailed probability distributions associated with non-Gaussian slip models. However, it is these latter slip models that likely more realistically describe observed slip distributions (Lavallée et al. 2006).

Continued acquisition of focused seismological and oceanographic observations can constrain the range of possible edge wave interactions. For example, additional work into limiting the parameter space of stochastic slip models for subduction zone earthquakes, particularly with regard to the stability parameter α (Eq. 6), can have a direct influence on which edge wave triads are most likely (Geist 2016). In addition, recent observations and analysis of current data (e.g., Bricker et al. 2007; Sobarzo et al. 2012; Benjamin et al. 2016) are able to distinguish dominant edge wave modes. The ADCP observations of the 2010 Chile tsunami by Sobarzo et al. (Sobarzo et al. 2012) in particular is able to not only identify three dominant frequencies that correspond to edge waves but also several other periods of energy (their

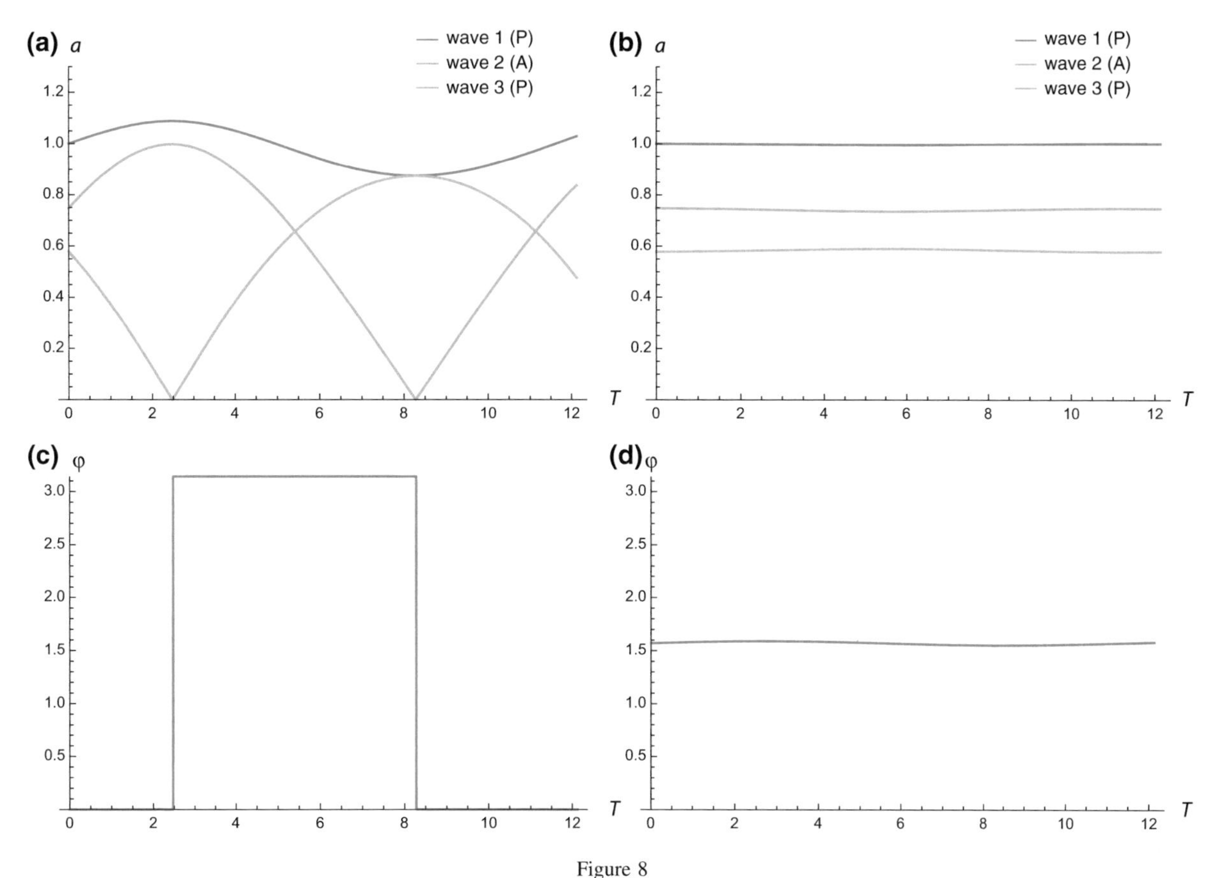

Figure 8
Variation of amplitude (**a**, **b**) and composite phase φ (**c**, **d**) for a 003 triad using a Cauchy source model, with a larger difference in amplitudes in waves 2 and 3 than in Fig. 8. **a**, **c** $\varphi(0) = 0^{+}$. **b**, **d** $\varphi(0) = \pi/2$

Fig. 7) that approximately satisfy the triad criteria (Eq. 9).

6. Summary

The range of interaction among tsunami edge waves in a resonant triad has been established using dynamical phase and non-zero initial amplitudes for all three waves. As indicated in a previous study (Geist 2016), a stochastic slip model based on different probability distribution results in a wide range of possible edge wave modes that can be generated by a continental subduction zone earthquake. In contrast to this previous study where the interaction was demonstrated using analytic functions that require an assumption of zero amplitude for one of the waves of the triad, a numerical solution to the governing differential equations was used in this study to establish the effect of dynamical phase and non-zero initial amplitude for all three waves. Results indicate a complex relationship between the level of interaction among the different waves and various initial conditions: combined phase φ, amplitude, and relative frequency among each of the waves. Conditions can be present where there is essentially no interaction among the waves to cases where an edge wave mode with initially small amplitude can grow to be as large or larger than the other modes in the triad. The effect of dynamical phase is most pronounced when the frequencies and amplitudes of the waves are of a similar order of magnitude. Continued analysis of edge wave from oceanographic observations, as well as constraining possible slip models for subduction zone earthquakes will help focus future numerical experiments.

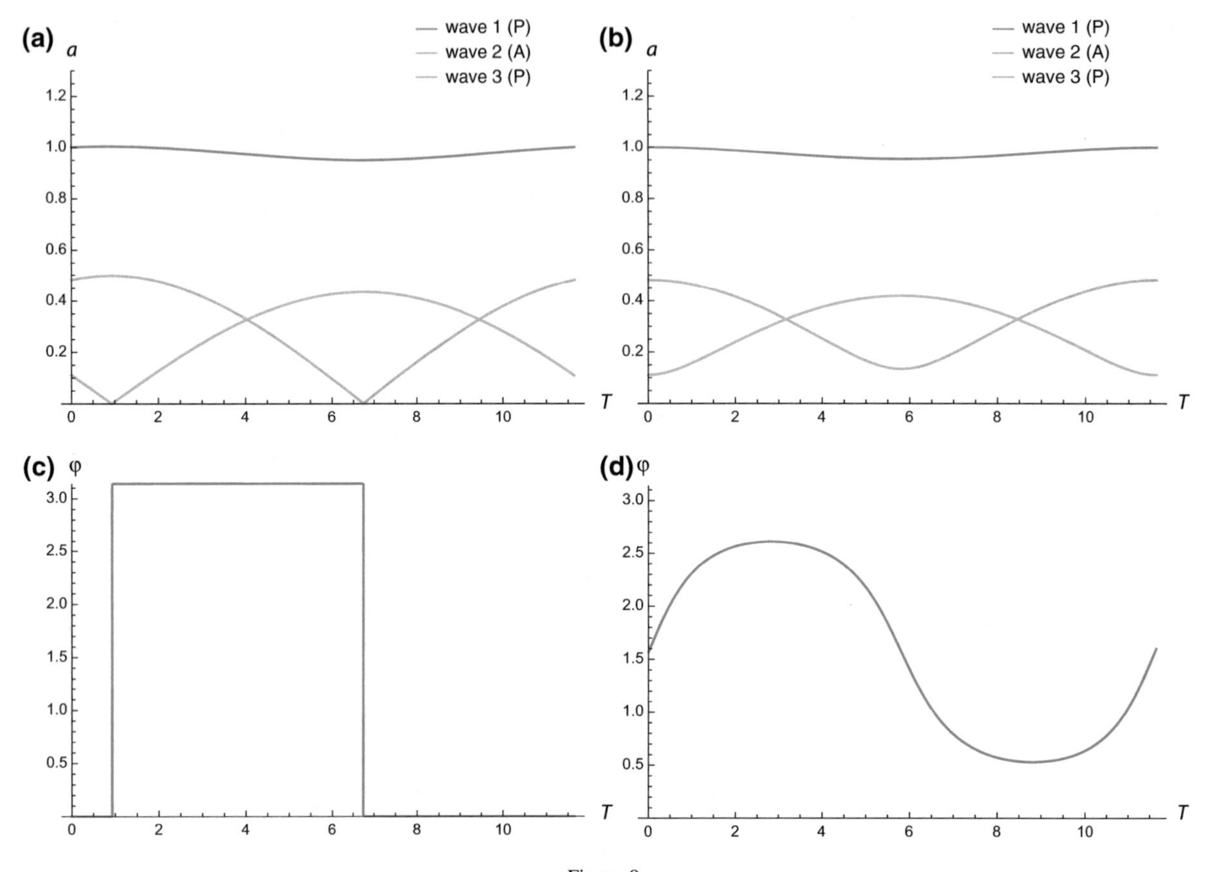

Figure 9
Variation of amplitude (**a**, **b**) and composite phase φ (**c**, **d**) for a 003 triad using a Cauchy source model, with a larger difference in amplitudes in waves 2 and 3 than in Fig. 8. **a**, **c** $\varphi(0) = 0^+$. **b**, **d** $\varphi(0) = \pi/2$

Acknowledgements

The author very much appreciates the constructive comments received on the manuscript by Efim Pelinovsky, Kenny Ryan, and an anonymous reviewer, as well as the thoughtful insights of Editor Alexander Rabinovich.

References

Abe, K., & Ishii, H. (1987). Distribution of maximum water levels due to the Japan Sea tsunami on 26 May 1983. *Journal of the Oceanographical Society of Japan, 43,* 169–182.

Alber, M. S., Luther, G. G., Marsden, J. E., & Robbins, J. M. (1998). Geometric phases, reduction and Lie–Poisson structure for the resonant three-wave interaction. *Physica D: Nonlinear Phenomena, 123,* 271–290. https://doi.org/10.1016/S0167-2789(98)00127-4.

Andrews, D. J. (1980). A stochastic fault model 1. Static case. *Journal of Geophysical Research, 85,* 3867–3877.

Armstrong, J. A., Bloembergen, N., Ducuing, J., & Pershan, P. S. (1962). Interactions between light waves in a nonlinear dielectric. *Physical Review, 127,* 1918–1939.

Benjamin, L. R., Flament, P., Cheung, K. F., & Luther, D. S. (2016). The 2011 Tohoku tsunami south of Oahu: High-frequency Doppler radio observations and model simulations of currents. *Journal of Geophysical Research: Oceans, 121,* 1133–1144. https://doi.org/10.1002/2015JC011207.

Bretherton, F. P. (1964). Resonant interactions between waves. The case of discrete oscillations. *Journal of Fluid Mechanics, 20,* 457–479.

Bricker J. D., Munger S., Pequignet C., Wells J. R., Pawlak G., & Cheung K. F. (2007). ADCP observations of edge waves off Oahu in the wake of the November 2006 Kuril Islands tsunami. *Geophysical Research Letters, 34.* https://doi.org/10.1029/2007gl032015.

Bustamante M. D., & Kartashova E. (2009a). Dynamics of nonlinear resonances in Hamiltonian systems. *Europhysics Letters, 85.* https://doi.org/10.1209/0295-5075/1285/14004.

Bustamante M. D., & Kartashova E. (2009b). Effect of the dynamical phases on the nonlinear amplitudes' evolution. *Europhysics Letters, 85.* https://doi.org/10.1209/0295-5075/1285/34002.

Carrier, G. F. (1995). On-shelf tsunami generation and coastal propagation. In Y. Tsuchiya & N. Shuto (Eds.), *Tsunami: Progress in prediction, disaster prevention and warning* (pp. 1–20). Dordrecht: Kluwer.

Catalán, P. A., Aránguiz, R., González, G., Tomita, T., Cienfuegos, R., González, J., et al. (2015). The 1 April 2014 Pisagua tsunami: Observations and modeling. *Geophysical Research Letters, 42,* 2918–2925. https://doi.org/10.1002/2015GL063333.

Cortés, P., Catalán, P. A., Aránguiz, R., & Bellotti, G. (2017). Tsunami and shelf resonance on the northern Chile coast. *Journal of Geophysical Research: Oceans, 122,* 7364–7379. https://doi.org/10.1002/2017JC012922.

Cree, W. C., & Swaters, G. E. (1991). On the topographic dephasing and amplitude modulation of nonlinear Rossby wave interactions. *Geophysical and Astrophysical Fluid Dynamics, 61,* 75–99. https://doi.org/10.1080/03091929108229037.

Dubinina, V. A., Kurkin, A. A., Pelinovsky, E. N., & Poloukhina, O. E. (2006). Resonance three-wave interactions of Stokes edge waves. *Izvestiya, Atmospheric and Oceanic Physics, 42,* 254–261.

Dubinina, V. A., Kurkin, A. A., & Polukhina, O. E. (2008). On the nonlinear interactions in triads of edge waves on the sea shelf. *Physical Oceanography, 18,* 117–132.

Fujima, K., Dozono, R., & Shigemura, T. (2000). Generation and propagation of tsunami accompanying edge waves on a uniform shelf. *Coastal Engineering Journal, 42,* 211–236.

Geist, E. L. (2009). Phenomenology of tsunamis: Statistical properties from generation to runup. *Advances in Geophysics, 51,* 107–169.

Geist, E. L. (2012). Near-field tsunami edge waves and complex earthquake rupture. *Pure and Applied Geophysics.* https://doi.org/10.1007/s00024-00012-00491-00027.

Geist, E. L. (2016). Non-linear resonant coupling of tsunami edge waves using stochastic earthquake source models. *Geophysical Journal International, 204,* 878–891. https://doi.org/10.1093/gji/ggv489.

Golovachev, E. V., Kochergin, I. E., & Pelinovsky, E. N. (1992). The effect of the Airy phase during propagation of edge waves. *Soviet Journal of Physical Oceanography, 3,* 1–7.

Greenspan, H. P. (1956). The generation of edge waves by moving pressure distributions. *Journal of Fluid Mechanics, 1,* 574–592.

Hasselmann, K. (1966). Feynamn diagrams and interaction rules of wave–wave scattering processes. *Reviews of Geophysics, 4,* 1–32.

Herrero, A., & Bernard, P. (1994). A kinematic self-similar rupture process for earthquakes. *Bulletin of the Seismological Society of America, 84,* 1216–1228.

Hindmarsh, A. C. (1983). ODEPACK, a systematized collection of ODE solvers. In R. S. Stepleman (Ed.), *Scientific computing.* Amsterdam: North-Holland.

Hsieh, W. W., & Mysak, L. A. (1980). Resonant interactions between shelf waves, with applications to the Oregon Shelf. *Journal of Physical Oceanography, 10,* 1729–1741.

Ishii, H., & Abe, K. (1980). Propagation of tsunami on a linear slope between two flat regions. Part I edge wave. *Journal of Physics of the Earth, 28,* 531–541.

Kajiura, K. (1972). The directivity of energy radiation of the tsunami generated in the vicinity of a continental shelf. *Journal of the Oceanographical Society of Japan, 28,* 260–277.

Kartashova, E. (2011). *Nonlinear resonance analysis: Theory, computation, applications.* Cambridge: Cambridge University Press.

Kartashova, E., & L'vov, V. S. (2007). Model of intraseasonal oscillations in Earth's atmosphere. *Physical Review Letters, 98,* 198501.

Kartashova, E., & L'vov, V. S. (2008). Cluster dynamics of planetary waves. *EPL (Europhysics Letters), 83,* 50012.

Kartashova, E., Raab, C., Feurer, C., Mayrhofer, G., & Schreiner, W. (2008). Symbolic computation for nonlinear wave resonances. In E. Pelinovsky & C. Kharif (Eds.), *Extreme ocean waves* (pp. 95–126). Dordrecht: Springer Netherlands.

Kaup, D. J., Reiman, A., & Bers, A. (1979). Space–time evolution of nonlinear three-wave interactions. I. Interaction in a homogeneous medium. *Reviews of Modern Physics, 51,* 275–309.

Kenyon, K. E. (1970). A note on conservative edge wave interactions. *Deep-Sea Research, 17,* 197–201.

Kirby J. T., Putrevu U., & Özkan-Haller H. T. (1998). Evolution equations for edge waves and shear waves on longshore uniform beaches. In: Proceedings of 26th International Conference on Coastal Engineering. ASCE, Copenhagen, Denmark, pp. 203–216.

Kurkin, A., & Pelinovsky, E. (2002). Focusing of edge waves above a sloping beach. *European Journal of Mechanics B/Fluids, 21,* 561–577.

Lavallée, D., Liu, P., & Archuleta, R. J. (2006). Stochastic model of heterogeneity in earthquake slip spatial distributions. *Geophysical Journal International, 165,* 622–640.

Lavallée, D., Miyake, H., & Koketsu, K. (2011). Stochastic model of a subduction-zone earthquake: Sources and ground motions for the 2003 Tokachi-oki, Japan, earthquake. *Bulletin of the Seismological Society of America, 101,* 1807–1821.

Leblond, P. H., & Mysak, L. A. (1978). *Waves in the ocean.* Amseterdam: Elsevier.

Longuet-Higgins, M. S., & Gill, A. E. (1967). Resonant interactions between planetary waves. *Proceedings of the Royal Society of London. Series A. Mathematical and Physical Sciences, 299,* 120–144. https://doi.org/10.1098/rspa.1967.0126.

Lynch, P. (2003). Resonant Rossby wave triads and the swinging spring. *Bulletin of the American Meteorological Society, 84,* 605–616. https://doi.org/10.1175/bams-84-5-605.

Lynch, P. (2009). On resonant Rossby–Haurwitz triads. *Tellus A, 61,* 438–445. https://doi.org/10.1111/j.1600-0870.2009.00395.x.

Lynch, P., & Houghton, C. (2004). Pulsation and precession of the resonant swinging spring. *Physica D: Nonlinear Phenomena, 190,* 38–62. https://doi.org/10.1016/j.physd.2003.09.043.

Mai P. M., & Beroza G. C. (2002). A spatial random field model to characterize complexity in earthquake slip. *Journal of Geophysical Research, 107.* https://doi.org/10.1029/2001jb000588.

Mei, C. C., Stiassnie, M., & Yue, D. K.-P. (2005). *Theory and applications of ocean surface waves. Part 2: Nonlinear aspects.* Singpore: World Scientific.

Miller, G. R., Munk, W. H., & Snodgrass, F. E. (1962). Long-period waves over California's continental borderland Part II. Tsunamis. *Journal of Marine Research, 20,* 31–41.

Monserrat, S., Vilibic, I., & Rabinovich, A. B. (2006). Meteotsunamis: Atmospherically induced destructive ocean waves in the tsunami frequency band. *Natural Hazards and Earth System Sciences, 6,* 1035–1051.

Munk, W., Snodgrass, F. E., & Carrier, G. F. (1956). Edge waves on the continental shelf. *Science, 123,* 127–132.

Munk, W., Snodgrass, F. E., & Gilbert, F. (1964). Long waves on the continental shelf: An experiment to separate trapped and leaky modes. *Journal of Fluid Mechanics, 20,* 529–554.

Okada, Y. (1985). Surface deformation due to shear and tensile faults in a half-space. *Bulletin of the Seismological Society of America, 75,* 1135–1154.

Pelinovsky, E., Polukhina, O., & Kurkin, A. (2010). Rogue edge waves in the ocean. *European Physical Journal Special Topics, 185,* 34–44.

Rabinovich, A. B., Candella, R. N., & Thomson, R. E. (2011). Energy decay of the 2004 Sumatra tsunami in the World Ocean. *Pure and Applied Geophysics, 168,* 1919–1950.

Reznik, G. M., Piterbarg, L. I., & Kartashova, E. A. (1993). Nonlinear interactions of spherical Rossby modes. *Dynamics of Atmospheres and Oceans, 18,* 235–252. https://doi.org/10.1016/0377-0265(93)90011-U.

Saito, T., Inazu, D., Tanaka, S., & Miyoshi, T. (2013). Tsunami coda across the Pacific Ocean following the 2011 Tohoku-Oki earthquake. *Bulletin of the Seismological Society of America, 103,* 1429–1443. https://doi.org/10.1785/0120120183.

Slunyaev, A., Didenkulova, I. I., & Pelinovsky, E. (2011). Rogue waters. *Contemporary Physics, 52,* 571–590.

Snodgrass, F. E., Munk, W. H., & Miller, G. R. (1962). Long-period waves over California's continental borderland Part I. Background spectra. *Journal of Marine Research, 20,* 3–30.

Sobarzo, M., Garcés-Vargas, J., Bravo, L., Tassara, A., & Quiñones, R. A. (2012). Observing sea level and current anomalies driven by a megathrust slope-shelf tsunami: The event on February 27, 2010 in central Chile. *Continental Shelf Research, 49,* 44–55. https://doi.org/10.1016/j.csr.2012.09.001.

Toledo, B. A., Chian, A. C. L., Rempel, E. L., Miranda, R. A., Muñoz, P. R., & Valdivia, J. A. (2013). Wavelet-based multifractal analysis of nonlinear time series: The earthquake-driven tsunami of 27 February 2010 in Chile. *Physical Review E, 87,* 022821.

Vela, J., Pérez, B., González, M., Otero, L., Olabarrieta, M., Canals, M., et al. (2014). Tsunami resonance in Palma Bay and Harbor, Majorca Island, as induced by the 2003 western Mediterranean earthquake. *The Journal of Geology, 122,* 165–182. https://doi.org/10.1086/675256.

Vennell, R. (2010). Resonance and trapping of topographic transient ocean waves generated by a moving atmospheric disturbance. *Journal of Fluid Mechanics, 650,* 427–443.

Wilhelmsson, H., Stenflo, L., & Engelmann, F. (1970). Explosive instabilities in the well-defined phase description. *Journal of Mathematical Physics, 11,* 1738–1742. https://doi.org/10.1063/1.1665320.

Yamazaki Y., & Cheung K. F. (2011). Shelf resonance and impact of near-field tsunami generated by the 2010 Chile earthquake. *Geophysical Research Letters, 38.* https://doi.org/10.1029/2011gl047508.

Yankovsky A. E. (2009). Large-scale edge waves generated by hurricane landfall. *Journal of Geophysical Research, 114.* https://doi.org/10.1029/2008jc005113.

(Received December 29, 2017, revised January 29, 2018, accepted January 31, 2018)

Pure Appl. Geophys.

https://doi.org/10.1007/s00024-018-1766-4

Pure and Applied Geophysics

On the Resonant Behavior of a Weakly Compressible Water Layer During Tsunamigenic Earthquakes

CLAUDIA CECIONI[1] and GIORGIO BELLOTTI[1]

Abstract—Tsunamigenic earthquakes trigger pressure waves in the ocean, given the weak compressibility of the sea water. For particular conditions, a resonant behavior of the water layer can occur, which influences the energy transfer from the sea-bed motion to the ocean. In this paper, the resonance conditions are explained and analyzed, focusing on the hydro-acoustic waves in the proximity of the earthquake area. A preliminary estimation of the generation parameters (sea-bed rising time, velocity) is given, by means of parametric numerical simulations for simplified conditions. The results confirm the importance of measuring, modeling, and interpreting such waves for tsunami early detection and warning.

Key words: Tsunami, hydro-acoustic waves, submarine earthquake.

1. Introduction

Fast sea-bed motions, such as those triggered by underwater earthquakes, generate both free-surface waves (tsunami) and low-frequency pressure waves (hydro-acoustic waves). Early studies on tsunami evolution in weakly compressible water have been carried out by Miyoshi (1954), Sells (1965), Kajiura (1970), and Yamamoto (1982). Later, analytical studies have solved the potential fluid problem in weakly compressible water, with fast rising motion of the sea bed (Smith 2015; Stiassnie 2010; Kadri and Stiassnie 2012; Eyov et al. 2013; Hendin and Stiassnie 2013; Kadri 2015) and including the effect of an elastic sea bottom (Panza et al. 2000; Balanche et al. 2009; Maeda and Furumura 2013).

Indeed, few measurements exist of hydro-acoustic waves generated by submarine earthquakes. Among these, we can cite the hydrophone measurements of the 2004 Sumatra event, analyzed by Okal et al. (2007); the measurements of the 2003 Tokachi-Oki earthquake (Japan) analyzed and compared with numerical simulations by Bolshakova et al. (2011), Nosov et al. (2007) and Nosov and Kolesov (2007); and the underwater pressure records of the 2012 Haida Gwaii earthquake (Canada), numerically reproduced by Abdolali et al. (2015a). Since full three-dimensional modeling in large geographical regions is computationally demanding, Sammarco et al. (2013) derived a depth-integrated equation for the mechanics of generation and propagation of hydro-acoustic waves, that has been conveniently applied over large-scale domains (Cecioni et al. 2015; Abdolali et al. 2014). Later, Abdolali et al. (2015b, c) extended the equation for porous bottom.

These analytical and numerical studies indicate that a fast sea-bed motion triggers pressure waves that propagate in the water layer; in situ acoustic measurements during submerged earthquake confirmed the possibility to record the generated hydro-acoustic waves. To use these acoustic signals as tsunami precursors, it is necessary to correlate the features of the measured hydro-acoustic wave with the earthquake parameters, as pointed out by Chierici et al. (2010), Levin and Nosov (2009), and Cecioni et al. (2014). This paper focuses on particular conditions of the generation mechanism that produce a resonant behavior of the water layer, revealing that the energy of the hydro-acoustic waves does not monotonously increase with the sea-bed velocity. This non-monotonic behavior of hydro-acoustic wave amplitude, first noted by Nosov (1999), is here mathematically explained. We point out that it is important to consider this behavior for interpretation of hydro-

[1] Engineering Department, Roma TRE University, via Vito Volterra 62, 00146 Rome, Italy. E-mail: claudia.cecioni@uniroma3.it

acoustic measurements. The paper presents parametric numerical simulations, aimed to analyze the correlation between the generation mechanism and the hydro-acoustic waves generated in the near field, i.e., close to the source.

The next section describes the mathematical problem of wave generation and propagation in a weakly compressible fluid. The following section presents the results of parametric computations, where the hydro-acoustic wave features are correlated with the generation mechanism. Finally, conclusions are given.

2. *Modeling Hydro-Acoustic Waves*

Fluid motion in a weakly compressible ocean, generated by sea-bed movement, is described in the framework of linearized theory, by the following governing equation and boundary conditions:

$$\begin{cases} \Phi_{tt} - c_s^2 \nabla^2 \Phi - c_s^2 \Phi_{zz} = 0 & \\ \Phi_{tt} + g\Phi_z = 0 & \text{at } z = 0 \\ \Phi_z + \nabla h \cdot \nabla \Phi + h_t = 0 & \text{at } z = -h(x,y,t), \end{cases} \tag{1}$$

where $\Phi(x,y,z,t)$ is the fluid velocity potential, ∇ and ∇^2 are, respectively, the gradient and the Laplacian in the horizontal plane x, y, with subscripts of independent variables denoting partial derivatives, and c_s is the celerity of sound in water (1500 m/s) and g is the gravitational acceleration (9.81 m/s^2). The system of Eq. (1) is valid for small amplitude waves and bottom motion, compared to the water depth. The wave generation is forced by the bottom boundary condition. The water depth $h(x, y, t)$ is defined as

$$h(x,y,t) = h_b(x,y) - \zeta(x,y,t), \tag{2}$$

where h_b is the fixed seafloor and ζ represents the sea-floor dislocation, due to an underwater earthquake.

The solution of the problem (1) can be analytically found by expanding a series of orthogonal functions, $f_n(z)$, the classic eigenfunctions of the constant depth homogeneous problem, which are locally valid even for the assumption of a mild slope sea bed ($\nabla h_b \ll kh_b$):

$$f_n(z) = \frac{\cosh[\beta_n(h+z)]}{\cosh(\beta_n h)}, \tag{3}$$

where β_n are the roots of the dispersion relation:

$$\beta_n = \begin{cases} n = 0 & \beta_n = \beta_0 & \omega^2 = g\beta_0 \tanh(\beta_0 h) \\ n \geq 1 & \beta_n = i\bar{\beta}_n & \omega^2 = -g\bar{\beta}_n \tan(\bar{\beta}_n h), \end{cases} \tag{4}$$

where overbar indicate the complex conjugate.

Indeed, the velocity potential in a compressible fluid is given by the summation of an infinite number of wave modes n; for each mode, the horizontal wavenumber is defined as

$$k_n = \sqrt{\frac{\omega^2}{c_s^2} + \beta_n^2}, \quad n = 0, 1, \ldots \tag{5}$$

Considering the relative magnitude of the two terms ω^2/c_s^2 an β_n^2, for each frequency ω, the wavenumbers could be real or imaginary. Therefore, a finite number of progressive modes and an infinite number of evanescent modes are solutions of (1). Conventionally, the wavenumber k_0 corresponds to the progressive gravity wave, while k_n with $n \geq 1$ corresponds to the progressive and evanescent hydro-acoustic waves.

Moreover, for each hydro-acoustic mode, the frequency $\frac{\omega^2}{c_s^2} = \beta_n^2$ represents the lower limit of the progressive wavenumbers. This cut-off frequency, ω_n, is given by approximating $\bar{\beta}_n h_b$ as $(n - 1/2)\pi$:

$$\omega_n = \frac{(2n-1)\pi c_s}{2h_b}, \tag{6}$$

and it corresponds to the natural frequency of the water layer with a depth of h_b.

Most research dealing with hydro-acoustic wave generation (Yamamoto 1982; Nosov 1999; Hendin and Stiassnie 2013; Kadri and Stiassnie 2012) model the sea-bed motion as a vertical displacement ζ that occurs in a finite time τ, with a constant velocity ζ_t. In the time domain, a constant sea-bed velocity ζ_t can be expressed by a rectangular function as follows:

$$\zeta_t = \frac{\zeta_0}{\tau} \mathrm{rect}\left(\frac{t}{2\tau}\right) \quad t \geq 0, \tag{7}$$

where ζ_0 is the final deformation of the sea-floor. In the frequency domain, its Fourier transform is

$$\tilde{\zeta}_t = \int_{-\infty}^{\infty} \zeta_t e^{-i\omega t} dt = \int_{-\tau/2}^{\tau/2} \frac{\zeta_0}{\tau} e^{-2\pi i f t} dt$$
$$= \frac{\zeta_0}{\pi f \tau} \frac{e^{\pi f \tau} - e^{-\pi f \tau}}{2i} = \zeta_0 \mathrm{sinc}(\pi f \tau). \tag{8}$$

The sinc function attains zero values for frequencies equal to

$$f = \frac{m}{\tau} \quad m = 1, 2, \ldots \tag{9}$$

At these frequencies, the source term is zero when the cut-off frequencies at which hydro-acoustic waves propagate (6) coincide with those of Eq. (9) and the earthquake energy is not transferred to the water layer as an acoustic wave. These conditions arise when τ is equal to m/f_n.

A numerical example is reported here to explain when these resonance conditions occur. Five numerical simulations have been performed solving the fluid velocity potential (problem 1) in a two-dimensional domain (x–z) and modeling the earthquake as a vertical rising displacement 10 km wide. For all five simulations, the water depth is constant over the entire domain and is equal to 1500 m, while the sea-bed velocity time series are different. Figure 1 shows on the left column the frequency spectra of five sea-bed velocity time series, and on the right column, the pressure signals generated from the corresponding five source terms at the sea bed 5 km away from the earthquake epicenter. The earthquake is modeled as a vertical rising displacement with constant velocity, ζ_t = 1 m/s, but with a sea-bed displacement time interval of τ = 1, 2, 4, 6, and 8 s in plot (a) to plot (e), respectively.

The vertical dashed lines represent the natural frequencies of the first six modes, which are invariant in all of the subplots given the constant water depth. The natural frequency of the first hydro-acoustic mode, f_1, is equal to 0.25 Hz. The plots (c) and (e) are related to a bottom motion with $\tau = 4$ s and $\tau = 8$ s, respectively, and, therefore, equal to $1/f_1$ and $2/f_1$. It can be noted that, for these sea-bed rising times, the source term function has zero values at the natural frequencies of the water layer. Under these conditions, the source term does not transfer energy to the eigen frequencies of the water layer dynamic system, as can be seen from the hydro-acoustic waves simulated at the sea bed (right column plots).

3. Numerical Simulations

Numerical computations for simplified conditions have been performed to investigate the resonant behavior cited in the previous section. All of the computations reported hereinafter have been carried out solving problem (1) by means of the finite-element method, for a two-dimensional (x–z) vertical domain. The earthquake is modeled as a constant velocity dislocation of a finite area along the x-axis, representing the dip dimension of the fault projected to the surface. The origin of the reference frame is at the undisturbed free surface and above the earthquake epicenter. For the simulations presented, the mesh is composed by triangular elements with a maximum size of 100 m; the time modeled is 100 s with a Δt = 0.1 s.

For the first set of parametric simulations, a fixed water depth is assumed $h_b(x) = 1.5$ km and the sea-bed dislocation has an horizontal extension of 30 km; the sea-bed rising time, τ, is varied, keeping constant the maximum sea-bed displacement ζ_0. The pressure signals generated by these parametric simulations have been analyzed for the time series extracted at the sea bottom 15 km away from the earthquake epicenter.

In Fig. 2, each point represents the spectral energy of the pressure signal for the parametric simulations. The energy E is obtained by integrating the density spectrum of the pressure signal over the frequency. It can be noted that the hydro-acoustic energy presents relative minimum values when the sea-bed rising time is exactly equal or a multiple of the natural period of the first hydro-acoustic mode, 4 s in this case.

As already noted by Nosov (1999) and confirmed from the results plotted in Fig. 2, the hydro-acoustic wave energy does not decrease monotonously with increasing τ, due to the modeling of the forcing term as a rectangular function, Eq. (7).

Let us now consider an earthquake model, where 10 km of sea bed rises vertically with constant velocity of 1 m/s up to ζ_0 = 1 m, during a time interval of τ = 1 s. When such earthquake occurs in an ocean 3 km deep, the generated hydro-acoustic waves will present the maximum oscillations at the natural frequencies of each acoustic propagating

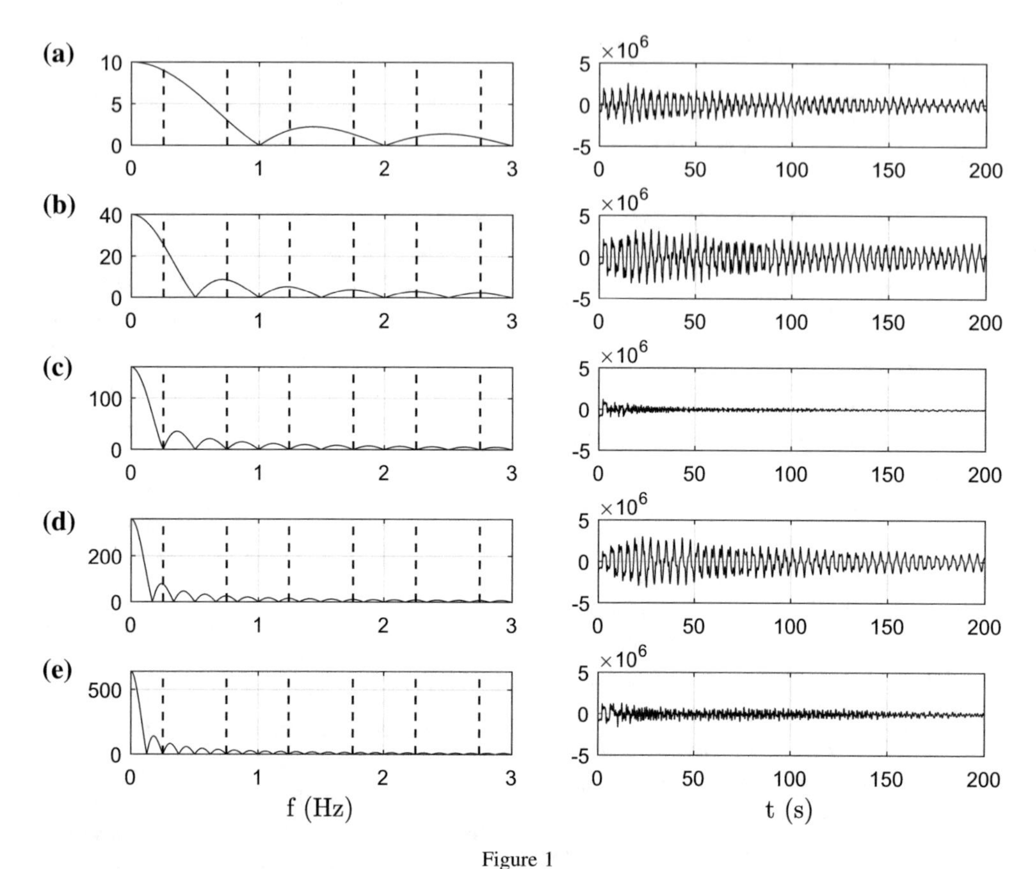

Figure 1
Left column: absolute values of the spectral density (m^2/s^2) of the forcing term $\zeta_t = \zeta_0/\tau$ for $t \leq \tau$, varying the sea bed rising time interval (τ = 1, 2, 4, 6, and 8 s, respectively, from **a–e**) while keeping constant the sea-bed velocity ζ_t = 1 m/s. The vertical dashed lines represent the first six natural frequencies of the water layer, 1.5 km deep. Right column: pressure signals (Pa) generated from the corresponding five source terms at the sea-bed 5 km away from the earthquake epicenter

mode, given by $f_n = (2n-1)c_s/4h_b$. The hydro-acoustic wave generated by this sea-bed movement is shown in Fig. 3. In the upper panel, the absolute values of the source term in the frequency domain are shown along with the small circles that represent the cut-off frequencies. In the lower panel, the continuous line represents the pressure spectrum measured at the sea bed 5 km away the epicenter. It is noted that the maximum values of this signal occur at the cut-off frequencies.

The dashed line in the lower panel indicates the envelope of the relative maxima of the pressure spectrum and is found to resemble the shape of the source term spectrum (upper panel). The results suggest that the modulation of the pressure spectra can provide information on the nature of the source term.

Further investigation on how to correlate the hydro-acoustic signal with the velocity of the sea-bed motion follows. For the same water depth h_b = 3 km, and the same 10 km of moving sea bottom, different kinematics of the wave generation mechanism are tested. In particular, keeping constant the rising time τ = 1 s, the rising velocity and the final sea-bed displacement are varied. The results, in terms of pressure spectrum at the bottom 5 km away from the earthquake epicenter, are presented in Fig. 4. Note that the *y*-axis is a logarithmic scale. From the darker to the lighter gray, the lines present the results for rising velocities equal to 1, 3, 5, 7, and 9 m/s, respectively. Since τ and h_b are identical in all five simulations, the pressure spectra represent all the same shape; however, as expected, their magnitude increases with increasing sea-bed velocity.

Furthermore, 40 simulations have been carried out for constant water depth $h_b = 3$ km, and an earthquake horizontal dimension of 10 km. The

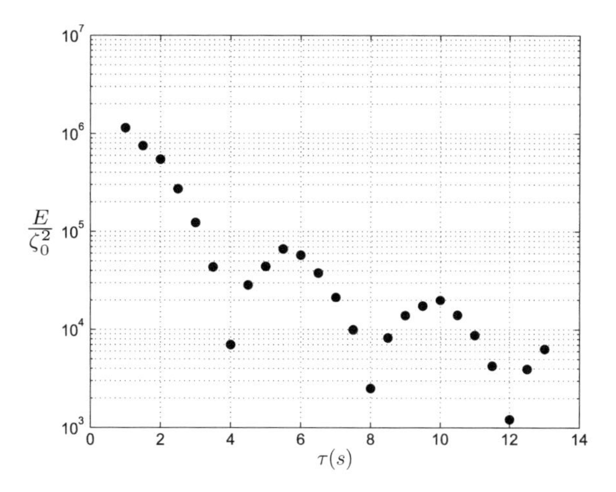

Figure 2
Spectral energy (E) of the pressure signal at the sea bottom 15 km from the epicenter, over the squared sea-bed vertical displacement (ζ_0) versus the sea-bed rising time τ. Numerical results of problem (1) are applied to a vertical sea section of constant depth equal to 1.5 km

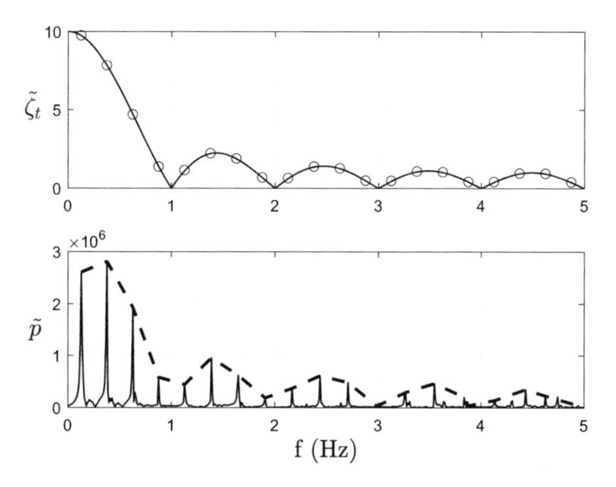

Figure 3
Upper panel: spectrum of the source term, modeled as a vertical displacement with a velocity of 1 m/s and a duration $\tau = 1$ s. The circles indicate the cut-off frequencies for the first 20 acoustic modes. Lower panel: spectrum of the pressure signal at the sea bottom (h_b = 3 km) 5 km away from the epicenter, the dashed line joins its maximum values

earthquake displacement and velocity are varied: for each fixed values of τ, from 1 s to 8 s, different values of sea-bed maximum displacement are tested ($\zeta_0 = 1, 3, 5, 7,$ and 9 m).

Figure 5 reports the integral of the energy density spectrum of the pressure signal at the sea bottom 5 km away from the epicenter, plotted against the sea-bed rising velocity. In the figure, the lines join the

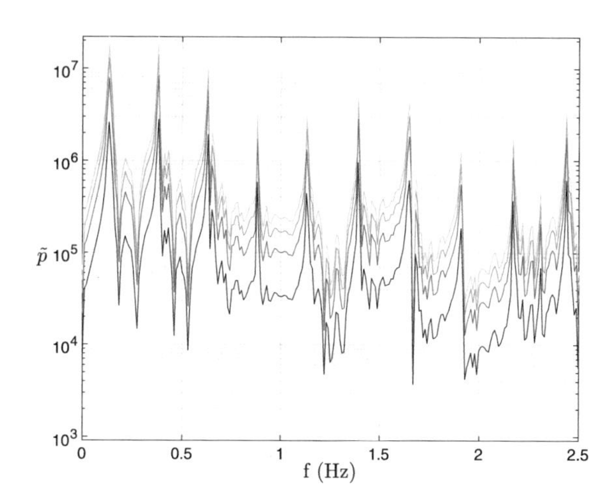

Figure 4
Spectrum of the pressure signal at the sea bottom (h_b = 3 km) 5 km from the earthquake epicenter. From the darker to the lighter gray, the lines represent the results for rising velocities equal to 1, 3, 5, 7, and 9 m/s, respectively, while τ is kept constant in all simulations equal to 1 s

values of the hydro-acoustic wave energy generated by sea-bed motion with the same value of τ. It is clear that the energy grows with increasing sea-bed rising velocity, for any value of τ. However, given a fixed sea-bed velocity, the energy increases with τ from τ = 1 s up to τ = 4 s (continuous lines); for sea-bed rising time from $\tau = 4$ s up to τ = 8 s (dashed lines), the energy decreases and it reaches a minimum value when τ = 8s, which corresponds to the natural period

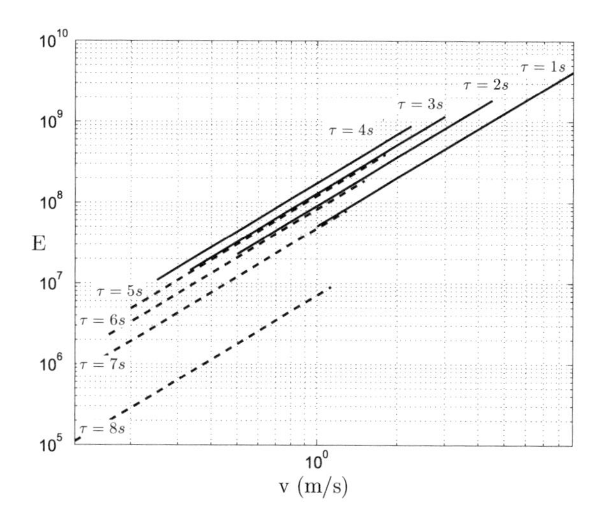

Figure 5
Lines join the values of the hydro-acoustic spectral energy varying with the sea-bed rising velocity for constant values of $\tau = 1, 2, 3, 4, 5, 6, 7,$ and 8 s

of the first hydro-acoustic mode. It can be concluded that there exist some conditions (resonance behavior) that amplify the energy transferred to the water layer ($\tau = 4$ s) and other conditions in which it is reduced ($\tau = 8$ s).

4. Conclusions

Results of a numerical investigation on hydro-acoustic wave features generated by fast sea-bed movements, such as those due to submarine earthquakes, have been presented in this paper. This analysis concerns the behavior of a weakly compressible water layer in the proximity of the generation area (near field). Explanation of the possible occurrence of resonance is given, showing which generation kinematics conditions produce an energy transfer from the sea-bed motion to the elastic oscillations of the water layer. The resonance condition arises for upward motion of the sea bed at constant velocity and for a finite duration τ, where τ is equal or a multiple of the natural period of the water layer.

The present study on the resonance behavior of the water layer in the near field allows identification of frequencies that carry significant energy content through hydro-acoustic waves. In this sense, wave propagation models can consider only those frequency values in the computations. The work of Sammarco et al. (2013) for example, presents a derivation of the depth-integrated equation, that is solved by dividing the spectrum into frequency bands and propagating each harmonics separately.

The results of this research highlight the importance of modeling and measurement of hydro-acoustic waves for innovative tsunami early warning detection and warning.

References

Abdolali, A., Cecioni, C., Bellotti, G., & Sammarco, P. (2014). A depth-integrated equation for large scale modeling of tsunami in weakly compressible fluid. *Coastal Engineering Proceedings, 1*(34), 736–748.

Abdolali, A., Cecioni, C., Bellotti, G., & Kirby, J. T. (2015a). Hydro-acoustic and tsunami waves generated by the 2012 Haida Gwaii earthquake: Modeling and in situ measurements. *Journal of Geophysical Research: Oceans, 120*(2), 958–971.

Abdolali, A., Cecioni, C., Kirby, J. T., Sammarco, P., Bellotti, G., & Franco, L.,. (2015b). Numerical modeling of low frequency hydro-acoustic waves generated by submarine tsunamigenic earthquake. In *International ocean and polar engineering conference*. International Society of Offshore and Polar Engineers.

Abdolali, A., Kirby, J. T., & Bellotti, G. (2015c). Depth integrated equation for hydro-acoustic waves with bottom damping. *Journal of Fluid Mechanics, 766*, R1.1–R1.12. https://doi.org/10.1017/jfm.2015.37.

Balanche, A., Guennou, C., Goslin, J., & Mazoyer, C. (2009). Generation of hydroacoustic signals by oceanic subseafloor earthquakes: A mechanical model. *Geophysical Journal International, 177*(2), 476–480.

Bolshakova, A., Inoue, S., Kolesov, S., Matsumoto, H., Nosov, M., & Ohmachi, T. (2011). Hydroacoustic effects in the 2003 Tokachi-oki tsunami source. *Russian Journal of Earth Science, 12*, ES2005(1–14).

Cecioni, C., Bellotti, G., Romano, A., Abdolali, A., & Sammarco, P. (2014). Tsunami early warning system based on real-time measurements of hydro-acoustic waves. *Procedia Engineering, 70*(C), 11–320.

Cecioni, C., Abdolali, A., Bellotti, G., & Sammarco, P. (2015). Large-scale numerical modeling of hydro-acoustic waves generated by tsunamigenic earthquakes. *Natural Hazards and Earth System Sciences, 15*(3), 627–636.

Chierici, F., Pignagnoli, L., & Embriaco, D. (2010). Modeling of the hydroacoustic signal and tsunami wave generated by seafloor motion including a porous seabed. *Journal of Geophysical Research: Oceans., 115*, C03015. https://doi.org/10.1029/2009JC005522.

Eyov, E., Klar, A., Kadri, U., & Stiassnie, M. (2013). Progressive waves in a compressible-ocean with an elastic bottom. *Wave Motion, 50*(5), 929–939.

Hendin, G., & Stiassnie, M. (2013). Tsunami and acoustic-gravity waves in water of constant depth. *Physics of Fluids, 25*(8), 086103(1–20).

Kadri, U., & Stiassnie, M. (2012). Acoustic-gravity waves interacting with the shelf break. *Journal of Geophysical Research: Oceans, 117*, C03035. https://doi.org/10.1029/2011JC007674.

Kadri, U. (2015). Acoustic-gravity waves interacting with a rectangular trench. *International Journal of Geophysics, 2015*, 806834. https://doi.org/10.1155/2015/806834.

Kajiura, K. (1970). Tsunami source, energy and the directivity of wave radiation. *Bulletin of the Earthquake Research Institute, 48*, 835–869.

Levin, B., & Nosov, M. (2009). *Physics of tsunamis*, vol. XI, pp. 327. Springer: Netherlands. https://doi.org/10.1007/978-1-4020-8856-8.

Maeda, T., & Furumura, T. (2013). FDM simulation of seismic waves, ocean acoustic waves, and tsunamis based on tsunami-coupled equations of motion. *Pure and Applied Geophysics, 170*(1–2), 109–127.

Miyoshi, H. (1954). Generation of the tsunami in compressible water (part I). *Journal of Oceanographical Society of Japan, 10*, 1–9.

Nosov, M. (1999). Tsunami generation in compressible ocean. *Physics and Chemistry of the Earth, Part B: Hydrology, Oceans and Atmosphere, 24*(5), 437–441.

Nosov, M., Kolesov, S. V., Denisova, A. V., Alekseev, A. B., & Levin, B. (2007). On the near-bottom pressure variations in the region of the 2003 Tokachi-Oki tsunami source. *Oceanology, 47*(1), 26–32.

Nosov, M., & Kolesov, S. (2007). Elastic oscillations of water column in the 2003 Tokachi-oki tsunami source: In-situ measurements and 3-D numerical modelling. *Natural Hazards and Earth System Science, 7*(2), 243–249.

Okal, E. A., Talandier, J., & Reymond, D. (2007). Quantification of hydrophone records of the 2004 Sumatra tsunami. In: K. Satake, E.A. Okal, J.C. Borrero (Eds.) *Tsunami and Its Hazards in the Indian and Pacific Oceans*, Pageoph Topical Volumes. Birkhäuser Basel. https://doi.org/10.1007/978-3-7643-8364-0_3.

Panza, G. F., Romanelli, F., & Yanovskaya, Tatiana B. (2000). Synthetic tsunami mareograms for realistic oceanic models. *Geophysical Journal International, 141*(2), 498–508.

Sammarco, P., Cecioni, C., Bellotti, G., & Abdolali, A. (2013). Depth-integrated equation for large-scale modelling of low-frequency hydroacoustic waves. *Journal of Fluid Mechanics, 722*. https://doi.org/10.1017/jfm.2013.153.

Sells, C. C. L. (1965). The effect of a sudden change of shape of the bottom of a slightly compressible ocean. *PhilosophicalTransactions of the Royal Society of London. Series A, Mathematical and Physical Sciences, 258*(1092), 495–528.

Smith, J. A. (2015). Revisiting oceanic acoustic gravity surface waves. *Journal of Physical Oceanography, 45*(12), 2953–2958.

Stiassnie, M. (2010). Tsunamis and acoustic-gravity waves from underwater earthquakes. *Journal of Engineering Mathematics, 67*(1–2), 23–32.

Yamamoto, T. (1982). Gravity waves and acoustic waves generated by submarine earthquakes. *International Journal of Soil Dynamics and Earthquake Engineering, 1*(2), 75–82.

(Received June 29, 2016, revised December 29, 2017, accepted January 4, 2018)

Pure Appl. Geophys.

https://doi.org/10.1007/s00024-018-1773-5

Pure and Applied Geophysics

Simulation of a Dispersive Tsunami due to the 2016 El Salvador–Nicaragua Outer-Rise Earthquake (M_w 6.9)

Yuichiro Tanioka,[1] Amilcar Geovanny Cabrera Ramirez,[2] and Yusuke Yamanaka[3]

Abstract—The 2016 El Salvador–Nicaragua outer-rise earthquake (M_w 6.9) generated a small tsunami observed at the ocean bottom pressure sensor, DART 32411, in the Pacific Ocean off Central America. The dispersive observed tsunami is well simulated using the linear Boussinesq equations. From the dispersive character of tsunami waveform, the fault length and width of the outer-rise event is estimated to be 30 and 15 km, respectively. The estimated seismic moment of 3.16×10^{19} Nm is the same as the estimation in the Global CMT catalog. The dispersive character of the tsunami in the deep ocean caused by the 2016 outer-rise El Salvador–Nicaragua earthquake could constrain the fault size and the slip amount or the seismic moment of the event.

Key words: Tsunami numerical simulation, the Boussinesq equations, outer-rise earthquake.

1. Introduction

A large earthquake (M_w 6.9) occurred at the outer rise of the Middle America trench off El Salvador and Nicaragua on November 24, 2016 (Fig. 1). The Global CMT solution shows that the mechanism of the earthquake was a normal fault type (strike = 127°, dip = 50°, rake = − 89°) with a centroid depth of 12 km and the seismic moment of 3.16×10^{19} Nm (M_w 6.9). A tsunami warning was issued along the coast of Nicaragua by the Nicaraguan Institute of Territorial Studies (INETER). A small tsunami was generated by the earthquake and was detected by only one ocean bottom pressure sensor, DART 32411 in Fig. 1. No damage was reported along the coast.

A large underthrust earthquake (M_w 7.3) occurred on August 27, 2012 near the 2016 earthquake (Fig. 2). The slip distribution of the 2012 earthquake was estimated using teleseismic body waves by Ye et al. (2013) and determined that a large slip occurred at the plate interface near the trench. Ye et al. (2013) also estimated a low radiated seismic energy ratio to the seismic moment (E_r/M_0) of 1.85×10^{-6} and suggested that this earthquake was a tsunami earthquake. Borrero et al. (2014) also concluded that the earthquake was a tsunami earthquake by analyzing the tsunami heights along the coast. Because the mechanism of the 2016 earthquake occurred in the outer-rise was a normal fault type, the 2016 event could be triggered by the 2012 earthquake which released the stress at the plate interface (Fig. 2).

Previously, Christensen and Ruff (1988) suggested that in the coupled subduction zone, including Central America, tensional outer-rise earthquakes occurred after large underthrust earthquakes at the plate interface. The most recent significant sequence is the occurrence of the 2007 outer-rise Kurile earthquake (M_w 8.0) after the 2006 underthrust Kurile earthquake (M_w 8.3) (Fujii and Satake 2008; Tanioka et al. 2008). Because the outer-rise event occurred at shallow depth with a high angle fault, the tsunami contains much higher frequency component than those tsunamis caused by typical underthrust earthquakes. Toh et al. (2011) shows that the far-field observed tsunami generated by the 2007 Kurile outer-rise earthquake had a dispersive character, but the that generated by the 2006 Kurile underthrust earthquake did not.

[1] Institute of Seismology and Volcanology, Hokkaido University, N10W8 Kita-ku, Sapporo, Hokkaido 060-0810, Japan. E-mail: tanioka@sci.hokudai.ac.jp
[2] Nicaraguan Institute of Territorial Studies, Managua, Nicaragua.
[3] Department of Civil Engineering, The University of Tokyo, Tokyo, Japan.

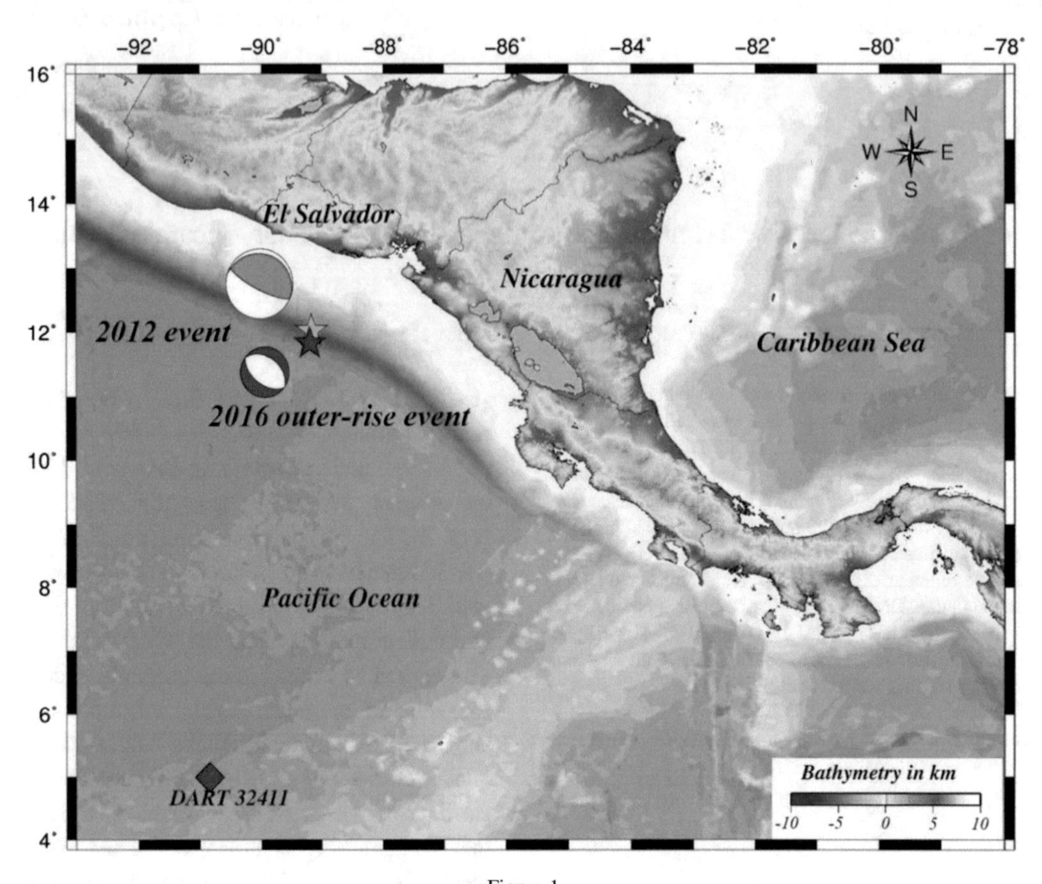

Figure 1
A map of the tsunami computational area with the epicenters (stars) and mechanisms of the 2012 and 2016 El Salvador–Nicaragua earthquakes. A diamond shows the location of the ocean bottom pressure sensor, DART 32411

A dispersive character of tsunamis caused by intraplate earthquakes was studied using three-dimensional tsunami simulation method by Saito and Furumura (2009). They concluded that the observed dispersive tsunami generated by the 2004 off Kii peninsula earthquake (M_w 7.4), intraplate earthquake occurred near the Nankai trough, was well modeled by the three-dimensional Navier–Stokes simulation and equally well modeled by the two-dimensional linear Boussinesq simulation but not by the linear long wave simulation. This indicates that intraplate earthquakes, including outer-rise events occurred near a trench, easily generate a short-wavelength deformation in a deep ocean and dispersive tsunamis.

The tsunami generated by the 2016 El Salvador–Nicaragua outer-rise earthquake had a dispersive character (Fig. 3). In this paper, the dispersive tsunami is modeled using appropriate tsunami numerical simulation methods. The source parameters of the earthquake are extracted from a dispersive character of the observed tsunami waveform.

2. Data and Fault Model

A tsunami waveform in Fig. 3 is obtained from the original observed waveform at an ocean bottom pressure sensor, DART 32411 shown in Fig. 1, by eliminating the tide estimated using the polynomial fitting technique. The sampling rate of the waveform is 15 s. A dispersive tsunami is clearly shown in Fig. 3 at about 60 min after the origin time of the earthquake.

To calculate the initial condition of the tsunami simulation, the Global CMT mechanism of the earthquake, strike = 127°, dip = 50°, and

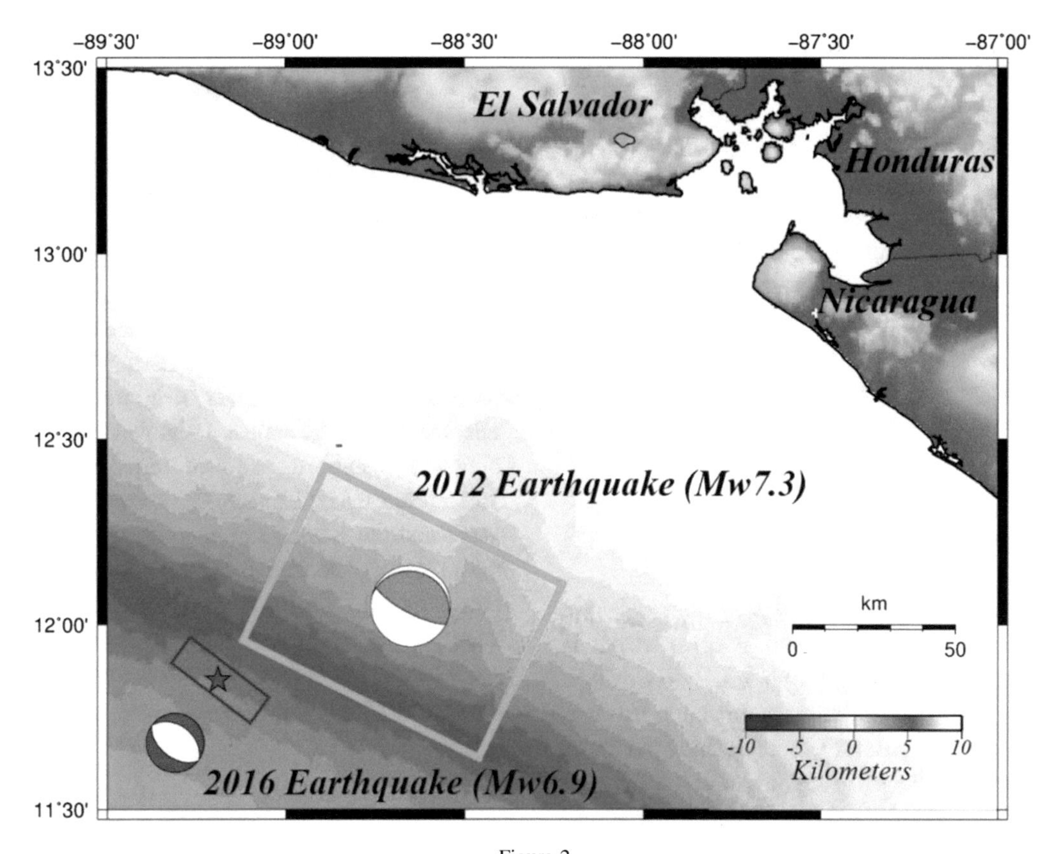

Figure 2
A large slip area (a green rectangular) of the 2012 underthrust earthquake estimated by Ye et al. (2013) and the fault model (a red rectangular) of the 2016 outer-rise earthquake estimated in this study

rake = − 89°, and a centroid depth of 12 km are used as the fault parameters of the earthquake. Because the mechanism is the pure normal fault type with a dip angle of about 45°, the calculated initial conditions are almost the same for two fault planes of the CMT solution (Fig. 2). Therefore, we arbitrary chose one of those fault planes. The fault length, L, and width, W, are calculated from scaling relationships for normal fault earthquakes described by Blaser et al. (2010). The corresponding relationships are $\log_{10} L = -1.61 + 0.46 M_w$ for the fault length and $\log_{10} W = -1.08 + 0.34 Mw$ for the fault width. The fault length and width is calculated to be 36.6 and 18.4 km, respectively, for the moment magnitude of 6.9. The vertical ocean bottom deformation due to the earthquake is computed from the above fault model using the equations of Okada (1985). This vertical deformation is used as the tsunami initial surface deformation.

3. *Tsunami Numerical Simulation Methods*

To compute a dispersive tsunami numerically, the linear Boussinesq equations are needed to be solved

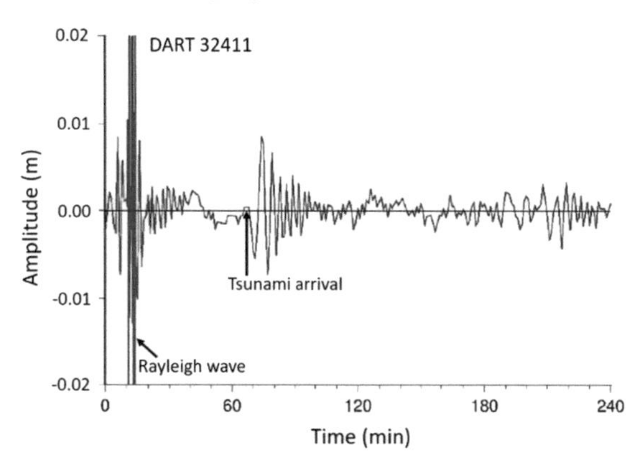

Figure 3
A tsunami waveform at the ocean bottom pressure sensor, DART 32411 in Fig. 1. In this waveform, the tidal components have been eliminated from the original waveforms

with an implicit scheme (e.g., Saito and Furumura 2009; Tanioka 2000). However, Imamura et al. (1990) developed a simple technique to use a numerical dispersion error caused by a finite difference scheme as a dispersion term in the Boussinesq equations. Using that technique, a dispersive tsunami is computed by solving the linear long wave equations instead of the linear Boussinesq equations. The numerical dispersion for the finite difference scheme is related to a grid spacing (Δx) and a time step (Δt). Then, the condition to describe a dispersion term with a numerical dispersion is called Imamura number, Im,

$$\mathrm{Im} = \frac{\Delta x}{2d}\sqrt{1 - c_0^2\left(\frac{\Delta t}{\Delta x}\right)^2} \approx 1, \quad (1)$$

where d is the ocean depth and $c_0 = \sqrt{gd}$ is the phase velocity. In this study, the average ocean depth is about 3 km, and a time step (Δt) is 2 s. By choosing a grid spacing (Δx) of 3 arc-minutes, Imamura number, Im, becomes 0.93. Therefore, first, we solve the linear long wave equations using a finite difference scheme with a staggered grid system of 3 arc-minutes grid spacing to model a dispersive tsunami. The 3 arc-minutes grid bathymetry data are made from the General Bathymetry Chart of the Oceans (GEBCO) dataset of 30 arc-seconds grid spacing.

Next, for more accurate simulation of a dispersive tsunami, the linear Boussinesq equations are numerically solved using a finite difference scheme with a staggered grid system (shown in Yamanaka et al. 2017). The governing equations are as follows (Goto 1991):

$$\frac{\partial \eta}{\partial t} + \frac{1}{R\cos\lambda}\left[\frac{\partial}{\partial \lambda} M\cos\lambda + \frac{\partial N}{\partial \theta}\right] = 0, \quad (2)$$

$$\frac{\partial M}{\partial t} + \frac{gh}{R}\frac{\partial \eta}{\partial \lambda} = -fN + \frac{1}{R}\left[\frac{\partial}{\partial \lambda}\left\{\frac{h^3}{3}F_1 + \frac{h^2}{2}F_2\right\} - \frac{\partial h}{\partial \lambda}\left\{\frac{h^2}{2}F_1 + hF_2\right\}\right], \quad (3)$$

$$\frac{\partial N}{\partial t} + \frac{gh}{R\cos\lambda}\frac{\partial \eta}{\partial \theta} = fM + \frac{1}{R\cos\lambda}\left[\frac{\partial}{\partial \theta}\left\{\frac{h^3}{3}F_1 + \frac{h^2}{2}F_2\right\} - \frac{\partial h}{\partial \theta}\left\{\frac{h^2}{2}F_1 + hF_2\right\}\right], \quad (4)$$

$$F_1 = \frac{1}{R\cos\lambda}\left\{\frac{\partial^2}{\partial t\partial \lambda}(u\cos\lambda) + \frac{\partial^2 v}{\partial t\partial \theta}\right\}, \quad (5)$$

$$F_2 = \frac{1}{R\cos\lambda}\left\{\frac{\partial}{\partial t}\left(u\cos\lambda\frac{\partial h}{\partial \lambda}\right) + \frac{\partial}{\partial t}\left(v\frac{\partial h}{\partial \theta}\right)\right\}, \quad (6)$$

where η is the water surface elevation, M (= uh) and N (= vh) are the fluxes in the λ and θ directions, u and v are the velocities in the λ and θ directions, h is the still ocean depth, g is the acceleration due to gravity, R is the radius of the Earth, f is the Coriolis parameter, λ is the latitude, θ is the longitude, and t is the time. The linear long wave equations are obtained using the above equations with an assumption that F_1 and F_2 are zero. The linear long wave and the Boussinesq equations were discretized based on the concepts of Goto et al. (1997) and Saito et al. (2014). The General Bathymetry Chart of the Oceans (GEBCO) dataset of 30 arc-seconds grid spacing was used for this tsunami numerical simulation.

4. Results

The dispersive tsunami is numerically computed using the simple method developed by Imamura et al. (1990) with the 3 arc-minutes grid spacing, and using the Boussinesq equations with the 30 arc-seconds grid spacing. For comparison, the tsunami is computed using the linear long wave equations with the 30 arc-seconds grid spacing. First, a slip amount of 1 m is assumed in those simulations to compare dispersive characters of tsunami waveforms. In Fig. 4, the tsunami waveform observed at DART 32411 in Fig. 1 is compared with three computed tsunami waveforms. The tsunami computed using the linear long wave equations does not explain the dispersive character of the observed tsunami waveform, and arrives at the station faster than the observed one. The tsunami computed using the simple method with Imamura number (Imamura et al. 1990) explains the dispersive tsunami, and arrives at the station at the same time as the observed one. The tsunami computed using the Boussinesq equations well explains the dispersive tsunami, and arrives at the station at the same time as the observed one.

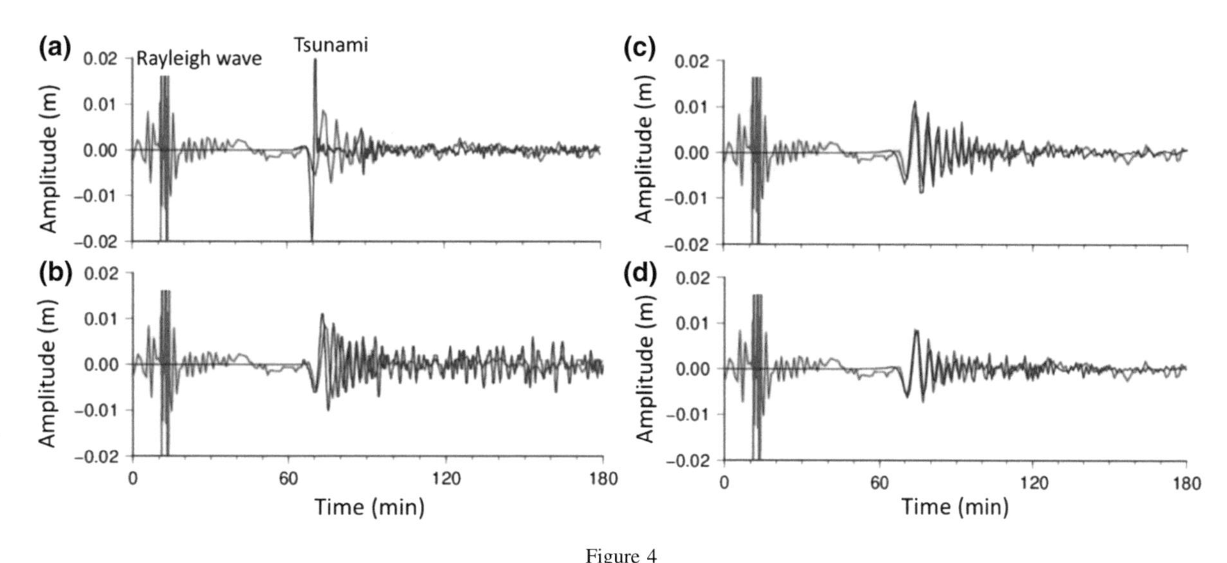

Figure 4
Comparisons of the observed and computed tsunamis, **a** computed tsunami using the linear long wave equations, **b** computed tsunami using Imamura number (Imamura et al. 1990), **c** computed tsunami using the linear Boussinesq equations, and **d** computed tsunami using the linear Boussinesq equations with the estimated slip amount of 0.74 m

Next, the slip amount of 0.74 m is calculated by comparing the amplitudes of the first negative and next large positive observed waves with the computed ones using the Boussinesq equation (Fig. 4d). The observed dispersive tsunami waveform is well explained by the computed one. The seismic moment is calculated to be 3.37×10^{19} Nm (M_w 6.9) by assuming a rigidity of 6.7×10^{10} N/m^2 which is a typical amount for outer-rise earthquakes (White et al. 1992; Shaw 1994). This seismic moment is similar to 3.16×10^{19} Nm (M_w 6.9) estimated in the Global CMT catalog.

Finally, we try to find the effect of the fault size on the dispersive character of tsunami and the seismic moment (or the slip amount). Previously, the fault length of 36.6 km and the fault width of 18.4 km are calculated using the scaling relationship for normal fault earthquakes described by Blaser et al. (2010). In this analysis, we keep the ratio of the fault length (L) and width (W) to be L:W = 2:1. Then, the fault size (L and W) is varied (L:W = 20 km:10 km, 30 km:15 km, 40 km:20 km, 50 km:25 km, and 60 km:30 km). The other fault parameters (strike, dip angle, rake, and the central location) are the same as those of the previous simulation. Tsunamis are again numerically computed using the linear Boussinesq equations. Figure 5 shows the comparisons of the observed tsunami waveform with the computed ones from the above five different fault sizes. The slip amount of each fault model is estimated by comparing the amplitudes of the first large positive and the next negative observed waves with the computed ones. Root-mean-square (RMS) residuals for observed and computed waveforms between 60 and 100 min after the origin time of the earthquake are calculated for five fault models and shown in Fig. 5. The computed tsunami from the fault model with a fault length of 20 km has more dispersive character than the observed tsunami as shown in Fig. 5a. The computed tsunamis from the fault models with fault lengths of 50 and 60 km have less dispersive character than the observed one (Fig. 5d, e). Finally, the computed tsunami from the fault model with a fault length of 30 km well explains the dispersive character of the observed tsunami (Fig. 5b). The RMS residual of 0.00178 m is the smallest for the computed tsunami from the fault model with a fault length of 30 km. Figure 6 shows a relationship of the seismic moment calculated from the slip amount of each fault model to the fault length (L). The calculated seismic moment of 3.16×10^{19} Nm for the best fault model with a fault length of 30 km is the same as the estimation in the Global CMT catalog. Figure 7 shows two snapshots of tsunami propagations using both the linear Boussinesq equations and the linear long wave equations at 30 min after the origin

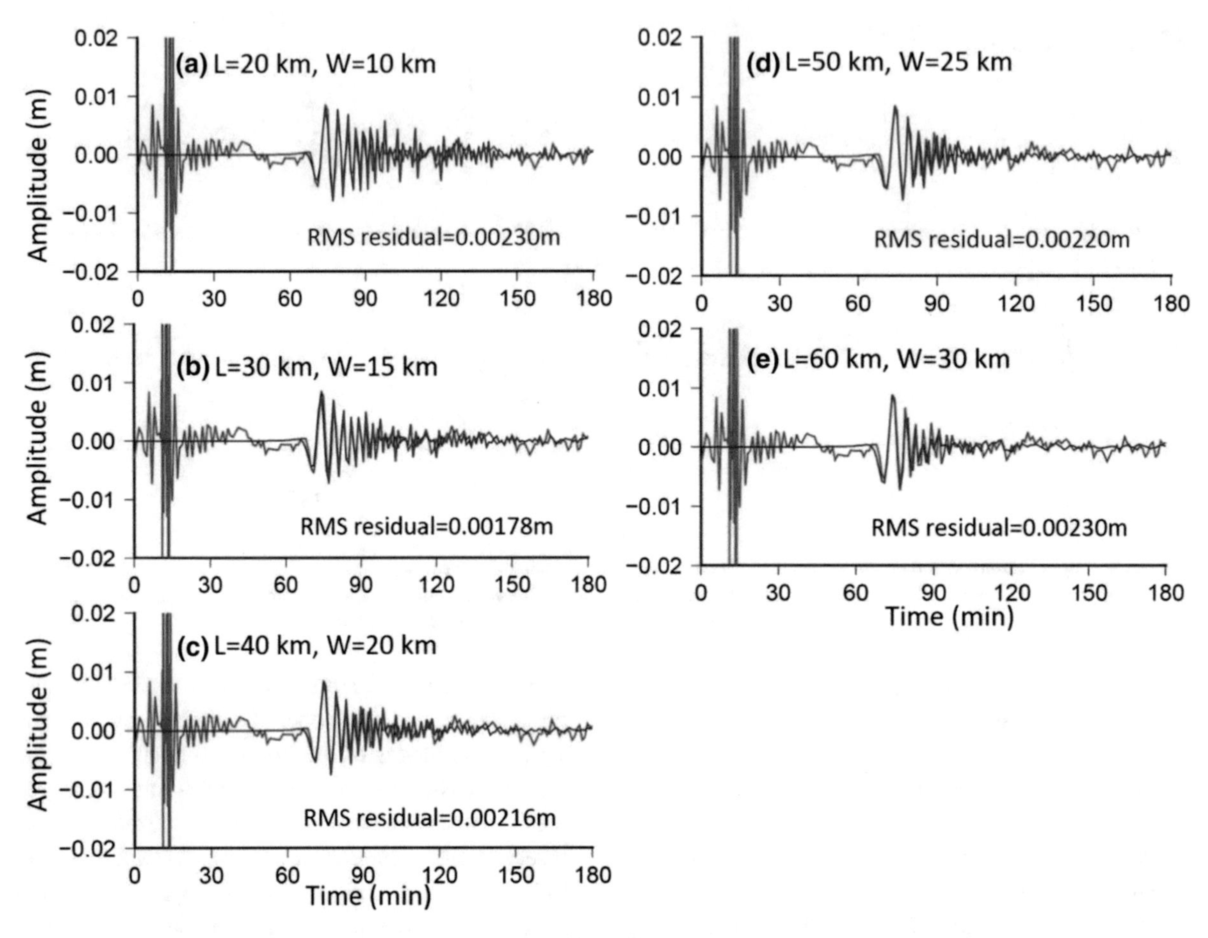

Figure 5
Comparisons of the observed and computed tsunamis using the linear Boussinesq equations, **a** for the fault length of 20 km and width of 10 km, **b** for the fault length of 30 km and width of 15 km, **c** for the fault length of 40 km and width of 20 km, **d** for the fault length of 50 km and width of 25 km, and **e** for the fault length of 60 km and width of 30 km. The RMS residual of observed and computed waveforms between 60 and 100 min for each fault model is also shown

time of the earthquake. The dispersive character of tsunami in all direction in the deep ocean is clearly seen in the snapshot for the liner Boussinesq simulation, but that is not seen in the snapshot for the linear long wave simulation.

5. Conclusions

The dispersive tsunami observed at the ocean bottom pressure sensor, DART 32411 in Fig. 1 was well simulated using the linear Boussinesq equations. The technique developed by Imamura et al. (1990) to use a numerical dispersion error caused by a finite different scheme for a dispersion term in the

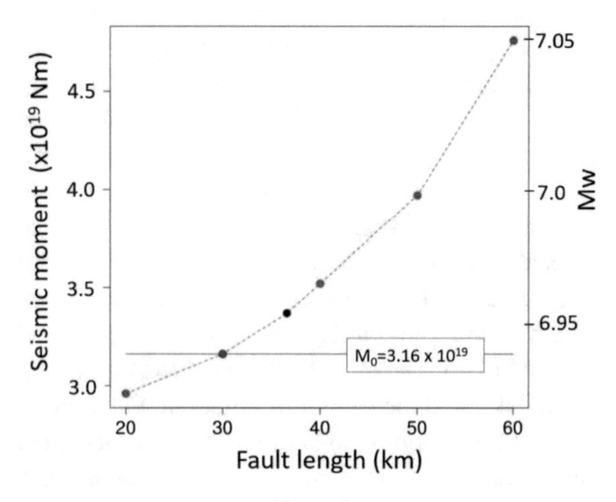

Figure 6
A relationship of the seismic moment calculated from the slip amount of each fault model to the fault length (L)

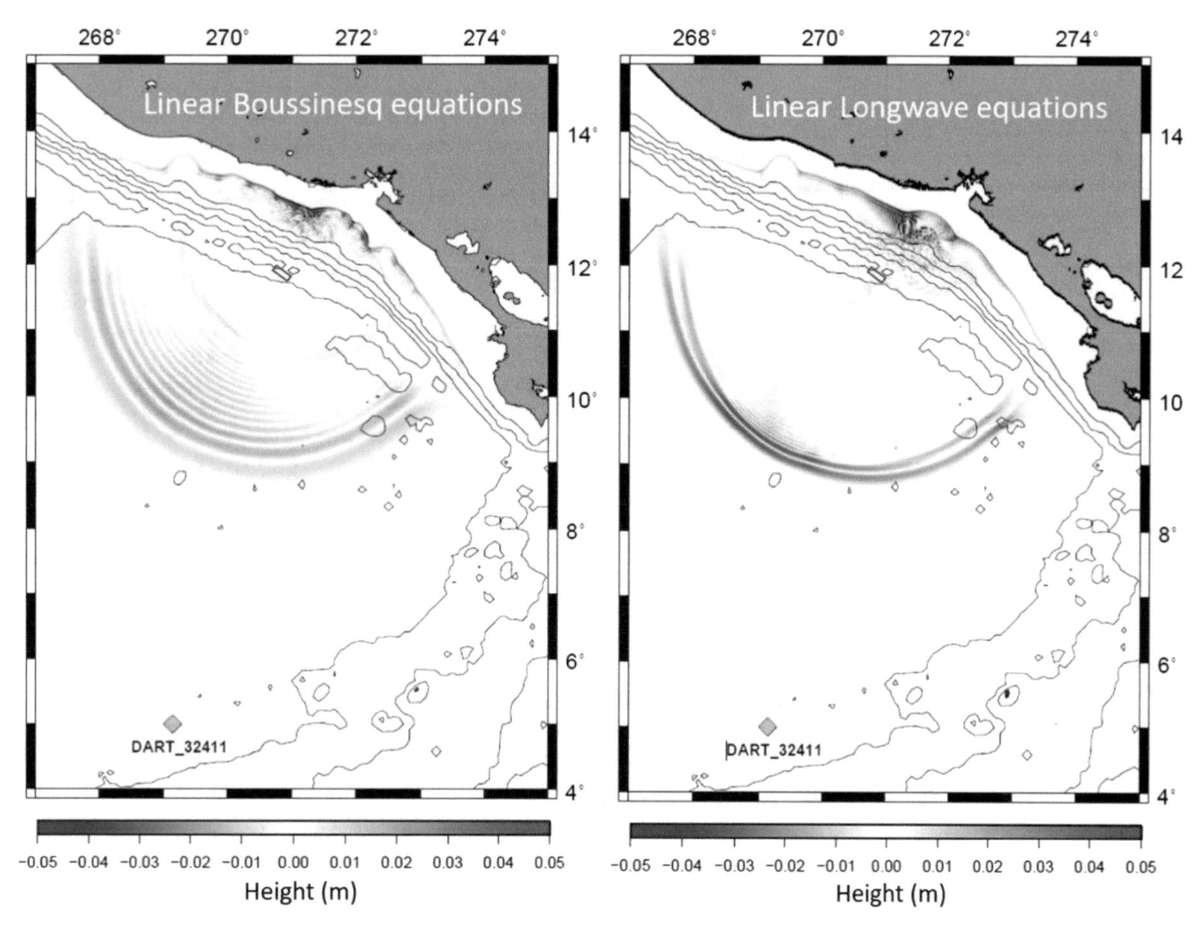

Figure 7
Two snapshots of tsunami propagations using both the linear Boussinesq equations (left) and the linear long wave equations (right) at 30 min after the origin time of the earthquake. Contours show the ocean depth. The contour interval is 1000 m. Red rectangles are the estimated fault model of the 2016 outer-rise earthquake

Boussinesq equations was also worked. From the dispersive character of tsunami waveform, the fault length and width was estimated to be 30 and 15 km, respectively (shown in Fig. 2). The seismic moment was calculated to be 3.16×10^{19} Nm which is the same as the estimation in the Global CMT catalog. The dispersive tsunami observed in the deep ocean caused by the 2016 outer-rise El Salvador–Nicaragua earthquake constrained the fault size and the slip amount or the seismic moment of the event.

Acknowledgements

Comments from two reviewers were helpful to improve the manuscript. Seitz' family provided a comfortable place to complete the manuscript. This study was supported by the Ministry of Education, Culture, Sports, Science and Technology (MEXT) of Japan, under its Earthquake and Volcano Hazards Observation and Research Program and JSPS KAKENHI Grant Number 16H01838.

References

Blaser, L., Krüger, F., Ohrnberger, M., & Scherbaum, F. (2010). Scaling relations of earthquake source parameter estimates with special focus on subduction environment. *Bulletin of the Seismological Society of America, 100*(6), 2914–2926.

Borrero, J. C., Kalligeris, N., Lynett, N. P., Fritz, H. M., Newman, A. V., & Convers, J. A. (2014). Observations and modeling of the August 27, 2012 earthquake and tsunami affecting El Salvador and Nicaragua. *Pure and Applied Geophysics, 171,* 3421–3435.

Christensen, D. H., & Ruff, L. J. (1988). Seismic coupling and outer rise earthquakes. *Journal of Geophysical Research: Solid Earth, 93*(B11), 13421–13444.

Fujii, Y., & Satake, K. (2008). Tsunami sources of the November 2006 and January 2007 great Kuril earthquakes. *Bulletin of the Seismological Society of America, 98*(3), 1559–1571.

Goto, C. (1991). Numerical simulation of the trans-oceanic propagation of tsunami. *Report of the Port and Harbour Research Institute, 30,* 4–19. **(in Japanese)**.

Goto, C., Ogawa, Y., & Shuto, N. (1997). Numerical method of tsunami simulation with the leap-frog scheme. IUGG/IOC Time Project Manuals Guides 35, UNESCO.

Imamura, F., Shuto, N., & Goto, C. (1990). Study on numerical simulation of the transoceanic propagation of tsunamis—part 2 characteristics of tsunami propagating over the Pacific Ocean, Zisin. *Journal of the Seismological Society of Japan, 43,* 389–402.

Okada, Y. (1985). Surface deformation due to shear and tensile faults in a half-space. *Bulletin of the Seismological Society of America, 75*(4), 1135–1154.

Saito, T., & Furumura, T. (2009). Three-dimensional simulation of tsunami generation and propagation: Application to intraplate events. *Journal of Geophysical Research, 114,* B02307. https://doi.org/10.1029/2007JB005523.

Saito, T., Inazu, D., Miyoshi, T., & Hino, R. (2014). Dispersion and nonlinear effects in the 2011 Tohoku-Oki earthquake tsunami. *Journal of Geophysical Research: Oceans.* https://doi.org/10.1002/2014jc009971.

Shaw, P. R. (1994). Age variations of oceanic crust Poisson's ratio: Inversion and a porosity evolution model. *Journal of Geophysical Research: Solid Earth, 99*(B2), 3057–3066.

Tanioka, Y. (2000). Numerical simulation of far-field tsunamis using the linear Boussinesq equation—the 1998 Papua New Guinea Tsunami. *Papers in Meteorology and Geophysics, 51,* 17–25. https://doi.org/10.2467/mripapers.51.17.

Tanioka, Y., Hasegawa, Y., & Kuwayama, T. (2008). Tsunami waveform analyses of the 2006 underthrust and 2007 outer-rise Kurile earthquakes. *Advances in Geosciences, 14,* 129–134.

Toh, H., Satake, K., Hamano, Y., Fujii, Y., & Goto, T. (2011). Tsunami signal from the 2006 and 2007 Kurile earthquakes detected at a seafloor geomagnetic observatory. *Journal of Geophysical Research, 116,* B02104. https://doi.org/10.1029/2010JB007873.

White, R. S., McKenzie, D., & O'Nions, R. K. (1992). Oceanic crustal thickness from seismic measurements and rare earth element inversions. *Journal of Geophysical Research: Solid Earth, 97*(B13), 19683–19715.

Yamanaka, Y., Tanioka, Y., & Shiina, T. (2017). A long source area of the 1906 Colombia–Ecuador earthquake estimated from observed tsunami waveforms. *Earth, Planets and Space, 69,* 163. https://doi.org/10.1186/s40623-017-0750-z.

Ye, L., Lay, T., & Kanamori, H. (2013). Large earthquake rupture process variations on the Middle America megathrust. *Earth and Planetary Science Letters, 381,* 147–155.

(Received November 24, 2017, revised December 30, 2017, accepted January 4, 2018)

Pure Appl. Geophys.

https://doi.org/10.1007/s00024-017-1746-0

Pure and Applied Geophysics

Ray Tracing for Dispersive Tsunamis and Source Amplitude Estimation Based on Green's Law: Application to the 2015 Volcanic Tsunami Earthquake Near Torishima, South of Japan

Osamu Sandanbata,[1] Shingo Watada,[1] Kenji Satake,[1] Yoshio Fukao,[2] Hiroko Sugioka,[3] Aki Ito,[2] and Hajime Shiobara[1]

Abstract—Ray tracing, which has been widely used for seismic waves, was also applied to tsunamis to examine the bathymetry effects during propagation, but it was limited to linear shallow-water waves. Green's law, which is based on the conservation of energy flux, has been used to estimate tsunami amplitude on ray paths. In this study, we first propose a new ray tracing method extended to dispersive tsunamis. By using an iterative algorithm to map two-dimensional tsunami velocity fields at different frequencies, ray paths at each frequency can be traced. We then show that Green's law is valid only outside the source region and that extension of Green's law is needed for source amplitude estimation. As an application example, we analyzed tsunami waves generated by an earthquake that occurred at a submarine volcano, Smith Caldera, near Torishima, Japan, in 2015. The ray-tracing results reveal that the ray paths are very dependent on its frequency, particularly at deep oceans. The validity of our frequency-dependent ray tracing is confirmed by the comparison of arrival angles and travel times with those of observed tsunami waveforms at an array of ocean bottom pressure gauges. The tsunami amplitude at the source is nearly twice or more of that just outside the source estimated from the array tsunami data by Green's law.

Key words: Ray tracing, tsunami propagation, dispersion, tsunami source, linear shallow-water waves, Green's law.

1. Introduction

Long-period tsunami waves are often approximated as shallow-water waves, whose velocities are determined only by water depth (e.g., Satake 1987). In reality, however, tsunami waves are dispersive, in other words, their velocities depend on not only water depth, but also wave frequency. Even for tsunamis induced by larger earthquakes, dispersion effects are not negligible for long-distance propagation (Glimsdal et al. 2006; Hossen et al. 2015). The dispersion effects are more significant for high-frequency or shorter waves as shown in real tsunami data; tsunami waves from a compact source such as induced by steep-dipping intraplate earthquakes or landslide are often dispersive (e.g., Baba et al. 2009; Saito et al. 2010; Glimsdal et al. 2006; Tappin et al. 2014). A very careful study by Glimsdal et al. (2013) summarized the importance of the dispersion effects on different types of tsunamis.

Ray tracing is commonly used in seismology and was applied to tsunami waves by Woods and Okal (1987) and Satake (1988). Their methods based on the eikonal equations can be used to investigate the bathymetry effects on tsunami propagation, such as ray refractions and travel times. However, their methods assume the shallow-water wave approximation that disregards the dispersion effects. Because tsunami velocities depend on frequency in reality, the ray paths or wavefronts should also depend on frequency. Lin et al. (2015) mapped tsunami phase and group travel times over different frequency ranges by array data analysis and showed that clear dispersion can be seen in a deep ocean area. To treat dispersive tsunamis, the ray-tracing method needs to be extended to dispersive waves.

Tsunami ray tracing has also been used to estimate source amplitude in combination with Green's law, which is derived from the conservation law of energy flux (e.g., Wiegel 1970; Murty 1977). Green's law relates tsunami amplitudes at two locations on the same ray path, and sometimes is used to estimate an initial tsunami amplitude at the source (e.g., Abe

[1] Earthquake Research Institute, The University of Tokyo, Tokyo, Japan. E-mail: osm3@eri.u-tokyo.ac.jp

[2] Japan Agency for Marine-Earth Science and Technology (JAMSTEC), Kanagawa, Japan.

[3] Kobe University, Kobe, Japan.

1973; Fukao and Furumoto 1975; Satake 1988). However, it may not be appropriate to connect amplitudes inside and outside the source region, because half of the potential energy of the initial sea-surface displacement shall be transformed into kinematic energy and tsunami energy propagates away in different directions out of the source.

The aim of this paper is to propose two improved methods for tsunami analysis, and to apply them to a real tsunami event. The first one is a ray tracing method extended to dispersive waves. Our application shows clear dispersion effects on wave propagations, and its validity is confirmed through comparison with observations. The second one is an extended method of Green's law to estimate the source amplitude. We argue that Green's law may not be applied inside the source region by showing analytical solutions of the linear shallow-water wave equations, and suggest taking into account the amplitude reduction rates near the source region to preliminarily estimate the source amplitude. As an example, we apply these improved methods to tsunamis caused by an earthquake in 2015 at a submarine volcano named Smith Caldera, near Torishima, Japan.

2. *The 2015 Torishima Earthquake Near Smith Caldera*

An earthquake occurred on May 1, 2015 (UTC), at a shallow part of a submarine volcanic body named Smith Caldera, along the Izu–Bonin arc (Fig. 1). Hereafter we call it "the 2015 Torishima earthquake". The focal mechanism reported in the Global CMT Catalog is composed of dominant compensated-linear-vector-dipole (CLVD) components (e.g., Aki and Richards 1980; Kanamori et al. 1993; Shuler et al. 2013a, b). Although the seismic moment magnitude of this earthquake is reported as a moderate value of M_w 5.7 by the U.S. Geological Survey, large tsunami waves were observed at islands on the arc, for example, a wave amplitude of 0.5 m at Hachijo Island, 180 km away from the epicenter (reported by Japan Meteorological Agency), which is unusually large for a tsunami from an earthquake with M less than 6. Because of its efficiency of tsunami generation, this earthquake may be regarded as a tsunami earthquake, as named by Kanamori (1972).

Tsunami waves from the 2015 Torishima earthquake were observed by an array of ocean bottom pressure (OBP) gauges deployed 100 km northeast away from the epicenter (Fig. 1b) (Fukao et al. 2016). The array consists of ten OBP stations distributed in a shape of a triangle, with a station distance of 10 km (Fig. 1b). Tsunami waves at all stations have similar waveforms characterized by a crest of about 2 cm amplitude. The data are of high quality, because the array was located relatively close to the epicenter and at deep oceans where near-shore effects are weak, hence direct information of the tsunami source can be inferred once the propagation effects are properly evaluated.

In this paper, we focus on tsunami propagation from the 2015 Torishima earthquake. As shown later in Sect. 3, the observed waves are strongly dependent on wave frequency, which is comparable to our simulated results by our ray tracing. In Sect. 4, the observed amplitude of 2 cm is used to estimate the source amplitude using extended Green's law. The mechanism of the earthquake is discussed in a separate paper (Fukao et al. in revision).

3. *Ray Tracing for Dispersive Tsunami*

3.1. *Mapping Dispersive Tsunami Velocity*

3.1.1 *Iterative Calculation of Dispersive Tsunami Velocity*

The phase velocity of dispersive tsunamis C is given by two forms of

$$C = \frac{\omega}{k} = \frac{2\pi f}{k}, \tag{1}$$

and

$$C = \frac{\omega}{k} = \sqrt{\frac{g}{k}\tanh kD} = \sqrt{\frac{g\lambda}{2\pi}\tanh\frac{2\pi D}{\lambda}}, \tag{2}$$

where ω, f and D represent angular frequency, frequency, water depth, respectively, and k, λ and g are wavenumber, wavelength and gravitational

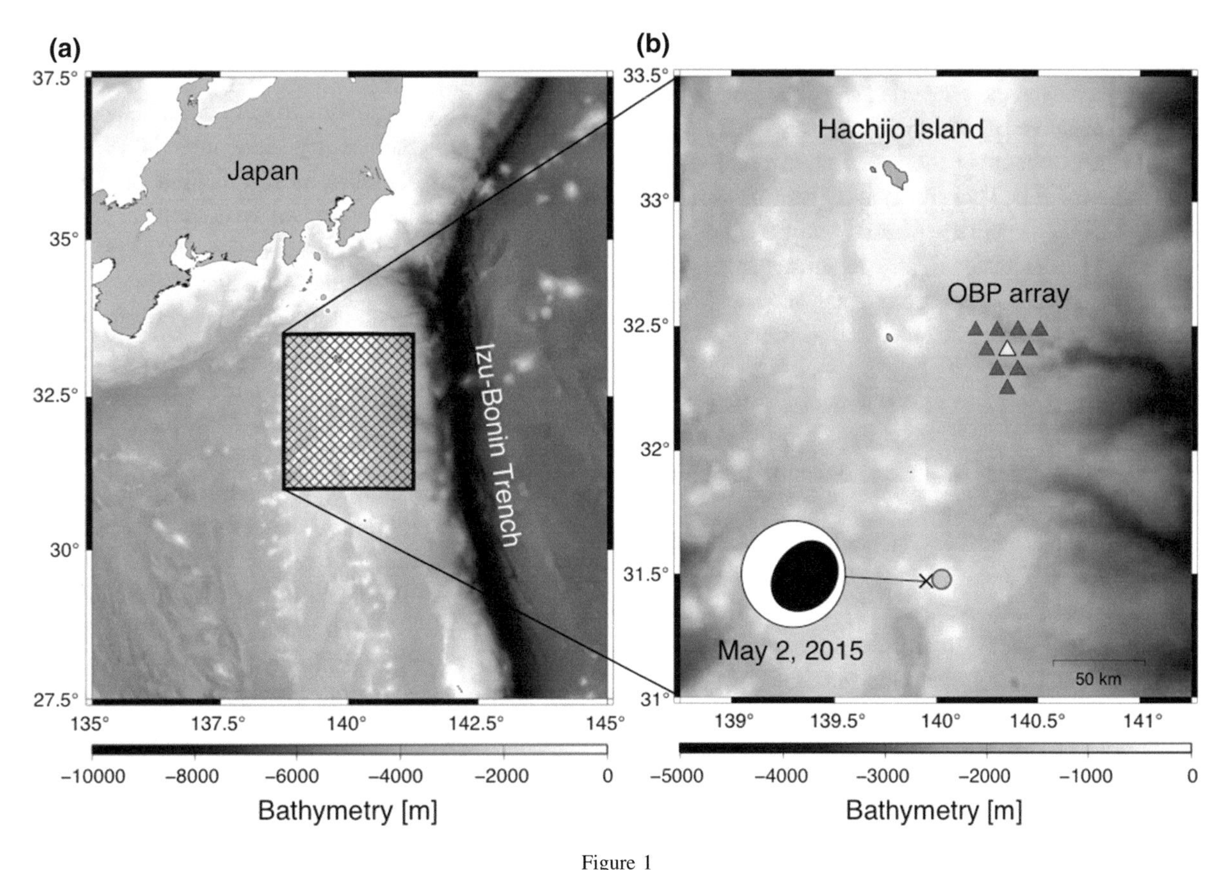

Figure 1

a Bathymetry of a southern ocean area of Japan along Izu–Bonin Trench. **b** Epicenter of the Torishima earthquake is shown in the black cross sign with its focal mechanism (data from the Global CMT Catalog). The rim geometry of Smith Caldera is approximatedly indicated by the red circle. Ocean bottom pressure (OBP) gauges are indicated by red or yellow triangles, and the yellow one is A05 station

acceleration, respectively (e.g., Lamb 1932; Dingemans 1997).

For ray tracing of dispersive tsunamis, phase velocity $C(f, D)$ for a given pair of frequency f and depth D is needed. However, C may not be calculated directly with Eqs. (1) or (2), because C has an unknown variable of wavenumber k. Therefore, to calculate wavenumber $k(f, D)$ for a given pair of frequency f and depth D, we first solve the combining form of Eqs. (1) and (2),

$$\frac{2\pi f}{k} = \sqrt{\frac{g}{k}\tanh kD}. \tag{3}$$

Once we have $k(f, D)$, phase velocity $C(f, D)$ can be calculated with Eqs. (1) or (2).

Because Eq. (3) may not be solved analytically, we solved it iteratively using a recursive algorithm originally developed by Watada (2013). We started iterative calculations,

$$k_n = \frac{2\pi f}{C_{n-1}}, \quad n = 1, 2, 3, \ldots \tag{4}$$

where

$$C_n = \sqrt{\frac{g}{k_n}\tanh k_n D} \quad n = 1, 2, 3, \ldots \tag{5}$$

with an initial value given by

$$C_0 = \sqrt{gD}. \tag{6}$$

The calculations were repeated until the dimensionless condition for convergence is satisfied,

$$\frac{|k_n - k_{n-1}|}{k_n} < \Delta c \frac{k_n}{2\pi f} = \frac{\Delta c}{C_{n-1}}, \tag{7}$$

where Δc is set 0.0001 m/s in this study, which sufficiently achieves the convergence of the values.

The recursive algorithm for iterative solution is equivalent to obtaining the intersection point of the two curves of Eqs. (1) and (2) at a given frequency f and depth D. A schematic illustration of the recursive algorithm to calculate $k(f, D)$ is shown in Fig. 2a, where f and D is 20.0 mHz and 1000 m, respectively. The x-component of the intersection point, $k(f, D)$, yields the solution of Eq. (3), and it is easily confirmed that the two curves have only one intersection point by showing the function $2\pi f/k - \sqrt{(g/k)}\tanh kD$ monotonically decreases as wavenumber k increases.

Using the iterative calculation, wavenumber $k(f, D)$ for a given pair of f and D can be computed, which yields phase velocity $C(f, D)$ of dispersive tsunamis with Eqs. (1) or (2). Examples of dispersion curves calculated by the algorithm are shown in Fig. 2b, in comparison with the phase velocities of shallow-water waves, which are constant for all frequencies.

3.1.2 Velocity Maps of Dispersive Tsunamis

Phase velocity maps at different frequency can be created from bathymetry data by the recursive algorithm. To demonstrate the method of mapping dispersive tsunami velocities, we used the bathymetry data JTOPO30, which contains 30 arcsec grid bathymetry data available from the Marine Information Research Center of the Japan Hydrographic Association (http://www.mirc.jha.or.jp/products/JTOPO30/). Figure 1b shows the calculation area including Smith Caldera and OBP array stations along the Izu–Bonin trench. The ocean area is characterized by a steep slope of seafloor from shallow areas on the Izu–Bonin ridge toward deep oceans along the trench.

We mapped the velocities at nine fixed frequencies from 2.0 to 10.0 mHz, with an interval of 1.0 mHz, among which those at 2.0, 5.0 and 10.0 mHz are shown in Fig. 3b–d. The frequency range is comparable to the dominant frequency range of observed tsunamis of the 2015 Torishima earthquake (Fukao et al. 2016). For comparison, the velocity map calculated with the linear shallow-water wave velocity, $C = \sqrt{gD}$, is also shown in Fig. 3a.

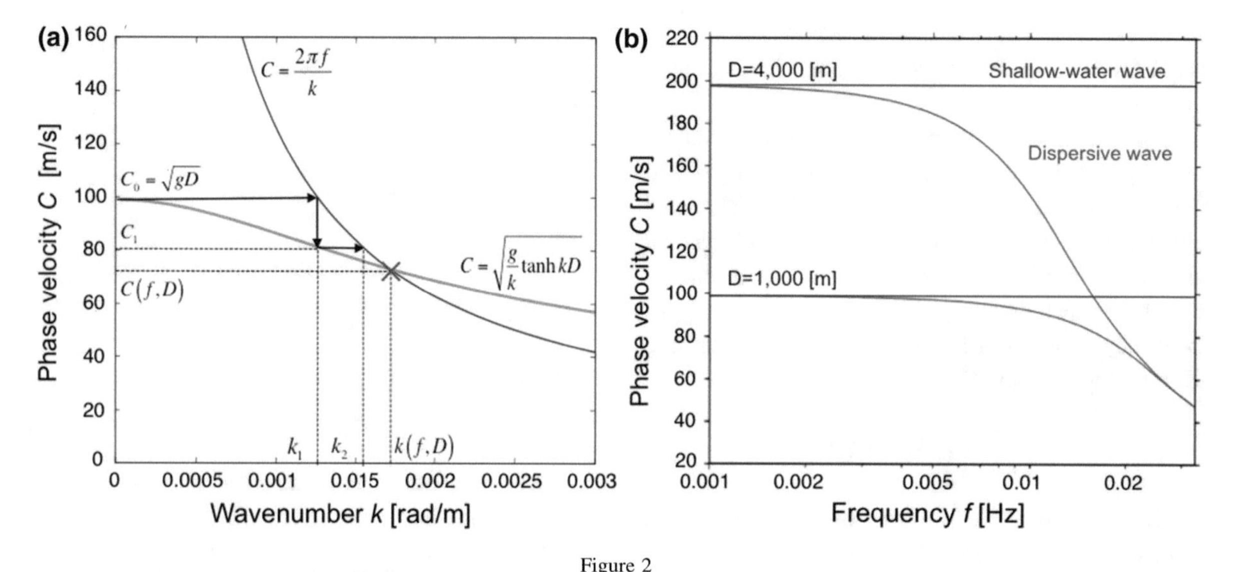

Figure 2

a Schematic illustration of the recursive algorithm to solve Eq. (3), where a fixed frequency f is 20 mHz and a fixed depth D is 1000 m. The green and purple curves represent Eqs. (1) and (2), respectively. The red cross sign is the intersection point of the two curves. **b** The phase speed curves of dispersive waves calculated by the iterative algorithm (red) and the shallow-water waves (blue) at depths of 1000 and 4000 m

A significant phase velocity contrast can be seen at lower- and higher-frequency waves particularly at deep oceans along the trench. Phase velocities at lower-frequency increase as the water depth increases (2.0 mHz: Fig. 3b), while velocities at higher-frequency are less sensitive to the water depth so that the velocity map tends to be homogeneous despite the large horizontal variation of the bathymetry (10.0 mHz: Fig. 3d).

3.2. *Ray Tracing*

Ray paths at different fixed frequencies, including the dispersion effects, are traced by numerical integrations. The ray equations of seismic surface waves in isotropic media are based on the eikonal equations, which are derived from the solution to the 2D Helmholtz wave equation (e.g., Wielandt 1993; Shearer 1999). The ray equations are expressed in general coordinates as

$$\frac{\mathrm{d}x_i}{\mathrm{d}s} = Cn_i, \tag{8}$$

$$\frac{\mathrm{d}n_i}{\mathrm{d}s} = -\frac{1}{C^2}\frac{\partial C}{\partial x_i}, \tag{9}$$

where C is the phase velocity, and x_i and n_i the ith components of location and slowness vector of ray path, respectively, and ds represents line element of the ray (e.g., Aki and Richards 1980; Julian 1970; Yomogida and Aki 1985). For the application to tsunami waves, it is useful to change the forms into the spherical coordinates;

$$\frac{\mathrm{d}\theta}{\mathrm{d}s} = \frac{1}{R}\cos\zeta, \tag{10}$$

$$\frac{\mathrm{d}\phi}{\mathrm{d}s} = \frac{1}{R}\frac{\sin\zeta}{\sin\theta}, \tag{11}$$

$$\frac{\mathrm{d}\zeta}{\mathrm{d}s} = \frac{\sin\zeta}{R}\frac{1}{C}\frac{\partial C}{\partial\theta} - \frac{\cos\zeta}{R\sin\theta}\frac{1}{C}\frac{\partial C}{\partial\phi} - \frac{\sin\zeta\cot\theta}{R}, \tag{12}$$

where θ and ϕ are colatitude and longitude of the ray at each time step, R is the radius of the earth, 6371 km in this study, ζ is the ray direction measured counter-clockwise from the south, and ds represents line element of the ray, which is calculated by $\mathrm{d}s = C \cdot dt$.

To trace ray paths at different frequencies, the ray equations, Eqs. (10)–(12), were integrated by the fourth-order Runge–Kutta method with a time step dt of 1.0 s. For phase velocity data, the 30 arcsec grid phase velocity maps at each frequency are used (Fig. 3). Their gradient fields were calculated by the central difference scheme after applying a spatial Gaussian filtering as explained in the following paragraph. Interpolation was carried out by the weighted mean using four neighboring points, which yields velocity and its gradient fields continuous along the rays.

Cares must be given to horizontal scales of velocity gradients, because the ray equations are valid only if

$$\frac{\delta' C}{C/\lambda} \ll 1, \tag{13}$$

where $\delta' C$ is the change in the gradient of velocity over a wavelength λ (Officer 1974). In other words, horizontal scales of velocity heterogeneities must be much longer than a wavelength. Equation (13) is the condition for the eikonal equation, which yields the ray equations (Eqs. 10–12), to be a good approximation of the wave equations. Otherwise, the ray directions may change abruptly at small-scale heterogeneities of velocities. To avoid the problem, we smoothed the velocity gradient maps by applying the Gaussian filter. Because wavelengths are different for different frequencies, the spatial parameter or standard deviation of the Gaussian function is set at each frequency. We defined the spatial parameter at each frequency $\sigma(f)$ in the following relationship;

$$\sigma(f) = \frac{\lambda_{\mathrm{ref}}(f)}{2} = \frac{C(f, D_{\mathrm{ref}})}{2f}, \tag{14}$$

where $C(f, D_{\mathrm{ref}})$ is the phase velocity at fixed frequency f and at reference depth D_{ref}, which is calculated by the recursive algorithm explained in Sect. 3.1. Although wavelengths also change at different depth and frequency, we simply assume the constant reference depth of 1000 m following a roughly average depth in the area between the epicenter and the OBP array. After trying different values of $\sigma(f)$, i.e., $\lambda_{\mathrm{ref}}(f)/4$ and $\lambda_{\mathrm{ref}}(f)$, we chose the value defined by Eq. (14), because it shows good balances between smoothing smaller-scale and preserving larger-scale bathymetry heterogeneities compared to scales of the reference wavelength at each frequency. For shallow-water waves, the

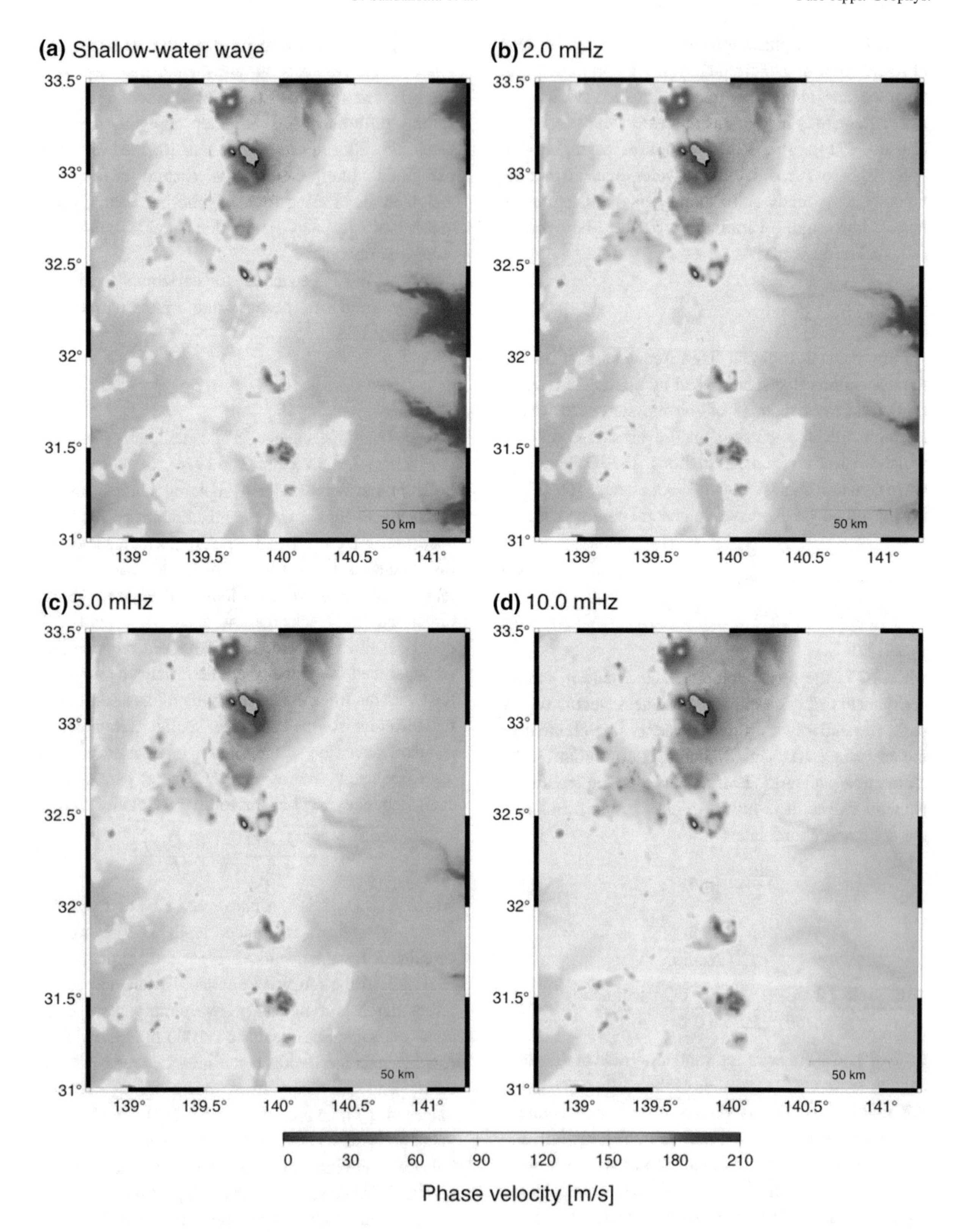
(a) Shallow-water wave
(b) 2.0 mHz
(c) 5.0 mHz
(d) 10.0 mHz
33.5°
33°
32.5°
32°
31.5°
31°
139°
139.5°
140°
140.5°
141°
50 km
0
30
60
90
120
150
180
210
Phase velocity [m/s]

◀Figure 3
Phase velocity maps for the region shown in Fig. 1b for **a** shallow-water wave and dispersive waves at a fixed frequency of **b** 2.0 mHz, **c** 5.0 mHz, and **d** 10.0 mHz

reference wavelength is defined by $\sqrt{gD_{\mathrm{ref}}}/f_{\mathrm{SWW}}$, where f_{SWW} is 2.0 mHz, because waves at 2.0 mHz have a wavelength of around 49 km at 1000 m depth, which is long enough for the shallow-water approximation. The velocity field itself has not been smoothed, to evaluate travel times more precisely.

It may be confusing that we use a phase velocity, but not group velocity for the ray equations. In this study we trace rays to visualize ray paths or phase travel times at each fixed frequency. Therefore, the line element $\mathrm{d}s$ is calculated by $\mathrm{d}s = C \cdot \mathrm{d}t$ in Eqs. (10)–(12). The same ray tracing equation can be used to trace a wave packet of dispersive tsunamis by using the group velocity instead of the phase velocity for the line element (e.g., Yomogida and Aki 1985; Červený 2001).

3.3. *Application to Tsunamis Caused by 2015 Torishima Earthquake*

The ray tracing for dispersive tsunamis was applied to the 2015 Torishima earthquake. In each phase velocity map created in Sect. 3.1, ray paths were traced from a point source at the center of Smith Caldera on the basis of tsunami source analysis by Fukao et al. (2016). The paths were emitted radially with the angle interval of 2° to retain clear visualization, and stopped if a ray approaches low velocity zones slower than 30 m/s, to avoid unrealistic rays at sharp velocity changes near coast on islands.

The results show that ray trajectories vary depending on frequencies (Fig. 4). First, ray paths at higher-frequency tend to focus along the shallow ridge running from south to north, because higher-frequency waves are more affected by small-scale horizontal variations of bathymetry (Fig. 4c, d). On the other hand, lower-frequencies or shallow-water waves are less affected by small structures on the seafloor, so that rays travel in a smooth way (Fig. 4a, b).

What is noted here is the strong dispersion, in particular, at the deep sea area close to the trench running along the north–south direction in the east of the calculation area. Although travel times on rays northward or southward has little difference at every frequency, there is a significant discrepancy in travel time in deep water. It can be seen easily if we compare wavefronts at the OBP array at different frequencies. First, travel times are smaller at lower-frequency. The tsunami travel times to A05 station at 2.0, 5.0 and 10.0 mHz are around 850, 880 and 950 s, respectively. Second, the slowness direction of wavefront at the array also changes from northwestward to northward, as the frequency becomes higher (see wavefront around the array at travel time of 1000 s in Fig. 4b–d). This frequency-dependent trend is attributed to the strong dispersion effect at the deep oceans close to the Izu–Bonin trench. In other words, lower-frequency waves travel faster in deep sea, and have a higher contrast in velocity between the shallow ridge and the deep trench, which results in significant wavefront refractions.

To measure the travel time and propagation direction at each frequency, Fukao et al. (2016) picked up phase-peak arrival times from band-pass filtered waves. Using the travel times at ten stations (Fig. 1b), they determined the slowness vector (Sx, Sy) and travel times T_0 at A05 at each frequency by the least squares method. Their phase analysis revealed that phase travel times to A05 station are longer at higher-frequency (Fig. 5a: blue line), and that slowness directions, $\theta = \tan^{-1}\frac{S_y}{S_x}$, deviate from the great-circle path from the epicenter (the array was located in the direction of around 80° measured counter-clockwise from east) (Fig. 5b: blue line).

We compare slowness directions θ and travel times T_0 at A05 station determined by our ray tracing method, with their observation (Fig. 5). We can see both travel times and slowness directions of our ray tracing are consistent with their analysis. This confirms the validity of our method to examine the dispersion effects on tsunami propagation as well as the bathymetry effects.

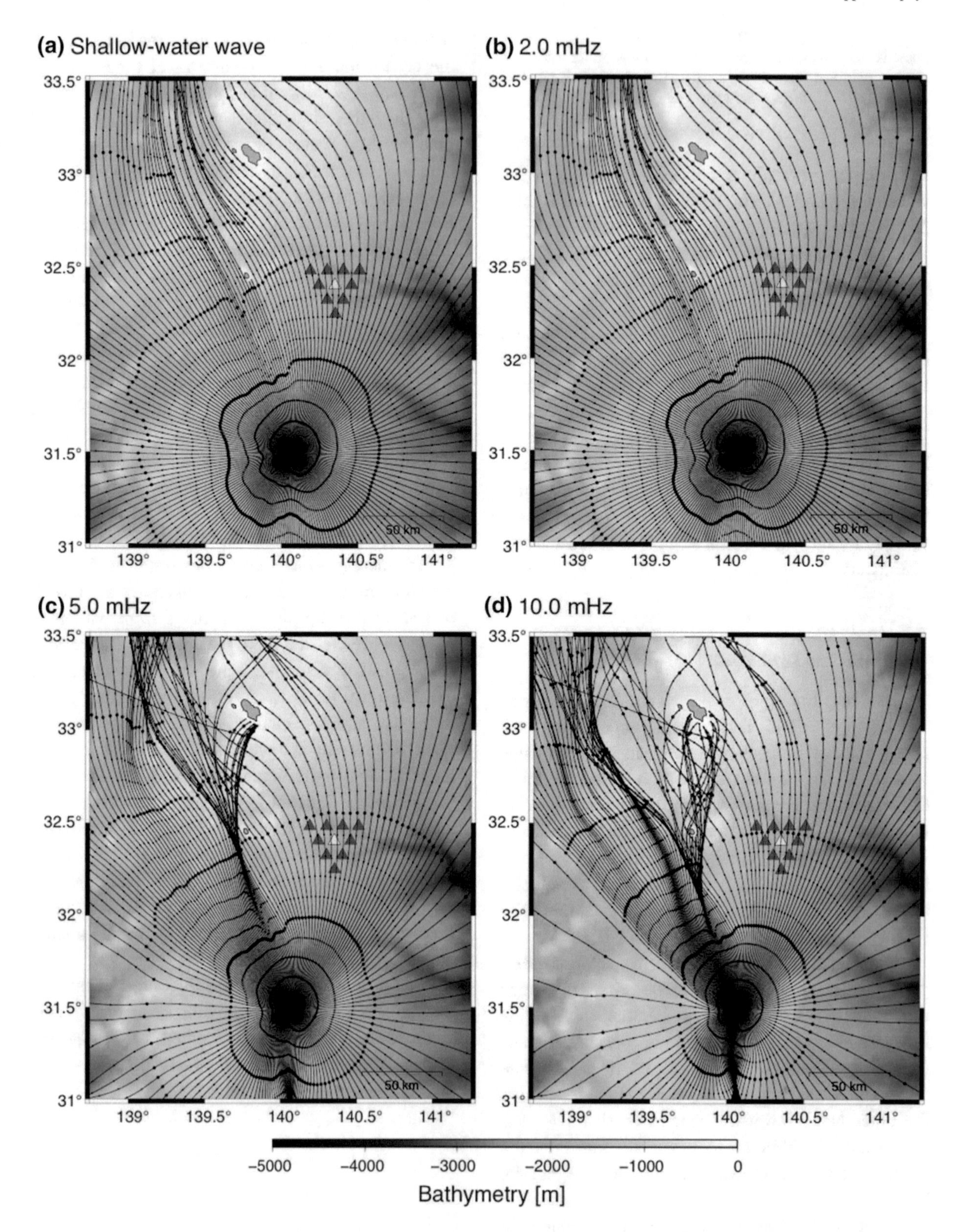
(a) Shallow-water wave
(b) 2.0 mHz
(c) 5.0 mHz
(d) 10.0 mHz
33.5°
33°
32.5°
32°
31.5°
31°
139°
139.5°
140°
140.5°
141°
50 km
−5000
−4000
−3000
−2000
−1000
0
Bathymetry [m]

◀Figure 4
Ray paths of **a** shallow-water waves and dispersive waves at a fixed frequency of **b** 2.0 mHz, **c** 5.0 mHz, and **d** 10.0 mHz. Ray paths radiate from the center of Smith Caldera with the interval of 2°. Small and large dots are plotted on paths every 100 and 500 s, respectively. Red or yellow triangles indicate the OBP stations, and the yellow one indicates A05 station

4. Source Amplitude Estimation

4.1. Green's Law

Green's law is based on the conservation of energy flux of propagating wavefront, and used to estimate the amplitude change of waves traveling in water of variable depth (e.g., Wiegel 1970; Murty 1977, Witting, 1981). Assuming a shallow-water wave without dissipation and reflection, wave amplitudes at different points along a ray path are related to each other by Green's law in the following formula;

$$A_2 = b\left(\frac{D_1}{D_2}\right)^{\frac{1}{4}} A_1, \qquad (15)$$

where A and D are wave amplitude and depth, respectively, and the subscripts 0 and 1 indicate two points along the ray trajectory. b is the refraction coefficient, which is defined as

$$b = \sqrt{\frac{W_1}{W_2}}, \qquad (16)$$

where W is the distance between two neighboring rays, which can be obtained from the ray tracing of shallow-water waves.

In previous studies, Green's law has been used to roughly estimate the tsunami source amplitude. For example, Abe (1973) and Fukao and Furumoto (1975) obtained the refraction coefficients from refraction diagrams, and estimated the average amplitude of initial sea-surface uplift from observed peak-to-peak wave heights or inundation heights at the coast. Satake (1988) computed the refraction coefficient by the ray tracing method and related the amplitude of a circular uplift source with wave heights at the coast. However, their analysis did not consider the tsunami behavior near the source region, so that there are possible biases in their analyses.

4.2. Solutions of Linear Shallow-Water Wave Equations for Constant Depth

In order to examine tsunami behavior when it is propagating away from the source region, we solve the linear shallow-water wave equation expressed as

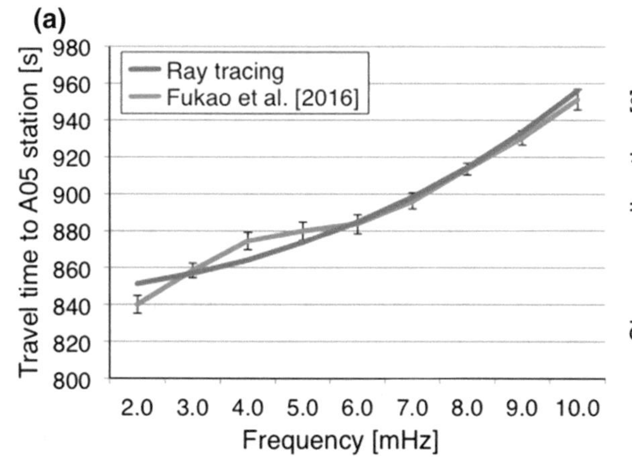

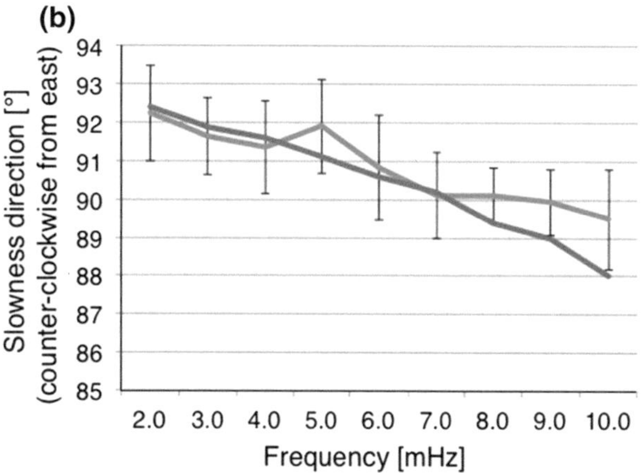

Figure 5
Comparison of plane-wave approximation results of **a** travel times at A05 station and **b** slowness directions between the ray tracing method (red) and the phase analysis of observed waves by Fukao et al. (2016) (blue). The A05 station of the OBP array is located in the great circle direction of around 80° measured counter-clockwise from the east

$$\nabla^2\eta(\boldsymbol{r},t) - \frac{1}{c^2}\frac{\partial^2\eta(\boldsymbol{r},t)}{\partial t^2} = 0, \quad (17)$$

where $\eta(\boldsymbol{r},t)$ is sea-surface height at space $\boldsymbol{r}$ and time t, and c is the shallow-water wave velocity, $\sqrt{gD}$. Depth is assumed constant here, because we focus on general properties when tsunami travels out of the source region, disregarding the bathymetry effect. For our aim of preliminary estimation of initial sea-surface displacement based on linear shallow-water waves, we employed the equation of linear shallow-water waves, Eq. (17), instead of dispersive waves (e.g., Kervella et al. 2007). The initial conditions of sea-surface profile without flow are given in the following formulas;

$$\eta(\boldsymbol{r},t)|_{t=0} = \eta_0(\boldsymbol{r}), \quad (18)$$

$$\left.\frac{\partial\eta(\boldsymbol{r},t)}{\partial t}\right|_{t=0} = 0. \quad (19)$$

In the following we examine plane waves in a one-dimensional space and cylindrically symmetric waves in a two-dimensional space. From the solutions, we show that amplitudes reduce when waves travel out of the source region, which implies the invalidity of Green's law inside the source. We also calculate A_{out}, the reduction rates in amplitude of outgoing waves at the source boundary relative to the initial amplitude within the source area. The rate A_{out} will be used to estimate source amplitude in Sect. 4.3.

4.2.1 One-Dimensional Problem: Plane Wave

We first consider the one-dimensional case. One-dimensional analysis can be applied when we consider an initial tsunami profile perpendicular to the strike direction of the fault of large earthquakes, because their wavefront has strong directivity and can be approximated as plane waves.

In this case, the wave equation Eq. (17) and the initial conditions Eqs. (18) and (19) are rewritten in the one-dimensional coordinate as

$$\frac{\partial^2\eta(x,t)}{\partial^2 x} - \frac{1}{c^2}\frac{\partial^2\eta(x,t)}{\partial t^2} = 0, \quad (20)$$

and

$$\eta(x,t)|_{t=0} = \eta_0(x), \quad (21)$$

$$\left.\frac{\partial\eta(x,t)}{\partial t}\right|_{t=0} = 0, \quad (22)$$

respectively, where x is the coordinate in space.

The solution of Eqs. (20)–(22) is analytically derived in the following formula:

$$\eta(x,t) = \frac{1}{2}\eta_0(x+ct) + \frac{1}{2}\eta_0(x-ct). \quad (23)$$

The two terms in Eq. (23) represent two waves that travel in the opposite directions. These two waves have the same profile as the initial profile, but the amplitudes are only a half of their initial amplitude in the source. This suggests that Green's law is not applied to amplitude inside the source area, because energy flux is distributed equally to the two waves when waves go out of the source, and that the reduction rate A_{out} is 0.5 for plane waves.

4.2.2 Two-Dimensional Problem: Cylindrical Wave

We next consider a cylindrically symmetric source. Tsunami events are sometime caused by a circular source such as volcanic eruptions or crustal deformation of submarine volcano, which may be classified into this case. Similar two-dimensional problems were solved by Kajiura (1970) and Carrier and Yeh (2005). We take an approach similar to Carrier and Yeh (2005) and investigate tsunami amplitude behaviors in the two simple cases of column and Gaussian uplifts.

In two-dimensional polar coordinates, the wave equation Eq. (17) is rewritten in the following formula;

$$\left(\frac{\partial}{\partial r^2} + \frac{1}{r}\frac{\partial}{\partial r}\right)\eta(r,t) - \frac{1}{c^2}\frac{\partial^2\eta(r,t)}{\partial t^2} = 0. \quad (24)$$

The initial conditions Eqs. (18) and (19) are expressed as

$$\eta(r,t)|_{t=0} = \eta_0(r), \quad (25)$$

$$\left.\frac{\partial\eta(r,t)}{\partial t}\right|_{t=0} = 0. \quad (26)$$

To solve the problem, the 0th order Hankel transformation is very useful. Definitions and important properties of the transformation are briefly summarized here for this problem. More details of the Hankel transformation can be found in a textbook such as Piessens (2000) and Arfken et al. (2013).

Let $f(r)$ be an arbitrary function for $r \geq 0$, the 0th order Hankel transformation of $f(r)$ is defined as

$$\hat{f}(s) = \mathcal{H}_0\{f(r)\} = \int_0^\infty rf(r)J_0(sr)\mathrm{d}r, \qquad (27)$$

where $\hat{f}(s)$ is the transform of $f(r)$ by the operator $\mathcal{H}_0$ with the transform variable s and $J_0(r)$ is the 0th order Bessel function. The inverse Hankel transformation is defined by

$$f(r) = \mathcal{H}_0^{-1}\{\hat{f}(s)\} = \int_0^\infty s\hat{f}(s)J_0(sr)\mathrm{d}s. \qquad (28)$$

When function $f(r)$ satisfies $\lim_{r\to\infty} f(r) = 0$, the transformation has an important property as follows;

$$\mathcal{H}_0\left\{\left(\frac{\partial}{\partial r^2} + \frac{1}{r}\frac{\partial}{\partial r}\right)f(r)\right\} = -s^2\mathcal{H}_0\{f(r)\} = -s^2\hat{f}(s). \qquad (29)$$

This property is useful to solve cylindrically symmetric problems including the Laplacian operator, because it enables us to transform a partial differential equation into an ordinary differential equation.

Now, let us solve the wave equation for cylindrically symmetric waves. Taking the 0th order Hankel transformation, Eq. (24) is rewritten as

$$s^2\hat{\eta}(s,t) + \frac{1}{c^2}\frac{\partial^2\hat{\eta}(s,t)}{\partial t^2} = 0, \qquad (30)$$

where $\hat{\eta}(s,t)$ is the transform of $\eta(r,t)$ by the operator $\mathcal{H}_0$. The general solution of Eq. (30) can be obtained easily in the form;

$$\hat{\eta}(s,t) = A(s)\cos(sct) + B(s)\sin(sct). \qquad (31)$$

The functions $A(s)$ and $B(s)$ are still unknown so that they need to be determined by the initial conditions. Taking the transformation on Eqs. (25) and (26), the initial conditions in space s are derived in the following formulas;

$$\hat{\eta}(s,t)|_{t=0} = \mathcal{H}_0\{\eta_0(r)\} = \hat{\eta}_0(s), \qquad (32)$$

$$\left.\frac{\partial\hat{\eta}(s,t)}{\partial t}\right|_{t=0} = \left.\frac{\partial}{\partial t}[\mathcal{H}_0\{\eta(r,t)\}]\right|_{t=0} = \left.\frac{\partial\hat{\eta}_0(s)}{\partial t}\right|_{t=0} = 0. \qquad (33)$$

Equations (31)–(33) yield the functions,

$$A(s) = \hat{\eta}_0(s), \qquad (34)$$

$$B(s) = 0. \qquad (35)$$

Hence Eq. (31) becomes

$$\hat{\eta}(s,t) = \hat{\eta}_0(s)\cos(sct). \qquad (36)$$

This is the transform of $\eta(r,t)$ by the 0th order Hankel transformation.

Finally, we obtain the solution by taking the 0th order inverse Hankel transformation on Eq. (36);

$$\eta(r,t) = \mathcal{H}_0^{-1}\{\hat{\eta}(s,t)\} = \int_0^\infty s\hat{\eta}(s,t)J_0(sr)\mathrm{d}s = \int_0^\infty s\hat{\eta}_0(s)\cos(sct)J_0(sr)\mathrm{d}s. \qquad (37)$$

Equation (37) yields the solutions for an arbitrary cylindrically symmetric initial sea-surface profile without initial flow.

Here, we assume two simple initial sea-surface profiles without flow. First, let us give $\eta_0(r)$ a column-shaped uplift with a unit amplitude;

$$\eta_0(r) = h(a-r), \qquad (38)$$

where $h(r)$ is the Heaviside step function, and a is radius of the column. The 0th order Hankel transform of Eq. (38) is

$$\hat{\eta}_0(s) = \mathcal{H}_0\{\eta_0(r)\} = \mathcal{H}_0\{h(a-r)\} = \frac{a}{s}J_1(as), \qquad (39)$$

where $J_1(r)$ is the 1st order Bessel function (see Table 9.1 in Piessens (2000)). By substituting Eq. (39) into Eq. (37), the solution for the column-shaped profile yields;

$$\eta(r,t) = a\int_0^\infty J_1(as)\cos(sct)J_0(sr)\mathrm{d}s. \qquad (40)$$

As another example, we take a Gaussian-shaped uplift with a unit amplitude of one at the peak expressed by

$$\eta_0(r) = \exp\left(-\frac{2r^2}{a^2}\right). \quad (41)$$

The 0th order Hankel transform of Eq. (41) is rewritten in the following form (Table 9.1 in Piessens (2000));

$$\hat{\eta}_0(s) = \mathcal{H}_0\{\eta_0(r)\} = \mathcal{H}_0\left\{\exp\left(-\frac{2r^2}{a^2}\right)\right\} = \frac{a^2}{4}\exp\left(-\frac{a^2s^2}{8}\right), \quad (42)$$

By substituting Eq. (42) into Eq. (37), the solution for the Gaussian initial profile is

$$\eta(r,t) = \frac{a^2}{4}\int_0^\infty s\exp\left(-\frac{a^2s^2}{8}\right)\cos(sct)J_0(sr)\mathrm{d}s. \quad (43)$$

In Fig. 6, we show sea-surface profiles at various elapsed times of cylindrically symmetric waves that initiated from the column and Gaussian uplift models, by numerically integrating Eqs. (40) and (43), respectively. The constant depth D and gravitational acceleration g are assumed as 800 m and 9.81 m/s^2, respectively. For both models, the spatial parameter a is set as 4 km, and we here regard it as the source size.

As waves travel out of the source region, amplitudes are reduced from the initial profile for both cases. Thus, cylindrically symmetric waves have a discrepancy in amplitude between inside and outside the source region in a way similar to 1-D plane waves. The reduction rate A_{out} was estimated by comparing the amplitudes at the center and at the edge of the source. It significantly depends on the initial shape; A_{out} is 0.5 and 0.25, for the column and Gaussian models, respectively (Fig. 6). Therefore, to estimate the source amplitude of a cylindrically symmetric profile, we need to consider the amplitude reduction rate depending on its shape.

Let us consider the valid range of Green's law for this problem. Now, assuming cylindrically symmetric waves over a constant depth, $A(r)$, amplitude at r, can be estimated from $A(a)$, the amplitude on the source boundary, by the following formula derived from Eq. (15);

$$A(r) = \sqrt{\frac{a}{r}}A(a), \quad (44)$$

because the refraction coefficient is given by Eq. (16) in the following form;

$$b = \sqrt{\frac{W(a)}{W(r)}} = \sqrt{\frac{4\pi a}{4\pi r}} = \sqrt{\frac{a}{r}}. \quad (45)$$

The estimated amplitude is shown in Fig. 6 with gray dashed lines. It is found that amplitude decay obeys Green's law only outside the source region. Therefore,

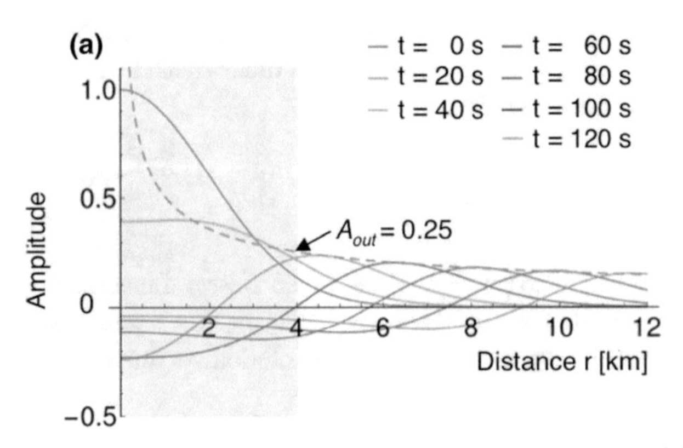

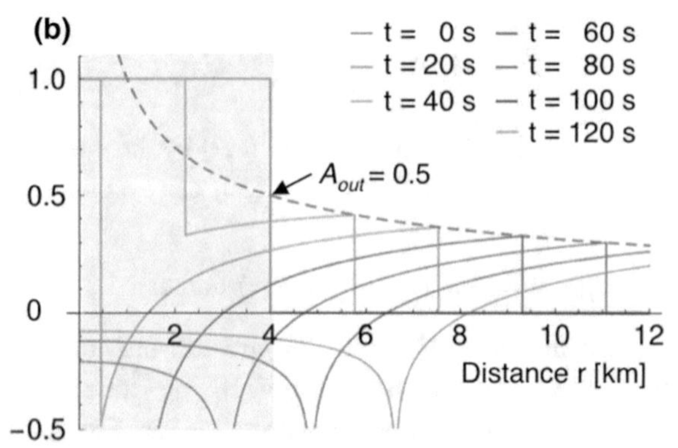

Figure 6
Sea surface profiles calculated by numerical integration at different elapsed times with initial conditions of **a** Gaussian-shaped uplift given by $\eta_0(r) = \exp(-2r^2/a^2)$ and **b** column-shaped uplift given by $\eta_0(r) = h(a-r)$, with a radius a of 4 km. The blue area represents the source region, and the gray dashed line represents the estimated amplitude (see text). Vertical unit is arbitral linear scale

we conclude that source amplitude can not be estimated directly from waves traveling outside the source by Green's law, for neither cylindrically symmetric waves nor plane waves over a constant depth.

4.3. Extension of Green's Law for Source Amplitude Estimation

We apply the extended method of Green's law for source amplitude estimation to the 2015 Torishima earthquake. First, we estimate the amplitude on the source boundary from the amplitude of station A05 using Eq. (15);

$$A_{\mathrm{BD}} = b\left(\frac{D_{\mathrm{OBP}}}{D_{\mathrm{BD}}}\right)^{\frac{1}{4}} A_{\mathrm{OBP}}, \tag{46}$$

where the subscripts BD and OBP indicate the location of the source boundary and A05 station, respectively. The refraction coefficient b is given by

$$b = \sqrt{\frac{W_{\mathrm{OBP}}}{W_{\mathrm{BD}}}}. \tag{47}$$

Using the ray tracing result of shallow-water waves (Fig. 4a), the refraction coefficient b is computed as 4.7. The ray paths for 2–10 mHz tsunami waves deviate from the long-wave path. We note that the refraction coefficients using those paths range from 4.5 to 4.9, and the variation does not affect our amplitude estimation. The depth at the source boundary and A05 station are 800, and 1762 m, respectively. According to Fukao et al. (2016), the amplitude at A05 station, or A_{OBP}, is 2 cm, and the source region estimated by the back projection method is a circular shape comparable to the caldera size with a radius of 4 km. With these numbers, the amplitude at the boundary A_{BD} is estimated to be around 11.4 cm from Eq. (46).

Using the reduction rates A_{out} of 0.5 and 0.25, we finally obtain the initial amplitude of around 22.8 and 45.6 cm, for the initial sea-surface uplift models with column and Gaussian shapes, respectively. This indicates that without considering the reduction effect near the source, the source amplitude may be underestimated by 50–75%.

We note that our estimates are based on the shallow-water wave approximation, which shall underestimate the source amplitude because wave heights also decay due to the dispersion, dissipation and reflection during propagations (e.g., Witting, 1981). In addition, the method is still incomplete because complex profiles of the initial conditions or uneven bathymetry near the source region may not be considered. However, this kind of preliminary estimation will be useful before a more precise analysis using numerical simulations over real bathymetry data is conducted, particularly when sufficient observations are unavailable

5. Discussion and Conclusions

We extended the ray tracing method to dispersive tsunamis. The ray tracing for dispersive waves is advantageous to examine propagations of wavefronts at different frequencies separately, and computationally costs less compared with direct numerical simulation of the dispersive waves. This enables us to investigate ray trajectories that may propagate in different ways from each other, which can not be visualized by other methods, such as the finite difference method of tsunami wave equations. We also reported that quantitative comparison of our ray-tracing results with measurements of tsunami phases over a dense OBP array is useful for a tsunami source study (Fukao et al. 2016).

In the procedure of mapping phase velocities, we filtered the velocity gradient fields on the basis of the ad hoc choice of a spatial filtering scale, which possibly affects the ray paths. Considering the validity condition of the ray equations, Eq. (13), the traced ray paths may be less reliable at shallow parts, because wavelength may be short and comparable to velocity heterogeneity scales. On the other hand, ray paths at deep water may be more reliable, because wavelengths are so long at deep water that rays are insensitive to small-scaled heterogeneities. For the range we analyzed, our choice of the filtering did not largely change the calculated travel time and the ray trajectory, which enable us to evaluate quantitatively the frequency dependency. Although cares must be given to the variation of ray trajectories and travel times due to different filtering scale, the ray tracing method for dispersive waves is useful to investigate general properties of propagation, particularly at deep oceans.

We also investigated the validity of Green's law by solving the linear shallow-water wave equations. The solutions revealed that as waves go out of the source region tsunami amplitudes change. By taking the amplitude reduction effect into account, Green's law was extended to estimation of the amplitude of assumed initial sea-surface profile. The method still has a limitation of linear shallow-water waves, or ignoring wave reflections, energy dissipation and dispersion effects, so that estimated amplitude may be underestimated. However, our method extending Green's law with the reduction rate in amplitude near tsunami source is useful for a preliminary source estimate, which yields a source amplitude depending on the assumed initial sea-surface profile, in particular, for tsunami events with insufficient observation. We suggest a tsunami source model of the 2015 Torishima event based on the ray-tracing analysis and tsunami simulations by the finite difference method for real bathymetry in a separate paper, Fukao et al. (in revision).

Acknowledgements

We used the bathymetry data, JTOPO30, provided by the Marine Information Research Center of the Japan Hydrographic Association. The earthquake information data are obtained from the Global Centroid Moment Tensor catalog. The information of tsunami at Hachijo Island is reported in Japanese by Japan Meteorological Agency (http://www.jma.go.jp/jma/press/1505/13a/1504recently-eq.pdf). We generated figures with the Generic Mapping Tools (GMT) data processing and package and Gnuplot. We also used Mathematica to analyze and plot the solutions of the wave equations. The simulation parts are supported by the JSPS KAKENHI Grant number JP17J02919. The observation was supported by JSPS KAKENHI Grant number 25247074. We appreciate to T. Maeda, R. Geller and T. Furumura for providing valuable comments. We thank four anonymous reviewers whose comments helped us to improve our paper.

References

Abe, K. (1973). Tsunami and mechanism of great earthquakes. *Physics of the Earth and Planetary Interiors, 7,* 143–153. https://doi.org/10.1016/0031-9201(73),90005-8.

Aki, K., & Richards, P. G. (1980). *Quantitative seismology: Theory and methods*. New York: W. H. Freeman.

Arfken, G. B., Weber, H. J., & Harris, F. E. (2012). *Mathematical methods for physicists: A comprehensive guide* (7th ed.). Waltham: Academic Press. https://doi.org/10.1016/b978-0-12-384654-9.00031-1.

Baba, T., Cummins, P., Thio, H., & Tsushima, H. (2009). Validation and joint inversion of teleseismic waveforms for earthquake source models using deep ocean bottom pressure records: A case study of the 2006 Kuril megathrust earthquake. *Pure Applied Geophysics, 166*(1–2), 55–76. https://doi.org/10.1007/s00024-008-0438-1.

Carrier, G. F., & Yeh, H. (2005). Tsunami propagation from a finite source. COMPUTER MODELING IN ENGINEERING AND SCIENCES, 10(2), 113. http://www.techscience.com/doi/10.3970/cmes.2005.010.113.pdf.

Červený, V. (2001). *Seismic ray theory*. Cambridge: Cambridge University Press. https://doi.org/10.1017/cbo9780511529399.

Dingemans, M. W. (1997). *Water wave propagation over uneven bottoms, part 1, linear wave propagation* (vol. 13). *Advanced series on ocean engineering*. River Edge: World Science Publishing.

Fukao, Y., & Furumoto, M. (1975). Mechanism of large earthquakes along the eastern margin of the Japan Sea. *Tectonophysics, 26,* 247–266. https://doi.org/10.1016/0040-1951(75),90093-1.

Fukao, Y., Sugioka, H., Ito, A., Shiobara, H., Sandanbata, O., Watada, S., & Satake K. (2016). 2015 Volcanic tsunami earthquake near Torishima island: Array analysis of ocean bottom pressure gauge records. Abstract [NH43B-1853] Presented at 2016 Fall Meeting, AGU, San Francisco, CA, 12–16 Dec.

Glimsdal, S., Pedersen, G., Atakan, K., Harbitz, C. B., Langtangen, H. P., & Lovholt, F. (2006). Propagation of the Dec. 26, 2004, Indian Ocean Tsunami: Effects of dispersion and source characteristics. *International Journal of Fluid Mechanics Research, 33*(1), 15–43. https://doi.org/10.1615/interjfluidmechres.v33.i1.30.

Glimsdal, S., Pedersen, G., Harbitz, C., & Løvholt, F. (2013). Dispersion of tsunamis: Does it really matter? *Natural Hazards and Earth Systems Sciences, 13,* 1507–1526.

Hossen, M., Cummins, P., Dettmer, J., & Baba, T. (2015). Tsunami waveform inversion for sea surface displacement following the 2011 Tohoku earthquake: Importance of dispersion and source kinematics. *Journal of Geophysical Research: Solid Earth, 120*(9), 6452–6473. https://doi.org/10.1002/2015JB011942.

Julian, B. R., (1970). Ray tracing in arbitrarily homogeneous media, Technical Note 1970-45, Lincoln Laboratory, Massachusetts Institute of Technology.

Kajiura, K. (1970). Tsunami source, energy and the directivity of wave radiation. Bull. Earthq. Res. Inst. Univ. 48, 835–869. http://ci.nii.ac.jp/naid/120000871347/.

Kanamori, H. (1972). Mechanisms for tsunami earthquakes. *Physics of the Earth and Planetary Interiors, 6,* 346–359. https://doi.org/10.1016/0031-9201(72),90058-1.

Kanamori, H., Ekström, G., Dziewonski, A., Barker, J. S., & Sipkin, S. A. (1993). Seismic radiation by magma injection: An anomalous seismic event near Tori Shima, Japan. *Journal of Geophysical Research, 98*(B4), 6511–6522. https://doi.org/10.1029/92JB02867.

Kervella, Y., Dutykh, D., & Dias, F. (2007). Comparison between three-dimensional linear and nonlinear tsunami generation models. *Theoretical and computational fluid dynamics, 21*(4), 245–269. https://doi.org/10.1007/s00162-007-0047-0.

Lamb, H. (1932). *Hydrodynamics* (6th ed.). Cambridge: Cambridge University Press.

Lin, F.-C., Kohler, M. D., Lynett, P., Ayca, A., & Weeraratne, D. S. (2015). The 11 March 2011 Tohoku tsunami wavefront mapping across offshore Southern California. *Journal of Geophysical Research: Solid Earth, 120,* 3350–3362. https://doi.org/10.1002/2014JB011524.

Murty, T. S. (1977). *seismic sea waves—tsunamis*. Ottawa: Department of Fisheries and Environment.

Officer, C. B. (1974). *Introduction to theoretical geophysics*. New York: Springer.

Piessens, R. (2000). *The Hankel transform, the transforms and applications handbook* (2nd ed.). Boca Raton: CRC Press. https://doi.org/10.1201/9781420036756.ch9.

Saito, T., Satake, K., & Furumura, T. (2010). Tsunami waveform inversion including dispersive waves: The 2004 earthquake off Kii Peninsula, Japan. *Journal of Geophysical Research, 115,* B06303. https://doi.org/10.1029/2009JB006884.

Satake, K. (1987). Inversion of tsunami waveforms for the estimation of a fault heterogeneity: Method and numerical experiments. *Journal of Physics of the Earth, 35*(3), 241–254. https://doi.org/10.4294/jpe1952.35.241.

Satake, K. (1988). Effects of bathymetry on tsunami propagation: Application of ray tracing to tsunamis. *Pure and Applied Geophysics, 126*(1), 27–36. https://doi.org/10.1007/BF00876912.

Shearer, P. (1999). *Introduction to seismology*. Cambridge: Cambridge University Press.

Shuler, A., Ekström, G., & Nettles, M. (2013a). Physical mechanisms for vertical-CLVD earthquakes at active volcanoes. *Journal of Geophysical Research: Solid Earth, 118,* 1569–1586. https://doi.org/10.1002/jgrb.50131.

Shuler, A., Nettles, M., & Ekström, G. (2013b). Global observation of vertical-CLVD earthquakes at active volcanoes. *Journal of Geophysical Research: Solid Earth, 118,* 138–164. https://doi.org/10.1029/2012JB009721.

Tappin, D. R., Grilli, S. T., Harris, J. C., Geller, R. J., Masterlark, T., Kirby, J. T., et al. (2014). Did a submarine landslide contribute to the 2011 Tohoku tsunami? *Marine Geology, 357*(1), 344–361. https://doi.org/10.1016/j.margeo.2014.09.043.

Watada, S. (2013). Tsunami speed variations in density–stratified compressible global oceans. *Geophysical Research Letters, 40,* 4001–4006. https://doi.org/10.1002/grl.50785.

Wiegel, R. L. (1970). *Earthquake Engineering*. Englewood Cliffs: Prentice-Hall.

Wielandt, E. (1993). Propagation and structural interpretation of nonplane waves. *Geophysical Journal International, 113,* 45–53. https://doi.org/10.1111/j.1365-246X.1993.tb02527.x.

Witting, J. M. (1981). A note on Green's law. *Journal of Geophysical Research, 86*(C3), 1995–1999. https://doi.org/10.1029/JC086iC03p01995.

Woods, M., & Okal, E. (1987). Effect of variable bathymetry on the amplitude of teleseismic tsunamis: A ray–tracing experiment. *Geophysical Research Letters*. https://doi.org/10.1029/GL014i007p00765.

Yomogida, K., & Aki, K. (1985). Waveform synthesis of surface waves in a laterally heterogeneous Earth by the Gaussian Beam Method. *Journal of Geophysical Research, 90*(B9), 7665–7688. https://doi.org/10.1029/JB090iB09p07665.

(Received April 30, 2017, revised October 28, 2017, accepted December 1, 2017)

Pure Appl. Geophys.

https://doi.org/10.1007/s00024-017-1744-2

Pure and Applied Geophysics

Tsunami Wave Run-up on a Vertical Wall in Tidal Environment

Ira Didenkulova[1,2] and Efim Pelinovsky[2,3,4,5]

Abstract—We solve analytically a nonlinear problem of shallow water theory for the tsunami wave run-up on a vertical wall in tidal environment. Shown that the tide can be considered static in the process of tsunami wave run-up. In this approximation, it is possible to obtain the exact solution for the run-up height as a function of the incident wave height. This allows us to investigate the tide influence on the run-up characteristics.

Key words: Wave run-up, sea wall, tsunami, tide, nonlinear shallow water theory.

1. Introduction

The tsunami wave run-up on the coast in most areas of the World Ocean occurs in the presence of a tide, the height of which can be significant. Tsunami approaching during the high tide aggravates the catastrophe, so that the tsunami penetrates long distances deep into the coast. On the contrary, at the low tide even a strong tsunami may remain unnoticed. That is why in international practice it is customary to measure tsunami characteristics "without a tide", subtracting the latter from the tide gauge, or simply reducing the wave height in observations. And the tide itself can be pre-calculated with a very good accuracy, so the errors associated with tidal fluctuations can be minimized.

Meanwhile, the tsunami impact on the coast in the tidal environment is a complex nonlinear problem, and the tsunami and tidal wave superposition alone is insufficient. The significant role of the tide in the process of tsunami wave generation by landslides is discussed in Kulikov et al. (1998) and Thomson et al. (2001). During the high water the coastal slope becomes thin, and at low tide it can lose stability and slide into the water, generating tsunami. Wave interaction with the tide affects the coastal zone ecology, in particular, the mixing of the fresh groundwater with the salt water (Xin et al. 2010). Numerical simulation of tsunami wave transformation at different tidal levels in one of the Spanish settlements demonstrates that the maximum drawdown height strongly depends on the tidal level, while the maximum run-up height is not so influenced by the tide (Lima et al. 2010). Tidal variations can also affect tsunami when it approaches the bay, strengthening or weakening them depending on the tidal phase (Behera et al. 2011). Tsunami penetration into rivers also depends on the tidal level, and numerical experiments emphasize the important role of tsunami interaction with the tide and river runoff (Tolkova et al. 2015).

To our knowledge, there are no analytical solutions describing the tidal influence on the characteristics of tsunami waves at the coast. In this study, we solve the nonlinear shallow water equations for wave run-up on a vertical wall at tidal environment and provide formulas for tsunami run-up height in the presence of tide.

The paper is organized as follows. The mathematical model and considered geometry is briefly presented in Sect. 2. The properties of tidal flow near the vertical barrier (sea wall, cliff, etc.) are discussed in Sect. 3. The tsunami run-up on a vertical wall is studied in Sect. 4. Main results are summarized in Conclusion.

[1] Department of Marine Systems, Tallinn University of Technology, Tallinn, Estonia. E-mail: ira.didenkulova@msi.ttu.ee
[2] Nizhny Novgorod State Technical University n.a. R. Alekseev, Nizhny Novgorod, Russia.
[3] Institute of Applied Physics, Nizhny Novgorod, Russia.
[4] National Research University, Higher School of Economics, Moscow, Russia.
[5] Special Research Bureau for Automation of Marine Researches, Yuzhno-Sakhalinsk, Russia.

2. Mathematical Model and Geometry of the Problem

Let us consider tsunami wave run-up on a vertical wall in the basin of constant depth in the presence of tide. The analysis is carried out in the framework of the 1D nonlinear shallow water theory considering normal propagation of tsunami waves to the beach:

$$\frac{\partial \eta}{\partial t} + \frac{\partial}{\partial x}[(h+\eta)u] = 0, \tag{1}$$

$$\frac{\partial u}{\partial t} + u\frac{\partial u}{\partial x} + g\frac{\partial \eta}{\partial x} = 0, \tag{2}$$

where η is the sea level displacement, u is the flow velocity averaged over the water depth, g is the gravity acceleration, and h is the water depth assumed to be constant. The x-axis is directed to the shore. On the right, the pool is limited by a vertical wall located at the point $x = 0$ (Fig. 1), so the boundary condition at the wall is

$$u\,(0,\ t) = 0. \tag{3}$$

3. The Tide at the Vertical Wall

To begin with, let us discuss the tidal wave properties in the basin described (Fig. 1) within the linear shallow water theory framework. The basin depth is assumed large enough to neglect nonlinear effects for the tide. For simplicity, let us confine to one tidal harmonic. Then the tide is described by

$$\eta_{\text{tide}}(x,\ t) = A\cos\left(\frac{\Omega x}{c}\right)\sin\,(\Omega t), \tag{4}$$

$$u_{\text{tide}}(x,\ t) = -\frac{Ag}{c}\sin\left(\frac{\Omega x}{c}\right)\cos(\Omega t), \tag{5}$$

satisfying the boundary condition Eq. (3) at the wall. Here A is the tide amplitude, Ω is the tidal wave frequency, and c is the wave celerity

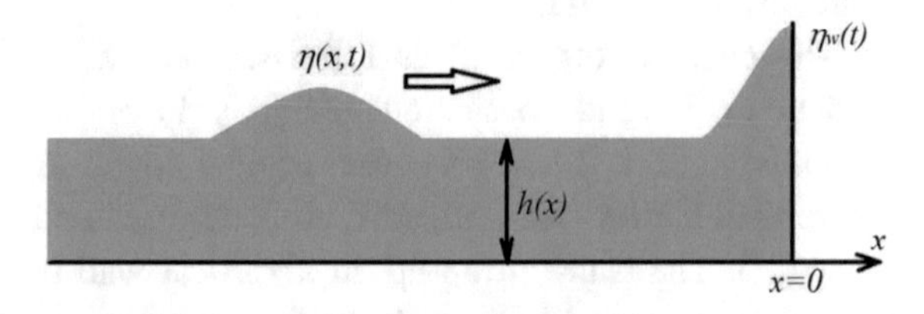

Figure 1
The geometry of the problem

$$c = \sqrt{gh}. \tag{6}$$

The tidal wave frequency Ω is small as compared with the tsunami wave frequency, its period is a 24-h multiple. Tidal harmonics (overtones) with a 6-h period are often observed in the coastal zone (Nekrasov 1975). The tsunami wave propagates over the typical shelf within an hour, so, in the first approximation, the tide can be considered static when examining the tsunami wave run-up on the shore.

Let us assume that the shelf length is L, and a tsunami wave approaching the shore is measured exactly at the edge of the shelf at the point $x = -L$. Then the condition that the tidal level does not change during the tsunami wave propagation is

$$\frac{\Omega L}{c} \ll 1. \tag{7}$$

In this case, as is follows from Eq. (5), the tidal current velocity is small (at the wall it is equal to zero) and can be neglected. In the same approximation, the tidal height is uniform along the shelf and can be replaced by

$$\eta_{\text{tide}}\,(x,t) \approx \eta_{\text{tide}}\,(t_0), \tag{8}$$

for $-L \le x \le 0$ and $t_0 \le t \le t_0 + 2L/c$, where t_0 is the time moment corresponding to the tsunami wave arrival at the shelf. As a result, in the first approximation, it can be assumed that the tide changes only the basin depth, which remains unchanged during the tsunami wave propagation and run-up on the shore. Obviously, the same result remains also for the nonlinear theory that allows considering tide static during tsunami wave propagation.

4. Tsunami Wave Interaction with a Vertical Wall

The static tide approximation allows solving the shallow water equations Eqs. (1–2) with constant coefficients for tsunami waves, replacing the unperturbed water depth, h by the modified one, taking into account the tidal influence ($h + \eta_{\text{tide}}$). Such a problem for waves in a basin of constant depth has already been solved in our works (Pelinovsky et al. 2008; Didenkulova and Pelinovsky 2011). Its solution is based on the transition from physical variables to Riemann invariants:

$$I_{\pm} = u \pm 2\left[\sqrt{g(h+\eta)} - \sqrt{gh}\right], \tag{9}$$

which are preserved along the characteristics, whose slopes in the (x, t) plane are determined by the velocities

$$c_{\pm} = \pm\sqrt{gh} + \frac{3}{4}I_{\pm} + \frac{1}{4}I_{\mp}. \tag{10}$$

We include the constant term $\sqrt{gh}$ in the definition of the Riemann invariants Eq. (9), so that each invariant presents the corresponding wave. For instance, in the wave propagating to the right $I_{+} \neq 0$ and $I_{-} = 0$.

Far from the wall, the incident wave is written through the Riemann invariants as follows:

$$\begin{aligned} u &= 2\left[\sqrt{g(h+\eta)} - \sqrt{gh}\right], \quad I_{-} = 0, \\ I_{+} &= 4\left[\sqrt{g(h+\eta)} - \sqrt{gh}\right]. \end{aligned} \tag{11}$$

Following the boundary condition Eq. (3), the invariant at the wall is equal to

$$I_{+} = 2\left[\sqrt{g(h+\eta_{w})} - \sqrt{gh}\right]. \tag{12}$$

From the invariant conservation during the wave propagation process, follows the connection between the incident wave $\eta\,(t)$ and the water level oscillations at the wall $\eta_{w}\,(t)$:

$$\frac{\eta_{w}(t)}{h} = 4\left[1 + \frac{\eta(t-\tau)}{h} - \sqrt{1 + \frac{\eta(t-\tau)}{h}}\right], \tag{13}$$

where τ is the wave travel time from the initial position $x = -L$ to the wall. Note that due to the nonlinearity of the problem, the wave travel time τ depends not only on the distance, but also on the parameters of the wave field, and therefore, is hard to define. However, if considering only maximum values of the wave field, Eq. (13) is simplified and does not contain τ any more:

$$\frac{R}{h} = 4\left[1 + \frac{H_0}{h} - \sqrt{1 + \frac{H_0}{h}}\right], \tag{14}$$

where R is the maximum run-up height, and H_0 is the tsunami amplitude when approaching the wall, see details in Pelinovsky et al. (2008) and Didenkulova and Pelinovsky (2011).

Further, returning to the original problem of tsunami propagation in tidal environment, we should replace the water depth, h by $[h + \eta_{\text{tide}}\,(t)]$. As a result, the tsunami wave height at the wall is

$$\frac{H_{w}}{h+\eta_{\text{tide}}} = 4\left[1 + \frac{H_0}{h+\eta_{\text{tide}}} - \sqrt{1 + \frac{H_0}{h+\eta_{\text{tide}}}}\right]. \tag{15}$$

In particular, if the tsunami wave height is small, then the approximated expression can be used

$$\frac{H_{w}}{H_0} \approx 2 + \frac{H_0}{2(h+\eta_{\text{tide}})}. \tag{16}$$

Equation (15) for normalized variables $H_{w}/(h + \eta_{\text{tide}})$ and $H_0/(h + \eta_{\text{tide}})$ is shown in Fig. 2 and demonstrates the monotonic growth of tsunami run-up height on the wall as a function of the incident wave amplitude. For comparison we also plot the corresponding result obtained within the linear theory, which implies that tsunami run-up height on the wall is equal to twice incident wave amplitude.

Thus, the tsunami run-up height increases at low tide and decreases at high tide. From this follows, that the tidal component exclusion from observation data, as is done in nowadays practice, does not provide universal and independent of the tide run-up characteristics.

For tsunami prevention and mitigation it is important to know the maximum water level at the

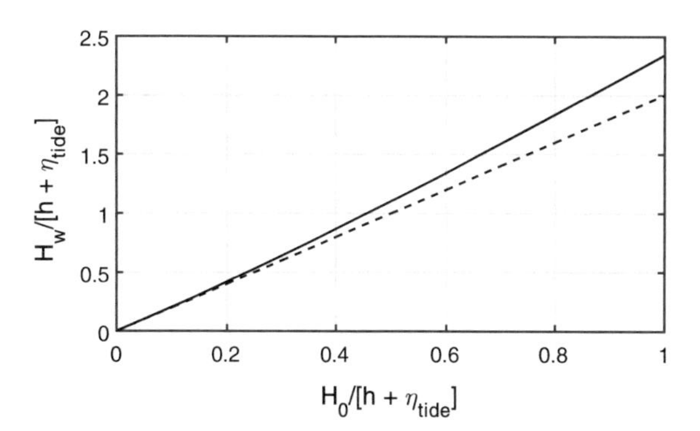

Figure 2
Tsunami run-up height on the wall versus incident wave amplitude: solid and dashed lines correspond to solutions obtained within nonlinear and linear theories, respectively

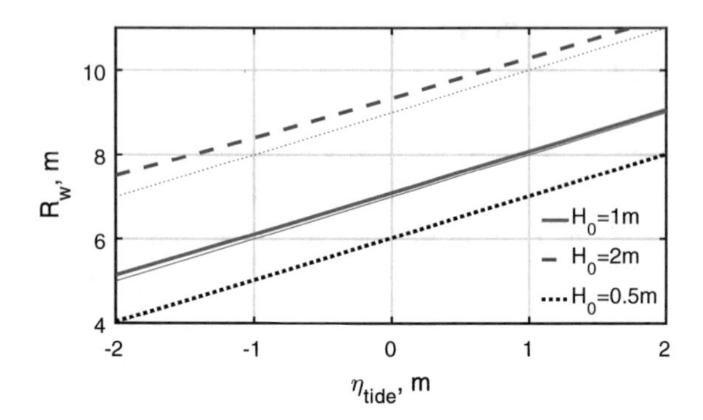

Figure 3
Water height at the wall versus the tidal level; the water depth, h is 5 m; sick and thin lines correspond to solutions obtained within nonlinear and linear theories, respectively

wall, which can be calculated explicitly using Eq. (15)

$$R_w = h + \eta_{tide} + H_w = (h + \eta_{tide})\left[5 + 4\frac{H_0}{h + \eta_{tide}} - 4\sqrt{1 + \frac{H_0}{h + \eta_{tide}}}\right]. \tag{17}$$

For estimates, let us take the water depth $h = 5$ m, the tsunami wave amplitude $H_0 = 1$ m and the tidal range from − 2 to + 2 m. In this case, the water level at the wall varies from 5.14 m (low tide) to 9.07 m (high tide), and the wave height above the tidal level varies from 2.14 to 2.07 m, indicating the nonlinear process of wave transformation near the wall. Figure 3 shows the dependence of water height at the wall versus the tidal level for three different amplitudes of the incident wave $H_0 = 0.5$, 1 and 2 m; the water depth, h is 5 m. By thin lines in Fig. 3, we also plot the corresponding outcome of linear superposition, which demonstrate the effect of the nonlinearity. It is natural that the effect of the nonlinearity is larger for waves of larger amplitude.

5. Conclusion

We present the exact solution to the nonlinear shallow water equations for tsunami wave run-up on a vertical wall in the presence of a static tide, the characteristics of which do not change during the process of tsunami run-up on the wall.

An explicit expression for the tsunami wave height at the wall is obtained. It is maximal for the low tidal level. Thus, the known procedure for excluding the pre-calculated tide from the observations does not completely eliminate the influence of tide on the wave run-up characteristics. Therefore, we recommend to the field survey campaigns to provide also the tidal level at the moment of tsunami arrival.

Note, that the tide–tsunami interaction influences the distance where tsunami wave breaks. Without tide this process was studied in Stoker (1957), Zahibo et al. (2008) and Didenkulova (2009) and these results can also be applied to the case of tide–tsunami interaction. In particular, the wave will break earlier at low tide than at high tide. Hence, these breaking effects will reduce tsunami run-up height stronger at low tide than at high tide.

In our analysis, the amplitude of the incident wave is independent of the tidal level. However, it may be incorrect for a given earthquake with the corresponding deep ocean tsunami characteristics, which are available from seismic data and deep water pressure gauges, respectively. These characteristics are not significantly affected by the tide. If one assumes that the deep ocean amplitude is fixed and that there is a gradual slope up to the basin Green's law may be invoked, as an estimate at any rate. Then

$$H_0 = H_a\left(1 + \frac{\eta_{tide}}{h}\right)^{-1/4}, \tag{18}$$

where H_a is the amplitude for $\eta_{tide} = 0$. For $h = 5$ m and tidal range of ± 2 m this gives an amplitude difference in the order of 20%, which is larger than the effect of the nonlinear coupling in the run-up formula Eq. (17), and should also be accounted in applications.

The developed theory is also valid for waves in narrow bays, if the geometry in Fig. 1 is supplemented by an arbitrary cross-section. As shown in Didenkulova and Pelinovsky (2011), Eq. (14) is universal for all such basins, and therefore, the corresponding Eqs. (15) and (17) are also universal for waves in the bays of various cross-sections.

Acknowledgements

This work is supported by the state programme of the Ministry of Education and Science of the Russian Federation (project No. 5.5176.2017/8.9), Russian President Grant MD-6373.2016.5, PUT1378 and RFBR Grants (16-55-52019 and 17-05-00067).

References

Behera, M. R., Murali, K., & Sundar, V. (2011). Effect of the tidal currents at the amphidromes on the characteristics of an N-wave-type tsunami. *Proceedings of the Institution of Mechanical Engineers Part M: Journal of Engineering for the Maritime Environment, 225,* 43–59.

Didenkulova, I. (2009). Nonlinear long-wave deformation and runup in a basin of varying depth. *Nonlinear Processes in Geophysics, 16,* 23–32.

Didenkulova, I., & Pelinovsky, E. (2011). Rogue waves in nonlinear hyperbolic systems (shallow-water framework). *Nonlinearity, 24,* R1–R18.

Kulikov, E. A., Rabinovich, A. B., Fine, I. V., Bornhold, B. D., & Thomson, R. E. (1998). Tsunami generation by landslides at the Pacific coast of North America and the role of tides. *Oceanology, 38,* 323–328.

Lima, V. V., Miranda, J. M., Baptista, M. A., Catalao, J., Gonzalez, M., Otero, L., et al. (2010). Impact of a 1755-like tsunami in Huelva, Spain. *Natural Hazards and Earth System Sciences, 10,* 139–148.

Nekrasov, A. V. (1975). *Tidal waves in marginal seas.* Lenigrad: Hydrometeoizdat. **(In Russian)**.

Pelinovsky, E., Kharif, C., & Talipova, T. (2008). Large-amplitude long wave interaction with a vertical wall. *European Journal of Mechanics-B/Fluids, 27,* 409–418.

Stoker, J. J. (1957). *Water waves: the mathematical theory with applications.* New York: Interscience Publishers Inc.

Thomson, R. E., Rabinovich, A. B., Kulikov, E. A., Fine, I. V., & Bornhold, B. D. (2001). On numerical simulation of the landslide-generated tsunami of November 3, 1994 in Skagway Harbor Alaska. In G. T. Hebenstreit (Ed.), *Tsunami Research at the End if a Critical Decade. Advances in Natural and Technological Hazards Research* (Vol. 18, pp. 243–282). Dordrecht: Kluwer.

Tolkova, E., Tanaka, H., & Roh, M. (2015). Tsunami observations in rivers from a perspective of tsunami interaction with tide and riverine flow. *Pure and Applied Geophysics, 172,* 953–968.

Xin, P., Robinson, C., Li, L., Barry, D., and Bakhtyar, R (2010) Effects of wave forcing on a subterranean estuary. Water Resources Research, 46, Article W12505.

Zahibo, N., Didenkulova, I., Kurkin, A., & Pelinovsky, E. (2008). Steepness and spectrum of nonlinear deformed shallow water wave. *Ocean Engineering, 35,* 47–52.

(Received March 28, 2017, revised October 19, 2017, accepted December 1, 2017)

Pure Appl. Geophys.

https://doi.org/10.1007/s00024-018-1804-2

Pure and Applied Geophysics

Implications on 1 + 1 D Tsunami Runup Modeling due to Time Features of the Earthquake Source

M. Fuentes,[1] S. Riquelme,[2] J. Ruiz,[1] and J. Campos[1]

Abstract—The time characteristics of the seismic source are usually neglected in tsunami modeling, due to the difference in the time scale of both processes. Nonetheless, there are just a few analytical studies that intended to explain separately the role of the rise time and the rupture velocity. In this work, we extend an analytical 1 + 1 D solution for the shoreline motion time series, from the static case to the kinematic case, by including both rise time and rupture velocity. Our results show that the static case corresponds to a limit case of null rise time and infinite rupture velocity. Both parameters contribute in shifting the arrival time, but maximum runup may be affected by very slow ruptures and long rise time. Parametric analysis reveals that runup is strictly decreasing with the rise time while is highly amplified in a certain range of slow rupture velocities. For even lower rupture velocities, the tsunami excitation vanishes and for larger, quicker approaches to the instantaneous case.

Key words: Tsunami, seismology, runup.

1. Introduction

The study of tsunamis from analytical approaches has been treated for decades (e.g. Kajiura 1970; Carrier and Greenspan 1958; Synolakis 1987; Kânoğlu 2004; Madsen and Schaffer 2010; Fuentes et al. 2013; Fuentes 2017). However, just a few analytic studies involve explicit time characteristics of the generation process with applications to tsunamis triggered by earthquakes (Hammack 1973; Dutykh and Dias 2007; Todorovska and Trifunac 2001; Dutykh and Dias 2009). The assumption of neglecting temporal effects comes from the fact that tsunami propagation velocities are by far slower than the rupture (Kajiura 1981). Kajiura (1981) demonstrated that there was equivalence between generating a tsunami from a static seafloor displacement or from a dynamic one, because of the much lower velocity of the tsunami propagating through the source. Since then, this term has not been a subject of importance in tsunami analytical modeling.

The evidence of a variety of slow earthquakes (Ide et al. 2007) suggests at least examining and testing this hypothesis. The 2004 Sumatra–Andaman megathrust earthquake has been studied by many authors investigating the relation between earthquake and tsunami generation (Ammon et al. 2005; Lay et al. 2005; Stein and Okal 2005). These studies showed that there is a slow component on the rupture towards the north, according to Lay et al. (2005). The main excitation of the tsunami was located around 500 km from the epicenter source occurring within the first 500 s. Then, the rupture shows a slow component in the next 700–800 km and the Bengal bay brings special attention because there are large runup heights, but no high-frequency earthquake was radiated, and buildings did not show signs of damage (Lay et al. 2005). This slow component of the rupture may play an important role in the tsunami generation.

Effects of tsunami amplification along the Japanese coastline were observed by Imai et al. (2010) by modeling delayed ruptures along the Nankai trough where they caused the sub-faults to have a temporal delay resulting in the worst case scenario. This approach could be thought as the use of a variable rupture velocity. Satake et al. (2013) performed a Multi time-window inversion to resolve the time history of the slip of the 2011 Tohoku-Oki earthquake, with constant rupture velocity. They observed

[1] Department of Geophysics, Faculty of Physical and Mathematical Sciences, University of Chile, Santiago, Chile. E-mail: mauricio@dgf.uchile.cl

[2] National Seismological Center, Faculty of Physical and Mathematical Sciences, University of Chile, Santiago, Chile.

that the best delayed slip model better explained the coastal tsunami heights than the instantaneous slip model, which overestimated them. Fukutani et al. (2016) studied the uncertainties in tsunami estimations due to kinematic rupture parameters. In particular, they considered the rupture origin location and velocity along the strike direction. These kinds of studies are practical for probabilistic tsunami hazard assessments.

Regarding the analytical approach, Todorovska and Trifunac (2001) studied the problem of a hump seafloor deformation with constant rupture velocity and instantaneous source (no rise time). They found that wave amplification occurs when rupture velocity is comparable with tsunami velocity. The attributed mechanism of amplification was wave focusing. On the other hand, Dutykh and Dias (2007) modeled a seafloor deformation with a finite rise time, but simultaneously along the fault plane (instantaneous rupture). Saito and Furumura (2009) included the source duration to obtain a criterion, in terms of the fault dimensions and depth, for discriminating when a dynamic generation should be preferred over a static initial condition. Saito (2013) incorporated the rise time to evaluate the influence of dynamic tsunami generation over the ocean-bottom pressure evolution. This is useful when using in combination with ocean-bottom pressure gauges to estimate the water displacement. Nevertheless, all those studies were applied in a constant depth ocean, with the 2 + 1 D linear potential theory. In this work, we trade-off one spatial dimension for bathymetry complexity. We consider a sloping beach model, which is suitable for earthquakes in subduction zones that triggers near-field tsunamis.

Geist (1998) reviewed the effects of rise time, rupture velocity, and dynamic overshooting. He used the definition of t^* as a dimensionless number from Hammack (1973). If this $t^* \ll 1$, then the static displacement is transferred immediately to the seafloor. The other extreme $t^* \gg 1$ is rare. He found that the runup decreases if the rise time increases, and the spatial variation of the rise time has almost no effects on the tsunami generation. Rise time value is between 1 and 20 s for subduction earthquakes (Geist 1998). The rupture time of the source is limited basically by two physical parameters: L and V_r, where L is the total rupture length and V_r is the rupture velocity. The source duration and the rupture time are affected depending on the rupture mode: unilateral and bilateral rupture propagation. Along-strike tsunami has a faster velocity in the ocean-ward direction because the bathymetry is deeper in that direction. Rupture modes are also important in tsunami directivity. It was demonstrated by Bouchon (1980) that, in the near field, there is a dynamic overshoot in the vertical displacement field to up-dip propagation of the rupture for an inverse fault. If the lower layer has low velocity value then the overshoot acquires an oscillatory behavior. Geist (1998) simulated this effect using the formula

$$u_z(x,t) = u_{0z}\left(1 - \cos(\omega_0 t)e^{-\frac{t}{t_R}}\right),$$

which is a ramp-like function including oscillations. Given that, the tsunami occurs in a much longer period of time than t_R.This should have a very small effect in the tsunami generation.

In this study, we will include rise time and rupture time into analytical modeling in order to study the amplification on the tsunami generation and runup.

2. Mathematical Formulation

We consider the forced linear shallow water equation in a sloping beach (Fig. 1):

$$\eta_{tt} - \alpha g(x\eta_x)_x = \eta_{0tt}, \tag{1}$$

where $\eta(x,t)$ is the water surface elevation, $\alpha =: \tan(\beta)$ is the slope of the beach, g is the gravity

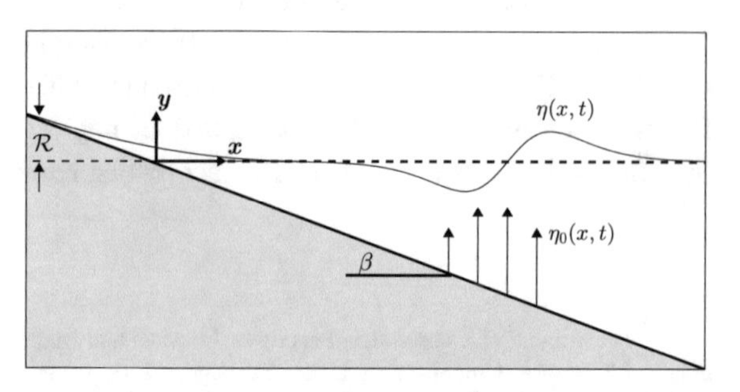

Figure 1
Sketch of the sloping beach domain and variables of the 1 + 1 D model

acceleration, and $\eta_{0tt}(x,t)$ is the forcing term applied to the sea bottom.

Tuck and Hwang (1972) solved Eq. (1) by using Hankel and Laplace transforms. The solution is

$$\eta(x,t) = \int_{-\infty}^{\infty} \mathrm{J}_0(2k\sqrt{x}) \int_{-\infty}^{\infty} \mathrm{J}_0(2k\sqrt{\xi})\eta_{1tt}(\xi,t) * \sin(\sqrt{\alpha g}kt)\mathcal{H}(t)\mathrm{d}\xi\mathrm{d}k,$$

where

$$\eta_1(x,t) = \eta_0(x,t) + [\eta(x,0) + t\eta_t(x,0)]\mathcal{H}(t),$$

$\mathcal{H}(t)$ is the Heaviside step function, $J_0(\cdot)$ is the zero-order cylindrical Bessel function, and $*$ denotes the convolution product in time.

2.1. Source Time-Dependent Solution

To model a non-instantaneous tsunami generation, we set null initial conditions, $\eta(x,0) = \eta_t(x,0) = 0$, and $\eta_0(x,t) = \zeta_0(x)T(x,t)$, where $\zeta_0(x)$ is the final shape of the seafloor deformation and $T(x,t) \in [0,1]$ is a temporal description on how $\zeta_0(x)$ is performed.T is directly related with the temporal description of the seismic source function. We define

$$T(x,t) = \left[\frac{t-t_\mathrm{V}}{t_\mathrm{R}}\mathcal{H}(t_\mathrm{R}+t_\mathrm{V}-t) + \mathcal{H}(t-t_\mathrm{R}-t_\mathrm{V})\right] \mathcal{H}(t-t_\mathrm{V}),$$

where t_R is the rise time, $t_\mathrm{V}(x) = \frac{|x-x_\mathrm{R}|}{V_\mathrm{r}}$ is the *rupture time* at location x for a bilateral rupture propagating with a constant velocity V_r, starting from an origin x_R (Fig. 2). For simplicity, we consider t_R constant.

Computing T_{tt} in the sense of the distributions, we get

$$T_{tt} = \frac{1}{t_\mathrm{R}}[\delta(t-t_\mathrm{V}) - \delta(t-t_\mathrm{V}-t_\mathrm{R})].$$

Defining

$$\mathcal{M}(\zeta,s)(t) = \frac{2}{\sqrt{\alpha g}}\int_0^{\infty}\frac{\mathcal{H}(\alpha g[t-s(\xi)]^2-4\xi)}{\sqrt{\alpha g[t-s(\xi)]^2-4\xi}}\zeta(\xi)\mathrm{d}\xi,$$

and following the same procedure as in Fuentes (2017) with each term, the approximated shoreline

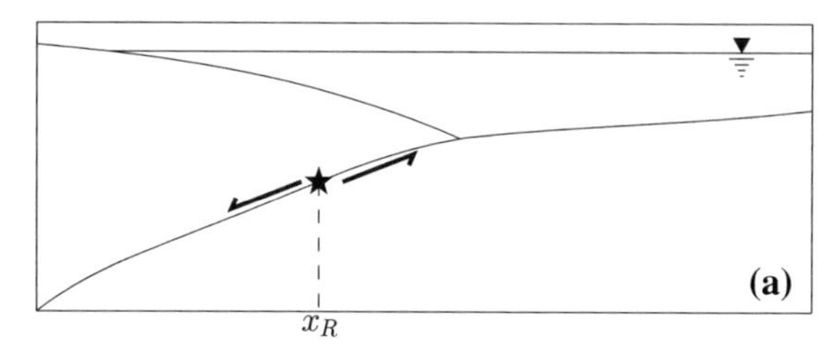

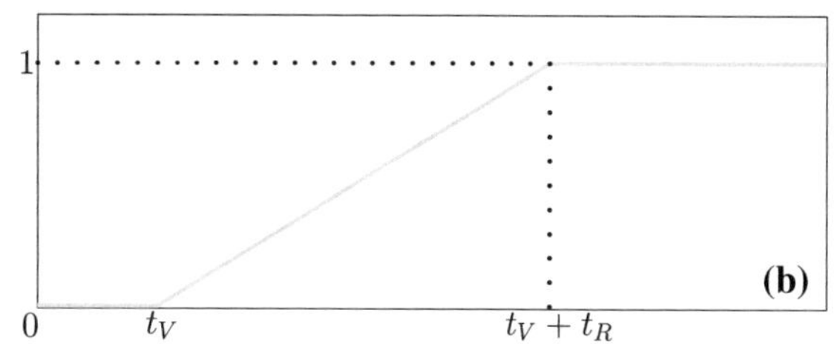

Figure 2
a Scheme of realistic setting of a subduction zone (Not to scale). The rupture origin location is denoted by x_R. **b** Source time function $T(x,t)$ for a given x

motion provided by the linear theory, $\eta_\mathrm{S}(t) =: \eta(0,t)$, can be written as

$$\eta_\mathrm{S}(t) = \frac{1}{t_\mathrm{R}}[\mathcal{M}(\zeta_0,t_\mathrm{V})(t) - \mathcal{M}(\zeta_0,t_\mathrm{V})(t-t_\mathrm{R})]. \quad (2)$$

From there, we can also obtain expressions for quantifying individual effects of t_R and V_r

$$\lim_{t_\mathrm{R}\to 0}\eta_\mathrm{S}(t) = \partial_t\mathcal{M}(\zeta_0,t_\mathrm{V})(t), \quad (3)$$

$$\lim_{V_r\to\infty}\eta_\mathrm{S}(t) = \frac{1}{t_\mathrm{R}}[\mathcal{M}(\zeta_0,0)(t) - \mathcal{M}(\zeta_0,0)(t-t_\mathrm{R})], \quad (4)$$

where Eqs. (3) and (4) are for null rise time and infinite rupture velocity, respectively.

Note that

$$\lim_{\substack{t_\mathrm{R}\to 0\\ V_\mathrm{r}\to\infty}}\eta_\mathrm{S}(t) = \lim_{\substack{V_\mathrm{r}\to\infty\\ t_\mathrm{R}\to 0}}\eta_\mathrm{S}(t) = \frac{2}{\sqrt{\alpha g}}\partial_t\int_0^{\infty}\frac{\mathcal{H}(\alpha g t^2-4\xi)}{\sqrt{\alpha g t^2-4\xi}}\zeta_0(\xi)\mathrm{d}\xi = \frac{1}{2}\frac{\partial}{\partial t}\left\{t\int_0^1\frac{\zeta_0\left(\frac{1}{4}\alpha g t^2 y\right)}{\sqrt{1-y}}\mathrm{d}y\right\}, \quad (5)$$

which is the same as the solution for the static case derived by Fuentes (2017).It is also clear to observe that

$$\lim_{V_r \to 0} \eta_S(t) = \lim_{t_R \to \infty} \eta_S(t) = 0.$$

2.2. The Seabed Deformation

As mentioned, the final shape of the initial condition is denoted by $\zeta_0(x)$. In tsunami modeling, the Okada's equations are widely used to compute the static deformation due to a finite fault in 3D (Okada 1985). It provides a solution for the problem of static deformation in an elastic half-space. Freund and Barnett (1976) solved the 2D problem of surface deformation and their solution can also handle non-uniform slip distributions.

2.2.1 Uniform Slip

We take a pure dip-slip (rake angle is 90°) fault of width W, length L, dip angle δ, slip U in a medium of Poisson ratio ν.The fault is oriented parallel to the coast, so the strike angle can be fixed, in our case, to 180° due to the axis orientation chosen (Fig. 3a). In the case of a pure dip-slip fault, the slip vector reduces to $(U_1, U_2, U_3) = (0, U, 0)$.

Both studies used different coordinate system, and then they will be reoriented to our reference system.

Placing the lower corner of the fault in $\left(-\frac{L}{2}, 0, -d\right)$ and letting L tends to infinity, the Okada's solution for the seabed displacements can be reduced to the following expressions:

$$U_z(x) = u_z\left(\frac{x}{d}\right) - u_z\left(\frac{x - W\cos(\delta)}{d - W\sin(\delta)}\right), \tag{6a}$$

$$U_h(x) = u_h\left(\frac{x}{d}\right) - u_h\left(\frac{x - W\cos(\delta)}{d - W\sin(\delta)}\right). \tag{6b}$$

Originally, the retained variable should be y, but we renamed it as x to keep consistency, and it represents the variable along the half-space. u_z and u_h are the dislocations defined as

$$u_z(\bar{x}) = \frac{U}{\pi}\left[\sin(\delta)\arctan(\bar{x}) - \frac{\cos(\delta) - \bar{x}\sin(\delta)}{1 + \bar{x}^2} + (3 - 8\nu)\delta\sin(\delta)\right], \tag{7a}$$

$$u_h(\bar{x}) = \frac{U}{\pi}\left[\cos(\delta)\arctan(\bar{x}) - \frac{\sin(\delta) + \bar{x}\cos(\delta)}{1 + \bar{x}^2} + \sin(\delta) + \delta\cos(\delta) - 2(1 - 2\nu)\delta\sin(\delta)\{2\delta\tan(\delta) + 1\}\right]. \tag{7b}$$

($\bar{x}$ denotes a ratio between the horizontal distance and depth of the fault endpoint) which are the same expressions obtained by Freund and Barnett (1976) for uniform slip and corrected Madariaga (2003). It must be noted that constant terms in the previous equations were ignored by other authors. However, it is not quite important because the final displacement U_i is the difference of dislocations u_i at the endpoints

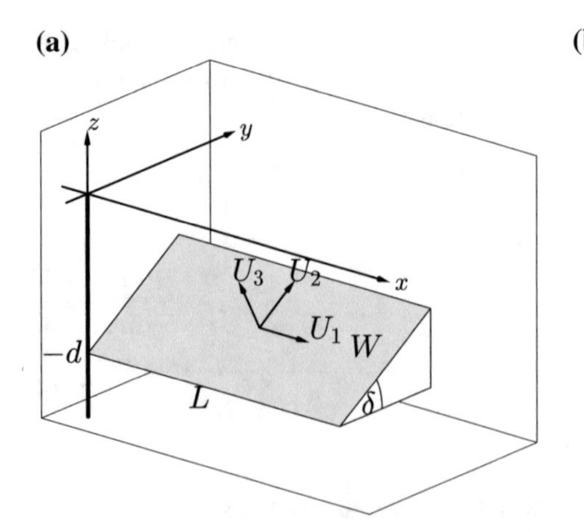

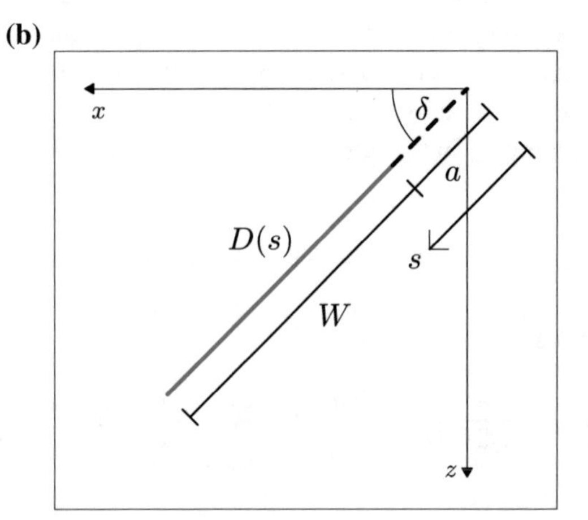

Figure 3
a Geometry and spatial orientation of the fault plane [adapted from Okada (1985)]. **b** Profile of a dip-slip fault with slip distribution $D(s)$. [adapted from Freud and Barnett (Freund and Barnett 1976)]

of the fault ($i = z, h$). Here, we will keep them for completeness.

In the worst case, the fault breaks up to the free surface of the half-space. It is well-known that a singularity is produced, but it is still possible to retain a closed form. This is obtained by letting d tends to $W\sin(\delta)^+$,

$$U_z(x) = \frac{U}{\pi}\left[\sin(\delta)\arctan\left(\frac{x}{d}\right) - \frac{d[d\cos(\delta) - x\sin(\delta)]}{x^2 + d^2} - C_1(x - W\cos(\delta))\right],$$

$$U_h(x) = \frac{U}{\pi}\left[\cos(\delta)\arctan\left(\frac{x}{d}\right) - \frac{d[x\cos(\delta) + d\sin(\delta)]}{x^2 + d^2} - C_2(x - W\cos(\delta))\right],$$

where $d = W\sin(\delta)$,

$$C_1(z) =: \frac{\pi}{2}\sin(\delta)\text{sgn}(z) + \cos(\delta)\mathbf{1}_{\{0\}}(z),$$

$$C_2(z) =: \frac{\pi}{2}\cos(\delta)\text{sgn}(z) + \sin(\delta)\mathbf{1}_{\{0\}}(z),$$

sgn(z) is the sign function and $\mathbf{1}_A(z)$ is the indicator function.

Following Tanioka and Satake (1996), the complete vertical displacement of the ocean bottom corresponds to the vertical component $U_z(x)$ plus the horizontal advection contribution. Also, we must translate the solutions to our coordinate system. Defining the trench axis location at x_0 and $x_e =: x_0 - W\cos(\delta)$,, we finally obtain

$$\zeta_0(x) = U_z(x - x_e) + \alpha U_h(x - x_e). \tag{8}$$

We can observe, when the term αU_h is neglected, that the maximum slip transformed into vertical displacement occurs in the singularity (Ward 2011); thus

$$\frac{\zeta_0(x_0)}{U} = \sin(\delta)\left(1 - \frac{\delta}{\pi}\right).$$

The optimal dip angle that maximizes the uplift is 63.76° giving 58% of slip converted into vertical displacement.

2.2.2 Non-uniform Slip

We call s to the local variable along the fault and

$$u_z(x,s) = \frac{1}{\pi}\left[\sin(\delta)\arctan\left(\frac{x - s\cos(\delta)}{s\sin(\delta)}\right) + \frac{sx\sin^2(\delta)}{x^2 - 2sx\cos(\delta) + s^2}\right], \tag{9a}$$

$$u_h(x,s) = \frac{1}{\pi}\left[\cos(\delta)\arctan\left(\frac{x - s\cos(\delta)}{s\sin(\delta)}\right) + \frac{[s - x\cos(\delta)]s\sin(\delta)}{x^2 - 2sx\cos(\delta) + s^2}\right]. \tag{9b}$$

Then, for a general slip distribution $D(s)$ and a fault of width W buried at a distance a from the origin, the final displacements are

$$U_z(x) = \int_a^{a+W} u_z(x,s)D'(s)\mathrm{d}s, \tag{10a}$$

$$U_h(x) = \int_a^{a+W} u_h(x,s)D'(s)\mathrm{d}s. \tag{10b}$$

Freund and Barnett (1976) proposed, as example, the following normalized slip distribution:

$$D'(r) = \begin{cases} \frac{12}{q^3}(q - r)r, & r < q \\ \frac{12}{(1-q)^3}(r - 1)(r - q), & r \geq q \end{cases}, \tag{11}$$

where $r = \frac{s-a}{W}$ and $q \in (0, 1)$.

In our coordinate system, $a = \frac{d}{\sin(\delta)} - W$ and the final vertical displacement is

$$\zeta_0(x) = U_z(x - x_e) + \alpha U_h(x - x_e), \tag{12}$$

with $x_e = x_0 + a\cos(\delta)$.

3. Numerical Tests

3.1. A Classical $M_w = 8.0$ Earthquake

In this section, we will set $\delta = 20°$ being a typical value in the Chilean subduction zone (Hayes et al. 2012). Once the dip angle is fixed, U and W are free to be chosen. Nonetheless, to keep realistic earthquake fault scales and to diminish the number of parameters, we use the scaling laws from Blaser et al. (2010), assuming a constant rigidity of the medium = 30GPa. Then,

$$\log(W) = -1.86 + 0.46M_w,$$

$$\log(U) = -3.15 + 0.47M_w,$$

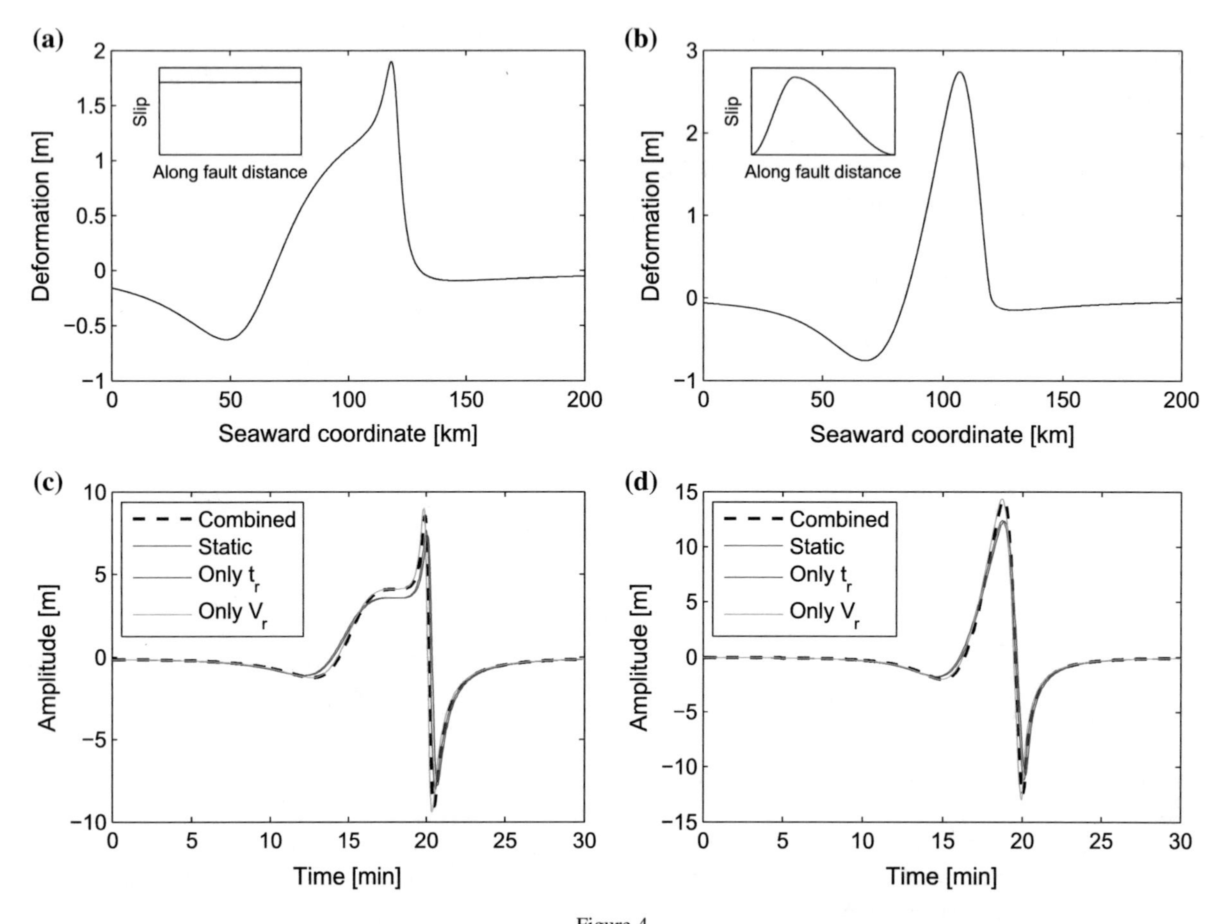

Figure 4
a Vertical displacement for a $M_w = 8.0$ earthquake (Eqs. 6a–8) computed at the sea bottom. Inner plot depicts the normalized uniform slip distribution used. **b** Vertical displacement for a $M_w = 8.0$ earthquake (Eqs. 9a, 9b, 10a, 10b, 12) computed at the sea bottom. Inner plot depicts the normalized skewed ($q = 0.3$) slip distribution used (Eq. 11). **c** Shoreline motion induced by **a** for different cases, with $V_r = 2$ km/s, $x_R = 80$ km (asymmetric bilateral rupture) and $t_R = 15$ s (Eqs. 2–5). **d** Shoreline motion induced by **b** for different cases, with $V_r = 2$ km/s, $x_R = 80$ km (asymmetric bilateral rupture) and $t_R = 15s$ (Eqs. 2–5)

where M_w is the moment magnitude, W is in kilometers, and D is in meters. This gives $W \approx 66$ km and $D \approx 4.1$ m.

Figure 4 displays the shoreline motion due to a typical $M_w = 8.0$ earthquake for a case with uniform slip and another with non-uniform slip distribution concentrated up-dip. One can see a clear difference in the deformation profile when assuming uniform or non-uniform slip. For the values chosen in the modeling, the results show that in terms of the shoreline motion, it presents larger peak-to-peak amplitudes when considering the combined effect of t_R and V_r than in the static case. In the case of the uniform slip fault, the amplification is 11% higher when comparing a combined effect of rupture velocity and rise time with a static initial condition. The case of non-uniform slip presents a 15% of amplification. These calculations were made with values indicated in the caption of Fig. 4.

In Figs. 5 and 6, we display the complete variation of the maximum runup relative to the static case for models with rupture velocity and rise time for four modes of rupture, for two slip distributions: uniform and heterogeneous (Fig. 4). Figure 6 shows that the amplification becomes very important for low rupture velocities (of the order of the tsunami velocity). For very slow ruptures, the rise-time has little affect on the maximum runup. For larger rupture

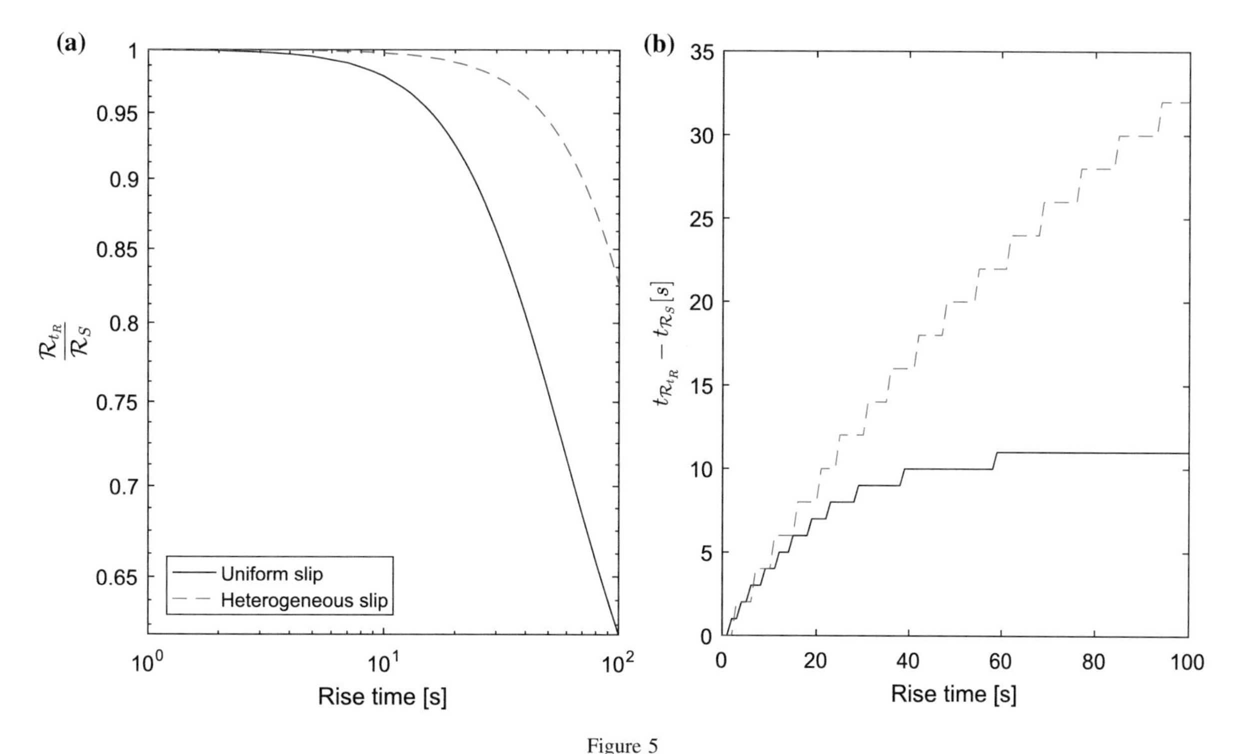

Figure 5
Effect of the rise time in the runup height and time shift for an instantaneous rupture. **a** Variation of the normalized runup (with respect to the static case) with variable rise time for homogeneous and heterogeneous slip distributions. The heterogeneous slip distribution was taken from eq. (11) with $q = 0.3$. **b** Same as **a** for the time shift relative to the peak runup time of the static case

velocities, when rise-time increases, the maximum runup tends to be in the order of the value of the static case. The complete behavior of these parameters will be examined in detail in the next section.

3.2. Parametric Analysis

There are many parameters involved in the whole problem of the tsunami runup; nonetheless, we focus on those aspects related to the temporal evolution of the source. Since the analytical solution is dependent on the initial deformation profile, we keep fixed that shape by using the same as one set in the previous subsection.

3.2.1 Rise Time

We call $\mathcal{R}_{t_\mathrm{R}}$ the maximum runup induced by an initial condition applied with a rise time t_R, and t_1 the time where it is attained.

$$\mathcal{R}_{t_\mathrm{R}} = \frac{1}{t_\mathrm{R}}[\mathcal{M}(\zeta_0, t_\mathrm{V})(t_1) - \mathcal{M}(\zeta_0, t_\mathrm{V})(t_1 - t_\mathrm{R})].$$

Similarly, we call $\mathcal{R}_0$ the maximum runup in the case of instantaneous rise of the seafloor deformation and t_0 the time where it is attained.

$$\mathcal{R}_0 = \partial_t \mathcal{M}(\zeta_0, t_\mathrm{V})(t_0).$$

By virtue of the mean value theorem, there exist $\bar{t} \in (t_1 - t_\mathrm{R}, t_1)$ such that

$$\mathcal{R}_{t_\mathrm{R}} = \partial_t \mathcal{M}(\zeta_0, t_\mathrm{V})(\bar{t}).$$

Since $\mathcal{R}_0$ is the maximum of the time series,

$$\mathcal{R}_{t_\mathrm{R}} \leq \mathcal{R}_0$$

which is true for any rise time. This result also shows that the rise time introduces a time delay in the runup. Figure (5) confirms the decreasing behavior of the ratio $\mathcal{R}_{t_R}/\mathcal{R}_0$ with increasing rise time (Fig. 5a) and also the trend that follows the time shift of the maximum runup relative to the static case (Fig. 5b). This test was performed with infinite rupture velocity,

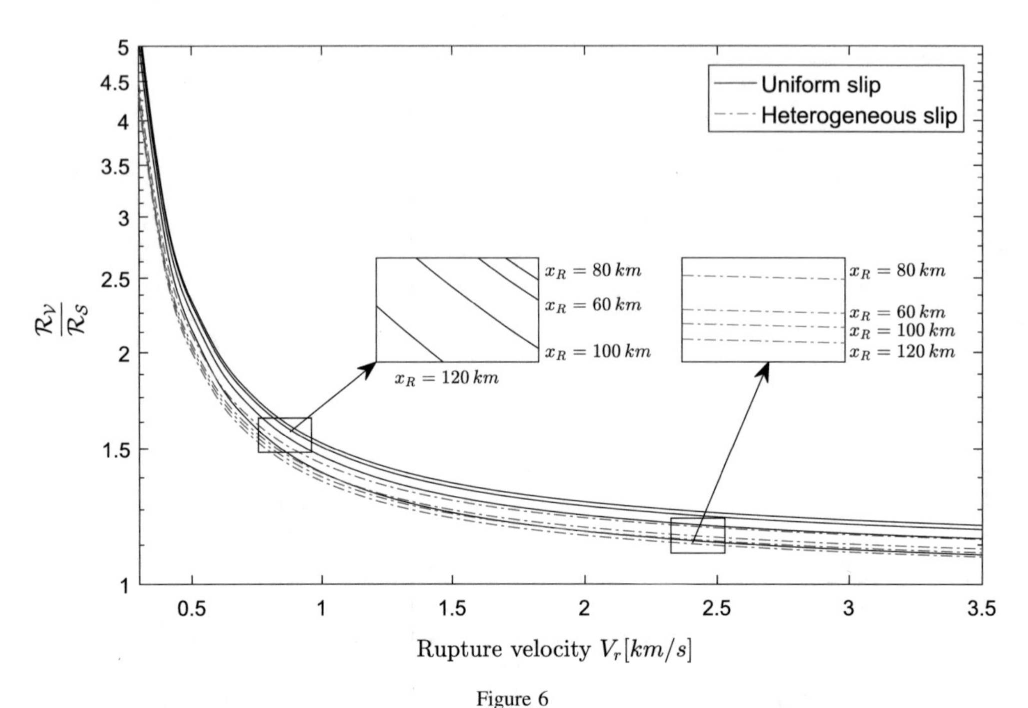

Figure 6
Amplification of the runup due to the rupture velocity considering uniform and non-uniform slip distributions. Each curve is associated with a different rupture origin location x_R. Both initial shapes were considered from subfigures (4a, b) with null rise time

which makes the origin location of the rupture x_R irrelevant.

3.2.2 Rupture Velocity

In this case, an analytical conclusion from the equations is more complicated to obtain. Nevertheless, we examine a wide range of rupture velocities where amplification is observed. Again, the results are compared with respect to the static case, and the rise time is set to zero in all these simulations.

Figure 6 shows the behavior of the amplification of the runup, relative to the static case, as a function of the rupture velocity. The variation with the rupture origin location reveals that the maximum amplification occurs for a bilateral rupture that initiates at the middle of the fault. Also, the amplification is greater for a rupture propagating downdip (updip origin) instead the opposite.

3.2.3 Combined Effect of V_r and t_R

As a summary, Figs. (7, 8) present the whole variation of the runup in terms of the rupture velocity and rise time. For instance, level curve "1" represents the isocontour where the effects of the rupture velocity and rise time are perfectly compensated. Nevertheless, as it can be deduced from the previous subsection, the worst-case scenario lies in the zone of low rupture velocities (0.3–0.5 km/s) and null rise time. Lower velocities were also tested (0.1–0.3 km/s), but numerical treatments for the integration become quite complex, even though the runup is still amplified, as founded by Todorovska and Trifunac (2001). In general, for regular earthquakes, the rupture velocity varies around 2.0–2.5 km/s and the rise time about 1–20 s (Kanamori and Brodsky 2004; Geist, 1998). This means that according to our model, the time dependency of the earthquake source is responsible for 15–20% of the amplification.

4. Discussion and Conclusions

Since continuous GPS and broadband seismometer observations have increased, seismologists have detected and observed several types of slow earthquakes or non-regular earthquakes such as slow slip

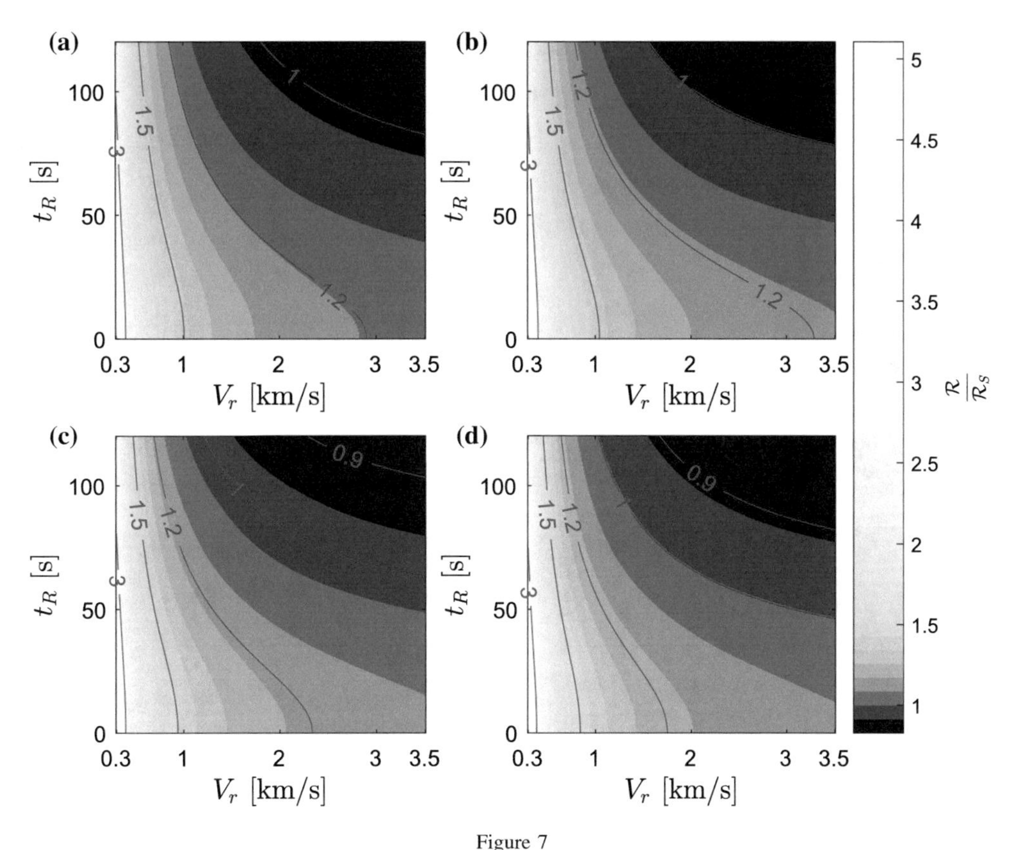

Figure 7
Color map of the maximum runup relative to the static case from Fig. 4a for four different rupture origins. **a** $x_R = 60$ km (unilateral up-dip). **b** $x_R = 80$ km (asymmetric bilateral). **c** $x_R = 100$ km (asymmetric bilateral). **d** $x_R = 120$ km (unilateral down-dip). Some level curves are displayed in red for visual guidance

events (SSE), episodic and tremors, slip earthquakes (ETS) (e.g. Miller et al. 2002; ; Beroza and Ide 2011), silent earthquakes (Kanamori and Stewart 1979), and tsunami earthquakes (Kanamori 1972) in certain zones around the world. A key question is: Do giant thrust tsunamigenic earthquakes produce slow rupture (0.1–0.5 km/s) velocities? There is no strong evidence of observations of such slow earthquake rupture velocity on tsunamigenic events in subduction zones. However, this is related to observational capabilities rather than no existence of such phenomena.

Large tsunamigenic earthquakes often produce large aftershocks immediately after the mainshock, then instruments are still reverberating for many hours. This complicates our observations of such small wave amplitudes that carry information on slow earthquake rupture velocity. Examples of this are the giant earthquakes of Valdivia 1960 and Sumatra 2004, both ruptures are very complex. The moment rate functions for both events have been subject of study (Kanamori and Cipar 1974; Lay et al. 2005; Ammon et al. 2005). These tsunamigenic earthquakes have shown slow rupture velocities; however, due to instrumental limitations, it is not well understood nor how slow they were, even the rupture area for both events were not determined until months or years after the mainshock (Barrientos and Ward 1990; Stein and Okal 2005).

For the giant 2004 Sumatra earthquake some runup observations are still not well explained, not just because of the limited resolution of the bathymetry or topography. During this event the Bay of Bengal did not experience structural damage and intensities were documented at levels I and II, suggesting the presence of a very slow rupture

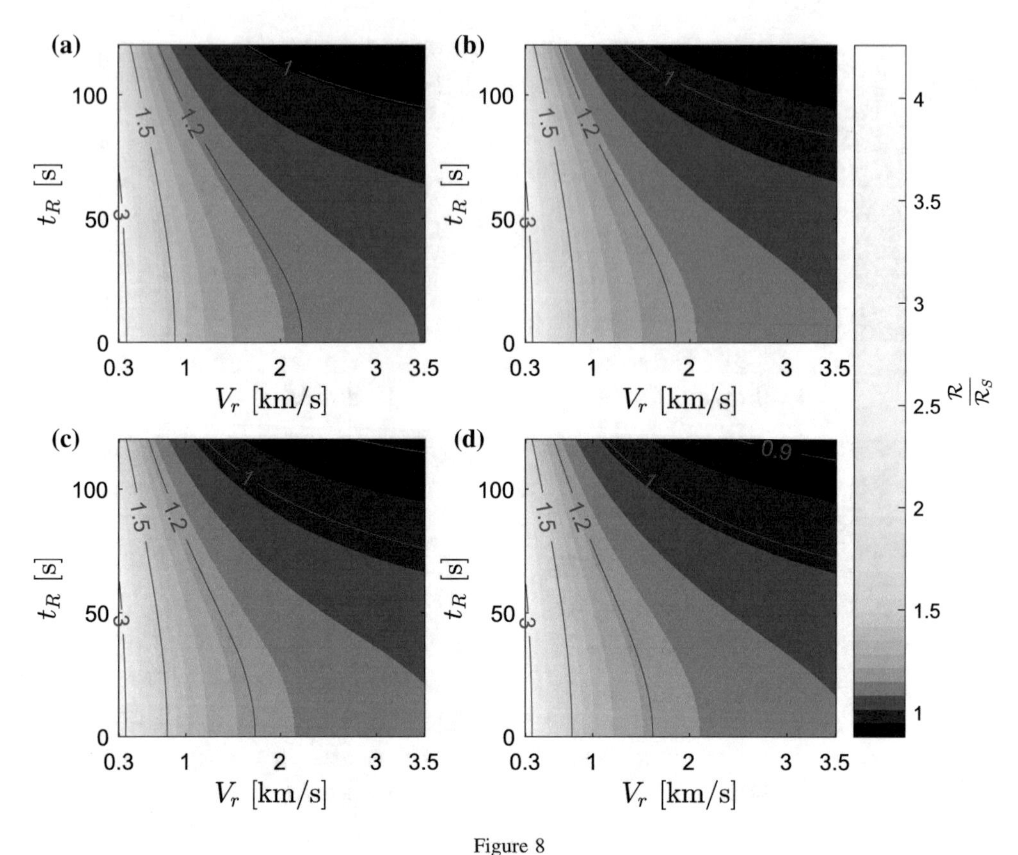

Figure 8
Color map of the maximum runup relative to the static case from Fig. 4b for four different rupture origins. **a** $x_R = 60$ km (unilateral up-dip). **b** $x_R = 80$ km (asymmetric bilateral). **c** $x_R = 100$ km (asymmetric bilateral). **d** $x_R = 120$ km (unilateral down-dip). Some level curves are displayed in red for visual guidance

component considering the inundation reached there (Lay et al. 2005). In this work, we isolate the effect of earthquake rupture velocity, rise time, and runup due to a rupture propagating along the dip direction. We computed the runup amplification due to a very slow moment rate in thrust earthquakes along the dip direction. We are aware of other amplification runup effects such as resonance in bays, shelf resonance, and edge waves. These effects are not taken into account in our analytical formulation; instead, we explore the effects on the runup due to source parameters controlling the rupture kinematic. The tsunami amplitude is larger with a slower rupture velocity (0.1–0.5 km/s) than with a regular one (2.0–2.5 km/s).

Due to the difficult equation treatment, we are still limited to obtain a solution including the source time function in the 2 + 1 D approximation. Nonetheless, with this 1 + 1 D solution we can still capture some overall features of the seismic source, and it has the advantage over previous studies that consider a sloping beach bathymetry rather than a flat ocean and the novelty of combining two temporal parameters at the same time.

We have demonstrated using a simple source model that time evolution of slip can play an important role in the tsunami modeling and its consequent runup. It cannot be neglected for some cases, especially when rupture velocities around 0.1 – 0.5 km/s can amplify the runup up to five times compared to the static case. This suggests that the system ocean-earth resonates with the tsunami wave periods (tsunami phase velocity rounds 0.2 km/s).

Mega-large earthquakes that present slow rupture velocities, which have generated enormous tsunamis in the near-field, might be conditioned by this slow rupture velocity component. This could be an explanation for the low earthquake intensity and the abnormal runup heights in Bengal Bay during the 2004 M_w 9.3 Sumatra earthquake. Further studies taking into account the combined effects of ruptures in the strike and dip directions are necessary to quantify accurately the amplification factors.

Acknowledgements

This work was entirely funded by the Programa de Riesgo Sísmico.

References

Ammon, C. J., Ji, C., Thio, H. K., Robinson, D., Ni, S., Hjorleifsdottir, V., et al. (2005). Rupture process of the 2004 Sumatra–Andaman earthquake. *Science, 308*(5725), 1133–1139.

Barrientos, S. E., & Ward, S. N. (1990). The 1960 Chile earthquake: inversion for slip distribution from surface deformation. *Geophysical Journal International, 103*(3), 589–598.

Beroza, G. C., & Ide, S. (2011). Slow earthquakes and nonvolcanic tremor. *Annual Review of Earth and Planetary Sciences, 39,* 271–296.

Blaser, L., Krüger, F., Ohrnberger, M., & Scherbaum, F. (2010). Scaling relations of earthquake source parameter estimates with special focus on subduction environment. *Bulletin of the Seismological Society of America, 100*(6), 2914–2926.

Bouchon, M. (1980). The motion of the ground during an earthquake: 2. The case of a dip slip fault. *Journal of Geophysical Research: Solid Earth, 85*(B1), 367–375.

Carrier, G. F., & Greenspan, H. P. (1958). Water waves of _nite amplitude on a sloping beach. *Journal of Fluid Mechanics, 4*(01), 97–109.

Dutykh, D., and Dias, F., (2007). Water waves generated by a moving bottom. In Tsunami and Nonlinear waves. In Tsunami and Nonlinear waves, 65-95. Springer Berlin Heidelberg.

Dutykh, D., & Dias, F. (2009). Tsunami generation by dynamic displacement of sea bed due to dip-slip faulting. *Mathematics and Computers in Simulation, 80*(4), 837–848.

Freund, L. B., & Barnett, D. M. (1976). A two-dimensional analysis of surface deformation due to dip-slip faulting. *Bulletin of the Seismological Society of America, 66*(3), 667–675.

Fuentes, M. (2017). Simple estimation of linear 1 + 1 D long wave run-up. *Geophysical Journal International, 209*(2), 597–605.

Fuentes, M., Ruiz, J., & Cisternas, A. (2013). A theoretical model of tsunami runup in Chile based on a simple bathymetry. *Geophysical Journal International, 196*(2), 986–995.

Fukutani, Y., Anawat, S., & Imamura, F. (2016). Uncertainty in tsunami wave heights and arrival times caused by the rupture velocity in the strike direction of large earthquakes. *Natural Hazards, 80*(3), 1749–1782.

Geist, E. L. (1998). Local tsunamis and earthquake source parameters. *Advances in Geo-physics, 39,* 117–209.

Hammack, J. L. (1973). A note on tsunamis: their generation and propagation in an ocean of uniform depth. *Journal of Fluid Mechanics, 60*(04), 769–799.

Hayes, G. P., Wald, D. J., and Johnson, R. L. (2012). Slab 1.0: A three-dimensional model of global subduction zone geometries. *Journal of Geophysical Research: Solid Earth, 117*(B1), B01302. https://doi.org/10.1029/2011JB008524.

Ide, S., Beroza, G. C., Shelly, D. R., & Uchide, T. (2007). A scaling law for slow earthquakes. *Nature, 447*(7140), 76–79.

Imai, K., Satake, K., & Furumura, T. (2010). Ampli_cation of tsunami heights by delayed rupture of great earthquakes along the Nankai trough. *Earth, Planets and Space, 62*(4), 427–432.

Kajiura, K. (1970). Tsunami source, energy and the directivity of wave radiation. *Bulletin of the Earthquake Research Institute, 48,* 835–869.

Kajiura, K. (1981). Tsunami energy in relation to parameters of the earthquake fault model. *Bulletin of the Earthquake Research Institute, 56,* 415–440.

Kanamori, H. (1972). Mechanism of tsunami earthquakes. *Physics of the Earth and Planetary Interiors, 6*(5), 346–359.

Kanamori, H., & Brodsky, E. E. (2004). The physics of earthquakes. *Reports on Progress in Physics, 67*(8), 1429.

Kanamori, H., & Cipar, J. J. (1974). Focal process of the great Chilean earthquake May 22, 1960. *Physics of the Earth and Planetary Interiors, 9*(2), 128–136.

Kanamori, H., & Stewart, G. S. (1979). A slow earthquake. *Physics of the Earth and Planetary Interiors, 18*(3), 167–175.

Kânoğlu, U. (2004). Nonlinear evolution and runup-rundown of long waves over a sloping beach. *Journal of Fluid Mechanics, 513,* 363–372.

Lay, T., Kanamori, H., Ammon, C., Nettles, M., Ward, S., Aster, R., et al. (2005). The great Sumatra–Andaman earthquake of 26 December 2004. *Science, 308*(5725), 1127–1133.

Madariaga, R. (2003). Radiation from a finite reverse fault in a half space. *Pure and Applied Geophysics, 160,* 555–577.

Madsen, P. A., & Schaffer, H. A. (2010). Analytical solutions for tsunami runup on a plane beach: single waves, N-waves and transient waves. *Journal of Fluid Mechanics, 645,* 27–57.

Miller, M. M., Melbourne, T., Johnson, D. J., & Sumner, W. Q. (2002). Periodic slow earthquakes from the Cascadia subduction zone. *Science, 295*(5564), 2423.

Okada, Y. (1985). Surface deformation due to shear and tensile faults in a half-space. *Bulletin of the Seismological Society of America, 75,* 1135–1154.

Saito, T. (2013). Dynamic tsunami generation due to sea-bottom deformation: Analytical representation based on linear potential theory. *Earth, Planets and Space, 65*(12), 1411–1423.

Saito, T., & Furumura, T. (2009). Three-dimensional tsunami generation simulation due to sea-bottom deformation and its interpretation based on the linear theory. *Geophysical Journal International, 178*(2), 877–888.

Satake, K., Fujii, Y., Harada, T., & Namegaya, Y. (2013). Time and space distribution of coseismic slip of the 2011 Tohoku earthquake as inferred from tsunami waveform data. *Bulletin of the Seismological Society of America, 103*(2B), 1473–1492.

Stein, S., & Okal, E. (2005). Speed and size of the Sumatra earthquake. *Nature, 434*(7033), 581–582.

Synolakis, C. E. (1987). The runup of solitary waves. *Journal of Fluid Mechanics, 185,* 523–545.

Tanioka, Y., & Satake, K. (1996). Tsunami generation by horizontal displacement of ocean bottom. *Geophysical Research Letters, 23*(8), 861–864.

Todorovska, M. I., & Trifunac, M. D. (2001). Generation of tsunamis by a slowly spreading uplift of the sea oor. *Soil Dynamics and Earthquake Engineering, 21*(2), 151–167.

Tuck, E. O., & Hwang, L.-S. (1972). Long wave generation on a sloping beach. *Journal of Fluid Mechanics, 51,* 449–461.

Ward, S. (2011). Tsunami. In: Gupta H.K. (eds) Encyclopedia of Solid Earth Geophysics. Encyclopedia of Earth Sciences Series. Springer, Dordrecht.

(Received June 22, 2017, revised January 20, 2018, accepted February 7, 2018)

Pure Appl. Geophys.

DOI 10.1007/s00024-017-1687-7

Pure and Applied Geophysics

A Collaborative Effort Between Caribbean States for Tsunami Numerical Modeling: Case Study CaribeWave15

SILVIA CHACÓN-BARRANTES,[1] ALBERTO LÓPEZ-VENEGAS,[2] RÓNALD SÁNCHEZ-ESCOBAR,[3] and NÉSTOR LUQUE-VERGARA[4]

Abstract—Historical records have shown that tsunami have affected the Caribbean region in the past. However infrequent, recent studies have demonstrated that they pose a latent hazard for countries within this basin. The Hazard Assessment Working Group of the ICG/CARIBE-EWS (Intergovernmental Coordination Group of the Early Warning System for Tsunamis and Other Coastal Threats for the Caribbean Sea and Adjacent Regions) of IOC/UNESCO has a modeling subgroup, which seeks to develop a modeling platform to assess the effects of possible tsunami sources within the basin. The CaribeWave tsunami exercise is carried out annually in the Caribbean region to increase awareness and test tsunami preparedness of countries within the basin. In this study we present results of tsunami inundation using the CaribeWave15 exercise scenario for four selected locations within the Caribbean basin (Colombia, Costa Rica, Panamá and Puerto Rico), performed by tsunami modeling researchers from those selected countries. The purpose of this study was to provide the states with additional results for the exercise. The results obtained here were compared to co-seismic deformation and tsunami heights within the basin (energy plots) provided for the exercise to assess the performance of the decision support tools distributed by PTWC (Pacific Tsunami Warning Center), the tsunami service provider for the Caribbean basin. However, comparison of coastal tsunami heights was not possible, due to inconsistencies between the provided fault parameters and the modeling results within the provided exercise products. Still, the modeling performed here allowed to analyze tsunami characteristics at the mentioned states from sources within the North Panamá Deformed Belt. The occurrence of a tsunami in the Caribbean may affect several countries because a great variety of them share coastal zones in this basin. Therefore, collaborative efforts similar to the one presented in this study, particularly between neighboring countries, are critical to assess tsunami hazard and increase preparedness within the countries.

Key words: tsunami inundation modeling, CaribeWave exercise, NEOWAVE tsunami model, RIFT tsunami model, international modeling group.

1. Introduction

According to the Historical Tsunami Database of the National Centers for Environmental Information (NCEI) of NOAA, 112 tsunami have affected the Caribbean basin from 1498 to 2017 (NGDC/WDS 2017). The most remarkable tsunami events within the Caribbean occurred in 1530 Venezuela, 1867 US Virgin Islands, 1900 Venezuela, 1918 Puerto Rico, and 2017 Colombia, with maximum heights of 7.3, 15.2, 10, 6.1, and 8 meters, respectively. In 1755, a tsunami originated from offshore Lisbon affected most of the Antilles, but not continental land. These numbers are in contrast to the vast amount of tsunami records observed at the Pacific basin, and their maximum heights. Accordingly, tsunami research at the Caribbean basin has been much scarcer and recent than at the Pacific Ocean. Tsunami have only recently been recognized as a risk for the Caribbean region, both for the Antilles and for the continental land (e.g., López et al. 2015; Lander et al. 2002; Hayes et al. 2014; Roger et al. 2013; Feuillet et al. 2011; ten Brink et al. 2008; Parsons and Geist 2008; Zahibo et al. 2003).

Concerns of tsunami affecting the Caribbean basin have risen following the aftermath of the Indonesian 2004 tsunami, when tsunami were recognized as a worldwide risk. The Caribbean basin has

[1] RONMAC and SINAMOT, Departamento de Física, Universidad Nacional de Costa Rica, P. O. Box 86-3000, Heredia, Costa Rica. E-mail: silviachaconb@gmail.com
[2] Department of Geology, University of Puerto Rico-Mayagüez, Call Box 9000, Mayaguez 00681-9000, Puerto Rico.
[3] Complejo Naval del Morro-Centro de Investigaciones Oceanográficas e Hidrográficas del Pacífico, DIMAR, Barrio 20 de Julio, Tumaco, Nariño, Colombia.
[4] Instituto de Geociencias, Universidad de Panamá, Campus Central Octavio Mendez Pereira, Urbanización El Cangrejo, Vía Simón Bolívar (Transísmica) con la intersección de la Vía Manuel Espinoza Batista y José de Fábrega, Ciudad de Panamá, Panamá.

considerable tsunami vulnerability due to its reduced size, the large number of small, low relief islands, and coastal development and densification (population and infrastructure). The Caribbean region is home for about 160 million people, counting Central America and Northern South America. Tourism in the region increases steadily according to the Caribbean Tourism Organization (CTO 2015), reaching a 6.3% growth during the first semester of 2015 for a total 14.3 million people visiting 28 CTO member states. Due to its tropical weather the Caribbean is a favorite tourist destination year round, thus increasing the floating population at any given time.

In 2005, the Intergovernmental Coordination Group of the Early Warning System for Tsunamis and other Coastal Hazards for the Caribbean and Adjacent Regions (ICG/CARIBE-EWS) as a subsidiary body of the IOC-UNESCO was established. It provides assistance on tsunami risk reduction to Member States in the Caribbean Region (IOC/UNESCO 2017). As a result, considerable advances in tsunami preparedness have been achieved in the region during the past 12 years, including increase on the coverage of seismic and sea level stations, tsunami evacuation maps, Tsunami Ready Recognition Program, and many others. For example, real-time seismic stations went from 10 to 100 reducing earthquake location time to less than a minute and sea level stations went from 5 to 76 dropping tsunami detection time from 3 h to between 5 and 30 min. ICG/CARIBE-EWS has formed four working groups for specific technical issues.

The Working Group 2 (WG2) focuses on Tsunami Hazard Assessment and it is in charge of defining potential tsunami sources and its consequences. Quantification of tsunami threat and determination of regions of higher tsunami threat is not straightforward. Since 2013, IOC/UNESCO with the support of WG2 organized several Experts Meeting on Tsunami Sources; at these meetings several worst-case scenario sources were defined at some regions within the Caribbean Basin. Sources from 2013 meeting were published in IOC/UNESCO (2013) and sources from 2016 meetings were presented in the American Geophysics Union Fall Meeting 2016 (Chacón-Barrantes et al. 2016).

The WG2 also has endeavored in quantifying the effects of potential tsunami sources through the application of tsunami numerical modeling. As a first approach to fulfill this task, the WG2 has intended to create a modeling system capable of simulating a tsunami caused by an earthquake within the Caribbean Sea. The modeling system should simulate the tsunami propagation through the entire basin and the inundation at selected locations where Member States have established tsunami inundation grids.

The goal of this paper is to perform a first approach in establishing a tsunami-modeling platform by the setup of a main propagation grid and several tsunami inundation grids for selected coastal regions of Colombia, Costa Rica, Panamá, and Puerto Rico. Here, this platform considered a hypothetical tsunami source within the Caribbean basin employed for the tsunami exercise CaribeWave15, to provide the states with an additional input for the exercise. In the future, the modeling platform can be used with other tectonic sources to simulate different scenarios, and can be extended to other Caribbean states for them to nest their inundation grids as well. As mentioned before, results can be used to increase tsunami preparedness as well as to spark collaborations and capacity building throughout the Caribbean.

The modeling platform presented in this study was performed under a unique collaborative effort between the four mentioned states, sharing bathymetric data and their seismic and tsunami expertise to perform joint research on tsunami propagation and inundation.

2. *Tectonic Setting*

The location of the Caribbean plate is primarily defined by the presence of two subduction zones along the eastern and western boundaries, while lateral motions predominantly define motion along the north and south boundaries. The North America and South America plates bound most of the Caribbean plate along the north, south, and eastern edges, while the Cocos plate subducts beneath the Caribbean plate's western boundary (Fig. 1). An uncommon thickening of the Caribbean crust during its eastward motion over the Galápago's hotspot has been suggested as the primary cause for the enriched oceanic crust inability to

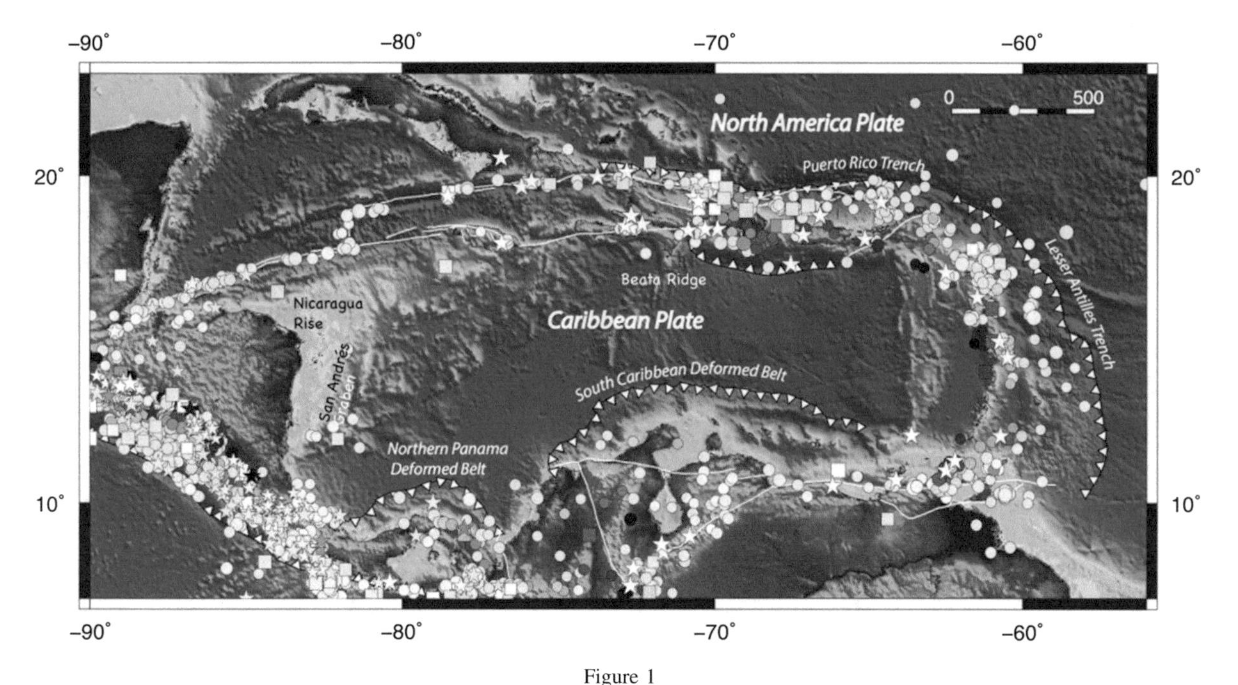

Figure 1
Map of the Caribbean region showing main tectonic features and seismicity. Subduction zones and/or prominent thrust tectonic environments (Puerto Rico Trench, Lesser Antilles Trench, Middle America Trench, Southern Caribbean Deformed Belt, Northern Panamá Deformed Belt) are depicted with white serrated lines. Stars show historical seismicity from the Global Earthquake Model/Global Historical Earthquake Catalog (2015) (1000–1903) and from Benito et al. (2012). Squares are events from the Centennial catalog (1900–1962) of Engdahl and Villaseñor (2002). Circles are seismic events from the NEIC/USGS Catalog (1960–2015). Symbols are color coded according to their depth: white = 0 km, dark red = 200 km

subduct along the eastern edge of the plate, therefore both normal oceanic crust of American plates have succumbed to subduction where the plate dips gently towards the west to form the Lesser Antilles Arc, a young volcanic island arc that has existed since the Eocene. With an average of 2 cm/yr of horizontal displacement (Jordan 1975; Dixon et al. 1998; López et al. 2006; DeMets et al. 2010), the motion of the Caribbean plate is slow and at times has been suggested that its motion is negligible with respect to the mantle reference frame (Müller et al. 1999). The fact that slow contracting motion results in mostly frontal collision along the eastern edge, and that shear zones have developed along the northern and southern boundaries, the region have had limited tsunamigenic events in the short lifespan of western civilization in the Caribbean (1492 to present). Nonetheless, the active nature of the collisional zones is still capable of generating lower frequency damaging events.

The Southwestern Caribbean is characterized by an active collision of the Central America Arc with western South America, which have resulted in the Atrato and Romeral suture zones since the Eocene. Continued convergence between North America and South America began to deform the Costa Rica-–Panamá arc resulting in the North Panamá Deformed Belt (NPDB), and overriding of Colombian basin below South America (Lu and MacMillan 1983; Case et al. 1990). GPS studies (Trenkamp et al. 2002; Bennett et al. 2014) have revealed that the arc is being thrusted eastward to override the Caribbean crust, making the NPDB a pseudo-subduction zone capable of generating significant earthquakes. Along the NPDB at least 12 major events have been registered since 1798 (Fig. 2), and 5 of them have caused tsunami, being the largest the 1882 east Panamá M_{W} 7.9 (Mendoza and Nishenko 1989; Camacho and Viquez 1993; Camacho et al. 2010) and the 1991 Limón M_{W} 7.7 (Plafker and Ward 1992; Camacho 1994). The CaribeWave15 scenario has been chosen as a strictly hypothetical source with the objective of producing the highest yield tsunami without compromising realistic fault parameters for the region of the NPDB.

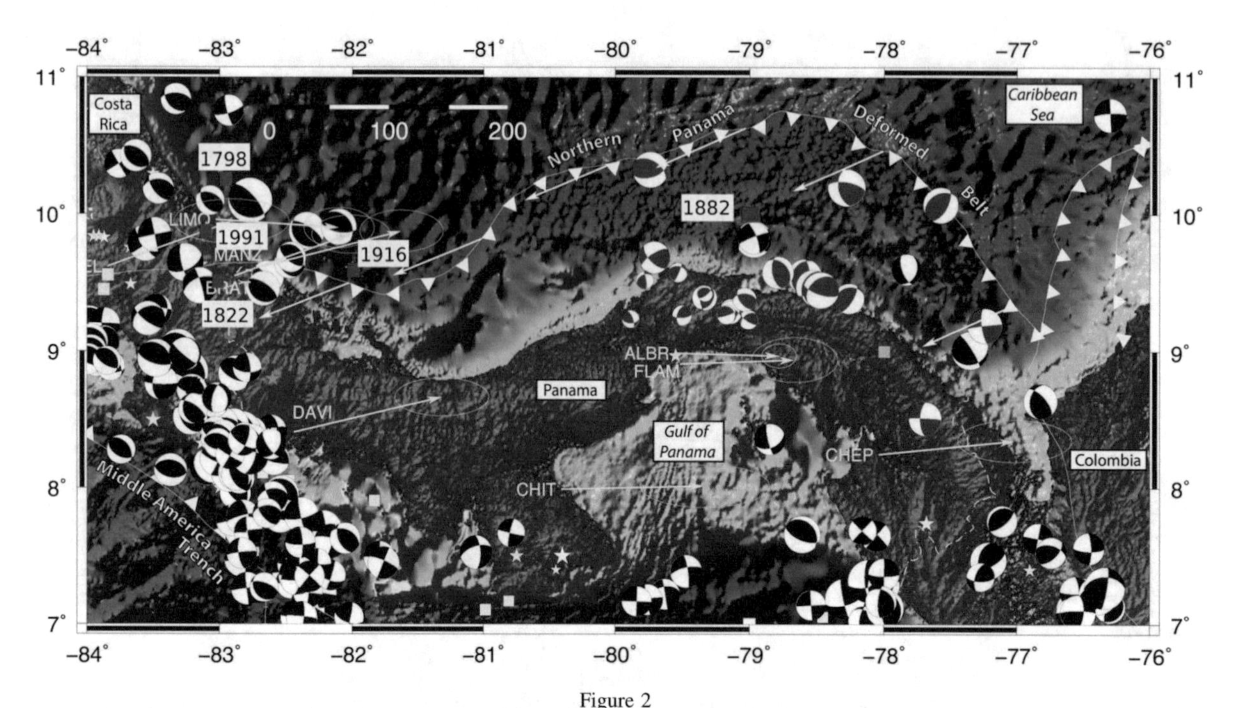

Figure 2
Tectonic setting of the Costa Rica-Panamá arc. Focal mechanisms with black compressional quadrant are from the Global Centroid Moment Tensor (2015) catalog, whereas red ones are from Camacho et al. (2010). White arrows indicate linear velocities predicted at selected locations along the Northern Panamá Deformed Belt using the Caribbean-Nazca pole of rotation from DeMets et al. (2010). Yellow arrows show observed velocities with their 95% confidence ellipsoid of GPS sites in Costa Rica and Panamá as published by Trenkamp et al. (2002). Stars and squares indicate historical events as described in Fig. 1

2.1. *Case Study*

The scenario employed to test the modeling system corresponds to the CaribeWave15 exercise carried out on March 23, 2015. Due to the small tsunami incidence at the Caribbean basin, CaribeWave exercises are performed annually to strengthen member states tsunami preparedness not necessarily following realistic scenarios. The 2015 scenario consisted in the simulation of a M_W 8.5 earthquake at the NPDB.

The Pacific Tsunami Warning Center (PTWC), the service provider for the ICG/CARIBE-EWS, distributed during the exercise a new set of enhanced products in an experimental stage, with the goal of providing detailed information to act as decision support tools. These new products include plots of Energy Forecast (maximum tsunami amplitude at deep ocean), Coastal Forecast (maximum tsunami heights at 1 m depth) and Forecast Polygons, as well as KMZ files with the Coastal Forecast points to be displayed by Google Earth and a Table of Forecast Statistics for Regional Polygons containing results for both coastal and offshore forecast points. The new enhanced products started their experimental phase at the Pacific basin in April 2013 and have been used officially there since October 2015. They have provided an acceptable real-time tsunami height forecast for that basin after the occurrence of several moderate and large tsunami since 2013. For the Caribbean basin, the experimental phase of the new enhanced products began in October, 2015, and they were officially adopted in March, 2016. However, these new products have not been verified there due to the lack of observed tsunamis in the Caribbean region since 2010.

3. *Study Regions*

Each state involved in this work defined its own study region. The island of San Andrés, situated 200 km off the east coast of Nicaragua was chosen as the case study for Colombia. In the eventuality of a tsunami within the Caribbean basin, the island of San Andrés, due to its location, coastal low relief and the number of permanent ($\sim$ 75,000) and floating population ($\sim$ 60,000), is highly vulnerable to tsunami threats, particularly during high-season (January, June, July, and December). The case study for Costa Rica was the center and south Caribbean coast, focusing on Limón city, as this region has records of tsunamis in the past. The south Caribbean coast of Costa Rica has a buoyant small-scale touristic industry consisting in many small hotels (less than three stories) distributed along approximately 40 km of coastline. The main road runs alongshore for the most part of the coast, and few dirt roads and pathways run inland and uphill. Although the relief might allow horizontal evacuation, the dense vegetation represents a handicap along most of this coastline. Unfortunately, there is a lack of fine-scale bathymetry along this region; consequently, it was not possible to build a high-resolution grid here. The only harbors at the Caribbean coast of Costa Rica are located at Limón city, which encloses also a hospital directly at the shoreline and a population of about 61,000 people (INEC 2011). About 114,000 people arrived in cruises during the touristic season of 2014–2015 (ICT 2015), and finer bathymetry is available there, although out of date. There are no evacuation maps or inundation maps, and no evacuation routes have been defined at any location of the Caribbean coast of Costa Rica; despite it was affected by moderate tsunamis in 1991 and 1822 (Camacho and Víquez 1993).

At Panamá, the chosen communities are located at lowlands close to the coast, and there is one school directly at the shoreline. At both communities, the population is less than 500 people but concentrated over a small area. Half of the buildings are small houses. There are no evacuation or inundation maps: this work is an approach to analyze possible scenarios for those communities toward the elaboration of tsunami evacuation maps and to make the communities aware about the risk and possible preparedness measures. The chosen communities are within the most exposed locations for tsunamis generated at the North Panamá Deformed Belt (NPDB).

In Puerto Rico, the chosen region of detail for tsunami inundation was Mayagüez in the western part of the island. This region was based upon the relation between the location of the source and the directivity of the maximum wave energy towards the north-eastern part of the Caribbean. Therefore, the city of Mayagüez was chosen to estimate and compare arrival times and inundation values for the only far-field location in this study. The city of Mayagüez has a population of 89,000, of which the majority live near the coast. The last tsunami affecting the island was in October 11, 1918, and back then the most affected area was the west, where Mayagüez was the most developed city subjected to the highest casualties and destruction.

4. Methods

4.1. Source Description

The NPDB is the most likely source of tsunami for the Southwestern Caribbean region. The CaribeWave15 scenario used two segments in this location that follow the arcuate morphology of the Panamá Arc; one oriented towards the ESE, and another oriented towards the WSW. It is worth noting that this scenario is not considered an entirely plausible source given that GPS studies of Trenkamp et al. (2002) and Bennett et al. (2014) suggest full thrust mechanism for the eastern segment, whereas predominant strike-slip component is expected along the western segment because its strike is sub-parallel to the motion of the GPS-derived velocities. Figure 2 shows linear velocities (white arrows) predicted at the NPDB using the Caribbean-Nazca Euler vector from DeMets et al. (2010), where compression is suggested for the eastern segment, whereas sub-parallel motion is predicted at the western segment. However, for exercise purposes a worst-case scenario was defined, capable of generating the highest tsunami amplitudes possible for the fault parameters and thus producing the most significant effects along the coast. With this goal, the CaribeWave15 Task Team considered both segments with thrust mechanisms rupturing together for a total seismic moment (M_0) of 7.1 $\times$ 10^{28} dyne-cm, producing a M_W 8.5 event.

The composite configuration of the fault area consists of a 120 km x 40 km western segment and a

Table 1

Source parameters: Sub-faults parameters given at CaribeWave15 Participants Handbook (IOC/UNESCO 2015) and PTWC Enhanced Products for CaribeWave15 Exercise (PTWC 2015)

	Participants handbook		PTWC
	Subfault east	Subfault west	
Latitude	10.2°N	9.87°N	10.3°N
Longitude	− 79°E	− 80.07°E	− 78.8°E
Depth	5 km	5 km	15 km
Strike	120°	71°	Unknown
Dip angle	40°	40°	Unknown
Rake	90°	90°	Unknown
Length	182 km	120 km	Unknown
Width	60 km	40 km	Unknown
Mean slip	13 m	9 m	Unknown
Shear modulus	3×10^{11} dyne/cm^2	3×10^{11} dyne/cm^2	Unknown
Seismic moment	4.87×10^{28} dyne-cm	1.43×10^{28} dyne-cm	Unknown

182×60 km eastern segment. Strikes of the fault planes are 77° and 120° for the western and eastern segments, respectively, with a common dip of 40° towards the SE and NE for the western and eastern segments, respectively. Table 1 shows the parameters of the two segments used as source for the simulation as defined in the CaribeWave15 Participants Handbook (IOC/UNESCO 2015). A PTWC Enhanced Products Supplement to the Participants Handbook (PTWC 2015) is available at the CaribeWave15 website (NOAA 2015) in which the PTWC-enhanced products are explained in detail and includes the plots of Energy Forecast (Fig. 7a, PTWC 2015) and Coastal Forecast (Fig. 7b, PTWC 2015) together with isochrones for tsunami arrival, for a M_W 8.5 that were sent during the exercise. However, it does not contain the fault parameters employed, only the epicenter coordinates and seismic moment magnitude (Table 1). The depth given at the PTWC Supplement is 15 km as opposite as the 5-km depth established at the Participants Handbook.

4.2. Numerical Model

We carried out all of our simulations using the tsunami numerical package NEOWAVE-2D (Yamazaki et al. 2011). NEOWAVE-2D is a depth-integrated numerical model that solves the incompressible Navier–Stokes equations with the option to consider either a hydrostatic or non-hydrostatic approach (Yamazaki et al. 2011). The model is implemented in finite differences and includes two-way nesting in several levels. NEOWAVE-2D includes the generation of the tsunami initial condition caused by deformation of the sea bottom calculated with the Okada model (Okada 1985) either static or kinematic multiple fault planes. For this study, the initial condition was calculated using the static sea surface deformation option with the fault parameters for two sub-faults, in hydrostatic mode. The simulations goal did not justify the employment of the non-hydrostatic mode, which is computationally more expensive. NEOWAVE-2D was chosen mainly due to three main reasons: first, it has been validated and verified for tsunami propagation and inundation following Synolakis et al. (2008); second, it is based on finite differences which serve the purpose to share the settings within several states; and third, it is freely available. Please see Yamazaki et al. (2011) for details and additional information regarding the numerical model NEOWAVE-2D.

4.3. Grids Setup and Bathymetric Data

A main grid was defined covering the Caribbean Sea with 1 arc-minute resolution, hereafter referred to as grid one (G1). The bathymetry of the grid was obtained after merging the grid at 1 arc-minute resolution provided by Puerto Rico Seismic Network (PRSN) covering east Caribbean Sea and the grid at 0.996 arc-minute resolution of the Western Caribbean provided by the Maritime General Direction of Colombia (DIMAR). Both grids contain good quality

Table 2

Grid parameters

Location	Size	dx	dt (s)	CFLmax
Caribbean G1	1741 × 958	60 arc-second (~ 1853 m)	3	0.4665
Colombia G2C	376 × 376	15 arc-second (~ 462 m)	1	0.5506
Colombia G3C	232 × 226	5 arc-second (~ 154 m)	0.5	0.5859
Costa Rica G2CR	316 × 271	12 arc-second (~ 371 m)	1.5	0.5939
Costa Rica G3CR	301 × 241	3 arc-second (~ 93 m)	0.5	0.5082
Costa Rica G4CR	361 × 361	1 arc-second (~ 31 m)	0.25	0.3951
Panamá G2P	166 × 226	12 arc-second (~ 371 m)	1.0	0.4665
Panamá G3P	169 × 277	3 arc-second (~ 93 m)	0.5	0.442
Panamá G4P	361 × 361	1 arc-second (~ 31 m)	0.5	0.388
Puerto Rico G2PR	896 × 596	12 arc-second (~ 371 m)	0.5	0.3405
Puerto Rico G3PR	397 × 997	3 arc-second (~ 93 m)	0.25	0.4626
Puerto Rico G4PR	601 × 541	1 arc-second (~ 31 m)	0.125	0.3103
Puerto Rico G5PR	541 × 676	1/3 arc-second (~ 10 m)	0.125	0.1783

bathymetric data around the mentioned states and were merged at about 72.6°W. It also incorporated bathymetric data from digitized nautical charts offshore south Costa Rica. The size, spacing, time step, and maximum Courant–Friedrichs–Laurent condition of G1 are detailed in Table 2 and its extent is shown in Fig. 3a, together with the location of the second-level grids nested by Colombia (G2C), Costa Rica (G2CR), Panamá (G2P) and Puerto Rico (G2PR) to G1.

A total of two grids were used to compute inundation values for the island of San Andrés. Grid two (G2C), with a spatial grid resolution of 15 arc-seconds (~ 462 m) was nested into G1. Grid three (G3C), with a spatial grid resolution of 5 arc-seconds (~ 154 m) was nested into G2C. The bathymetry and extent of both grids are shown in Fig. 3e, and their main parameters are detailed in Table 2. Data from digitized nautical charts were merged with 30 arc-seconds bathymetry data from GEBCO (IOC et al. 2003) and high-resolution LiDAR topography. The nautical charts employed were identified as 044, 200, and 201 from the National Hydrographic Service of the Oceanographic and Hydrographic Caribbean Research Center of DIMAR (CIOHC from the Acronym in Spanish), with scales ranging from 1:1,000 to 1:250,000. LiDAR data is 30-cm resolution in both horizontal and vertical directions, also from DIMAR.

Three grids were nested for Costa Rica and also for Panamá. In both cases, the grids have resolutions of 12, 3, and 1 arc-second, which correspond to approximately 370.6, 92.6, and 30.9 m, respectively. For Costa Rica, the nested grids cover the southern coast and focus on Limón city for flooding (Fig. 3c and Table 2). The bathymetry for those grids was built after merging GEBCO 30 arc-seconds bathymetry with digitized nautical charts 28048 and 28051 from the Defense Mapping Agency Hydrographic/Topographic Center of the United States of America. Those nautical charts are the only source of bathymetry data in that region, having resolutions of 1:145,290 and 1:15,000 and their data was obtained from surveys performed in 1938 and 1987, respectively. Unfortunately these data do not represent the current bathymetry, as the 1991 Limón M_W 7.7 earthquake changed the nearshore bathymetry and topography; Plafker and Ward (1992) and Denyer et al. (1995) measured coastal uplift up to 1.5 m associated to that earthquake. The topography was obtained from SRTM data with 30-m resolution (Jarvis et al. 2008).

For Panamá, the nested grids cover the central Caribbean coast, the region known as "Costa Arriba de Colón", focusing on the towns of Viento Frío and Palenque, with extent and bathymetry shown in Fig. 3d and parameters detailed in Table 2. The bathymetric data at Panamá were obtained after digitized nautical charts 26065 and 26066 from the Defense Mapping Agency Hydrographic/Topographic Center of the United States of America at 1:75,000. Also data from the deep-water nautical

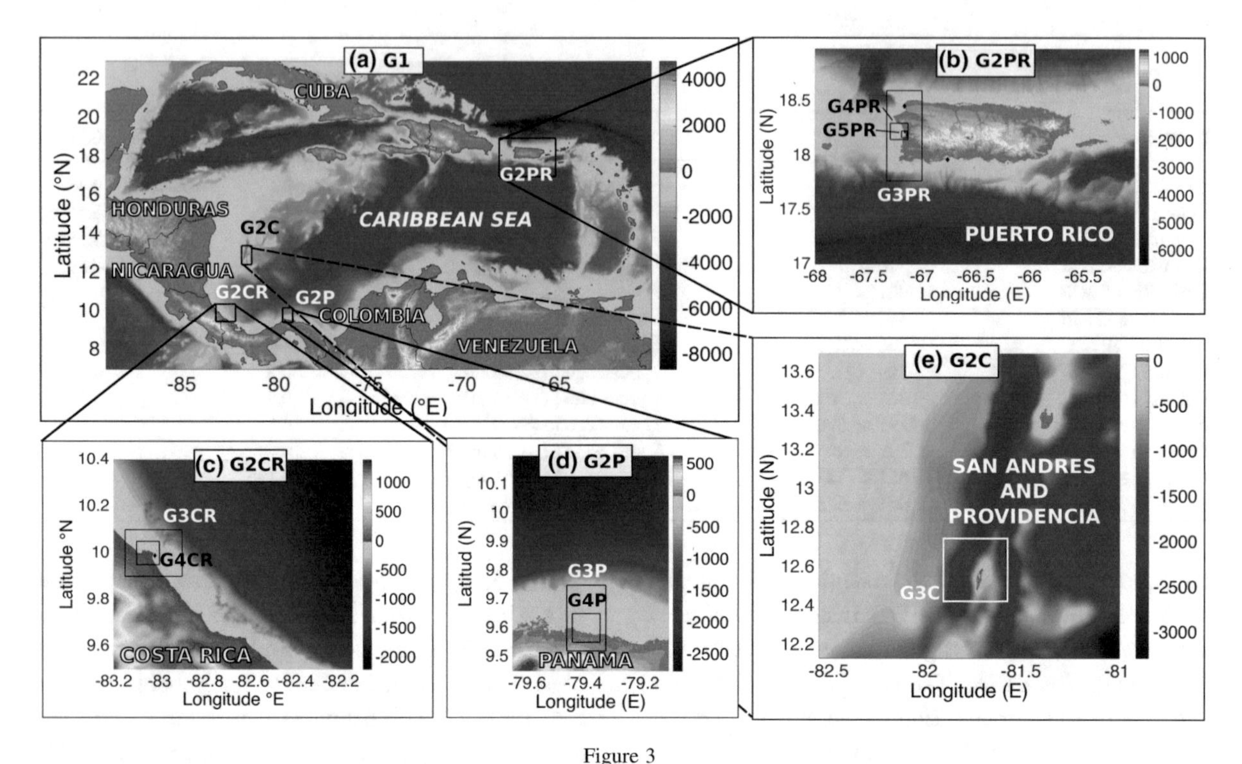

Figure 3

Domain dimensions of bathymetry grids used in this study. **a** Bathymetry of the first grid (G1) and location of nested second-level grids for **b** Puerto Rico (G2PR), **c** Costa Rica (G2CR), **d** Panamá (G2P) and **e** San Andrés Island (G2C). Nesting of grids three, four and five, when applies, are shown in **b–e** by thick black lines. Black diamonds show the position of tidal gauges

chart 4120 covering the Caribbean Sea from the General Maritime Direction of Colombia (DIMAR) at 1:400,000 were employed. Topographic data was obtained from a Digital Elevation Model (DEM) from NextMap Intermap (Intermap Technologies 2010), which is a compilation of SRTM, ASTER y GEOTOPO data.

The grids employed in the simulations for Puerto Rico were created from data obtained through the National Center for Environmental Information (NCEI, formerly NGDC) from the US National Oceanic and Atmospheric Administration (NOAA). The 1 arc-second MHW DEM grid available from the NCEI website was used to generate grids two (G2PR), three (G3PR) and four (G4PR), whereas the 1/3 arc-second resolution grid available was used to generate grid five (G5PR) (Table 2). G2PR and G3PR were degraded to 12 and 3 arc-seconds, respectively. G4PR retained its native resolution but was cropped to the domain shown in Fig. 3b for the greater Mayagüez area. Similarly, G5PR was cropped to generate a domain for the downtown coastal region of the city of Mayagüez.

5. Results and Discussion

5.1. Co-Seismic Deformation

The seismic deformation for the exercise scenario consists of two fault segments divided between east and west. Two input files corresponding to the two fault segments were ingested into NEOWAVE to perform the Okada static sea-floor deformation shown in Fig. 4a. The resulting deformation area has a dimension of 191 km in length and 67 km in width for the eastern segment and 124 km in length, 45 km in width for the western segment. With these dimensions, the computed dislocation area is slightly larger for both segments (compare 180 × 60 km for the eastern segment, and 120 × 60 km for the western segment). A maximum deformation of 7.4 m

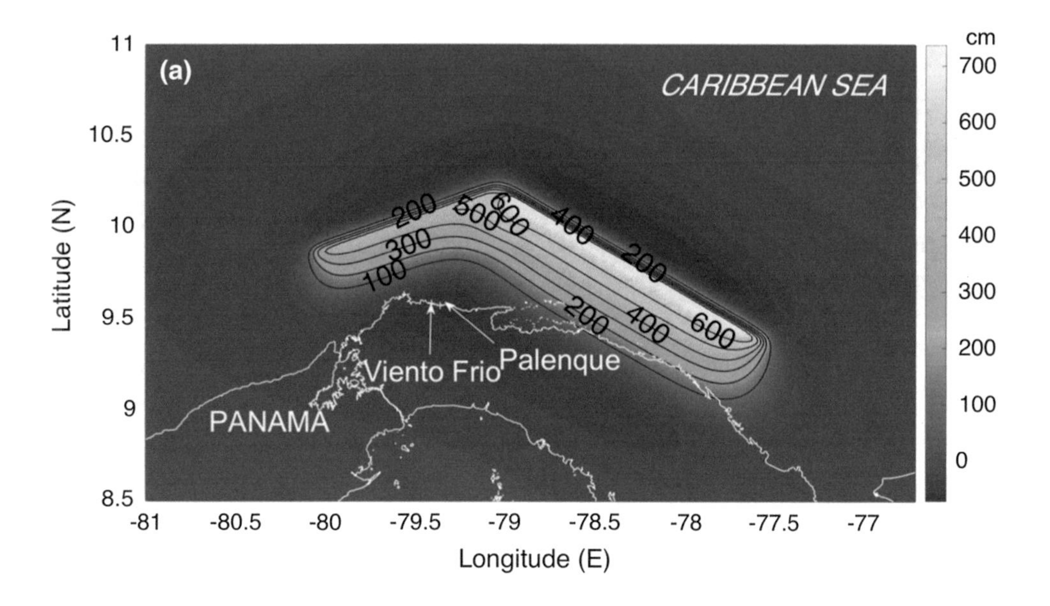

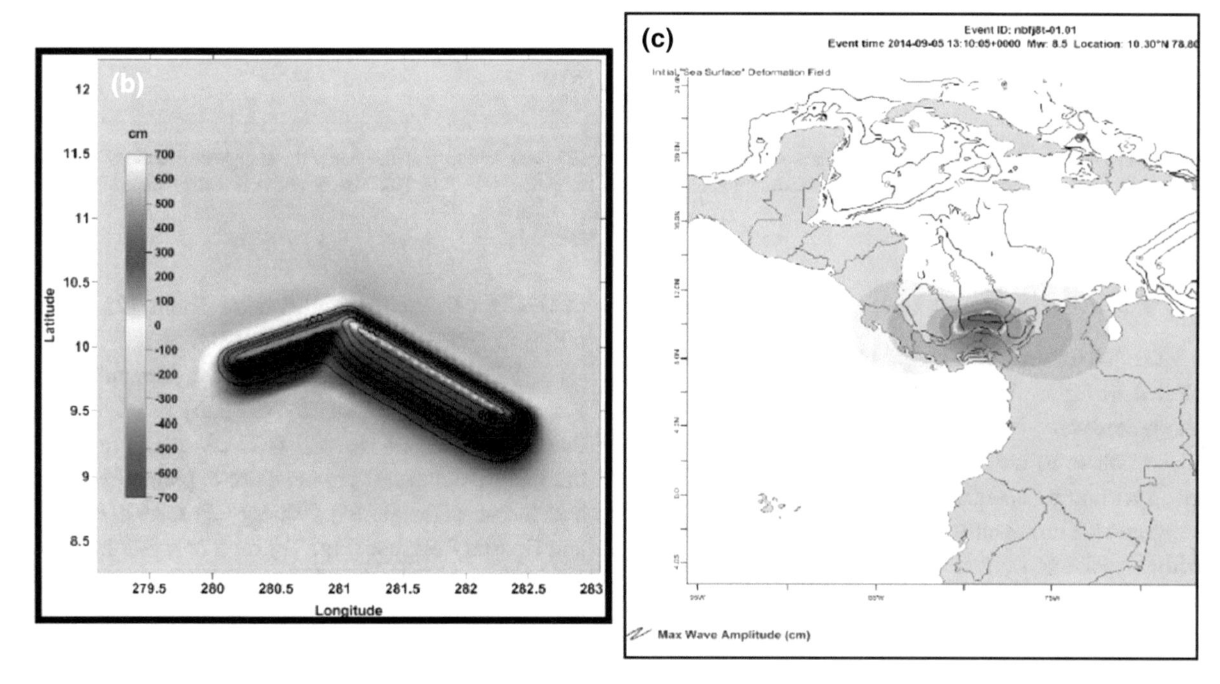

Figure 4
a Co-seismic deformation using Okada (1985) dislocation model. Contours and colors are in centimeters. Viento Frío and Palenque (Panamá) are shown with white crosses. **b** Fig. III-2 from Participants Handbook (IOC/UNESCO 2015). Combined source dislocation from the two faults used in this scenario in ATFM and RIFT. **c** Fig. III-3 from Participants Handbook (IOC/UNESCO 2015). SIFT source dislocation for this scenario

occurs along the shallow portion of the eastern segment, on a location near the junction of the two segments, whereas the maximum deformation along the western segment is 6.3 m. The distribution resembles the deformation depicted in Figure III-2 from the CaribeWave15 Participant's handbook (Fig. 4b) that has a maximum co-seismic uplift of about 7 m.

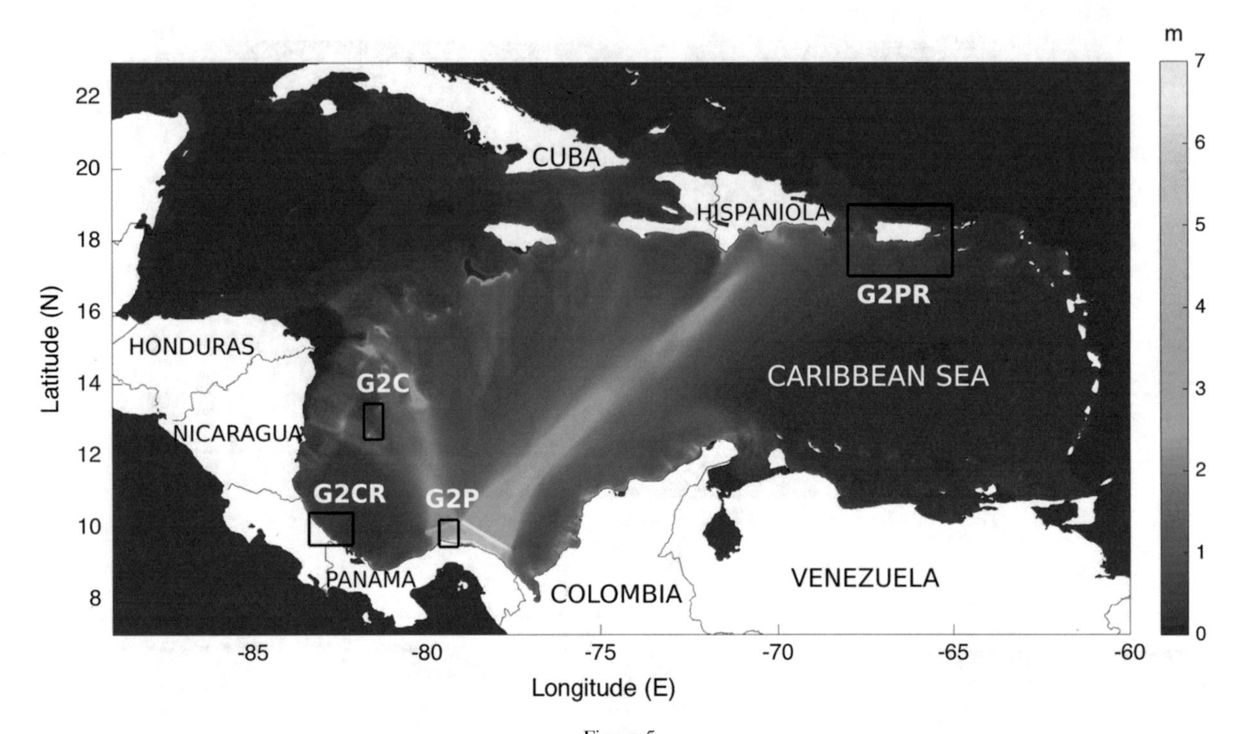

Figure 5
Maximum tsunami height at grid one (G1). The color scale is in meters and is saturated to visualize better the energy distribution. The scale shown reaches a maximum of 3 m, while the maximum tsunami height computed is 13.7 m. Thick black lines show the extent of grid two (G2) for the four states

5.2. Tsunami Propagation Over the Caribbean Basin

The maximum wave height obtained here at G1 is plotted in Fig. 5; the color scale is saturated but the maximum wave height within the Caribbean was of 13.7 m close to the source, which is consistent with the maximum co-seismic uplift and the source magnitude. The directivity of the tsunami is coherent with the seismic source, as there are two directions of maximum energy propagation. The strongest energy beam was directed towards Dominican Republic, also funneled by Beata Ridge, this agreed with the eastern segment presenting a larger slip than the western segment. The second energy beam was directed towards Cayman Islands and San Andrés Island. Hess Escarpment and San Andres Graben refracted the wave directing it to San Andres Archipelago and Nicaragua Rise. Both directions of maximum energy are also visible on the snapshots shown in Fig. 6, obtained with NEOWAVE. The wave front arrives within few minutes to G2P for Panamá, then to G2CR for Costa Rica at about 25 min (Fig. 6a), shortly after to G2C for Colombia, and it takes about 125 min to arrive to G2PR for Puerto Rico (Fig. 6b).

A PTWC Enhanced Products Supplement to the Participants Handbook (PTWC 2015) is available at the CaribeWave15 website (NOAA 2015) in which the PTWC-enhanced products are explained in detail and includes the plots of Energy Forecast (Fig. 7a) and Coastal Forecast (Fig. 7b) for a M_W 8.5 that were sent during the exercise. However, these plots are not the same as the plots of Energy Forecast (Figure III-5, IOC/UNESCO 2015) and Coastal Forecast (Figure III-7, IOC/UNESCO 2015) available at the Participants Handbook. The maximum tsunami height plotted in the Energy Forecast at the PTWC Supplement and distributed during the exercise was 6.76 m, smaller than the co-seismic uplift obtained for a M_W 8.5, which has no physical sense. Also, for this plot the beam of energy directed to northeast is smaller than the beam directed to northwest, which is inconsistent with the source, as the eastern segment released almost four times more energy than the

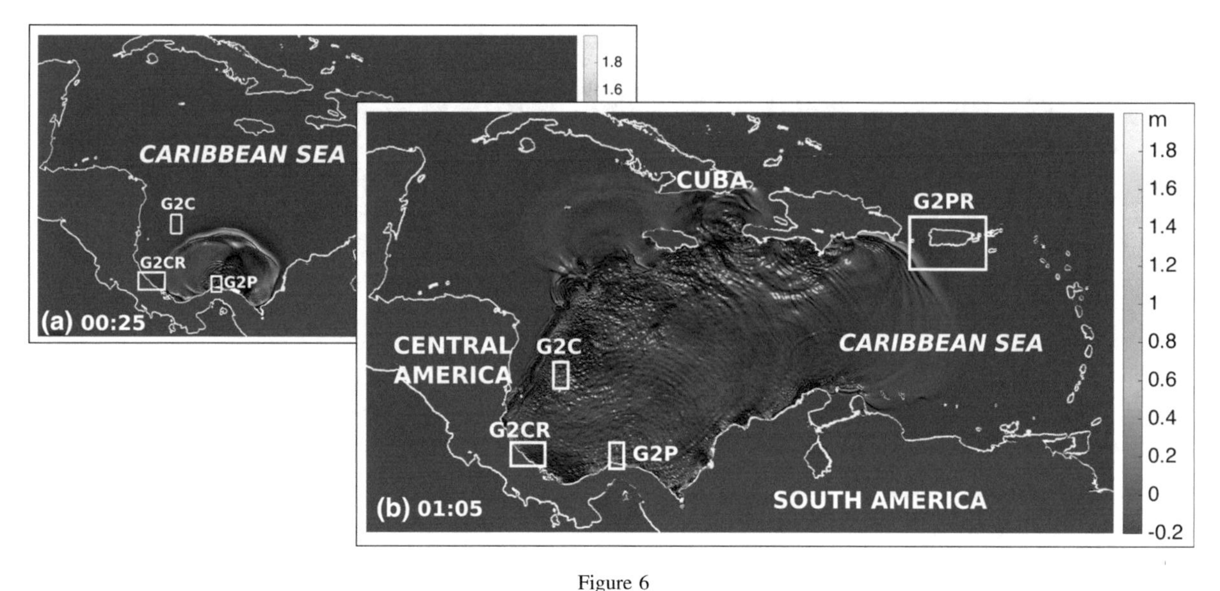

Figure 6
Propagation of wave in grid one (G1), **a** at 25 min arriving to grid two for Costa Rica (G2CR) and soon to arrive to grid two for Colombia (G2C); and **b** at 125 min, arriving to grid two for Puerto Rico (G2PR). Thick white lines show the extent of grid two for the four states. Tsunami heights given in meters

western segment. Additionally, there is no information on the source parameters employed by PTWC as mentioned in Sect. 4.1 and Table 1. On the other hand, the directionality in the Energy Forecast figure in the Participant's Handbook (Figure III-5 at IOC/UNESCO 2015) is consistent with the source; still maximum wave amplitude of 7.6 m is indicated for a M_W 8.16.

5.3. *Model Results at Tidal Gauges*

5.3.1 *Tsunami Heights*

Figure 8 shows the location of existing tide gauges at the four participating states and Fig. 9 shows numerical marigrams at those gauges. The largest values, corresponding to a maximum tsunami height of 7.2 m, 13 min after the earthquake were computed at El Porvenir, a tide gauge located within the deformation region and where water started to raise its level immediately after the earthquake. This gauge is located outside G2P, therefore the modeling results were obtained at G1.

For Puerto Limón, the water started to recede 24 min after the earthquake and started to rise 13 min after, reaching a maximum height of 0.9 m, 46 min after the earthquake, this tidal gauge is located within Limón harbor, which was not resolved by the model resolution. For San Andrés tidal gauge, the water started to recede 40 min after the earthquake and 48 min after the earthquake reached a maximum of 2.6 m.

For the case of Puerto Rico, while the Mayagüez site was computed with the high-resolution 1/3 arc-second grid (G5PR), marigrams for Peñuelas and Aguadilla were computed using G2PR (12 arc-second) and G3PR (3 arc-second), respectively. The maximum wave height among the three gauges was computed at the Mayagüez tidal gauge of 1.4 m and corresponded to the first elevated wave arriving 158 min ($t = 2.633$ h) after the onset of the tsunami (Fig. 9). A peculiar resonance with periodicity of 24 min was observed in this tidal gauge, implying the Mayagüez bay is susceptible to increased tsunami hazards and a prolonged effect can be observed for multiple hours. The fact that the Mayagüez basin is a shallow platform may be responsible for the phenomenon that needs to be identified and characterized better for tsunami hazard mitigation. On the other hand, tide gauges at Aguadilla (north of Mayagüez) and Peñuelas (east of Mayagüez along the southern coast) showed a more typical tsunami wave height

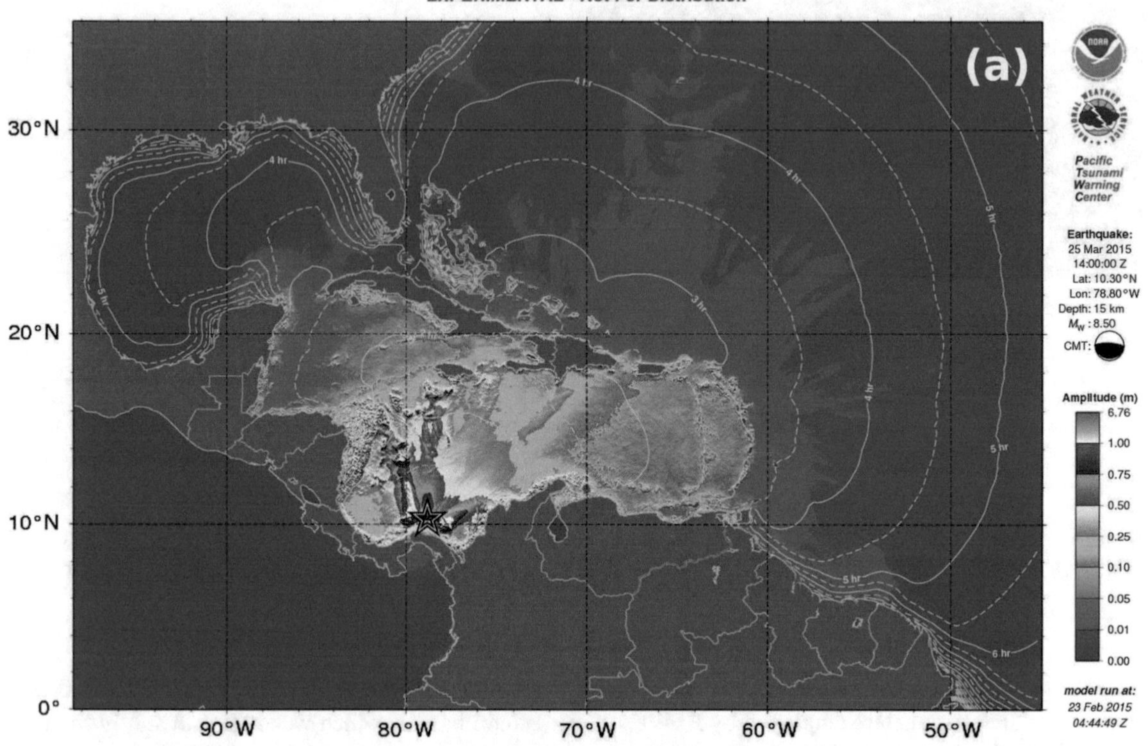
PTWC Energy Forecast
EXPERIMENTAL - Not For Distribution
(a)
30°N
20°N
10°N
0°
90°W
80°W
70°W
60°W
50°W
Pacific Tsunami Warning Center
Earthquake:
25 Mar 2015
14:00:00 Z
Lat: 10.30°N
Lon: 78.80°W
Depth: 15 km
M_W : 8.50
CMT:
Amplitude (m)
6.76
1.00
0.75
0.50
0.25
0.10
0.05
0.01
0.00
model run at:
23 Feb 2015
04:44:49 Z

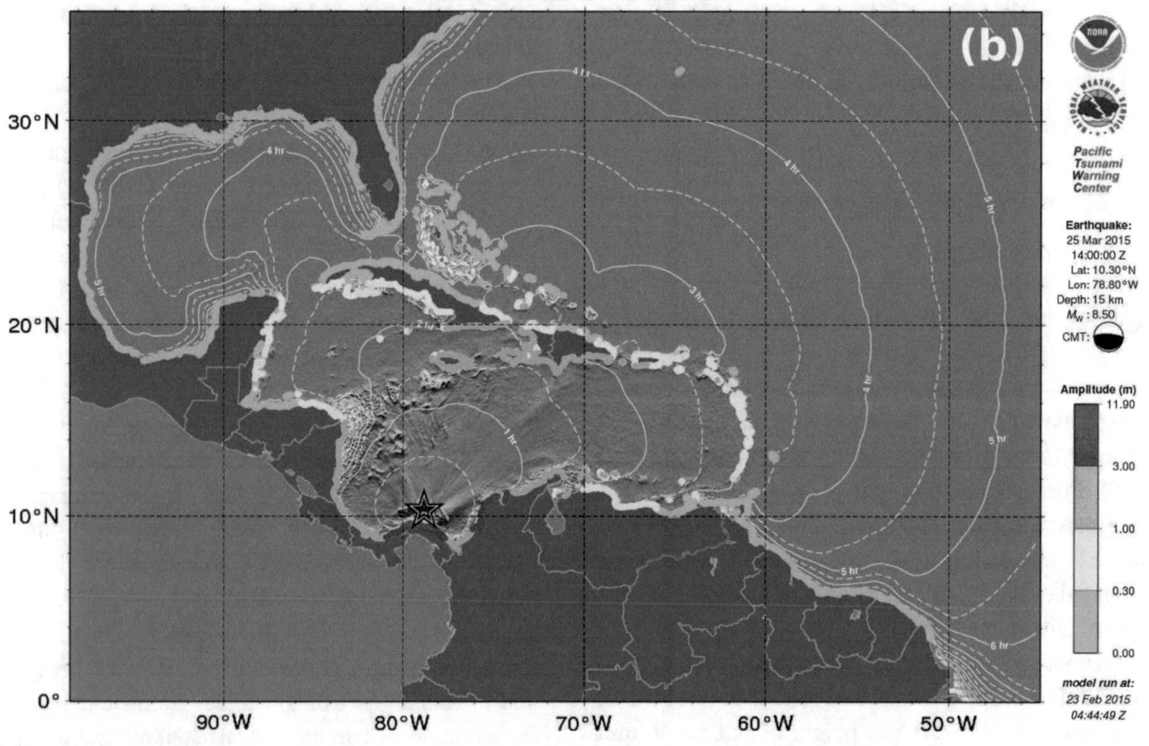
PTWC Coastal Forecast
EXPERIMENTAL - Not For Distribution
(b)
30°N
20°N
10°N
0°
90°W
80°W
70°W
60°W
50°W
Pacific Tsunami Warning Center
Earthquake:
25 Mar 2015
14:00:00 Z
Lat: 10.30°N
Lon: 78.80°W
Depth: 15 km
M_W : 8.50
CMT:
Amplitude (m)
11.90
3.00
1.00
0.30
0.00
model run at:
23 Feb 2015
04:44:49 Z

◀Figure 7
a Energy Forecast and **b** coastal forecast products distributed by PTWC during CaribeWave15 Exercise (PTWC 2015)

history and energy dissipation. The tide gauge at Peñuelas recorded sea withdrawal at 122 min ($t = 2.033$ h), whereas at Aguadilla and Mayagüez occured at 134 min ($t = 2.233$ h) and 141 min ($t = 2.35$ h), respectively. Once again, the shallow nature of the Mayagüez basin results in a slower propagation that enables the wave to arrive faster to the northwestern tip of Puerto Rico before it does at Mayagüez. The Aguadilla tide gauge showed a maximum wave height of 0.9 m and a high frequency oscillation of less than a meter in amplitude. In Peñuelas the maximum height was 1.2 m observed at 141 min after the onset of the tsunami ($t = 2.35$ h). The second most prominent peak reached 0.4 m at 172 min ($t = 2.867$ h) followed by a uniform oscillation that fluctuates between 0.2 and -0.2 m. All travel times were calculated from synthetic marigrams, both for first arrival and arrival of maximum heights.

PTWC provides Coastal Forecast in three formats: a figure, a KMZ file for Google Earth, and a Table of Statistics. The values of Coastal Forecast are estimated using RIFT model results offshore and applying Green's Law to estimate tsunami coastal amplitudes at nearshore points with 1-m depth (PTWC 2015). The KMZ file of the PTWC-enhanced products allows to analyze the forecasted coastal amplitudes at each forecast point along the affected coasts. Two sets of tsunami messages were sent during the exercise: one named "CaribeWave 15 Message" corresponding to the PTWC Messages included in Annex VI of the Participants Handbook and the other one named "CaribeWave 15 Experimental Products" corresponding to the Text Products included in the Supplement. Within both sets of messages, simulated tsunami maximum heights recorded at tidal gauges obtained using a combination

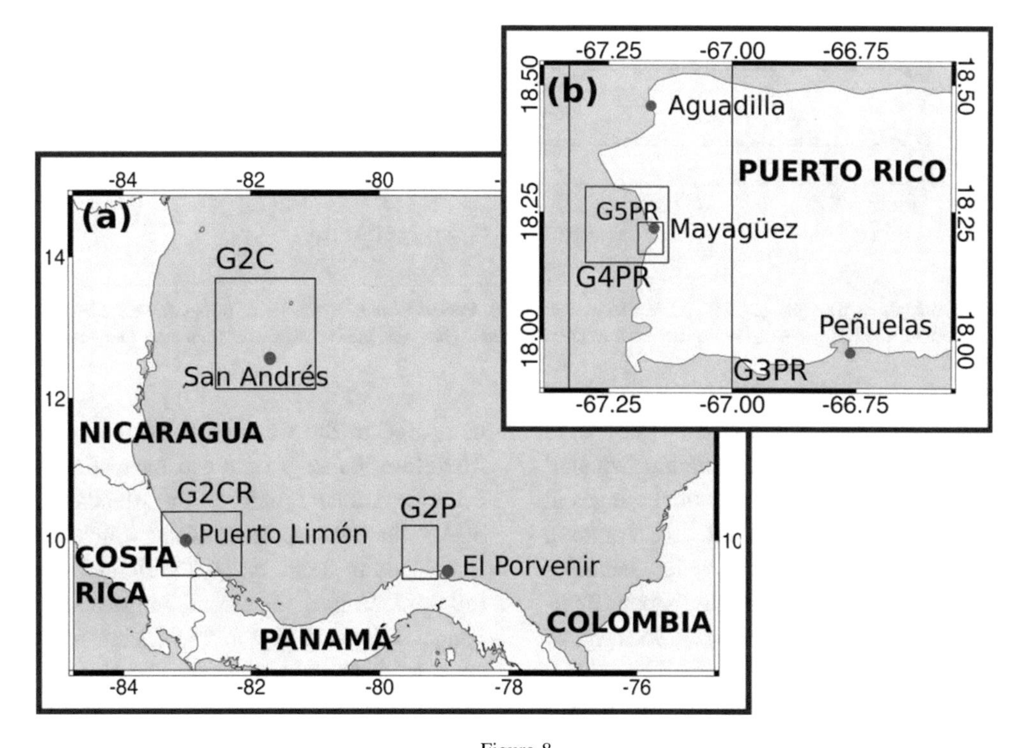

Figure 8
Map location of tidal gauges within the Caribbean region. **a** Tidal gauges San Andrés, Puerto Limón and El Porvenir. **b** Tidal gauges at Puerto Rico: Aguadilla, Mayagüez and Peñuelas

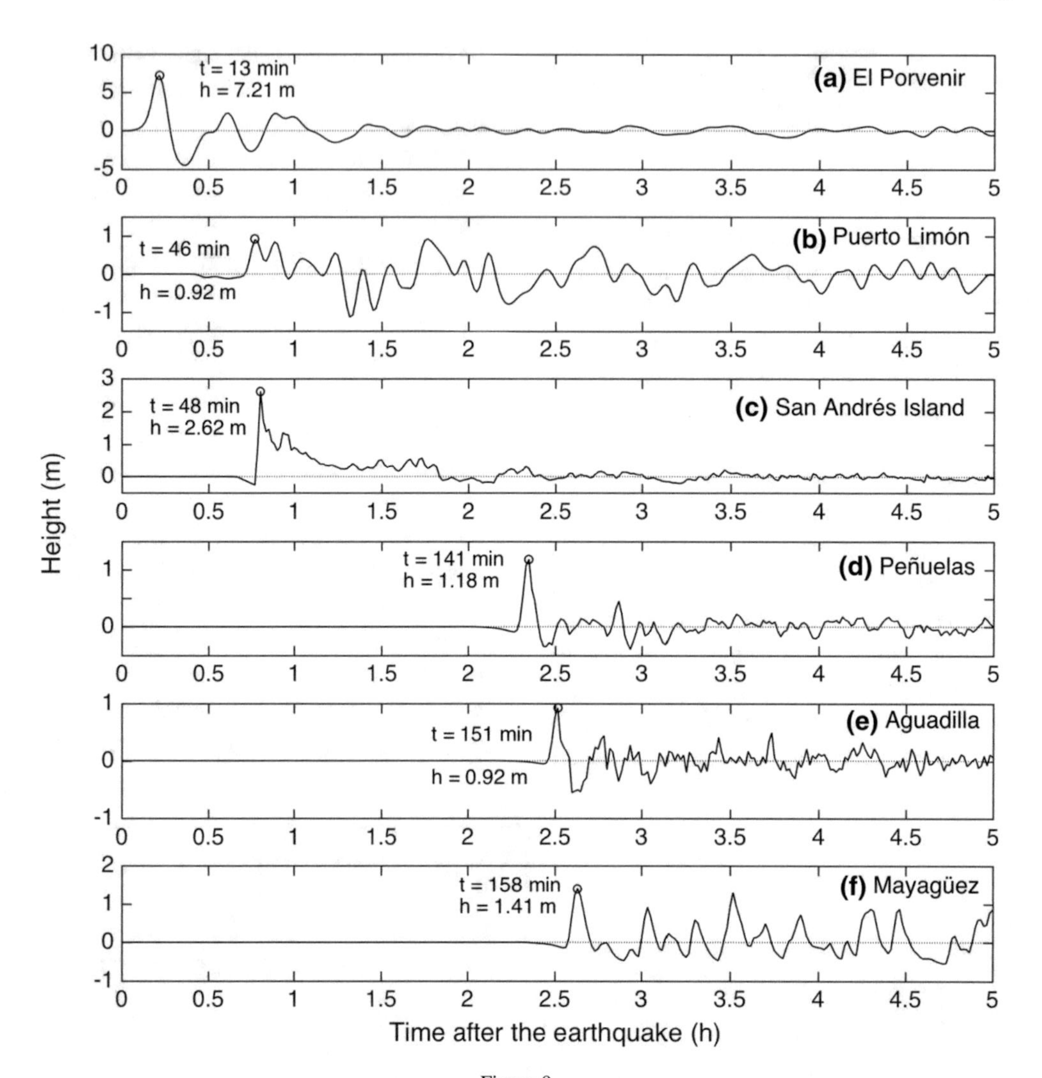

Figure 9

Synthetic tsunami records at tidal gauges at: **a** El Porvenir, Panamá, **b** Puerto Limón, Costa Rica, **c** San Andrés Island, Colombia and **d** Peñuelas, **e** Aguadilla, and **f** Mayagüez, Puerto Rico. All were computed at the finest grid available for each case. Gauge locations in Fig. 8

of RIFT models results and Green's Law were provided. Figure 10 compares maximum tsunami heights obtained with NEOWAVE at tidal gauges at El Porvenir (Panamá), San Andrés (Colombia), Puerto Limón (Costa Rica) and Mayagüez, Peñuelas and Aguadilla (Puerto Rico) with the heights indicated in both sets of tsunami messages, location of the gauges in Fig. 8.

Exercise products were obtained with Green's Law, which assumes (1) that the coastline is linear and exposed to the open ocean, (2) that there are no significant wave reflections and no dissipation by turbulence, and (3) that the bathymetry varies slowly compared to the wavelength of the tsunami waves. Therefore, Green's Law can overestimate or underestimate tsunami heights on islands smaller than 30-km diameter approximately, which is the case of El Porvenir (approx. 0.5 km) and San Andrés (approx. 13 km). For San Andrés, also the tidal gauge is located in shallow waters in front of the coral reef, which dissipates tsunami energy. Green's Law values are also not reliable in harbors and bays, as in Puerto Limón; this harbor is not well solved even by G4CR considered here. The purpose of comparing results obtained with non-compatible methodologies (Green's Law versus inundation

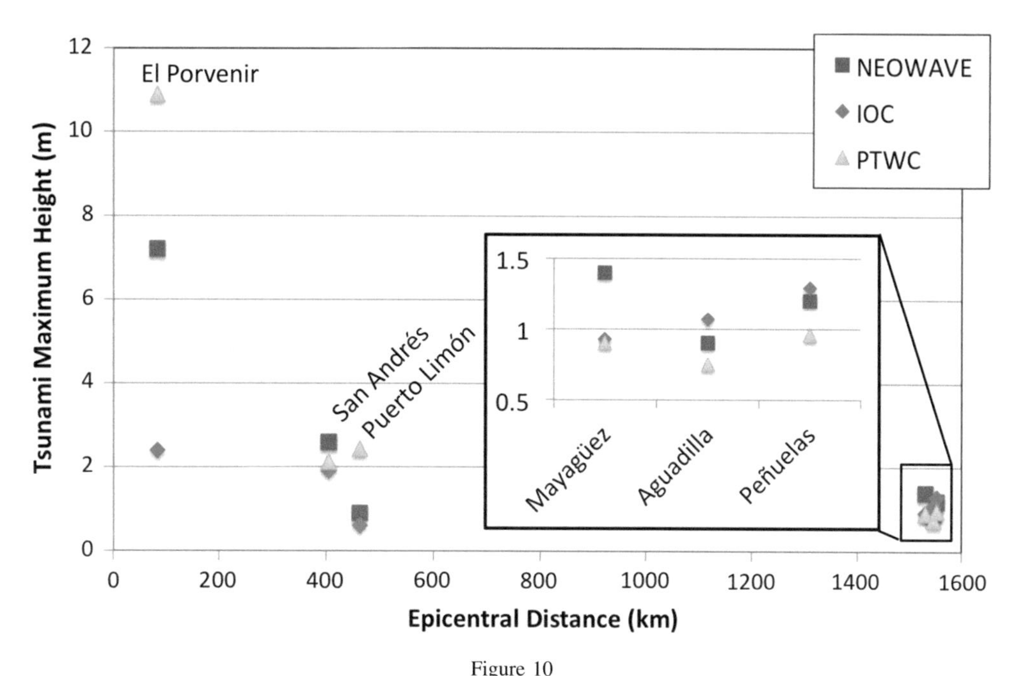

Figure 10
Comparison of tsunami maximum heights obtained with NEOWAVE (red squares) and given by the Participant's Handbook (IOC, blue diamonds) and the Supplement (PTWC, green triangles)

modeling) was to provide orientation on how to interpret the enhanced products distributed by PTWC during a real event. However, even when the parameters employed at the Participants Handbook and PTWC Supplement should have been the same as the ones employed here, the plots of energy forecast presented there indicate that it might not be the case, hindering the comparison's purpose.

5.3.2 Tsunami Arrival Times

Estimated arrival times provided for the exercise were calculated considering a point source instead of finite faults, which increased largely the arrival time in the nearfield, but yielded comparable values for Puerto Rico in contrast to a finite fault (Figs. 11, 12). For El Porvenir, Panamá, Participants Handbook predicted ETA of 24 min, but as the location is within the co-seismic deformation region the tsunami arrived immediately, with the maximum tsunami height arriving 13 min after the earthquake. For Puerto Limón, and Costa Rica, the Participants Handbook and the Supplement predicted arrival times of 54 and 56 min (Fig. 7), respectively, but NEOWAVE resulted in the tsunami arriving 24 min after the earthquake, and the maximum was obtained 46 min after the earthquake. For San Andrés, Colombia, the first arrival occurred 42 min after the earthquake but it was not indicated in any of the exercise publications, it could only be inferred from Fig. 7, being between half hour and an hour. The maximum tsunami height arrived 46 min after the earthquake at NEOWAVE simulation, being in good agreement with the PTWC-Enhanced Products value of 48 min, but being much faster than the Participants Handbook value of 143 min. For Puerto Rico ETA of first tsunami wave published in the Participant's Handbook for Aguadilla are in good agreement but a discrepancy of 6 min exists for Mayagüez. However, while the PTWC Enhanced Product arrival time of the maximum wave at Mayagüez is in agreement with our simulations, the estimated value published by the Participants Handbook occurs 54 min later, clearly overestimating the arrival of the wave that would produce maximum damage at Mayagüez. Unfortunately, values for either Peñuelas or Aguadilla are missing from the publication and thus a direct comparison cannot be made. Travel time

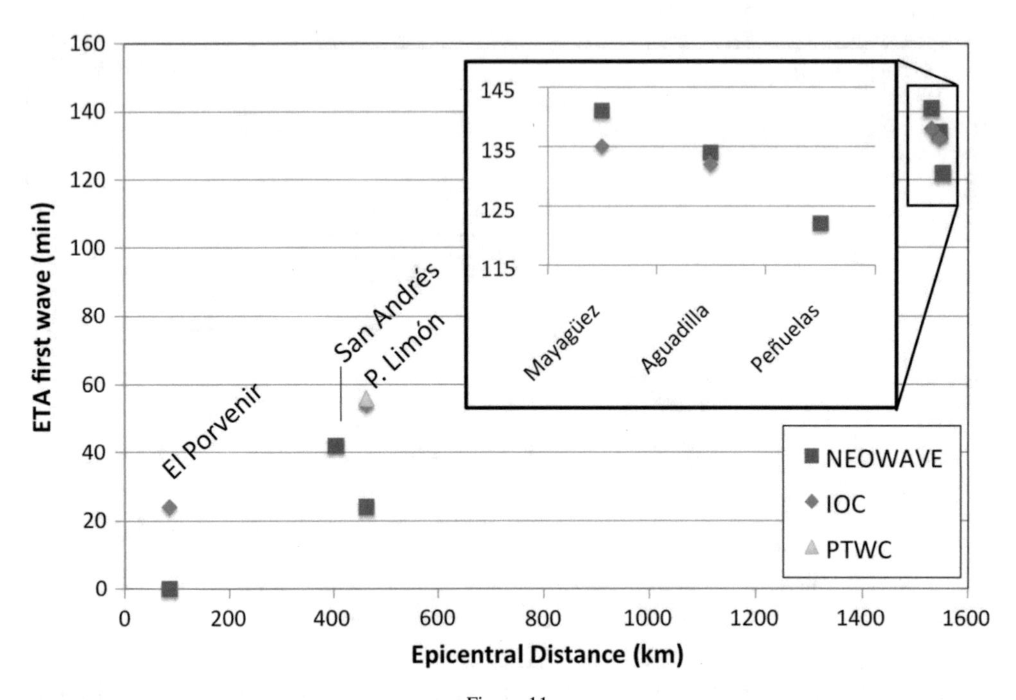

Figure 11
Comparison of estimated time of arrival (ETA) of the first tsunami wave obtained with NEOWAVE (red squares) and given by the Participant's Handbook (IOC, blue diamonds) and the Supplement (PTWC, green triangles)

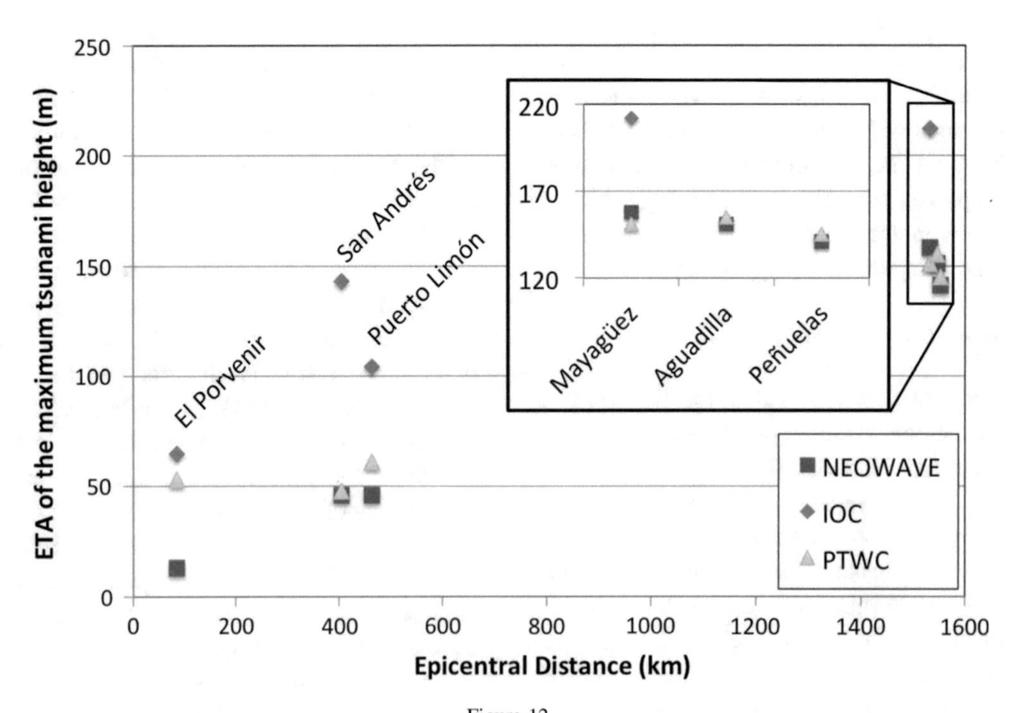

Figure 12
Comparison of estimated time of arrival (ETA) for the maximum tsunami heights obtained with NEOWAVE (red squares) and given by the Participant's Handbook (IOC, blue diamonds) and the Supplement (PTWC, green triangles)

contours in Fig. 7 show arrival times of about 2.5 h, larger than obtained here.

5.4. *Tsunami Inundation*

Regarding the results obtained here, a maximum coastal tsunami height of 21.4 m was calculated using nested grids and inundation modeling with NEOWAVE. This maximum height is about twice the maximum resulted at G1 using rigid wall approach. The discrepancy can be explained by the limitations on the rigid wall approach and the coarse resolution of G1 that does not resolve some coastal features influencing the tsunami heights. NEOWAVE tsunami inundation results obtained here were not compared with coastal forecast provided for the exercise due to uncertainties on the source employed to calculate those products, as mentioned above.

The authors are aware that the source considered here does not represent accurately the tectonic potential of the region and the modeling results were meant to complement the exercise products. However, the analysis of the inundation results for each state can assist identifying the parts of the coasts more exposed to tsunami threat for similar sources.

5.4.1 Colombia

According to Fig. 6a, the tsunami arrived to San Andrés and Providence archipelago from the southeast. The maximum tsunami height obtained for San Andrés Island (G3C) was 5.36 m over high-tide level, as shown in Fig. 13b. The effect of the coral reef in mitigating the tsunami is also evident on the maximum tsunami heights obtained for grid two (Fig. 13a), through a reduction of up to 60% on the wave heights before the tsunami arrives to the coast. The highest tsunami heights were concentrated downtown San Andrés, at the highest vulnerability zone: the region where the stores and hotels are concentrated, with tsunami heights between 2 and 4 m.

The earthquake very likely would not be felt there, consequently disaster management authorities must assure the proper mechanisms to disseminate the warning timely. The tsunami arrived approximately after 40 min, also requiring the population to be properly prepared to ensure a successful speedy evacuation. The inundation simulated here using NEOWAVE is coherent with the geomorphology of San Andrés Island, particularly with the coral reef acting as a barrier.

5.4.2 Costa Rica

For Costa Rica, the tsunami arrived from the east (Fig. 6a). The maximum tsunami runup within G2CR was 2.42 m around Cahuita, which resulted as partially flooded by the tsunami (Fig. 14a). The maximum tsunami heights resulting for G4CR indicate that tsunami heights over 1 m inundated the region where the Hospital Tony Facio is located (Fig. 14b). The Limón Harbor presented slightly smaller tsunami heights, but Moín Harbor presented tsunami heights of almost 2 m. The airport region presented wave heights over 2 m and along the region covered by G4CR the tsunami had its highest heights at Portete of about 2.3 m. The hospital and the airport are key facilities in emergency response if an earthquake and/or a tsunami strike the Caribbean coast of Costa Rica. More specifically, the hospital is located at a dead-end street at which there is also an elementary school, obstructing even more a tsunami evacuation.

The earthquake Mercalli intensity in Costa Rica for this scenario did not reach V (moderate), according with the Shake Map distributed at the CaribeWave15 Participants Handbook, which would not trigger self-evacuation. At the same time, the estimated time of arrival is less than half hour, hindering the chances to issue a timely evacuation order. Even when we are considering a non-realistic source, the tsunami evacuation plan for the south Caribbean coast in Costa Rica must consider a similar case and build capacities on the population to behave accordingly without waiting for a warning or an evacuation order. Detailed calculations of tsunami runup and penetration were performed only for Limón but we recommend it to be extended to the south Caribbean coast of Costa Rica, with a more realistic source, to improve tsunami planning and preparedness.

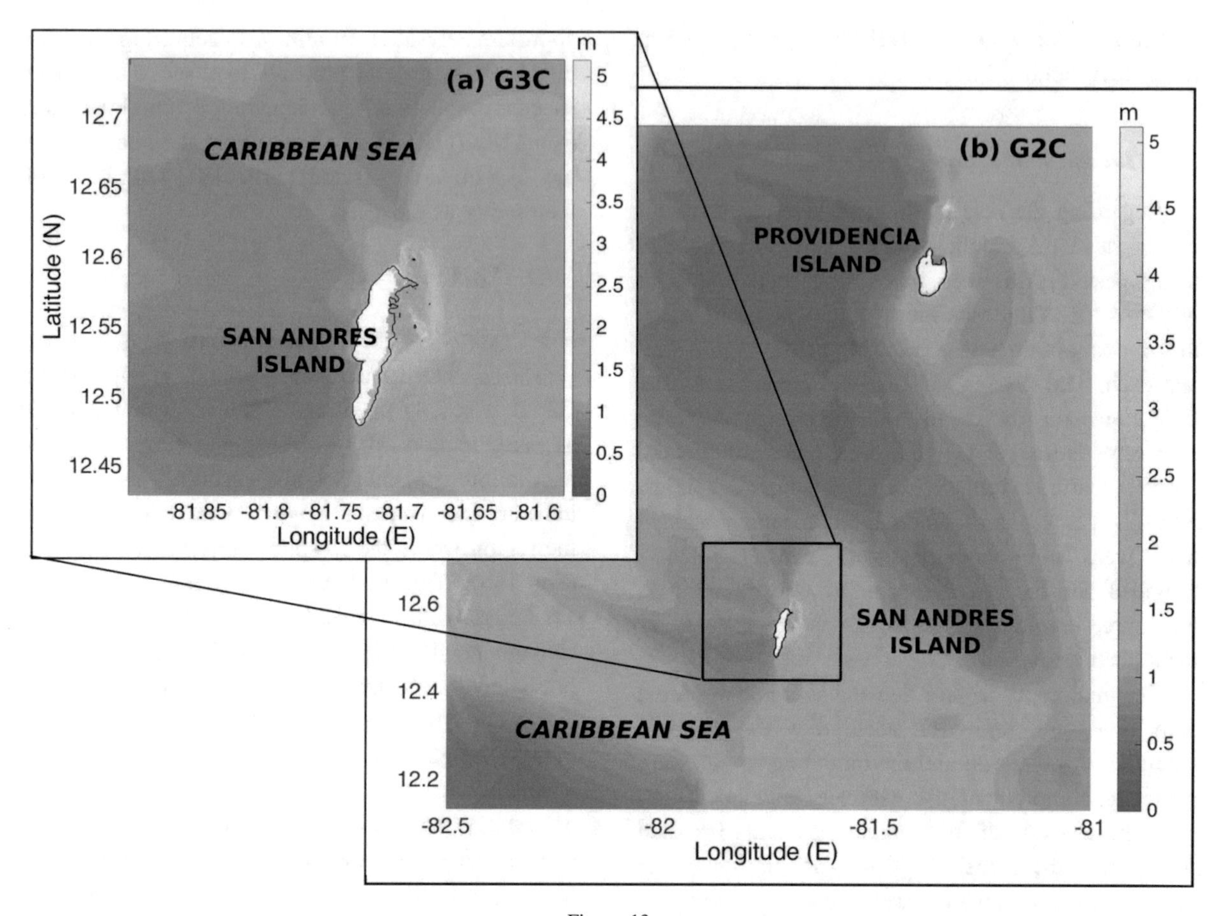

Figure 13
Maximum wave height for Colombia **a** grid three (G3C) and **b** grid two (G2C). Tsunami heights in meters. Extension of G2C and G3C in Fig. 3e

5.4.3 Panamá

The study region at Panamá is right in front of the source region, and it has plain, low relief. Accordingly, maximum tsunami heights of 21.4 m were obtained with NEOWAVE in G4P; and about 20 m and 15 m were obtained at Palenque and Viento Frío, respectively (Fig. 15), with arrival times of about 15 min. Here the Shake Map predicted Mercalli Intensity of VI (strong), which very likely would trigger a self-evacuation if people were prepared.

Only a few highlands might be available for evacuation during such an event. Due to extremely short arrival time, local population must be properly educated on how to react and where to go in case of a strong earthquake. Tsunami evacuation maps and proper signalization are recommended, particularly due to recent touristic development in the region. It is worth noting that a more realistic, plausible tectonic source should be simulated to compare values obtained in this study to evaluate whether vertical evacuation structures need to be considered to mitigate tsunami risk at these potentially high vulnerable regions.

The coastal morphology of the "Costa Arriba de Colón" in Panamá, and south Caribbean coast of Costa Rica might amplify tsunami runup and penetration in some places beyond the results obtained here. However, a better quantification and understanding of those effects can only be reached through numerical studies relying on higher resolution bathymetric and topographic data for both countries. The nautical charts employed to build the Costa Rica and Panamá grids are the only bathymetric data available

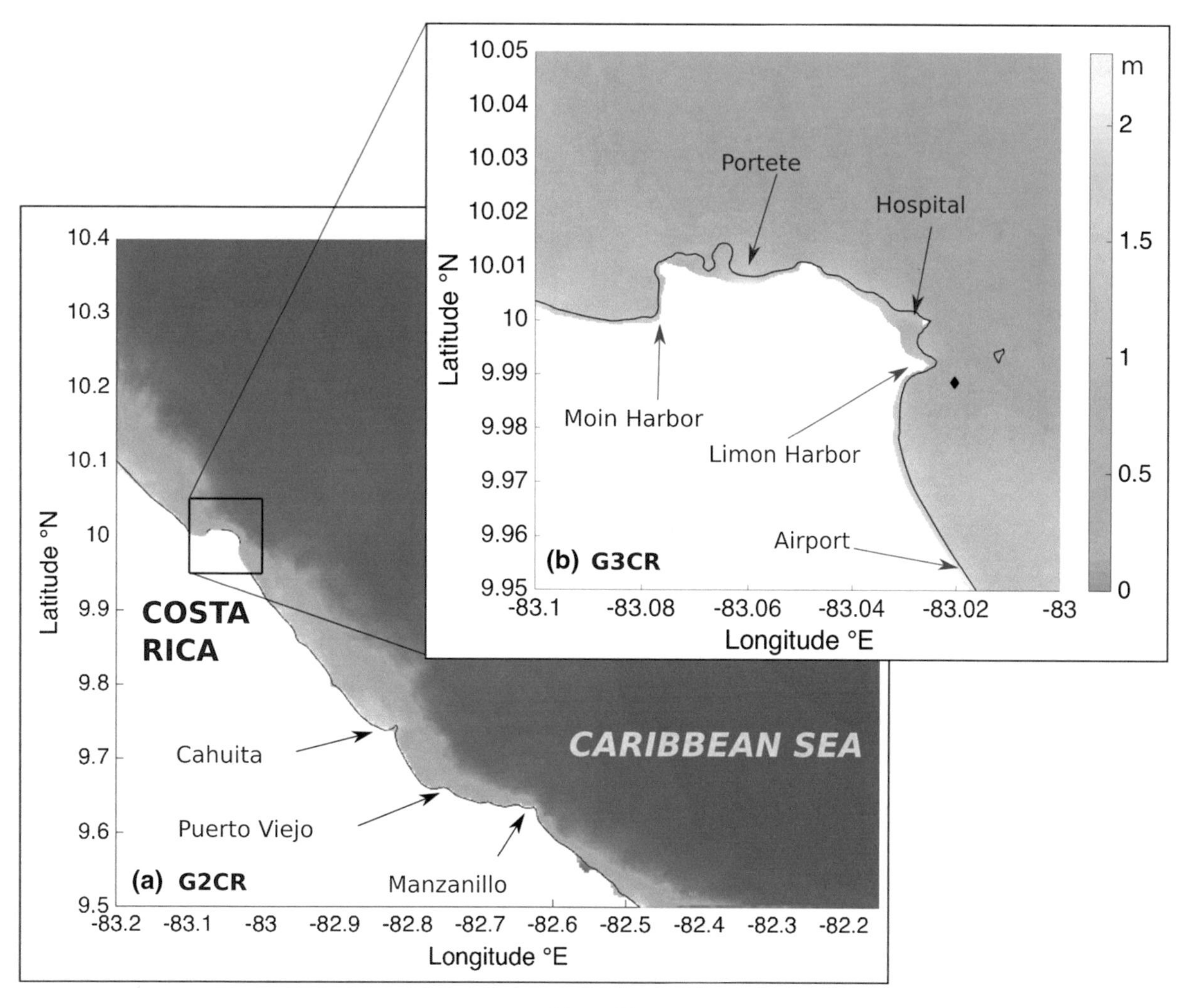

Figure 14
Maximum wave height for Costa Rica at **a** grid two (G2CR) and **b** grid four (G4CR). Tsunami heights in meters. Both **a**, **b** have the same colormap. Extension of G2CR and G3CR in Fig. 3c

for that region, although they do have neither the resolution nor the date desired. In the case of Costa Rica, the 1991 Limón earthquake changed drastically the coastal morphology (Plafker and Ward 1992; Denyer et al. 1995), and those changes are not present in the nautical charts.

5.4.4 Puerto Rico

For Puerto Rico, the tsunami arrived from the southwest (Fig. 6b) and maximum tsunami height within G5PR was 2.6 m, a few meters offshore from the Mayagüez shoreline. The maximum runup on the same grid was 1.5 m, only inundating 350 m from the shoreline and thus did not reach the downtown area of the city (located 1.5 km from the shore). On the other hand, inundation along the coast to the south, although not particularly severe, affected communities with vertical evacuation needs (San José, Guanajibo, etc.). Similarly, there are low-lying areas south of Mayagüez (as it can be seen from Fig. 16a) that are affected by the inundation, suggesting locations suitable for paleotsunami studies. Elsewhere along the western coast of the island, the maximum values indicate noticeable inundation and runup ($\sim$ 1 m) in Añasco (north of Mayagüez—Fig. 16b), Boquerón (Fig. 16c near 18 N, $-$67.17E), and less than 1 m along the southern coast of Puerto

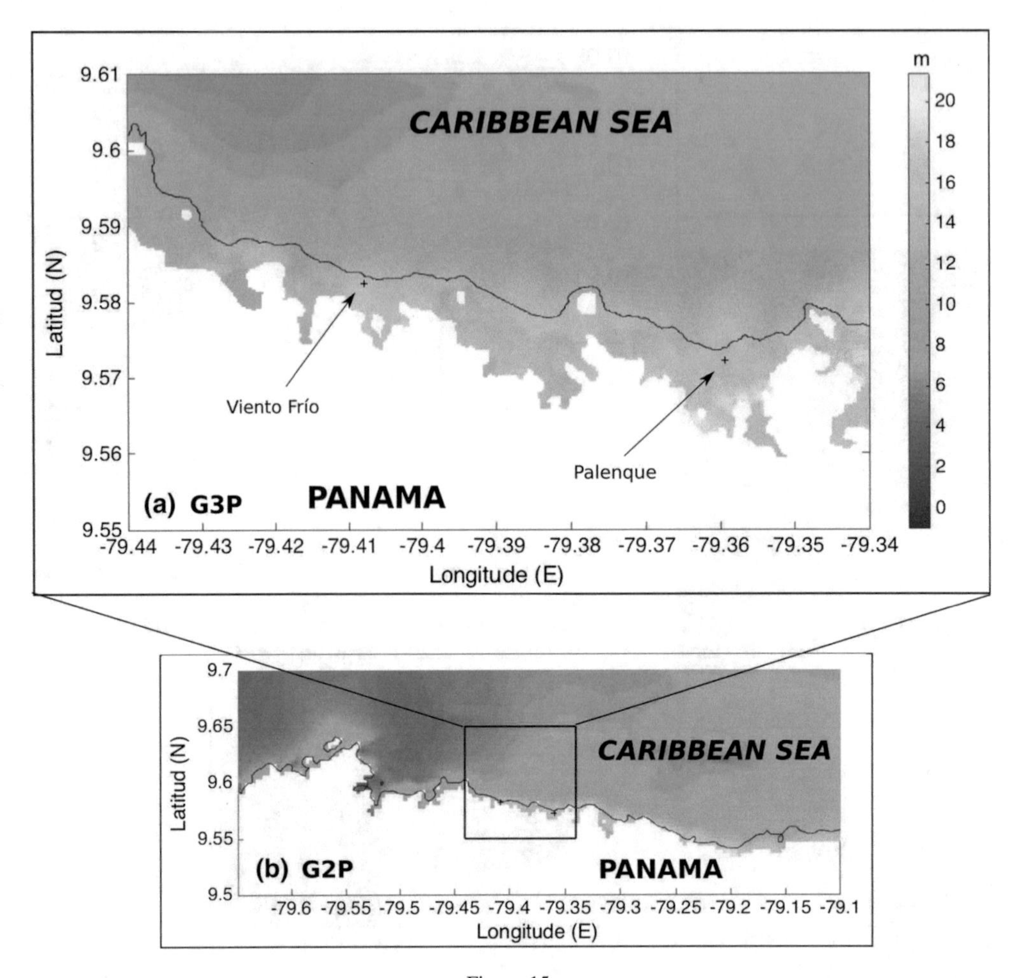

Figure 15
Maximum wave height for Panamá in meters for grid four (G4P) (top) and grid two (G2P) (bottom). Black crosses represent the locations of Viento Frío (left) and Palenque (right). Extension of G2P, G3P and G4P in Fig. 3d

Rico (Fig. 16c). For Puerto Rico, tsunami arrival times were over two hours, much larger than for the other three locations, ensuring enough time for evacuation. Also, Puerto Rico is the only one of the four states having Tsunami Ready Recognition for all locations along its coastline. Tsunami Ready Recognition means that the community has evacuation maps, properly marked evacuation routes and protocols to disseminate warnings among other measures.

6. Conclusions

The tsunami simulated in this study employed a seismic scenario that consisted of a hypothetical M_W 8.5 earthquake along two segments of the Northern Panamá Deformed Belt (NPDB). This scenario was employed for the tsunami exercise CaribeWave15. The resulting tsunami affected the coasts of immediate countries and effects were quantified for selected locations in Colombia (San Andrés Island), Costa Rica, Panamá, and Puerto Rico. We used the tsunami numerical model NEOWAVE (Yamazaki et al. 2011) to simulate the co-seismic deformation and the resulting tsunami. We set up a main propagation grid at which each state nested their inundation grids, two for Colombia, three for each Costa Rica and Panamá and four for Puerto Rico.

The maximum tsunami height obtained here within the Caribbean basin using rigid wall approach,

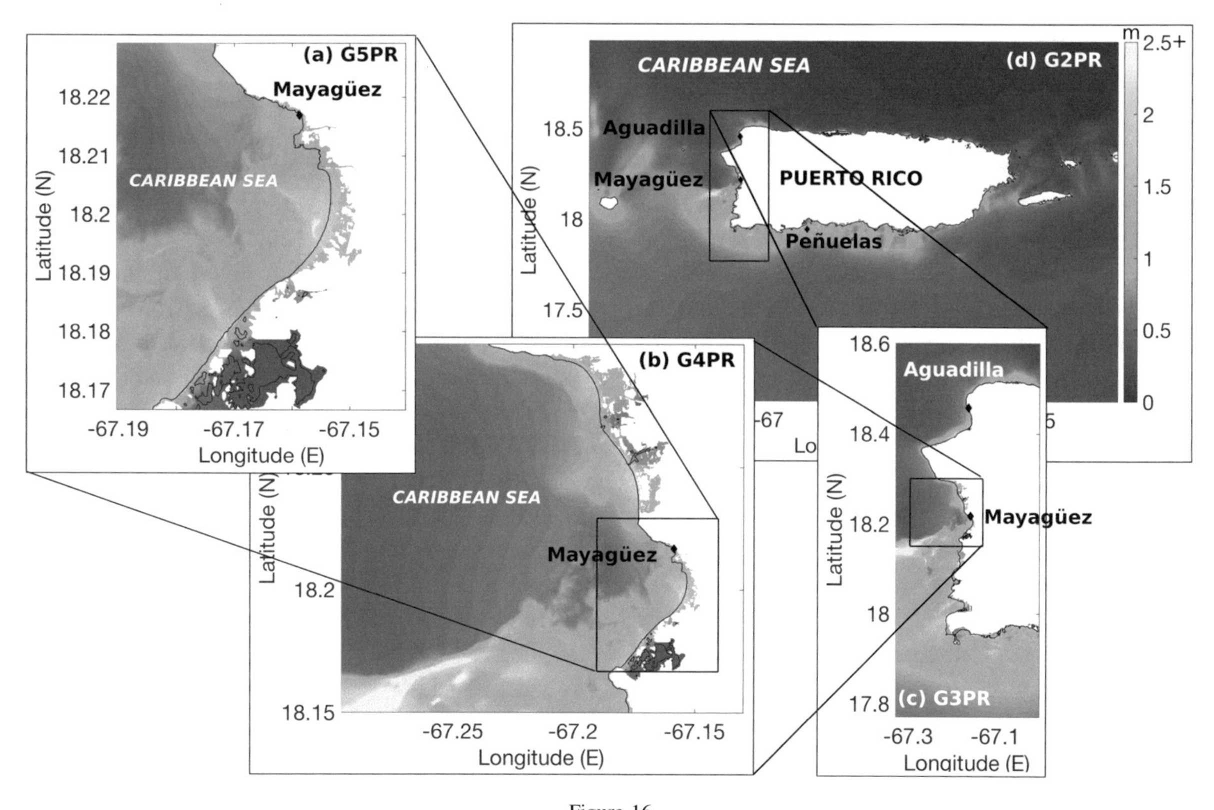

Figure 16
Maximum wave height for Puerto Rico in meters for **a** grid five (G5PR), **b** grid four (G4PR), **c** grid three (G3PR) and **d** grid two (G2PR). Black diamonds show the position of tidal gauges. Extension of G2P, G3P, G4P and G5P in Figs. 3b and 8b

was of 13.7 m at Panamá. The tsunami energy propagation directions agreed with the source and compared well with the results of the AFTM and RIFT propagation models described in the CaribeWave15 Participants Handbook (IOC/UNESCO 2015). However, there were differences on energy forecast, coastal forecast, estimated arrival times for the first arrival and the maximum height at tidal gauges depending on which product version for the exercise were compared to: Participants Handbook or Supplement. The values calculated using NEOWAVE were either smaller or larger depending on the location. According to the simulations performed here, this event would yield wave heights of over 5 m at San Andrés Island, up to 2.4 m at Costa Rica, more than 20 m in Panamá and around 1 m at Puerto Rico.

The coral reef fringing San Andrés Island mitigated the tsunami for the central west coast of the island, as the wave height decreased by 60%. Nonetheless, we observed average tsunami heights of 3.5 m occurring in the northwest side of the island, where most of its hotels, commerce, and economic development occurs, hence posing an important challenge for emergency managers. We find that for this scenario, the combination of the observed 40 min of wave arrival time plus the elevation and extension of the island allows for carrying out a successful evacuation plan, if the population is properly prepared. Therefore, it is recommended to generate tsunami mitigation and evacuation plans to reduce the vulnerability of the island.

For Costa Rica, the chosen scenario presents a challenge on terms of tsunami evacuation. An earthquake triggering a tsunami at the NPDB very likely would not represent a shaking strong enough to prompt a self-evacuation but the tsunami would arrive in less than half hour, with maximum heights over 2 m. It is recommended to increase tsunami preparedness in the region to consider similar scenarios, particularly considering the recent touristic development, possible evacuation routes and the topography and dense vegetation of the region.

Palenque and Viento Frío in Panamá are within the deformation region. Horizontal evacuation of those communities could be particularly difficult because few places exhibit topographic heights in excess of 10–15 m and our simulations yielded tsunami heights that reached up to 20 m. Although the CaribeWave15 scenario portrayed in this study is not entirely realistic, real tsunami like that of 1882 (M_W 7.9) might yield similar estimated arrival times and maximum heights for this region.

The south Caribbean coast of Costa Rica has increased considerably its small-scale touristic activity in recent years. The towns at "Costa Arriba de Colón" region of Panamá are still fishermen towns, however, plans for large-scale development are in progress. Consequently, detailed tsunami inundation simulations would be very useful to increase tsunami preparedness at both regions. However, the ability to produce detailed results is hampered by the currently available bathymetry and topography resolution.

Puerto Rico is the farthest of the study sites in this research. The shallow platform of the Mayagüez basin results in a slight delay of the first wave arrival compared to the tide gauge located in Aguadilla farther up north of Mayagüez. On the other hand, a comparison of maximum wave heights indicates that Mayagüez receives the highest wave amplitude (1.40 m) of the three study sites in Puerto Rico. This may be associated to the high-resolution bathymetry grid employed at this location (G5PR for Mayagüez, G3PR for Aguadilla, and G2PR for Peñuelas). Our results indicate that the geometry of the NPDB source chosen for the exercise does not have a catastrophic impact to the western coast of Puerto Rico as it is observed in the near field sites. However, it is interesting that a resonance of 24 min was observed at the Mayagüez tide gauge, indicating that further studies should be conducted to characterize the tsunami hazard posed by the morphology of the Mayagüez basin.

Green's Law employed at PTWC new enhanced products provides only a rough estimate of tsunami shoaling, as it does not account for bathymetric features that can increase or decrease tsunami heights. This might have happened for locations as San Andrés Island, Puerto Limón, Palenque and Viento Frío. Unfortunately, a detailed comparison could not be performed because of inconsistencies of the exercise products with the source parameters given by the Participants Handbook.

The research presented in this study is the result of an international collaboration between four Caribbean states. By joining data and resources, a platform was created that can enhance future research collaborations among more states. This platform used NEOWAVE, a robust, well-established tsunami numerical model that allows non-hydrostatic approach, heterogeneous faulting with the same or different initiation times, among other features. However, a similar platform employing a simpler and more accessible numerical model, although equally verified and validated, could be used in the future for the benefit of low technical-level users. NEOWAVE required a relatively high level of technical expertise to operate, and this most certainly would be a limiting factor for any Caribbean state without the appropriate resources to perform their own simulations. Employing a more user-friendly platform, such as ComMIT (NOAA 2016) as an alternative numerical model, could facilitate a larger number of states within the Caribbean to generate their own simulations and use it as a tool for warning and planning purposes. Additionally, having reliable bathymetric data would enhance tsunami hazard assessments within the Caribbean Sea.

Acknowledgments

Part of this research was done with a grant of the "Fondo de Apoyo para el Fortalecimiento de Alianzas Estratégicas para el Desarrollo de Proyectos Colaborativos Internacionales", of the Universidad Nacional of Costa Rica. Authors would like to appreciate the valuable suggestions and technical help provided by Y. Yamazaki and J. Horrillo. Also, special thanks go to the reviewers that made this manuscript a valuable contribution to the Caribbean tsunami community, particularly to Jean Rogers and Francois Schindele.

References

Benito, M. B., Lindholm, C., Camacho, E., Climent, A., Marroquín, G., Molina, E., et al. (2012). A new evaluation of seismic hazard for the central america region. *Bulletin of the Seismological Society of America, 102*(2), 504–523.

Bennett, R. A., Spinler, J. C., Compton, K., Rockwell, T. K., & Gath, E. (2014). Global positioning system constraints on active crustal deformation in Central Panamá. *Seismological Research Letters, 85*(2), 278–283. doi:10.1785/0220130177.

Camacho, E. (1994). El tsunami del 22 de abril de 1991 en Bocas del Toro, Panamá, Review Geologists American Central Vol. esp. Terremoto de Limón, 61–64, doi:10.15517/rgac.v0i0.13422.

Camacho, E., Hutton, W., & Pacheco, J. F. (2010). A new look at evidence for a Wadati-Benioff zone and active convergence at the North Panamá deformed belt. *Bulletin of the Seismological Society of America, 100*(1), 343–348.

Camacho, E., & Víquez, V. (1993). Historical seismicity of the north Panama deformed belt. *Review Geologists American Central 15,* 49–64. doi:10.15517/rgac.v0i15.13238.

Caribbean Tourist Organization (CTO) (2015), Latest Statistics 2015, July 29 2015, 5 pp. Available at http://www.onecaribbean.org/wp-content/uploads/Lattab15_FINAL.pdf.

Case, J. E., Shagam, R., and Giegengack, R. F. (1990). Geology of the northern Andes: an overview, in The Caribbean Region, the Geology of North America. In: G. Dengo, Case, J. (Eds.). The Geological Society of America, 1990 vol. H., pp. 177–200.

Chacón-Barrantes, S., López-Venegas, Alberto; Macías, Jorge; Zamora, Natalia; Moore, Christopher; Llorente, Miguel. Numerical Simulation of Several Tectonic Tsunami Sources at the Caribbean Basin. (2016) AGU Fall Meeting, Abstract NH43A-1799. Available at http://abstractsearch.agu.org/meetings/2016/FM/NH43A-1799.html

DeMets, C., Gordon, R. G., & Argus, D. F. (2010). Geologically current plate motions. *Geophysical Journal International, 181*(1), 1–80.

Denyer, P., Arias, O., & Personius, S. (1995), Efecto tectónico del terremoto de Limón, *Review Geologists American Central* Vol. esp. Terremoto de Limón, 39-52. doi:10.15517/rgac.v0i0.13394.

Dixon, T. H., Farina, F., DeMets, C., Jansma, P., Mann, P., & Calais, E. (1998). Relative motion of the Caribbean plate and associated boundary zone deformation based on a decade of GPS observations. *Journal of Geophysical Research, 103,* 15157–15182.

Engdahl, E.R., and Villaseñor, A. (2002). Global seismicity: 1900–1999. In: Lee, W.H.K., Kanamori, H., Jennings, P.C., Kisslinger (Eds.). International Handbook of Earthquake and Engineering Seismology. Academic Press pp. 665–690.

Feuillet, N., Beauducel, F., & Tapponnier, P. (2011). Tectonic context of moderate to large historical earthquakes in the Lesser Antilles and mechanical coupling with volcanoes. *Journal of Geophysical Research, 116,* B10308. doi:10.1029/2011JB008443.

Global Centroid Moment Tensor. (2015). Catalog Search. [Available online at http://www.globalcmt.org/CMTsearch.html]. [Accessed 2015]

Global Earthquake Model/Global Historical Earthquake Catalogue. (2015). Catalog search. [Available online at http://www.emidius.eu/GEH/download/GEM-GHEC-v1.kmz]. [Accessed 2015].

Hayes, G. P., McNamara, D. E., Seidman, L., & Roger, J. (2014). Quantifying potential earthquake and tsunami hazard in the Lesser Antilles subduction zone of the Caribbean region. *Geophysical Journal International, 196,* 510–521. doi:10.1093/gji/ggt385.

ICT (2015). Anuario Estadístico de Turismo 2015. Instituto Costarricense de Turismo (ICT).

INEC (2011). X Censo Nacional de Población y VI de Vivienda. Instituto Nacional de Estadística y Censos de Costa Rica (INEC).

IOC, IHO and BODC (2003) Centenary Edition of the GEBCO Digital Atlas, published on CD-ROM on behalf of the Intergovernmental Oceanographic Commission and the International Hydrographic Organization as part of the General Bathymetric Chart of the Oceans; British Oceanographic Data Centre, Liverpool.

IOC/UNESCO (2013) Earthquake and Tsunami Hazard in Northern Haiti: Historical Events Earthquake and Tsunami Hazard in Northern Haiti: Historical Events.

IOC/UNESCO (2015). Exercise Caribe Wave/Lantex 15 Participant Handbook. Intergovernmental Oceanographic Commission Technical Series 118.

IOC/UNESCO (2017). Caribbean Home. Intergovernmental Oceanographic Commission Tsunami Programme website. [Available online at http://www.ioc-tsunami.org/index.php?option=com_content&view=article&id=9&Itemid=15&lang=en] [Accessed 2015].

Intermap Technologies. (2010). Product Handbook & Quick Start Guide v.4.4.3. [Available online at http://www.intermap.com/hs-fs/hub/395294/file-1384266545-pdf/pdf/brochures/INTERMAP_Product_Handbook-web.pdf] [Accessed 2015].

Jarvis, A., Reuter, H.I., Nelson, A., and Guevara, E. (2008). Hole-filled SRTM for the globe Version 4, http://srtm.csi.cgiar.org. Accessed 2015.

Jordan, T. H. (1975). The present-day motions of the Caribbean plate. *Journal of Geophysical Research, 80,* 4433–4439.

Lander, J., Whiteside, L., & Lockridge, P. (2002). A Brief History of Tsunamis in the Caribbean Sea. *Science of Tsunami Hazard, 20*(1), 57–94.

López AM, Chacón-Barrantes S, Zamora N, et al. (2015). Tsunamis from Tectonic Sources along Caribbean Plate Boundaries. In: AGU Fall Meeting Abstracts. San Francisco, p T11E – 2942

López, A. M., Stein, S., Dixon, T. H., Sella, G., Jansma, P. E., Weber, J., et al. (2006). Is there a Northern Lesser Antilles Forearc block? *Geophysical Research Letters.* doi:10.1029/2005GL025293.

Lu, R. S. and McMillan, K. J. Multichannel seismic survey of the Colombia basin and adjacent margin, in Stories in continental margin geology. In: Watkins, J. S. and Drake, C. L. (Eds.) Amer. Assoc. Petr. Geolog., 1983, Memoir 34, pp. 395–410.

Mendoza, C., & Nishenko, S. (1989). The North Panama earthquake of 7 September 1882. *The Bulletin of the Seismological Society of America, 79,* 1264–1269.

Müller, R. D., Royer, J. Y., Cande, S. C., Roest, W. R., & Maschenkov, S. (1999). New constraints on the Late Cretaceous/Tertiary plate tectonic evolution of the Caribbean, Caribbean Basins. *Sedimentary Basins of the World, 4,* 39–55.

NEIC/USGS. (2015). Earthquake Archives. http://earthquake.usgs.gov/earthquakes/search/. Accessed 2015.

NGDC/WDS. (2017). Global Historical Tsunami Database. http://www.ngdc.noaa.gov/hazard/tsu_db.shtml. Accessed 2015.

NOAA. (2015). CARIBE WAVE/LANTEX 2015 website. http://www.srh.noaa.gov/srh/ctwp/?n=caribewave2015. Accessed 2015.

NOAA. (2016). Community Model Interface for Tsunami (ComMIT). NOAA Center for Tsunami Research. http://nctr.pmel.noaa.gov/ComMIT/. Accessed 2016.

Okada, Y. (1985). Surface deformation due to shear and tensile faults in a half-space. *The Bulletin of the Seismological Society of America, 75*(4), 1135–1154.

Parsons, T., & Geist, E. L. (2008). Tsunami probability in the Caribbean Region. *Pure and Applied Geophysics, 165*, 2089–2116. doi:10.1007/s00024-008-0416-7.

Plafker, G., & Ward, S. (1992). Backarc thrust faulting and tectonic uplift along the Caribbean Sea coast during the April 22, 1991 Costa Rica earthquake. *Tectonics, 11*(4), 709–718.

PTWC. (2015). PTWC enhanced products for CaribeWave/LANTEX 2015 Panama Scenario (Supplement to the CaribeWave/LANTEX 2015 Handbook). [Available online]. http://www.srh.noaa.gov/images/srh/ctwp/PTWCEnhancedProductsCARIBE_WAVE2015Final.pdf. Accessed 2015.

Roger, J., Dudon, B., & Zahibo, N. (2013). Tsunami hazard assessment of Guadeloupe Island (F.W.I.) related to a megathrust rupture on the Lesser Antilles subduction interface. *Natural Hazards and Earth Systems Sciences, 13*, 1169–1183. doi:10.5194/nhess-13-1169-2013.

Synolakis, C. E., Bernard, E. N., Titov, V. V., Kânoğlu, U., & González, F. I. (2008). Validation and verification of Tsunami numerical models. *Pure and Applied Geophysics, 165*(11–12), 2197–2228. doi:10.1007/s00024-004-0427-y.

ten Brink US, Twichell D, Geist EL, et al (2008) Evaluation of Tsunami Sources with the Potential to Impact the U.S. Atlantic and Gulf Coasts—a Report to the Nuclear Regulatory Commission: U.S. Geological Survey Administrative Report.

Trenkamp, R., Kellogg, J. N., Freymueller, J. T., & Mora, H. P. (2002). Wide plate margin deformation, southern Central America and northwestern South America, CASA GPS observations. *Journal of South America Earth Sciences, 15*, 157–171.

Yamazaki, Y., Cheung, K. and Zowalik, Z. (2011). Depth-integrated, non-hydrostatic model with grid nesting for tsunami generation, propagation, and run-up, International Journal for Numerical Methods in Fluids. 67 (December 2010), 2081–2107, doi:10.1002/fld.2485.

Zahibo, N., Pelinovsky, E., Yalciner, A. C., et al. (2003). The 1867 Virgin Island Tsunami. *Natural Hazards and Earth Systems Sciences, 3*, 367–376.

(Received August 30, 2016, revised October 1, 2017, accepted October 6, 2017)

Pure Appl. Geophys.

DOI 10.1007/s00024-017-1713-9

Pure and Applied Geophysics

Coastal Amplification Laws for the French Tsunami Warning Center: Numerical Modeling and Fast Estimate of Tsunami Wave Heights Along the French Riviera

A. Gailler,[1] H. Hébert,[1] F. Schindelé,[1] and D. Reymond[2]

Abstract—Tsunami modeling tools in the French tsunami Warning Center operational context provide rapidly derived warning levels with a dimensionless variable at basin scale. A new forecast method based on coastal amplification laws has been tested to estimate the tsunami onshore height, with a focus on the French Riviera test-site (Nice area). This fast prediction tool provides a coastal tsunami height distribution, calculated from the numerical simulation of the deep ocean tsunami amplitude and using a transfer function derived from the Green's law. Due to a lack of tsunami observations in the western Mediterranean basin, coastal amplification parameters are here defined regarding high resolution nested grids simulations. The preliminary results for the Nice test site on the basis of nine historical and synthetic sources show a good agreement with the time-consuming high resolution modeling: the linear approximation is obtained within 1 min in general and provides estimates within a factor of two in amplitude, although the resonance effects in harbors and bays are not reproduced. In Nice harbor especially, variation in tsunami amplitude is something that cannot be really assessed because of the magnitude range and maximum energy azimuth of possible events to account for. However, this method is well suited for a fast first estimate of the coastal tsunami threat forecast.

Key words: Amplification law, tsunami warning system, coastal tsunami forecast.

1. Introduction

In the French tsunami Warning Center (CENALT) framework, the modeling tools in deep ocean currently implemented provide no-dimension tsunami warning at basin scale only (i.e., a red/orange/yellow color scale is used to express the watch/advisory/information warning levels). For now, coastal wave heights values are obtained from real time tide gage records exclusively (Schindelé et al. 2015). The forecast system includes a propagation database consisting of more than 2000 pre-computed tsunamis scenarios in the western Mediterranean. Each scenario represents a unit tsunami from an Mw 6.7 earthquake. These scenarios are linearly scaled to produce events from Mw 5.8 to 7.9, with respect to rupture zone coherency in its geodynamic context (Gailler et al. 2013). Together with the forecasting system, another operational tool based on real time computing is implemented as part of the CENALT framework. This second simulation tool works also in deep ocean only. It relies on the same numerical method, but takes advantage of multiprocessor approaches and more realistic seismological parameters, once the focal mechanism is estimated. Both simulation systems thus provide tsunami offshore maps only (i.e., they do not take into account the coastal response to tsunami arrival), which means evaluations compared to tide gage observations are unsuitable. Working with deep ocean propagation modeling enables application of only the properties of the linearity of the physics of tsunami generation and propagation in the ocean (because amplitudes are very small compared to the water depth).

A detailed modeling of the response in bays and harbors must be performed over an increasingly fine grid. That means introducing reduced time steps, and the use of a fully non-linear code. The problem can be approached from nested grids computation (i.e., bathymetric grids of increasing resolution toward the coast; e.g., Heinrich et al. 1998, Hébert et al. 2007, Gailler et al. 2015), but remains too time consuming in an operational context. This limitation is enhanced by the fact that the Mediterranean French coasts can be hit by near-field and regional tsunami, so warning

[1] CEA, DAM, DIF 91297 Arpajon, France. E-mail: audrey.gailler@cea.fr

[2] Laboratoire de Géophysique, CEA, DASE, PO BOX 640, 98713 Papeete, Tahiti, French Polynesia.

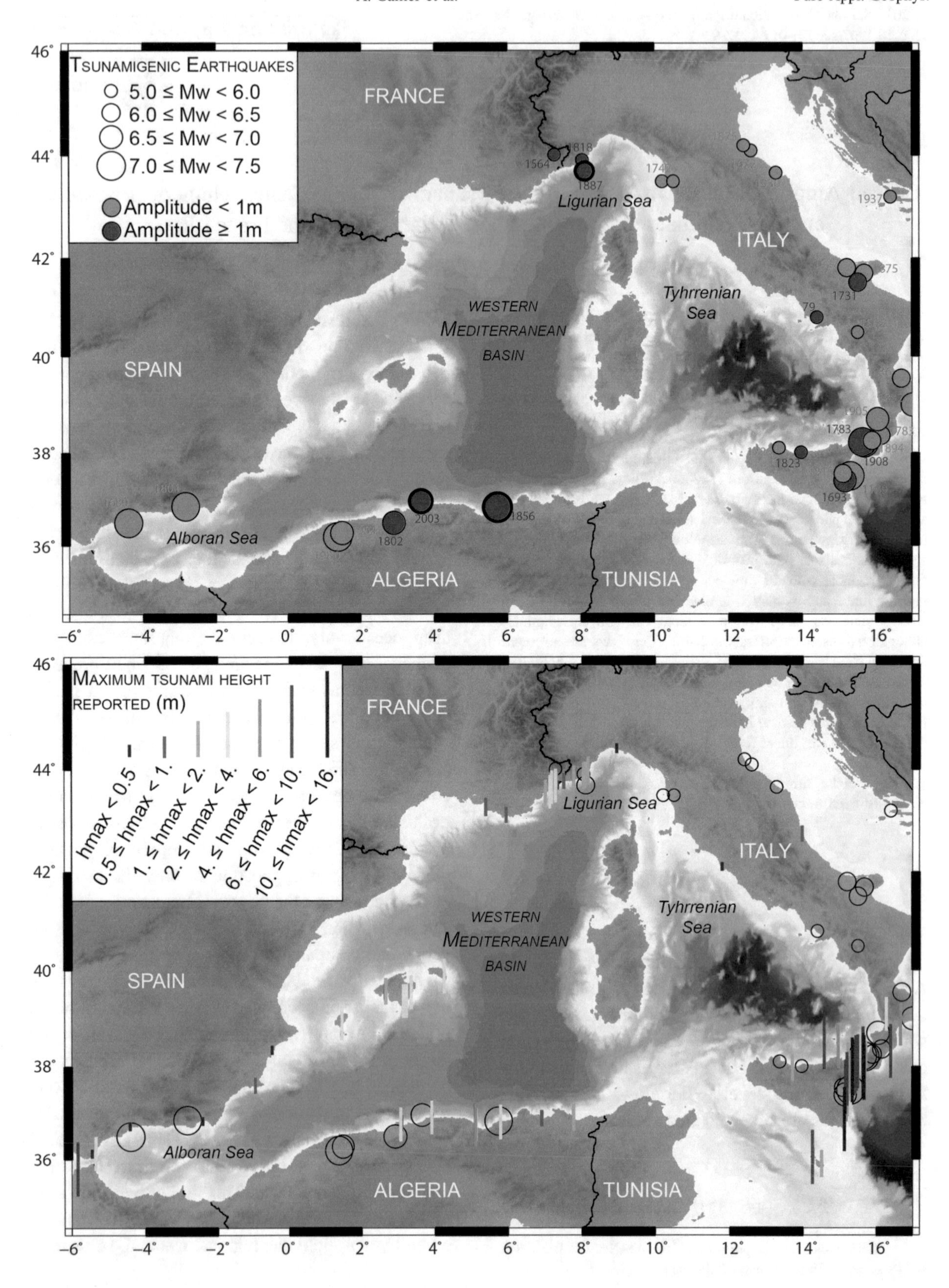
Tsunamigenic Earthquakes
5.0 ≤ Mw < 6.0
6.0 ≤ Mw < 6.5
6.5 ≤ Mw < 7.0
7.0 ≤ Mw < 7.5
Amplitude < 1m
Amplitude ≥ 1m
FRANCE
SPAIN
ITALY
ALGERIA
TUNISIA
Ligurian Sea
Tyhrrenian Sea
Western Mediterranean Basin
Alboran Sea
1564
1818
1887
2003
1856
1802
1731
1783
1908
1693
1823
1937
Maximum tsunami height reported (m)
hmax < 0.5
0.5 ≤ hmax < 1.
1. ≤ hmax < 2.
2. ≤ hmax < 4.
4. ≤ hmax < 6.
6. ≤ hmax < 10.
10. ≤ hmax < 16.
46°
44°
42°
40°
38°
36°
−6°
−4°
−2°
0°
2°
4°
6°
8°
10°
12°
14°
16°

◀Figure 1
[top] Historical tsunamigenic earthquakes in the western Mediterranean basin (from Gailler et al. 2013). Circles are located at the epicentral position. Circle size is proportional to the earthquake magnitude. Circle color indicates if the tsunami wave amplitude was higher or lower than 1 m. Circles highlighted in black show the location of the historical events used in this study. [bottom] Maximum tsunami height reported along the western Mediterranean coastline, due to the historical events shown on top

messages in the area must be issued in less than 70 min.

The challenge is to obtain rapid estimates of forecast coastal tsunami wave heights. Many efforts have been made in this regard since the 1990s (e.g., Synolakis 1991; Titov 2009; Hayashi 2010; Greenslade et al. 2014; Park et al. 2015; Riquelme et al. 2015).

A simple and computationally very fast method to provide maximum wave heights estimates at coastal locations uses a classic linear amplification law named Green's law (Green 1838). This must be considered as a rough approximation, as the complex response of a particular harbor does not follow Green's law exactly. However, this approach was successfully tested by Hayashi (2010) in the case of the 1896 Meiji Sanriku tsunami, adding proportional constants to the amplification equation. The complex coastal response due to sharp seafloor and bathymetry/topography transitions could be approached using an empirical correction factor, a function of the coastal configuration. This has been shown in recent studies performed for the French Polynesian sites, for which numerous tsunami records are available (Reymond et al. 2012; Jamelot and Reymond 2015).

Based on the tools developed at the Polynesian Center of tsunami warning (CPPT), this study proposes to test this new method using a transfer function derived from the Green's law empirically modified (Reymond et al. 2012; Jamelot and Reymond 2015) to take into account the specific character of various amplification sites (bays, harbors, steep slope areas,...). For French Polynesia, the number of tsunami records since 1960 (up to 35 observations per site) is consistent and allows an accurate evaluation of the method (Reymond et al. 2012). This method does not reproduce the resonance effects in harbors and bays, but is suitable for a fast forecast in the CPPT operational context with a standard error globally less than a factor of 2, especially for observations above 10 cm.

In the western Mediterranean Sea, several historical earthquakes (magnitude < 7.5) are known to have triggered tsunamis (Fig. 1). However, the moderate generated waves are poorly documented and the observation database is very sparse compared to the large amount of information available for Pacific coastlines (e.g., Gusiakov et al. 1997, 2005; Iwabuchi et al. 2008; National Geophysical Data Center/World Data Service (NGDC/WDS): Global Historical Tsunami Database. National Geophysical Data Center, NOAA. https://doi.org/10.7289/V5PN93H7).

Historical tsunami impact along the French coastline is not well known and wave height records and observations are almost nonexistent. In 1887, a strong earthquake in the Genoa Gulf, off Imperia (Italy) (magnitude 6.2–6.5), triggered a tsunami that caused several floodings along the French Riviera, from Antibes to Menton (Eva and Rabinovich 1997). Up to 2 m maximum wave heights were reported along the Ligurian coasts (e.g., Larroque et al. 2012). A recent study of Ioualalen et al. (2014) reevaluated the maximum amplitude of the tide gage record located in Genoa harbor to about 10–12 cm (peak-to-trough). The historical tsunamigenic earthquakes triggered in the Messina Straits in Italy (1693, 1783, 1908) generated large withdrawals and flooding in the Sicily-Calabria area (e.g., Tinti et al., 1999) but did not impact the French coastline. The North Algerian seismically active margin hosts several possibilities of strong submarine earthquakes able to produce important tsunamis in the western Mediterranean Sea (Kherroubi et al. 2009) (Fig. 1). The key event that stimulated earthquake-induced tsunami awareness in the area was the Boumerdès-Zemmouri thrust fault earthquake (Mw 6.9) in Algeria in May 2003. Quantitative observations along the French coasts were available through two tide gage records (one of good quality in Sète, and an undersampled one in Nice; Alasset et al. 2006; Hébert and Alasset 2003). The amplitudes did not exceed 10–60 cm, but significant eddies and sea withdraws were reported in several small marinas in southern France (Sahal et al.

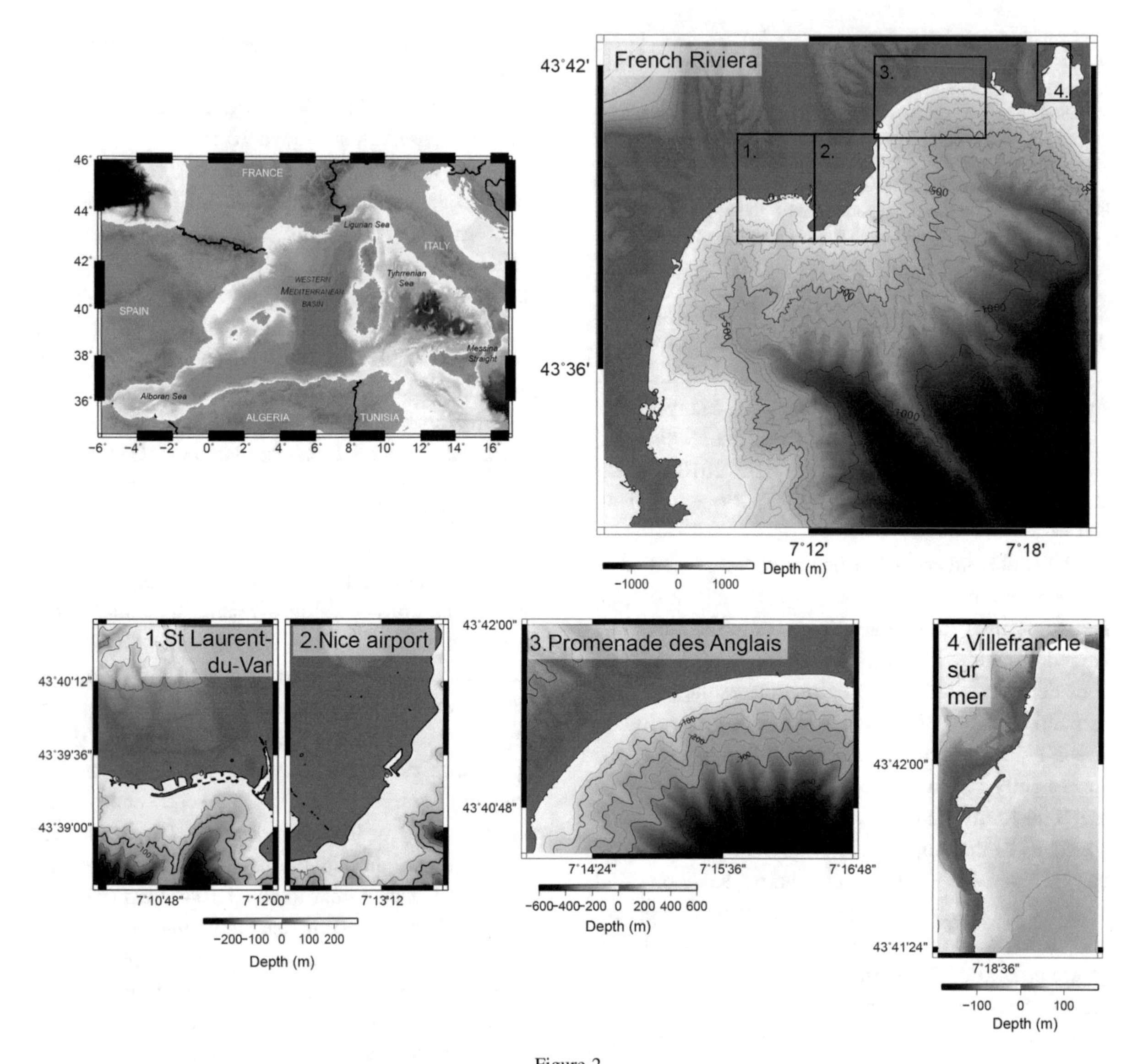

Figure 2
[top left] Nice test-site location; [top right] detailed view of the French Riviera indicating the location of the high resolution grids; [bottom] 5 m resolution bathy-topo grids used in this study

2009). Even if historical earthquakes in the northern Algerian margin capable of triggering tsunamis in the western Mediterranean Sea are not well known, the sequence of events could go back to the Algiers earthquake in 1365 (Yelles-Chaouche 1991). The strongest tsunamigenic earthquakes include the $M \sim 7.1$ 1856 Djijelli (Roger and Hébert, 2008) and $M \sim 7.3$ 1980 El Asnam events (Roger et al. 2011). The 1980 El Asnam earthquake was followed by a small tsunami recorded on several tide gages along the SE Spanish coast with amplitude of oscillations (peak to trough) of 48 cm. The location of the rupture zone (mainly inland) and the azimuth of the main axis of tsunami propagation meant the tsunami was unlikely to impact the French coasts. The 1856 Djijelli tsunami is less documented. However, recent tsunami modeling of this event shows that the azimuth of the main axis of tsunami propagation amplifies effects towards the French coasts (Roger and Hébert 2008), especially along the French Riviera.

In this work, we focus on the distribution of the tsunami amplitude for 4 zones along the French Riviera around the Nice region, from west to east: St Laurent-du-Var, Nice airport, the Promenade des Anglais and the bay of Villefranche-sur-mer (Fig. 2). Nine historical and synthetic earthquakes located in the two major tsunamigenic zones in the western Mediterranean which could impact most of the French coastlines are considered. Regarding these nine sources, rapid tsunami heights forecasts are calculated based on amplification laws, then compared to nested grids simulation results, the aim being to develop a new operational method for tsunami warning centers.

2. Method and Limitations

The coastal amplification of a tsunami can be described relying on tsunami energy conservation, as proposed in Green's law (1838). The tsunami speed c, following the non-dispersive approximation, is written: $c = \sqrt{gh}$ where h is the water depth and g the gravity.

Considering that the square of the energy, which is proportional to the tsunami amplitude η multiplied by its speed, is conserved, we can deduce that the product $\eta h^{1/4}$ must remain constant close to the coasts. This implies that the run-up, represented by η_2 taken at depth h_2, must follow:

$$\eta_2 = \eta_1 \cdot \left(\frac{h_1}{h_2}\right)^{1/4}. \quad (1)$$

This formula, named pure Green's law, provides an approximation of the wave heights at the coastline, but does not give the theoretical run-up defined as the elevation reached by the maximum inundation due to the tsunami. It expresses the conservation of wave energy flux, to extend the gridded wave field into the harbor at depth h_2, with respect to a nearby deep-water grid point at depth h_1.

This constitutes a drastic approximation, as the complex response of a particular harbor or an off-shore relief does not follow Green's law exactly. No new specific amplification law has been developed since Synolakis' initiative in 1991 (Synolakis 1991). So we choose the Reymond et al. (2012) and Jamelot and Reymond (2015) approach that involves a transfer function derived from the Green's law. The authors have studied and established that for the French Polynesian sites, for which numerous tsunami records are available, the specific response in harbors and bays can be approached using an empirical correction factor β, a function of the coastal configuration (eq. [2]). The associated operational tool developed by the CPPT provides, in less than 5 min, a forecast distribution of the tsunami amplitude for 30 coastal sites located in French Polynesia. This method does not take into account the local resonance effects, but has shown its ability in terms of CPPT fast evaluation of the coastal tsunami risk, in particular for the Chilean event of September, 2015.

For a set of values (see next section), we thus use a transfer function based on Green's law empirically modified to take into account the special amplifying sites depending on their geometry or bathymetric slope. This transfer function is applied directly to the maximum water heights distribution calculated in deep ocean (i.e., deep ocean model) at depth h_1. The operation is carried out on 5 m resolution bathy-topo grids (built from the Litto 3D© project, Fig. 2) in order to produce tsunami wave heights forecast maps as close as possible to the coastline.

The approximation η_2 at depth h_2 is calculated from empirical amplification laws following four levels of depth (Reymond et al. 2012; Jamelot and Reymond 2015):

- If $h_2 \geq h_1$; η_2 is directly extracted from the deep ocean model (η_1 in the 2 min bathy grid)
- If $h_E \leq h_2 < h_1$; the pure Green's law is applied on the deep ocean model, i.e.: $\eta_2 = \eta_1.(h_1/h_2)^{1/4}$
- If $h_0 \leq h_2 < h_E$; the transfer function derived from the Green's law is applied on the deep ocean model. This expression provides a linear approximation of the response between the depth h_E at the entrance of a bay or harbor and the target point at minimal depth h_0, where β is the attenuating or amplifying factor as a function of the studied site configuration:

Table 1

Characteristics of the nine seismic sources used in this study

Scenario	ID	Magnitude	Rupture zone parameters										
			nb of source segments	Lon (°)	Lat (°)	Depth (km)	Slip (m)	Strike (°)	Dip (°)	Rake (°)	Length (km)	Width (km)	Rock rigidity (N/m2)
Ligure 1887 (Eva and Rabinovich 1997)	L	6.5	1	8.3	43.80	10.2	0.45	71	85	90	45	10	35.E9
Boumerdes 2003 (Semmane et al. 2005)	S	7.2	128	3.3419									4.0139
36.663 37.12	2.4	22.9	0.007 3.5	54	47	90	4						(*128)
4 (*128) 3.8E+10		2.2E+10											
Boumerdes 2003 (from Yelles et al. 2004) inflated	Y	7.5	1	3.565	36.803	12.	2.5	60	42	84	85	35	30.E9
Djijelli 1856 (Roger and Hébert 2008)	J7.1	7.1	3	5.476	36.950	7.0	1.0	75	40	90	25	20	4.5E+10
				5.736	37.080	7.0	1.5	85	40	90	37	20	4.5E+10
				6.150	37.178	7.0	1.5	75	40	90	44	20	4.5E+10
Djijelli 1856 inflated	J7.8	7.8	3	5.476	36.950	7.5	6.0	75	40	90	25	22	4.5E+10
				5.736	37.080	7.5	6.0	85	40	90	37	22	4.5E+10
				6.150	37.178	7.5	6.0	75	40	90	44	22	4.5E+10
CASST-413 (fault parameters from the	413-	65 35.E9	6.5	1	6.95	37.29	4.	0.45	91	50	90	45	10
CENALT pre-computed tsunami database)	413-	70 35.E9	7.0	1	6.95	37.29	7.	1.5	91	50	90	42	18
413-75	7.5	1	6.95	37.29	12.	2.6	91	50	90	85	29	35.E9	
413-75 s	7.5	1	6.95	37.29	9.	8	91	50	90	33	22	35.E9	

$$\eta_2 = \eta_1 \cdot \left[1 + \frac{(h_E - h_2)}{h_E}\right] \cdot \left(\frac{h_1}{h_2}\right)^{1/4} \quad (2)$$

- If $0 \leq h_2 < h_0$; the maximum water height η_2 obtained at depth h_0 (eq. [2]) is continued to the coastline (depth = 0).

The main limitation of the Green's law method is that it requires a large database of previous observations for a given coastal area, to define the empirical parameters of the correction equation. As no such data (i.e., historical tide gage records of significant tsunamis) are available for the western Mediterranean basin, a synthetic dataset is calculated for both synthetic and well-known historical tsunamigenic earthquakes in the area (Table 1, Fig. 3).

3. Chosen Sources and Tested Parameters

In the western Mediterranean basin, the tsunamigenic earthquake areas can be divided into four major zones: the North Algerian margin, the Ligurian Sea, the Tyrrhenian Sea with the Messina Strait, and the Alboran Sea. Only the first two zones can generate a tsunami threat along the French Mediterranean coastlines. Hence eight sources localized along the Algerian margin and one in the Ligurian Sea are chosen for this study, based on historical events and synthetic worst-case scenarios. These events cover the range of source location, magnitude and azimuth most representative of the main tsunami sources for the French Riviera area. So most of the reasonable tsunami impacts along the French Riviera are represented within these nine scenarios. Among these nine

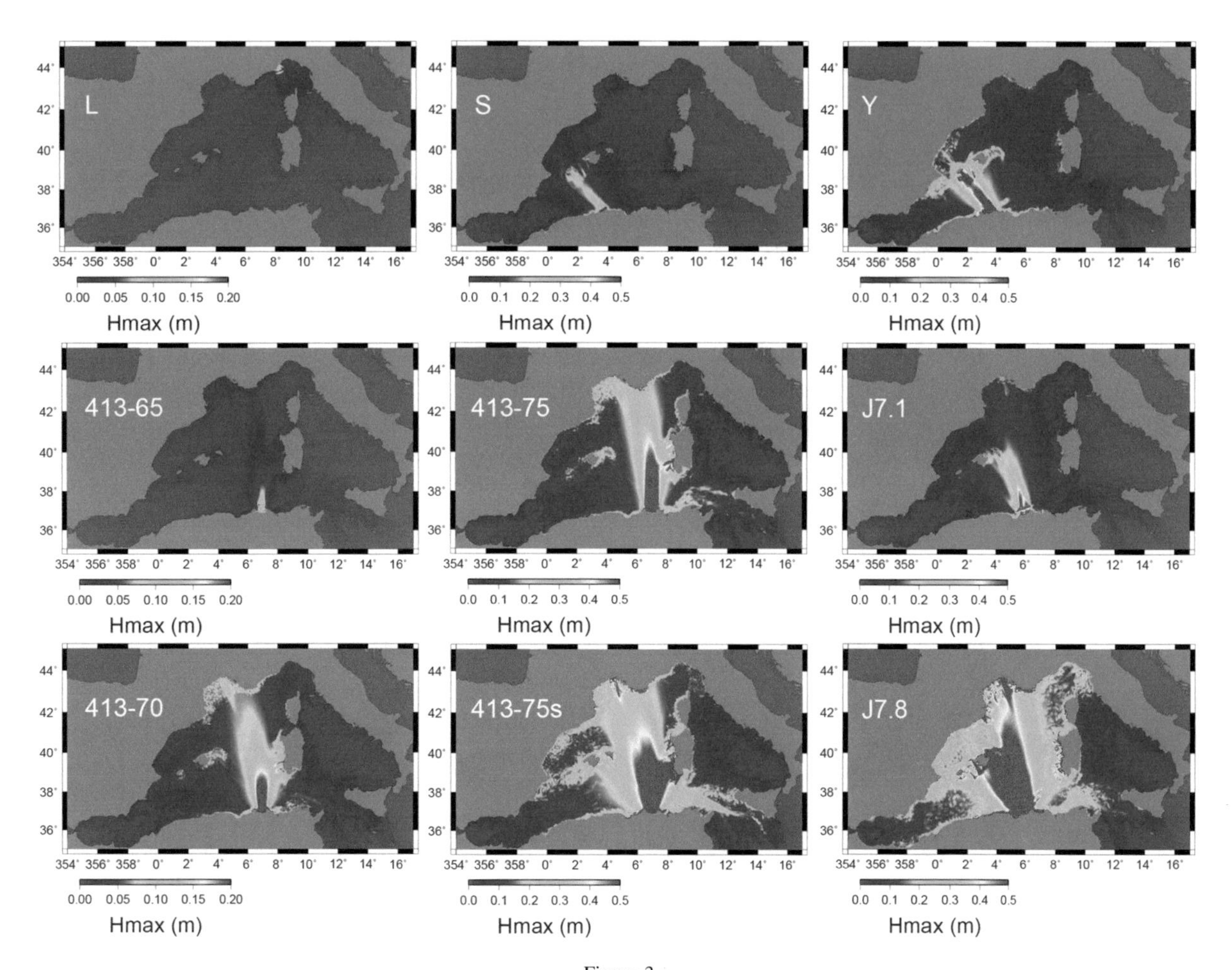

Figure 3
Maximum water heights calculated for each of the nine seismic sources of the study (results in the 2 min resolution mother grid, after 3 h of tsunami propagation)

events, three are based on historical events (1856, 1887 and 2003, Fig. 1):

- The Boumerdès earthquake (2003), presented by the Semmane et al. (2005) source, magnitude 7.2 (128 segments).
- A source consisting of one segment only, for which the rupture parameters are consistent with the 2003 Boumerdès earthquake (Yelles et al. 2004), but where the magnitude is artificially increased to 7.5.
- The Djijelli earthquake (1856), presented by the Roger and Hébert (2008) source, magnitude 7.1 (three segments). The fault strike trends ENE–WSW.
- A source similar to the 1856 Djijelli earthquake, with a magnitude value increased artificially to 7.5 using a 6 m slip.
- Four sources for which the parameters come from the No. 413 fault of the CENALT pre-computed scenario database (Gailler et al. 2013; Schindelé et al. 2015), considering magnitudes 6.5, 7.0 and 7.5 with a moderate slip (2.6 m) and 7.5 with a high slip (8 m). The No. 413 fault is located in the 1856 Djijelli earthquake area but uses on E–W strike.
- The last one, a source located in the Ligurian Sea, corresponding to the 1887 Imperia earthquake, estimated magnitude 6.5 (Eva and Rabinovich 1997).

For each of these nine seismic sources, the synthetic dataset is obtained through accurate numerical tsunami propagation and inundation modeling by

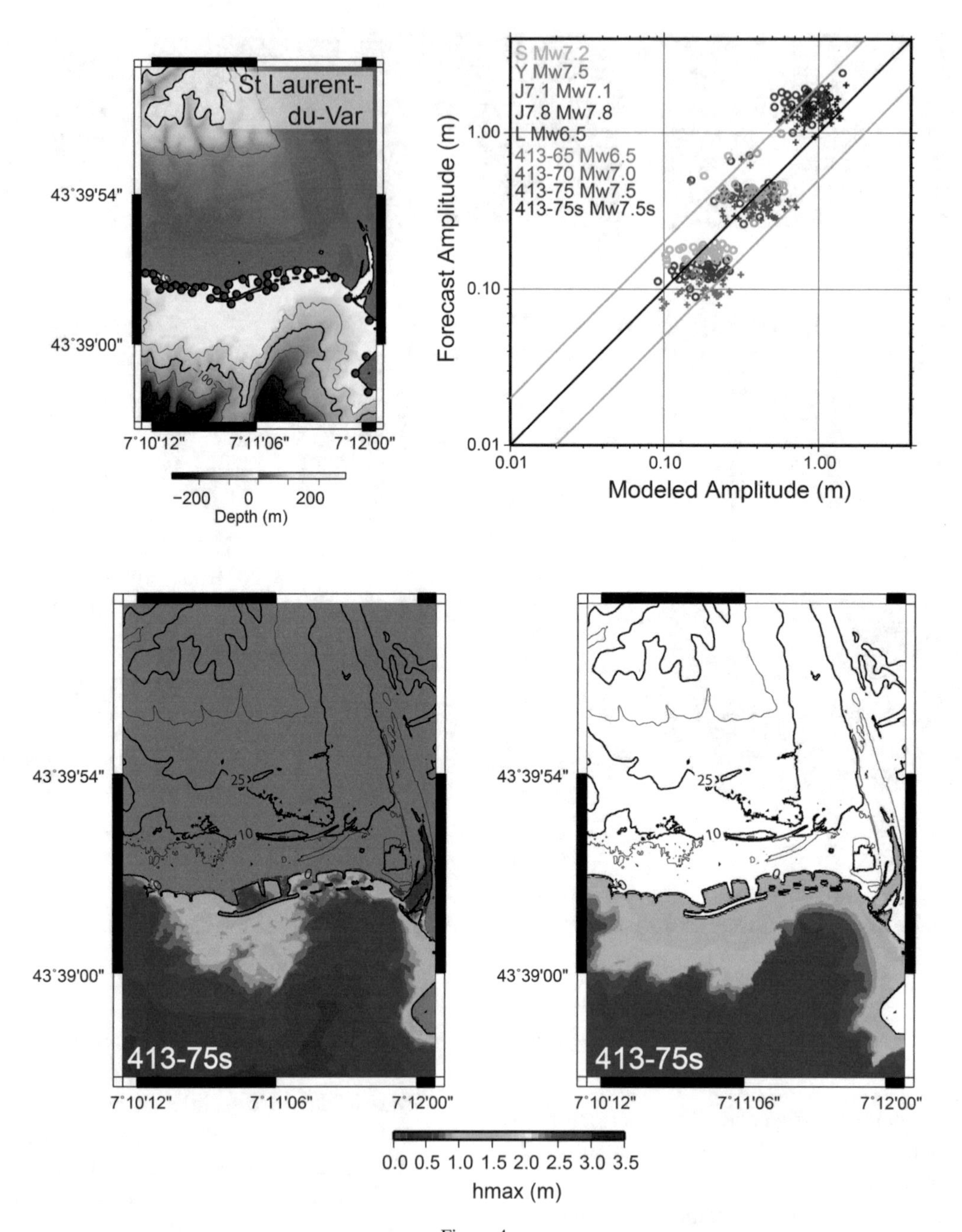

Figure 4
[top, right] Comparison of the tsunami heights estimated from amplification laws for each of the nine seismic sources tested with the results provided by the 5 m resolution nested grids simulation, at each red dot along the coast of St Laurent-du-Var [map on the left]. [bottom] Comparison of maximum water height maps obtained for St Laurent-du-Var in the case of CENALT database source No. 413, at magnitude 7.5, strong slip (Table 1) in the 5 m resolution nested grids simulation (left), with extrapolation given by amplification laws (right)

using several nested bathymetric grids characterized by coarse resolution over deep water regions and an increasingly fine resolution close to the shores. The synthetic data calculated for the higher resolution grids (25 and 5 m, Fig. 2) are then used as reference (i.e., "synthetic observation") to approximate the empirical parameters of the correction equation.

The empirical approximation methodology presented above is applied to the modeling results obtained on a coarse bathymetry grid (2 min)

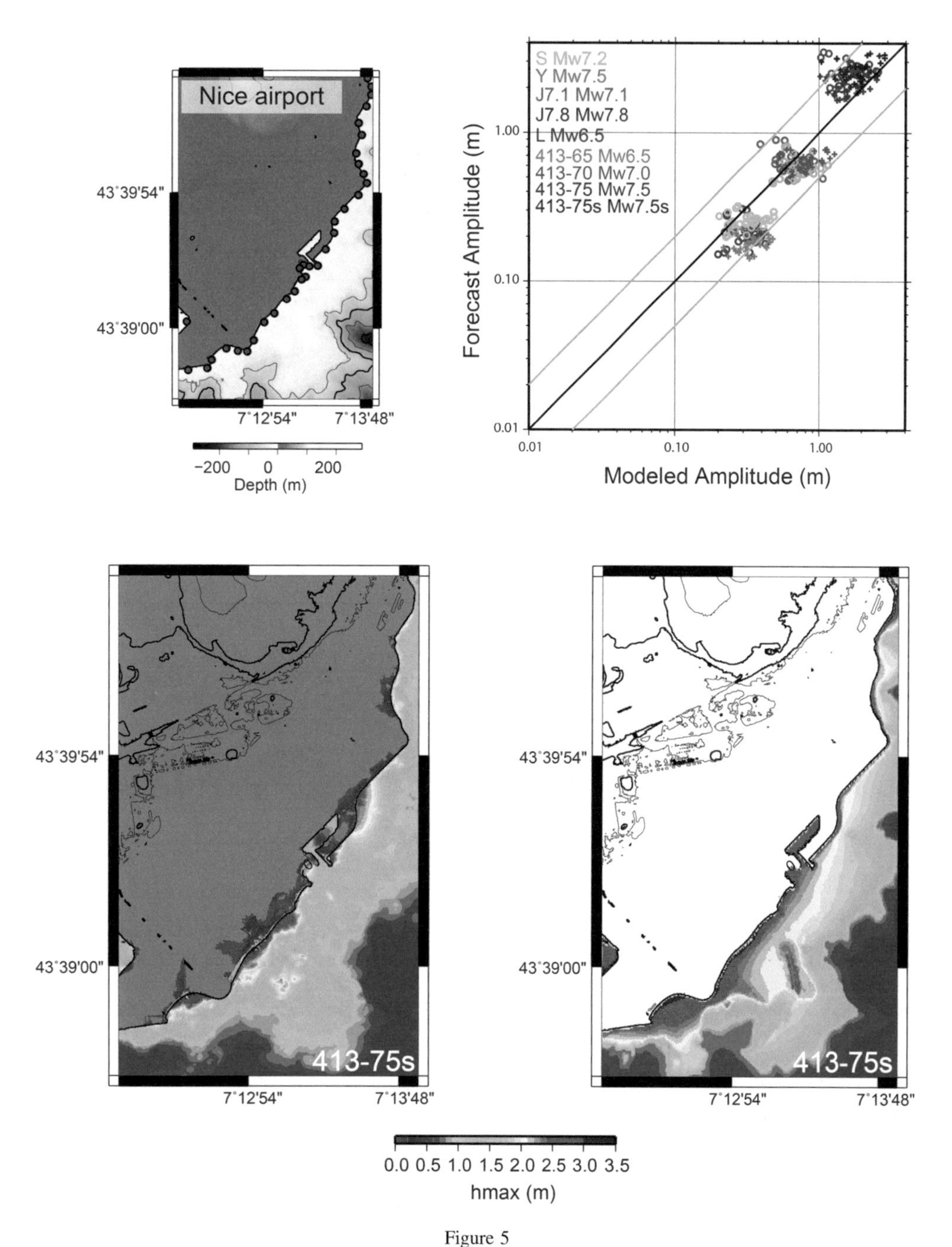

Figure 5
[top, right] Comparison of the tsunami heights estimated from amplification laws for each of the nine seismic sources tested with the results provided by the 5 m resolution nested grids simulation, at each red dot along the coast of Nice airport [map on the left]. [bottom] Comparison of maximum water height maps obtained for Nice airport in the case of CENALT database source No. 413, at magnitude 7.5, strong slip (Table 1) in the 5 m resolution nested grids simulation (left), with extrapolation given by amplification laws (right)

computed by the fast simulation tool currently implemented in the CENALT. The obtained values are compared with the values computed in the high resolution synthetic dataset, in order to evaluate to what extent the simple approach by amplification laws can explain these synthetic data used as

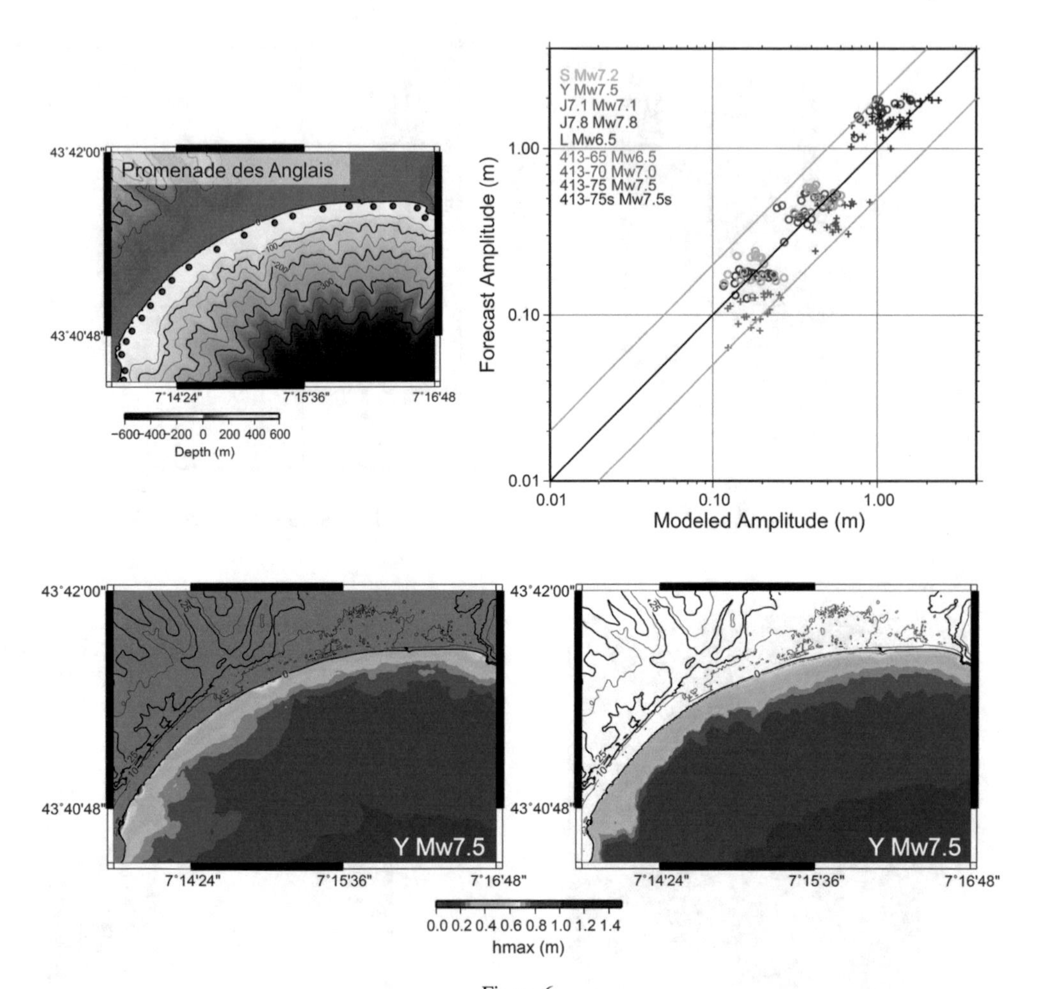

Figure 6

[top, right] Comparison of the tsunami heights estimated from amplification laws for each of the nine seismic sources tested with the results provided by the 5 m resolution nested grids simulation, at each red dot along the Promenade des Anglais [top left]. [bottom] Comparison of maximum water height maps obtained along the Promenade des Anglais in the case of the Boumerdès source increased to magnitude 7.5 (Table 1) in the 5 m resolution nested grids simulation (left), with extrapolation given by amplification laws (right)

reference. The objective being to obtain a good fit between deep ocean models after empirical coastal amplification and accurate models obtained from nested grids after time-consuming runs (1 h per run using 72 processors).

In order to achieve this goal, several input parameters of the Green's law and the transfer function have been tested for the Nice test site:

(i) The water depth h_1 from which the maximum water heights η_1 coming from the basin scale modeling (2 min grid) must be clipped and extrapolated using the Green's law (values of h_1 tested: 1000, 500, 200, 100 m).
(ii) The depth h_E from which the transfer function is applied (values of h_E tested: 200, 100, 75, 50, 20 m).
(iii) The minimum depth h_0 from which the calculation of empirical amplification can be considered as non-relevant (values of h_0 tested: 10, 5, 2 m).
(iv) The value of the $\boldsymbol{\beta}$ factor, which varies as a function of the target area configuration. Typically, slightly negative values of β are clues of attenuating sites, e.g., protected by very steep slopes. In contrast, positive values of β are an indicator of amplifying sites, e.g., narrow bays and gentle slopes.

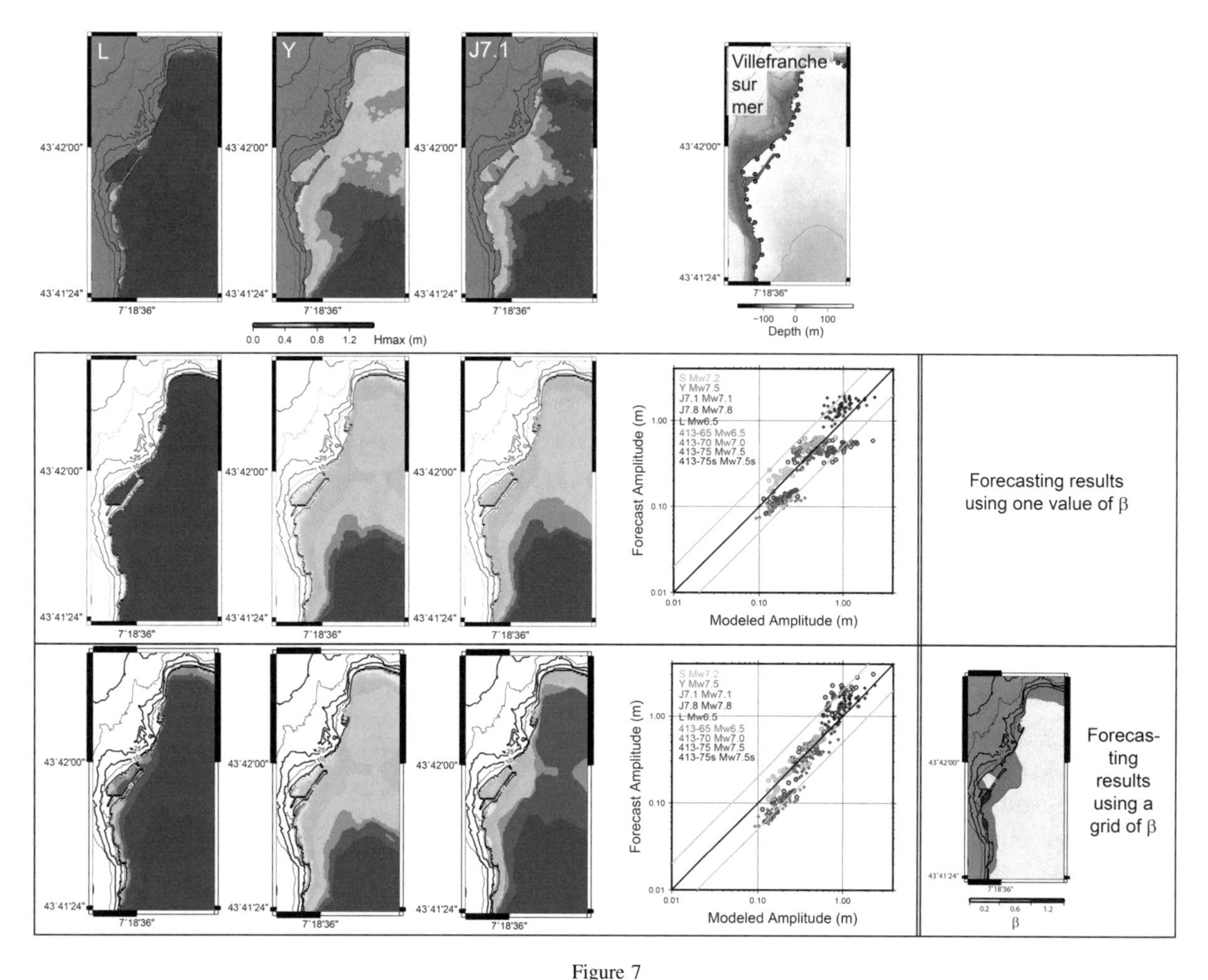

Figure 7
Comparison of maximum water height maps obtained in the bay of Villefranche-sur-mer in the case of one Ligurian (L, Mw = 6.5) and two Algerian (*Y*, Mw = 7.5; *J*7.1, Mw = 7.1) sources (Table 1) in the 5 m resolution nested grids simulation (top), with extrapolation given by amplification laws with a constant $\beta = 0.8$ (middle) and an envelope of regionalized β (bottom). Values of regionalized β as a function of more amplifying or attenuating specific local geometries are shown on the bottom-right map. The graphs in the middle compare the tsunami heights estimated from amplification laws (middle—constant β, bottom—envelop of β) for each of the nine seismic sources tested with the results provided by the 5 m resolution nested grids simulation, at each red dot along the coast of Villefranche-sur-mer bay [map on top right]

(v) The grid resolution on which the calculated amplification using empirical laws is performed. For a comparison, the computation was carried out on two grid levels: the 25 m « baie de Anges » grid, and the three grids of 5 m resolution focused on St Laurent-du-Var airport, Promenade des Anglais, and Villefranche-sur-mer (Fig. 2). The best results are obtained for the calculations on the 5 m grids.

Thus, the pure Green's law (1) has been applied to the models given at isobaths h_1 1000, 500, 200 and 100 m, with an extrapolation towards h_E depths of 200, 100, 75, 50 and 20 m. The transfer function (2) has been applied from h_E isobaths of 200, 100, 75, 50 and 20 m, with an extrapolation towards h_0 depths of 10, 5, and 2 m. Among the combinations of tested parameters, the most relevant for the four zones of the Nice test site are the following: $h_1 = 500$ m; $h_E = 50$ m; $h_0 = 5$ m. The methodology presented in Sect. 2 thus becomes:

(1) if $h_2 \geq 500$ m $\Rightarrow \eta_2 = \eta_1$ (directly extracted from the deep ocean model)

(2) if $50 \leq h_2 < 500$ m $\Rightarrow \eta_2 = \eta_1 \cdot \left(\frac{h1}{h2}\right)^{1/4}$
(3) if $5 \leq h_2 < 50$ m $\Rightarrow \eta_2 = \eta_1 \cdot \left[1 + \beta\frac{(hE-h2)}{hE}\right] \cdot \left(\frac{h1}{h2}\right)^{1/4}$
(4) if $0 \leq h_2 < 5$ m $\Rightarrow \eta_2 = \eta_2$ at 5 m depth

4. Results

On each of the three 5 m resolution bathy-topo grids of the Nice test site, the pure Green's law eq. (1) was applied to the maximum water height coarse grid model (2 min grid) from depth $h_1 = 500$ m, with an extrapolation toward depth $h_E = 50$ m. Then, between 50 and 5 m, the transfer function (2) using the β factor (linked to the coastal configuration, zone-dependent) was applied. The comparison is done with tsunami values of maximum water heights obtained in the detailed nested grids simulation for the same seismic sources (Table 1) in the high resolution 5 m grids. Values for beta are obtained once a good fit is found on the Modelled Amplitude/Forecast Amplitude regression curve for each site. We present the results from west to east for the sites of St Laurent-du-Var, Nice airport, Promenade des Anglais and Villefranche-sur-mer.

4.1. St Laurent-du-Var

A β value of 0.2 along St Laurent-du-Var beach (westward 7.202°E) is applied. This slightly positive value expresses the weakly amplifying role of this small open bay.

Figure 4 shows that for all the compared coastal points, almost all estimates are accurate to within a factor of 2. The mean error is − 0.17 m for the nine tested events, with a standard deviation of ± 0.34 m.

The least satisfying results are obtained for the strongest magnitude source (Djijelli Mw 7.8). The prediction based on linear approximations overestimates results compared to the multigrid calculation for $M \geq 7.5$ events. In the cases of small magnitude sources (i.e., 6.5), the prediction results tend to underestimate the values obtained from the nested grids calculation.

Comparison of the maximum water height predicting maps with the nested grids models (Fig. 4, left) shows that the inundation is not predicted clearly, but the coastal values show a good agreement, with a slight overestimation using the empirical laws in the case of source CENALT No. 413 (Table 1) at magnitude 7.5. The largest differences are obtained in the small harbor of St Laurent-du-Var, protected by a breakwater and at the mouth of the river to the east. Here the forecast method tends to overestimate the maximum wave heights.

4.2. Nice Airport

A β value of 1.2 along the Nice airport runway (eastward 7.202°E) is applied. This higher value expresses the strongest amplification related to the large amount of backfill material built on the sea.

Most of the estimates are beyond a factor of 2 for all the coastal points compared (Fig. 5). The mean error and associated standard deviation are − 0.15 and ± 0.50 m, respectively, consistent with the values obtained for the St Laurent-du-Var area.

The prediction based on linear approximations appears to overestimate results compared to the multigrid calculation for $M \geq 7.5$ events, whereas the prediction in the cases of sources of small magnitude tend to underestimate the values obtained from the nested grids calculation.

The comparison of the maximum water height prediction maps with the nested grids models (Fig. 5, left) in the case of source CENALT No. 413 at magnitude 7.5 shows that the inundation is not predicted clearly, but the coastal values show good agreement, with a slight overestimation using the empirical laws.

4.3. Promenade des Anglais

The bathymetric-topographic variations along the Promenade des Anglais being relatively homogeneous, a β value of 0.4 is applied, expressing the amplified response linked to the slightly concave shape of the beach.

Figure 6 shows that for all the coastal points compared, the estimate is accurate to within a factor of 2 in amplitude. The mean error is 0.09 m for the nine tested events, with a standard deviation of ± 0.26 m.

The least satisfying results are obtained for the strongest magnitude source (Djijelli Mw 7.8). The prediction based on linear approximations provides almost systematically slightly overestimated results compared to the multigrid calculation for $M \geq 7.5$ events. In the cases of small magnitude sources (i.e., 6.5), the prediction results tend to underestimate the values obtained from the nested grids calculation. Regarding the comparison of the maximum water height maps (Fig. 6, left), several small local amplifications modeled in nested grids are not reproduced by the linear approximation prediction, e.g. westward the Promenade.

4.4. *Villefranche-sur-mer*

For the site of Villefranche-sur-mer, two approaches are tested:

(1) A single value of β for the whole grid is applied: $\beta = 0.8$ (Fig. 7, middle), expressing the globally amplified response of this narrow bay.
(2) A map of regionalized values of β is applied, as a means to amplify specific local geometries (e.g., breakwaters, inner harbor,...). Amplifying areas, such as the elbow-shaped breakwater protecting the harbor, are associated with $\beta = 1.2$. For less amplifying zones, such as the inner portions of the bay and harbor and the SW coast, β is set to 0.6. Finally, a β of 0.2 is attributed to areas showing weakest amplification (e.g., middle of the harbor) (Fig. 7, bottom).

Comparison of the maximum water height maps (Fig. 7) shows a good agreement between the predictions and the nested grids models, independent of the β strategy employed. Nevertheless, the approach using an envelope of β values clearly reproduces the local amplifications modeled by nested grids with better accuarcy.

Figure 7 reveals that for all the coastal points compared, the estimate is generally accurate to within a factor of 2. The points seem to be slightly more focused along $y = x$ line with the regionalized β approach. However, the mean error/standard deviation of 0.11 ± 0.32 m is greater than the statistical values of 0.04 ± 0.31 m given by predictions using a constant β for the whole grid. The predictions for the lowest magnitude sources are once again characterized by a slight underestimate of the nested grids models. The tendency to overestimate the strong magnitude sources is however less visible for this site.

Further research on this site and others is necessary to determine whether or not the strategy of using β envelopes enhances the prediction accuracy.

4.5. *Discussion*

Table 2 collates the β values used in formula (2) for the Nice area and the one defined for French Polynesia by the CPPT (Jamelot and Reymond 2015). The same trend is found: the more amplifying the sites are, the more the associated β value is positive. These values remain strongly site-dependent: linked to the coastal shape, the associated water depths, the extent of the continental shelf, the continental shelf/abyssal plain geometry transition, etc. The reference points used (18 tide gage observations of real events for French Polynesia, against the nested grids results for the Nice test site) also have a strong influence on the β value calculation at each site.

Table 2

Comparison of the β values used in formula (2) for the Nice area and the values defined for French Polynesia by the CPPT (Jamelot and Reymond 2015)

French Riviera			French Polynesia		
Site	Type	*b*	Site	Type	*b*
St Laurent-du-Var	Beach	0.2	Marquesas Bay	Shallow bay	1.4
Nice airport	Seaside backfill	1.2	Papeete (Tahiti, Society Islands)	Reef barrier	− 0.35
Promenade des Anglais	Large and deep bay	0.4	Faaa airport (Tahiti, Society Islands)	Lagoon	− 0.35
Villefranche-sur-mer	Narrow and deep bay	0.2–1.2	Rangiroa (Tuamotu archipelago)	Atoll	− 0.4
			Rurutu (Austral Islands)	Harbor	0

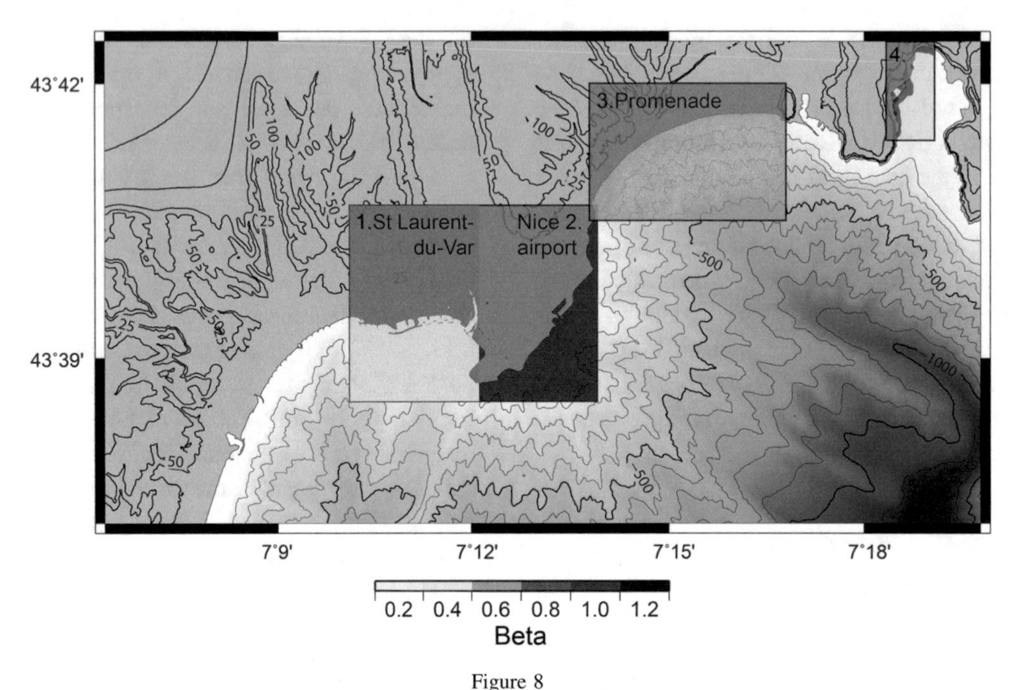

Figure 8

Values of regionalized β as a function of more amplifying or attenuating specific local geometries along the French Riviera determined from this study [(1) St Laurent-du-Var; (2) Nice airport; (3) Promenade des Anglais; (4) Villefranche-sur-mer]

Along the French Riviera, it should be noted that the β factor can vary from 1.2 to 0.2 between neighboring sites (Fig. 8). From west to east, we have a weakly amplifying site (St Laurent-du-Var), then a strongly amplifying one (Nice airport), then once again a weakly amplifying one (Promenade des Anglais). For the eastern site (Villefranche-sur-mer), the complex response of the bay due to its bathymetry and coastal variations leads to a regionalization of the amplifying factors within the bay, as tsunami wave behavior differs from the inner portion of the bay to either side of the breakwater protecting the harbor. These different amplifying factors along the French Riviera constitute an important result which provides explanations about the variability of coastal tsunami heights along the coastlines. This is due to the variability of submarine slopes, the shape and size of harbors, bays and beaches, and the interferences with possible offshore infrastructures (e.g., breakwaters and backfills).

This work shows that the pure Green's law is not able to give satisfactory estimates for all coastal points, since strongly amplifying sites can neighbor weakly amplifying ones. Including a site-dependent amplification factor provides better forecast approximation. The transfer function based on Green's law applied here could be modified, with additional corrective factors to improve the estimates at low and strong magnitudes.

In the operational scope, the CENALT should be able to provide rapid coastal estimates of tsumani heights at different scales to the civil protection authorities, for example department scale grids associated to the actual forecast points, as well as higher resolution warning maps focusing on amplifying harbors, beaches and bays, such in the same way as presented in this work.

5. Conclusion

On the basis of nine historical and synthetic sources tested in this work, a fast prediction tool of coastal tsunami amplitude based on a transfer function derived from Green's law has been developed for the western Mediterranean French coasts. The first encouraging results from the Nice test site show a good agreement with high resolution modeling: the linear approximation provides accurate estimates within a factor of 2 very quickly (within 1 min). A comparison of the β values used in formula (2) for the Nice area with the one defined for French Polynesia by the CPPT is presented in the following appendix.

The extrapolation by empirical laws does not allow us to consider local amplification close to shore other than bathymetric effects. The approach using regionalized values of β instead of a constant value needs to be studied further. A way to compute the theoretical run-up should also be explored.

Acknowledgements

The authors wish to acknowledge with thanks the Editor Alexander Rabinovich along with Fred Stephenson, Gillian Royle, Julien Roch and two anonymous reviewers for their efforts and comments which contributed to a substantial improvement of the first draft manuscript. This work has received funding from the EU FP7 project ASTARTE, Assessment, Strategy and Risk Reduction for Tsunamis in Europe grant No. 603839 (Project ASTARTE).

References

Alasset, P. J., Hébert, H., Maouche, S., Calbini, V., & Meghraoui, M. (2006). The tsunami induced by the 2003 Zemmouri earthquake (Mw = 6.9, Algeria): modelling and results. *Geophysical Journal International, 166*(1), 213–226.

Eva, C., & Rabinovich, A. B. (1997). The February 23, 1887 tsunami recorded on the Ligurian coast, western Mediterranean. *Geophysical Research Letters, 24*(17), 2211–2214.

Gailler, A., Calais, E., Hébert, H., Roy, C., & Okal, E. (2015). Tsunami scenarios and hazard assessment along the northern coast of Haiti. *Geophysical Journal International, 203*(3), 2287–2302.

Gailler, A., Hébert, H., Loevenbruck, A., & Hernandez, B. (2013). Simulation systems for tsunami wave propagation forecasting within the French tsunami warning center. *Natural Hazards and Earth System Sciences, 13,* 2465–2482.

Green, G. (1838). On the motion of waves in a variable canal of small depth and width. *Transactions of the Cambridge Philosophical Society, 6,* 457–462.

Greenslade, D. J., Annunziato, A., Babeyko, A. Y., et al. (2014). An assessment of the diversity in scenario-based tsunami forecasts for the Indian Ocean. *Continental Shelf Research, 79,* 36–45.

Gusiakov, V. K., Marchuk, A. G., & Osipova, A. V. (1997). Expert tsunami database for the Pacific: motivation, design, and proof-of-concept demonstration. In G. Hebenstreit (Ed.), *Perspectives on Tsunami Hazard Reduction* (pp. 21–34). Netherlands: Springer.

Hayashi, Y. (2010). Empirical relationship of tsunami height between offshore and coastal stations. *Earth, Planets and Space, 62*(3), 269–275.

Hébert, H., & Alasset, P. J. (2003). The tsunami triggered by the 21 May 2003 Algiers earthquake. *EMSC/CSEM Newsletter, 20,* 10–12.

Hébert, H., Sladen, A., & Schindelé, F. (2007). Numerical modeling of the great 2004 Indian Ocean tsunami: focus on the Mascarene Islands. *Bulletin of the Seismological Society of America, 97*(1A), S208–S222.

Heinrich, P., Schindele, F., Guibourg, S., & Ihmlé, P. F. (1998). Modeling of the February 1996 Peruvian tsunami. *Geophysical Research Letters, 25*(14), 2687–2690.

Ioualalen, M., Larroque, C., Scotti, O., & Daubord, C. (2014). Tsunami mapping related to local earthquakes on the French-Italian Riviera (western Mediterranean). *Pure and Applied Geophysics, 171*(7), 1423–1443.

Iwabuchi, Y., Sugino, H., Ebisawa, K., Imamura, F. (2008). Reviewed database of the tsunami run-up data on the documentation in Japan. In Proceedings of the 14th World Conference on Earthquake Engineering, Paper ID (pp. 15–0036).

Jamelot, A., & Reymond, D. (2015). New tsunami forecast tools for the French Polynesia tsunami warning system. Part II: numerical modelling and tsunami height estimation. *Pure and Applied Geophysics, 172*(3–4), 805–819.

Kherroubi, A., Déverchère, J., Yelles, A., et al. (2009). Recent and active deformation pattern off the easternmost Algerian margin, western Mediterranean Sea: new evidence for contractional tectonic reactivation. *Marine Geology, 261*(1), 17–32.

Larroque, C., Scotti, O., & Ioualalen, M. (2012). Reappraisal of the 1887 Ligurian earthquake (western Mediterranean) from macroseismicity, active tectonics and tsunami modelling. *Geophysical Journal International, 190*(1), 87–104.

Park, H., Cox, D. T., & Petroff, C. M. (2015). An empirical solution for tsunami run-up on compound slopes. *Natural Hazards, 76*(3), 1727–1743.

Reymond, D., Okal, E. A., Hébert, H., & Bourdet, M. (2012). Rapid forecast of tsunami wave heights from a database of pre-computed simulations, and application during the 2011 Tohoku tsunami in French Polynesia. *Geophysical Research Letters, 39*(11), L11603. https://doi.org/10.1029/2012GL051640.

Riquelme, S., Fuentes, M., Hayes, G. P., & Campos, J. (2015). A rapid estimation of near-field tsunami runup. *Journal of Geophysical Research: Solid Earth, 120*(9), 6487–6500.

Roger, J., & Hébert, H. (2008). The 1856 Djijelli (Algeria) earthquake and tsunami: source parameters and implications for tsunami hazard in the Balearic Islands. *Natural Hazards and Earth System Sciences, 8*(4), 721–731.

Roger, J., Hébert, H., Ruegg, J. C., & Briole, P. (2011). The El Asnam 1980 October 10 inland earthquake: a new hypothesis of tsunami generation. *Geophysical Journal International, 185*(3), 1135–1146.

Sahal, A., Roger, J., Allgeyer, S., Lemaire, B., Hébert, H., Schindelé, F., et al. (2009). The tsunami triggered by the 21 May 2003 Boumerdes-Zemmouri (Algeria) earthquake: field investigations on the French Mediterranean coast and tsunami modelling. *Natural Hazards and Earth System Sciences, 9*(6), 1823–1834.

Schindelé, F., Gailler, A., Hébert, H., et al. (2015). Implementation and challenges of the Tsunami Warning System in the western

Mediterranean. *Pure and Applied Geophysics, 172*(3–4), 821–833.

Semmane, F., Campillo, M., & Cotton, F. (2005). Fault location and source process of the Boumerdes, Algeria, earthquake inferred from geodetic and strong motion data. *Geophysical Research Letters, 32*(1), L01305. https://doi.org/10.1029/2004GL021268.

Synolakis, C. E. (1991). Tsunami runup on steep slopes: how good linear theory really is. *Natural Hazards, 4*(2–3), 221–234.

Tinti, S., Armigliato, A., Bortolucci, E., & Piatanesi, A. (1999). Identification of the source fault of the 1908 Messina earthquake through tsunami modelling Is it a possible task? *Physics and Chemistry of the Earth, Part B: Hydrology, Oceans and Atmosphere, 24*(5), 417–421.

Titov, V. V. (2009). Tsunami forecasting. In A. Robinson & E. Bernard (Eds.), *The Sea* (Vol. 15, pp. 371–400). Cambridge: Harvard University Press. **(ISBN 0674031733, 9780674031739)**.

Titov, V. V. (2009). Tsunami forecasting. In A. Robinson & E. Bernard (Eds.), *The Sea* (Vol. 15, pp. 371–400). Cambridge: Harvard University Press. **(ISBN 0674031733, 9780674031739)**.

Yelles-Chaouche, A. (1991). Coastal Algerian earthquakes: a potential risk of tsunamis in Western Mediterranean–Preliminary investigation. *Science of Tsunami Hazards, 9*(1), 47–54.

Yelles, K., Lammali, K., Mahsas, A., Calais, E., & Briole, P. (2004). Coseismic deformation of the May 21st, 2003, $M_w = 6.8$ Boumerdes earthquake, Algeria, from GPS measurements. *Geophysical Research Letters, 31*(13), 13610. https://doi.org/10.1029/2004GL019884.

(Received May 16, 2017, revised October 24, 2017, accepted October 26, 2017)

Pure Appl. Geophys.

https://doi.org/10.1007/s00024-018-1824-y

Pure and Applied Geophysics

Evaluating the Effectiveness of DART® Buoy Networks Based on Forecast Accuracy

DONALD B. PERCIVAL,[1,2] DONALD W. DENBO,[3,4] EDISON GICA,[3,4] PAUL Y. HUANG,[5] HAROLD O. MOFJELD,[3,4] MICHAEL C. SPILLANE,[3,4] and VASILY V. TITOV[3]

Abstract—A performance measure for a DART® tsunami buoy network has been developed. DART® buoys are used to detect tsunamis, but the full potential of the data they collect is realized through accurate forecasts of inundations caused by the tsunamis. The performance measure assesses how well the network achieves its full potential through a statistical analysis of simulated forecasts of wave amplitudes outside an impact site and a consideration of how much the forecasts are degraded in accuracy when one or more buoys are inoperative. The analysis uses simulated tsunami amplitude time series collected at each buoy from selected source segments in the Short-term Inundation Forecast for Tsunamis database and involves a set for 1000 forecasts for each buoy/segment pair at sites just offshore of selected impact communities. Random error-producing scatter in the time series is induced by uncertainties in the source location, addition of real oceanic noise, and imperfect tidal removal. Comparison with an error-free standard leads to root-mean-square errors (RMSEs) for DART® buoys located near a subduction zone. The RMSEs indicate which buoy provides the best forecast (lowest RMSE) for sections of the zone, under a warning-time constraint for the forecasts of 3 h. The analysis also shows how the forecasts are degraded (larger minimum RMSE among the remaining buoys) when one or more buoys become inoperative. The RMSEs provide a way to assess array augmentation or redesign such as moving buoys to more optimal locations. Examples are shown for buoys off the Aleutian Islands and off the West Coast of South America for impact sites at Hilo HI and along the US West Coast (Crescent City CA and Port San Luis CA, USA). A simple measure (coded green, yellow or red) of the current status of the network's ability to deliver accurate forecasts is proposed to flag the urgency of buoy repair.

Key words: Aleutian Islands tsunami sources, Buoy network performance measure, Crescent City CA, DART® data inversion, Hilo HI, Network assessment, Port San Luis CA, South American tsunami sources, Tsunameter, Tsunami buoys, Tsunami forecasts, Tsunami simulation, Tsunami source estimation.

[1] Applied Physics Laboratory, University of Washington, Box 355640, Seattle, WA 98195-5640, USA. E-mail: dbp@apl.washington.edu

[2] Department of Statistics, University of Washington, Box 354322, Seattle, WA 98195-4322, USA.

[3] NOAA/Pacific Marine Environmental Laboratory, 7600 Sand Point Way NE, Seattle, WA 98115, USA.

[4] Joint Institute for the Study of the Atmosphere and Ocean, University of Washington, Seattle, WA 98195-5672, USA.

[5] National Tsunami Warning Center, National Weather Service, Palmer, AK 99645, USA.

1. Introduction

Tsunamis are potentially devastating disasters for coastal regions worldwide. The loss of life due to tsunamis can be greatly reduced when the propagation time of a tsunami is long enough to allow for timely issuance of warnings. Beginning in the late 1940s soon after a devastating tsunami that hit Hawaii in 1946, the National Oceanic and Atmospheric Administration (NOAA) and its predecessors established warning systems to provide tsunami forecasts to the United States and other nations (Whitmore 2009). The research presented here is part of an ongoing effort to improve the speed and accuracy of such forecasts through enhanced observational capabilities.

Prompted in part by the scale of destruction and unprecedented loss of life following the 2004 Indian Ocean tsunami, NOAA has deployed 39 Deep-ocean Assessment and Reporting of Tsunamis (DART®) buoys (to date, eight countries other than the United States have set out additional buoys). These buoys can observe the passage of a tsunami and relay data related to its arrival time and amplitude in near-real time. These data are used by the Short-term Inundation Forecast for Tsunamis (SIFT) application developed at the NOAA Center for Tsunami Research (NCTR); for details, see Gica et al. (2008) and Titov (2009). SIFT combines DART® data with precomputed geophysical models to estimate so-called unit source coefficients (Percival et al. 2011). As outlined in Sect. 2, these coefficients are used to forecast wave amplitudes at locations in the open ocean outside of impact sites (coastal regions of particular interest, e.g., Hilo HI). In turn forecasts of these wave amplitudes play a critical role in forecasting coastal inundation at impact sites.

In this article we consider the problem of how to evaluate the effectiveness of a DART® buoy network in providing accurate and timely forecasts of open-ocean wave amplitudes. The purpose of the network is not merely to detect tsunamis (for which instruments much less complex than DART® buoys could be deployed), but, more importantly, to collect data that can be used to provide accurate inundation forecasts in a timely manner—hence we focus our evaluation of the network on the quality of the forecasts rather than just on the simpler task of detection. For an existing network the goal is to devise a simple measure of effectiveness that can be monitored over time and used to raise an alarm if the network deteriorates. For an alteration to an existing network (typically through adding, decommissioning or relocating a buoy), the goal is to quantify the effect of the change in a simple manner. The challenge in both cases is that the effectiveness of wave amplitude forecasts is predicated on a number of interacting factors. The desire for a simple measure is at odds with the inherent complexity of the forecast process. What is needed is a useful measure that does not oversimplify. Here, we propose a measure that makes certain simplifications but nonetheless captures key factors limiting the accuracy of forecasted wave amplitudes. Our approach uses a succession of summaries of simulated forecasts. (An ideal measure would make use of actual rather than simulated forecasts, but there is simply not enough actual data for realistic evaluation of an existing network, much less a proposed network.)

The effort described here is not the first to consider the evaluation of a network designed to contribute to tsunami warnings. Greenslade and Warne (2012) is the first work to evaluate an existing heterogeneous network (i.e., one consisting of both tide gauges and DART® buoys) for its effectiveness for tsunami warnings. Their evaluation is done purely in terms of travel times. Our work focuses on just DART® buoys and uses accuracy of forecasted wave amplitudes as the evaluation criterion while paying nominal attention to travel times. Greenslade and Warne (2012) note that network design is closely related to network evaluation and point to Spillane et al. (2008) on the design of DART® buoy networks. The current effort builds upon the latter in using some of the same tools (e.g., the propagation database described in Gica et al. (2008)), but does network evaluation in a manner substantively different from network design.

2. *Construction of Open-Ocean Wave Amplitude Forecasts in Practice*

Here we give an overview of how open-ocean wave amplitude forecasts are made in practice using SIFT. Generation of these forecasts presumes that a tsunami-causing earthquake originates from within one or more unit sources, which are 100 km by 50 km sections of fault planes in tsunamigenic regions. Figure 1 shows locations of a subset of 74 predesignated unit sources tiling the region where the Pacific plate is being subducted within the Aleutian–Alaska subduction zone. The sources chosen are in 2 rows (`b` and `a`) and 37 columns (`001–037`). Row `b` is adjacent to the plate boundary, while row `a` is a continuation of the descending slab to greater depths (we do not make use of additional rows used to represent deeper Aleutian–Alaska earthquakes). The expanded view in the figure shows rows `b` and `a` along with 3 of the columns (`025`, `026` and `027`). We refer to the unit source in, e.g., row `b` and column `026` as `ac026b`. The figure also shows the locations of three impact sites (Hilo (HIL), Crescent City (CCY) and Port San Luis (PSL)) and the current locations of 14 DART® buoys (labeled clockwise by *U*, *V*, ..., *Z*, 1, 2, ..., 8; the expanded view includes buoy 46403, which is labeled as `6`). There are 39 DART® buoys in all, but these 14 are germane for handling events arising from Aleutian–Alaska earthquakes in that, as called for by standard operating procedures, their data could be used as the basis for issuing a warning at least 3 h in advance to at least one of the three impact sites (Whitmore et al. 2008).

A geophysical model is precomputed for each pairing of one of the 74 unit sources either with a buoy or with a location on the gridded open-ocean

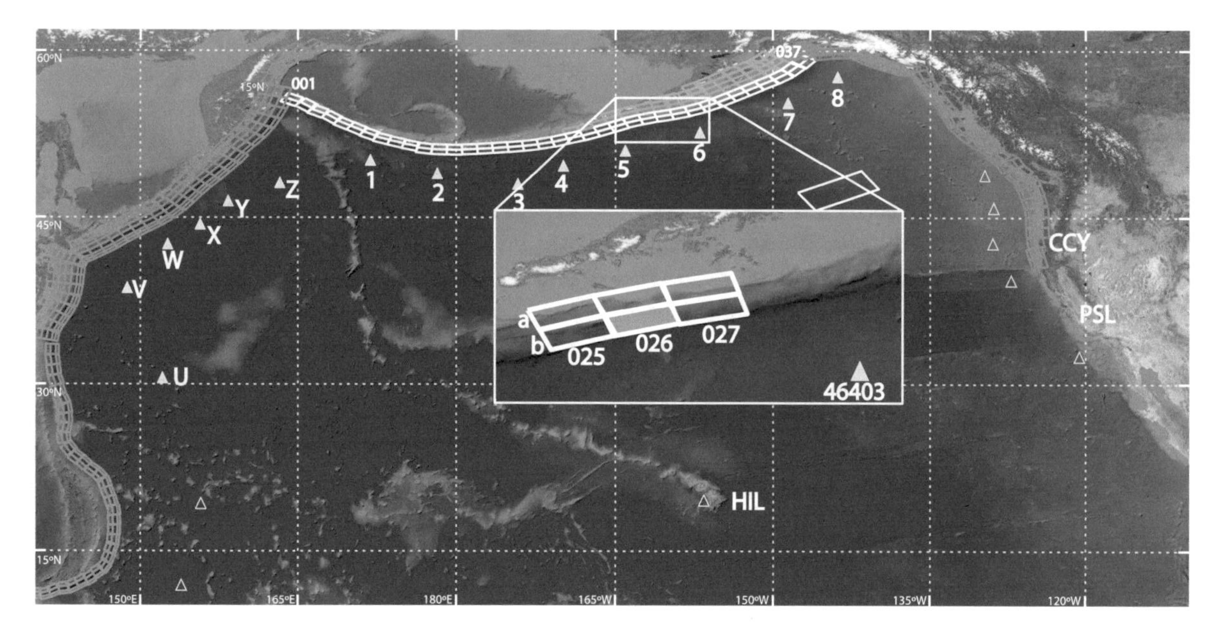

Figure 1
Locations of 74 unit sources in Aleutian–Alaskan subduction zone (small white rectangles, with an expanded view of six of them), DART® buoys in the Northern Pacific (triangles, 14 of which are labeled by `U–Z` and `1–8`) and three impact sites (HIL marks Hilo, while CCY and PSL show where Crescent City and Port San Luis are). The unit sources are arranged in two rows denoted by `a` (upper row) and `b` (lower) and in 37 columns labeled as, from left to right, `001–037`. Going clockwise, the World Meteorological Organization designations for the 14 buoys are (`U`) 21413, (`V`) 21418, (`W`) 21401, (`X`) 21419, (`Y`) 21402, (`Z`) 21416, (`1`) 21415, (`2`) 21414, (`3`) 46413, (`4`) 46408, (`5`) 46402, (`6`) 46403, (`7`) 46409 and (`8`) 46410. The expanded view shows central unit source `ac026b` and the five unit sources abutting it, along with the location of the nearby buoy 46403

boundary of the inundation model for a particular impact site [see Gica et al. (2008) for details about these models, which are stored in a propagation database maintained at NCTR]. The model predicts what would be observed over time at a deep-water location if a moment magnitude $M_W = 7.5$ reverse fault earthquake were to originate from within a particular unit source. Figure 2 shows examples of predictions for buoy 46403 due to an earthquake from either unit source `ac026b` or one of five abutting it (for use later on, these predictions are marked by $\mathbf{g}_1$, ..., $\mathbf{g}_6$). If, as is usually the case, the moment magnitude of the earthquake is different from 7.5, then, relying on the linearity of tsunami waves in deep water, the predictions are adjusted by a nonnegative multiplicative factor α known as a source coefficient (the assumption of a reverse fault earthquake imposes the nonnegativity constraint). Earthquakes can span more than one unit source, in which case the model is taken to be a linear combination of individual unit source models, with the coefficients in the linear combination, say α_k, $k = 1, \ldots, K$, all constrained to be nonnegative.

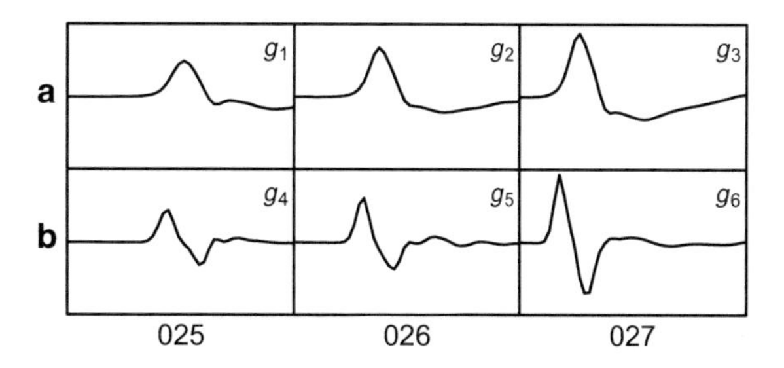

Figure 2
Models for what would be observed at DART® buoy 46403 from an earthquake originating at the center of one of six unit sources in the Aleutian–Alaskan subduction zone (the 100 km by 50 km sections of the ocean corresponding to these unit sources are highlighted in Fig. 1). The top row shows the models for unit sources `ac025a`, `ac026a` and `ac027a` (labeled as $\mathbf{g}_1$, $\mathbf{g}_2$ and $\mathbf{g}_3$); the bottom, for `ac025b`, `ac026b` and `ac027b` ($\mathbf{g}_4$, $\mathbf{g}_5$ and $\mathbf{g}_6$). The time span for each displayed model is 45 min, and the span starts 10 min after the start of the earthquake. The vertical axes all range from −4.1 cm to 4.1 cm

During the passage of a tsunami, a DART® buoy collects and transmits a time series of bottom pressure (BP) measurements. Each measurement represents the BP averaged over a 1-min span. The spans are non-overlapping, and the measurements occur once per minute, thus forming what we refer to as a 1-min stream of data. Once a sufficient stretch of this stream has been collected by the first buoy that detects the tsunami event, the source coefficients α_k are estimated using constrained least squares (Percival et al. 2014). With the passage of time, more 1-min data from this and potentially other buoys become available, and the estimates $\hat{\alpha}_k$ of the source coefficients can be reevaluated and possibly refined. In operational tsunami forecasting, the estimated source coefficients are applied to a sub-region of the propagation database to provide the boundary forcing fields for the real-time inundation model run with detailed simulation of waves, currents and inundation for the impact site (Gica et al. 2008); however, in the simulation study discussed in the next section, we avoid use of the inundation model by choosing an adjacent deep-water location for each impact site of interest. This approach allows us to efficiently generate a large numbers of simulations by linear combinations of table look-ups.

3. *Evaluation of Open-Ocean Wave Amplitude Forecasts via Simulation*

In theory the best way to evaluate the effectiveness of a network of DART® buoys would be to compare coastal inundation forecasts formulated during actual tsunami events with the actual inundations. While such comparisons are routinely done as a vital check on current operational procedures, use of these alone is not sufficient for network evaluation. A meaningful assessment would require not only multiple events, but also ones treated in the same manner (no change in the network of buoys, no equipment upgrades, use of the same methodology for forming the forecasts, etc.). Even if conditions for generating forecasts could be fixed, the amount of time needed to collect data for a reliable evaluation would be inordinate given the rate at which tsunamis occur. In addition, if we were to insist upon using actual events to assess effectiveness, changes to a network could not be assessed until long after the changes had been put into place. For planning purposes, we need the ability to assess the effectiveness of any proposed changes. We thus entertain basing our assessments on simulated forecasts. The simulated forecasts assume that the DART® buoys are an inherent and vital component of the forecast process (in the Appendix we briefly consider forecasts based on seismic information alone).

As noted in the previous section, the run-up model for generating inundation forecasts at a particular impact site requires its initial conditions be set by wave amplitudes in the open ocean at a nearby location. The quality of the forecasts generated by the run-up model depends critically on these initial conditions, which are provided by forecasts of open-ocean wave amplitudes. Rather than assessing a network via the quality of inundation forecasts directly, we can do so indirectly via the quality of open-ocean wave amplitude forecasts. In the simulation studies discussed below, this indirect approach is computationally attractive because calculation of run-up models is time-consuming. This approach, however, ignores errors in forecasts that might arise from imperfect bathymetry and/or topography.

With the simplification of dealing with open-ocean wave amplitude forecasts rather than inundation forecasts, there are four tasks at hand to simulate and evaluate the forecasts of interest. Here we glean the key points in our handling of the four tasks, with full details being provided in the Appendix (Sects. 7.1, 7.2, 7.3, 7.4).

The first task is to play the role of nature (Sect. 7.1). We must simulate a tsunami signal as it passes by a particular buoy along with what we will regard as the wave amplitudes presumed to occur outside of an impact site of interest. We assume the signal originates within a unit source that we refer to as the central unit source and that is surrounded by five other unit sources (in the expanded view of Fig. 1, the central unit source is `ac026b`). For each of the six unit sources, we have a geophysical model for what would be observed at a particular buoy for a tsunami generating event originating at the center of the unit source. The signal is a weighted average of the six models, with the weights set by picking at random a location within the central unit source. This location is used to define a new unit source that is constructed from pieces of the original six unit sources, with the sizes of the pieces determining the weights. We also have a geophysical model for what would be observed in the open ocean outside of an impact site of interest. We apply the same weights to these models to form the wave amplitudes presumed to occur in the open ocean.

The second task is to play the role of nature's observer (Sect. 7.2). We must simulate the data that a particular DART® buoy would collect during a tsunami event. We do so by adding actual tidal fluctuations and background noise recorded under ambient conditions by the buoy (if archived data are not available for a particular buoy, we use data from a suitable surrogate).

The third task is to use the simulated data to estimate source coefficients, which in turn are used to forecast the wave amplitudes at an impact site (Sect. 7.3). Estimation is accomplished in terms of a linear regression model that, in addition to the source coefficients, has coefficients designed to compensate for tidal fluctuations. The source coefficients are associated with the geophysical models, but, to mimic what happens in practice, we permit potential mismatches between the tsunami signal and its assumed model. One last consideration is how much data to use for estimating the coefficients. This is settled by determining the location of the first peak in the signal and including data spanning from 1 hr prior to this peak to 21 min after the peak (this assumes that the peak occurs at least 1 hr after the start of the earthquake generating the tsunami signal—if not, then the first data point coincides with the start of the earthquake). We estimate the coefficients in the linear regression model using least squares. We then use the estimated source coefficients to form forecasted wave amplitudes.

The fourth and final task is to evaluate the quality of the forecasted wave amplitudes by comparing them with the presumed amplitudes using an appropriate metric (Sect. 7.4). To do so, we repeat the first three tasks 1000 times, resulting in 1000 forecasted amplitudes to be compared with 1000 presumed amplitudes. We evaluate how well the forecasts do by looking at the difference between the maximum forecasted and presumed amplitudes and using these differences to form a root-mean-square error, which is the basis for the network assessment discussed in the next section.

4. *Network Assessment*

In the previous section, we described our approach for assessing the effectiveness of a particular buoy in forecasting wave amplitudes in the open ocean outside of an impact site of interest when an earthquake-generating tsunami arises from within a particular central predesignated unit source. The approach takes into account major factors influencing the quality of the forecasts (including tides, background noise, uncertainties in the signal model and amount of available buoy data). In this section, we turn our attention to network assessment. To focus our discussion, we consider how well the existing network of DART® buoys does in responding to tsunami events originating first from the Aleutian Islands (Sect. 4.1) and then from South America (Sect. 4.2).

4.1. First Case Study: Aleutian Islands

Figure 1 shows the locations of 74 predesignated unit sources tiling the Aleutian–Alaskan subduction zone, the 14 DART® buoys closest to this zone [8 along the Aleutian Islands (labeled `1–8`) and 6 to the west (`U–Z`)] and three impact sites (Hilo, Crescent City and Port San Luis—these are labeled as HIL, CCY and PSL). To evaluate this network of buoys, we consider all possible triads involving one buoy, one impact site and one central unit source selected from `ac002b`, ..., `ac036b` (there are thus $14 \times 3 \times 35 = 1470$ triads in all). Figure 3 shows root-mean-square errors (RMSEs) of forecasted maximum wave amplitudes in the open ocean outside of Hilo [see Eq. (6) in the Appendix for the definition of RMSE]. The RMSEs are shown only for those buoy/unit source combinations such that, as called for by standard operating procedures, a warning to Hilo could be issued at least 3 h in advance of the arrival of the tsunami (when this criterion is applied to all three impact sites, the number of relevant triads shrinks from 1470 down to 835). The RMSEs for the 14 buoys are plotted versus unit source. Let us concentrate on DART® buoy 46403 [also used as an

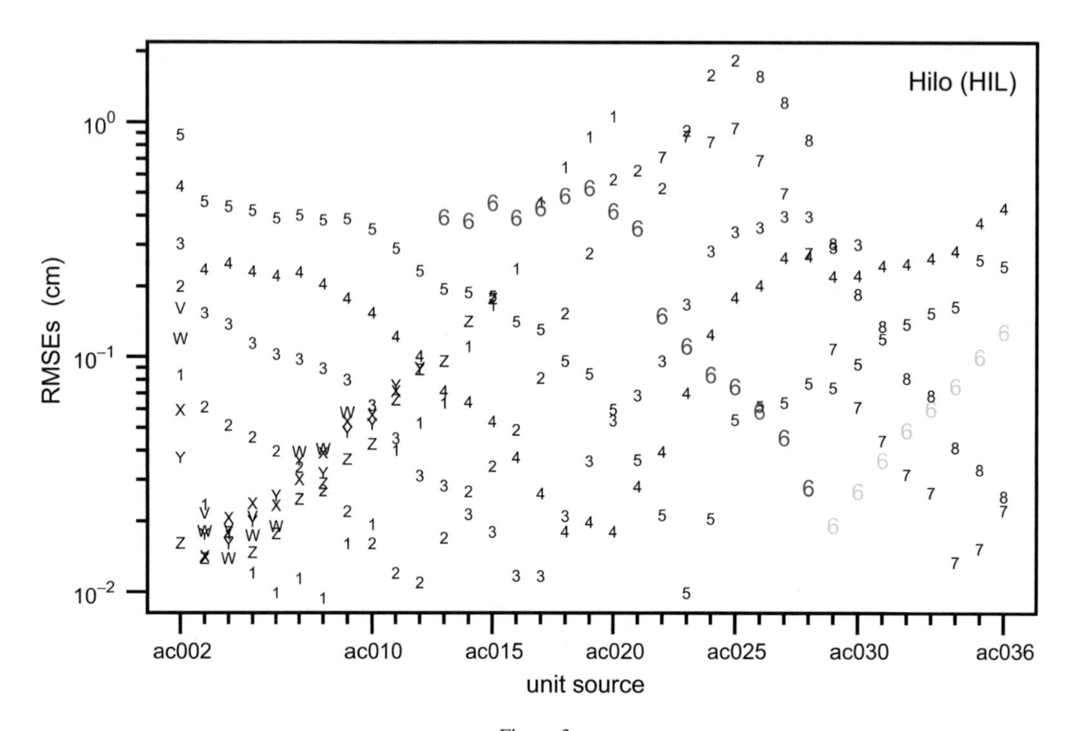

Figure 3
RMSEs for forecasted maximum wave amplitudes in the open ocean outside of Hilo based on 14 buoys (labeled as indicated in Fig. 1) in combination with central unit sources ac002b–ac036b from the Aleutian–Alaskan subduction zone. The RMSEs for buoy 46403 (labeled as 6) are colored, with red indicating its pairing with ac026b

example in Figs. 1 and 2 and throughout the Appendix (Sect. 7)]. This buoy's label is 6, and the only non-black labels in Fig. 3 are the 6s. One is colored red, and its location in the figure tells us that, for a $M_W = 7.5$ earthquake originating within unit source ac026b, the RMSE over 1000 forecasts at Hilo based on buoy 6 is equal to 5.9×10^{-2} cm (this RMSE is also displayed by the 4 in the 21 min category in Fig. 18, which is discussed in Sect. 7.4 of the Appendix). The RMSEs for ac002b, ..., ac012b are not shown because a tsunami originating from one of these unit sources would not arrive at buoy 6 soon enough to issue a timely warning to Hilo.

For assessing the ability of a network of buoys to provide good forecasted maximum wave amplitudes, individual RMSEs are of not as much interest as the best, i.e., lowest, RMSE that can be achieved across all buoys for an event originating from within a particular unit source. The second best RMSE is also of interest because a comparison of this with the best RMSE tells us how much degradation in forecasted amplitude there would be if the buoy with the best RMSE were to become inoperative. Figure 4a simplifies Fig. 3 by showing just the first (blue) and second best (red) RMSEs. There is a prominent pinching pattern between the two sets of RMSEs. The unit sources where the pinches occur are ones for which the distances from the source to the two associated buoys are approximately equal. For events originating from pinch locations, if the buoy with the best RMSE were to drop out, there would be little impact on the ability of the network to forecast wave amplitudes outside of Hilo because of the existence of another buoy that performs almost as well. Parts b and c of Fig. 4 are similar to a, but now use Crescent City and Port San Luis as impact sites rather than Hilo. Pinching patterns are evident again, with the locations of the pinches occurring close to the same unit sources as for Hilo. With some exceptions, the first and second best RMSEs are also associated with the same buoys. These results indicate that there is some degree of commonality across impact sites that are at transoceanic distances from the unit sources.

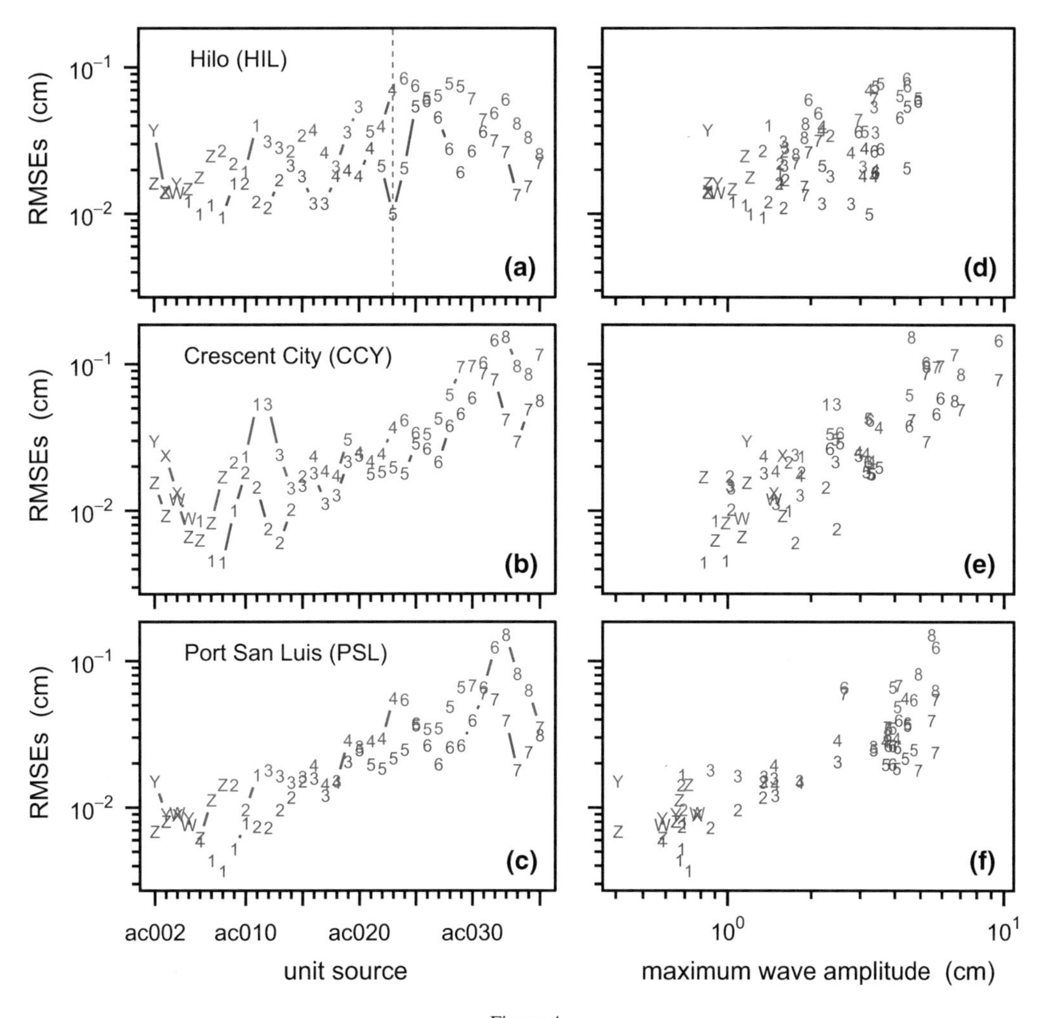

Figure 4
Best (blue) and second best (red) RMSEs for forecasted maximum wave amplitudes in the open ocean outside of, from top to bottom rows, Hilo, Crescent City and Port San Luis. The RMSEs are plotted versus central unit sources `ac002b`, ..., `ac036b` in the left-hand column and, in the right-hand, versus maximum wave amplitudes. Plot **a** shows RMSEs extracted from Fig. 3, and the vertical dashed line crosses the RMSEs for unit source `ac023b` in combination with buoys `5` and `4`

In addition to pinching patterns, there are particularly prominent upward trends in Fig. 4b, c for Crescent City and Port San Luis. A glance at Fig. 1 shows that Hilo is more centrally located with respect to the Aleutian–Alaskan unit sources than the California sites, both of which are off to one side. The travel times from the 35 central unit sources to the California sites both have a nearly monotonic decay proceeding from `ac002b` to `ac036b`, whereas those for Hilo decrease from `ac002b` to `ac021b` and then increase along the remaining 15 unit sources. As travel time increases, the maximum amplitude of a tsunami signal should tend to decrease. Presuming an inverse relationship between travel time and signal amplitude, the prominent upward trends at the California sites suggest that RMSE increases as signal amplitude increases. To explore this suggestion, the right-hand column of Fig. 4 shows the RMSEs of the forecasted maximum wave amplitudes for each of the impact sites versus the maximum wave amplitudes outside of the three impact sites. The maximum wave amplitudes were extracted from a propagation data base of predictions linking the center of each central unit source with the open ocean location outside of the impact site. Crescent City and Port San Luis have prominent upward trends, but Hilo, less so, which is the same set of patterns displayed in the left-hand column. This qualitative comparison offers support for the hypothesis that the upward trends in the left-hand column are due at least in part to a positive relationship between RMSE and signal amplitude.

To quantify how well the existing network of DART® buoys can respond to events emanating from the Aleutian–Alaskan subduction zone, let us focus on forecasting maximum wave amplitudes based on single buoys and consider what happens if one buoy were to drop out. As a measure of both the contribution of individual buoys to the network and the overall strength of the network, consider the maximum relative increase in RMSE across all 35 central unit sources due to a particular buoy becoming inoperative. Suppose, for example, that buoy 5 becomes inoperative. Focusing on Hilo, Fig. 4a indicates that, if a tsunami were to emanate from unit source `ac023b` (corresponding to the dashed vertical line), loss of 5 would result in a 7 fold increase in the best available RMSE because we would then have to rely on buoy 4 (the RMSEs would also increase at the California sites, but less so). For Hilo, buoy 5 also has the best RMSE for three additional unit sources (`ac022b`, `ac024b` and `ac025b`), but the relative increases in RMSE due to 5 being gone are all smaller than that for `ac023b`. For events emanating from any of the remaining 31 unit sources, the loss of 5 would not cause a deterioration in the ability of the network to forecast Hilo's wave amplitudes.

Figure 5 quantifies how the network is affected when a single buoy becomes inoperative. For every buoy that can be used to deliver a warning at least 3 h in advance for a particular impact site due to an event arising from at least one of the 35 central unit sources, we examine the ratio of two particular RMSEs for each unit source. The first RMSE is the best available RMSE after elimination of the buoy in question, and the second is the best RMSE when all DART® buoys are operational. We then concentrate on just the largest such ratio across all unit sources. This ratio reflects a worst-case scenario in a clear-cut manner and is arguably preferable to a more complicated measure that attempts to take into account degradations across all unit sources. The left panel of Fig. 5 shows these RMSE ratios for 14 buoys and at Hilo. The smallest possible ratio is unity, which is indicated by the left-most red dashed line. A ratio of unity occurs when a buoy's RMSEs are never the best at any of the 35 unit sources. Such a buoy does not participate in the blue portion of Fig. 4a and typically is not in close proximity to at least one unit source (see Fig. 1). Under a worst-case scenario, loss of these buoys would not degrade any forecasts. By this measure, buoys with an RMSE ratio of unity are of secondary importance to Hilo for responding to events arising from the Aleutian Islands (they are primarily intended to handle events arising elsewhere). When the ratio is greater than unity, the network deteriorates at least to some degree when the buoy in question is out of commission. For these buoys, Fig. 5 indicates the unit source linked to the largest proportional increase in RMSE. The largest RMSE ratio occurs when buoy 5 (indicated by a solid blue circle) is not available to handle an event arising from unit source `ac023b`. Thus, assuming a worst-case scenario, loss of buoy 5 would cause the largest degradation in the network's ability to forecast maximum wave amplitudes at Hilo. The middle and right panels of Fig. 5 show corresponding results for Crescent City and Port San Luis. For a buoy whose ratios are greater than unity for multiple impact sites, the associated unit sources are either the same or quite close to one another. With the exceptions of buoy 2, the RMSE ratios for the two California sites are remarkably similar.

Figure 6 explores the effect on network performance when two buoys drop out. As in Fig. 5, there are three panels, one for each impact site. Each panel has 15 red dotted horizontal lines, which breaks the panel up into 14 subplots, one for each of the relevant buoys. Consider the bottommost subplot in the left-hand panel, which is for buoy U in combination with Hilo. There are 13 dots in the subplot, each indicating the maximum ratio of two particular RMSEs across the 35 unit sources. The first RMSE is the best remaining after U drops out along with one of the other 13 buoys, and the second is the best RMSE when all 14 buoys are present (going from top to bottom in the subplot, the dots are associated with

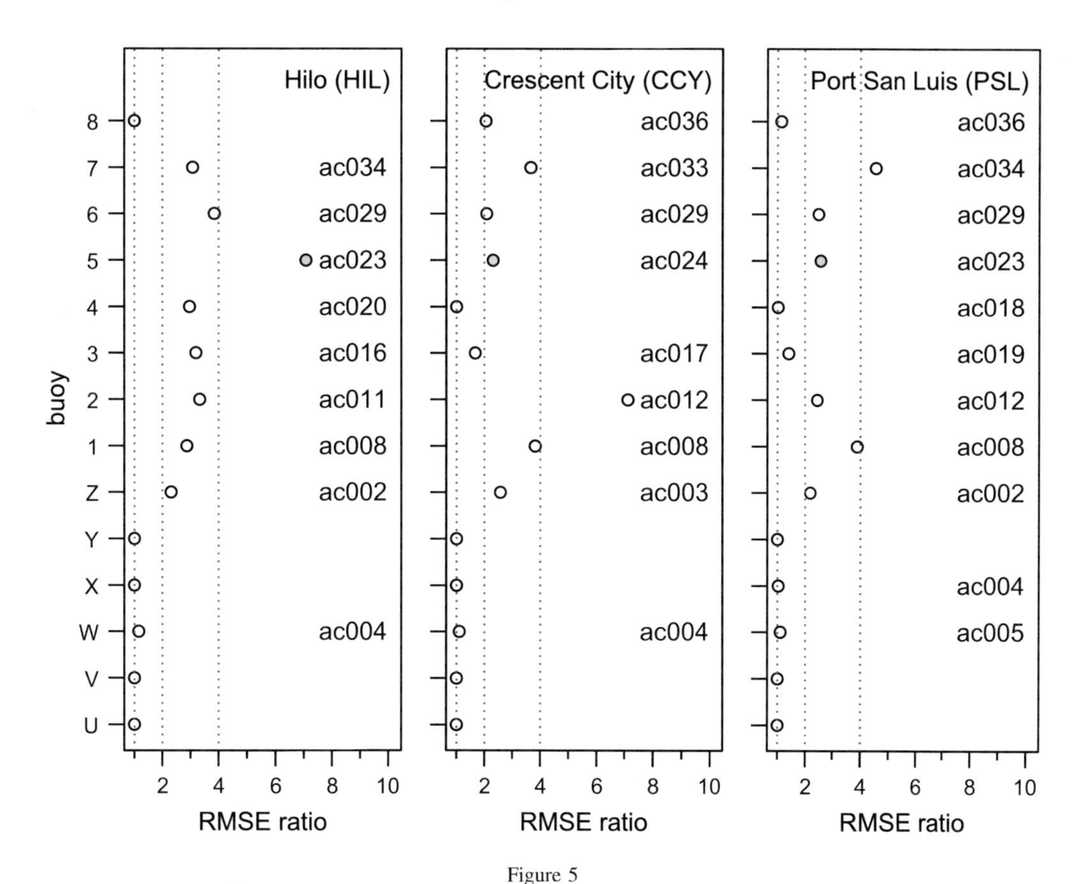

Figure 5
Effect on performance of DART® buoy network across 35 unit sources and at three impact sites due to one-by-one removal of the 14 buoys depicted in Fig. 1. For each buoy, the largest increase in RMSE over the 35 unit sources is shown relative to the best available RMSE when that buoy is dropped as compared to when the full network of buoys is available. The case where buoy 5 is dropped is highlighted with colored circles. The one corresponding to Hilo is the change in RMSE indicated by the vertical dashed line in Fig. 4a. The three red dashed lines mark RMSE ratios of 1, 2 and 4. These ratios are used to define a green, yellow and red network status—see Fig. 7 and the discussion pertaining to it

buoys 8–V). The blue dot is the RMSE ratio when both U and 5 are inoperative (the blue dots in the other subplots also indicate ratios in which 5 is involved). Because buoy U never has the best RMSE for any of the 35 unit sources, the ratios merely reflect what happens when the second buoy drops out. As a result, the dots in the subplot have same pattern as the topmost 13 circles in the left-hand panel of Fig. 5. Thus, when examining any of the other 13 subplots, we can compare it to the bottommost subplot to ascertain of how much the leave-out-two pattern deviates from the leave-out-one pattern. The same holds for the middle and right-hand panels for Crescent City and Port San Luis.

The subplots for buoys V–8 in Fig. 6 are organized in a manner similar to that for U. Consider the subplot for buoy 5 in combination with Hilo. Because any ratio involving 5 is represented by a blue dot, all of the dots in this subplot are blue (in each of the other 13 subplots for Hilo, there is exactly one blue dot). With two exceptions, the ratios all have a value of 7.1. The identical ratios come about because dropping 5 leads to the largest proportional increase in RMSE across all 35 unit sources (see Fig. 4a). The two exceptional ratios are 9.7 and 11.2. A dot in a given subplot representing the RMSE ratio for a particular pair of buoys must be replicated in one other subplot. The dot for 9.7 is replicated in the subplot for 6, and the one for 11.2 has a duplicate in the subplot for 4. Hence, under the worst-case scenario, the concurrent loss of 5 and 4 would lead to an order of magnitude increase in RMSE. This

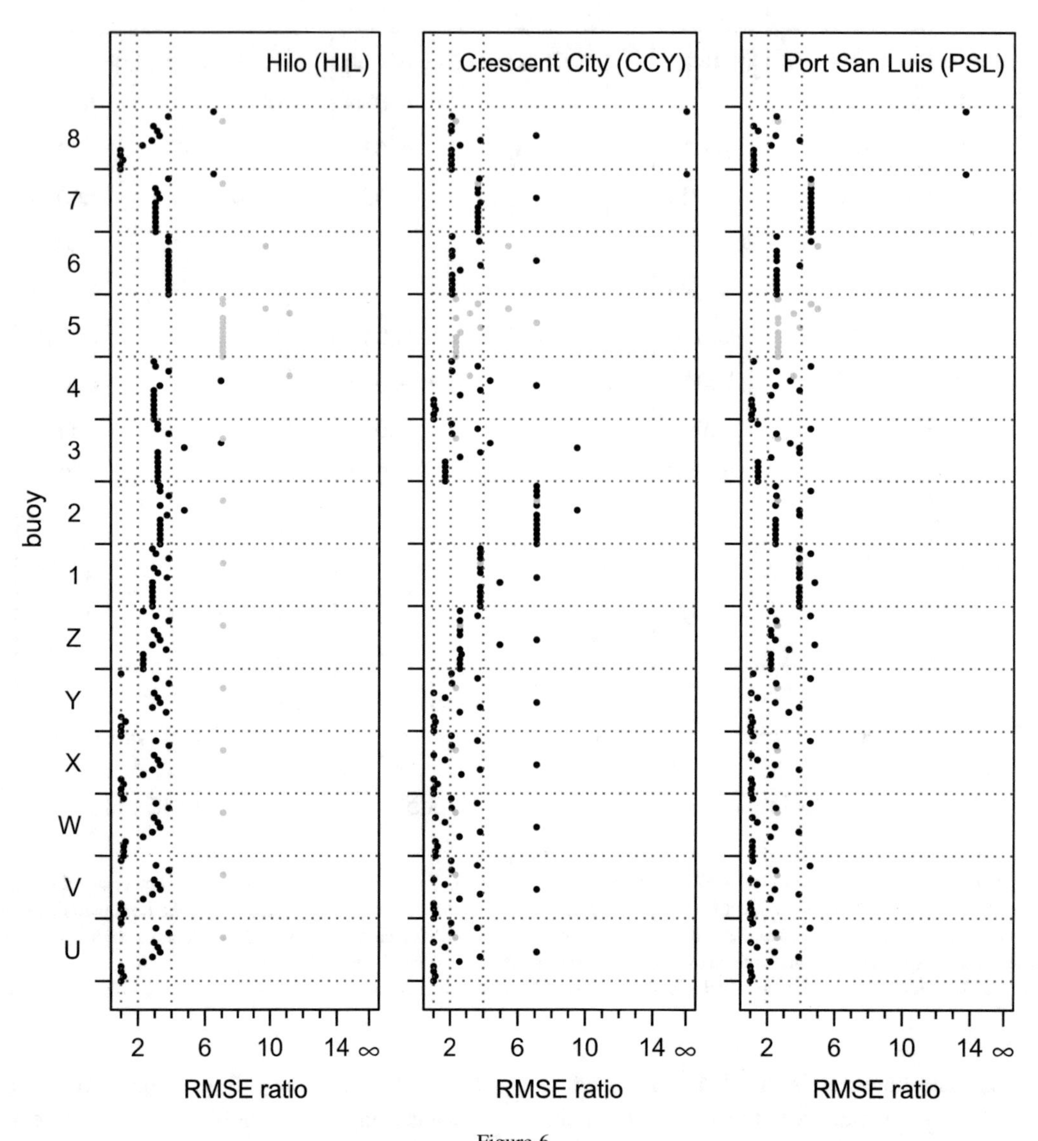

Figure 6
Effect on performance of DART® buoy network across 35 unit sources and at three impact sites due to concurrent dropout of two of 14 buoys depicted in Fig. 1. For each pair of buoys, the largest increase in RMSE over the 35 unit sources is shown relative to the best available RMSE when both buoys are dropped as compared to when the full network of buoys is available. The RMSE ratios are shown as dots. The blue dots show cases for which buoy 5 is one of the dropped buoys. The three red vertical dashed lines mark RMSE ratios of 1, 2 and 4. These ratios are used to define a green, yellow and red network status—see Fig. 7 and the discussion pertaining to it

scenario involves a tsunami-generating earthquake occurring in unit source `ac023b`, for which the third best RMSE is for buoy `6`. The second worst-case scenario would be to lose `5` and `6` for an event in `ac028b`, for which the third best RMSE is for buoy `4`.

The subplots for eight buoys (`Z–7`) in the left-hand panel of Fig. 6 all have ratios greater than unity. These same buoys also have ratios greater than unity under the leave-out-one scenario (left-hand panel of Fig. 5). If one of these buoys were to be inoperative, loss of a second buoy from this set of eight could further compromise the network's ability to forecast maximum wave amplitudes at Hilo. The maximum ratios in all eight subpanels are associated with blue dots, which means, once any buoy amongst these eight other than 5 becomes inoperative, the worst-case scenario is to lose 5 also (as noted previously, the worst case overall is the loss of 5 and 4 together). If any of the 14 buoys were to becomes inoperative, a study of the corresponding subplot would tell us how much the ability of the network to forecast Hilo wave amplitudes would deteriorate if another buoy were to become inoperative before the first one is returned to operational status.

The middle and right-hand panels of Fig. 6 show leave-out-two RMSE ratios for the two California impact sites. For Crescent City (middle panel), the worst-case scenario occurs if buoys 7 and 8 were inoperative during an event originating in either unit source ac034b, ac035b or ac036b (see Fig. 4b). These two buoys are the only ones that can provide a warning 3 h in advance to Crescent City. Because there is no other buoy to rely on, we can consider the RMSE ratio to be infinite. The second worst-case scenario occurs when buoys 2 and 3 drop out for an event from ac012b leading to the RMSE ratio of 9.5, with 1 providing backup. After these worst cases, there are a number of RMSE ratios equal to 7.1, all of which involve buoy 8 and unit source ac012b.

Turning now to Port San Luis (right-hand panel of Fig. 6), the RMSE ratio for the worst-case scenario is 13.7, which occurs when 7 and 8 are inoperative and is associated with ac034b (see Fig. 4c). With these two buoys gone, buoy 6 has the best RMSE amongst the remaining buoys satisfying the 3-h warning time constraint. The second and third largest RMSE ratios are 4.9 and 4.8. The latter is of interest because the two buoys involved are 1 and Z, and the unit source is ac008b (this is in agreement with Fig. 4c). With both of these buoys gone, buoy W has the best remaining RMSE. Note that, in this case, the RMSE-based metric for network evaluation differs from one based solely on travel time since the travel times between unit source ac008b and buoys X and Y are both shorter than that for W (see Fig. 1). A possible explanation is the orientation of ac008b.

We can also explore the effect on network performance when three buoys drop out. For Hilo, the worst-case scenario is for buoys 4, 3 and 5 to be unavailable when an event arises from unit source ac020b. Buoy 6 then has the best RMSE amongst the remaining buoys satisfying the 3-h warning time constraint, and the associated RMSE ratio raises to 23.0 (as compared to 11.2 when two buoys—5 and 4—drop out during an event arising from unit source ac023b). For Crescent City, we have already noted that loss of 7 and 8 means that none of the remaining buoys can provide a 3-h warning for events from either ac034b, ac035b or ac036b. With the additional loss of 6, events from ac031b, ac032b or ac033b are also problematic because of lack of a fourth buoy that can provide a 3-h warning. There are similar concerns for ac029b and ac030b when 5, 6 and 7 are jointly inoperative. Turning to Port San Luis, the worst-case scenario involves buoys 7, 8 and 6; the associated unit source is ac034b; and backup comes from buoy 5 with an RMSE ratio of 42.2. The buoys and unit sources match up well with the leave-out-two case.

To help assess the effect of one, two or three buoys becoming inoperative, we introduce a simple indicator of the current status of the network reflecting its ability to forecast maximum wave amplitudes at a given impact site. The indicator has three levels:

1. if the RMSE ratio is less than 2, we award the network a 'green' status (the network is healthy);
2. a ratio between 2 and 4 indicates a 'yellow' status (the network has deteriorated somewhat); and
3. a ratio greater than 4 is a 'red' status (there is serious deterioration)

(the values 2 and 4 are placeholders for our demonstration, but are arbitrary and subject to change when used in practice). In Fig. 5, there are vertical dotted lines at 2 and 4 in each of the plots—these delineate the boundaries between the three levels. Considering the plot for Hilo, we see that the network would still have a green status even if any one of 6 buoys (U to Y or 8) out of a total of 14 (i.e., 43%) were to drop out by itself; on other hand, loss of 5 would result in a red status (7%), while losing any of the remaining 7 buoys would give a yellow status (50%). Figure 7a depicts these percentages, and plots (b) and (c) show corresponding results for the two California sites. The middle and bottom rows of Fig. 7 shows the percentages when two or three buoys drop out together. Not surprisingly, as more buoys drop out, the percentage of cases with red status increases for a given impact site. Note that three buoys can drop out

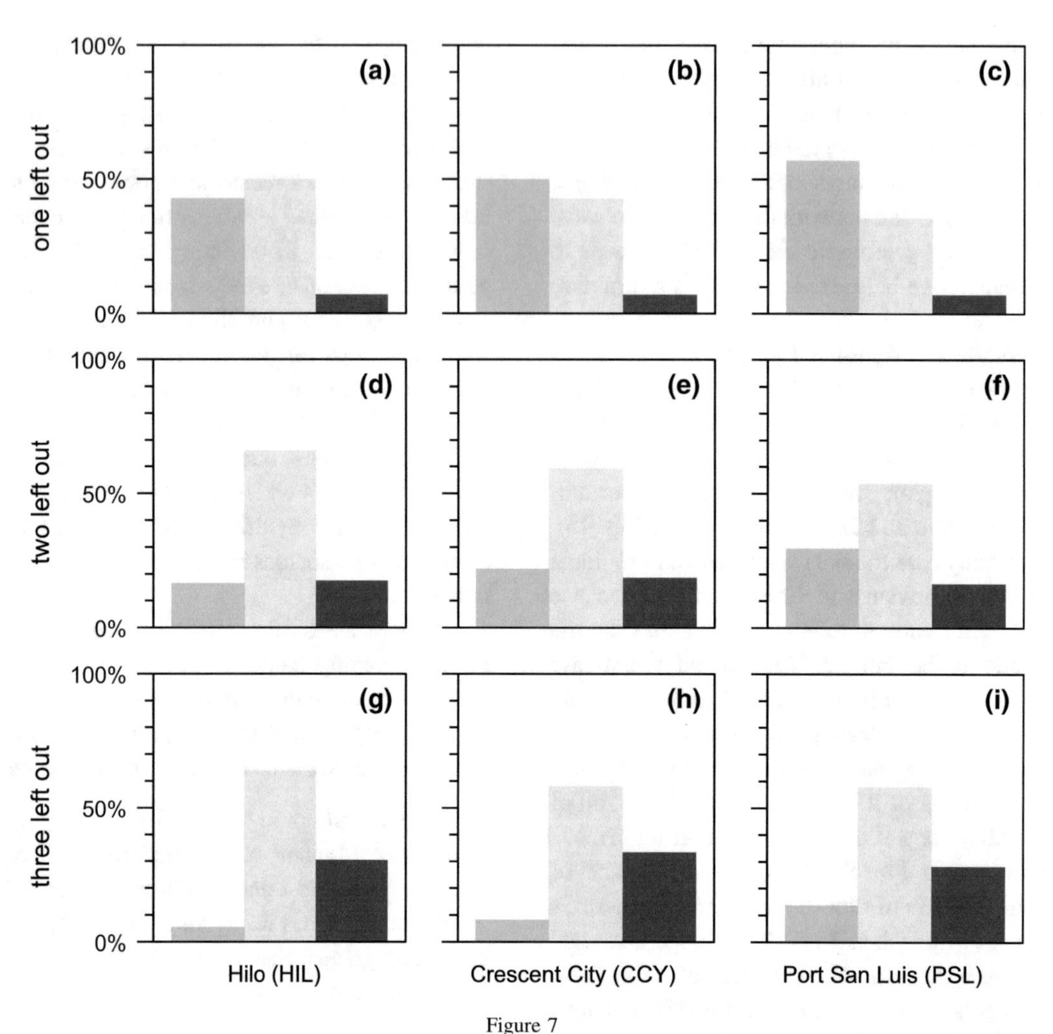

Figure 7
Percentages of cases that would result in a green, yellow or red network status for Hilo, Crescent City and Port San Luis (left to right columns) due to 1, 2 or 3 of 14 DART® buoys becoming inoperative (top to bottom rows, respectively)

but there can still be a green rating at all three impact sites for events arising from Aleutian Islands (but presumably not elsewhere). On the other hand, dropout of the single buoy `7` can raise a red alarm for Port San Luis. Once a particular buoy has dropped out, we can assess the effect of loss of one of the other buoys, thus giving an indication of the urgency of returning the buoy to normal operation. This exercise would address the question of how close the array is to falling into a red status.

As a concluding example, the National Data Buoy Center (`ndbc.noaa.gov`) reported that six of the 14 buoys we have been considering were inoperative at the time of writing (early May 2017), namely, `U`, `V`, `W`, `X`, `Y` and `1`. Loss of `1` alone is enough to raise a yellow alarm for all three impact sites (see Fig. 5). This loss coupled with the loss of any one of the other five buoys does not increase the alarm level for the three sites (see Fig. 6). With `1` gone, the alarm status for Hilo would switch from yellow to red only if buoy `5` were to become inoperative; for Crescent City, if `Z` or `2` become inoperative; and for Port San Luis, if `Z` or `7` become inoperative (again, see Fig. 6).

4.2. *Second Case Study: South America*

In the preceding case study we concentrated on evaluating an existing network of DART® buoys.

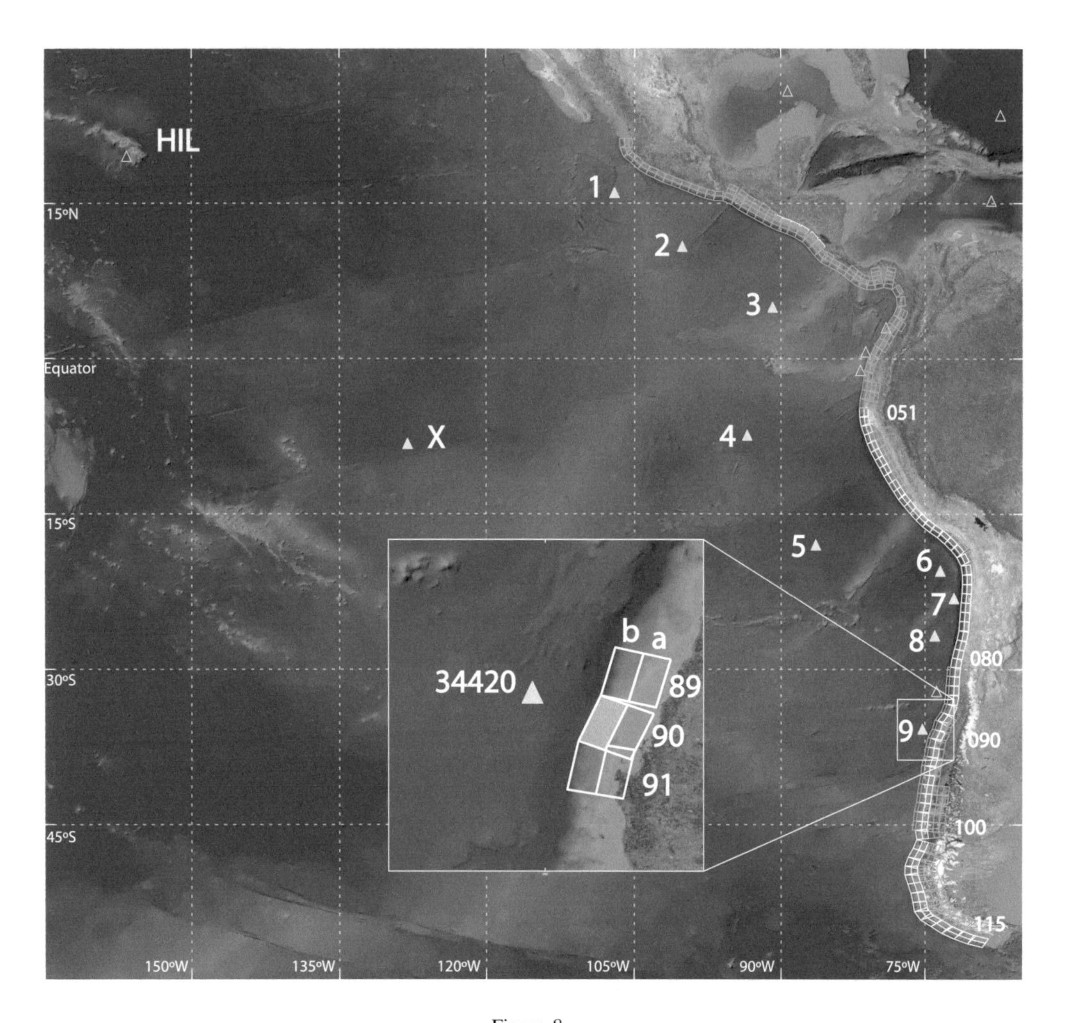

Figure 8
Locations of 130 unit sources along the West Coast of South America, DART® buoys (triangles, ten of which are labeled by `1–9` and `X`) and the Hilo impact site (HIL). The unit sources are arranged in two rows denoted by `a` (eastern) and `b` (western) and in 65 columns denoted by, from top to bottom, `051–115`. Going clockwise from the top, the World Meteorological Organization designations for the ten buoys are (`1`) 43412, (`2`) 43413, (`3`) 32411, (`4`) 32413, (`5`) 32412, (`6`) 32401, (`7`) 32403, (`8`) 32402, (`9`) 34420 and (`X`) 51406. The expanded view shows central unit source `cs090b` and the five unit sources surrounding it, along with the nearby buoy 34420

Here we illustrate another use for our proposed methodology, namely, to assess the impact of extending an existing network by adding a new buoy.

Figure 8 is a map of the West Coast of South America showing the locations of 130 predesignated unit sources, nine operational DART® buoys (labeled `1–9`) and one decommissioned buoy (`X`). These buoys can provide forecasts both for Southern Hemisphere impact sites and for Northern ones such as Hilo (indicated by HIL on the map; we do not report on results for Crescent City and Port San Luis because they are quite similar to those for Hilo). Merely to exercise our proposed methodology, we presume buoys `1–8` to be a preexisting network and consider the benefit to Hilo of adding buoy `9`. In addition we explore a hypothetical question: what would have been the effect on Hilo if, rather than augmenting the network by adding buoy `9` (a coastal buoy), buoy `X` (an open-ocean buoy) had been reactivated instead?

Figure 9 shows RMSEs of forecasted maximum wave amplitudes at Hilo for tsunami events arising from South America. The requirement that a warning to Hilo must be issued at least 3 h in advance of the

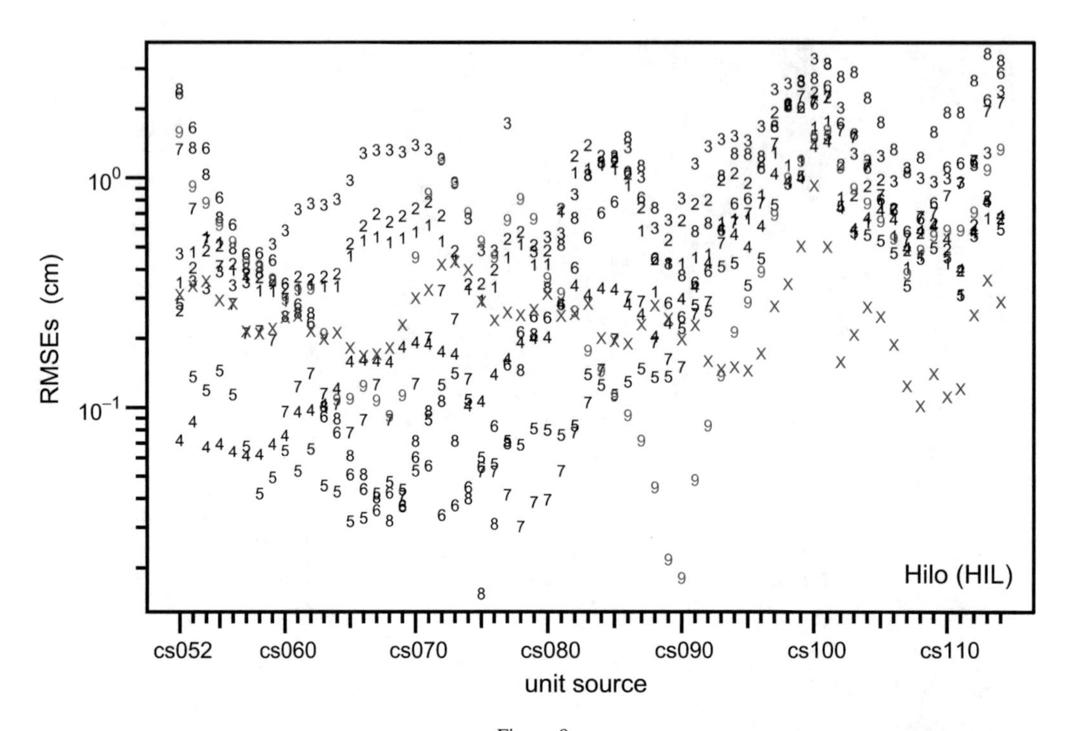

Figure 9
RMSEs for forecasted maximum wave amplitudes in the open ocean outside of Hilo based on ten buoys (labeled as indicated in Fig. 8) in combination with central unit sources `cs052b–cs114b` along the West Coast of South America

arrival of the tsunami is satisfied for all buoy/unit source combinations with two exceptions: buoy `1` with `cs109b`, and the same buoy with `cs114b`. The RMSEs for buoys `9` and `X` are colored, respectively, red and blue. For the network consisting of buoys `1–8` (black symbols in Fig. 9), there is an increase in RMSE of approximately an order of magnitude going from the northern-most central unit sources (`cs052b–cs080b`) to the southern-most (`cs100b–cs114b`). This pattern is in contrast to Fig. 3 for the Aleutian–Alaskan subduction zone, where the RMSEs are not strikingly different across all unit sources. Augmenting the network of eight buoys with coastal buoy `9` leads to a decrease in best RMSE by a factor to two or more for unit sources `cs087b–cs094b` (the unit source closest to `9` is `cs090b`—see Fig. 8); however, although there is some improvement in best RMSE from unit source `cs085b` up to `cs097b`, there is none elsewhere. The augmented network thus does not improve forecasts of wave amplitudes at Hilo for events arising from the southern-most unit sources.

Suppose now that we augment the `1–8` network by adding the open-ocean buoy `X` rather than buoy `9`. We would now have a factor of two or more decrease in best RMSE for unit sources `cs093b–cs114b` (`cs100b` is an exception—there is still a noticeable decrease, but less than a factor of two; in addition, there are small improvements in best RMSEs for unit sources `cs091b` and `cs092b`). Thus, in contrast to augmenting the network with buoy `9`, augmenting with `X` improves the response of the network in handling events events from the southern-most unit sources. The `X` augmentation improves the best RMSE at 24 unit sources, while the `9` augmentation, at 12 unit sources; however, the drastic improvement offered by the `9` augmentation at unit sources close to `cs090` is not matched anywhere by the `X` augmentation.

As is evident from Fig. 8, a quicker warning to Hilo from an event originating from a unit source near the tip of South America is possible using data from buoy `9` rather than `X`. Consider unit source `cs109b` as an example. As indicated by Fig. 9, the three smallest RMSEs for `cs109b` are associated with buoys X, 5 and 9 and have values of, respectively, 0.14, 0.49 and 0.56 cm. The corresponding travel times from cs109b to these buoys are 9.8, 5.9 and 3.0 h, whereas the travel time from cs109b to Hilo is 15.4 h. Use of data from any of the three buoys should result in a warning issued well within the 3 h called for by standard operating procedures. Buoys 5 and 9 yield forecasts of approximately the same quality, so adding 9 to the 1 to 8 network offers no improvement in RMSE. The cleaner signal appearing at X offers a decrease in RMSE by a factor of 3.5 over the one at 5. If we assume a wait of 1 hr after the arrival of the signal to capture data 21 min past the first peak, warnings based on 5 and 9 could be issued 8.5 and 11.4 h in advance, whereas the corresponding time for buoy X is 4.6 h. Adding X to the existing network improves the RMSE (and hence the quality of the warning) at the expense of decreasing the warning time by 3.9 h; on the other hand, adding 9 offers no improvement in RMSE but increases the warning time by 2.9 h. If we think of 5 as providing an initial assessment of the event sufficiently early, adding X opens up the possibility of a later—but still timely—higher-quality assessment whereas adding 9 does not.

In terms of improving forecasts for Hilo, augmenting the 1 to 8 network by reactivating the open-ocean buoy X is arguably preferable to going with coastal buoy 9; however, there are many other impact sites of importance, each of which would need to be evaluated in a similar manner, and the individual evaluations would need to be combined to come up with an objective evaluation of which buoy is the preferable augmentation (how best to combine the evaluations is a subject for future research). Our intent is not to take issue with the current placement of South American buoys, but rather to demonstrate how our methodology can be used to evaluate variations of an existing network.

5. *Discussion*

Four aspects of our method for evaluating buoy networks merit additional discussion. First, as described in Sect. 3 and discussed in more detail in the Appendix (Sect. 7), our approach is based on certain simplifications, including (but not limited to)

1. tsunami-generating events that come from within a single unit source;
2. mismatches between tsunami events and their associated models solely due to uncertainty in source location; and
3. use of data from a single DART® buoy to forecast wave amplitudes at an impact site even though tsunami events are routinely observed at multiple buoys (i.e., we do not combine together data from several buoys to generate forecasts as is done during actual tsunami events).

With regard to 1, we are in effect assuming that highly localized tsunami events are useful for assessing the effectiveness of a network of buoys even though events originating from regions spanning more than a single unit source occur more typically. A valid concern with this simplification is that use of tsunami events that are large, complex and/or spread out over an area larger than that trapped by a single unit source might lead to a substantially different network evaluation. While we hypothesize that this is not the case, it is a hypothesis that should be addressed in future research. With regard to 2, while there are other potentially important causes for mismatches besides the one we have chosen to focus on, we are assuming that, while including these additional causes might lead to increased RMSEs, the pattern of RMSEs across unit sources would not change to such a degree that network assessment would be substantially different. With regard to 3, we are assuming that forecasts based on each buoy separately can assess the effectiveness of the overall network and that using two or more buoys simultaneously won't lead to a significantly different evaluation even though use of data from multiple buoys for generating forecasts is more typical. Going beyond these three simplifying assumptions introduces levels of complexity that would complicate the simulation procedures needed for the network evaluation. In addition, assumption 3 provides a way of evaluating the contribution to the network of each buoy individually in a simple manner.

Second, our performance measure is predicated on tsunami events coming from a specific subduction zone (e.g., Aleutian–Alaskan) and heading toward a specific impact site (e.g., Hilo). We have not addressed the question of how to take the measures for different subduction zones and different impact sites and combine them together. For example, combining measures for multiple subduction zones and Hilo would lead to an evaluation of how well Hilo is serviced by the global buoy network (and would likely reveal that, while buoys U, V and W in Fig. 1 proved to be of limited use to Hilo for events arising from the Aleutian–Alaskan subduction zone, they are vital for events arising in the Western Pacific such as the disastrous March 2011 Japan event). On the other hand, combining measures for a particular subduction zone and multiple impact sites would tell us how well the buoy network offers protection globally for events arising from that particular zone. Combined measures such as these are certainly of interest, but how best to do so is outside the scope of this article.

Third, we have concentrated on impact sites that are distant from the unit sources generating the tsunami events. For these sites insisting on a 3-h warning time is appropriate. For sites that are close to a particular unit source (of which there are many along the West Coast of South America), a more stringent warning time (e.g., 30 min) must unfortunately be entertained. A mixture of impact sites—some distant and some close to a given unit source—adds a degree of complexity (beyond the scope of this article) for assessing how well a tsunami event arising from such a source is handled by a buoy network.

Fourth, our work on the Aleutian–Alaska subduction zone points to an interesting potential practical method for quantifying uncertainties in forecasts. Figure 10 shows RMSEs at the three impact sites versus maximum wave amplitude at the buoys. For a given buoy, Fig. 10 shows that the lowest RMSEs are associated with the highest maximum wave amplitudes. That is, buoys that are positioned well to sample the strongest portion of the main tsunami beam yield the best forecasts in terms of RMSEs. The relationship between the maximum wave amplitudes at the buoys and the RMSEs is roughly linear on a log/log scale, with a fairly strong sample correlation of -0.88. This suggests that the RMSE in forecasts at impact sites can be usefully assessed from the maximum wave amplitude at the buoys. This opens up the possibility of assigning a quality measure to forecasts at impact sites based on a simple summary of the data collected at the buoy.

6. *Summary and Conclusions*

We have developed a performance measure that quantifies the impact on forecasts of open-ocean wave amplitudes near an impact site due to loss of one or more buoys in a network of DART® buoys. The measure is based on a simulation procedure that takes into account key factors contributing to uncertainty in the forecasts. One factor is the model for the tsunami signal, which is never known perfectly in practice, in part due to uncertainty in the location of the tsunami-generating earthquake. We mimic this uncertainty by taking a predetermined central unit source and relocating it through random selection of a new location for its center. This relocation leads to perturbations both in the tsunami signal as it appears at DART® buoys and in the open-ocean wave amplitudes. The perturbations are codified by source coefficients, which serve as weights for forming the perturbed signal and wave amplitudes through linear combinations of signals and wave amplitudes associated with the central unit source and ones abutting it.

Tidal fluctuations and background noise recorded at the buoys are two additional factors contributing to forecast uncertainty. We mimic these through the use of archived DART® data recorded under ambient conditions either by the buoy of interest or by a surrogate operating under similar oceanographic conditions. The simulated tsunami signal and randomly selected archived data are added together to form simulated buoy data. These data are the response in a regression model for estimating source coefficients, which in turn lead to estimates of wave amplitudes. Discrepancies between the estimated and presumed coefficients translate into errors in forecasted wave amplitudes. These estimation errors are due in part to tidal fluctuations and background noise in the simulated buoy data, but are also influenced by the amount of available data and by a mismatch between the regression model and the presumed model for the signal, both of which we mimic in our simulations.

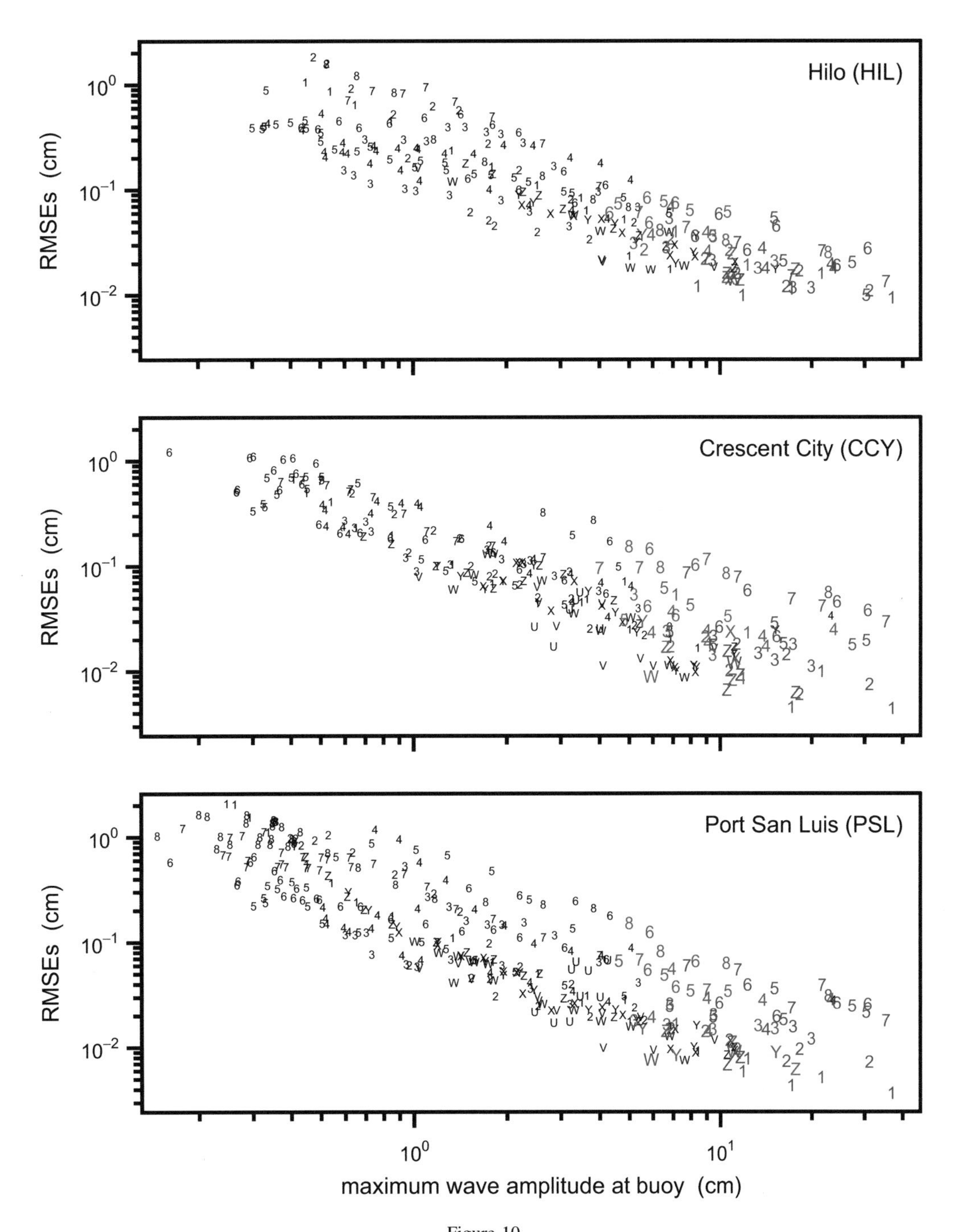

Figure 10
As in the right-hand column of Fig. 4, but now with the RMSEs of the maximum wave amplitudes forecasted at the open ocean outside of the impact sites plotted versus maximum wave amplitudes at the DART® buoys rather than plotted versus maximum wave amplitudes in the open ocean outside of the impact sites. While only the best RMSEs (colored by blue symbols) and second best (red) are shown in Fig. 4, both these and all remaining RMSEs (black) are shown here

A thousand replications of our simulation procedure for a given unit source/buoy combination gives a thousand forecasts of open-ocean wave amplitudes at as many impact sites as desired. To summarize the quality of the forecasts, we compare the thousand maximum values of the forecasted amplitude to the corresponding presumed maximum values by forming a root-mean-square error (RMSE). The RMSEs for a network of buoys in conjunction with a set of unit sources relevant for a particular impact site form the basis for network evaluation. The evaluation consists of dropping one or more buoys from the network and determining what change there is in the minimum RMSE for each unit source. We propose to use the maximum change in minimum RMSE as a metric for measuring the health of the buoy network as one or more buoys become inoperative. We also propose a simple green/yellow/red indicator of network health as a management tool.

Currently we have prototyped our procedure for evaluating the effectiveness of DART® buoy networks in the `R` language (Ihaka and Gentleman 1996; R Development Core Team 2010). Using the `R` code as guidance, current plans are to encapsulate the methodology we have described into a tool that would serve multiple purposes. We propose to use the tool to go beyond our Aleutian Island and South American case studies to do a comprehensive study of the existing network of 39 buoys. This study might identify unit sources for which the current network configuration is particularly sensitive at certain impact sites to the dropout of buoys. We could also use the tool to assess the impact of adjusting the network by relocating, adding or removing buoys. Subject to additional research, it should be possible to augment the tool to investigate the use of tide gauges to help mitigate the loss of buoys. Despite the simplifications we have made, we contend that our proposed methodology is an effective and realistic way to assess network effectiveness and that a tool built around this methodology would be valuable for managing networks of DART® and other types of tsunami buoys.

Acknowledgements

This work was funded by the Joint Institute for the Study of the Atmosphere and Ocean (JISAO) under NOAA Cooperative Agreement No. NA15OAR4320063 and is JISAO Contribution No. 2714. This work is also Contribution No. 4507 from NOAA/Pacific Marine Environmental Laboratory. The authors thank Peter Dahl for discussion on a running example.

7. Appendix: Details on Simulations and Evaluation of Forecasts

Here we give details about how we have chosen to simulate and evaluate open-ocean wave amplitude forecasts. This material is summarized in Sect. 3, where we noted four tasks to be carried out, which are described in Sects. 7.1, 7.2, 7.3, 7.4. We illustrate the four tasks by focusing on a representative triad consisting of one predesignated central unit source (`ac026b`), one DART® buoy (46403) and one impact site (Hilo HI).

7.1. Simulation of Tsunami Signal and Open-Ocean Wave Amplitudes

Our first task is to simulate a tsunami event, which manifests itself as a tsunami signal that passes by a particular DART® buoy, say, 46403 (Fig. 1 shows the location of this buoy). We denote this signal by $\boldsymbol{t}$. A simple way to specify $\boldsymbol{t}$ would be to use one of the precomputed models predicting what 46403 should see if the tsunami event were to originate from within a particular predesignated unit source, say, `ac026b` (shown in Fig. 1). We would thus just set $\boldsymbol{t}$ equal to $\boldsymbol{g}_5$ depicted in Fig. 2; however, we have chosen *not* to adopt this simple approach because an important factor impacting open-ocean wave amplitude forecasts is modeling of the tsunami signal $\boldsymbol{t}$ (this is the subject of Sect. 7.3). Because we cannot model the signal perfectly in an actual tsunami event, there is a danger of generating unrealistic simulated forecasts if we were to assume the signal to be identical to its model. One of many sources of mismatch is uncertainty in the location of the source within the subduction zone. Figure 11 illustrates the effect of relocating unit source `ac026b`, producing a relocated 100 km by 50 km source that overlaps (in this example) with three additional predesignated sources. To create the relocated source, we picked a random point (the red diamond) within `ac026b` and used this point as the center for the relocated unit source (blue rectangle). We can now compute a model predicting

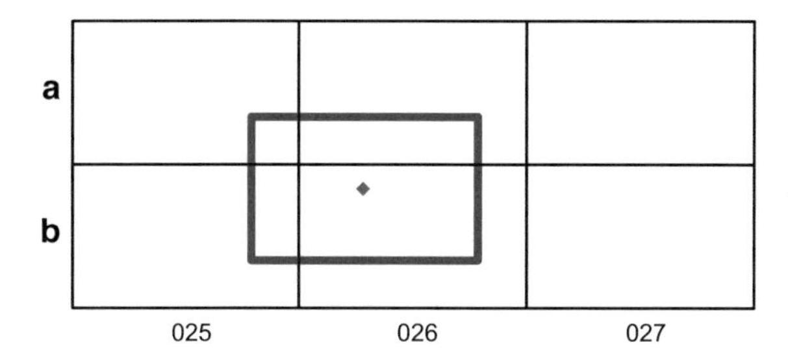

Figure 11
Six rectangles (black lines) representing 100 km by 50 km sections of ocean associated with predesignated unit sources ac025a, ac026a and ac027a (top row) and ac025b, ac026b and ac027b (bottom) in the Aleutian–Alaskan subduction zone (these are highlighted in Fig. 1). The red diamond within the rectangle for the central unit source ac026b defines the center of a rectangle for a relocated unit source (blue lines). The relocated rectangle intersects the rectangles associated with ac025a, ac026a, ac025b and ac026b (these are associated with models $\boldsymbol{g}_1$, $\boldsymbol{g}_2$, $\boldsymbol{g}_4$, $\boldsymbol{g}_5$ shown in Fig. 2)

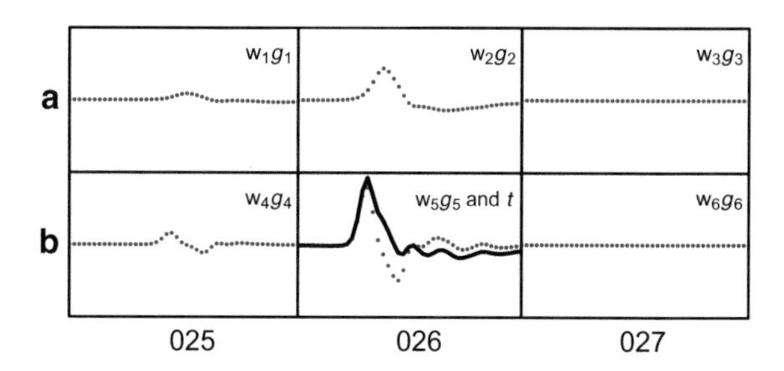

Figure 12
Tsunami signal $\boldsymbol{t}$ (black curve, middle bottom plot) formed from linear combination of models $\boldsymbol{g}_1, \ldots, \boldsymbol{g}_6$ as per Eq. 1 with $w_1 \doteq 0.28$, $w_2 \doteq 1.04$, $w_3 = 0$, $w_4 \doteq 0.57$, $w_5 \doteq 2.11$ and $w_6 = 0$ (the $\boldsymbol{g}_k$ models are shown in Fig. 2). The blue dotted curves show $w_k\boldsymbol{g}_k$, and the sum of all six of these is equal to $\boldsymbol{t}$ (because $w_3 = w_6 = 0$, models $\boldsymbol{g}_3$ and $\boldsymbol{g}_6$ do not contribute to $\boldsymbol{t}$). The time spans are the same as in Fig. 2, but now the vertical axes range from −6.6 cm to 6.6 cm

what would be observed at buoy 46403 if an earthquake were to originate from the randomly relocated unit source. This new model will differ from the precomputed models shown in Fig. 2. If we take this new model to be our tsunami signal $\boldsymbol{t}$ and if we were then to entertain modeling $\boldsymbol{t}$ using a subset of the models shown in Fig. 2, then the constructed signal need not be exactly equal to its model. Randomly choosing a set of locations within a central unit source (ac026b in this example) then leads to an ensemble of relocated sources that reflects the uncertainty in source location.

Because it is time consuming to generate geophysical models, we make an assumption of linearity and construct the tsunami signal $\boldsymbol{t}$ using a linear combination of precomputed models. For the example shown in Fig. 11, the signal would be a linear combination of the models for unit sources ac025a, ac026a, ac025b and ac026b (the central unit source). These models are labeled in Fig. 2 as $\boldsymbol{g}_1$, $\boldsymbol{g}_2$, $\boldsymbol{g}_4$ and $\boldsymbol{g}_5$. The weights w_k in the linear combination are based on geometry and are dictated by the degree of intersection of the relocated source with the predesignated sources—in this example, $\boldsymbol{g}_5$ (corresponding to ac026b) gets the most weight because it intersects the most, while $\boldsymbol{g}_1$ (corresponding to ac025a) gets the least nonzero weight. Figure 12 illustrates the construction of the N-dimensional column vector $\boldsymbol{t}$ in terms of the six vectors $\boldsymbol{g}_k$ shown in Fig. 2:

$$\boldsymbol{t} = \sum_{k=1}^{6} w_k \boldsymbol{g}_k \tag{1}$$

(black curve, lower middle plot). Note that, since neither $\boldsymbol{g}_3$ nor $\boldsymbol{g}_6$ are involved in constructing $\boldsymbol{t}$, the weights w_3 and w_6 are zero. Past experience with actual tsunami events suggests that the weights should not be arbitrarily set if we want to regard the simulated event as originating from a 100 km by 50 km region ((Hanks and Kanamori, 1979; Papazachos et al., 2004)). We use the normalization $\sum_k w_k = 4$ in all of our simulations because a larger setting than this would result in a tsunami signal whose magnitude would be more realistically associated with multiple unit sources. For the example shown in Fig. 11, we have $w_1 \doteq 0.28$, $w_2 \doteq 1.04$, $w_4 \doteq 0.57$ and $w_5 \doteq 2.11$.

We can construct additional tsunami signals by selecting other points at random within the rectangle for the central unit source ac026b. Figure 13 breaks the rectangle for ac026b up into four quadrants of equal size (labeled as I, II, III and IV). The specific models $\boldsymbol{g}_k$ that are used to construct the tsunami signal depend upon which quadrant the randomly selected point falls in—these are listed in Fig. 13 for the four quadrants. The random pick in Fig. 11 falls in quadrant II, and hence, as previously noted, the signal is constructed using $\boldsymbol{g}_1$, $\boldsymbol{g}_2$, $\boldsymbol{g}_4$ and $\boldsymbol{g}_5$. No matter into which quadrant the random pick falls, the model $\boldsymbol{g}_5$ is always included with a weight w_5 satisfying $1 \leq w_5 \leq 4$, and this weight is close to 4

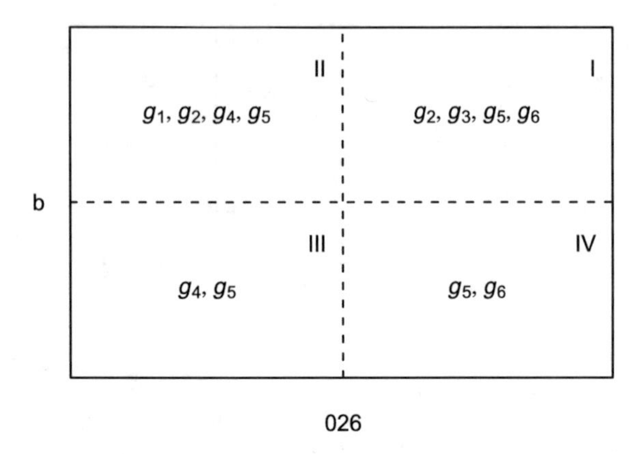

Figure 13
Division of rectangle for central unit source ac026b into four quadrants (I, II, III and IV). A point chosen at random within this rectangle is equally likely to fall in each of the four quadrants. The models $\boldsymbol{g}_k$ used to construct the tsunami signal $\boldsymbol{t}$ depend upon which quadrant the random pick falls in—these models are listed in the middles of the quadrants. The model $\boldsymbol{g}_5$ for ac026b is always part of the constructed tsunami signal. The weight w_k assigned to each model depends upon the location of the random pick within the quadrant, with the weight for $\boldsymbol{g}_5$ increasing as the random pick gets closer to the center of the rectangle

when the random pick is close to the center of the rectangle. The upper two quadrants involve three models in addition to $\boldsymbol{g}_5$, whereas the lower quadrants involve just one (either $\boldsymbol{g}_4$ or $\boldsymbol{g}_6$). The characteristics of the constructed tsunami signals will thus depend upon the randomly selected quadrant and where the randomly selected point occurs within the quadrant. No matter where the random pick ends up, we can write the resulting constructed tsunami signal as per Eq. (1), with the understanding that either two or four of the w_ks will be zero.

We can also construct tsunami signals by focusing on a unit source other than ac026b; however, for simplicity, we will use as a central unit source only those having a setup similar to ac026b, namely, that they are in row b and are surrounded by five other unit sources. This restriction for central unit sources simplifies computer code somewhat and arguably should not adversely impact the evaluation of the effectiveness of a buoy network (a hypothesis that is yet to be investigated).

Given a relocated unit source, there are two methods for generating the corresponding open-ocean wave amplitudes $\boldsymbol{h}$ outside of an impact site of interest. The more accurate method is via high-resolution model runs, which are time consuming; the less accurate makes use of lower-resolution runs already available in a precomputed propagation data base, which has the advantage of being easy to extract. We tested both methods and found that results based on the propagation data base differed little from those based on high-resolution model runs. We have thus elected to use the propagation data base approach, for which we take the presumed wave amplitudes to be

$$\boldsymbol{h} = \sum_{k=1}^{6} w_k \boldsymbol{h}_k, \tag{2}$$

where the weights w_k are identical to those in Eq. 1, while $\boldsymbol{h}_k$ is the forecast of the wave amplitudes in the open ocean that would occur from an event occurring in the same predesignated unit source associated with $\boldsymbol{g}_k$ (note, however, that, while $\boldsymbol{h}_k$ depends upon the location of the unit source and the location in the open ocean, it does not depend upon *any* of the buoy locations).

As an example, Fig. 14 shows wave amplitudes $\boldsymbol{h}_k$ at an open-ocean location outside of Hilo associated with the same six predesignated unit sources considered in Figs. 1 and 2. The left-hand plot of Fig. 15 shows $\boldsymbol{h}$ formed using the weights stated in the caption to Fig. 12 (the right-hand plot is explained in Sect. 7.3).

7.2. Simulation of 1-Min Stream

Our second task is to simulate data as it would be recorded at a DART® buoy. Due both to the complexity of recorded tsunami data and the way in which these data are used to create wave amplitude forecasts, simulations of observations and forecasts that are fully realistic are difficult and time consuming to create. The approach we take is to impose simplifications that nonetheless retain the salient factors limiting the ability of a buoy network to contribute to accurate forecasts. As a starting point, we assume the 1-min stream recorded at a particular buoy is given by

$$\bar{\boldsymbol{y}} = \boldsymbol{x} + \boldsymbol{t} + \boldsymbol{\epsilon}, \tag{3}$$

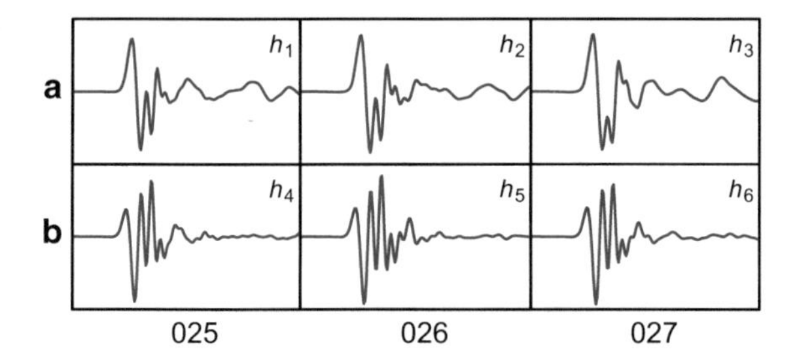

Figure 14
Models $\boldsymbol{h}_1,\ldots,\boldsymbol{h}_6$ for the wave amplitudes in the open ocean outside of Hilo from an earthquake originating from either the predesignated central unit source ac026b (corresponding to $\boldsymbol{h}_5$) or one of five adjacent unit sources ac025a (corresponding to $\boldsymbol{h}_1$), ac026a ($\boldsymbol{h}_2$), ac027a ($\boldsymbol{h}_3$), ac025b ($\boldsymbol{h}_4$) or ac027b ($\boldsymbol{h}_6$) located in the Aleutian–Alaskan subduction zone (Fig. 1 shows the locations of the 100 km by 50 km sections of the ocean associated with these unit sources). The time span for each displayed model is 3 h, and the span starts at 4 h from the beginning of the earthquake. The vertical axes all range from −1.5 cm to 1.5 cm

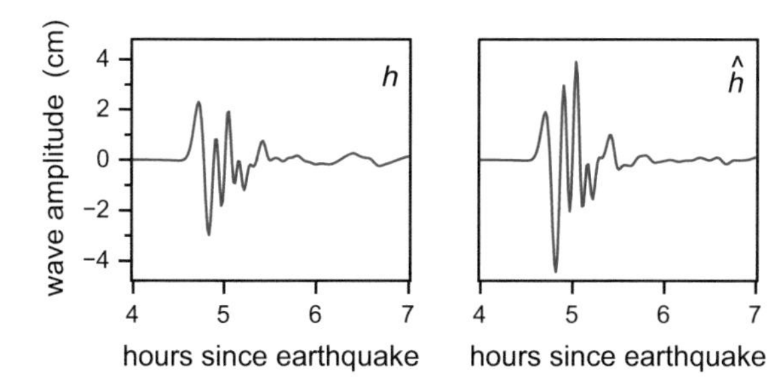

Figure 15
Open-ocean wave amplitudes $\boldsymbol{h}$ outside of Hilo as dictated by relocated unit source of Fig. 11 (left-hand plot) and associated forecasted wave amplitudes $\hat{\boldsymbol{h}}$ (right-hand). The amplitudes $\boldsymbol{h}$ are given as per Eq. 2 with the same setting for w_k as listed in the caption to Fig. 12. The amplitudes $\hat{\boldsymbol{h}}$ are given as per Eq. 5 with $\mathcal{K} = \{4, 5\}$, $\hat{\alpha}_4 \doteq 1.33$ and $\hat{\alpha}_5 \doteq 2.10$

where $\bar{\boldsymbol{y}}$ is a vector containing N consecutive values from the 1-min stream; $\boldsymbol{x}$ represents tidal fluctuations; $\boldsymbol{t}$ is the tsunami signal constructed as per Eq. (1); and $\boldsymbol{\epsilon}$ is background noise. Tidal fluctuations and background noise in the DART® data are two factors that can adversely impact inundation forecasts, so it is important to handle these realistically. Rather than simulating these factors, we can use archived DART® data that were recorded under ambient conditions and retrieved during routine servicing of the buoy. The tsunami signal in Eq. (3) is missing during ambient conditions, so random samples from historical data can serve to generate $\boldsymbol{x} + \boldsymbol{\epsilon}$ (for details about the sampling procedure, see Sect. 3, (Percival et al., 2015)). A complication is that not all currently deployed buoys have associated archived data (and proposed buoys certainly don't). To handle such buoys, we use data from a surrogate buoy with good matching oceanographic conditions.

Figure 16 shows a simulated 1-min stream $\bar{\boldsymbol{y}}$ constructed as per Eqs. (1) and (3). The top plot shows the tsunami signal $\boldsymbol{t}$ (this is the same as the black curve in middle bottom plot of Fig. 12). The middle plot shows the sum of tidal fluctuations $\boldsymbol{x}$ and background noise $\boldsymbol{\epsilon}$, which were obtained from archived data for DART® buoy 46403. The bottom plot shows $\bar{\boldsymbol{y}}$, which is the sum of the time series in the top and middle plots.

7.3. *Estimation of Source Coefficients and Forecasting of Open-Ocean Wave Amplitudes*

With simulated 1-min stream $\bar{\boldsymbol{y}}$ in hand, the third task is to estimate the source coefficients α_k. In practice factors limiting the accurate estimation of α_k include tidal fluctuations, background noise, an insufficient amount of data and imperfect knowledge about the underlying tsunami signal. We assume a linear regression model of the form

$$\bar{\boldsymbol{y}} = \mu \boldsymbol{1} + \beta_1 \boldsymbol{c} + \beta_2 \boldsymbol{s} + \sum_{k \in \mathcal{K}} \alpha_k \boldsymbol{g}_k + \boldsymbol{e}, \tag{4}$$

where $\boldsymbol{1}$ is an N-dimensional vector of ones associated with the regression coefficient μ; $\boldsymbol{c}$ is a vector with elements $\cos(\omega n \Delta)$, $n = 0, 1, \ldots, N-1$, and is associated with β_1—here ω is the tidal frequency M2 and $\Delta = 1$ min; $\boldsymbol{s}$ is associated with β_2 and is similar to $\boldsymbol{c}$ but its elements are given by $\sin(\omega n \Delta)$; $\mathcal{K}$ is a subset of $\{1, 2, \ldots, 6\}$ that specifies the $\boldsymbol{g}_k$ to be used to model the tsunami signal; and $\boldsymbol{e}$ is a vector of stochastic errors assumed to have zero mean (note that this vector is *not* taken to be the same as $\boldsymbol{\epsilon}$ in Eq. (3)). As discussed in Percival et al. (2015), the coefficients μ, β_1 and β_2 and their associated vectors serve to model the tidal fluctuations in Eq. (3) in a simple—but statistically efficient—manner that is superior to other methods for handling the tides including filtering. Modeling of the signal $\boldsymbol{t}$ is handled through specification of $\mathcal{K}$. We consider four protocols for setting $\mathcal{K}$, which are of interest because they

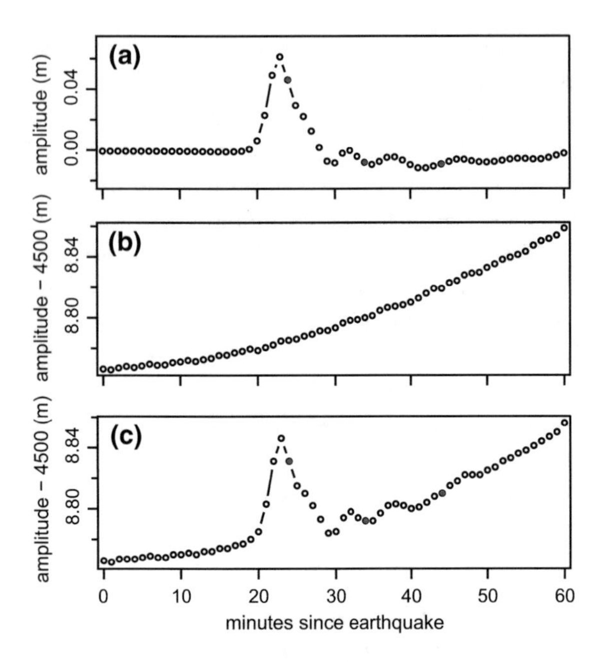

Figure 16
Construction of a simulated 1-min stream $\bar{\boldsymbol{y}}$. Plot **(a)** is the simulated tsunami signal—this is the same as black curve in middle bottom plot of Fig. 12. The three red circles indicate data occurring 1, 11 and 21 min after the signal's peak value. Plot **(b)** shows tidal fluctuations and background noise, i.e., $\boldsymbol{x} + \boldsymbol{\epsilon}$, recorded by DART® buoy 46403 under ambient conditions. Plot **(c)** is $\bar{\boldsymbol{y}}$, which is the sum of the time series in plots (**b**, **c**)

make different assumptions about what is known about the random pick and because they lead to different degrees of mismatch between the constructed signal and its model.

(a) The *single unit source protocol* ('1 protocol' for brevity) sets $\mathcal{K}$ to be $\{5\}$; i.e., the model for the signal is $\alpha_5 \boldsymbol{g}_5$. With probability one, this protocol yields an incorrect signal model in the sense that the constructed signal consists of $\boldsymbol{g}_5$ combined with either one or three additional $\boldsymbol{g}_k$s, whereas our model for it involves just $\boldsymbol{g}_5$. This protocol in effect assumes we only know within which predesignated unit source the earthquake occurred (i.e., we know nothing about the random pick).
(b) The *two unit sources protocol* ('2 protocol') assumes we have limited information about the random pick, namely, whether the earthquake originates from the left-hand or right-hand side of the rectangle (the random pick in Fig. 11 is left-handed). For a left-hand pick, we set $\mathcal{K}$ to $\{4, 5\}$, and we set it to $\{5, 6\}$ for a right-hand pick. For both settings, there are two $\boldsymbol{g}_k$s in the model. With this protocol, there is a 50% chance of having a mismatch between the constructed signal and its model, and the nature of the mismatch is that the model has two too few unit sources.
(c) The *four unit sources protocol* ('4 protocol') also assumes we know the left- or right-hand location of the earthquake. Now we set $\mathcal{K}$ to $\{1, 2, 4, 5\}$ for a left-hand pick and to $\{2, 3, 5, 6\}$ otherwise. There is again a 50% chance of having a mismatch, but now the nature of the mismatch is that there are two too many unit sources.
(d) The *matched protocol* sets $\mathcal{K}$ so that it contains *exactly* the indices used in forming the constructed signal; i.e., we use the same set of $\boldsymbol{g}_k$s both to form the signal and to model it so there is no mismatch. This protocol essentially presumes knowledge of the quadrant in which the random pick falls.

After selection of the protocol $\mathcal{K}$, the regression model of Eq. (3) is now fully specified. The model involves three coefficients for the tidal model (μ, β_1 and β_2) and either one, two or four source coefficients α_k for the signal model. We estimate the regression coefficients using constrained least squares; i.e., the estimated coefficients $\hat{\mu}$, $\hat{\beta}_1$, $\hat{\beta}_2$ and $\hat{\alpha}_k$ are those minimizing

$$\left\| \bar{\boldsymbol{y}} - \mu \boldsymbol{1} - \beta_1 \boldsymbol{c} - \beta_2 \boldsymbol{s} - \sum_{k \in \mathcal{K}} \alpha_k \boldsymbol{g}_k \right\|_2^2 \text{ subject to } \alpha_k \geq 0 \text{ for all } k,$$

where $\|\boldsymbol{x}\|_2^2 = \sum_n x_n^2$ is the squared Euclidean norm of a vector $\boldsymbol{x}$ with elements x_n. The nonnegativity constraint on each source coefficient is critical because, without it, there is nothing to prevent the least square estimate of α_k from being negative, which would render it inconsistent with the type of generating event we are assuming (a reverse fault earthquake).

Another important factor that impacts the quality of the inundation forecasts is the amount of data N available for estimating the source coefficients. In general, more data in $\bar{\boldsymbol{y}}$ improves the quality of the estimated coefficients, which, in turn, improves inundation forecasts. To study the effect of N, we consider three settings intended to loosely mimic what would be available during different stages of an ongoing tsunami event. These settings are based on the location of the first peak of the constructed signal $\boldsymbol{t}$ (alternatively we could use the first peak of the noisy data $\bar{\boldsymbol{y}}$, but, to avoid issues that arise in automatic detection of a peak possibly distorted by noise, we have chosen to use the constructed signal instead). The first setting is 1 min past the first peak in the signal, and the second and third settings are 11 and 21 min past the peak. The three settings are illustrated in Fig. 16a, c using red circles. For this particular example, the settings correspond to 24, 34 and 44 min after the start of the earthquake, and we would estimate the coefficients based on placing the corresponding first $N = 25$, 35 or 45 simulated buoy measurements in (c) into $\bar{\boldsymbol{y}}$. For tsunami signals other than this example, we identify the first peak and the same related three points, but only use a maximum of 60 min worth of data prior to the first peak in cases where this peak occurs more than 1 hr after the start of the earthquake (this mimics the amount of the 1-min stream available during an actual event).

As an example of how the sample size influences source coefficient estimation, let us focus on the simulated DART® buoy data $\bar{\boldsymbol{y}}$ shown in Fig. 16c and set $\mathcal{K}$ as per the 2 protocol. Because the random pick is left-handed, the model for the tsunami signal becomes $\alpha_4 \boldsymbol{g}_4 + \alpha_5 \boldsymbol{g}_5$. There is thus a mismatch between the model and signal since the latter makes use of $\boldsymbol{g}_1$ and $\boldsymbol{g}_2$ in addition to $\boldsymbol{g}_4$ and $\boldsymbol{g}_4$. Table 1 shows the estimated source coefficients for the three sample sizes along with the presumed weights w_4 and w_5. The estimates improve with increasing N in the sense that $\hat{\alpha}_k$ gets closer to w_k.

The forecasted wave amplitudes are

$$\hat{\boldsymbol{h}} = \sum_{k \in \mathcal{K}} \hat{\alpha}_k \boldsymbol{h}_k. \tag{5}$$

For the 2 protocol with a left-handed pick, these amplitudes are given by

$$\hat{\boldsymbol{h}} = \hat{\alpha}_4 \boldsymbol{h}_4 + \hat{\alpha}_5 \boldsymbol{h}_5,$$

where $\boldsymbol{h}_4$ and $\boldsymbol{h}_5$ are depicted in Fig. 14. Use of the estimates for α_k corresponding to $N = 45$ in Table 1

Table 1

Estimated source coefficient $\hat{\alpha}_4$ and $\hat{\alpha}_5$ for three sample sizes N along with presumed weights w_4 and w_5

k	w_k	$\hat{\alpha}_k$		
		$N = 25$	$N = 35$	$N = 45$
4	0.57	88.92	1.63	1.33
5	2.11	2.51	2.34	2.10

leads to the forecasted wave amplitudes shown in the right-hand plot of Fig. 15.

To summarize, estimates of the source coefficients α_k are imperfect in practice due to major factors including, inter alia,

1. an imperfect tidal model;
2. background noise;
3. a limited amount of data;
4. a mismatch between the assumed model and the actual tsunami signal; and
5. seismic noise, which is potentially important, but, because of timing, often does not come into play.

Our simulation study takes into account 1–4, but ignores seismic noise (which the next generation of DART® buoys is designed to suppress entirely).

7.4. Evaluation of Forecasted Open-Ocean Wave Amplitudes

Our final task is to quantify how well the forecasted wave amplitudes $\hat{\boldsymbol{h}}$ match the presumed amplitudes $\boldsymbol{h}$. We considered three metrics. The first is the maximum cross-correlation between $\hat{\boldsymbol{h}}$ and either $\boldsymbol{h}$ or a lagged versions thereof; the second is the squared difference between the maximum amplitudes in $\hat{\boldsymbol{h}}$ and $\boldsymbol{h}$; and the third is the squared difference between the maximum 4-h 'energies' in $\hat{\boldsymbol{h}}$ and $\boldsymbol{h}$, where the energies in question are the sum of squares of data over all possible 4-h stretches contained within $\hat{\boldsymbol{h}}$ or $\boldsymbol{h}$. The latter two metrics are of more operational interest than maximum cross-correlation, which is also the most sensitive of the three to innocuous misalignments in time between $\hat{\boldsymbol{h}}$ and $\boldsymbol{h}$. In tests to date, we have found that use of either maximum wave amplitudes or maximum energies leads to evaluations of network effectiveness that are qualitatively similar. Since maximum wave amplitudes are less time consuming to compute, we stick with these in all that follows. For the example shown in Fig. 15, we have $\max\{\boldsymbol{h}\} \doteq 2.3$ cm and $\max\{\hat{\boldsymbol{h}}\} \doteq 3.9$ cm, and the metric is $(\max\{\hat{\boldsymbol{h}}\} - \max\{\boldsymbol{h}\})^2 \doteq 2.6$ cm^2.

A more thorough assessment of how well we can forecast wave amplitudes outside of Hilo requires repeating what is set forth in the previous three subsections. We do so by generating 1000 different tsunami signals $\boldsymbol{t}$ as per Eq. 1. The signals are associated with 1000 independent random picks from within unit source ac026b. Each pick serves as a center for a relocated unit source whose location dictates the weights w_k used to form both $\boldsymbol{t}$ and the presumed open-ocean wave amplitudes $\boldsymbol{h}$ of Eq. 2. Each signal $\boldsymbol{t}$ is then added to an independently drawn random sample of data recorded under ambient conditions by an appropriate DART® buoy. This addition forms the simulated 1-min stream $\bar{\boldsymbol{y}}$ as per Eq. 3. For each realization, we then compute twelve sets of estimated source coefficients $\hat{\alpha}_k$ corresponding to twelve formulations of the linear model of Eq. 4. The formulations involve the four protocols for setting $\mathcal{K}$ in combination with the three ways of determining the amount of data N to be placed in $\bar{\boldsymbol{y}}$. For a given formulation, we use the source coefficient estimates $\hat{\alpha}_k$ to generate—as per Eq. 5—forecasted wave amplitudes $\hat{\boldsymbol{h}}$ outside of the chosen impact site (Hilo in our representative triad). We then compare the peak values in $\boldsymbol{h}$ and $\hat{\boldsymbol{h}}$.

Figure 17 shows forecasted maximum amplitudes versus corresponding presumed amplitudes for 1000 realizations of our representative triad (central unit source ac026b, DART® buoy 46403 and a location in the open ocean outside of the impact site Hilo). Each realization corresponds to a single point in each of the four plots. In addition to the 1000 points, each plot has a diagonal line indicating where a point would fall if the forecasted and presumed amplitudes were in perfect agreement. For all four plots, we set N such that 21 min worth of data past the peak value in the signal $\boldsymbol{t}$ is used to estimate the source coefficients and other parameters in the regression model (Eq. 4). Plots (a) to (d) correspond to, respectively, the 1, 2, 4 and matched protocols $\mathcal{K}$ for fully specifying the model.

The colors of the points in Fig. 17 indicate the quadrant in which the random pick for a particular realization fell. Of the 1000 picks, 245 were in quadrant I (colored red); 247, in II (green); 268, in III (blue); and 240, in IV (black). This distribution is not inconsistent with each pick being equally likely to fall in any of the four quadrants. As noted in Fig. 13, picks in quadrants I and II involve linear combinations of four $\boldsymbol{g}_k$s. For the 1 and 2 protocols, forecasted amplitudes are thus based on mismatched models. The mismatch is due to relevant $\boldsymbol{g}_k$s being left out of the regression model. As a result, the red and green points in plots (a) and (b) have prominent scatter off the diagonal line. Plots (c) and (d) indicate that the forecasts are markedly better when there is no model mismatch, as occurs in the 4 and matched protocols (note that the red and green points in (c) and (d) are necessarily identical). In the absence of model mismatch, the remaining inaccuracies in the forecasts are mainly due to background noise and the tidal component.

By contrast, picks in quadrants III and IV (blue and black points) correspond to linear combinations of two $\boldsymbol{g}_k$s. There is a model mismatch with the 1 and 4 protocols, but not for the 2 and matched protocols (note that the blue and black points in (b) and (d) are necessarily identical). Interestingly, when the mismatch is due to one less $\boldsymbol{g}_k$ in the 1 protocol, there are highly structured deviations from the diagonal line. On the other hand, when the mismatch is due to two extraneous $\boldsymbol{g}_k$s, as happens with the 4 protocol, the scatter about the diagonal line visually does not increase much over that of the 2 protocol (no mismatch).

We can summarize how well forecasted maximum wave amplitudes match up with the presumed amplitudes by computing a root-mean-square error (RMSE) for each scatterplot. Letting $\max\{\hat{\boldsymbol{h}}_l\}$ and $\max\{\boldsymbol{h}_l\}$ represent the forecasted and presumed amplitudes for the lth realization, this error is defined as

$$\mathrm{RMSE} = \left[\frac{1}{1000}\sum_{l=1}^{1000}\left(\max\{\hat{\boldsymbol{h}}_l\} - \max\{\boldsymbol{h}_l\}\right)^2\right]^{1/2}. \tag{6}$$

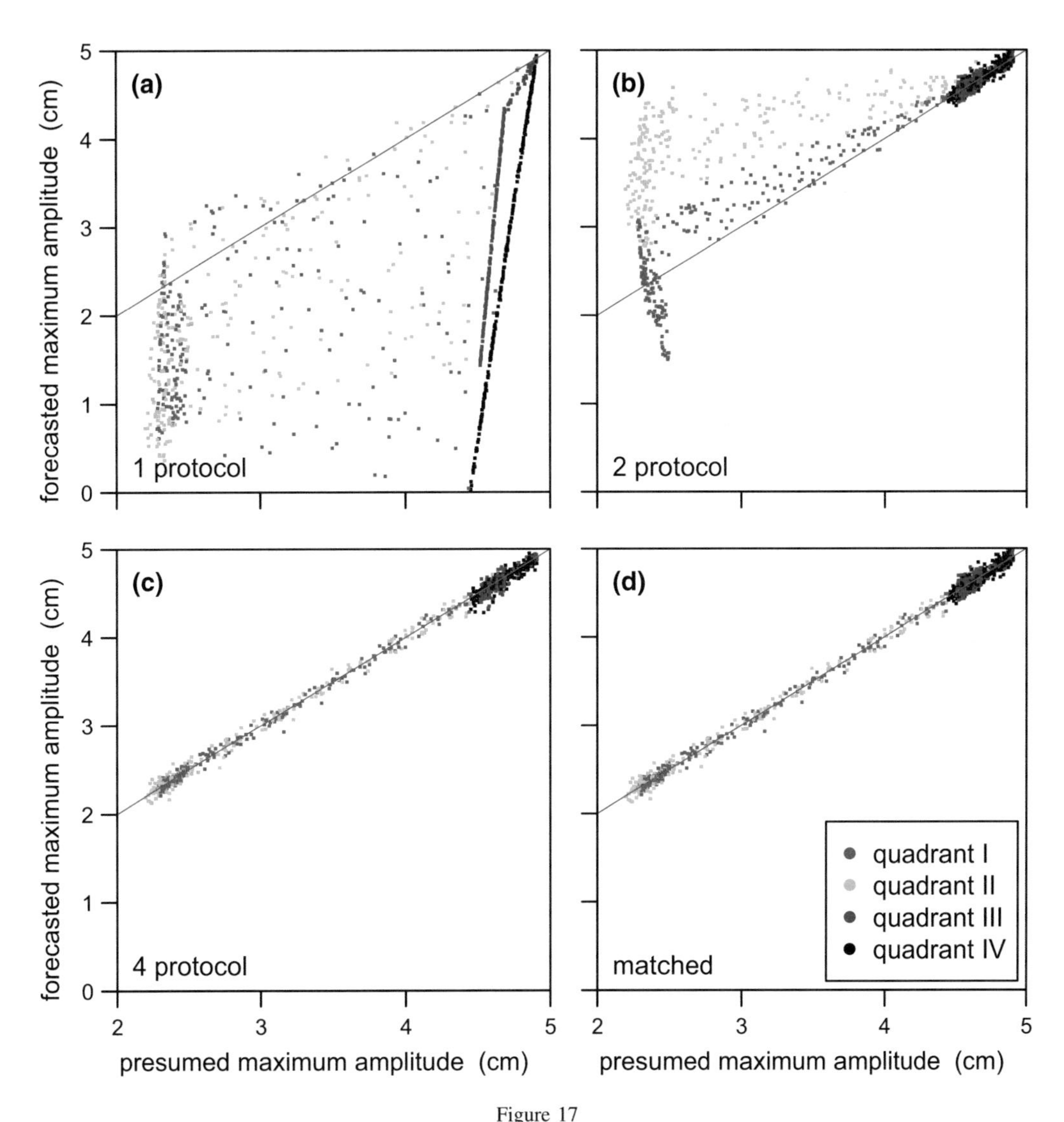

Figure 17
Scatterplots of forecasted versus presumed maximum wave amplitudes for 1000 realizations of representative triad (central unit source `ac026b`, buoy 46403 and impact site Hilo). Plots **a**–**d** correspond to the 1, 2, 4 and matched protocols for fully specifying the regression model of Eq. (4). Each realization is based on a random pick that has an equal chance of falling within one of the four quadrants shown in Fig. 13, with the quadrant determining which models $\boldsymbol{g}_k$ receive nonzero weights w_k in constructing the signal $\boldsymbol{t}$ via Eq. (1). As per the legend on plot (**d**), the colors of the points in the scatterplots indicate into which quadrant the random pick fell. The diagonal line in each plot shows where a point would fall if the forecasted and presumed amplitudes were in perfect agreement. The amount of data N used in the regression model is dictated by 21 min past the first peak in the constructed signal $\boldsymbol{t}$

Figure 18 shows the RMSEs for the four protocols in combination with three settings N for the amount of data used to estimate the source coefficients (dictated by up to either 1, 11 or 21 min past the first peak in the signal $\boldsymbol{t}$). The four RMSEs corresponding to the scatterplots in Fig. 17 are shown above the '21' label on the horizontal axis. In addition this figure has a dashed line showing the RMSE for a so-called seismic solution, for which a forecast is generated using $4\boldsymbol{g}_5$. This procedure assumes idealized seismic information about the tsunami event, namely, that the generating earthquake occurs within the central unit source `ac026b` and that the magnitude of the earthquake suggests setting the source coefficient to

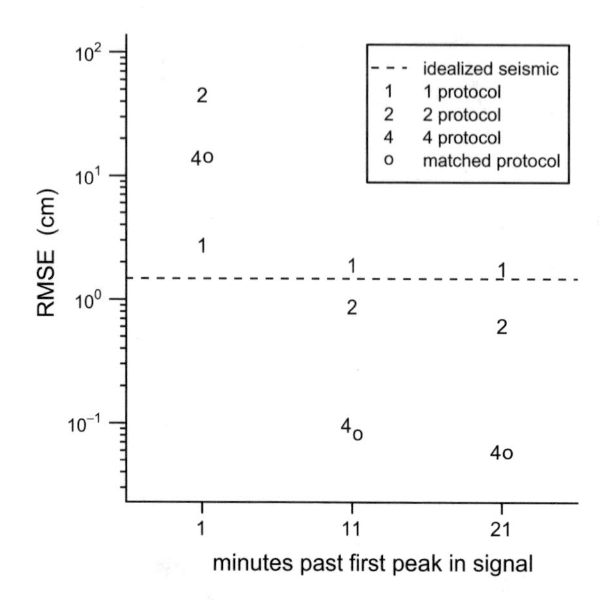

Figure 18
Root-mean-square errors (RMSEs) for forecasted versus presumed maximum wave amplitudes for 1000 realizations of representative triad (central unit source ac026b, buoy 46403 and impact site Hilo). There are twelve RMSEs in all (each computed as per Eq. (6)) corresponding to selecting one of the four protocols (the 1, 2, 4 and matched protocols) in combination with three different amounts of data (dictated by 1, 11 and 21 min past the first peak in each constructed signal $\boldsymbol{t}$) to be placed into $\bar{\boldsymbol{y}}$ in Eqs. (3) and (4). The RMSEs corresponding to the four scatterplots of Fig. 17 are plotted above the '21' label on the horizontal axis. The horizontal dashed line indicates the RMSE for forecasted amplitudes based on idealized seismic information only (and not any simulated data collected by buoy 46403)

4, which corresponds to our standard normalization $\sum_k \alpha_k = 4$ for the relocated unit source. The seismic-based forecast is of interest because it is arguably the best that can achieved without any information supplied by data from DART® buoy 46403.

For all four protocols for formulating the regression model, the RMSEs decrease as more data are collected by the DART® buoy. The decrease is minimal for the 1 protocol, but substantial for the other three protocols (in particular, there is a drop of more than two orders of magnitude for the 4 and matched protocols). The 1 protocol always involves a mismatch between the signal and its model. For the other protocols, there is at least a 50% chance of no mismatch. The poor performance of the 1 protocol points out the need for an adequate regression model. On the other hand, there is little difference in the RMSEs for the 4 and matched protocols even though the former involves mismatches in about 50% of the realizations, whereas the latter involves none. In contrast to the 1 protocol, the nature of the mismatch with the 4 protocol is too many $\boldsymbol{g}_k$s rather than too few. This result suggests that having extraneous $\boldsymbol{g}_k$s in the model (overfitting) does not significantly impact the forecasted wave amplitudes.

The seismic-based forecast outperforms the buoy-based forecasts either when there is an insufficient amount of data or when the regression model is always inadequate due to missing $\boldsymbol{g}_k$s (the 1 protocol). When there is too little data, forecasts deteriorate considerably as the complexity of the regression model increases (i.e., more coefficients must be estimated). For our representative triad, there is some value to be gained by waiting the extra 10 min from 11 min past the first peak to 21 min (the RMSEs drop by 60% to 70%). For other triads involving different central unit sources and different buoys, the gain can be more substantial.

For the network assessment discussed in Sect. 4, we concentrate on the 4 protocol with N selected as dictated by 21 min after the first peak in the signal. This choice is close to the best RMSE for most—but not all—triads and should reflect the contribution of a particular buoy to wave amplitude forecasts at an impact site once enough data have been collected and once an adequate regression model has been identified.

References

Gica, E., Spillane, M.C., Titov, V.V., Chamberlin, C.D., and Newman, J.C. (2008), Development of the forecast propagation database for NOAA's Short-term Inundation Forecast for Tsunamis (SIFT). NOAA Technical Memorandum OAR PMEL–139, p. 89. http://nctr.pmel.noaa.gov/pubs.html

Greenslade, D. J. M., & Warne, J. O. (2012). Assessment of the effectiveness of a sea-level observing network for tsunami warning. *Journal of Waterway, Port, Coastal, and Ocean Engineering, 138*(3), 246–255.

Hanks, T. C., & Kanamori, H. (1979). A moment-magnitude scale. *Journal of Geophysical Research, 84*(B5), 2348–2350.

Ihaka, R., & Gentleman, R. (1996). R: a language for data analysis and graphics. *Journal of Computational and Graphical Statistics, 5*, 299–314.

Papazachos, B. C., Scordilis, E. M., Panagiotopoulos, D. G., Papazachos, C. B., & Karakaisis, G. F. (2004). Global relations between seismic fault parameters and moment magnitude of earthquakes. *Bulletin of the Geological Society of Greece*, *36*, 1482–1489.

Percival, D. B., Denbo, D. W., Eblé, M. C., Gica, E., Huang, P. Y., Mofjeld, H. O., et al. (2015). Detiding DART® buoy data for real-time extraction of source coefficients for operational tsunami forecasting. *Pure and Applied Geophysics*, *172*(6), 1653–1678.

Percival, D. B., Denbo, D. W., Elbé, M. C., Gica, E., Mofjeld, H. O., Spillane, M. C., et al. (2011). Extraction of tsunami source parameters via inversion of DART® buoy data. *Natural Hazards*, *58*(1), 567–590.

Percival, D. M., Percival, D. B., Denbo, D. W., Gica, E., Huang, P. Y., Mofjeld, H. O., et al. (2014). Automated tsunami source modeling using the sweeping window positive elastic net. *Journal of the American Statistical Association*, *109*(506), 491–499.

R Development Core Team (2010) http://www.r-project.org/

Spillane, M.C., Gica, E., Titov, V.V., and Mofjeld, H.O. (2008), Tsunameter network design for the U.S. DART® arrays in the Pacific and Atlantic Oceans. NOAA Technical Memorandum OAR PMEL–143, p. 165. http://nctr.pmel.noaa.gov/pubs.html

Titov, V. V. (2009). Tsunami forecasting. In E. N. Bernard & A. R. Robinson (Eds.), *The sea, volume 15: Tsunamis* (Vol. 15, pp. 371–400). Cambridge: Harvard University Press.

Whitmore, P. M. (2009). Tsunami warning systems. In E. N. Bernard & A. R. Robinson (Eds.), *The sea, volume 15: Tsunamis* (pp. 401–442). Cambridge: Harvard University Press.

Whitmore, P., Benz, H., Bolton, M., Crawford, G., Dengler, L., Fryer, G., et al. (2008). NOAA/West coast and Alaska Tsunami Warning Center Pacific Ocean response criteria. *Science of Tsunami Hazards*, *27*(2), 1–21.

(Received May 18, 2017, revised January 21, 2018, accepted February 26, 2018)

Pure Appl. Geophys.

DOI 10.1007/s00024-017-1660-5

Pure and Applied Geophysics

Numerical Procedure to Forecast the Tsunami Parameters from a Database of Pre-Simulated Seismic Unit Sources

César Jiménez,[1,2] Carlos Carbonel,[1] and Joel Rojas[1]

Abstract—We have implemented a numerical procedure to forecast the parameters of a tsunami, such as the arrival time of the front of the first wave and the maximum wave height in real and virtual tidal stations along the Peruvian coast, with this purpose a database of pre-computed synthetic tsunami waveforms (or Green functions) was obtained from numerical simulation of seismic unit sources (dimension: 50×50 km^2) for subduction zones from southern Chile to northern Mexico. A bathymetry resolution of 30 arc-sec (approximately 927 m) was used. The resulting tsunami waveform is obtained from the superposition of synthetic waveforms corresponding to several seismic unit sources contained within the tsunami source geometry. The numerical procedure was applied to the Chilean tsunami of April 1, 2014. The results show a very good correlation for stations with wave amplitude greater than 1 m, in the case of the Arica tide station an error (from the maximum height of the observed and simulated waveform) of 3.5% was obtained, for Callao station the error was 12% and the largest error was in Chimbote with 53.5%, however, due to the low amplitude of the Chimbote wave (<1 m), the overestimated error, in this case, is not important for evacuation purposes. The aim of the present research is tsunami early warning, where speed is required rather than accuracy, so the results should be taken as preliminary.

Key words: Tsunami, numerical simulation, forecasting.

1. Introduction

According to historical seismicity (Silgado 1978; Beck and Nishenko 1990), many destructive earthquakes have occurred in Peru. Some of which have generated tsunamis that have affected ports and coastal towns, for instance, the remarkable tsunami of Callao 1746 in central Peru (Dorbath et al. 1990; Jiménez et al. 2013; Mas et al. 2014a) and the tsunami of southern Peru in 1868, both megathrust earthquakes with a magnitude around M_w 9.0. In the twentieth century, there were the tsunamigenic earthquakes of 1960 in Lambayeque (M_w 7.6), 1966 in northern Lima (M_w 8.1), 1974 in southern Lima (M_w 8.1) (Beck and Ruff 1989), 1996 in Chimbote (M_w 7.5) and 1996 in Nazca (M_w 7.7). According to instrumental seismicity (after 1960), only in the twenty-first century there have been two major tsunamigenic earthquakes in Peru: Camana in 2001 (M_w 8.4) (Adriano et al. 2016) and Pisco in 2007 (M_w 8.1) (Jiménez et al. 2014). Accordingly, regarding tsunami hazard, the main issue in Peru is the near-field tsunami event, where the tsunami arrival time is in the range of 15–60 min (Mas et al. 2014b). Therefore, it is necessary to develop a system to forecast, in the shortest possible time, the tsunami parameters from a database of pre-simulated seismic unit sources.

The "Centro Nacional de Alerta de Tsunamis" from Peru (CNAT in Spanish, corresponding to Peruvian Tsunami Warning Center) is the official representative to international institutions such as the Pacific Tsunami Warning Center (PTWC), responsible for issuing information bulletins, alert or alarm in case of occurrence of tsunamis for near-field or far-field origin. For near-field events, it is not convenient to compute a numerical tsunami simulation immediately after the occurrence of an earthquake because this process may take several hours and the arrival time is in order of the minutes. For this purpose, it is required to process quickly and accurately the seismic information (hypocenter location and magnitude) provided by the Instituto Geofísico del Perú (IGP) for earthquakes generated in the near field (local) or by USGS (United States Geological Survey) for earthquakes generated in the far field. To accelerate the processing of information is necessary to implement

[1] Laboratorio de Física de la Tierra, Universidad Nacional Mayor de San Marcos, Av. Germán Amézaga N° 375, Lima 1, Peru. E-mail: cjimenezt@unmsm.edu.pe

[2] Centro Nacional de Alerta de Tsunamis, Dirección de Hidrografía y Navegación, Jr. Roca 118, Callao, Peru.

some numerical procedure to process such information automatically from a database of pre-computed tsunami scenarios.

1.1. State of the Art

As research antecedents, Gica et al. (2008), implemented a database of pre-calculated seismic events to forecast tsunami parameters (arrival time, maximum tsunami height) with seismic unit sources of 50 × 100 km^2 using a bathymetry of 4 arc-min resolution. This system is used by the National Oceanic and Atmospheric Administration (NOAA) and this pre-computed database serves as input to an intermediate resolution model along the coast. On the other hand, Greenslade et al. (2011) implemented a model to forecast tsunamis from a database of seismic scenarios, which is used by the Tsunami Warning Center in Australia for buoys at offshore (not for tidal gauges on the coast), the bathymetry resolution is 4 arc-min and the size of the seismic source is 100 × 100 km^2.

Jiménez (2010) presented a numerical procedure to determine the occurrence of tsunamis based on hypocentral parameters and magnitude. Also, an algorithm to calculate quickly the arrival time of the first tsunami wave (based on a linear trajectory followed by the tsunami wave according to actual bathymetry) was implemented. That research was the starting point for the next step: the calculation of the tsunami wave height from tsunami numerical modeling.

In the present paper, the results of Jiménez (2010) are extended to calculate the arrival time of the front of the first tsunami wave and the maximum wave height (of the whole computation time) for real and virtual coastal tide gauge stations. This is based on a database or catalogue of synthetic tsunami waveforms (or Green functions) obtained by tsunami numerical simulation of unitary seismic sources of dimensions 50 × 50 km^2, unitary dislocation with a focal mechanism based on Global Centroid Moment Tensor (CMT) catalogue and a 5 km depth for the shallower end of the subfault that is near the trench. The bathymetry has a grid resolution of 30 arc-sec (approximately 927 m). Hence, this paper covers important and interesting work being undertaken to develop a tsunami scenario database for Peru. It will provide a tool that can be used to determine if tsunami warnings are needed for parts of the Peruvian coastline.

2. Data

2.1. Bathymetry Data

The tsunami wave speed and directivity depend on bathymetry; therefore, a good description of bathymetry is important to conduct numerical simulation of propagation of tsunamis. The global bathymetry model is taken from GEBCO 30 (2017), which has a grid resolution of 30 arc-sec or approximately 927 m. The fine bathymetry in coastal zones has been updated from marine soundings provided by the Directorate of Hydrography and Navigation (DHN) of the Peruvian Navy. Figure 1 shows the bathymetry of the whole Pacific Ocean, the seismicity for 2015 and the principal subduction zones.

2.2. Focal Mechanism Parameters and Seismicity

The focal mechanism parameters (azimuthal or strike angle θ, dip angle δ, and rake angle λ) are taken from historical averages of the catalogue of the Global Centroid Moment Tensor (CMT) for a given subduction zone, (http://www.globalcmt.org, last accessed: December 2015). In some cases, the strike angle is taken according to the orientation of the trench or parallel to the coast, also the rake or dislocation angle is taken as 90° to obtain a maximum vertical seafloor deformation. The upper side depth of the subfault near the trench is taken as 5 km; according to dip angle, the intermediate depth subfault is set to 17–21 km and the depth of the subfault close to coast is around 30 km.

To define the computational grid, it is necessary to know the location of the subduction zones in the Pacific Ocean. From seismicity data and focal mechanism parameters is possible to deduce the seismic subduction zones along the Pacific Ocean. These subduction zones contain: Chile, Peru to Central America up to northern Mexico, Cascadia zone, Alaska, Aleutians islands, Kuril Islands, Japan, Marianas trench, Tonga and Kermadec (Fig. 1).

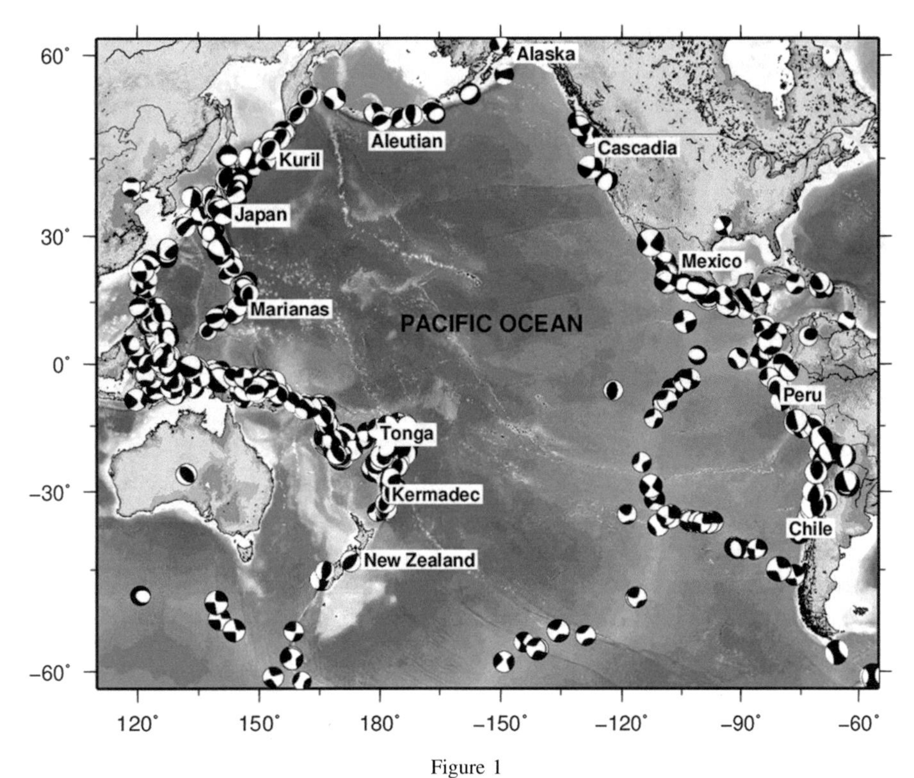

Figure 1

Bathymetry of the Pacific Ocean from GEBCO 30. The *circles* represent the focal mechanism of seismic events ($M_w > 5.0$) of 2015 from Global CMT catalogue. The tags represent the principal subduction zones

3. Methodology

The methodology could be summarized in the following form: First, the coseismic deformation must be calculated and used as the initial condition for the tsunami wave propagation. Later, a database or catalogue of synthetic tsunami waveforms was implemented from tsunami numerical simulation of unitary seismic sources.

3.1. Coseismic Deformation

The initial condition for tsunami propagation is calculated using the deformation model of Okada (1992) for an elastic, homogeneous and isotropic semi-infinite medium. Seismic sources are considered of dimensions $50 \times 50\ \mathrm{km}^2$, unitary dislocation (slip $= 1$ m) and depth of the upper side of the shallower unit source is 5 km; the depths of the deeper unit sources will depend on the dip angle and the width (50 km, in this case) of the unit source. The focal mechanism parameters are taken from the mean values of the CMT catalogue and the seismicity of the corresponding subduction zone (from southern Chile to northern Mexico).

We assume that the generation of coseismic deformation of the seafloor produces an immediate and identical deformation on the ocean surface (due to the incompressibility of the fluid). As the ocean is a fluid, the deformation does not hold its shape and the initial displacement radiates away from the source creating a tsunami wave that propagates in all directions (Okal 2009). Figure 2 shows an example of deformation field.

3.2. Tsunami Numerical Modeling

A modified version of TUNAMI-F1 (IOC-Intergovernmental Oceanographic Commission 1997) numerical model was used. This code was written in Fortran. The algorithm corresponds to a linear model in a regular grid-system using spherical coordinates based on shallow water theory. The differential equations (linear momentum conservation and mass conservation or continuity equation) are integrated numerically using the method of "leap frog" in a finite differences scheme. A numerical stability condition (CFL condition) of less than 0.8 was used to avoid numerical instabilities.

The initial conditions are based on the deformation theory by Okada (1992). The boundary conditions along the coastline (for the linear model) represent a vertical wall, so we suggest the use of a correction factor for the synthetic tsunami waveform of each tidal station (to approximate a non-linear system).

The numerical code is written in Fortran and runs in a Linux system. The digital processing of input data and output results was made in Matlab. The graphical user interface (GUI) was implemented also in Matlab. The final application runs under Windows or Linux.

3.3. Unitary Seismic Sources and Green Functions

The set of unitary seismic sources and tsunami waveforms form the database of the model. The number of unitary seismic sources is proportional to the geographical extent of the database. Therefore, the chances of predicting the tsunami parameters of an earthquake generated somewhere in the Pacific Ocean will be greater.

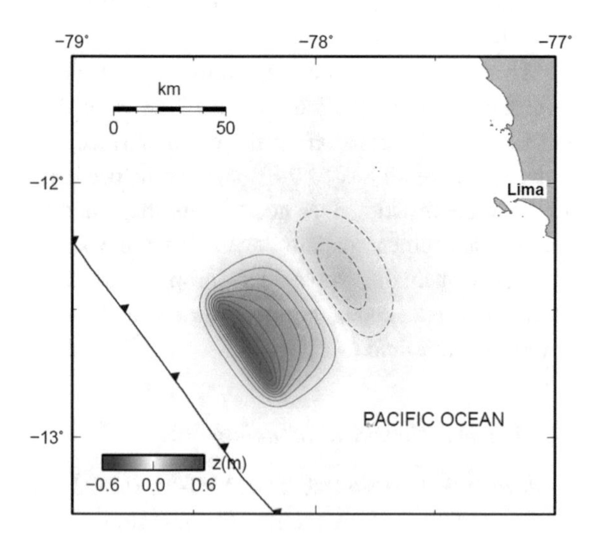

Figure 2
Example of coseismic deformation field of a unitary seismic source located offshore Lima. The *red color* represents the uplift and *blue color* represents subsidence. The *barbed lines* represent the Peruvian trench

Taking into account the seismicity of the Peruvian coast and adjacent countries, the computational domain was initially chosen from Antofagasta (Chile) to Ecuador. For the first stage, this area has been divided into 130 "unit seismic sources" of dimensions 50×50 km^2. The idea is to simulate the propagation of tsunami waveforms for each seismic unit source and calculate the Green functions G (t) or synthetic mareograms at each port and coastal town in Peru, where real and virtual tide gauge stations have been selected. Currently and in a second stage, 465 seismic unit sources have been modeled, from southern Chile to northern Mexico.

3.4. Virtual Tidal Stations

There are 17 numerical tide gauge station within the computational area, many of them correspond with the actual location of real tidal stations. These virtual stations correspond to ports and coastal towns (Fig. 3). The geographical coordinates and the empirical correction factor of each station are shown in Table 1.

A major aspect of the database is the development of correction factors for each individual tidal station. These correction factors account for the amplification or attenuation of the tsunami waves in shallow waters due to non-linear processes that cannot be simulated using linear model. These correction factors are determined in an empirical way as the rate of the maximum observed tsunami height over the maximum simulated tsunami height.

3.5. Governing Equations of the Numerical Model

All the equations are highly dependent on earthquake magnitude Mw and epicenter location. According to scaling laws between the dimensions of a fault plane or seismic source (of rectangular geometry) and the magnitude of the earthquake for subduction zones (Papazachos et al. 2004), we have the basic empirical relations

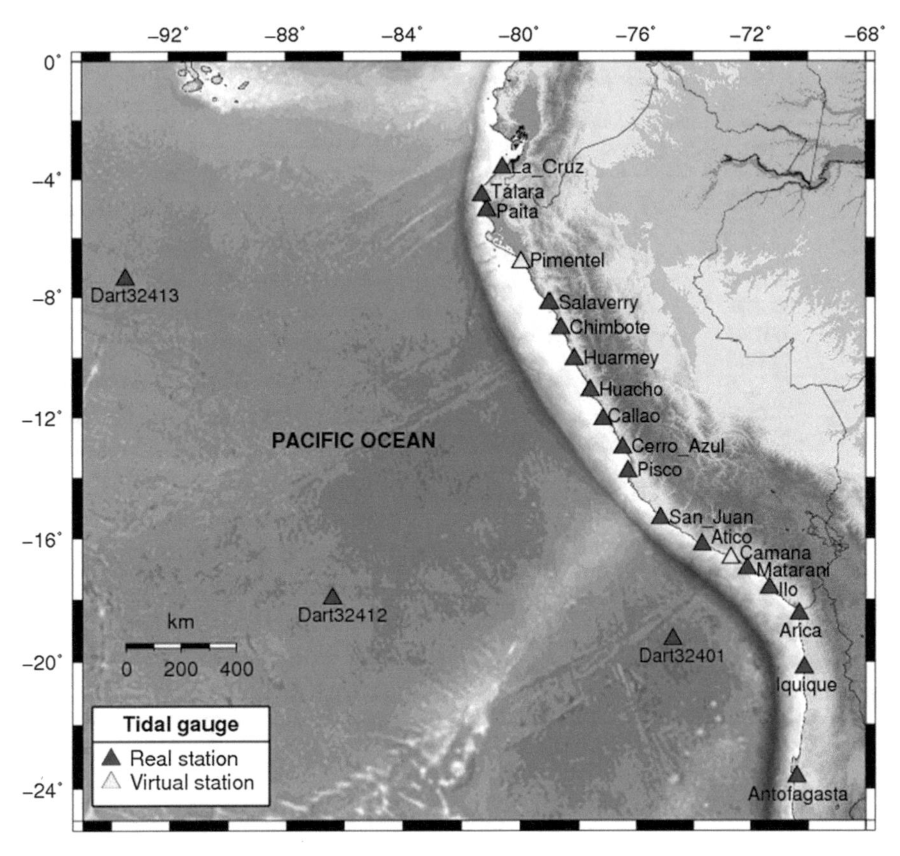

Figure 3
Location of virtual and real tide gauge station to be considered in the numerical modeling

$$\log(L) = 0.55M - 2.19 \tag{1}$$

$$\log(W) = 0.31M - 0.63, \tag{2}$$

where L is the length (in km) of the seismic source, W is the width (in km) of the seismic source, and M is the moment magnitude (M_w). The seismic moment defined by (Aki 1966) is

$$M_0 = \mu LWD, \tag{3}$$

where μ is the mean rigidity modulus (4×10^{10} N/m^2) of the rupture zone and D is the dislocation or slip of the seismic source. The relationship between seismic moment (in N $\times$ m) and the moment magnitude is given by the relation,

$$\log(M_0) = 1.5M_w + 9.1. \tag{4}$$

The model output (tsunami waveform) is the linear combination of the synthetic tsunami waveforms or Green functions G (t) corresponding to the seismic unit sources contained in the rupture geometry (Fig. 4). The number of seismic unit sources is given by

$$N = \frac{LW}{L_0 W_0}, \tag{5}$$

where $L_0 = 50$ km, $W_0 = 50$ km, corresponding to the dimensions of the unitary seismic source. Therefore, for an earthquake of M_w 7.0 only one seismic source is required, but for an earthquake of M_w 8.0, 5 seismic sources are required (Table 2). The criterion in selecting the deeper or shallower unit seismic source is based on an algorithm that takes into account the nearest distance from the epicenter. Also, an amplification factor is required: the dislocation or mean slip D. To linearize the model, an empirical correction factor K_i for each coastal tidal station is required. The signal of the tsunami f_i (t) in i-th tidal station is given by

Table 1

Location of real and virtual tidal stations used in the numerical simulation

N	Station	Lat (°)	Lon (°)	K_i
1	La Cruz	−03.6337	−80.5876	0.74
2	Talara	−04.5751	−81.2827	0.88
3	Paita	−05.0837	−81.1077	0.94
4	Pimentel	−06.8396	−79.9423	0.75
5	Salaverry	−08.2279	−78.9818	0.70
6	Chimbote	−09.0763	−78.6128	0.46
7	Huarmey	−10.0718	−78.1616	1.05
8	Huacho	−11.1218	−77.6162	0.95
9	Callao	−12.0689	−77.1667	0.80
10	Cerro Azul	−13.0253	−76.4808	0.82
11	Pisco	−13.8061	−76.2919	0.65
12	San Juan	−15.3556	−75.1603	0.60
13	Atico	−16.2311	−73.6944	1.00
14	Camaná	−16.6604	−72.6838	0.40
15	Matarani	−17.0009	−72.1088	0.80
16	Ilo	−17.6445	−71.3486	0.79
17	Arica	−18.4758	−70.3232	0.70

The parameter K_i represents the empirical correction factor that takes into account the linearity

$$f_i(t) = DK_i \sum_{j=1}^{N} G_{ij}(t), \tag{6}$$

where G_{ij} (t) is the Green function (synthetic tsunami waveform) for the i-th tidal station due to the j-th seismic unit source.

4. Results

4.1. Graphical User Interface: "Pre-Tsunami"

With the information of hypocentral parameters, conditions of tsunami generation and algorithm of pre-computed seismic unit sources, we have implemented an application with a graphical user interface developed in Matlab programming language. The input data are the hypocentral parameters: magnitude M_w, hypocentral depth (this is used for discriminating the sources greater than 60 km depth), geographical latitude and longitude of the epicenter and origin time. The outputs are the parameters of seismic source geometry and a diagram of the likely rupture geometry on a map.

The subroutine "Arrival Time" calculates the arrival times from the dependence of wave velocity with respect to the bathymetry. The subroutine "Catalogue" calculates the arrival time of the front of the first wave and the maximum wave height of the tsunami (Fig. 5) in ports and coastal towns of Peru, from a database of synthetic tsunami waveforms or Green functions. The time of calculation is very fast, in the order of a few seconds due to the pre-calculated synthetic tsunami waveforms.

The subroutine "Automatic IGP" loads the hypocentral parameters from the website of the seismological service of IGP (Instituto Geofísico del Perú, in spanish) to avoid the manual introduction of these parameters in the case of some big earthquake in the near field reported by the Peruvian seismological service of IGP.

The graphical user interface of Pre-Tsunami application is quite friendly and easy to use by the operator of the tsunami warning and constitutes an important computational tool to forecast and decision making for issuing bulletins, alert or alarm in case of occurrence of tsunamis. The last version (V4.6) was updated in August 2016 (Fig. 6).

4.2. Output of the Application Pre-Tsunami

The output of the model will be useful to forecast tsunamis in the near, regional and far field. The output parameters are: the arrival time of the front of the first wave and the maximum height of the tsunami wave. These results will be useful for issuing the alert or alarm of the tsunami. For example, for a hypothetical M_w 8.6 magnitude earthquake with epicenter offshore of Lima (Lat = −12.00°, Lon = −78.00°, Depth = 20 km, origin time: 00:00), according to the threshold of tsunami heights of PTWC, maximum evacuation ($H_{max} > 3$ m) would be declared for Huacho, Callao and Cerro Azul (Table 3).

4.3. Application to the Tsunami of Chile 2014 (M_w 8.2: USGS Preliminary Report)

A big earthquake and tsunami was generated in Iquique Chile on 01 April 2014 (Lay et al. 2014; An et al. 2014). The hypocentral parameters (longitude, latitude, depth and origin time) reported by NEIC-USGS (http://earthquake.usgs.gov/earthquakes/map/)

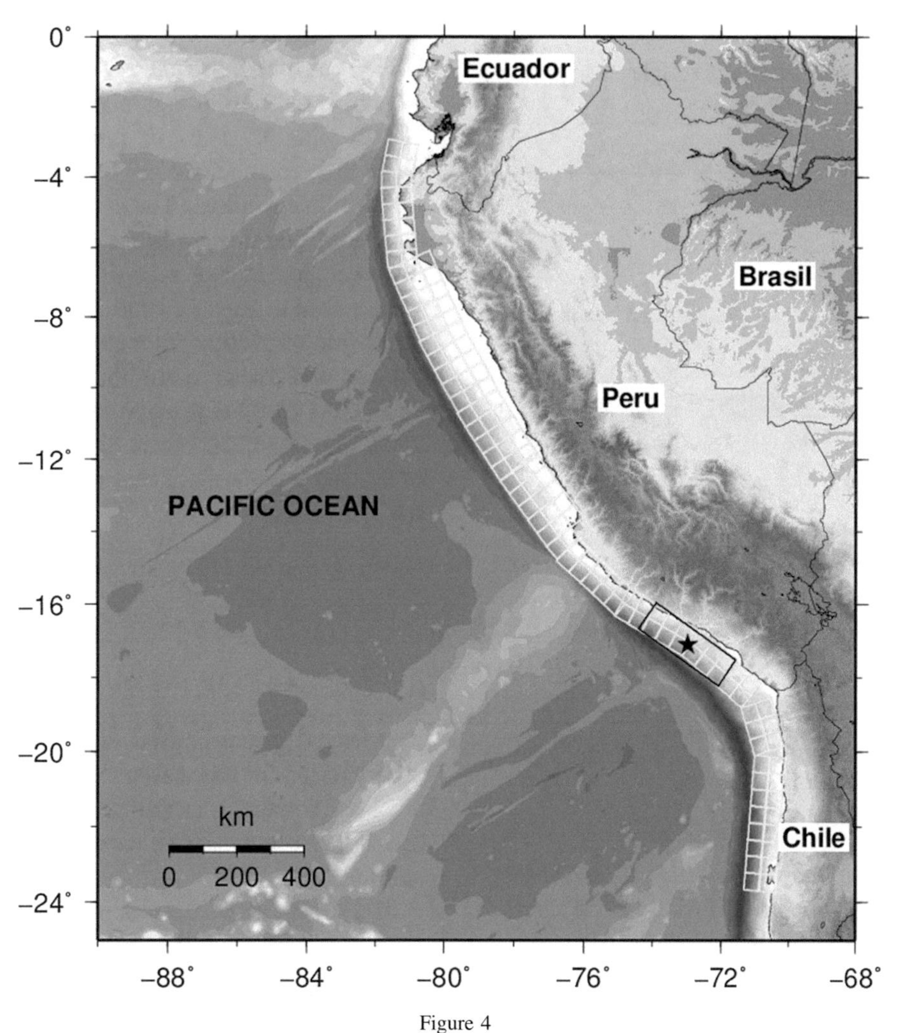

Figure 4
Location of unit seismic sources and an example of a major earthquake (*black rectangle*). The *black star* represents the epicentroid (gravity center of the seismic source)

were: lon = −70.817°, lat = −19.642°, z = 25 km, origin time: 23:46:46 UTC.

The acquisition of data from the National Tide Gauge Network of Peru (http://www.dhn.mil.pe) and Arica Station (tide gauge from Chilean network, available on: http://www.ioc-sealevelmonitoring.org) was performed. Digital signal processing was performed for these signals; a high-pass Butterworth filter with cut frequency: $f_c = 6.94 \times 10^{-5}$ Hz and a low-pass filter ($f_c = 0.0083$ Hz) was applied to remove components of long period (tides) and short period (surges); so, the maximum amplitude of the tsunami wave was calculated (Table 4). The calculation of the parameters of the simulated tsunami was conducted using the Pre-Tsunami application (version 4.6). This event has allowed calibrating the values of the correction factor K_i for each station.

The results show a very good correlation for stations with wave amplitude greater than 1 m, meanwhile for Pisco and Chimbote stations the correlation is poor. For the case of Arica station, the error was of 3.5% and Ilo's error was 5.7%. In the case of Callao station, the error was 12% and the largest error was in Pisco (34.8%) and Chimbote (53.5%). This does not mean that the method presented is flawed and not useful, rather it shows the limitations of the numerical method. With additional data from DART buoys and future

Table 2

Number of unit sources N required according to the magnitude of the earthquake

Magnitude (M_w)	N
7.0	1
7.5	2
8.0	5
8.5	12
9.0	32

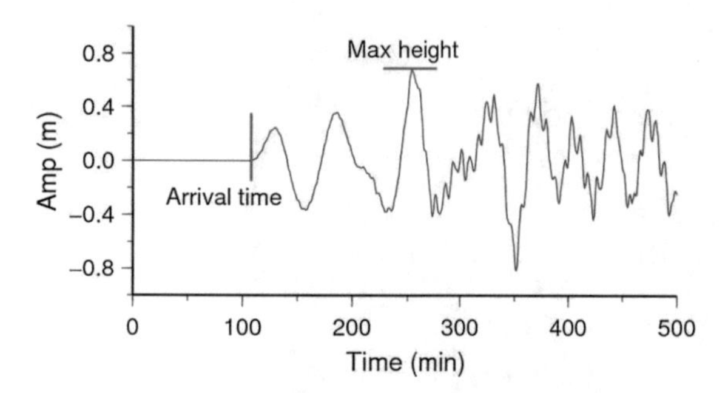

Figure 5
Diagram showing a synthetic tsunami waveform. The *vertical red line* represents the arrival time and the *horizontal red line*, the maximum tsunami height

tsunamis, updating and calibration of Pre-Tsunami application will be performed (i.e., the empirical correction factors of each tidal station), achieving on the future a greater reliability. Additionally, due to the low amplitude of the Pisco and Chimbote wave (<1 m), the overestimated error is not important for evacuation purposes.

4.4. *Projection of the Catalogue of Seismic Unit Sources*

The current addition in the database of the simulation is shown in Fig. 7, geographically, from southern Chile to northern Mexico with 465 seismic unit sources simulated. We hope to conduct the simulation of seismic unit sources for the entire Pacific Ocean corresponding to subduction zones of Pacific Seismic Ring. This implies a large computational effort with the simulation of more than 1000 seismic unit sources, for a computational grid around the Pacific Ocean with 30 arc-sec resolution or 1 min resolution for the far field.

The Directorate of Hydrography and Navigation (DHN) of the Peruvian Navy have acquired, at beginning of 2016, a high-performance computer or cluster with 32 cores or processors for scientific numerical calculations to conduct these tsunami numerical simulations. The computation time will be considerably reduced with the use of this high-performance computer: each seismic unit source simulation is computed in approximately 30 min for a computational grid of 3720 × 5760 elements, while this very computation lasted more than 3 h in a personal computer with microprocessor i7 of 3.40 GHz.

In the future, we expect to conduct the inversion of tsunami waveforms to obtain the distribution of the seismic source.

5. *Discussion*

5.1. *Limitations of the Model*

A linear model is used to simulate a non-linear phenomenon, for this reason we suggest the use of an empirical correction factor for each tidal station. This correction factor must be calibrated with every new tsunami event.

The model is strongly dependent on the seismic magnitude and epicenter location. In addition, the model takes the mean values of the focal mechanism CMT catalogue, if the variation of focal mechanism is great, the forecast will not be very good.

In most of the cases, if the epicenter does not match with the epicentroid, the results will be biased due to the position of the seismic source. It is difficult to obtain the geometry of the seismic source immediately after the occurrence of the earthquake; this issue can be resolved with the aftershock distribution, many hours later.

For the far field, the numerical model does not take into account the effects of dispersion, which is a non-linear effect. The model assumes homogeneous and regular deformation; the reality is that the seismic source can have a heterogeneous asperity distribution.

5.2. *Strengths of the Model*

The time that the application spends to provide the tsunami parameters is immediate, since it has a

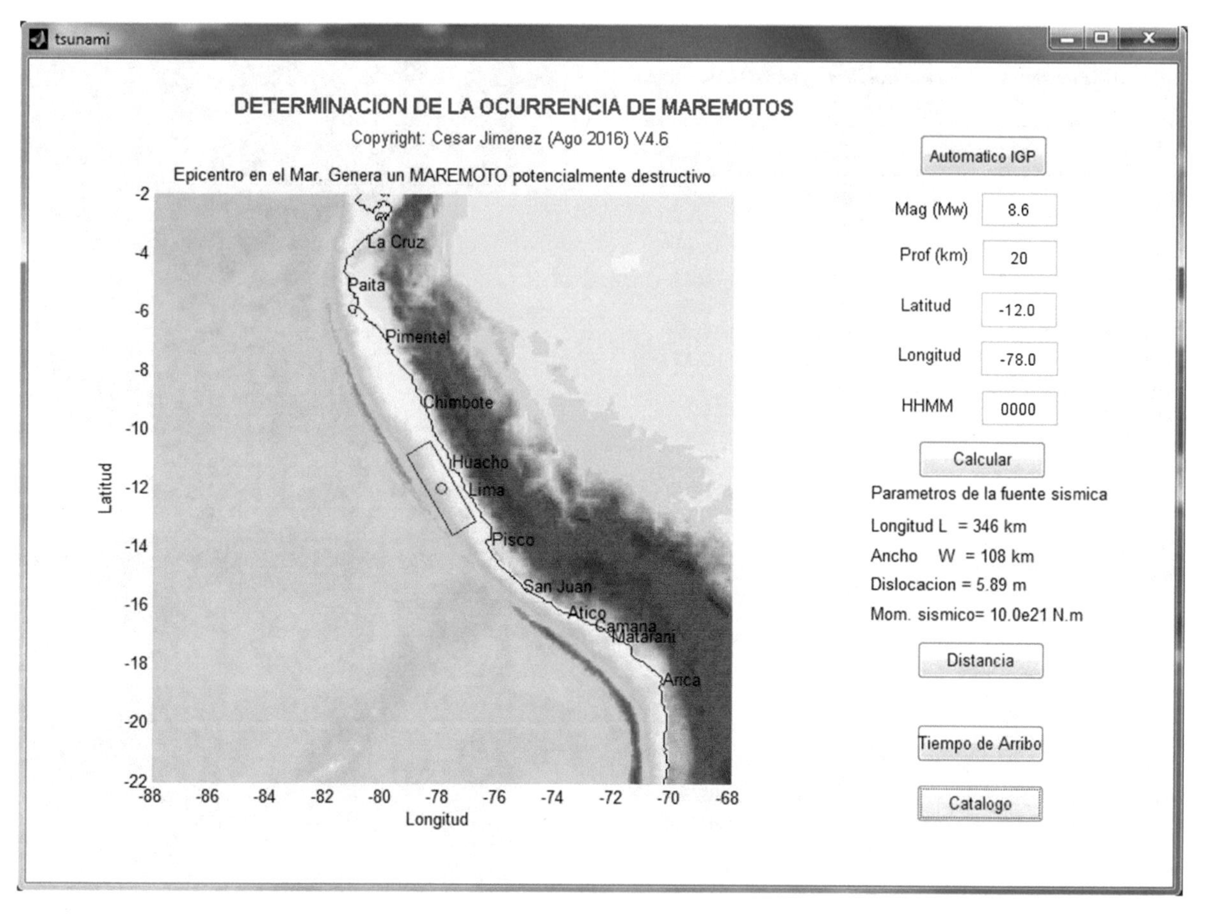

Figure 6
Graphical user interface pre-tsunami (Version 4.6)

database of simulations previously computed. The model works well when the epicenter matches with the epicentroid or gravity center of the rupture geometry, as the case of Chile earthquake of 2014.

This model is applicable for seismic events in the near and regional field, due to the reaction time is in the order of a few minutes to 1 h. On the other hand, for tsunamis in the far field, the forecaster has a greater reaction time, in the order of many hours.

The GUI (Graphical User Interface) is user friendly and easy to use by an operator (it is not necessary the operator be an expert in programming or numerical simulation). As a post-tsunami analysis, the Green functions of the model can be used to obtain the distribution of the seismic source, with an inversion calculation.

We have mentioned that this model requires the aftershock distribution to solve the location of the seismic source. Therefore, in a real application, the accuracy of this system without knowing the real rupture geometry is good, from the point of view of the tsunami early warning operator.

6. *Summary and Conclusions*

We have presented a numerical procedure to determine quickly the occurrence of a tsunami and to forecast some parameters: the arrival time of the front of the first tsunami wave and the maximum tsunami wave height in the different coastal cities. The application "Pre-Tsunami" is of practical use in the Peruvian Tsunami Warning Center.

The numerical model tries to reproduce a physical process (as is the process of interaction of tsunami waves near the coast) through a linear simulation

Table 3

Output of the numerical model: arrival time T_A and maximum wave height H_{max}

Port	Region	T_A	H_{max} (m)
La Cruz	N	01:54	0.26
Talara	N	01:05	0.12
Paita	N	01:15	0.52
Pimentel	N	01:42	0.95
Salaverry	N	01:34	2.70
Chimbote	C	01:11	1.39
Huarmey	C	00:22	2.32
Huacho	C	00:20	3.41
Callao	C	00:24	6.75
Cerro Azul	C	00:37	3.35
Pisco	C	00:40	2.10
San Juan	C	00:43	0.91
Atico	S	00:55	0.28
Camaná	S	01:15	0.28
Matarani	S	01:09	0.18
Ilo	S	01:22	0.40
Arica	S	01:39	0.59

Region: Northern (N), Central (C) and Southern (S)

Table 4

Comparison of the observed and simulated maximum tsunami wave heights at gauge sites for Chilean earthquake and tsunami of 01 April 2014

Station	Observed (m)	Simulated (m)	% Error
Arica	2.00	2.07	3.5
Ilo	1.05	1.11	5.7
Matarani	0.52	0.60	15.4
San Juan	0.45	0.51	13.3
Pisco	0.23	0.31	34.8
Callao	0.25	0.28	12.0
Chimbote	0.15	0.23	53.5
Salaverry	0.20	0.18	(−) 10.0
Paita	0.10	0.11	10.0

Negative values (−) indicate underestimation

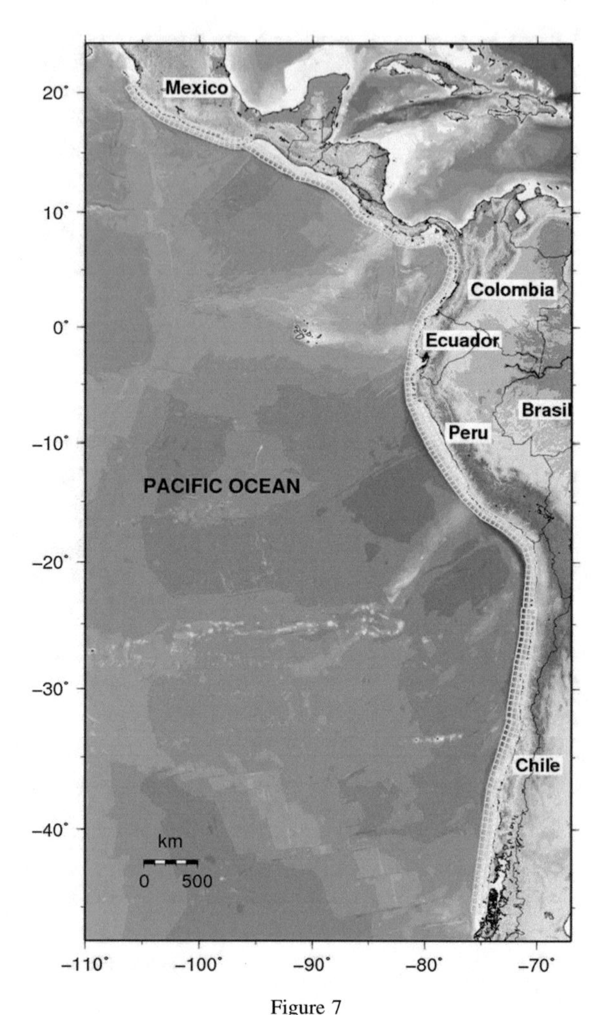

Figure 7
The small *yellow squares* represent the unitary pre-calculated seismic sources. The advance covers the subduction zone from southern (Chile) to northern Mexico with 465 seismic sources

model (as is the process of propagation of the tsunami) to use the principle of superposition of the Green functions. This could suggest the use of a "linearity correction factor" for each tidal station, which can be empirically calibrated and updated with each new tsunami event.

Despite the limitations of the model, the results are of practical use as a first approximation for purposes of fast estimation of tsunami parameters to issue the alerts and warnings of tsunamis in the near field, due to the speed with which the tsunami parameters are obtained.

The application of the model to the earthquake and tsunami in northern Chile 2014 (M_w 8.2) indicates a very good correlation between the maximum heights of observed and simulated waves for amplitudes greater than 1 m. In the case of tidal station of Arica (amplitude = 2.0 m), the error was 3.5%; for Ilo station (amplitude = 1.05 m), the error was 5.7%; Callao's error was 12%. The largest error was in Chimbote with 53.5%, however, due to the low amplitude of the Chimbote wave (<1 m), the overestimated error is not important for evacuation purposes. We must take into account that the aim of the research is tsunami early warning, where speed is required rather than accuracy.

Acknowledgements

This research was sponsored in part by the Vice-rectorate for Research and Postgraduate from "Universidad Nacional Mayor de San Marcos" (Group Research Project 2017) and by the "Direccion de Hidrografia y Navegacion" from Peruvian Navy. In addition, we acknowledge the anonymous reviewers.

REFERENCES

Adriano, B., Mas, E., Koshimura, S., Fujii, Y., Yanagisawa, H., & Estrada, M. (2016). Revisiting the 2001 Peruvian earthquake and tsunami impact along Camana beach and the coastline using numerical modeling and satellite imaging. In: Chapter I of Tsunamis and earthquakes in coastal environments Springer Int. doi:10.1007/978-3-319-28528-3_1.

Aki, K. (1966). Estimation of earthquake moment, released energy, and stress-strain drop from the G wave spectrum. *Bulletin Earthquake Research Institute, 44,* 73–88.

An, C., Sepúlveda, I., & Liu, P. L.-F. (2014). Tsunami source and its validation of the 2014 Iquique, Chile Earthquake. *Geophysical Research Letters, 41,* 3988. doi:10.1002/2014GL060567.

Beck, S., & Nishenko, S. (1990). Variations in the mode of great earthquake rupture along the Central Peru subduction zone. *Geophysical Research Letters, 17*(11), 1969–1972.

Beck, S., & Ruff, L. (1989). Great earthquakes and subduction along the Peru trench. *Physics of the Earth and Planetary Interiors, 57,* 199–224.

Dorbath, L., Cisternas, A., & Dorbath, C. (1990). Assessment of the size of large and great historical earthquakes in Peru. *Bulletin of Seismological Society of America, 80*(3), 551–576.

GEBCO 30. (2017). General Bathymetric Chart of the Oceans. Web: http://www.gebco.net. Accessed Nov 2016.

Gica, E., Spillane, E., Titov, V., Chamberlin, C., & Newman, J. (2008). Development of the forecast propagation database for NOAA's short-term inundation forecast for Tsunamis (SIFT). NOAA Technical Memorandum OAR PMEL-139, pp 89.

Greenslade, D., Allen, S., & Simanjuntak, M. (2011). An evaluation of tsunami forecasts from the T2 scenario database. *Pure and Applied Geophysics, 168,* 1137–1151.

IOC-Intergovernmental Oceanographic Commission. (1997). IUGG/IOC TIME project numerical method of tsunami simulation with the leap-frog scheme. UNESCO, Paris. http://www.jodc.go.jp/info/ioc_doc/Manual/122367eb.pdf. Accessed Jan 2016.

Jiménez, C. (2010). Software para determinación de ocurrencia de maremotos. *Boletín de la Sociedad Geológica del Perú., 104,* 25–31.

Jiménez, C., Moggiano, N., Mas, E., Adriano, B., Fujii, Y., & Koshimura, S. (2014). Tsunami waveform inversion of the 2007 Peru (Mw 8.1) earthquake. *Journal of Disaster Research, 8*(2), 266–273.

Jiménez, C., Moggiano, N., Mas, E., Adriano, B., Koshimura, S., Fujii, Y., et al. (2013). Seismic source of 1746 Callao Earthquake from tsunami numerical modeling. *Journal of Disaster Research, 9*(6), 954–960.

Lay, T., Yue, H., Brodsky, E., & An, C. (2014). The 1 April 2014 Iquique, Chile, Mw 8.1 earthquake rupture sequence. *Geophysical Research Letters, 41,* 3818–3825.

Mas, E., Adriano, B., Kuroiwa, J., & Koshimura, S. (2014a). Reconstruction process and social issues after the 1746 earthquake and tsunami in Peru: Past and present challenges after tsunami events. *InPost Tsunami Hazards, 44,* 97–109.

Mas, E., Adriano, B., Pulido, N., Jimenez, C., & Koshimura, S. (2014b). Simulation of tsunami inundation in Central Peru from future megathrust earthquake scenarios. *Journal of Disaster Research, 9*(6), 961–967.

Okada, Y. (1992). Internal deformation due to shear and tensile faults in a half space. *Bulletin of Seismological Society of America, 82*(2), 1018–1040.

Okal, E. (2009). Excitation of tsunamis by earthquakes, chapter V. In: E. Bernard, A. Robinson (Eds.), The sea: Tsunamis, volume 15 (pp. 137–177). London: Harvard University Press.

Papazachos, B., Scordilis, E., Panagiotopoulos, D., & Karakaisis, G. (2004). Global relations between seismic fault parameters and moment magnitude of Earthquakes. *Bulletin of the Geological Society of Greece, 36,* 1482–1489.

Silgado, E. (1978). Historia de los Sismos más notables ocurridos en el Perú (1513–1974). Boletín N° 3. Instituto de Geología y Minería. Lima, Perú.

(Received April 15, 2016, revised August 25, 2017, accepted August 29, 2017)

Pure Appl. Geophys.

https://doi.org/10.1007/s00024-018-1830-0

Pure and Applied Geophysics

Holocene Tsunamis in Avachinsky Bay, Kamchatka, Russia

Tatiana K. Pinegina,[1] Lilya I. Bazanova,[1] Egor A. Zelenin,[2] Joanne Bourgeois,[3] Andrey I. Kozhurin,[1,2] Igor P. Medvedev,[4] and Danil S. Vydrin[5]

Abstract—This article presents results of the study of tsunami deposits on the Avachinsky Bay coast, Kurile-Kamchatka island arc, NW Pacific. We used tephrochronology to assign ages to the tsunami deposits, to correlate them between excavations, and to restore paleo-shoreline positions. In addition to using established regional marker tephra, we establish a detailed tephrochronology for more local tephra from Avachinsky volcano. For the first time in this area, proximal to Kamchatka's primary population, we reconstruct the vertical runup and horizontal inundation for 33 tsunamis recorded over the past ~ 4200 years, 5 of which are historical events – 1737, 1792, 1841, 1923 (Feb) and 1952. The runup heights for all 33 tsunamis range from 1.9 to 5.7 m, and inundation distances from 40 to 460 m. The average recurrence for historical events is ~ 56 years and for the entire study period ~ 133 years. The obtained data makes it possible to calculate frequencies of tsunamis by size, using reconstructed runup and inundation, which is crucial for tsunami hazard assessment and long-term tsunami forecasting. Considering all available data on the distribution of historical and paleo-tsunami heights along eastern Kamchatka, we conclude that the southern part of the Kamchatka subduction zone generates stronger tsunamis than its northern part. The observed differences could be associated with variations in the relative velocity and/or coupling between the downgoing Pacific Plate and Kamchatka.

Key words: Kamchatka, subduction zone, Avachinsky Bay, Earthquake, tsunami deposits, tephrochronology, paleo-shoreline reconstruction, Avachinsky volcano.

Electronic supplementary material The online version of this article (https://doi.org/10.1007/s00024-018-1830-0) contains supplementary material, which is available to authorized users.

[1] Institute of Volcanology and Seismology FED RAS, 9 Piip Boulevard, Petropavlovsk-Kamchatsky 683006, Russia. E-mail: pinegtk@yandex.ru

[2] Geological Institute RAS, 7 Pyzhevsky Lane, Moscow, Russia.

[3] Earth and Space Sciences, University of Washington, Seattle, USA.

[4] P.P. Shirshov Institute of Oceanology RAS, 36 Nakhimovsky Prospekt, Moscow, Russia.

[5] Lomonosov Moscow State University, 1 Leninskie Gory, Moscow, Russia.

1. Introduction

For most coastal countries, historical tsunami records are quite short and are reliable only for the last several centuries. However, for establishing tsunami recurrence and intensity and for developing tsunami hazard zoning, a much longer catalog of tsunami data is required because catastrophic tsunamis occur infrequently, on a time scale of hundreds to a few thousand years (Bourgeois 2009). A reliable prediction of possible tsunami parameters at specific parts of the coast cannot rely on data from one or a few historical events.

At present, many scientists are studying prehistoric-tsunami (paleotsunami) deposits to restore the parameters of ancient tsunamis over thousands of years. Research groups have applied this approach since the 1990s, and it became widely used after the disastrous 2004 Indian Ocean tsunami. Methods applied, as in this paper, can produce detailed geological records of tsunamis back to about 6000 years ago, following the rapid early-to-middle Holocene rise in sea level (Woodroff and Horton 2005). Preservation of this record also depends on the presence of accumulative coastal settings and on the coastal tectonic regime.

In Russia, the Far Eastern (Pacific) coast is the most affected by tsunamis (Soloviev and Ferchev 1961; Soloviev 1972, 1978). The most dangerous tsunamigenic earthquakes in the Russian Far East occur along the Kurile-Kamchatka subduction zone, along the northwestern margin of the Pacific Ocean. The catalog of historical tsunamis for this region is short (first event 1737) and quite incomplete before the twentieth century (Zayakin and Luchinina 1987; Pinegina and Bazanova 2016). Even recent tsunamis (e.g., 1997 Kronotsky) have geographically limited data (Bourgeois and Pinegina 2018). Therefore, studies of tsunami deposits are a key source of reliable information about these hazardous events.

In this paper, we present new, detailed information about tsunami deposits for the past ~ 4200 years on the Avachinsky Bay coast, Kamchatka Peninsula (Fig. 1). To determine paleotsunami runups and inundations for different time intervals we apply corrections for vertical movement of the coast and for horizontal seaward progradation.

On most coasts along eastern Kamchatka, Holocene marine accumulative (wave-built) terraces as expressed in modern topography are typically not older than ~ 2000 years (Pinegina 2014), limiting paleotsunami data and analysis. However, partly because of intense search and study, we have established a > 4000-year record in the case herein. This study thus maximizes extension of the tsunami catalog for Kamchatka. This task is particularly important because the central coast of Avachinsky Bay (named Khalaktyrsky beach) is the recreational zone of Petropavlovsk-Kamchatsky city and adjacent settlements, and includes industrial facilities and agricultural land.

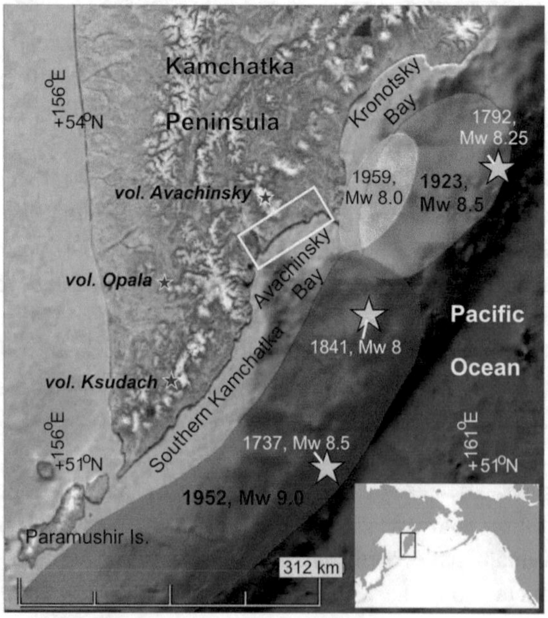

Figure 1
Location of local historical tsunamigenic earthquakes affecting the Avachinsky Bay coast (modified from Gusev 2004, 2006). Rupture areas for twentieth century earthquakes; pre-twentieth century earthquakes show estimated epicenters. See Fig. 2 for general map and features of Kamchatka. The white rectangle shows the position of Fig. 3

Avachinsky Bay is located on the southeast coast of Kamchatka Peninsula (Fig. 1). Petropavlovsk-Kamchatsky, the most populated city and the regional center (Fig. 2), is situated in its middle part, on the northern bank of Avachinsky Harbor.

2. *Review: Seismic and Volcanic Activity in the Avachinsky Bay Region from Historical Records and Previous Studies*

2.1. *Tsunamigenic Earthquakes*

By the beginning of the twentieth century, a world network of seismic stations was already capable of recording earthquakes of magnitude $M \geq 7.0$ from the Kamchatka region. The first seismic station in Kamchatka (in Petropavlovsk-Kamchatsky) was established in 1915. Since then two great earthquakes (04 Feb 1923 and 05 Nov 1952) along the Kamchatka subduction zone generated tsunamis that affected Avachinsky Bay (Fig. 1). Another large earthquake, on 05 Nov 1959 (Fig. 1) occurred at greater depth and caused a small tsunami of up to 2 m on the northern coast of Avachinsky Bay (Zayakin and Luchinina 1987; Gusev 2004). The 1960 Chilean tsunami did not exceed the active beach along Avachinsky Bay.

In 1959, the seismic station "Petropavlovsk" became involved in a tsunami warning system. The period of detailed seismological observation in Kamchatka and the Kurile Islands began in 1961. Since then, there have been several strong earthquakes accompanied by a tsunami: Ozernovskoe 1969, Kronotskoe 1997, and Central-Kuriles 2006 and 2007. However, on the coast of Avachinsky Bay, there were no noticeable tsunamis observed during this period—they were recorded only by the tide gauge in Petropavlovsk-Kamchatsky, and their amplitude did not exceed few cm. The information about parameters of significant historical tsunamis along Avachinsky Bay has been refined and significantly supplemented during the study of their deposits (Pinegina and Bazanova 2016; this study; Fig. 3).

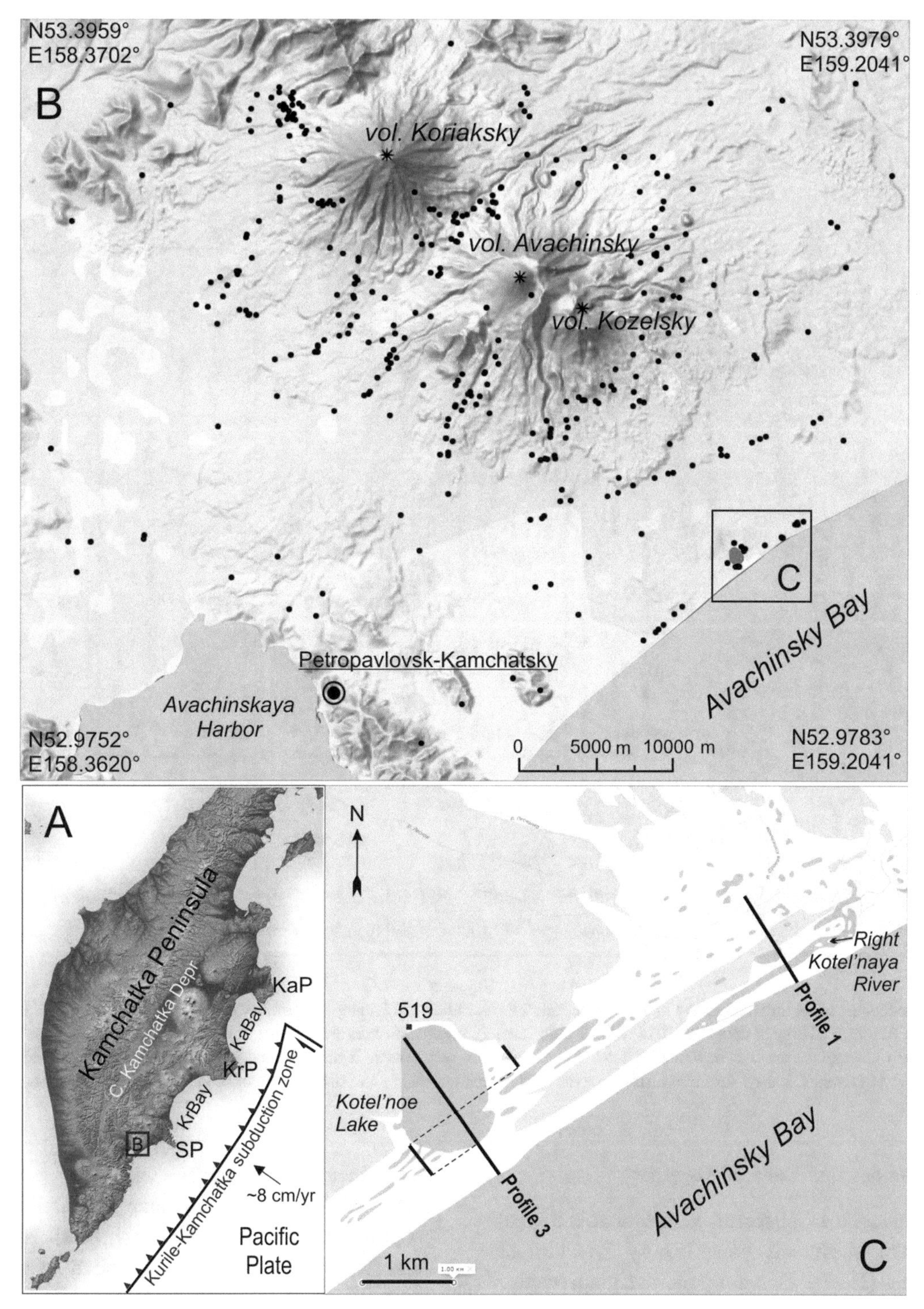

Figure 2
Field research area and setting: **a** General map of Kamchatka Peninsula with location of the field area (black box, **b**); sub-peninsulas and bays: *SP* Shipunsky Peninsula, *KrP* KrBay—Kronotsky Peninsula and Bay, *KaP* KaBay—Kamchatsky Peninsula and Bay; also Central Kamchatka Depression. **b** Observation points (black dots) in the vicinity of Avachinsky volcano where the geological sections of Holocene soil and peat were studied and sampled for tephrostratigraphy; **c** the site of detailed studies on the coast of Avachinsky Bay, locating profiles 1 and 3 and key peat section #519. Additional measurements and excavation descriptions were made along parallel profiles near profile 3, on either side of Kotel'noe Lake

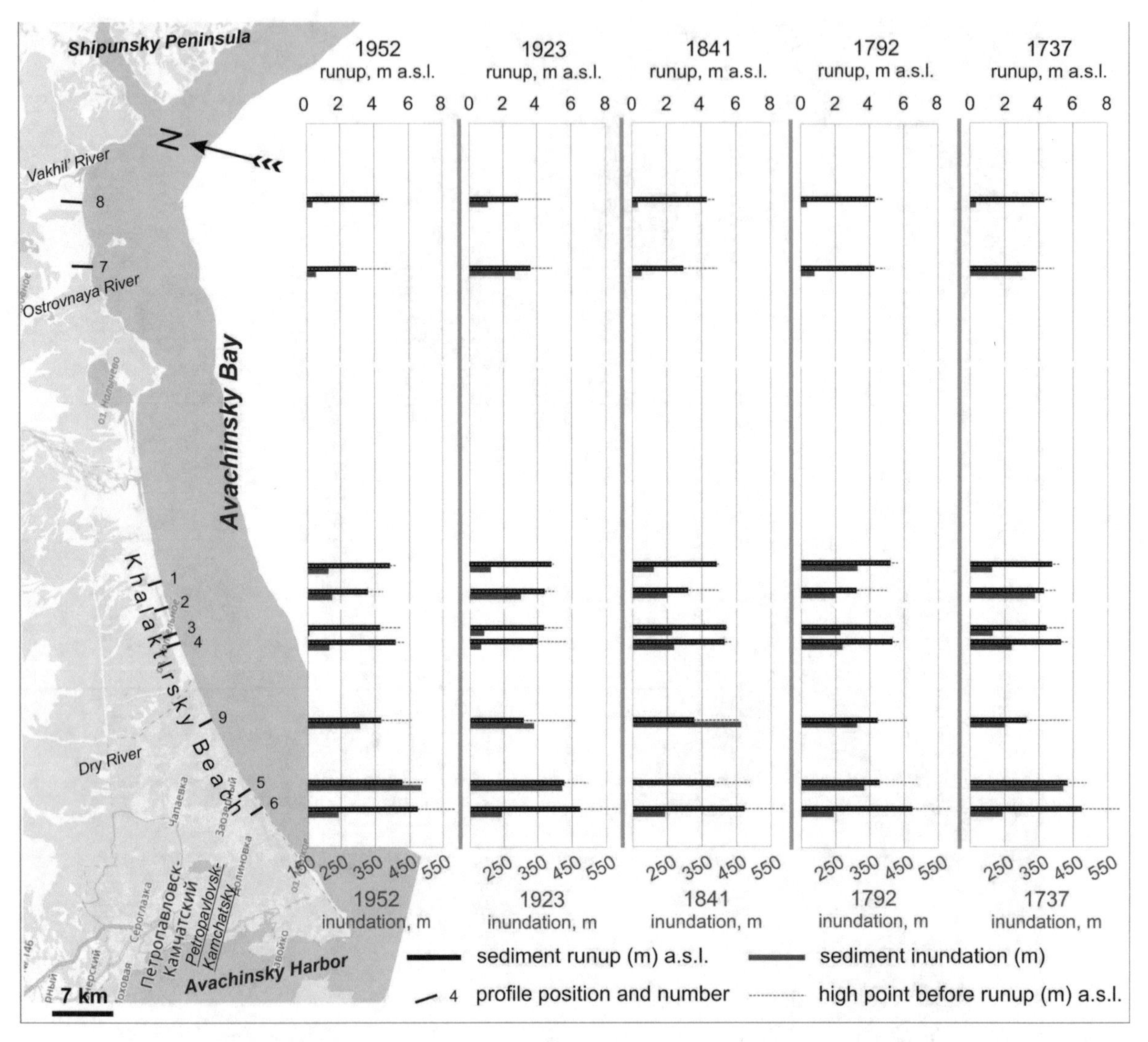

Figure 3
Tsunami-sediment runup, inundation, and high point before runup for historical tsunamis on the Avachinsky Bay coast (data in Pinegina and Bazanova 2016 with additional data from 2016 to 2017 field studies). Inundation axes start from 150 m for clarity of comparison between the events. The topographic map was downloaded from the cartographic web service Yandex Maps (https://yandex.ru/maps/). Modern coastal topography is used for these historical events; eighteenth century shoreline may have been 50 ± 25 m narrower

2.2. *Tsunamis and Tsunami Deposits*

A catalogue of historical tsunamis affecting the Kamchatka coast was compiled by Zayakin and Luchinina (1987), including pre-twentieth century tsunamigenic earthquakes in 1841, 1792 and 1737 (Figs. 1, 3). Strong Kuril-Kamchatka tsunamis have also been recorded all around the Pacific (NCEI database: https://www.ngdc.noaa.gov/hazard/hazards.shtml). The NCEI database indicates that Kamchatka tsunamigenic earthquakes with Mw $\geq$ 8.3 have caused dangerous transoceanic tsunamis. Due to geographical location, tsunamis from Kamchatka most strongly affect the Hawaiian Islands (see Table S1 in the Electronic Supplement). Far-field tsunami records supplement the limited Kamchatka data in the period of macroseismic observations.

The first investigation of tsunami deposits along the Avachinsky Bay coast was conducted in summer

2000. At that time, we described the deposits of 13 tsunamis in the last ~ 3500 years (Pinegina et al. 2002). At the time of this work the region's tephrostratigraphy was not studied in sufficient detail, so the age of some historical tsunamis was determined incorrectly. The paleotsunami history was also incomplete, and runup and inundation were presented without reconstructing the ancient shorelines. Newly updated data on historical tsunamis is presented in Pinegina and Bazanova (2016), and a review of Kamchatka paleotsunamis for the past 1500–2000 years is presented in Pinegina (2014).

2.3. *Ashfalls*

For stratigraphic markers in this study area, an area which we call "Kotel'ny", we have used three well-known tephras from large Holocene eruptions on southern Kamchatka (Braitseva et al. 1997a, b; Melekestsev et al. 1996) as well as additional, more local tephra from Avachinsky volcano. The geological sections in our field site include regional marker ashes from Ksudach volcano ($KSht_3$—AD 1907, KS_1—1800 ^{14}C BP) and Barany Amphitheatre crater, formed at the foot of Opala volcano (OP—1500 ^{14}C BP) (Fig. 1, Table S4). The field site, moreover, is located 22–23 km southeast of the active crater of Avachinsky volcano (Fig. 2), whose Holocene tephra have been repeatedly dispersed in the region.

The eruptive history of Avachinsky volcano for the past 12,000 years has been reconstructed by detailed geologic, stratigraphic and tephrochronological studies (Braitseva et al. 1998; Bazanova 2013). The list of known Avachinsky eruptions includes 156 events, recorded as deposits of ashfalls and pyroclastic flows, of which 118 tephra layers belong to the active Young Cone. Individual layers have been traced in sections along radial and circular profiles around the crater (Fig. 2b), with tephra identification confirmed by mineralogical and chemical analyses. Radiocarbon dating helped verify tephra correlations: 189 radiocarbon dates were used for stratigraphic and chronological reconstructions (Bazanova 2013 and earlier), of which 54 are reported herein (Table 1; Fig. 4). Isopach maps have been compiled for most Avachinsky fall units, and as a result its dispersal axes and areas have been determined (Bazanova et al. 2001, 2003, 2005).

In this study, at the Kotel'ny site, we take advantage of the long record of tephra from Avachinsky volcano in its several stages of eruptive history. The oldest parts of our sections generally include thin layers from the marginal zones of ashfall tephra associated with the so-called Andesite-stage eruptions (7250 – 3500 ^{14}C BP) (Braitseva et al. 1998). Juvenile rocks of this Andesite stage are mostly low-K andesite (white, yellow, light gray pumice, sometimes with dense gray fragments). Toward the end of the Andesite stage, two initial catastrophic eruptions of the active Young Cone of Avachinsky volcano occurred about 3500 and 3300 ^{14}C years BP and produced widespread fallout deposits (Bazanova et al. 2003), including at the Kotel'ny site.

Over the past 3800 years, the SSE oceanward sector of the volcano has remained one of the areas most exposed to ashfalls (Bazanova et al. 2001). Tephra units of the Young Cone are predominantly composed of juvenile black or dark gray and brown scoria of low- to medium-K basaltic andesite (Bazanova et al. 2003; Krasheninnikov et al. 2010), some with rare small lapilli of andesitic pumice. In historical time, Avachinsky ashfalls occurred at the Kotel'ny site in 1779, 1827, 1855 and 1945. The characteristics of these eruptions and the peculiarities of their tephra distribution on the coast of Avachinsky Bay have been considered in earlier publications (Melekestsev et al. 1994a, b; Pinegina and Bazanova 2016).

We have adopted a new system for indexing the eruptions of the Avachinsky volcano (as in Pinegina and Bazanova 2016). Early tephrochronological studies considered five paroxysmal eruptions and assigned them codes AV_1–AV_5 (Braitseva et al. 1997a, b). Later publications (Braitseva et al. 1998; Bazanova et al. 2003) proposed indexing style IAV_x for the Andesite-stage eruptions and $IIAV_x$ for the Young Cone stage, where the codes include the stage number and the sequence number (x) of the explosion (see Table S2). However, further work has enumerated about 150 eruptive events (this paper; Bazanova 2013) and revealed the flaws of using such indexing, which makes tephra identification, section description, text review and scientific discussion difficult, even for the authors. Therefore, for prehistoric eruptions of Avachinsky volcano, we index the

Table 1

Radiocarbon dates and ages of Avachinsky volcano eruptions

S/N	Tephra code[a]	Laboratory number[b, c]	^{14}C age (years before 1950 AD)	Average ^{14}C age[d]	Calibrated age AD/BC (probability 95.4%)[d]	Material for dating	Place of sampling
1	AV550	GIN-11369 (C)	540 ± 60	553 ± 46	AD 1299(1371)1440	Peat above the tephra	Kotel'noe Lake, the Pacific coast
		GIN-11370 (C)	570 ± 70			Peat under the tephra	Kotel'noe Lake, the Pacific coast
2	AV580	GIN-11370 (C)	570 ± 70	576 ± 46	AD 1296(1353)1427	Peat above the tephra	Kotel'noe Lake, the Pacific coast
		GIN-7804 (H′)	580 ± 60			Peat under the tephra	Kotel'noe Lake, the Pacific coast
3	AV750	GIN-8105 (H_1)	740 ± 110	732 ± 46	AD 1210(1272)1388	Soil above the tephra	Koriaksky volcano
		GIN-7803 (H′)	730 ± 50			Peat under the tephra	Kotel'noe Lake, the Pacific coast
4	AV800	GIN-7802 (H′)	800 ± 40		AD 1166(1233)1278	Peat under the tephra	Kotel'noe Lake, the Pacific coast
5	AV1000	GIN-13163 (C)	1020 ± 60	1006 ± 25	AD 985(1020)1147	Peat above the tephra	Zhupanov ridge, Shaibnaya River
		GIN-7801 (H′)	1000 ± 40			Peat above the tephra	Kotel'noe Lake, the Pacific coast
		GIN-7800 (H′)	1000 ± 40			Peat under the tephra	Kotel'noe Lake, the Pacific coast
		GIN-13164 (C)	1020 ± 70			Peat under the tephra	Zhupanov ridge, Shaibnaya River
6	AV1100	GIN-7800 (H′)	1000 ± 40	1081 ± 29	AD 894(965)1018	Peat above the tephra	Kotel'noe Lake, the Pacific coast
		GIN-7799 (H′)	1160 ± 40			Peat under the tephra	Kotel'noe Lake, the Pacific coast
7	AV1250	GIN-7798 (H′)	1250 ± 60	1257 ± 34	AD 670(738)868	Peat above the tephra	Kotel'noe Lake, the Pacific coast
		GIN-6934 (C)	1260 ± 40			Soil under the lahar deposit	Avachinsky volcano
8	AV1600[f]	IVAN-399 (H_1)	1620 ± 80	***1622 ± 45***	***AD 264(424)541***	Peat above the tephra	Petropavlovsk-Kamchatsky
		IVAN-593 (C)	1570 ± 110	1623 ± 44	AD 335(437)545	Peat under the tephra	Hot River, Nalichevo Valley
		GIN-7796 (H′)	1640 ± 60			Peat under the tephra	Kotel'noe Lake, the Pacific coast
9	AV2000	GIN-11375 (H_1)	1960 ± 40	1975 ± 34	BC 48(AD26)AD115	Peat above the tephra	Kotel'noe Lake, the Pacific coast
		GIN-11376 (C)	2010 ± 60			Peat under the tephra	Kotel'noe Lake, the Pacific coast

Table 1 *continued*

S/N	Tephra code[a]	Laboratory number[b, c]	^{14}C age (years before 1950 AD)	Average ^{14}C age[d]	Calibrated age AD/BC (probability 95.4%)[d]	Material for dating	Place of sampling
10	AV2400	GIN-11378 (H_1)	2380 ± 40	2388 ± 32	BC 729(464)396	Peat under the tephra	Kotel'noe Lake, the Pacific coast
		GIN-8516 (H_1)	2400 ± 50			Peat under the tephra	Left Avacha River
11	AV2450	GIN-11379 (H_1)	2460 ± 40		BC 761(603)415	Peat under the tephra	Kotel'noe Lake, the Pacific coast
12	AV2550	GIN-6931 (H_1)	2580 ± 60	2559 ± 34	BC 806(764)549	Soil above the tephra	Avachinsky volcano
		GIN-11381 (C)	2550 ± 40			Peat under the tephra	Kotel'noe Lake, the Pacific coast
13	AV2650	GIN-10639 (C)	2650 ± 70	2662 ± 46	BC 912(830)787	Tephra-containing peat	Vakhil' River
		GIN-6909 (C + H_1)	2670 ± 60			Tephra-containing soil	Avachinsky volcano
14	AV2700	GIN-7439 (C + H_1)	2720 ± 70	2711 ± 26	BC 906(858)811	Soil above the tephra	Avachinsky volcano
		IVAN-562	2710 ± 70			Charcoal from PFl[e] deposits	Avachinsky volcano
		GIN-11382 (H_1)	2710 ± 30			Peat under the tephra	Kotel'noe Lake, the Pacific coast
15	AV2800	GIN-11383 (H_1)	2790 ± 30		BC 1011(941)846	Peat under the tephra	Kotel'noe Lake, the Pacific coast
16	AV3100	GIN-11385 (C)	3100 ± 40		BC 1449(1354)1260	Peat under the tephra	Kotel'noe Lake, the Pacific coast
17	AV3300[g]	GIN-6929 (C)	***3300 ± 80***	***3279 ± 28***	***BC 1678(1523)1463***	Soil above the tephra	Avachinsky volcano
		GIN-6896	***3270 ± 40***	3276 ± 25	BC 1617(1558)1501	Charcoal from PFl deposits	Avachinsky volcano
		GIN-6897	***3280 ± 40***			Charcoal from PFl deposits	Avachinsky volcano
		GIN-11374 (H_1)	3260 ± 70			Peat under the tephra	Kotel'noe Lake, the Pacific coast

Table 1 *continued*

S/N	Tephra code[a]	Laboratory number[b, c]	^{14}C age (years before 1950 AD)	Average ^{14}C age[d]	Calibrated age AD/BC (probability 95.4%)[d]	Material for dating	Place of sampling
18	AV3500[g]	IVAN-708	***3440 ± 50***	***3510 ± 17***	***BC 1885(1877, 1840,***	Peat above the tephra	Sharomy vil., Kamchatka R.
		(C)	***3470 ± 120***	3508 ± 16	***1827, 1795, 1782) 1745***	Soil above the tephra	Uzon caldera
		IVAN-294	***3450 ± 40***		BC 1891(1824)1767	Wood buried by PFl deposits	Avachinsky volcano
		(C)	***3510 ± 50***			Wood buried by PFl deposits	Avachinsky volcano
		GIN-7129 (C)	***3460 ± 40***			Charcoal from PFl deposits Charcoal from PFl	Avachinsky volcano
		GIN-7128 (C)	***3510 ± 100***			deposits Charcoal from PFl deposits	Avachinsky volcano
		GIN-7134	***3570 ± 40***			Soil under PFl deposits	Avachinsky volcano
		IVAN-815	***3580 ± 70***			Soil under the tephra	Avachinsky volcano
		GIN-6056	***3560 ± 50***			Soil under the tephra	Avachinsky volcano
		GIN-7130 (C)	***3580 ± 90***			Peat under the tephra	Avachinsky volcano
		GIN-6361 (C)	***3570 ± 60***			Peat under the tephra	Kamchatka R., the Big Yar
		IVAN-843	3470 ± 70				Kotel'noe Lake, the Pacific coast
		(C + H_1)					
		IVAN-385					
		(H′)					
		GIN-11372					
		(C)					
19	AV3700	GIN-11371	3680 ± 70	3685 ± 68	BC 2284(2076)1892	Peat under the tephra	Kotel'noe Lake, the Pacific coast
		(C)	3750 ± 250			Soil under the tephra	Avachinsky volcano
		GIN-6917 (C)					
20	AV3800	GIN-6917 (C)	3750 ± 250	3811 ± 24	BC 2339(2247)2146	Soil above the tephra	Avachinsky volcano
		IVAN-561	3760 ± 60			Charcoal from PFl deposits	Avachinsky volcano
		IVAN-844	3750 ± 120			Soil under the tephra	Avachinsky volcano
		(C + H_1)	3800 ± 100			Soil under the tephra	Avachinsky volcano
		GIN-7440	3820 ± 30			Peat under the tephra	Zhupanovsky volcano, SE foot
		(C + H_1)	3860 ± 70			Soil under the tephra	Avachinsky volcano
		Le-7866					
		GIN-6916					

[a]Codes of Avachinsky volcano tephra include index AV and rounded average radiocarbon age (up to 50 years or up to 10 years for the intervals with a higher frequency of eruptions); for codes in previous publications see Table S2

[b]Radiocarbon dating from the bulk samples was carried out at the Geological Institute of the Russian Academy of Sciences, Moscow (GIN) and the Institute of Volcanology, Far-Eastern Branch of the Russian Academy of Sciences, Petropavlovsk-Kamchatsky (IVAN), and the Institute for the History of Material Culture of the Russian Academy of Sciences, Saint-Petersburg (Le)

[c]Alkaline extracts: C—cold (first); H_1—hot (second); H_2—hot (the third); H′—hot (in the case when only a hot extracts were used)

[d]Average radiocarbon and calibrated ages of tephra calculated with program OxCal v.4.24 (Bronk 2009) with the IntCal13 calibration curve (Reimer et al. 2013)

[e]PFl—pyroclastic flow

Bold italics—data published in: [f](Dirksen and Bazanova 2010); [g](Bazanova et al. 2003, 2016)

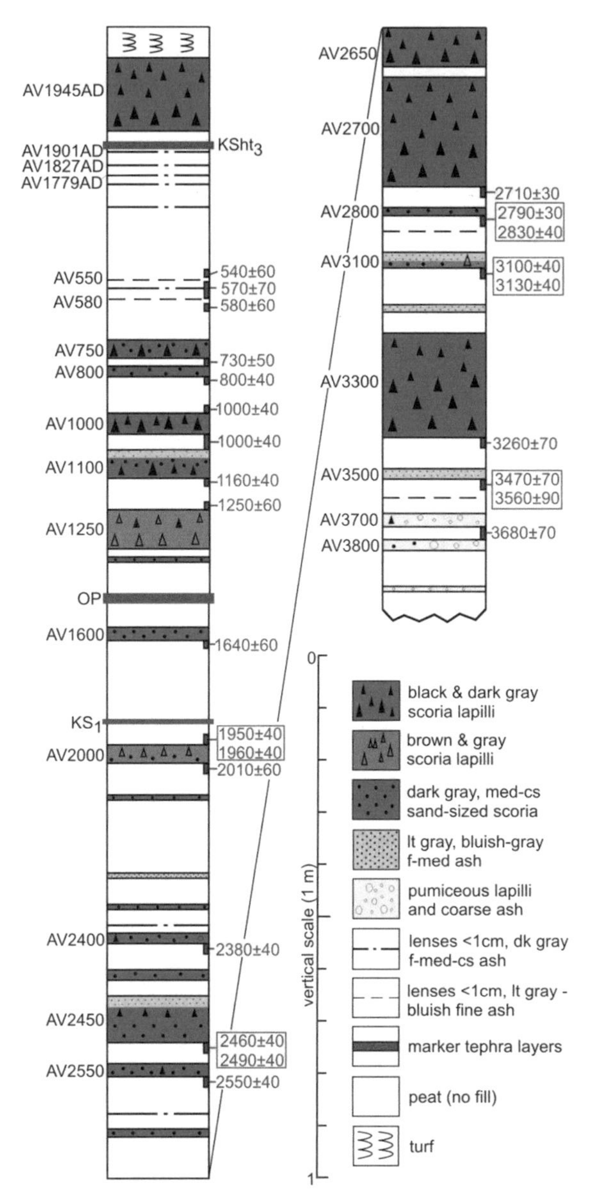

Figure 4
Key peat section near Kotel'noe Lake (site 519). Location of the section is shown in Fig. 2c and Fig. S2. Codes and ages of tephra layers as in Tables 1 and 2, and Tables S3 and S4. Radiocarbon dates are in red; dates on successive alkaline extractions from the same sample are shown in boxes (see Table 1)

tephra layers by volcano (AV) and a rounded ^{14}C age, which makes the tephra stratigraphy clearer. We summarize the new and old eruption codes in Table S2 to support comparison of this article to previous publications with different codes for the same eruptions.

3. Methods

To start, we interpreted satellite imagery to identify the oldest beach ridges preserved along the present coast of Avachinsky Bay. We used aerial imagery and declassified KH-9 Hexagon stereoscopic imagery (scanned films provided by the US Geological Survey) and imagery from cartographic web services (Google Earth, Yandex Maps, Bing Maps), where available, with resolution of ~ 4–6 m per pixel. We considered profiles 1 and 3, across the broadest accumulative marine terraces, as most promising for our studies because they preserve the oldest coastal accumulations (Fig. 2a, Fig. S1, Fig. S2). Older, lower beach ridges that have been partially buried by peat can be identified as linearly elongated vegetation "islands" clearly visible on multispectral images, even if they do not differ in height from the surrounding surface. The possibility to make excavations and coring operations in the wetlands depends on site accessibility and the level of groundwater, the latter elucidated by spectral signatures corresponding to different vegetation types. At times we used a pump to access levels under the water table. Outcrops along streamcuts through beach ridges are also a significant source of information, particularly where the channel is orthogonal to a beach ridge.

We measured nine topographic profiles perpendicular to the shoreline along ~ 70 km of Avachinsky Bay coastline (Fig. 3) using a Trimble M-3 total station with distance measurement accuracy of ± 3 mm/km and angular accuracy of ± 1″. The heights of all the points measured along profiles were corrected according to known heights from the Reference State Geodetic Network. The lengths of profiles 1 and 3 are about 1100 and 1500 m, respectively; profile 3 is a composite to include the area submerged by Kotel'noe Lake (Fig. 2c). The height of beach ridges along the profiles ranges from 3 to 5.5–6 m above sea level (a.s.l.). The width of the active beach in this area is about 70 m, with its upper limit (where vegetation becomes dense) at elevations of 4.5–5.0 m. The shape of the profiles is saw-toothed with a slight increase in elevation of the beach ridges towards the sea. Such a shape indicates a net relative-

Table 2

Reconstructed parameters of historical and prehistoric tsunamis on the Avachinsky Bay coast (along the profiles 1 and 3)

Tsunami sediments and tephras in the composite section	Tsunami parameters (*H*/*L*/*h*)						The ordinal number of prehistoric tsunamis for each profile (Fig. 7)		Composite numbering for two profiles (used for Figs. 8,9)
	Profile 1			Profile 3			Profile 1	Profile 3	
	H[a]	*L*[b]	*h*[c]	*H*	*L*	*h*			
Tsunami 1952[d]	5	210	4.7	4.3	150	5.5	Historical	Historical	1
AV1945AD									
Tsunami 1923[d]	5	210	4.7	4.4	190	5.5	Historical	Historical	2
$KSht_3$									
AV1855AD									
Tsunami 1841[d]	5	210	4.7	5.4	270	5.5	Historical	Historical	3
AV1827AD									
Tsunami 1792[d]	5.2	310	5.7	5.4	270	5.5	Historical	Historical	4
AV1779AD									
Tsunami 1737[d]	5	210	4.7	4.4	210	5.5	Historical	Historical	5
Tsunami ~ 1400–1700 AD[d]	3.5	460 ± 70	5.7	5	300 ± 50	5.7	1	1	6
AV550									
Tsunami	5.7	120 ± 20	5.5	4	150 ± 50	5.5	2	2	7
AV750									
Tsunami	5.7	120 ± 20	5.5				3		8
AV800									
Tsunami	5	140 ± 15	5.7	5.2	220 ± 20	5.7	4	3	9
Tsunami	1.9	285 ± 15	5.7				5		10
AV1000									
Tsunami				4.1	110 ± 20	5.4		4	11
AV1100									
Tsunami	3.9	315 ± 20	5.7	5	220 ± 20	5.7	6	5	12
AV1250									

Table 2

continued

Tsunami				4.4	150 ± 20	5.7		6	13
Tsunami	2.8	350 ± 20	5.7	5	220 ± 20	5.7	7	7	14
OP									
Tsunami				4.9	190 ± 20	5.7		8	15
AV1600									
Tsunami	3.7	315 ± 20	5.7	4.9	190 ± 20	5.7	8	9	16
KS_1									
Tsunami	3.4	175 ± 55	4.7 (5.2)	4.2	200 ± 40	5.7	9	10	17
AV2000									
Tsunami	3.4	175 ± 55	4.7 (5.2)	4.9	120 ± 40	5.7	10	11	18
AV2400									
Tsunami				4.2	200 ± 40	5.7		12	19
AV2450									
Tsunami	3.4	175 ± 55	4.7 (5.2)	4.9	130 ± 40	5.7	11	13	20
AV2550									
Tsunami				4.2	160 ± 80	5.7		14	21
Tsunami				4.2	160 ± 80	5.7		15	22
Tsunami				4.2	160 ± 80	5.7		16	23
AV2650									
Tsunami	2.5	365 ± 55	[e]				12		24
AV2700									
Tsunami	3.9 2.4	180 ± 55 670 ± 55	4.7 (5.2)[e]	4.2	120 ± 50	5.2	13	17	25
AV2800									
Tsunami	3.4	175 ± 55	4.7 (5.2)	4.4	560 ± 20	5.3[e]	14	18	26
Tsunami				4.4	560 ± 20	5.3[e]		19	27
AV3100									
Tsunami	3.2 2.4	265 ± 55 670 ± 55	4.7 (5.2)[e]	4.4	560 ± 20	5.3[e]	15	20	28
Tsunami	2.8	215 ± 55	4.7 (5.2)				16		29
AV3300									

Table 2

continued

Tsunami	2.9	425 ± 60	4.3	4.2	90 ± 90	5.3	17	21	30
AV3500									
Tsunami	3.2	196 ± 60	4.3[e]				18		31
	2.4	565 ± 60							
AV3700									
Tsunami				4.3	40 ± 40	4.7		22	32
AV3800									
Tsunami	2.8	365 ± 15	4.3				19		33

The left column contains only dated units of tephra. The dark gray fill of a cell means the presence of tephra in the excavations along the profile, light gray indicates the presence of tephra traces; unshaded implies the absence of tephra

[a]*H*, sediment runup elevation (m above sea level, corrected for paleotsunami cases

[b]*L*, sediment inundation limit (m from the shoreline, corrected for paleotsunami)

[c]*h*, maximum profile elevation between the shoreline and sediment inundation limit (m above sea level, corrected for paleotsunami)

[d]Published in Pinegina and Bazanova (2016)

[e]Tsunami deposits from river outcrop sections; the tsunami probably propagated up the Right Kotel'naya River

sea-level rise of no more than a few meters from the Late-Middle Holocene to the present.

Along profiles 1 and 3 we made 37 excavations to a depth of ~ 1 to ~ 3.5 m and described tephra layers and tsunami deposits (Figs. S1, S2). Excavations were made on or between almost every beach ridge; distances between excavations vary from 10–20 to 50–70 m. The depth of the pits depended on the thickness of soil mantling the relict (not active at present) beach ridges. Usually, we dug to a depth of 1–1.5 m below the soil-pyroclastic sequences into clean sand, which represents the ancient active beach. The upper age boundary of the time when a given beach ridge was no longer active is determined by the age of the oldest tephra at the base of the soil (as in Pinegina et al. 2013). Ages of tsunami deposits are bracketed by the ages of underlying and overlying tephras.

Peat sections (e.g., Fig. 4; Fig. S2) were studied in detail because peat's high rate of accumulation permits good preservation of even thin tephra and tsunami deposits. Moreover, peat can be used for radiocarbon dating. We cored the peat up to a depth of 4–5 m with a manual peat corer, and then the cores were described and sampled in detail.

3.1. Tephrochronology

Tephrochronological studies with radiocarbon dating provide comprehensive stratigraphic and chronological control for prehistoric events. This method has been successfully applied in Kamchatka, for example to determine the ages of landforms and of event deposits (many references). A primary goal of the Kotel'ny site study was to find and study a reference section with the maximum number of tephra layers. We succeeded in generating such a section from the peat bog on the NW side of Kotel'noe Lake (Site 519, Figs. 2c, 4).

The tephrostratigraphic framework in this reference section, where Avachinsky tephra layers predominate, goes back ~ 4200 years (i.e., the calibrated age BP of AV3800). The tephra layers were described by stratigraphic position and by appearance—thickness, grain size, grading and color of particles, lithic content, textural features, etc. They were identified/indexed by tracking and comparison with previously studied tephra in sections toward Kozelsky volcano (Fig. 2b). We collected peat samples for radiocarbon dating (Fig. 4), which we used for chronological reconstructions.

With this reference section, we made a layer-by-layer study of Holocene soil-pyroclastic sequences, including tsunami deposits, in coastal excavations along profiles 1 and 3 (Figs. S1, S2). The known and dated tephra horizons correlated from our reference section allowed us to trace tsunami deposits along and across beach ridges for mapping and for age estimation of each tsunami horizon.

3.2. Tsunami Deposits

In each excavation along a measured profile, we identified tsunami deposits, and described their thickness, grain size, stratification and grading; we took selected samples and photo-documented the section walls. Tsunami deposits in soil-pyroclastic sequences at the Kotel'ny site are generally thin sand layers (up to 20 cm), which tend to decrease landward in both layer thickness and grain size. The deposits consist mostly of the same material as the marine wave-built terrace and active beach—mainly dark, medium—to coarse-grained sand with some layers and patches of rounded gravel and pebbles. Once deposited, the tsunami layers are protected from erosion by dense coastal vegetation and are buried in the soil-pyroclastic sequence. If a tsunami leaves a deposit when there is snow cover on the coast, then as the snow melts, the deposit is lowered to the turf surface, where the deposit is protected from erosion by old and new vegetation (MacInnes et al. 2009a; Bourgeois and Pinegina 2018). We used standard criteria for tsunami-deposit identification along the Kuril-Kamchatka coastline, as described in previous papers (e.g., Pinegina et al. 2000; Bourgeois et al. 2006; Pinegina 2014). Here we will not describe these criteria in detail; we only note that the study sites are away from eolian, fluvial and storm influence.

It is possible to evaluate a minimum distance (sediment inundation, L) and a minimum height (sediment runup at inundation point, H) of past historical and prehistoric tsunamis from tsunami-deposit distribution (Fig. S3; also Bourgeois and Pinegina 2018). The most landward section with a given tsunami deposit approximates tsunami runup and inundation, limited by distance to the next landward section and also by the assumption that the tsunami transported sediment to its landward limit. To verify a pinch-out line, we try to control for non-preservation by examining more than one more section landward. The pinch-out elevation for a given tsunami deposit on a particular profile specifies a minimum wave height. If the tsunami crossed the beach-ridge plain orthogonally, the reconstructed tsunami height may also be bounded by the maximum elevation (h) of beach ridges between the

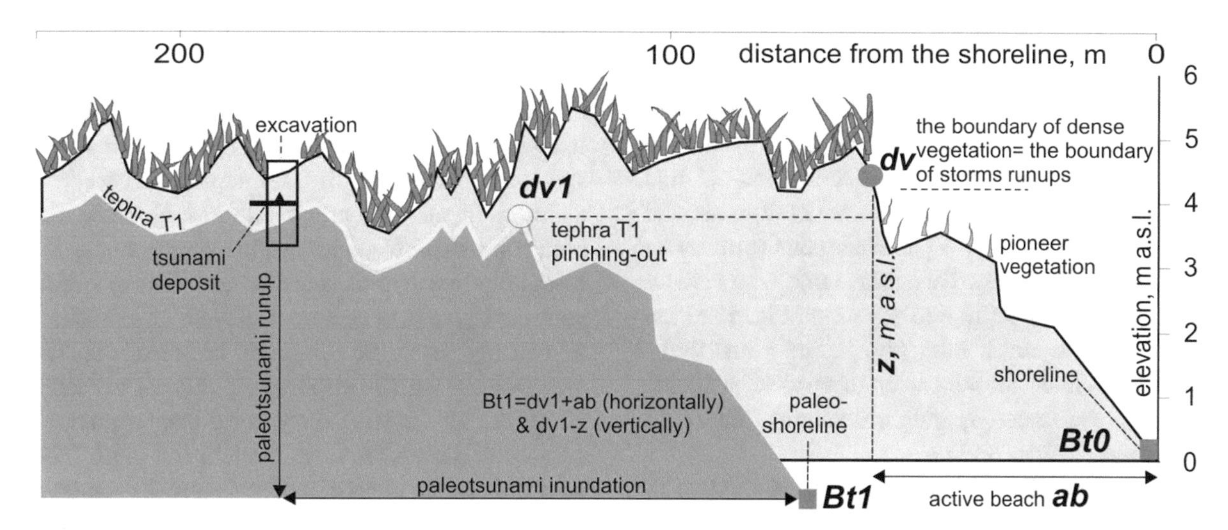

Figure 5

Reconstruction of a paleo-shoreline position (example of a fragment of profile 3 with a schematic tephra T1): dv beginning of dense vegetation (present day upper boundary of the active beach); dv1 upper boundary of the active beach at the time of T1 tephra fall; z elevation (m a.s.l.) of dv; ab distance between dv and modern shoreline Bt0 (blue square); Bt1 shoreline at the time of T1 tephra fall

shoreline and tsunami-deposit pinch-out (Fig. S3). Tsunami heights closer to the shoreline on a relatively flat plain will be higher than runup *H* in any case, to drive the tsunami to its landward limit *L* (Fig. S3).

Our calculation of tsunami runup and inundation may be somewhat underestimated because the tsunami flow can go farther landward than its sediments. Some recent tsunami studies have shown that actual (water) runup did not much exceed sediment runup in several cases (MacInnes et al. 2009a, b). However, studies of 2011 Tohoku tsunami deposits show that is some cases the difference between water runup and sediment runup can be significant (Goto et al. 2011; Abe et al. 2012).

3.3. *Estimation of Paleotsunami Size by Reconstructing Ancient Shoreline Position*

For accurate estimation of paleotsunami runup and inundation, it is necessary to reconstruct the paleoshoreline position at the time of tsunami propagation as well as subsequent changes in surface elevation. We use tephra stratigraphy and tephra mapping along measured topographic profiles to reconstruct paleoprofiles and paleoshoreline positions (as in Pinegina et al. 2013; Pinegina 2014; MacInnes et al. 2016).

Volcanic ash-falls accumulate and are preserved on a vegetated (marine-terrace) surface, but at the active beach, tephra will be washed away during the next storm. In a prehistoric case, the same holds true, so the landward limit of the active beach at the time of the ash-fall is the modern tephra pinch-out line (Fig. 5). For further calculations, we need to assume that storm conditions at a particular point (profile) on the coast have remained about the same, so a vertical shift from pinch-out line (e.g., dv1 on Fig. 5) to the modern active beach limit (dv) is due primarily to vertical tectonic movement or to eustatic sea-level change; the latter is minimal during the Late Holocene.

The calculation of the paleoposition relative to the modern shoreline (x; y = 0; 0) (Bt0 on Fig. 5) of any point on a profile at the time of a particular tephra fall comprises two steps. First, subtract the

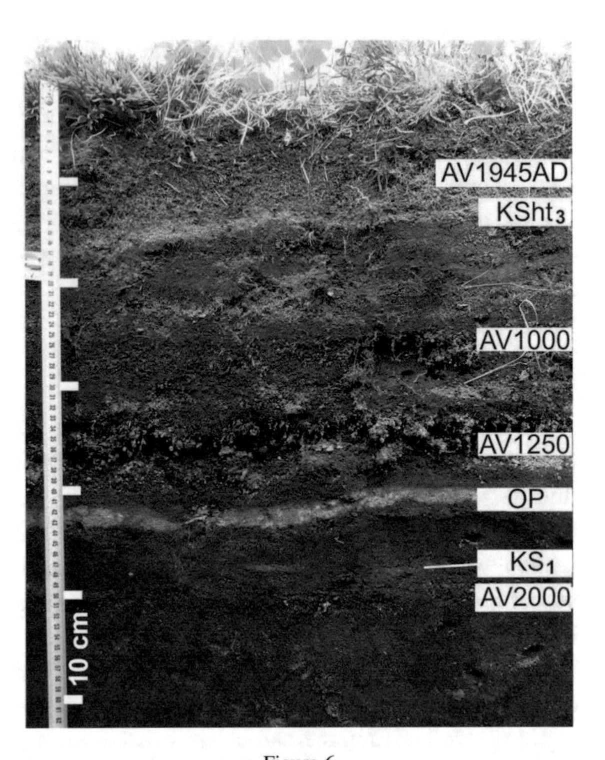

Figure 6
Section 727 (Figure S2, profile 3), where the regional marker tephras ($KSht_3$, OP, KS_1) are distinguished and easily recognized due to their bright color and individual features (Table S4). Also labeled are some of the most clearly visible (on this photo) horizons of Avachinsky tephra

(x; y) coordinates of tephra base at its pinch-out (dv1 on Fig. 5) from the (x; y) coordinates of the tephra base at a given excavation point (excavation in Fig. 5). Second, add the width of the modern active beach and the height (a.s.l.) of its upper boundary (dv on Fig. 5) to the coordinates calculated in the first step. If the point of pinch-out for a specific tephra (e.g., dv1 for tephra T1) was not exposed in an excavation, we would place this point on the profile between the seawardmost T1-containing excavation and the next seaward excavation without T1. The error of such an estimation is roughly half the distance between the two excavations, which in this study did not exceeded 60 m. We also correct for vertical displacement of paleoshorelines relative to modern sea level. That is, if the current elevation of the paleo-dv1 point is below the present dv, we add the difference (dv-dv1) between their elevations to the entire "old" (older than dv1) part of the profile.

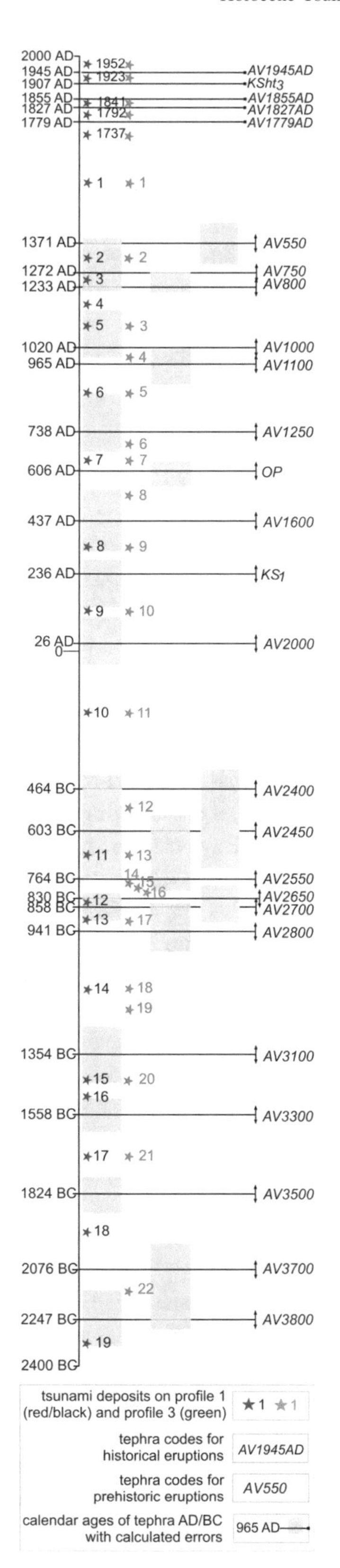

◀Figure 7
Composite chronological section from excavations along the Avachinsky Bay coast with positions of tephra and tsunami events. Tsunami are shown in the middle interval between two nearest tephras. Vertical axis indicates time in calendar years BC/AD. For tephra codes see Table 1 and Table S3. Tsunami numbers are same as in Table 2

4. Results and Discussion

During field surveys on the Avachinsky Bay coast, we excavated and described more than 100 geological sections on the Holocene accumulative marine terrace and on adjacent Late Pleistocene fluvioglacial and alluvial terraces (Fig. 2). Based on this work, we have created a tephrochronological framework, reconstructed the position of ancient shorelines, identified tsunami deposits and estimated tsunami recurrence rate and intensity.

4.1. Tephrochronology

We have obtained a reliable tephrochronological framework with radiocarbon dates for most of tephra horizons in our reference section, which has served as the basis for further stratigraphic and chronological comparisons. In the reference peat section (section 519, Fig. 2c) near Kotel'noe Lake, we identified 42 tephra layers from the past ~ 4200 years and generated radiocarbon dates for many of them (Fig. 4); some of these tephra, traced in other sectors of the volcano, were dated previously. Most (39) tephra are from Avachinsky volcano, including the four lowest, which belong to the final eruptions of the Andesite stage; the remaining 35 record activity of the Young Cone. We collected 26 ^{14}C samples from key peat section 519 and obtained calibrated ^{14}C ages (AD/BC) for most of the tephras (Fig. 4; Table 1, including prior dating). The even distribution of dates from the top of the peat section to the bottom, and the consistency of dates for each tephra unit, including dates from other sections (Table 1) support the reliability of the obtained dates. In addition to the reference peat section, we described 37 excavations and mapped tephra horizons along profiles 1 and 3 (Figs. S1, S2).

The strongest eruptions with a volume of pyroclastic material $\geq$ 0.1 km^3 are most clearly expressed at the Kotel'ny site and are reliably recognized; these include: AV1945AD, AV750, AV1000, AV1100, AV1600, AV2650, AV2700, AV3300. These fall units have distinct dispersal axes directed to the SE (most), SSE, S, and E (Bazanova et al. 2005). It is also easy to identify the tephra of some moderate eruptions (AV1250, AV2450). However, some thin tephra layers in the 519 peat section are hardly visible or were unrecognized in the soil-pyroclastic sequences along profiles 1 and 3. For detailed descriptions of the individual horizons of tephra in these sections see Table S3.

In the field area, tephras of the Avachinsky Young Cone stage are usually composed of black to brown scoria of various grain sizes, variable density, and almost uniform composition. The tephra from weak eruptions (e.g., AV550) or from the marginal ashfall zones of significant eruptions (e.g., AV3500) typically occur as thin layers or lenses of light gray and bluish-gray ash (Table S3).

Southern-Kamchatka-derived marker tephras from Ksudach volcano and the Barany Amphitheatre crater are consistently present in both peat and soil-pyroclastic sequences (Figs. S1, S2). They are easily distinguished by their light color in otherwise dark-brown sections (Fig. 6). The characteristics of these tephras are provided in Table S4.

4.2. *Historical Tsunami Deposits on the Coast of Avachinsky Bay*

Comparing available historical descriptions of tsunami heights on the Avachinsky Bay coast (Zayakin and Luchinina 1987) with sediment runups reconstructed in our studies, we conclude that they are roughly equal. The parameters for historical tsunamis on profiles 1–6 (Fig. 3) were determined earlier (Pinegina and Bazanova 2016); data for profiles 7–9 were obtained during fieldwork in 2016–2017 and are presented here for the first time. On the coast of Avachinsky Bay, we have identified deposits of five historical tsunamis (1737, 1792, 1841, 1923 and 1952). Their inundation did not exceed 480 m from the shoreline, with runup < 6.3 m a.s.l. (Fig. 3). On profiles 1 and 3, inundation of historical tsunamis did not exceed 310 m, with runup < 5.4 m a.s.l. (Table 2). The mean recurrence rate of these strong (> 4–5 m runup) tsunamis over almost 300 years is one event per 55–56 years.

Analysis of these historical earthquakes and tsunamis (Figs. 1, 3) shows that tsunami sources at Avachinsky Bay can be located on the subduction zone within or north or south of the bay. That is, earthquakes with Mw ~ 8–9 situated in south Kamchatka (e.g., 1952) and the southern part of Kronotsky Bay (e.g., 1923) may also generate dangerous tsunamis on the Avachinsky Bay coast.

4.3. *Paleotsunami*

Twenty-eight of the 33 tsunami deposits we have documented on profiles 1 and 3 are prehistoric (Table 2; Figs. S1, S2) and are shown as a composite section in Fig. 7. The oldest of these deposits lies just under the AV3800 tephra fall ($\sim$ 4250 calendar BP), and should be close to this age. Older tsunami deposits in the area are not preserved due to the absence of older marine terraces. We suppose that the main factors for this absence are the global Middle Holocene sea-level rise (Woodroff and Horton 2005) and an overall general net lowering of the central part of the Avachinsky Bay coast accomplished primarily by coseismic subsidence (Pinegina et al. 2015).

The runup heights for all tsunamis in profiles 1 and 3 (Table 2) vary from 1.9 to 5.7 m with a mean value of 4.1 m a.s.l. Tsunami inundation distances range from 40 to 460 m (most in $\sim$ 150–300 range), averaging 255 m, excluding tsunamis possibly propagated along the river (see Table 2 asterisk cases). The minimum highest point a tsunami had to exceed along a given topographic profile (h in Figure S3) ranges from 2.4 to 5.7 m with a mean of 5.15 m a.s.l. (Table 2).

From our data, we conclude that paleotsunami parameters for the past $\sim$ 4.2 millennia are comparable to historical ones. However, horizontal inundation in prehistoric cases may be underestimated if it occurred before erosional shoreline retreat following coseismic subsidence. We think this underestimation is generally not important because on the Avachinsky Bay coast, we have been able to identify only three events of coseismic subsidence in the past $\sim$ 4200 years (Pinegina et al. 2015), each event followed quickly by coastal progradation.

Because tsunami runup highly depends on coastal relief, we believe that horizontal inundation along lower, flatter profiles should be more informative for reconstructing tsunami intensity (strength, size). On steep slopes and cliffs close to the shoreline, the runup can be much greater, whereas on low-elevation, low-relief profiles, inundation will be greater. Moreover, even in the geographically limited case of our field area, the elevation of the coastal beach ridges in the southern part of the Khalaktirsky beach is slightly higher than on profiles 1 and 3. Respectively, the restored tsunami runups are also higher there (see Fig. 3). Thus, tsunami hazard analysis must take coastal topography into account, as well as both runup and inundation.

4.4. *Analysis of Data Loss Depending on the Age of Tsunami Deposits*

To evaluate the fidelity of long-term tsunami recurrence statistics, we need to estimate the quality of preservation of their deposits over a millennial time scale. This issue was considered in detail in a central Kuril Islands study (MacInnes et al. 2016). For our current study area, we tried to analyze tsunami-deposit preservation using various horizons of well-studied volcanic tephras as time benchmarks (Fig. 8).

A priori, we assume that the average recurrence interval of earthquakes and tsunamis on the millennial scale should be comparable for a given region because the seismic regime of the subduction zone varies on a longer time scale. Millennial intervals will

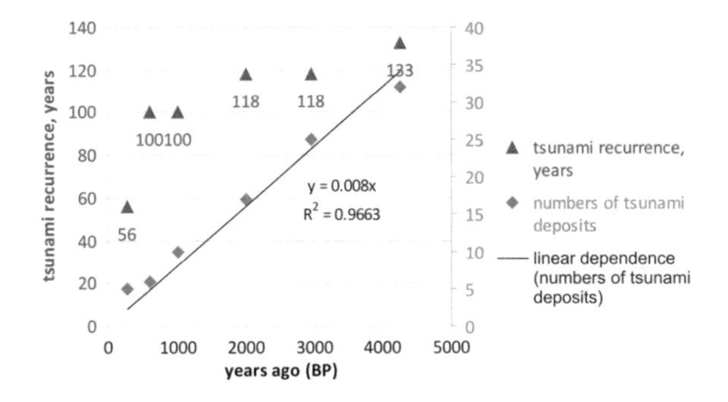

Figure 8
Analysis of tsunami deposit preservation and tsunami recurrence during the past ~ 4200 years along Avachinsky Bay. The time intervals are set by well-dated tephra

include seismic cycles of different events up to and including mega-earthquakes, such as AD 1952 and AD 1737 (Zayakin and Luchinina 1987). For our field site, the cumulative number of tsunami deposits for different time intervals (the past 4250, 2950, 2000, 1000, 600 and 279 years, based on calibrated tephra ages) is shown in Fig. 8. The cumulative number of deposits per time interval is well described by a linear relationship (R^2 close to 1). The average tsunami-deposit recurrence for the same time intervals (red numbers in Fig. 8) is more variable but for the last ~ 600, 1000, 2000 and 2950 years the average frequency of tsunamis was almost the same, at 1 event per 100–118 years. For the past 4250 years, the average tsunami-deposit frequency is reduced a little (1 event in 133 years), which could be due to poorer preservation, or to fewer sections studied. The shortest time interval (279 years, delimited by the 1737 tsunami deposit) includes only the historical period when tsunami deposits were emplaced every 56 years on average, twice as often as in other intervals. This pattern is common (e.g., MacInnes et al. 2016) and can be explained partly because tsunami chronology preceding the catalog tends to underestimate a number of events in cases of doubtful separation of tsunami deposits from each other in the sections, without a historical record to compare. Therefore, in evaluating tsunami frequency in the historical and prehistoric period, we postulate that we lose information for about half of the paleo-events. Because the largest events (such as 1952 and 1737) leave a more distinct and widespread deposit (Pinegina et al. 2003; MacInnes et al. 2010; Pinegina 2014) we would argue that the loss of paleoseismological information for weaker events is more typical, and that the geological traces of the largest tsunamis are typically well preserved.

4.5. *Analysis of Tsunami Recurrence Depending on Tsunami Intensity*

For tsunami hazard analysis, it is important to know not only general tsunami recurrence, but also the intensity (strength, size) of each individual tsunami, as well as the recurrence of different sizes of tsunamis. These data are necessary for long-term tsunami and earthquake prediction and for mapping

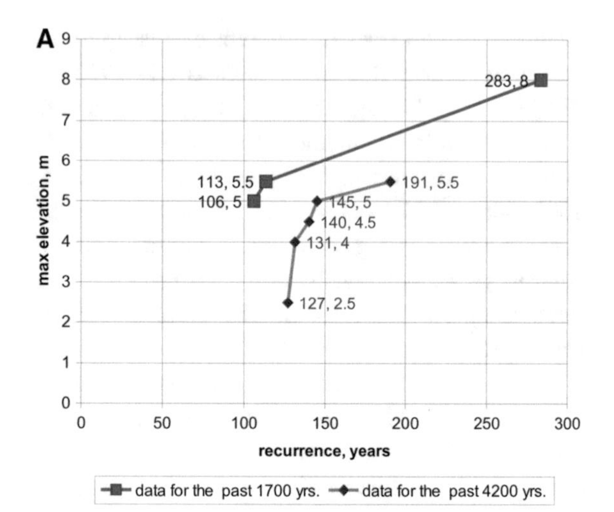

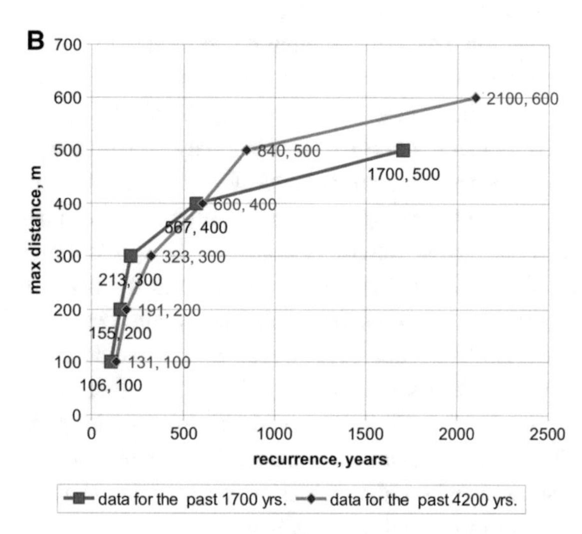

Figure 9
Cumulative graphs of tsunami recurrence for the Avachinsky Bay coast: **a** for exceeded elevations (sediment runup) and **b** for exceeded distances from shoreline (sediment inundation) based on all tsunami deposits present in sections for the past ~ 1700 years (Pinegina 2014) and for the past ~ 4200 years (this study)

tsunami hazard zones and making evacuation plans. We use cumulative graphs of tsunami-deposit runup and inundation distance versus tsunami-deposit recurrence (Fig. 9a, b). To generate higher statistical significance, we have used both newly obtained data and previously published data from 6 (of 9 on Fig. 3) profiles on the Avachinsky Bay coast for the past 1700 years (Pinegina 2014) and from 7 (of 9) profiles with historical events (Pinegina and Bazanova 2016), as well as profiles 1 and 3 (of 9) reported herein, for the past 4200 years.

The runup heights determined from 6 profiles for 1700 years (plus historical cases) are slightly higher than from the two profiles for 4200 years (Fig. 9a), likely because runup heights determined from tsunami deposits always depend on the actual heights existing on a coastal topographic profile. Thus, because profiles 1 and 3 are relatively low, the orange line on Fig. 9a is offset to the right, that is, it shows lower frequencies for tsunamis with runup greater than 5 m, even though in the same bay. Thus, we base our overall analysis on the more complete (though shorter time-scale) dataset (purple line). We conclude that tsunamis with a height of 5 m or more occur on the coast of Avachinsky Bay every 100 (106.5) years on average, while 8 m and higher events occur every 280 (283.6) years on average (purple line in Fig. 9a).

On the other hand, lower profiles permit longer inundation distances, which are shown by the shorter recurrence intervals of inundation of 500 m or more for the orange line in Fig. 9b. The scatter at shorter inundation distances and the differences between the profile sums (purple and orange) is likely not statistically significant. Tsunamis with an inundation distance $\geq$ 100 m occur every 100 years on average, and tsunamis with inundation $\geq$ 500 m occur every 840 years on lower profiles and every 1700 years, on average on higher profiles (Fig. 9b).

There is no direct relationship between graphs (a) and (b) in Fig. 9 for a number of reasons, including the coastal topography, as discussed above. Tsunami wavelength (and therefore, inundation) and tsunami height (and therefore, runup) are dependent on earthquake source parameters. Tsunami wavelength and wave period are largely set by source width, whereas initial tsunami height depends on source depth and slip amplitude (Cox and Machemehl 1986).

4.6. Discussion: Seismic Potential of the Kurile-Kamchatka Subduction Zone

Historical earthquakes causing significant tsunamis in Avachinsky Bay have included Mw > 8 events from the southern part of Kronotsky Bay (1923) to south Kamchatka (1737 and 1952) (Figs. 1, 3; Pinegina and Bazanova 2016). Inundation distances for all historical tsunamis on 9 topographic profiles varied from 180 to 480 m, and runup heights from 3.5 to 6.3 m. Along the northern coast of Avachinsky Bay tsunami intensity of the 1952 and 1737 mega-events (Mw ~ 9) was similar to the still-large but smaller events with magnitudes 8–8.5. Based on tsunami deposits preserved on the Avachinsky Bay coast, analysis of paleotsunami intensity over the past ~ 4200 years shows that prehistoric events were quite comparable to historical events (Table 2), including the great 1737 and 1952 cases.

In southern Kamchatka (south of Petropavlovsk-Kamchatsky) and the northern Kurile Islands, according to available geological and historical data, the tsunamis of 1737 and 1952 were significantly larger than in Avachinsky Bay, with runup heights of more than 15–20 m and inundation distances of over 1000 m (Krasheninnikov 1786; Zayakin and Luchinina 1987; MacInnes et al. 2010; Pinegina 2014). Moreover, investigations of paleotsunami deposits in this same region show that tsunami intensity (runup and inundation) for some paleo-events is comparable to 1952 and 1737 (Pinegina 2014). These data allow us to conclude that the southern segment of the Kamchatka subduction zone can generate stronger earthquakes and regionally larger tsunamis than from Avachinsky Bay northward, including Kamchatsky Bay (KaB on Fig. 2a; Bourgeois and Pinegina 2018). Avachinsky and Kamchatsky bays also have comparable recurrence intervals for runup and inundation (slightly shorter recurrence for long inundations in Kronotsky). A similar statistical analysis has yet to be applied to south Kamchatka data.

The possible causes of such distribution of seismicity do not yet have an accepted explanation, but the observed differences are almost certainly associated with variations of subduction-zone parameters along Kamchatka. For example, there is an apparent southward increase in coupling between the Pacific and Okhotsk plates along the subduction zone (Bürgmann et al. 2005). Bürgmann et al. (2005) believe that there is a clear correlation between/among currently locked segments of the subduction zone, locations of the strongest historical earthquake sources, positions of negative gravity anomalies, and seafloor relief. Their interpretation of satellite geodetic measurements data (GPS) suggests that in front of south Kamchatka there is a zone with the most significant plate coupling; therefore, large-scale slip occurs at times in this region, accompanied by earthquakes with Mw ~ 9 and their attendant tsunamis.

MacInnes et al. (2010) interpreted the presence of a high-slip region along southern Kamchatka based on modeling multi-segment displacement amplitudes for the source of the 1952 Kamchatka great earthquake and comparing those amplitudes to 1952 tsunami runup from historical records and tsunami-deposit data. They concluded that the tsunami runup was best explained by stronger deformation along south Kamchatka. Paleotsunami intensity along the south Kamchatka coast is also large (Pinegina 2014) and suggests millennial-scale consistency in subduction-zone behavior.

An oceanward shift of the northern edge of the subducted part of Pacific plate, or trench rollback, which begins at about the latitude of Shipunsky Peninsula (Lander and Shapiro 2007) may also explain weaker coupling between the plates along the northern subduction zone. The possibility of the so-called trench rollback has been demonstrated by physical and numerical modeling (Schellart et al. 2007; Stegman et al. 2006). The shift of the eastern part of the Central Kamchatka Depression toward the ocean (Kozhurin et al. 2008; Kozhurin and Zelenin 2017) also supports a hypothesis of slab retreat.

5. *Summary and Conclusion*

As a result of tephrostratigraphy studies and radiocarbon dating, we have constructed a composite tephrochronological section for the central part of Avachinsky Bay. This section covers the past ~ 4200 years and includes 28 horizons of tephra with determined ages. This combined chronological framework was the basis for reconstructing an age sequence for paleoseismic events imprinted in tsunami deposits. This part of our study has an independent importance as a set of reliable data for various other investigations. For example, our descriptions of sections allow estimating the frequency of tephra falls for volcanic hazard assessment.

Deposits of 33 tsunamis, including 5 of historical age (back to AD 1737), were identified along the Avachinsky Bay coast in excavations going back ~ 4200 years. For each we determined tsunami inundation distance and runup height using corrections for ancient shoreline positions. Runup heights from tsunami deposits are about 2–6 m, with typical inundation 150–450 m; some tsunamis left deposits at distances of more than 500–600 m, probably due to propagation upstream along a river. Clearly variations in these numbers are related not just to tsunami intensity but also to local topography.

The average tsunami recurrence in Avachinsky Bay is 56 years for the historical period (to AD 1737) and ~ 100–133 years for the past few millennia, showing a direct relationship between recurrence interval (in years) and tsunami size (in m, runup or inundation). From our data we conclude that about half the prehistoric events, mostly weaker ones, are missed in a chronology constructed from tsunami deposits, whereas deposits of the most significant tsunamis are expected to be preserved in the geological record. Obtained data show the important relationship between tsunami recurrence and runup, and tsunami recurrence and inundation. For example, tsunamis with runup $\geq$ 5 m occur along Avachinsky Bay once in 100 years on average, $\geq$ 8 m once in 280 years on average. Tsunamis with an inundation of $\geq$ 100 m occur every 100 years on average, $\geq$ 500 m every 840 years on average.

We conclude that the decrease in tsunami intensity along the Avachinsky Bay coast in comparison with south Kamchatka can be explained by different parameters of seismicity caused either by variations in plate coupling, or in plate convergence rate, or by both these factors. This conclusion is supported by the long-term consistency of historical and paleotsunami data at a given location along the subduction zone, and yet the variability from location to location from Kamchatsky Bay (Bourgeois and Pinegina 2018) to this study site (Avachinsky Bay) to southern Kamchatka (Pinegina 2014).

Judging from our paleoseismic analysis this field area compared to other studies along the subduction zone, the southern part of the Kamchatka subduction zone can generate stronger earthquakes and tsunamis than its northern part. This difference tends toward agreement of other analyses of the subduction zone but cannot distinguish a specific cause. The obtained data about tsunami history at Avachinsky Bay and other sites along the Kuril-Kamchatka subduction zone are important for the entire Pacific region. According to available data (tsunami catalog for the Hawaiian Islands), we can conclude that tsunamis from Kamchatka earthquakes with Mw $\geq$ 8.3 could be hazardous on remote Pacific coasts.

Acknowledgements

This work was supported by Grants from RFBR #15-05-02651, #18-05-00407 and 18-5-003 FEB RAS (to Pinegina). Expeditionary studies partly supported by Grants RFBR #16-05-00090, #15-05-05505, and RNF (project #14-50-00095).

Author contributions All authors participated in field work. TKP is primarily responsible for the tsunami-deposit analyses, radiocarbon calibration and writing the manuscript; LIB is primarily responsible for the tephrostratigraphy and writing the manuscript; EAZ made all geodetic measurements and calculations and helped prepare the manuscript; JB worked with all aspects of the manuscript.

Compliance with ethical standards

Conflict of interest The authors declare they have no conflicts of interest.

References

Abe, T., Goto, K., & Sugawara, D. (2012). Relationship between the maximum extent of tsunami sand and the inundation limit of the 2011 Tohoku-oki tsunami on the Sendai Plain, Japan. *Sedimentary Geology, 282,* 142–150.

Bazanova, L. I. (2013). 12,000-year of explosive activity at Avachinsky volcano. In Volcanism and related processes. Abstracts of the regional conference dedicate to Volcanologists Day (pp. 69–70). IVS FEB RAS, Petropavlovsk-Kamchatsky (**in Russian**).

Bazanova, L. I., Braitseva, O. A., Dirksen, O. V., Sulerzhitsky, L. D., & Danhara, T. (2005). Ashfalls from the largest Holocene Eruptions along the Ust'-Bol'sheretsk—Petropavlovsk-Kamchatsky traverse: sources, chronology, recurrence. *Journal of Volcanology and Seismology, 6,* 30–46. (**in Russian with English abstract**).

Bazanova, L. I., Braitseva, O. A., Melekestsev, I. V. & Puzankov, M. Y. (2001). Potential hazards from the Avachinsky volcano eruptions. In B. V. Ivanov (Ed.), *Geodynamics and volcanism of the Kurile-Kamchatka Island-Arc system* (pp. 390–407) (**in Russian with English abstract**).

Bazanova, L. I., Braitseva, O. A., Puzankov, M. Y., & Sulerzhitsky, L. D. (2003). Catastrophic plinian eruptions of the initial cone-building stage of the Young Cone of Avachinsky volcano (Kamchatka). *Vulkanologiya i Seismologiya (Journal of Volcanology and Seismology), 5,* 20–40. **(in Russian with English abstract)**.

Bazanova, L. I., Melekestsev, I. V., Ponomareva, V. V., Dirksen, O. V., & Dirksen, V. G. (2016). Late Pleistocene and Holocene volcanic catastrophes in Kamchatka and in the Kurile islands. Part 1. Types and classes of catastrophic eruptions as the leading components of volcanic catastrophism. *Journal of Volcanology and Seismology, 10*(3), 151–169.

Bourgeois, J. (2009). Geologic effects and records of tsunamis. In Robinson, A.R. & Bernard, E.N., eds., The Sea, Vol. 15, Tsunamis. Harvard University Press, p 53–91.

Bourgeois, J., & Pinegina, T. K. (2018). 1997 Kronotsky earthquake and tsunami and their predecessors, Kamchatka, Russia. *Natural Hazards and Earth System Sciences, 18,* 335–350.

Bourgeois, J., Pinegina, T., Ponomareva, V., & Zaretskaia, N. (2006). Holocene tsunamis in the southwestern Bering Sea, Russian Far East, and their tectonic implications. *Geological Society of America Bulletin, 118*(3/4), 449–463.

Braitseva, O. A., Bazanova, L. I., Melekestsev, I. V., & Sulerzhitsky, L. D. (1998). Largest Holocene eruptions of Avacha volcano, Kamchatka (7250–3700 ^{14}C years B.P.). *Volcanology and Seismology, 20*(1), 1–27.

Braitseva, O. A., Ponomareva, V. V., Sulerzhitsky, L. D., Melekestsev, I. V., & Bailey, J. (1997a). Holocene key-marker tephra layers in Kamchatka, Russia. *Quaternary Research, 47*(2), 125–139.

Braitseva, O. A., Sulerzhitsky, L. D., Ponomareva, V. V., & Melekestsev, I. V. (1997b). Geochronology of the greatest Holocene explosive eruptions in Kamchatka and their imprint on the Greenland glacier shield. *Transactions (Doklady) of the Russian Academy of Sciences. Earth Science Sections, 352*(1), 138–140.

Bronk, R. C. (2009). Bayesian analysis of radiocarbon dates. *Radiocarbon, 51*(1), 337–360.

Bürgmann, R., Kogan, M. G., Steblov, G. M., Hilley, G., Levin, V. E. & Apel, E. (2005). Interseismic coupling and asperity distribution along the Kamchatka subduction zone. *Journal of Geophysical Research: Solid Earth, 110*(B7). https://doi.org/10.1029/2005JB003648

Cox, J. C., & Machemehl, J. (1986). Overload bore propagation due to an overtopping wave. *Journal of Waterway, Port, Coastal, and Ocean Engineering, 112*(1), 161–163.

Dirksen, O. V., & Bazanova, L. I. (2010). An eruption of the veer cone as a volcanic event during the increase of volcanic activity at the beginning of the christian era. *Journal of Volcanology and Seismology, 4*(6), 378–384.

Goto, K., Chagué-Goff, C., Fujino, S., Goff, J., Jaffe, B., Nishimura, Y., et al. (2011). New insights of tsunami hazard from the 2011 Tohoku-oki event. *Marine Geology, 290*(1), 46–50.

Gusev, A. A. (2004). The schematic map of the source zones of large Kamchatka earthquakes of the instrumental epoch. In Gordeev E. I., Chebrov V. N. (Eds.), "*Complex seismological and geophysical researches of Kamchatka. To 25th Anniversary of Kamchatkan Experimental and Methodical Seismological Department*" (p. 445), Petropavlovsk-Kamchatsky (**in Russian**).

Gusev, A. A. (2006). Large earthquakes in Kamchatka: locations of epicentral zones for the instrumental period. *Journal of Volcanology and Seismology, 6,* 39–42. **(in Russian with English abstract)**.

Kozhurin, A. I., Ponomareva, V. V., & Pinegina, T. K. (2008). Active faulting in the south of Central Kamchatka. *Bulletin of Kamchatka Regional Association "Educational-Scientific Center". Earth Sciences, 12*(2), 10–27. **(in Russian with English abstract)**.

Kozhurin, A. I., & Zelenin, E. A. (2017). An extending island arc: the case of Kamchatka. *Tectonophysics, 706,* 91–102.

Krasheninnikov, S.P. (1786). Description of the land of Kamchatka: vol. 1 (p. 438). Imperial Academy of Sciences (**in Russian**).

Krasheninnikov, S., Portnyagin, M. & Bazanova, L. (2010). Chemical evolution of Avachinsky volcano (Kamchatka) during the Holocene. Geophysical Research Abstracts. European Geosciences Union 12, EGU2010-633-1.

Lander, A. V., & Shapiro, M. N. (2007). The Origin of the Modern Kamchatka Subduction Zone. In Eichelberger J. et al., eds., Volcanism and Subduction: The Kamchatka Region, 57–64.

MacInnes, B., Bourgeois, J., Pinegina, T. K., & Krchunovskaya, E. A. (2009a). Before and after: geomorphic change from the 15 November 2006 Kuril Island tsunami. *Geology, 37*(11), 995–998.

MacInnes, B., Krchunovskaya, E., Pinegina, T., & Bourgeois, J. (2016). Paleotsunamis from the central Kuril Islands segment of the Japan-Kuril-Kamchatka subduction zone. *Quaternary Research, 86*(1), 54–66.

MacInnes, B., Pinegina, T. K., Bourgeois, J., Razhigaeva, N. G., Kaistrenko, V. M., & Kravchunovskaya, E. A. (2009b). Field survey and geological effects of the 15 November 2006 Kuril tsunami in the Middle Kuril Islands. *Pure and Apply Geophysics, 166,* 9–36.

MacInnes, B. T., Weiss, R., Bourgeois, J., & Pinegina, T. K. (2010). Slip distribution of the 1952 Kamchatka great earthquake based on near-field tsunami deposits and historical records. *Bulletin of the Seismological Society of America, 100,* 1695–1709.

Melekestsev, I. V., Braitseva, O. A., Bazanova, L. I., Ponomareva, V. V., & Sulerzhitskiy, L. D. (1996). A particular type of catastrophic explosive eruptions with reference to the Holocene subcaldera eruptions at Khangar, Khodutka Maar, and Baraniy Amfiteatr volcanoes in Kamchatka. *Volcanology and Seismology, 18*(2), 135–160.

Melekestsev, I. V., Braitseva, O. A., Dvigalo, V. N., & Bazanova, L. I. (1994a). Historical eruptions of Avacha volcano, Kamchatka. Attempt of modern interpretation and classification for long-tern prediction of the types and parameters of future eruptions. Part 1 (1737–1909). *Volcanology and Seismology, 15*(6), 649–666.

Melekestsev, I. V., Braitseva, O. A., Dvigalo, V. N., & Bazanova, L. I. (1994b). Historical eruptions of Avacha Volcano, Kamchatka. Attempt of modern interpretation and classification for long-tern prediction of the types and parameters of future eruptions. Part 2 (1926–1991). *Volcanology and Seismology, 16*(2), 93–114.

National Centers for Environmental Information (NCEI, formerly NGDC), Natural Hazards Data, Images and Education, Tsunami and Earthquake databases [online]. https://www.ngdc.noaa.gov/hazard/hazards.shtml. Accessed 16 Mar 2018

Pinegina, T. K. (2014). Time-space distribution of tsunamigenic earthquakes along the pacific and bering coasts of Kamchatka: insight from paleotsunami deposits (p. 235). Doctor of

Geological Science dissertation, Institute of Oceanology RAS, Moscow (**in Russian**).

Pinegina, T. K., & Bazanova, L. I. (2016). New data on characteristic of historical tsunami on the coast of Avacha Bay (Kamchatka). *Bulletin of Kamchatka Regional Association "Educational-scientific center". Earth Sciences, 31*(3), 5–17. (**in Russian with English abstract**).

Pinegina, T. K., Bazanova, L. I., Melekestsev, I. V., Braitseva, O. A., Storcheus, A. V., & Gusyakov, V. K. (2000). Prehistorical tsunamis in Kronotskii Gulf, Kamchatka, Russia: a progress report. *Volcanology and Seismology, 22*(2), 213–226.

Pinegina, T. K., Bazanova, L. I., Zelenin, E. A. & Kozhurin, A. I. (2015). Identification of Holocene mega-earthquakes along the Kurile-Kamchatka subduction zone. In Chebrov, V. N. (Ed.), *Proceedings of 5th science and technology conference "Problems of complex geophysical monitoring of the Russian Far East"* (pp. 373–377), Petropavlovsk-Kamchatsky, Sept. 27–Oct. 3. Obninsk: GS RAS (**in Russian**).

Pinegina, T. K., Bourgeois, J., Bazanova, L. I., Braitseva, O. A. & Egorov, Y. O. (2002). Tsunami and analysis of tsunami risk at Khalatyrka beach region of Petropavlovsk-Kamchatsky, pacific coast of Kamchatka. In *Proceedings of the International workshop "Local tsunami warning and mitigation"* (pp. 122–131), Petropavlovsk-Kamchatsky, September 10–15, Moscow, Yanus-K.

Pinegina, T. K., Bourgeois, J., Bazanova, L. I., Melekestsev, I. V., & Braitseva, O. A. (2003). Millennial—scale record of Holocene tsunamis on the Kronotskiy Bay coast, Kamchatka, Russia. *Quaternary Research, 59,* 36–47.

Pinegina, T. K., Bourgeois, J., Kravchunovskaya, E. A., Lander, A. V., Arcos, M. E. M., Pedoja, K., et al. (2013). A nexus of plate interaction: segmented vertical movement of Kamchatsky Peninsula (Kamchatka) based on Holocene aggradational marine terraces. *The Geological Society of America Bulletin, 125*(9/10), 1554–1568.

Reimer, P. J., Bard, E., Bayliss, A., Beck, J. W., Blackwell, P. G., Bronk, R. C., et al. (2013). IntCal13 and Marine13 Radiocarbon age calibration Curves 0–50,000 years cal BP. *Radiocarbon, 55*(4), 1869–1888.

Schellart, W. P., Freeman, J., Stegman, D. R., Moresi, L., & May, D. (2007). Evolution and diversity of subduction zones controlled by slab width. *Nature, 446,* 308–311.

Soloviev, S. L. (1972). Recurrence of earthquakes and tsunamis in the Pacific Ocean. In *Proceedings of SakhKNII, 29. The waves of the tsunami* (pp. 7–47). Yuzhno-Sakhalinsk (**in Russian**).

Soloviev, S. L. (1978). Basic data on the tsunami on the Pacific coast of the USSR, 1737-1976. In Nayka, M. (Ed.), *The study of tsunamis in the open ocean* (pp. 61–136) (**in Russian**).

Soloviev, S. L. & Ferchev, M. D. (1961). A summary of data on the tsunami in the USSR. In *Bull. Council on Seismology, 9* (pp. 23–55). USSR Academy of Sciences (**in Russian**).

Stegman, D. R., Freeman, J., Schellart, W. P., Moresi, L., & May, D. (2006). Influence of trench width on subduction hinge retreat rates in 3-D models of slab rollback. *Geochemistry Geophysics Geosystems, 7*(3), 1–22.

Woodroff, S. A., & Horton, B. P. (2005). Holocene sea-level changes in the Indo-Pacific. *Journal of Asian Earth Sciences, 25*(1), 29–43.

Zayakin, Y. A. & Luchinina, A. A. (1987). Catalogue tsunamis on Kamchatka. In (p. 51). Obninsk: Vniigmi-Mtsd (**Booklet in Russian**).

(Received February 21, 2018, revised March 7, 2018, accepted March 8, 2018)

Pure Appl. Geophys.

https://doi.org/10.1007/s00024-018-1840-y

Pure and Applied Geophysics

Historical Tsunami Records on Russian Island, the Sea of Japan

N. G. Razjigaeva,[1] L. A. Ganzey,[1] T. A. Grebennikova,[1] Kh. A. Arslanov,[2] E. D. Ivanova,[1] K. S. Ganzey,[1] and A. A. Kharlamov[3]

Abstract—In this article, we provide data evidencing tsunamis on Russian Island over the last 700 years. Reconstructions are developed based on the analyses of peat bog sections on the coast of Spokoynaya Bay, including layers of tsunami sands. Ancient beach sands under peat were deposited during the final phase of transgression of the Medieval Warm Period. We used data on diatoms and benthic foraminifers to identify the marine origin of the sands. The grain size compositions of the tsunami deposits were used to determine the sources of material carried by the tsunamis. The chronology of historical tsunamis was determined based on the radiocarbon dating of the underlying organic deposits. There was a stated difference between the deposition environments during tsunamis and large storms during the Goni (2015) and Lionrock (2016) typhoons. Tsunami deposits from 1983 and 1993 were found in the upper part of the sections. The inundation of the 1993 tsunami did not exceed 20 m or a height of 0.5 m a.m.s.l. (0.3 above high tide). The more intensive tsunami of 1983 had a run-up of 0.65 m a.m.s.l. and penetrated inland from the shoreline up to 40 m. Sand layer of tsunami 1940 extend in land up to ~ 50 m from the present shoreline. Evidence of six tsunamis was elicited from the peat bog sections, the deposits of which are located 60 m from the modern coastal line. The deposits of strong historic tsunamis in the Japan Sea region in 1833, 1741, 1614 (or 1644), 1448, the XIV–XV century and 1341 were also identified on Russian Island. Their run-ups and inundation distances were also determined. The strong historic tsunamis appeared to be more intensive than those of the XX century, and considering the sea level drop during the Little Ice Age, the inundation distances were as large as 250 m.

Key words: Historical Tsunami, deposits, radiocarbon dating, Little Ice Age, Russian Far East, Sea of Japan.

[1] Pacific Geographical Institute, Far East Branch of the Russian Academy of Science, FEB RAS, Radio Street, 7, 690041 Vladivostok, Russia. E-mail: nadyar@tig.dvo.ru

[2] St.-Petersburg State University, St. Petersburg, Russia. E-mail: arslanovkh@mail.ru

[3] Shirshov' Institute of Oceanology RAS, Moscow, Russia. E-mail: harl51@mail.ru

1. Introduction

Tsunamis are among the peculiar and dangerous phenomena found on the coast of the Sea of Japan. Only large tsunamis with epicenters in the Pacific Ocean penetrate to the Sea of Japan, examples of which include Kamchatka (1952), Chile (1960), and Tohoku (2011) (Kurkin et al. 2004; Kaistrenko et al. 2011). In Primorye, tsunami 2011 was recorded by coastal tide gauges in Nakhodka (maximal height 15.4 cm), Preobrazhenie (21.5 cm) and Rudnaya Pristan (18.2 cm) (Kaistrenko et al. 2011; Shevchenko et al. 2014). However, the strongest tsunamis are usually caused by earthquakes with epicenters that are located on the Sea of Japan along a narrow line on the shelf area along the Japanese Islands. The seismicity pattern implies that the relative plate motion in the eastern margin of the Sea of Japan is primarily accommodated by large earthquakes with characteristic magnitudes of 7.5–7.7 and hypocenter depths to 35 km. The deformation and earthquakes have been attributed to the formation of an incipient subduction zone (Kobayashi 1983; Nakamura 1983). That zone is considered to be a boundary between the Eurasian (Amurean) and North American (Okhotsk) Plates, where strong tsunamigenic earthquakes have occurred periodically (Fukao and Furumoto 1975; Seno et al. 1996; Wei and Seno 1998). Wei and Seno (1998) interpreted the margin as a convergent plate boundary lying between the Amurian and Okhotsk Plates, along which the estimated convergence rates are 15 mm/year in the south and 8 mm/year in the north. Both east- and west-dipping active faults are known in the eastern margin (Tamaki and Honza 1985). The Japanese Islands, with the exception of eastern Hokkaido, are located on the eastern margin of the Eurasian Plate, under which both the Pacific and Philippine Sea Plates are subducted, and the relative motions of these three tectonic plates are believed to trigger large earthquakes in the region. The Philippine Sea Plate moves northwestward and is subducted beneath Southwest (SW) Japan along the Nankai Trough at a velocity of 4 cm/year (Seno et al. 1993), and the Pacific Plate moves to the west at a velocity of 9–10 cm/year (DeMets et al. 2010) and is subducted beneath NE Japan along the Japan Trench. The Pacific Plate is also subducted under the Philippine Sea Plate along the Izu–Ogasawara Trench. The E–W contraction causes a number of large inland crustal earthquakes, particularly along the back-arc region far from the plate subduction boundary (i.e., the Japan Trench). Takahashi (2017) suggested that the cause of inland crustal earthquakes along the Japan Sea side of NE Japan is the Philippine Sea Plate, not the Pacific Plate.

The predominant mechanism of these earthquakes is upthrow faults along the steeply dipping planes of faults (Gusiakov 2016). Historical tsunamis that reached the shores of Primorye are weaker than in Japan. Since 1940, only 5 notable tsunamis have been noted on the Primorye coast from 17 events that occurred within the basin of the Sea of Japan (Kurkin et al. 2004). Among those events, two strong tsunamis caused by underwater earthquakes (M 7.7 and 7.8) occurred in the northern Sea of Japan on May 26, 1983, and July 12, 1993. Their epicenters were located 30 km from the northwestern shore of Honshu and to the west and southwest of Hokkaido near Okushiri Island (Satake 1986; Tsuji et al. 1994; Ohtake 1995; Satake and Tanioka 1995; Tanioka et al. 1995). These tsunamis reached the coast of Primorye and were the largest of those historically recorded (Go et al. 1985; Gorbunova et al. 1997; Polyakova 2012). The tsunami of August 1, 1940 caused by underwater earthquakes (M 7.5) did not leave much evidence (Go et al. 1972; Soloviev and Go 1974).

Due to the short period of the tsunami observation, data on tsunami recurrence and the extent of their influence on shorelines are not sufficiently reliable. The only way to gain more information on large tsunamis from the past is to use geological methods to find signs of tsunamis manifested as sand layers in coastal sections. Studying tsunami deposits aids in estimating their extents and frequency, which are crucial for deciding whether they pose a danger to territories. Our first results on the eastern Primorye coasts were obtained from studies of the areas over which tsunami deposits are distributed and their chronological frameworks over the last 2.5 thousand years (Razjigaeva et al. 2014; Ganzey et al. 2015). The distribution, preservation and composition of the deposits of the 1983 and 1993 tsunamis have been studied in the eastern Primorye bays, which have different geomorphological structures (Ganzey et al. 2017).

The goal of this research is to reconstruct the historical tsunami occurrence times on the shores of Russian Island, which is included in the area of the city of Vladivostok. After being connected with the Muraviov-Amursky Peninsula by a bridge after APEC Summit 2012, Russian Island has continued to experience progressive development (Ganzei 2016; Baklanov et al. 2017). A significant number of city visitors travel to the beaches of the touristic seaside area, and tent camps appear in summer. In that regard, we must be aware of dangerous processes at the seaside.

2. *Data from Previous Observations of the Historical 1983 and 1993 Tsunamis on Russian Island and Near Vladivostok*

Immediately after the 1983 tsunami, observations were undertaken at the eastern part of Russian Island, where the run-up ranged from 0.5 (Paris Bay) to 1.5 m (Ayax and Zhidkov bays) and the inundation distance reached 36 m (Polyakova 2012). The tsunami began with a drop in sea level, low level lasted for 20 min, after which the sea level began to rise rapidly. The tsunami was manifested as a rapid tide. Eyewitnesses observed a strong current, which was especially noticeable between the sections of the floating pier, where the water was churning, as in a mountain river. The high level lasted for 10–15 min, after which the return flow began. Fluctuations in sea level lasted for approximately 1–2 h (Polyakova 2012). Within Vladivostok, the run-up heights varied from 0.66 m (Golden Horn Bay, mareographic station) to 3–4 m (Gornostai Bay), where the inundation distance reached 250–300 m, and the maximal run-up of 5–6 m was recorded at Tikhaya Bay (Go et al. 1985; Gorbunova et al. 1987; Polyakova 2012). The tsunami in 1983 in Vladivostok and nearby bays was more intense than the 1993 tsunami. No research was conducted on Russian Island after the 1993 tsunami. Data were obtained only in some particular coastal parts in Vladivostok, where the run-up height reached 0.3–0.9 m. On the coast of Ussuri Bay (Sobol and Gornostai bays), the run-up heights were 1.3–1.5 m, the maximum recorded level of which was 2.2 m in Chumaki Bay and on Engelm Cape, and the inundation distances were less than 10–15 m (Go et al. 1985; Gorbunova et al. 1987). During the observations on the coast immediately after the tsunami, its deposits were not studied.

3. *Materials and Methods*

The investigation of tsunami deposits on the eastern and southern parts of Russian Island was conducted in 2015–2016 (Fig. 1). The main research objects were the wetlands located beyond the areas of storm influence. Sections sampled by hand drilling and using a testing geoslicer were described from the shoreline inland. At some sites, pits were dug. The hypsometric positions of the sections were determined by line leveling. Corrections for tides were made. A summary of the tide gauge observations at the Vladivostok site, which is the site closest to Russian Island, is presented in a monograph (Mean Sea Level 2003). Irregular half-day tides are typical near Vladivostok, mean tidal amplitude is 0.2 m, and mean tropical tide range 0.26–0.32 m (http://esimo.oceanography.ru/tides/index.php?endsea=9&station1=5; Tidal Phenomena 2003).

In all the sections, we sampled sand layers to identify their origins. Peat layers were sampled at key sites to define the paleoecological status during the tsunamis. Samples from the intertidal zone, beach and storm ridges were used as potential material sources. All samples were stored in wet conditions. Grain size, diatom and foraminiferal analyses were among the analyses conducted on the deposits. To study grain size composition, the samples were sieved using sieves with γ steps and a weighed using a high-precision Sartorius balance.

The diatom analysis was carried out using standard techniques (The diatoms 1974). The samples were treated with a solution of hydrogen peroxide and washed with distilled water. For certain diatom preparations, a heavy liquid (mixture of H_2O:CoJ_2: KI = 1:1.5:2.25, density 2.4 g/sm^3) was used. Permanent preparations were made on 18x18 mm cover glasses and Elyashev aniline–formaldehyde resin with refraction indices n = 1.66–1.68. Diatoms were identified and counted under an oil immersion light microscope (Axioskop-Carl-Zeiss) at 1200× magnification. When possible, 200–300 frustules were counted per sample. Fragments containing more than half of a valve were included in the counts. The taxonomic compositions of the diatoms and their ecological characteristics were identified following Krammer and Lange-Bertalot (1988), Diatom (1950) and The diatoms (1992). The percentages of individual taxa were calculated from the total number of taxa. The ecological data given by the abovementioned authors enabled most of the taxa to be grouped accordingly based on their habitat, salinity, pH, and biogeography.

The foraminifera were studied under a binocular microscope at 100× magnification after the deposits were washed through a 0.063 mm sieve and dried immediately before identification. Because concentrations of foraminifera were low, the percentages were not calculated. The following references were used to identify the ecologies of the species: Foraminifera (1979), Preobrazhenskaya and Troitskaya (1996) and Sen Gupta (2002).

To elicit the tsunami ages, we performed the radiocarbon dating of the underlying peat at St. Petersburg State University (Table 1). The peat samples were treated with 2% HCl and NaOH to remove carbonates and foreign humic acids. The samples of well-decomposed peat were dated using hot 2% NaOH applied to the humic acid extract. The calibration of radiocarbon dates to calendrical dates was undertaken using the OxCal 4.2 program (https://c14.arch.ox.ac.uk; Bronk Ramsey 2008) with the IntCal13 calibration dataset (Reimer et al. 2013). We also used tephrastratigraphic data to determine the age of the coastal landform. We found volcanic ash in a soil section on ancient storm ridge and in a peat bog section near the lake. To identify the source of the tephra, a volcanic glass microprobe analysis was performed at the V.

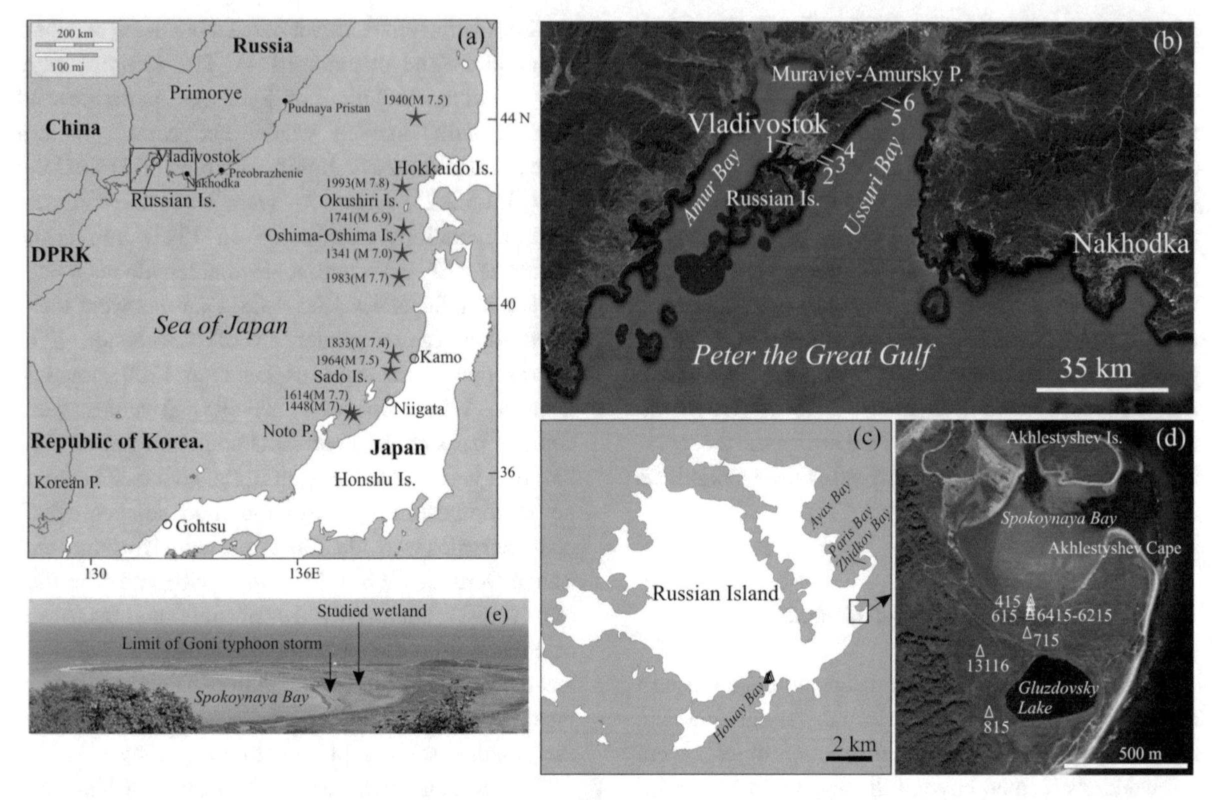

Figure 1

Map of the study area and positions of the studied sections. **a** The Sea of Japan with the epicenters of strong historical tsunamis (AD years) and earthquake magnitudes, **b** Peter the Great Gulf, 1 Golden Horn Bay, 2 Sobol Bay, 3 Tikhaya Bay, 4 Gornostai Bay, 5 Engelm Cape, 6 Chumaki Bay, **c** Russian Island with the position of Spokoynaya Bay, **d** position of the transect on the Spokoynaya Bay coast, **e** photo of the south coast of Spokoynaya Bay with the studied wetland and Akhlestyshev Cape

Table 1

Radiocarbon dates obtained from the peat section with tsunami deposits on the coast of Spokoynaya Bay, Russian Island

Sample number	Depth, m	Material dated	^{14}C age, BP	Calibrated age, 2σ	Laboratory index
1/615	0.16–0.17	Peat	$\delta^{14}C = 3.82 \pm 0.83\%$	1955–1957 (54.1%) 2006–2009 (41.3%)	LU-8031
4/615	0.36–0.37	«	330 ± 70	390 ± 80	LU-8036
5/615	0.42–0.43	«	540 ± 50	560 ± 40	LU-8037
6/615	0.58–0.59	*Zostera*	700 ± 60	650 ± 50	LU-8038

G. Khlopin Radium Institute, St. Petersburg. A total of 34 glass shards were measured.

4. The Geomorphological Setting

Unique conditions for deposits of tsunamis with small run-ups (up to 1–1.5 m) existed on the eastern coast of Russian Island at the coast of the closed Spokoynaya Bay, where a low wetland is located out of reach of the influence of large storms and river flood occurrences. The entrance from the bay is isolated by the small Akhlestyshev Island. The bay is shallow (less than 1 m in depth), and the bay was formed as a result of the formation of a double tombolo, which connected a small paleo-island (Cape

Akhlestyshev) with Russian Island (Fig. 1). A barrier landform separated the lagoon where Gluzdovsky Lake was later formed. The lake and bay are connected by a small channel. The height of the barrier landform on the open sea side is 4–5 m. This landform began to form at middle Holocene under sea level higher than in late Holocene and the present. The height of the ancient storm ridge on the inner side of the bay, which comprises gravel and pebbles, is 2.75 m. Within the bay, the modern storm ridge currently has a height of 0.7 m, and the beach (5–8 m width) only dries out during low tides. At the lower level, there is an almost horizontal tidal zone that is up to 10 m wide.

A low marine terrace that is flat and slightly inclined towards the sea (up to 1.3 m in height) is located between the modern and ancient storm ridges (Fig. 2). Its surface is a wetland that is overgrown by reeds, sedges and other moisture-loving herbs. A vast peatland (width up to 500 m) is developed around the lake.

A peat section near the Gluzdovsky Lake 13116 and behind the ancient storm ridge and soil profile 715 at the top of ancient storm ridge on the inner side of the bay include a thin tephra layer comprising green-light gray or yellow silt (Fig. 2). The volcanic glass is attributed to the Baitoushan Volcano. The tephra has a high-K (K_2O 5.55–6.54%) trachytic composition, which is typical of ash erupted from the Baitoushan Volcano on the China/North Korea border (Chen et al. 2016). The eruption occurred in the winter season of 946/947 AD (Xu et al. 2013). We correlated this volcanic ash with the widespread tephra B–Tm marker that is widely distributed over an extensive area (Machida 1999; Oppenheimer et al. 2017). This tephra is found on the coast of Eastern and South Primorye (Ganzey et al. 2015) and serves as a good marker of the Medieval Warm Period. No marine sands were found in the peat near the lake (sections 815 and 13116).

5. *Observations of the Effects of Extreme Storms on the Spokoynaya Bay Coast*

Among the difficult problems encountered when identifying tsunami deposits is finding differences between the evidence left by tsunamis and heavy storms or storm surges. In the closed Spokoynaya Bay, the impacts of storms are usually weak. In winter, the bay freezes, and winter storms do not penetrate there. In summer and early autumn, strong storms are usually caused by passing deep cyclones or tropical typhoons. In 2015–2016 in Primorye, including the Vladivostok area, two of the strongest typhoons, Goni in August 26, 2015, and Lionrock in August 31, 2016, collapsed in accompaniment with extreme storms. This gave us an opportunity to assess the range of penetration of storm waves inland and to observe the effects of their impacts on the shore. Observations conducted immediately after the storms showed that these two extreme storms were accompanied by surges. The maximal line of penetration of storm waves is well expressed by shafts of the seagrass *Zostera marina* and by marine garbage. During Typhoon Goni, the impacts of oncoming storm waves did not extend beyond the heights of the modern storm ridges. The inundation distance was up to 3–4 m and reached 8 m near the mouth of the channel connecting the lake and bay. During Typhoon Lionrock, the storm surge was more significant, and the coastal lowland was flooded up to 16 m. The storm surges were not highly turbulent, and sheets of deposits were not formed. In the flood zone, the grass was tilted toward the land. There was no evidence of sand accumulation or peat erosion. We suppose that sedimentological effects of strong storms in the past were similar to these two extreme storm surges.

6. *Tsunami Deposits: Distribution, Grain Size, and Microfossils*

The low wetland (up to 1.3 m high) situated between the modern and ancient storm ridges on the inner side of the bay was the most informative for finding traces of historical tsunami (Fig. 2). The sections comprise two units of marine and terrestrial origin.

At the bottom, there is marine fine silty sand (modes 0.1–0.125, 0.125–0.16 mm). The admixtures of silt reach 28.6%, and the material is moderately sorted. The number of diatoms in the sand is small,

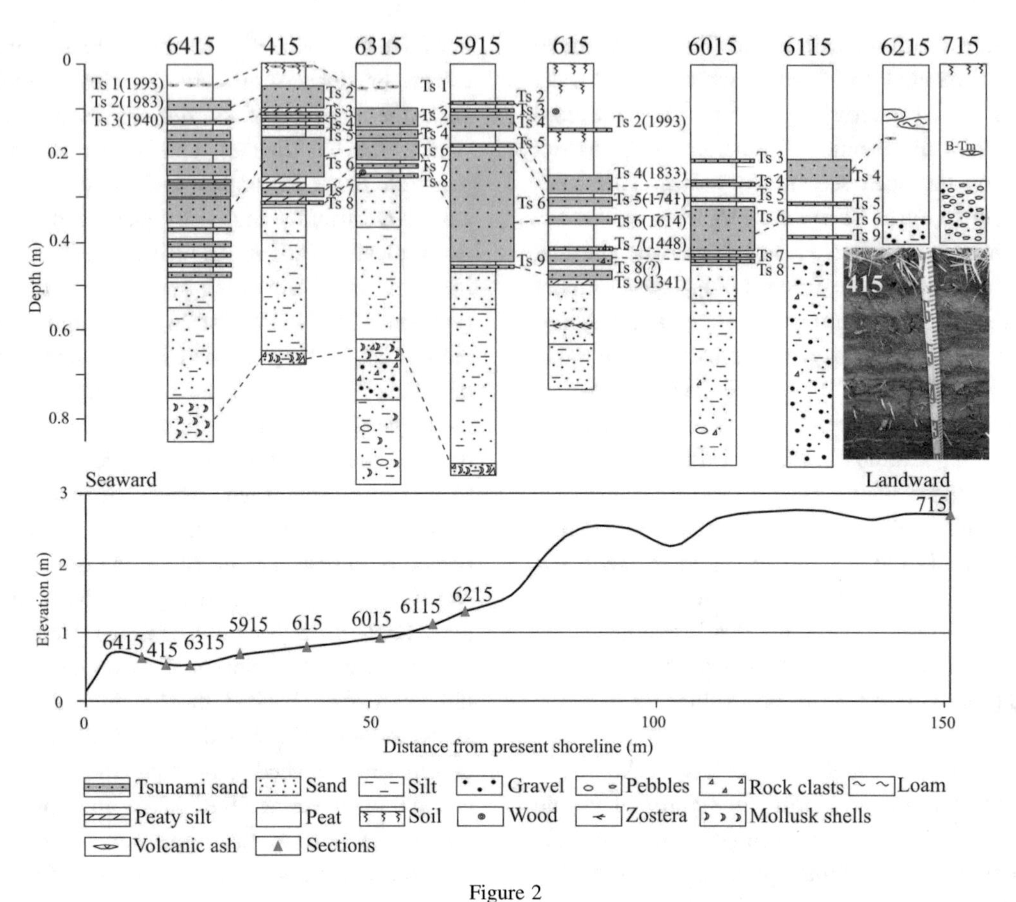

Figure 2
a Peat bog sections with tsunami sands on the coast of Spokoynaya Bay, **b** profile crossing the wetland; the gray triangles indicate the sections

and sublittoral planktonic species *Paralia sulcata* (80%) and epiphytic species *Cocconeis scutellum*, which are commonly found attached to macrophytes in the intertidal zone, are most frequent (Fig. 3). Many large frustules (*Arachnoidiscus ehrenbergii, Biddulphia biddulphiana, Odontella turgida,* and *Isthmia nervosa*) have a poor safety and are represented by fragments that can be connected with active turbulence of the tsunami flow (Dawson 2007). In deposits, there are rare benthic foraminifera (*Cribrostomoides jeffreysii* and *Jadammina macrescens*). In the layer with abundant fragments of mollusk shells, a few species of benthic foraminifera were found: *Haplophragmoides hancocki* from shallow epifauna and *Buccella frigida* and *Cribroelphidium asterineum*, which have a wide range of habitats that spread from coastal lagoons to the outer shelf (Sen Gupta 2002). The top of the marine unit comprised medium well-sorted sands of an ancient beach. The ^{14}C ages of the sands are 700 ± 60 BP and 650 ± 50 cal. yr. BP for LU-8038, which was obtained from a peat lens composed of *Zostera marina* that was found in beach sand. Closer to the barrier form, the base of the incisions is composed of gravel with small pebbles.

The peat and soil in the upper part of the sections include numerous layers of marine sands (Fig. 2). All of the sand layers have clear and abrupt lower contacts above peat or organic soil, and at times, an erosion base. The accumulation of material occurred on the uneven surface of the wetland. Inland fining is observed. The correlations of layers between different sections are not always clear. Some sand layers in sections located near the modern storm ridge were probably shaped at the times of extreme storms accompanied by wind-induced surges. It is assumed

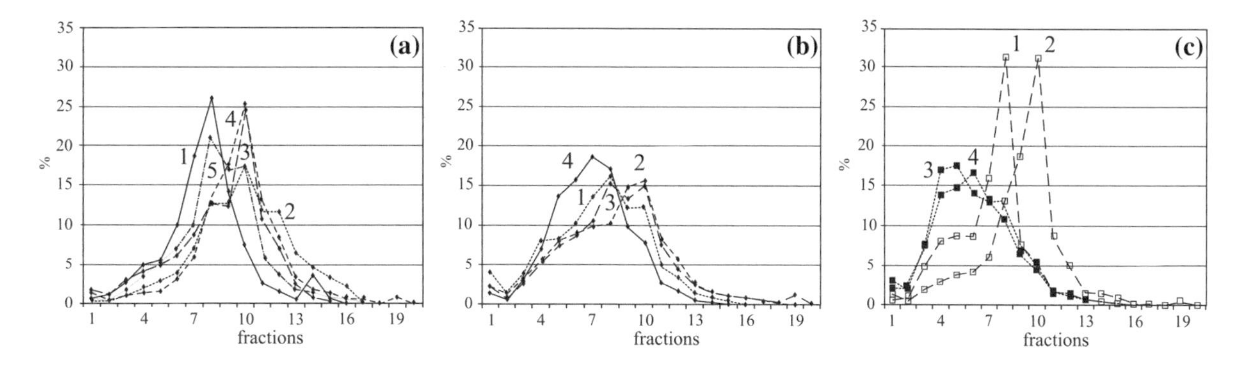

Figure 3
Grain-size distributions of the Spokoynaya Bay deposits: **a** tsunami deposits, which formed mainly due to supplies of beach sands (1 1993, 2 1983, 3 1833, 4 1614 or 1644, 5 XIV–XVI centuries); **b** tsunami deposits, which formed due to material supplied mainly from the bottom of the bay (1 1940, 2 1741, 3 1448, 4 1341); **c** beach sands (1 modern, 2 ancient) and inshore sands (3, 4). The horizontal scale shows the following grain-size fractions: 1 < 0.05; 2—0.05–0.063; 3—0.063–0.08; 4—0.08–0.1; 5—0.1–0.125; 6—0.125–0.16; 7—0.16–0.2; 8—0.2–0.25; 9—0.25–0.315; 10—0.315–0.4; 11—0.4–0.5; 12—0.5–0.63; 13—0.63–0.8; 14—0.8–1; 15—1–1.25; 16—1.25–1.6; 17—1.6–2; 18—2–3; 19—3–4; 20—4–5; 21 > 5 mm

that sections located beyond the storm ridge include sand layers of tsunamigenic origin.

In the tops of peat sections 6415, 415, and 6315, a layer (T_s 1) of light gray fine sand with admixed medium sand (thickness of $\leq$ 1 cm) was found. In places, the layer breaks up into separate lenses. This layer can be traced to 18 m from the high tide line (elevations up to 0.5 m). This deposit presumably appeared was deposited during the 1993 tsunami. In the same place, old logs were found along the modern shoreline. In contrast to a modern beach, different sizes of poorly sorted sands and tsunami deposits occur due to single and bimodal distribution curves (modes of 0.2–0.25 and 0.315–0.4 mm), and they include admixtures of silt (up to 15%) and are better sorted (Fig. 3). Material was primarily transferred from a beach and fine fractions—from the bottom of the bay which is proved with diatom analysis data. Twenty-seven species of marine and brackish water diatoms were found in the deposit (17.5–69.8% of the total content, maximally in section 6415), including sublittoral planktonic *Paralia sulcata, Hyalodiscus scoticus* and benthic *Aulacodiscus affinis, Diploneis smithii, Cocconeis scutellum, C. scutellum* var. *parva, Amphora marina, Opephora mutabilis, Grammatophora oceanica, Navicula cancellata* var. *retusa,* and *Tryblionella marginulata* in addition to fragments of pelagic *Thalassiosira decipiens, Thalassiosira* sp., and *Coscinodiscus* sp. Among the brackish water species, *Diploneis interrupta* (up to 63.9%) is dominant, *Planothidium hauckianum* (6.7%), *Fallacia pygmaea* (2.5%) are abundant, and *Cosmioneis lundstroemii, Fallacia cryptolyra, Rhopalodia musculus, Nitzschia vitrea, N. vitrea* var. *salinarum, N. sigma, Diploneis smithii* var. *pumila,* and *Halamphora coffeiformis* are also present. Freshwater species are more usual for the wetland; *Pinnularia lagerstedtii* (up to 25.9%), *Diploneis oblongella* (up to 21.2%), *Rossithidium nodosum* (5%), *Nitzschia amphibia, N. terrestris, N. palea, Eunotia bilunaris,* aerophils *Tryblionella debilis* and *Luticola mutica* and fresh-brackish water species *Cosmioneis pusilla* (15.3%), *Craticula halophila, Cocconeis pediculus* and others are present here. The presence of *Nitzschia vitrea* and *N. vitrea* var. *salinarum*, along with other marine and brackish water species in the upper part of the soil in section 615 (Fig. 4), may also evidence the penetration of this tsunami. In addition, agglutinated species of benthonic foraminifers *Jadammina macrescens* and *Miliammina fusca* were found in the sand, and tests were attached on dry seaweed remnants. These species belong to the euryhaline type and are well adapted to low and unstable salinities. They are widespread in the littoral zone, in marshes, intertidal environments (Sen Gupta 2002) and live in shallow lagoons and closed inlet of the Primorye (Foraminifera 1979; Annin 2000).

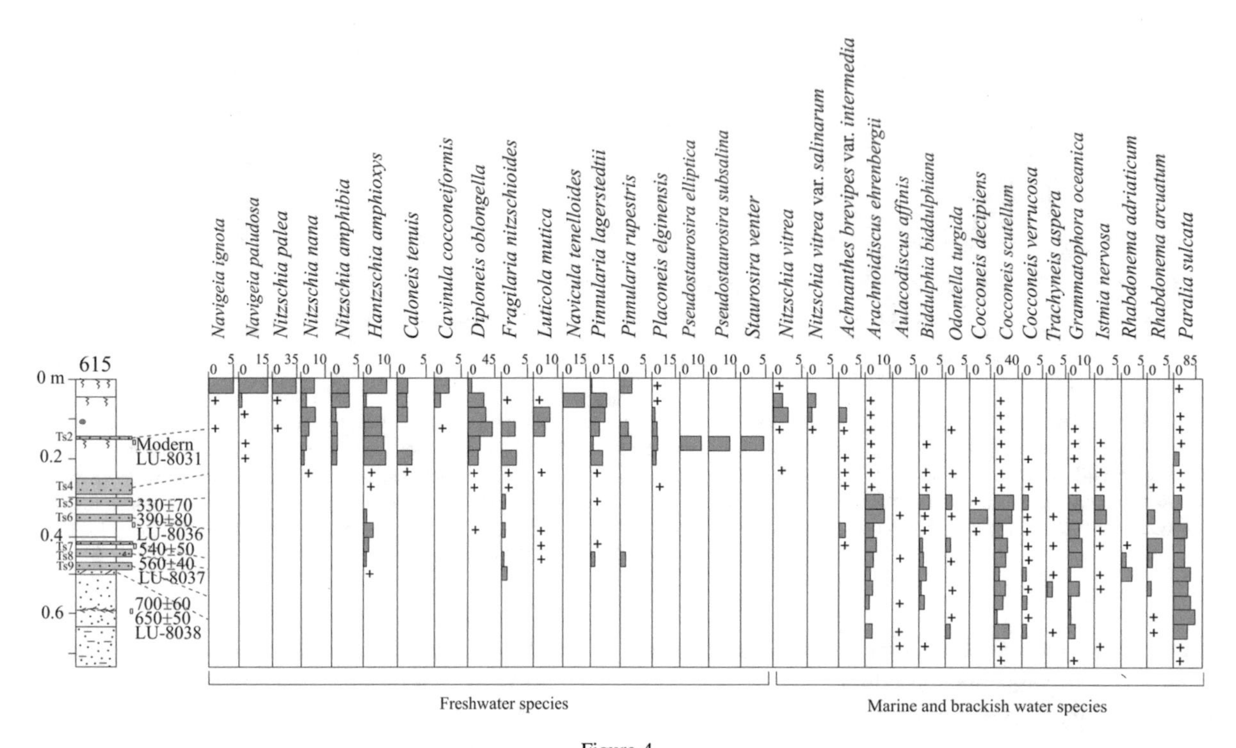

Figure 4
Distribution of diatoms in Spokoynaya Bay coastal wetland deposits (section 615). The diatom content is ≤ 1%, and rare frustules are shown by plus signs

Sand layer T_s 2 is well represented in its extension 40 m inland from the shoreline (heights up to 0.65 m) (section 615). At point 615, sometimes the layer breaks into two parts, which may include redeposited pieces of soil. The maximum thickness (5 cm) was reached behind the storm ridge (section 415), and in the back area, the thickness decreases to 1 cm (section 615). The deposit contains fine and medium sand with admixtures of coarser sand (to 22.6%), gravel (to 5.8%) and silt (to 6.7%). The grain size distribution curve is unimodal (mode at 0.315–0.4 mm) and sometimes bimodal, with modes from 0.2 to 0.25 mm. The main source of the material was the beach and the intertidal zone. The modern age of the deposit confirms the ^{14}C-date of LU-8031 in the underlying peat, which began forming after 1955 ($\delta^{14}C = 3.82 \pm 0.83\%$).

The marine origin of the sand was demonstrated using the data from the diatom analysis of 20 species of marine and brackish water diatoms (up to 79.6%, site 415). The marine species predominantly included *Paralia sulcata* (17.1%), which is typical for the gulf and bays, and benthic *Amphora marina* (2.9%), *Cocconeis scutellum* (1.9%), *Opephora mutabilis, Grammatophora angulosa, Amphora proteus, Trigonium arcticum, Mastogloia pumila, Cocconeis costata, Petroneis marina* were also found. Among the brackish water species, *Planothidium hauckianum*, *Rhopalodia musculus*, *Fallacia pygmaea*, *Diploneis interrupta,* and *Nitzschia vitrea* were found. Freshwater diatoms are mainly represented by species inhabiting ponds with middle or slightly high water pH, including *Amphora ovalis, Cocconeis placentula* var. *euglypta, Cosmioneis pusilla, Diploneis oblongella,* and *Navicula peregrina.*

In sand layer of T_s 2 in section 615, marine diatoms in addition to those in the diagram are represented by *Cocconeis scutellum* var. *parva, Planothidium hauckianum,* and *Rhopalodia musculus*. Fragments of pelagic *Rhizosolenia* sp. and *Coscinodiscus* sp. were also found. Among the freshwater, diatoms are species that reflect different degrees of moisture. Species typical of soils such as *Luticola mutica* and those able to live in weakly wet places, such as *Cavinula cocconeiformis, Tryblionella debilis,* and *Nitzschia amphibia*, were found. The soil underlying the sand includes large numbers of *Staurosira venter, Pseudostaurosira elliptica, P. subsalina,* and *Stauroforma exiguiformis,* which are widely distributed in shallow lakes with aquatic vegetation and in creeks and small rivers. These species were probably redeposited by tsunamis from the channel connecting the lake and sea.

Possibly, one of thin sand layers (T_s 3) in upper peat bogs in sections 6415, 415, 5915 and 6015 can be an evidence of tsunami in Primorye of 1940. Deposit is represented by fine sand with high content of silt (up to 16.4%). Mode 0.2–0.25 mm is well pronounced. The sources of the material were mainly from the beach, silt fraction supplied from intertidal zone and the bay floor. The deposit contains marine and brackish water diatoms (30%). Brackish water species dominate, including *Planothidium hauckianum, Diploneis interrupta, Cosmioneis lundstroemii, Fallacia cryptolyra, Halamphora coffeiformis, Nitzschia sigma.* Among marine species *Paralia sulcata* occurred in communities (colonies), implying weak turbulence, such as conditions existing during small tsunami run-up. Benthic *Cocconeis scutellum, C. scutellum* var. *parva,* and *Trachyneis aspera* were met too.

As a key site for reconstructions of more ancient historical tsunamis, we chose section 615, which is located 39 meters from shoreline. In the lower part of section 615, there are interlayers of grass peat, peaty silts and greenish-gray sands (thicknesses up to 5 cm) (Fig. 2). Eighty-eight species and varieties of freshwater diatoms and 55 species and varieties of marine and brackish water diatoms have been identified in the section. Only the most abundant taxa and indicative species were graphed on the diagram (Fig. 4). The marine origin of the sand layers in the peat was proven using the diatom analysis data. In addition, marine diatoms are abundant both in peat layers and between sand layers, which confirms the intense leakage of seawater through peat bogs that were not thick enough due to tsunami impact.

Layer T_s 4 (4 cm thick) is represented by fine-medium sand with coarse sand admixture (10.5%) and silt (9.5%). The deposits are well sorted, with a mode of 0.315–0.4 mm. The material was transported mainly from the beach, as confirmed by the low content of diatom frustules. The most common marine and brackish water diatoms are *Paralia sulcata* and benthic *Cocconeis scutellum*, and in addition to those in the diagram, *C. scutellum* var. *parva, C. distans,* and *Tryblionella plana* were found. Larger frustules are broken and poorly preserved. Among the freshwater diatoms, species living in various biotopes have been found, which indicates their allochthonous origin. *Diploneis oblongella*, which is typical for moist subaerial environments, *Luticola mutica* and *Hantzschia amphioxys*, which live under dry subaerial conditions, and arcto-boreal epiphyte *Fragilaria nitzschioides*, which prefers the less mineralized waters of mountain streams and small lakes, were observed together.

Moving downwards, layer T_s 5 (1–2 cm thickness) is composed of fine-medium moderately sorted sand (mode 0.315–0.4 mm) with coarse sand (9.8%), gravel (2.9%) and silt (12%). Apparently, the wave captured a large amount of material from inshore as well as from the beach. The sand includes mud clasts that are evidence of rapid deposition. Deposits include many diatom valves, and marine and brackish water diatoms make up 97% of the total composition. *Cocconeis scutellum* and *Paralia sulcata* dominate, and in addition to those in the diagram, sublittoral benthic *C. distans* (1.5%), *Grammatophora angulosa* (1.5%), *Diploneis smithii* and plankton *Actinoptychus senarius* are also found. Among freshwater *Fragilaria nitzschioides* and aerophilic *Pinnularia lagerstedtii*, which mainly inhabit low wet places, moist mosses were found.

The next layer (T_s 6) of coarse-medium sand has variable thicknesses (from 2 to 5 cm in section 615 to 25 cm in small depressions in section 5915). The lower contact of the layer is erosional and strongly wavy with pockets. Rip-up clasts of peat were found. The deposit is well sorted (mode 0.315–0.4 mm), with admixtures of coarser sand (13.6%), gravel (3.4%), and silt (3.7%). Its grain size characteristics are similar to those of the beach sand and tidal flat materials. The sand includes mud clasts (2 × 2 cm). Fifteen species of marine and brackish water diatoms were found in the deposits (total 90%) (Fig. 3). In addition to those indicated in the diagram, *Cocconeis scutellum* var. *parva* (3.5%), *C. costata, C. distans, C.*

pellucida, Diploneis littoralis and *Grammatophora angulosa* were found. Frustules of *Arachnoidiscus ehrenbergii* have poor preservation. Fragments of pelagic *Coscinodiscus* sp. and *C. asteromphalus* were also observed. Among the eight freshwater species, *Hantzschia amphioxys* inhabiting different biotopes, including shallow ponds and humid soils, and *Fragilaria nitzschioides*, which is typical for small lakes and small creeks, were found. The ^{14}C ages of 330 ± 70 BP and 390 ± 80 cal. year BP for LU-8036 were obtained from the underlying peat.

The lower part of the peat includes 3–4 sand layers (T_s 7, 8 and 9). Two thin layers (thicknesses of 3–8 mm) with sharp wavy lower boundaries of medium-fine well-sorted sand (mode 0.2–0.25 mm) with rare unrounded rock clasts (up to 2 cm) possibly correspond to one of the same events (T_s 7). Material was mainly supplied from the bay bottom, beach and tidal flat (Fig. 3). Rock clasts could be transferred from the capes there are bedrock outcrops (Triassic sandstones). They could not have been sourced from slopes because they are located far from the shore. There are 27 taxa of marine and brackish water diatoms (up to 97.1%) in the deposits. *Paralia sulcata* (up to 51.3%) and *Cocconeis scutellum* (up to 45.7%) dominate. In addition to those graphed on the diagram, benthic *Cocconeis scutellum* var. *parva* (3.8%), *Ctenophora pulchella* (1.1%) provide evidence of noticeable impact. Plankton *Actinoptychus senarius, Odontella aurita*, benthic *Cocconeis californica, C. distans, Diploneis chersonensis, D. smithii, Surirella fastuosa,* fragments of pelagic *Thalassionema nitzschioides, Thalassiosira* sp., and *Rhizosolenia* sp. are also present. Freshwater *Pinnularia eifelana, P. lagerstedtii, P. rupestris, Placoneis clementis* and other species typical of lake and wetland environments are also present. The ^{14}C ages of 540 ± 50 BP and 560 ± 40 cal. year BP for LU-8037 were obtained from the peat underlying the sand layer T_s 7.

In the studied section 615, sand layer at a depth of 0.43–0.45 m (T_s 8) is represented by medium-fine sand (mode 0.2–0.25 mm), and the grain size composition similar to that of the tidal flat zone deposits. Rare unrounded rock clasts (up to 1 cm) were found. Among the marine and brackish water diatoms, there were 15 taxa (95.5%), with a prevalence of associated *Paralia sulcata, Cocconeis scutellum* and *C. scutellum* var. *parva* (5.6%). In addition to those shown in the diagram, *Cocconeis distans* (1.1%), *Amphora proteus, C. decipiens, Diploneis dydima, Planothidium hauckianum* and a fragment of pelagic *Coscinodiscus* sp. are present. The consistency of freshwater diatoms remains the same. Fragments of marine crustaceans and benthic foraminifera were found in the sediment (2.2 number/g), including mainly *Jadammina macrescens* and rare *Miliammina fusca* attached to algae.

In studied section 615 in the lower part of the peat unit, a layer of fine sand denoted T_s 9 (2 cm thickness) (mode 0.16–0.2 mm) with silt (12.3%) was found. The material was primarily supplied from the bottom of the bay. Among the marine and brackish water diatoms, there were 11 taxa (98.7%). *Paralia sulcata* and *Cocconeis scutellum* dominate. In addition to those shown in the diagram, benthic *Catenula adhaerens, Cocconeis scutellum* var. *parva, C. distans, Planothidium hauckianum* were attributed to concomitant. The composition of the freshwater diatoms did not change. Fragments of marine crustaceans and benthic foraminifera (0.8 number/g) were found in the deposit, and all shells were attached to algae remains. Many *Jadammina macrescens* and single examples of *Cribrostomoides jeffreysii* (epifaunal species) freely moving on the surface of the sediment are often attached to algae and proliferate from shallow water to the outer shelf (Sen Gupta 2002).

On the south coast of Russian Island, we found modern tsunami deposits in Holuay Bay. The beach was deeply anthropogenically altered because sand from accumulative landforms was taken to construct a fortress in the early XX century. In the wetlands at the head of the bay, 100 m from the shoreline at the back of the storm ridge, a layer of poorly sorted sand with gravel was found under thin soil. Marine diatoms were encountered: neritic south-boreal *Thalassionema nitzschioides* and benthic *Cocconeis scutellum.*

7. Chronology of Tsunamis and Correlations with Known Historical Events of the Sea of Japan Region

The wetland between the modern and ancient storm ridges on the coast of the Spokoynaya Bay began to form during the Little Ice Age (Fagan 2000), as confirmed by radiocarbon data and the absence of tephra B–Tm, which was found in the soil profile at the top of the ancient storm ridge and in the peat bog near the lake. Marine units underlying the peat formed in the last minor transgression of the Medieval Warm Period (Korotky et al. 1997).

Geological data on the construction of the shores and shelf of Primorye indicate that the main tendency in the Pleistocene was slow tectonic subsidence (Korotky et al. 1980). At the end of the Pleistocene and during the Holocene, the rates of tectonic movements were incommensurable with sea level fluctuations. During late Pleistocene and the Holocene, the coast of Primorye is considered to have been tectonically stable, and there are marine deposits from minor Holocene transgressions lying above sea level (Korotky et al. 1997). GPS data show that crustal movements along the largest faults located to the east from Vladivostok do not exceed a few millimeters (Shestakov et al. 2006; Gerasimenko et al. 2016). The calculated vertical GPS velocities at sites belonging to the regional geodynamic network for the period between 1997 and 2009 indicate the nearly internal (between network sites) and external (with respect to the Eurasian tectonic plate, EUR) stability of the southeast region of Russia (Shestakov et al. 2011). Based on leveling measurements in Vladivostok city during the three epochs of 1941–1970–1987, the relative vertical velocities along the leveling profile for this period were quite small and alternated no more than 1 mm year^{-1} (Gerasimenko et al. 2015). This indicates that there is also a low level of present tectonic activity in the Primorsky region. Because no prominent periodic oscillations or secular variations in mean sea level were observed along the south Primorye coast (Mean Sea Level 2003), it follows that one can neglect the influence of contemporary crustal vertical movements on the mean sea level in the studied area.

On the coast of the Spokoynaya Bay, the upper boundary of the deposits of this transgression have elevations up to 0.39 m (beach facie) and up to 0.17 m (inshore facie), which indicate a sea level rise of $\sim$ 0.5 m above that at present. The medium well-sorted sands of ancient beaches were accumulating during the final phase of the transgression on the boundary with the Little Ice Age.

The ^{14}C-dates of 700 ± 60 BP and 650 ± 50 cal. year BP (1218-1399 AD) for LU-8038 are close to 1300 AD, a time when there were dramatic changes in the development of the environment in different regions (Nunn 2000). This date marks the time when the Little Ice Age began (Fagan 2000). Cooling was accompanied by a drop in sea level. There are no detailed data on the uneven course of sea level decline for the coast of Primorye. The radiocarbon dating of peat below the present sea level showed that the sea level was 0.5 m lower than at present (Korotky et al. 1997). The peat unit on the coast of Spokoynaya Bay began to form under the conditions of a minor regression. A sharp drop in sea level was the main reason for the change in the marine environments of sedimentation to continental ones. The shoreline shifted to a relatively modern location, and in place of the shallow part of the Spokoynaya Bay, there was low land. The accumulative landform in the northern part of the bay and the small island did not exist. The bay was more open. The presence of boreal–tropical benthic species of diatoms (*Rhabdonema adriaticum* and *Biddulphia biddulphiana*) in tsunami sands indicates that the shallow part of the bay was well warmed in the summer, even during cooling.

The lake was probably not connected to the sea during the Little Ice Age period—there were no channels from the lake, and deposits of this age almost do not contain any brackish water diatoms. The layers of marine sands at the continental parts of sections could have appeared because of high run-ups that were probably related to tsunami impact. Observations of modern extreme storms have shown that deposits of sand sheets do not form in the flood zone. The situation is very different from the shores of open and semi-enclosed bays subjected to strong and extreme storms (Ganzey et al. 2010). On closed bays of Russian Island, including Spokoynaya Bay, traces of storms are likely to be layers with an abundance of *Zostera marina*, which were found in the Little Ice Age beach sands and in the section of the modern storm ridge.

The assumption of the formation of layers of sea sands during coseismic subsidence seems unlikely. There are no geological data reflecting strong earthquakes in this region for the last millennium. The data on modern postseismic movements in South

Primorye based on GPS measurements after the Tohoku earthquake of 2011 also indicate slight displacements in the Vladivostok region from +0.7 to –7.5 mm per year, but coseismic vertical offsets were not detected (Gerasimenko et al. 2016). Movements of similar amplitude could not lead to periodic flooding of the low-lying parts of the coast of Russian Island.

The studied wetland could have been partially flooded during the tsunamis of 1983 and 1993. The two layers of sands, T_s 1 and T_s 2, in the top of the sections were probably deposited during these two tsunamis, which were the largest of the twentieth century. The highest run-ups (> 3–4 m) were observed on the coast of Eastern Primorye (Go et al. 1985; Gorbunova et al. 1997; Polyakova 2012), where we discovered sand sheets stretching for distances of up to 300 m onto land (Ganzey et al. 2015, 2017).

On the coast near Vladivostok, including Russian Island, the tsunami run-ups were minor. Usually, tsunami with small run-up is unlikely to leave deposits and their traces are difficult to find in coastal sections (Dawson and Shi, 2000). The observations on South Kuriles show that historical tsunamis of XX–XXI centuries with run-up less than 5 m did not form a extensive deposits (Razzhigaeva et al. 2007; 2017). In Eastern Primorye, the best defined sand sheets of tsunami 1983 and 1993 occur in the upper part of sections in costal lowland of bay shores, where the wave run-up was more than 3 m (Gansey et al. 2017). It was only the unique conditions for material deposition in the closed Spokoynaya Bay that allowed the deposits of these tsunamis to be found. The distribution of tsunami sands made it possible to reconstruct the inundation distance, which generally corresponds to the observations made immediately after tsunamis in nearby bays. The inundation distance for the 1993 tsunami was not more than 20 m, and the run-up height was 0.5 m a.m.s.l. Perhaps the penetration of the tsunami was greater than the zone in which sand was deposited, and evidence of the limit of tsunami run-up (up to 39 m from the shoreline) may be provided by the presence of brackish water and marine diatoms in soil (section 615). The more intensive tsunami of 1983 had a run-up 0.7 m a.m.s.l. and penetrated inland from the shoreline up to 50 m. On the south coast of Russian Island, the deposits were probably left behind by one of XX century tsunamis, the run-up of which exceeded 1 m.

One of the thin sand layers T_s 3 in the peat in sections 6415–415 located near the shoreline may be evidence of a tsunami in Primorye in 1940, which left poor information to assess and discuss (Go et al. 1972; Soloviev and Go 1974). The epicenter of the earthquake that caused the tsunami was located in the northeastern part of the Sea of Japan (Satake 1986). On the coast of Eastern Primorye, the manifestations of that tsunami were comparable with those of the events of 1983 and 1993. In Vladivostok, a tsunami was observed after 2 h, but there is no data on the altitude of sea level (Go et al. 1972). The tsunami run-up and inundation distance in Rudnaya Pristan was comparable to the tsunami of 1983 (Go et al. 1972; Kurkin et al. 2004). Sand layer of possible tsunami 1940 whish was found in peat of Russian Island can be traced at 0.7 m amsl and extend in land up to ~ 50 m from the present shoreline.

Tsunamis were quite frequent in the Sea of Japan from the XIV to XIX centuries, when, according to chronicles, as many as 14 strong events were observed (Soloviev and Go 1974; Hatori and Katayama 1977; NGDC/WDS Global historical tsunami database; HTDB/WLD..., 2015). At least four tsunamis were recorded at the coastal lowland of the Spokoynaya Bay, the deposits of which extend to 60 meters inland from the present shoreline. The scales of these tsunamis greatly exceeded those of the 1983 and 1993 tsunamis.

The T_s 4 deposits probably record the tsunami of December 7, 1833, which was caused by the Shonai-oki earthquake (7.4–7.8 M), the epicenter of which was located to the north-east of Sado Island. The tsunami was catastrophic in Japan, the run-up height on the Honshu Island coast near Niigata City reached 5–6 m; at Kamo City, the tsunami had a run-up height of 9 m high, and the waves spread to north to Hokkaido and south to the Noto Peninsula (Soloviev and Go 1974; Hatori and Katayama 1977; Iida 1984; NGDC/WDS Global historical tsunami database). On the east coast of Russian Island, the run-up height exceeded 1 m, and the inundation zone was > 50 m. Because the tsunami occurred in December, the tsunami's waves were aggravated by ice fragments, which increased erosion; this may be related to the poor preservation of diatom valves. Example of the tsunami with ice on the Far Eastern shores is the tsunami 2011 on the Shikotan Island, South Kuril (Kaistrenko et al. 2011; 2013). The ice floes enhanced the bottom erosion on shoals and destruction of low-lying coastal peatland played an important role in the transport and deposition of material (Razjigaeva et al. 2013).

Traces of a strong tsunami that occurred on August 29, 1741 are possibly captured on Russian Island (sand layer T_s 5); this tsunami was caused by a large underwater landslide during the eruption of the Oshima–Oshima Volcano, which is located on a small island to the west of Southern Hokkaido (Nishimura and Miyaji 1995; Satake 2007). That tsunami was the largest historical tsunami in the region and was observed throughout the western coast of Japan, especially in southwest Hokkaido (the height of the waves recorded in Chronicles was up to 13 m, and, according to oral reports, was up to 34 m), in northwest Honshu (the wave height was up to 6 m) and Sado Island (with wave heights of 5 m and up to 8 m from oral reports), and the Noto Peninsula (4 m); it caused powerful destruction and a large number of victims (Soloviev and Go 1974; Hatori and Katayama 1977; Satake 2007; NGDC/WDS Global historical tsunami database). The tsunami reached the Tsushima Strait coast, and near Gohtsu (Honshu Island), the run-up height was 2 m. On the east coast of the Korean Peninsula (320 km from the epicenter), the tsunami height reached 3–4 m (Satake 2007). The deposits of the tsunami of 1741 were found on the coast of the Eastern Primorye (Nishimura et al. 2014; Ganzey et al. 2015) and southwestern Hokkaido, where the run-up of this tsunami exceeded the tsunami of 1993 (Kawakami et al. 2017). The data for Russian Island confirm that the tsunami of 1741 was more intense event on Primorye coasts than the tsunamis of the twentieth century.

A XVII century tsunami left sand layer T_s 6. The tsunami run-up was at least 1.5 m, and the inundation zone reached 200 m (considering the lower sea level during the regression than that at present). The deposits of the tsunami, which were formed at approximately 400 cal. year BP, are well expressed on the coast of Eastern Primorye, where run-up is estimated to have been up to 5 m (Ganzey et al. 2015). In the XVII century, there were large tsunamis in the Sea of Japan (Niigata Prefecture) on November 26, 1614 (M 7.7), and October 18, 1644 (M 6.9, Akita Prefecture) (Soloviev and Go 1974; Iida 1984). It is believed that the tsunami produced by a 1611 earthquake was significant (Hatori and Katayama 1977). The first tsunami was caused by a strong earthquake (M 7.7), the epicenter of which was located to the southwest of Sado Island, and a second earthquake (M 6.9) occurred near the northwestern coast of Honshu. The tsunami caused great destruction and fatalities on the west coast of the Japanese islands, where wave heights reached 6 m (Soloviev and Go 1974; Hatori and Katayama 1977; NGDC/WDS Global historical tsunami database). The ^{14}C-dates of 330 ± 70 BP and 390 ± 80 cal. year BP (1435–1670 AD—93.2%) for LU-8036 from the underlying tsunami deposits peat confirm the conformity of the age of the sand to the historical tsunami.

The low part of the peat section 615 records three tsunamis from the XIV–XVI centuries. ^{14}C-date of 540 ± 50 BP and 560 ± 40 cal. year BP (1367–1442 AD) for LU-8037 was obtained from peat underlying sand layer T_s 7. Deposits from the same tsunami have probably been found on the coast of East Primorye. In some bays, well-defined sand layers left by an ~ 600 BP tsunami have been found, with ^{14}C-dates of 580 ± 80 BP and 600 ± 50 cal. year BP, (1274-1449AD) for LU-7104, 580 ± 40 BP and 600 ± 40 cal. year BP, (1207–1422 AD) for LU-6562, and 530 ± 50 BP and 580 ± 50 cal. year BP (1301–1449 AD) for LU-6558, and the tsunami run-up exceeded 4 m (Ganzey et al. 2015). It is difficult to attribute this tsunami on Russian Island to a well-known event. There are records in Japanese Chronicles regarding an intense tsunamigenic earthquake (M 7) with an epicenter near Sado Island in 1448 (Iida 1984; NGDC/WDS Global historical tsunami database, HTDB/WLD… 2015). There was probably another event on Primorye coast that is not reflected in the Japanese records. If the epicenter of the earthquake that caused the tsunami was near West Hokkaido, it could not have been very destructive to the inhabited coasts of Honshu Island. Sand layer T_s 8 was perhaps left by a XIV–XV century tsunami. A primary event possibly corresponding to an XIV century tsunami is recorded in sand layer T_s 9 in the peat bog in Spokoynaya Bay. One of the well-known and strong events in the Sea of Japan occurred in 1341 (October 31), for which the epicenter of the earthquake (*M* 7) was located near northwestern Honshu Island. The tsunami caused many victims on the west coast of Hokkaido and northern part of Honshu (Hatori and Katayama 1977; NGDC/WDS Global historical tsunami database). Perhaps that tsunami reached Russian Island.

Within the peatland near Gluzdovskoe Lake, the peat lying above the tephra does not contain marine diatoms, which shows that tsunamis have not impacted this part of the coastal lowland and indicates that tsunami waves during the last millennium did not exceed 2.7 m there.

It is hard to say which of the tsunamis that reached Russian Island were the strongest. There is uncertainty regarding the position of the coastline during the small regression in the Little Ice Age. There is no direct connection between the findings of pelagic diatom species and tsunami intensity. The relationship between marine planktonic, benthic and brackish water diatoms probably reflects changes in the configuration of the shoreline rather than changes in the run-up of the tsunami. It can be assumed that in the tsunami deposits formed in the more open bay (T_s 6–9) during the regression of the Little Ice Age, marine species, mainly represented by *Paralia sulcata*, predominate in the diatom composition. When the level began to rise, the closed shallow bay was formed. Under conditions of warming and increasing humidity, fresh water began to flow from the lake into the sea, and the bay became fresher. This led to the fact that in deposits T_s 1–3 of the tsunamis of 1983, 1993 and 1940, the contents of brackish water species increased. In all cases, the compositions of diatoms with abundances of sublittoral species in the tsunami deposits indicate that the transfer of material during the tsunamis occurred from shallow depths, which is confirmed by the presence in the tsunami deposits of benthonic foraminifera, which are typical for shallow water environments.

The broken diatom frustules in the tsunami deposits indicate events that were accompanied by more severe erosion (Grebennikova et al. 2002; Dawson 2006; Sawai et al. 2009). The seasons when the tsunamis occurred could have played a role in the depositional processes during the tsunamis. Tsunamis occurring in winter with ice could have significantly greater erosion capacity. The deposits of such tsunamis would include severely broken diatoms frustules. This is observed in the T_s 4 deposits of the 1833 tsunami, which occurred in December. Broken frustules of diatoms were also found in the T_s 6 deposits of the 1614 (1644) tsunami, along with other evidence of erosional processes during that tsunami.

The interpreted results of the grain size composition of the sand are also ambiguous. From the distribution curve shapes and sediment sorting, the material sources can be well identified. A comparison of the tsunami deposits of Spokoinaya Bay with other sedimentary deposits near the study area showed that the tsunami deposits are divided into two types (Fig. 3). The sources of deposits T_s 1, 2, 4, 6, and 8 of in 1993, 1983, 1833, 1614 (1644?), and XIV–XVI century tsunamis were mainly beach sands. Deposits Ts 3, 5, 7 and 9 of the tsunamis of 1940, 1741, 1448, and 1341 were mainly derived from the bay's bottom. However, the dependence is not always maintained such that a stronger tsunami redeposited material from the bottom of the bay and that the deposits of a weak tsunami would only be derived from the beach. The compositions of the deposits were controlled by the situations during material deposition: the duration that water remained in the inundation zone and intensities of the up-flow and return flow.

It can be assumed that the historical tsunamis in the Little Ice Age that occurred at lower sea levels and on shorelines far from the modern one exceeded the known tsunamis of the twentieth century in scale. The main argument is the deeper distribution of the deposits of those tsunami inlands, even at the present position of the shoreline. Only those tsunamis left traces of erosion in the wetland sections.

8. Conclusions

Detailed geological records of strong tsunamis that occurred in the last 600–700 years in the Sea of Japan region have been discovered on Russian Island on the coast of Spokoynaya Bay. The bay is a unique target for researching such events; it is a sedimentary trap because of its isolation from the effects of strong storms and has sufficient sandy material that leaves well-marked sand sheets in the inundation areas of tsunamis, and there is an absence of rivers or streams that can erode deposits during flooding. The tsunami deposits are localized in a narrow area along the coast, the width of which is less than 60 m. That fact indicates low tsunami run-up. Taking into account the decline in sea level during the Little Ice Age, the inundation zones and material accumulations could reach 200–250 m, and the tsunamis of that age were therefore larger events than those of the XX century. Deposits brought by tsunamis according to grain size characteristics are not complete analogs of beach, tidal flat and inshore sands but represent material combinations from different sources that are represented in different proportions, depending on the characteristics of the waves and erosion–accumulation processes. The largest tsunamis were accompanied by the erosion of the bottom of the bay and its entrance capes. Weaker tsunamis usually captured material from the beach and tidal zone. The particular characteristics of the tsunami deposits found on the coast of Spokoynaya Bay reflect that material from inshore sources was mainly redeposited. Therefore, the deposits include great amounts of marine and brackish water diatoms, benthic foraminifera and pieces of crustaceans. Although tsunamis may have small run-ups, they can endanger tourists and temporary structures located on the beach and low relief levels, which must be considered in the development of recreational land use planning and management for Russian Island.

Acknowledgements

The work was undertaken with financial support from the Russian Foundation for Basic Research (RFBR), Russia (grant # 15-05-00179), and the FEB RAS Program "Far East" (grant #18-5-003). The 2016 field work was also supported by the Russian Science Foundation (RSF) (grant # 14-50-00095). We would like to thank two anonymous reviewers for help and productive comments that significantly improved the manuscript.

References

Annin, V. K. (2000). Benthic foraminifera assemblages in Posiet Bay (Sea of Japan) and their habitats. *Oceanology, 40*(1), 830–838.

Baklanov, P. Ya., Avdeev, Yu. A., Romanov, M. T. (2017). A new phase in development of Vladivostok City and its agglomeration. *The Territory of New Opportunities. The Herald of Vladivostok State University of Economics and Service*, 9, 27–46. https://doi.org/10.24866/VVSU/2073-3984/2017-3/27-46 (**in Russian**).

Bronk Ramsey, C. (2008). Deposition models for chronological records. *Quaternary Science Reviews, 2*(1–2), 42–60. https://doi.org/10.1016/j.quascirev.2007.01.019.

Chen, X.-Y., Blockley, S. P. E., Tarasov, P. E., Xu Y. -G., McLean, D., Tomlinson, E. L., Albert, P. G., Liu, J. -Q., Müller, S., Wagner, M., & Menzies, M. A. (2016). Clarifying the distal to proximal tephrochronology of the Millennium (B-Tm) eruption, Changbaishan Volcano, northeast China. *Quaternary Geochronology*, 33, 61–75. 10.1016/j.quageo.2016.02.003.

Dawson, S., & Shi, S. (2000). Tsunami deposits. *Pure and Applied Geophysics, 157,* 875–897.

Dawson, S. (2007). Diatom stratigraphy of tsunami deposits: Examples from the 1998 Papua New Guinea tsunami. *Sedimentary Geology, 200,* 328–335. https://doi.org/10.1016/j.sedgeo.2007.01.011.

DeMets, C., Gordon, R. G., & Argus, D. F. (2010). Geologically current plate motions. *Geophysical Journal International, 181,* 1–80. https://doi.org/10.1111/j.1365-246X.2009.04491.

Diatom analysis. (1950). Leningrad: Gosgeolizdat.

Fagan, B. (2000). The little ice age. How climate made history 1300-1850. New-York: Basic Books.

Foraminifera of Far Eastern Seas of the USSR. (1979). V.I. Gudina (Ed.). Novosibirsk: Nauka.

Fukao, Y., & Furumoto, M. (1975). Mechanism of large earthquakes along the eastern margin of the Japan Sea. *Tectonophysics, 26,* 247–266.

Ganzei, K. S. (2016). Dynamics of land use (2007–2014) and future prospects for development of Russkii Island (Gulf of Peter the Great). *Geography and Natural Resources, 37*(3), 257–263. https://doi.org/10.1134/S1875372816030094.

Ganzey, L. A., Razjigaeva, N. G., Nishimura, Yu., Grebennikova, T. A., Kaistrenko, V. M., Gorbunov, A. O., Arslanov, Kh. A., Chernov, S. B., & Naumov, Yu. A. (2015). Deposits of historical and paleotsunamis on the coast of Eastern Primorye. *Russian Journal of Pacific Geology*, 9(1), 64–79. https://doi.org/10.1134/S1819714015010029.

Ganzey, L. A., Razjigaeva, N. G., Nishimura, Yu., Grebennikova, T. A., Gorbunov, A. O., Kaistrenko, V. M., et al. (2017). Deposits of the 1983 and 1993 tsunamis on the coast of Primorye. *Oceanology, 57*(4), 568–579. https://doi.org/10.1134/S0001437017040075.

Ganzey, L. A., Razzhigaeva, N. G., Kharlamov, A. A., & Ivelskaya, T. N. (2010). Extreme storms in 2006–2007 on Shikotan Island and their impact on the coastal relief and deposits. *Oceanology, 50*(3), 425–434. https://doi.org/10.1134/S0001437010030112.

Gerasimenko, M. D., Shestakov, N. V., Kolomiets, A. G., Gerasimov, G. N., Takahashi, H., Sysoev, D. V., et al. (2016). Vertical crustal movements of south of Primorsky krai and their relationship with the subduction zone geodynamic processes.

Geodesy and cartography, 3, 30–34. https://doi.org/10.22389/0016-7126-2016-909-3-30-34

Gerasimenko M. D., Shestakov N. V., Kolomiets A. G., Gerasimov G. N., Takahashi H., Sysoev D. (2015). Contemporary vertical movements of the southern Primorye region and their relation to subduction zone processes. In: B.W. Levin, O.N. Likhacheva (Eds). Proc. Int. Sci. Conf Geodynamical Processes and Natural Hazards. Lessons of Neftegorsk. Yuzhno-Sakhalinsk, May 26-30, 2015. V. 2, (pp. 42–45). Vladivostok, Dal'nauka. **(in Russian)**.

Go, Ch. N., Ivaschenko, A. I., Simonov, K. V., & Soloviev, S. L. (1985). Manifestations of Japan Sea tsunami May 26, 1983 on the coast of the USSR. In: E.N. Pelinovsky (Ed.), Tsunami rolling on coast (pp. 171–180). Gorky: Institute of Applied Physics RAS. **(in Russian)**.

Go, Ch. N., Leonidova, N. I., & Leonov, N.N. (1972). Some data on tsunami on August 1, 1940 in the Sea of Japan. In S.L. Soloviev, A.I. Ivaschenko, A.A. Poplavsky (Eds.) Tsunami Waves (pp. 279–283). Yuzhno-Sakhalinsk: Sakhalin Complex Scientific Research Institute USSR Academy of Science. **(in Russian)**.

Gorbunova, G. V., Didenko, G. V., D'yachenko, V. D., Nagornykh, T. V., Poplavskii, A. A., Poplavskaya, L. N., Kharlamov, A. A., & Shelepov, G. P. (1997). Study of July 12–13, 1993 tsunamic manifestations on the coast of the Primorsky Krai. In K.F. Sergeev (Ed.), Geodynamics of Tectonosphere of the Pacific-Eurasia Conjunction Zone. V. 8 (pp. 7–28), Yuzhno-Sakhalinsk: Institute of Marine Geology and Geophysics, FEB RAS.

Grebennikova, T. A., Razzhigaeva, N. G., Iliev, A. Ya., & Kaistrenko, V. M. (2002). Diatom analysis using for identification of paleotsunami deposits. In B. V. Levin & M. A. Nosov (Eds.), *Local tsunami: notice and risk decreasing* (pp. 19–31). Moscow: Yanus-K.

Gusiakov, V. K. (2016). Tsunamis on the Russian Pacific coast: history and current situation. *Russian Geology and Geophysics, 57*(9), 1259–1268. https://doi.org/10.1016/j.rgg.2016.08.011.

Hatori, T., & Katayama, M. (1977). Tsunami behavior and source areas of historical tsunamis in the Japan Sea. *Bulletin Earthquake Research Institute, 52*, 49–70. **(in Japanese)**.

HTDB/WLD (Historical Tsunami Database for the World Ocean), 2000 BC to present (2015). Novosibirsk: Tsunami Laboratory, ICMMG SD RAS http://tsun.sscc.ru/nh/tsunami.php. Accessed 04 March 2018.

Iida, K. (1984). *Catalog of tsunamis in Japan and its neighboring countries*. Toyota: Aichi Institute of Technology.

Kaistrenko, V., Razjigaeva, N., Kharlamov, A., & Shishkin, A. (2013). Manifestation of the 2011 Great Tohoku tsunami on the Coast of the Kuril Island: A Tsunami with Ice. *Pure and Applied Geophysics, 170*, 1103–1114. https://doi.org/10.1007/s00024-012-0546-9.

Kaistrenko, V. M., Shevchenko, G. V., & Ivelskaya, T. N. (2011). Manifestation of the Tohoku Tsunami of 11 March, 2011 on the Russian Pacific Ocean coast. *Problems Engineering Seismology, 38*(1), 41–64. **(In Russian)**.

Kawakami, G., Nishina, K., Kase, Y., Tajika, J., Hayashi, K., Hirose, W., Sagayama, T., Watanabe, T., Ishinaru, S., Koshinizu, K., Takahashi, R., & Hirakawa, K. (2017). Stratigraphic records of tsunami along the Japan Sea, southwest Hokkaido, northern Japan. *Island Arc, 26*, 1–18. https://doi.org/10.1111/iar.12197.

Kobayashi, Y. (1983). Beginning of plate "subduction". *Gekkan-Chikyu, 3*, 510–518. **(in Japanese)**.

Korotky, A. M., Grebennikova, T. A., Pushkar, V. S., Razzhigaeva, N. G., Volkov, V. G., Ganzey, L. A., et al. (1997). Climatic changes of the territory of South Far East at Late Pleistocene-Holocene. *Bulletin Far East Branch, Russian Academy of Science, 3*, 121–143. **(in Russian)**.

Korotky, A. M., Karaulova, L. P., Troitskaya, T. S. (1980). *Quaternary deposits of the Primorye. Stratigraphy and paleogeography*. Novosibirsk: Nauka. **(in Russian)**.

Krammer, K., & Lange-Bertalot, H. (1988). *Bacillariophyceae. Süßwasserflora von Mitteleuropa, Epithemiaceae, Surirellaceae*. Stuttgart: Gustav Fisher.

Kurkin, A. A., Pelinovskii, E. N., Choi, B. H., & Lee, J. S. (2004). A comparative estimation of the tsunami hazard for the Russian coast of the Sea of Japan based on numerical simulation. *Oceanology, 44*(2), 179–188.

Machida, H. (1999). The stratigraphy, chronology and distribution of distal marker-tephras in and around Japan. *Global and Planetary Change, 21*, 71–94.

Mean Sea Level. (2003). In: A. S. Vasiliev, F. S. Terziev, A. N. Kosarev (Eds.), *Hydrometeorology and Hydrochemistry of Seas. Vol. VIII. Japan Sea, Issue 1*. (pp. 270–272). St. Petersburg: Gidrometeoizdat. **(in Russian)**.

Nakamura, K. (1983). Possible nascent trench along the eastern Japan Sea as the convergent boundary between Eurasia and North American plates (in Japanese). *Bulletin Earthquake Research Institute, Tokyo University, 58*, 721–732.

NGDC/WDS Global historical tsunami database. National Geographical data center. http://www.ngdc.noaa.gov/hazard/tsu.shtml. https://doi.org/10.7289/V5PN93H7. Accessed 04 March 2018.

Nishimura, Yu., & Miyaji, N. (1995). Tsunami deposits from the 1993 southwest Hokkaido earthquake and the 1640 Hokkaido Komagatake eruption, northern Japan. *Pure and Applied Geophysics, 144*, 720–733.

Nishimura, Y., Razjigaeva, N., Ganzey, L., Grebennikva, T., Kaistrenko, V., Gorbunov, A., & Nakamura, Y. (2014). Insight of large tsunami recurrence around the Sea of Japan revealed by surveys of historical and pre-historical tsunami. Chiba. 2014 Japan Earth and Planetary Science Joint Meeting. Abstract. Chiba, Japan. p. MIS23–19.

Nunn, P. D. (2000). Environmental catastrophe in the Pacific Islands around AD 1300. *Geoarchaeology, 15*(7), 715–40. https://doi.org/10.1002/1520-6548(200010)15

Ohtake, M. (1995). A seismic gap in the eastern margin of the Sea of Japan as inferred from the time-space distribution of past seismicity. *The Island Arc, 4*, 156–165.

Oppenheimer, C., Wacker, L., Xu J., Galván, J. D., Stoffel, M., Guillet, S., Corona, C., Sigl, M., Cosmo, N. D., Hajdas, I., Pan, B., Breuker, R., Schneider, L., Esper, J., Fei, J. Hammond, J.O.S., & Büntgen, U. (2017). Multi-proxy dating the "Millennium Eruption" of Changbaishan to late 946 CE. *Quaternary Science Reviews, 158*, 164–171. https://doi.org/10.1016/j.quascirev.2016.12.024.

Polyakova, A. M. (2012). *Dangerous and especially dangerous hydrometeorological phenomena in the Northern Pacific and tsunami waves near the Primorye coast*. Vladivostok: Dalnauka. **(in Russian)**.

Preobrazhenskaya, T. V., & Troitskaya, T. C. (1996). *Foraminifera of Far East Seas. Part 1. Foraminifera of littoral of Lesser Kuril Arc*. Vladivostok: Dalnauka. **(in Russian)**.

Razjigaeva, N. G., Ganzey, L. A., Grebennikova, T. A., Ivanova, E. D., Kharlamov, A. A., Kaistrenko, V. M., et al. (2013). Coastal sedimentation associated with the Tohoku tsunami of 11 March 2011 in South Kuril Islands, NW Pacific Ocean. *Pure and Applied Geophysics, 170,* 1081–1102. https://doi.org/10.1007/s00024-012-0478-4.

Razjigaeva, N. G., Ganzey, L. A., Grebennikova, T. A., Ganzey, K. S., Nishimura, Y., Kaistrenko, V. M., et al. (2014). Chronology of tsunamis documented in sections of the coastal lowlands in East Primorye. *Doklady Earth Sciences, 459*(2), 1609–1612. https://doi.org/10.1134/S1028334X14120204.

Razzhigaeva, N. G., Ganzey, L. A., Grebennikova, T. A., Kharlamov, A. A., Arslanov, Kh A, Kaistrenko, V. M., et al. (2017). The Problem of Past Megatsunami Reconstructions on the Southern Kurils. *Russian Journal of Pacific Geology, 11*(1), 34–45. https://doi.org/10.1134/S1819714017010079.

Razzhigaeva, N. G., Ganzey, L. A., Grebennikova, T. A., Kharlamov, A. A., Il'ev, A. A., & Kaistrenko, V. M., (2007). Deposits of Shikotan earthquake 1994 tsunami. *Oceanology, 47*(4), 579–587.

Reimer, P. J., Bard, E., Bayliss, A., Beck, J. W., Blackwell, P. G., Bronk Ramzey, C., Buck, C. E., Cheng, H., Edwards, R. L., Friedrich, M., Grootes, P.M., Guilderson, T. P., Haflidason, H., Hajdas, I., Hatté, C., Heaton, T. J., Hoffmann, D. L., Hogg, A. G., Hughen, K. A., Kaiser, K. F., Kromer, B., Manning, S. W., Niu, M., Reimer, R. W., Richards, D. A., Scott, E. M., Southon, J. R., Staff, R.A., Turney, C., S. M., & van der Plicht, J. (2013). IntCal13 and marine 13 radiocarbon age calibration curves 0-50.000 years cal BP. *Radiocarbon, 55 (4),* 1869–1887. https://doi.org/10.2458/azu_js_rc.55.16947.

Satake, K. (1986). Re-examination of the 1940 Shakotan-oki earthquake and the fault parameters of the earthquakes along the eastern margin of the Japan Sea. *Physics of the Earth and Planetary Interiors, 43,* 137–147.

Satake, K. (2007). Volcanic origin of the Oshima-Oshima tsunami in the Japan Sea. *Earth Planet Space, 59,* 381–390. https://doi.org/10.1186/BF03352698.

Satake, K., & Tanioka, Y. (1995). Tsunami generation of the 1993 Hokkaido Nansei-oki earthquake. *Pure and Applied Geophysics, 144,* 803–821.

Sawai, Y., Jankaew, K., Martin, M., Prendergast, A., Choowong, M., & Charoentitirat, T. (2009). Diatom assemblages in tsunami deposits associated with the 2004 Indian Ocean tsunami at Phra Thong Island, Thailand. *Marine Micropaleontology, 73,* 70–79. https://doi.org/10.1016/j.marmicro.2009.07.003.

Sen Gupta, B. K. (2002). Foraminifera in marginal marine environments. In: K. Barun, B.K. Sen Gupta (Eds.), Modern Foraminifera (pp. 141–160). Dortrecht: Kluwer Academic Publishers.

Seno, T., Sakurai, T. S., & Stein, S. (1996). Can the Okhotsk plate be discriminated from the North American plate? *J. Geophysics Research, 101,* 11305–11351.

Seno, T., Stein, S., & Gripp, A. (1993). A model for the motion of the Philippine Sea Plate consistent with NUVEL–1 and geological data. *J. Geophysics Research, 98,* 17941–17948.

Shestakov, N. V., Gerasimenko, M. D., Kolomiets, A. G., Gerasimov, G. N., Gavrilov, A. A., Kasahara, M., & Kato, T. (2006). Last processing of geodynamics GPS measurement in Primorye. 5th Workshop on Subduction Processes emphasizing the Japan-Kuril-Kamchatka-Aleutian Arcs (JKASP-5). Abstract. July 6-14 Sapporo, Japan. p. 90–93.

Shestakov, N. V., Gerasimenko, M. D., Takahashi, H., M. Kasahara, M., Bormotov, V.A., Bykov, V.G., Kolomiets, A.G., Gerasimov, G.N., Vasilenko, N.F., Prytkov, A.S., Timofeev, V.YU., Ardyukov D.G. (2011). Present tectonics of the southeast of Russia as seen from GPS observations. Geophys. J. Int., 184, 529–540.

Shevchenko, G., Ivelskaya, T., & Loskutov, A. (2014). Characteristics of the 2011 Great Tohoku Tsunami on the Russian Far East Coast: deep-Water and Coastal Observations. *Pure and Applied Geophysics, 171*(12), 3329–3350. https://doi.org/10.1007/s00024-013-0727-1.

Soloviev, S. L., & Go, Ch N. (1974). *Catalogue of Tsunami on the Western Pacific Coast.* Moscow: Nauka. **(in Russian)**.

Tidal Phenomena. (2003). In: A. S. Vasiliev, F. S. Terziev, A. N. Kosarev (Eds.) Hydrometeorology and Hydrochemistry of Seas. Vol. VIII. Japan Sea, Issue 1, (pp. 262–265). St. Petersburg: Gidrometeoizdat. (**in Russian**).

Takahashi, M. (2017). The cause of the east-west contraction of Northeast Japan. *Bulletin Geological Survey Japan, 68*(4), 155–161. https://doi.org/10.9795/bullgsj.68.155.

Tamaki, K., & Honza, E. (1985). Incipient subduction and obduction along the eastern margin of the Japan Sea. *Tectonophysics, 119,* 381–406.

Tanioka, Y., Satake, K., & Ruff, L. (1995). Total analysis of the 1993 Hokkaido Nansei-oki earthquake using seismic wave, tsunami, and geodetic data. *Geophysics Research Letters, 22,* 9–12.

The diatoms of the USSR. (1974). *Fossil and recent.* Leningrad: Nauka.

The diatoms of the USSR. (1992). *Fossil and modern.* Leningrad: Nauka.

Tsuji, Y., Kato, K., Arai, K., & Ueda, K. (1994). Run-up height distribution of tsunami due to the southwest Hokkaido earthquake along coast of southwest Japan. *Kaiyo Monthly, 7,* 110–122.

Wei, D., & Seno, T. (1998). Determination of the Amurian plate motion, in Mantle Dynamics and Plate Interactions in East Asia. In M. F. J. Flower, S. Chung, C. Lo, & T. Lee (Eds.), *Mantle dynamics and plate interaction in East Asia, Geodynamics Series, 27* (pp. 337–346). Washington: AGU.

Xu, J., Pan, B., Liu, T., Hajdas, I., Zhao, B., Yu, H., et al. (2013). Climate impact of the Millennium eruption of Changbaishan volcano in China: insight from high-precision radiocarbon wiggle-match dating. *Geophysical Research Letters, 40*(1–6), 1–6. https://doi.org/10.1029/2012GL054246.

(Received October 9, 2017, revised March 6, 2018, accepted March 12, 2018)

Pure Appl. Geophys.

https://doi.org/10.1007/s00024-017-1745-1

Pure and Applied Geophysics

Airburst-Generated Tsunamis

Marsha Berger[1] and Jonathan Goodman[1]

Abstract—This paper examines the questions of whether smaller asteroids that burst in the air over water can generate tsunamis that could pose a threat to distant locations. Such airburst-generated tsunamis are qualitatively different than the more frequently studied earthquake-generated tsunamis, and differ as well from tsunamis generated by asteroids that strike the ocean. Numerical simulations are presented using the shallow water equations in several settings, demonstrating very little tsunami threat from this scenario. A model problem with an explicit solution that demonstrates and explains the same phenomena found in the computations is analyzed. We discuss the question of whether compressibility and dispersion are important effects that should be included, and show results from a more sophisticated model problem using the linearized Euler equations that begins to addresses this.

Key words: Tsunami, asteroid-generated airburst, shallow water equations, linearized Euler equations.

1. Introduction

In Feb. 2013, an asteroid with a 20 m diameter burst 30 km high in the atmosphere over Chelyabinsk, causing substantial local damage over a 20,000 km^2 region (Popova et al. 2013). The question arises what would be the effect of an asteroid that bursts over the ocean instead of land? The concern is that the atmospheric blast wave might generate a tsunami threatening populated coastlines far away.

There is little literature on airburst-generated tsunamis. Most of the literature on asteroids study the more complicated case of water impacts, where the meteorite splashes into the ocean (Weiss et al. 2006; Gisler et al. 2010; Gisler 2008). This involves much more complicated physics. The only reference we are aware of that relates to a blast-driven water wave is from the 1883 volcanic explosion of Krakatoa (Harkrider and Press 1967). The authors report a tide gauge in San Francisco registered a wave that could not be explained by a tsunami. There is also some analytic work in Kranzer and Keller (1959), where they derive asymptotic formulas for water waves from explosions and from initial cavities. There is more literature on meteo-tsunamis. These are also driven by air-pressure events and have similarities to our case, but occur in a different regime of air speed and water depth.

This paper studies the behavior of airburst-generated tsunamis, to better understand the potential threat. We compute the ocean response to a given (specified or pre-computed) atmospheric overpressure. The ocean response to this overpressure forcing, including the possible tsunami, is simulated. Typically, the shallow water equations are used for long-distance propagation, since they efficiently and affordably propagate waves over large trans-oceanic distances. Other alternatives, such as the Boussinesq equations, are much more expensive, and at least for the case of earthquake-generated tsunamis the difference seems to be small (Liu 2009).

In the first part of the paper, we present simulations under a range of conditions using the shallow water equations and the GeoClaw software package (GeoClaw 2017). We compute the ocean's response to an overpressure as calculated in Aftosmis et al. (2016). The overpressure was found by simulating the blast wave in air, and extracting the ground footprint. Roughly speaking, the blast wave model corresponds to the largest meteor that deposits all of its energy in the atmosphere without actually reaching the water surface. With this forcing, if there is no sizeable response, then we can conclude that that airbursts do not effectively transfer energy to the ocean, and there is little threat of distant inundation.

[1] Courant Institute, New York University, 251 Mercer St., New York City, NY 10012, USA. E-mail: berger@cims.nyu.edu

In general, our results using the shallow water equations suggest that airburst-generated tsunamis are too small to cause much coastal damage. Of course, depending on local bathymetry there could be an unusual response that is significant. For example, Crescent City, California is well known to be subject to inundation due to the configuration of its harbor and local bathymetry. However, we find that to generate a large enough response so that the water floods the coastline, the blast has to be so close that the blast itself is the more dangerous phenomenon. This is also the conclusion reached by Gisler et al. (2010) and Melosh (2003) for the case of asteroid water impacts.

In the second part of this paper, we study model problems to better understand and describe the phenomena we compute in the first part. The first model problem is based on the one-dimensional shallow water equations for which we can obtain an explicit closed form solution. It assumes a traveling wave form for the pressure forcing. Actual blast waves only approximately satisfy this hypothesis for a short time before their amplitudes decay. Nevertheless, the model explains several key features that we observe in the two-dimensional simulations. We observe a response wave that moves with the speed of the atmospheric forcing. There is also the gravity wave, or tsunami, moving at the shallow water wave speed c_w that is generated by the initial transient of the atmospheric forcing. We study in detail the response wave, or 'forced' wave, but the two are closely related. The analysis shows that the forced wave is proportional to the local depth of the water h at each location, a phenomena clearly seen in our computations. The model problem also allows us to assess the importance of nonlinear modeling. For most physical situations related to airburst tsunamis, the linear and nonlinear models give similar predictions.

In our final section, we assess the effect of corrections to the shallow water equations arising from compressibility and dispersion using a second model problem—the linearized Euler equations. Airbursts have much shorter time and length scales than the earthquakes that generate tsunamis, comparable to the acoustic travel time to the ocean floor. This leads to the question of whether compressibility of the ocean water could be a significant factor. In addition, airbursts have much shorter wavelengths, on the order of 10–20 km, at least for meteors with diameter less than 200 m or so. Recall that the shallow water model results from assuming long wavelengths and incompressibility of the water. Our results show that for airburst-generated tsunamis, dispersion can be significant but that compressibility is less so, suggesting interesting avenues for future work.

This work is an outgrowth of the 2016 NASA-NOAA Asteroid-generated Tsunami and Associated Risk Assessment Workshop. The workshop conclusions are summarized in Morrison and Venkatapathy (2017). Several other researchers also performed simulations, and videos of all talks are available online.[1]

2. *Two-Dimensional Simulations*

In this section, we present results from two sets of simulations. We use a 250MT blast, which roughly corresponds to a meteor with a 200 m diameter entering the atmosphere with a speed of 20 km/s. Generally speaking, this is the largest asteroid that would not splash into the water.[2] For each location, we did several simulations varying the blast locations with no meaningful difference in results, so we only present one representative computation in each set of simulations.

In the first set of results, we locate the blast in the Pacific about 180 km off the coast of near Westport, Washington. This spot was chosen since it is well studied by the earthquake-generated tsunami researchers due to its proximity to the M9 Cascadia fault (Petersen et al. 2002; Gica et al. 2014). By the time the waves reach shore, they have decayed and are under a meter high. Since they do not have the long length scales of earthquake tsunamis, we did not see any inundation on shore. In the second set of results, we move the location offshore to Long Beach,

[1] All presentations are available at https://tsunami-workshop.arc.nasa.gov/workshop2016/sched.php

[2] Initially, we used a blast wave corresponding to a 100MT blast, but since no significant response was found we do not include those results here. We also did simulations where we increased the 250MT pressure forcing by a factor of 1.2 with no change to the conclusions.

California, where there is significant coastal infrastructure, and has also been studied extensively in relation to earthquake tsunamis (Uslu et al. 2010). We place the blast approximately 30 km from shore, so that there is less time for the waves to decay. In all simulations, bathymetry is available from the NOAA National Center for Environmental Information website.

To perform these simulations, we use a model of the blast wave simulated in Aftosmis et al. (2016). The ground footprint for the overpressure was extracted, a Friedlander profile was fit to the data, and its amplitude as a function of time was modeled by a sum of Gaussians. The 250MT model is shown in Fig. 1, with a few of the profiles drawn in black to illustrate their form. The profiles start with the rise in pressure from the incoming blast, and are followed by the expected rarefaction wave (underpressure) some distance behind. Note that the maximum amplitude is over four atmospheres, but decays rapidly from its initial peak. In the model, the blast wave travels at a fixed speed of 391 m/s. If we take the sound speed at sea level to be 343 m/s, this corresponds to a Mach 1.14 shock.. This may be less accurate at early times. If the asteroid enters at a low angle of incidence, the blast wave travels more quickly when it first hits the ground. This would also lead to a more anisotropic response when see from the ground. Here, however, we assume that the blast wave is radially symmetric. We then use this model of the overpressure as a source term in our two-dimensional shallow water simulations using the software package `GeoClaw`.

`GeoClaw` is an open source software package developed since 1994 (LeVeque et al. 2011) for modeling geophysical flows with bathymetry using the shallow water equations. It is mostly used for simulations of tsunami generation, propagation and inundation. `GeoClaw` uses a well-balanced, second-order finite volume scheme for the numerics (LeVeque 1997; George 2008). Some of the strengths of `GeoClaw` include automatic tracking of coastal inundation, robustness in its handling of dry states, a local adaptive mesh refinement capability, and the automated setup that allows for multiple bathymetry input files with varying resolution. A bottom friction term is included using a constant Manning coefficient of 0.025. The results below do not include a Coriolis force, which we have found to be unimportant. There is no dispersion in the shallow water equations. In 2011, the code was approved by the US National Tsunami Hazard Mitigation Program (NTHMP) after an extensive set of benchmarks used to verify and validate the code (González 2012).

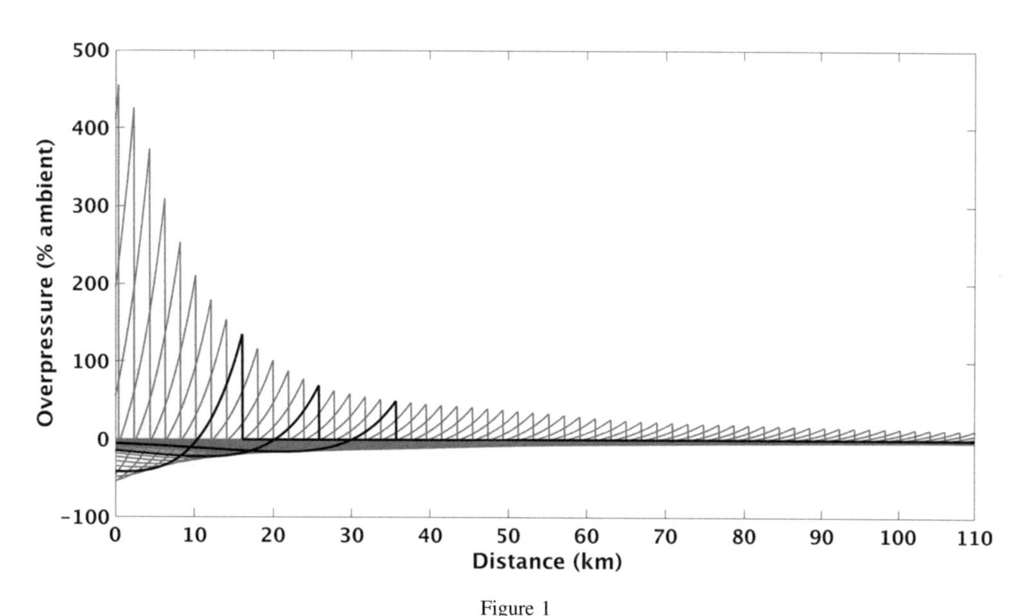

Figure 1
Ground footprint for 250MT blast wave overpressure as a function of distance from the initial blast. The curves are drawn every 5 s. A few of the curves are drawn in black to more clearly show a typical Friedlander profile

2.1. Westport Results

For this set of simulations, the 250MT blast is located at $-126.25°$ longitude and $46.99°$ latitude, about 30 km from the continental slope. The ocean is 2575 m deep at this spot. The blast location is about 180 km from shore. Many simulations were performed with different mesh resolutions. The finest grids used in the adaptive simulations had a resolution of 1/3 arc second. Three bathymetric data sets were used—a 1 min resolution covering the whole domain, a 3 s resolution nearer shore, and a 1/3 arc second bathymetry that included the shoreline itself.

In Fig. 2, we show a Hovmöller plot through the center of the blast location at a fixed latitude. On the left is the atmospheric overpressure for the first 300 s. This is the forcing that travels at 391 m/s. On the right is the amplitude of the water's response. Two waves traveling at different speeds are visible. A shallow water gravity wave travels with speed $c_w \approx \sqrt{gh}$ which at the blast location is 158 m/s. It is evident that the blast wave travels approximately twice as fast as the gravity wave. The blast wave reaches the edge of the graph in just over 150 s instead of the 300 s of the main water wave (in blue, since it is a depression). Also visible in the wave height plot is a wave that starts off in red and travels at the same speed as the blast wave, and whose amplitude decays more rapidly. Here, the color scale saturates below the maximum value in each plot so that smaller waves are visible.

Figure 3 shows the maximum amplitude found at any time in the simulation at that location. Note that the color bar is not linear in this plot, so that the different levels can more easily be seen. Nearest the blast location the maximum wave amplitude is over 10 m, but it decay rapidly. As the waves approach shore, the waves are amplified in a non-uniform way by the bathymetry. The coastline is outlined in black. We do not see any inundation of land, although admittedly at this resolution it would be hard to see.

Figure 4 shows the time history of wave heights through several gauge plots. The left plot shows seven gauges placed $0.1°$ apart (about 10 km at this latitude), starting about 1 km from the blast and at the same latitude. The gauge closest to the blast location has a maximum amplitude that reaches 5 m. Subsequent gauges show a very rapid decay in maximum amplitude. These positive elevation waves are the water's response to the blast wave overpressure, and travel at the same speed as the blast wave. Most of

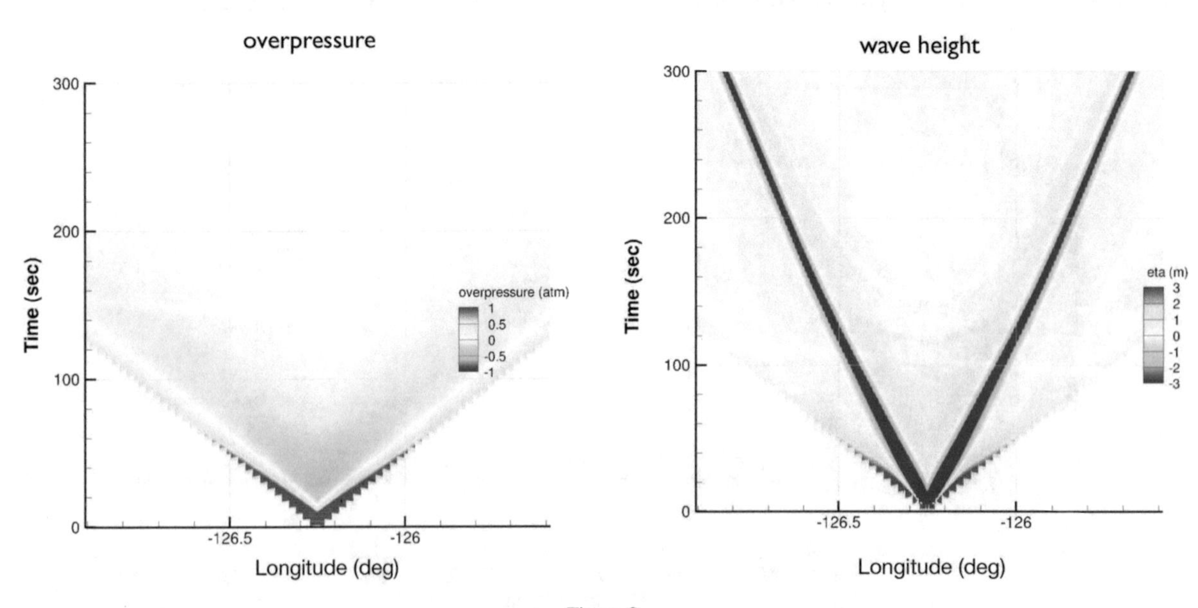

Figure 2
The Hovmöller plot shows the overpressure in atmospheres through the center of the blast location (left) and the wave height (right). The blast wave speed is approximately twice the gravity wave speed, and its amplitude decays more rapidly

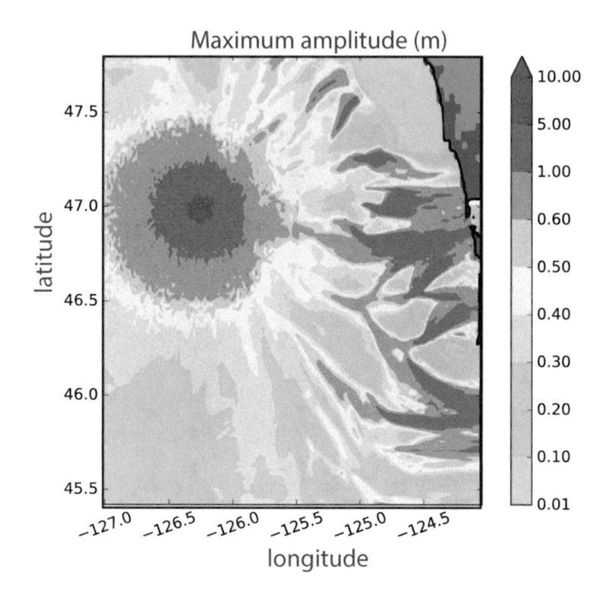

Figure 3
Maximum amplitude found between the blast location and the shoreline during the simulation. Note the color scale is not equally spaced

the ocean's response at this location appears as a depression, not an elevation. The negative amplitude wave travels at the gravity wave speed, $\sqrt{gh}$, where the water has depth h. It shows much less decay in amplitude. For example, looking at gauge 3 and 5, the peak amplitude decays from 2.7 to 0.72 m in about 50 s, whereas between 100 and 200 s, the trough decays from −5.3 to −4.1 m in about 100 s.

Figure 4 right shows gauges approaching the shore, at latitude 46.88° so that it lines up with the opening to Gray's Harbor. The gauges start about 100 km away from the blast. These are not equally spaced but are placed from 0.25 to 0.1° apart (from 25 to 10 km at this latitude), becoming closer as they approach shore and the bathymetry changes more rapidly. Shoaling is observed as the wave amplitudes increase, seen in gauges 17 and higher. The maximum elevation is between 0.5 and 1 m, and its duration is short, at least compared to earthquake-generated tsunamis.

Finally, Fig. 5 shows several close-ups of the region near shore. The waves are of uneven strength due to focusing from the bathymetry. The maximum amplitude is around 1 m. The sequence shows waves reflecting from the coastline but not flooding it. Some waves enter Grays Harbor, but they small amplitude and do not flood the inland area either.

2.2. *Long Beach Results*

For this set of experiments, we move the simulations to Long Beach, California. We locate the blast very close to shore so that the waves do not have time

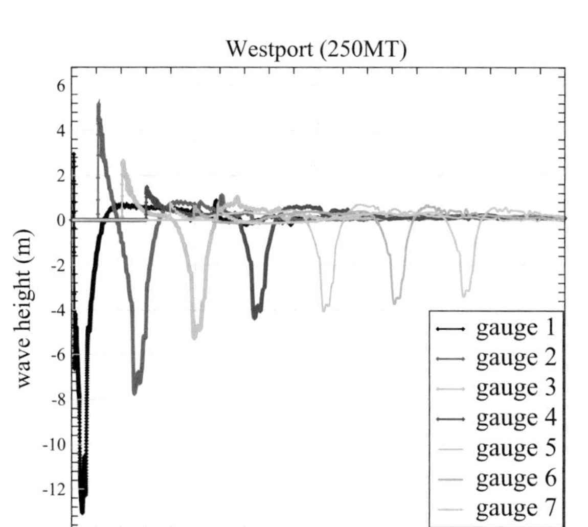

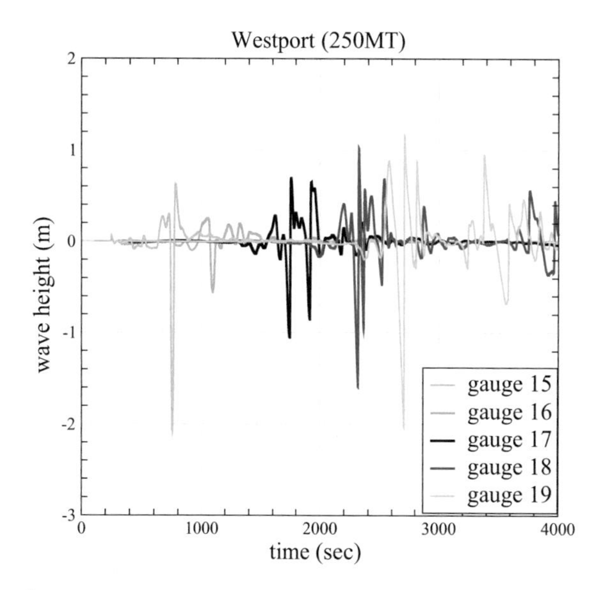

Figure 4
(Left) Gauges near blast location, every 0.1° starting 0.01° from blast. Gauges show rapid decay of maximum amplitude, much slower decay of maximum depressions. (Right) Gauges approaching shoreline show similar wave forms with decreasing maximum amplitude before shoaling increases it. These gauge locations are marked in Fig. 5

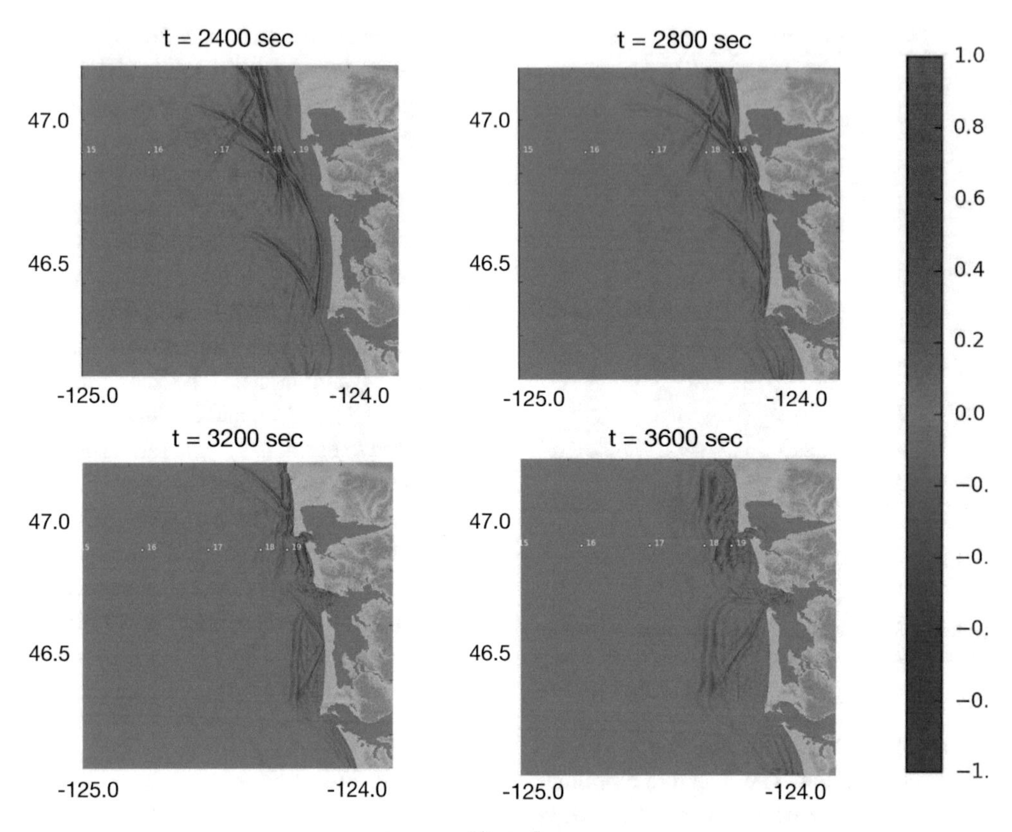

Figure 5
Zoom of waves approaching shoreline around Westport. No inundation is observed

to decay. Again we have detailed bathymetry at a resolution of 1/3 arc second between Catalina Island and Long Beach, and use a 1 min dataset outside of this region. The blast is located at -118.25° longitude and 33.41° latitude, where the ocean is 797 m deep. This is about 30 km from shore. Figure 6 shows the region where the blast is located, and a zoom of the Long Beach harbor where we will look for flooding.

Figure 7 shows the ocean response at several points in time. The black circle on the plots indicates the location of the air blast. In the plot marked at 25 s, note how the wave height in red that is closest to the blast location is not as circular as the air blast itself. It also decays faster than in the Westport computation. This will be explained by the model problem presented in the next section.

There is a breakwater that protects Long Beach. It reflects most of the waves that reach it, with only a small portion getting through the opening. Waves that reach the harbor go around the breakwater, and are reflected from the shoreline back into this region.

Figure 8 shows a plot of the maximum water amplitudes seen in the harbor area. We do see some overtopping of land, but it is very small. In several locations, it reaches 0.5 m, where the inlet exceeds its boundaries, and on the dock in the middle. The region with the largest accumulation is just outside the harbor before the breakwater, where the maximum amplitude seen is between 3 and 6 m. There is a steep cliff here, however, and the water does not propagate inland. Paradoxically, in other experiments where the blast was located closer to shore by a factor of 2, there was no overtopping. This can also be explained by our model problem in the next section.

3. *Shallow Water Model*

In this section, we present a one-dimensional model of the shallow water equations (SWE) that explains much of the behavior seen in the previous

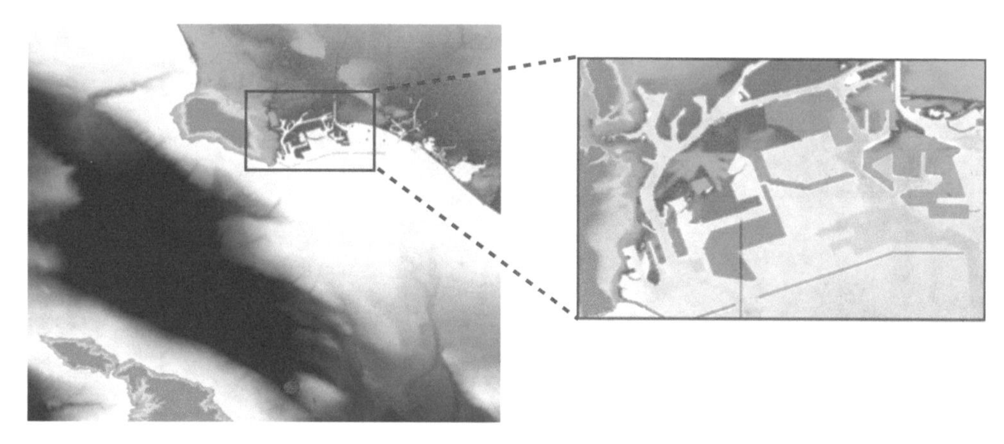

Figure 6
Location of airburst northeast of Catalina, and zoom of Long Beach shoreline

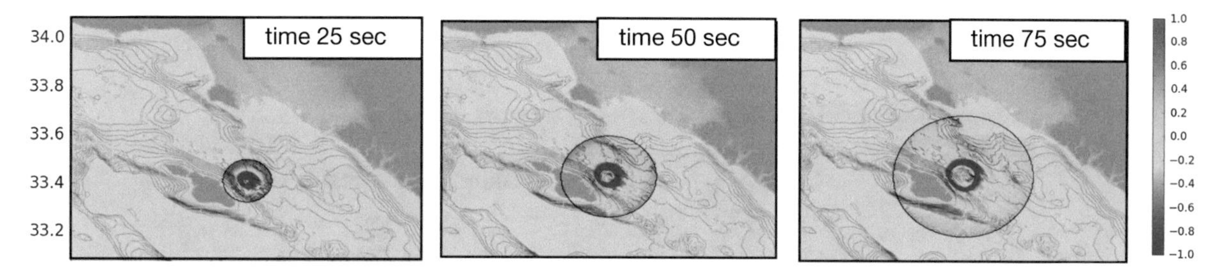

Figure 7
Snapshots at early times of blast wave and ocean waves in Long Beach simulation

examples. In the SWE, the atmospheric overpressure appears as an external forcing p_e in the momentum equation. In one space dimension, it is

$$h_t + (hu)_x = 0, \\ (hu)_t + (hu^2 + \frac{1}{2}gh^2)_x = \frac{-h p_{e_x}}{\rho_w}, \tag{1}$$

where h is the height of the water surface over the bottom, u is the depth-averaged velocity of the water in the x direction, g is gravity, and ρ_w is the density of water. The external pressure forcing is p_e, and its x derivative is p_{e_x}. The notation is illustrated in Fig. 9. We assume constant bathymetry in the model, $B(x,t) = 0$. See Vreugdenhil (1994) for these equations, or Mandli (2011) for a clear and complete derivation. In this section (and the next), the conclusions are in the last few paragraphs after the analysis.

3.1. Derivation and Analysis

As stated in the introduction, we simplify the pressure forcing by assuming it has the form of a traveling wave, and look for solutions h and u that are traveling waves too. This means they are functions only of the moving variable

$$m = x - st\ .$$

so that $\partial_t h = -s\,\partial_m h(x - st) = -s\,h_m$. Equation (1) becomes a pair of ordinary differential equations

$$-sh_m + (hu)_m = 0, \tag{2}$$

$$-s(hu)_m + (hu^2 + \frac{1}{2}gh^2)_m = \frac{-hp_{e_m}}{\rho_w}\ . \tag{3}$$

Equation (2) can be integrated to give $-sh + hu =$ const . We evaluate the constant by

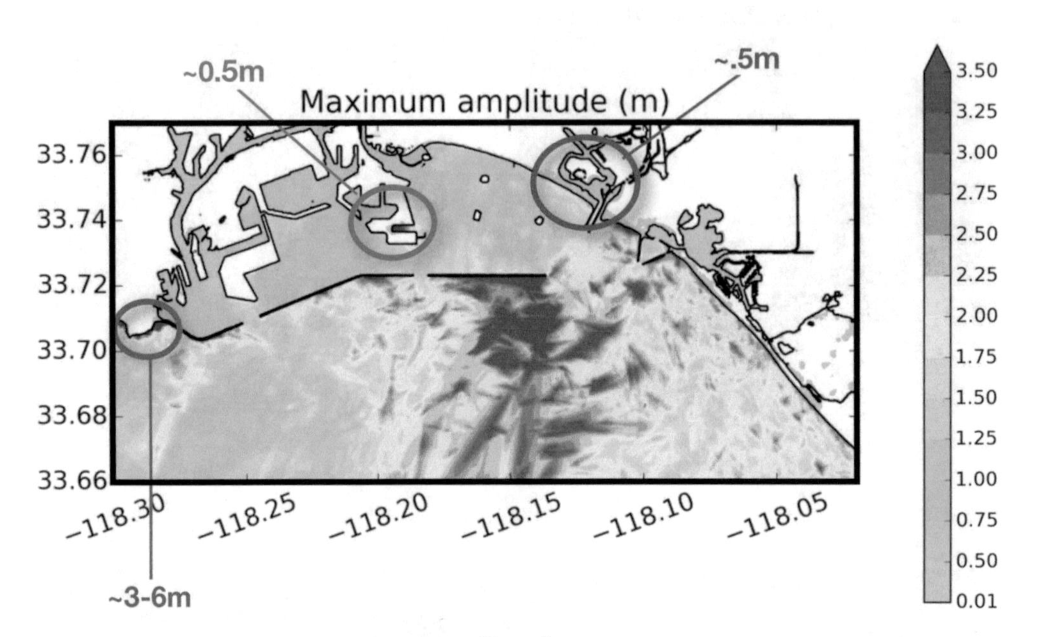

Figure 8
Maximum amplitude plot shows 0.5 m of water overtopping the dock and the riverbank

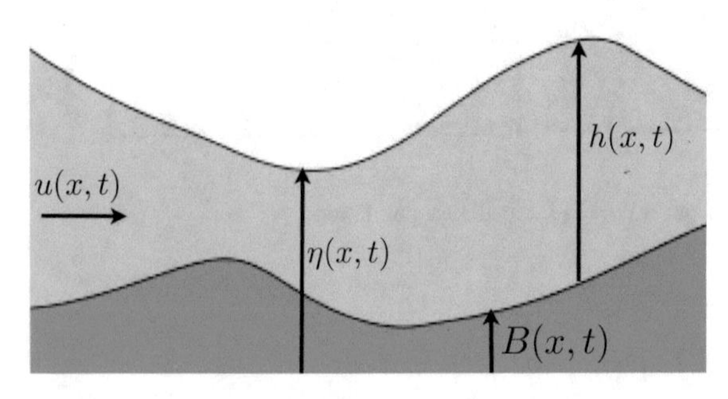

Figure 9
Illustration of variables for Eq. (1). In the general case, the bathymetry $B(x, t)$ varies in space and possibly time. Here, we take it to be constant

taking $m \to \infty$, where $u \to 0$ and $h \to h_0$, with h_0 the undisturbed water height. (We assume that the overpressure has localized support, and goes to zero as $m \to \infty$.) Therefore, $-sh + hu = -sh_0$. This may be re-written as:

$$u(m) = \frac{s(h(m) - h_0)}{h(m)} .$$

We use this to eliminate u from (3), which gives

$$-s\,(s(h - h_0))_m + \left(\frac{s^2(h - h_0)^2}{h}\right)_m + \left(\frac{1}{2}gh^2\right)_m = \frac{-hp_{em}}{\rho_w} .$$

After some algebra, this leads to

$$\frac{s^2}{2}\left(\frac{h_0^2}{h^2}\right)_m + g\,h_m = \frac{-p_{em}}{\rho_w} . \tag{4}$$

As before, this may be integrated exactly. Again we use the boundary conditions $h \to h_0$, $u \to 0$, and $p_e \to 0$ as $m \to \infty$. The result is

$$\frac{s^2}{2}\left(1 - \frac{h_0^2}{h(m)^2}\right) + gh_0\left(1 - \frac{h(m)}{h_0}\right) = \frac{p_e(m)}{\rho_w} . \tag{5}$$

To summarize, Eq. (5) is the water's response according to shallow water theory. The solution of the differential equation system (2) and (3) is an algebraic relation between the overpressure and the response height. It tells us that the water height at a point $m = x - st$ is determined by the overpressure at the same point.

To get a better feel for the behavior of the solution (5), we linearize it, writing $h(m) = h_0 + h_r(m)$ where $h_r(m)$ is the response height. The linearization uses the relation

$$\left(\frac{h_0}{h(m)}\right)^2 = \frac{h_0^2}{(h_0 + h_r)^2} \approx 1 - \frac{2h_r}{h_0} ,$$

which is valid when $h_r \ll h_0$. This is our case, since the change in wave height h_r is a number in meters where h_0 is typically measured in kilometers. The linearization of (5) is

$$h_r = \frac{h_0 \, p_e}{\rho_w (s^2 - c_w^2)} \, . \quad (6)$$

Figure 10 shows that the full response theory (5) and the linear approximation (6) are very close to each other for the parameters of interest. The plot uses a constant depth of $h_0 = 4$ km, and takes $\rho_w = 1025$ kg / m ^{3}. The maximum difference between the nonlinear and linear wave heights in Fig. 10 is half a meter, less than 2%, when the overpressure is five atmospheres.

To enumerate the consequences of the response predicted by (5) and (6), we observe

1. The response wave height h_r is linearly proportional to the depth h_0. A pressure wave over deep ocean has a stronger effect than a pressure wave over a shallower continental shelf. This explains why locating the blast in the Long Beach case closer to shore had less of an effect. If the distance offshore of the air blast from Long Beach is halved to 15 km, the ocean is only 90 m deep, resulting in approximately 1/10 the impact response. On the other hand, there is almost no difference in the decay rate of the shallow water waves before they reach shore.
2. If $s > c_w$, then p_e and h_r have the same sign. The response height is positive in regions of positive overpressure. This contradicts an intuition that positive overpressure would depress the water surface. This response is similar to the case of a forced oscillator in vibrational analysis. Consider for example $\ddot{x} = -x + A\cos(\omega t)$. The steady solution is

$$x(t) = \frac{A}{1 - \omega^2} \cos(\omega t) \, .$$

For $\omega > 1$, the response $x(t)$ has the opposite sign from the forcing $A\cos(\omega t)$. For pressure forcings with speeds slower than the water speed, the water response would be a depression, with h_r negative. This response is clearly seen in the all the simulations. The wave that travels at the speed of the blast wave is an elevation. However, in the Long Beach results, we can see that the response wave is not uniformly circular when the depth of the water changes rapidly. Note that if we take the speed of sound in air is 343 m/s, the water would have to be more than 12 km deep for the gravity wave speed to exceed the speed of the pressure forcing. So in all cases on earth we expect an elevation of the sea surface beneath the pressure wave.
3. The response is particularly strong when the forcing speed s is close to the gravity wave speed $c_w \approx 200$ m/s., (for $h_0 = 4$ km). In this case, we have a Proudman resonance (Proudman 1929; Monserrat et al. 2006). This is the regime for meteo-tsunamis, in basins whose depth leads to

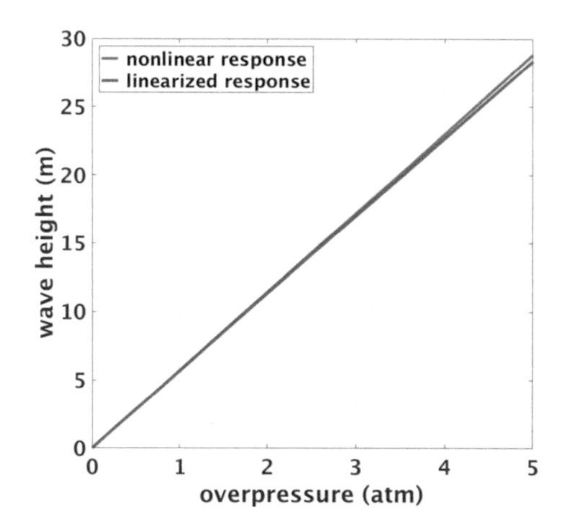

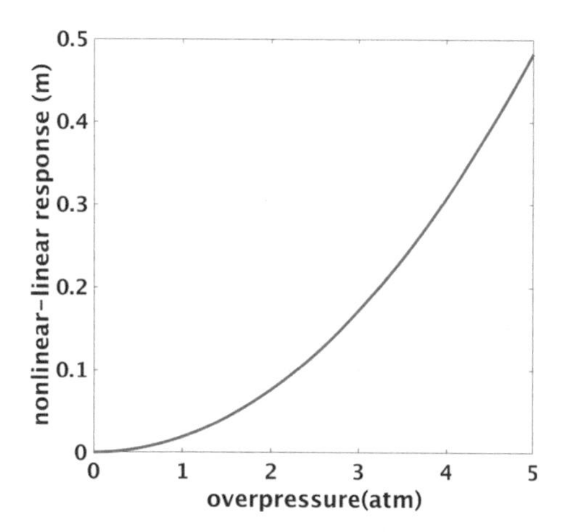

Figure 10
Wave height as a function of overpressure, for the nonlinear Eq. (5) and the linearized Eq. (6), using $h_0 = 4$ km and $s = 350$ m/s. The curves are very close. The right figure is a plot of their difference, which is on the order of a percent

gravity wave speeds that match the squall speeds. These speeds are much slower than the speed of sound in air.

3.2. Shallow Water Model Computations

This subsection presents solutions of the one dimensional model (1) computed using `GeoClaw`. These show the generation of freely propagating gravity waves along with the forced response wave described above. The sealevel and bathymetry are taken to be flat for $t \leq 0$, so the initial conditions for this simulation are $h = 0$ and $u = 0$ at $t = 0$. For $t \geq 0$, the pressure forcing is

$$\begin{aligned} p_e &= p_{\text{ambient}} \exp(-0.1(x-st)^2) \\ &= p_{\text{ambient}} \exp\left(-\frac{1}{2}\left(\frac{x-st}{\sqrt{5}}\right)^2\right), \end{aligned} \tag{7}$$

with p_{ambient} = 1 atm. One can check that the solution to the linearized equations is

$$\begin{aligned} h(x,t) = h_r(x-st) &- \left(\frac{s}{c_w}+1\right)\frac{h_r(x-c_w t)}{2} \\ &+ \left(\frac{s}{c_w}-1\right)\frac{h_r(x+c_w t)}{2}, \end{aligned} \tag{8}$$

The computed solution to the nonlinear equation is very close to this. It corresponds to the response wave moving to the right, and two gravity waves. Conservation of water requires that the sum of the amplitudes is zero:

$$1 - \frac{1}{2}\left(\frac{s}{c_w}+1\right) + \frac{1}{2}\left(\frac{s}{c_w}-1\right) = 0\,. \tag{9}$$

The gravity waves move at the shallow water speed $c_w = \sqrt{gh_0}$. The forced wave (what we have been calling the response wave) moves with speed s.

These initial conditions have a non-zero air blast pressure wave only at time $t > 0.0$. It is not a pure traveling wave, since the forcing p_e is zero before $t = 0.0$. The general solution to this linearized shallow water model is a combination of the inhomogeneous solution with forcing function p_e, which gives the response wave h_r, and the solution to the homogeneous problem with $p_e = 0$, which consists of left and right moving gravity waves of the same shape h_r but different amplitudes. The full solution is a linear combination of all three waves that satisfy the initial conditions. Equation (8) shows that the left-going tsunami wave (rightmost term) will have a smaller amplitude in absolute value for $s/c_w > 1$ than the right going wave (middle term on right hand side). Also, the latter will be a depression, since it has amplitude $-0.5 \cdot (1 + s/c_w))$.

Numerical results illustrating this are shown in Fig. 11 at time 250 s. Solid lines show the water wave heights, and dashed lines show the air overpressure profiles. In Fig. 11 left, the speed $s = 350$ m/s, somewhat larger than the speed of sound in air. For this case, since $s > c_w$, (for h_0 = 4 km, $c_w = \sqrt{gh_0} \simeq 198$ m/s), the forced wave height is positive since the overpressure is positive. Note that the gravity wave at the same point in time trails the pressure wave. The right-moving gravity wave is a depression, the left moving wave is a smaller elevation. Since this calculation is in one space dimension, the waves do not decay. In the two-dimensional shallow water equations, the gravity wave decays with the square root of distance. Also, the pressure blast wave and, therefore, the leading water response would both decay too.

By contrast, Fig. 11 right shows the water's response for an overpressure moving at 120 m/s, slower than the gravity wave ($s < c_w$). The tsunami waves travel at the same speed in both computations, but they have different amplitudes and signs. The tsunami wave is the opposite sign as the wave due to the pressure. This is consistent with conservation of mass.

To give a more complete picture, two more experiments with $s > c_w$ but different forcings are shown in Fig. 12. The figures on the left use a Gaussian pressure forcing but their magnitude is ramped up for the first 100 s. This results in quite a different-looking gravity wave. The right figure uses a typical Friedlander blast profile described in Sect. 2 for the overpressure, but keeping the amplitude constant at 1 atm. It looks similar to the Gaussian example above.

3.3. Breakdown of Smooth Solutions

In a bit of a digression, this subsection examines in a little more detail the nonlinear shallow water

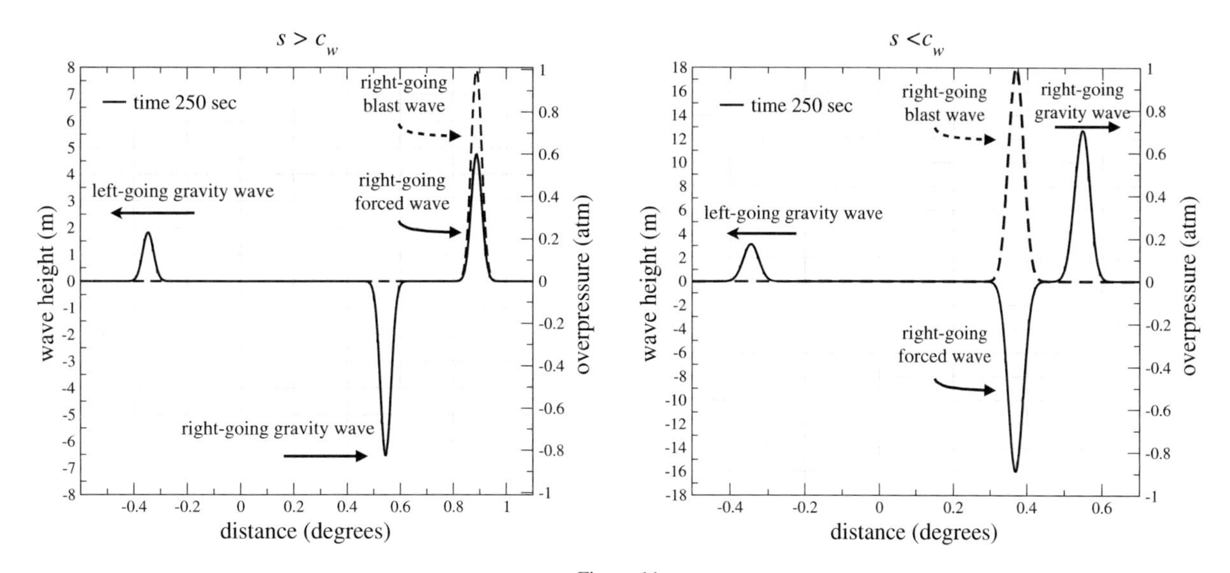

Figure 11
Numerical simulations showing wave heights h_r (left axis) and overpressure p_e (labeled on right axis). Left experiment uses $s > c_w$, on the right the pressure front is slower. On the left, the wave heights (solid line) under the pressure pulse (dashed line) are positive, as Eq. (6) predicts, and the tsunami wave trails the pressure wave. On the right, the pressure pulse is above a negative wave height (depression), and the tsunami leads the pressure wave

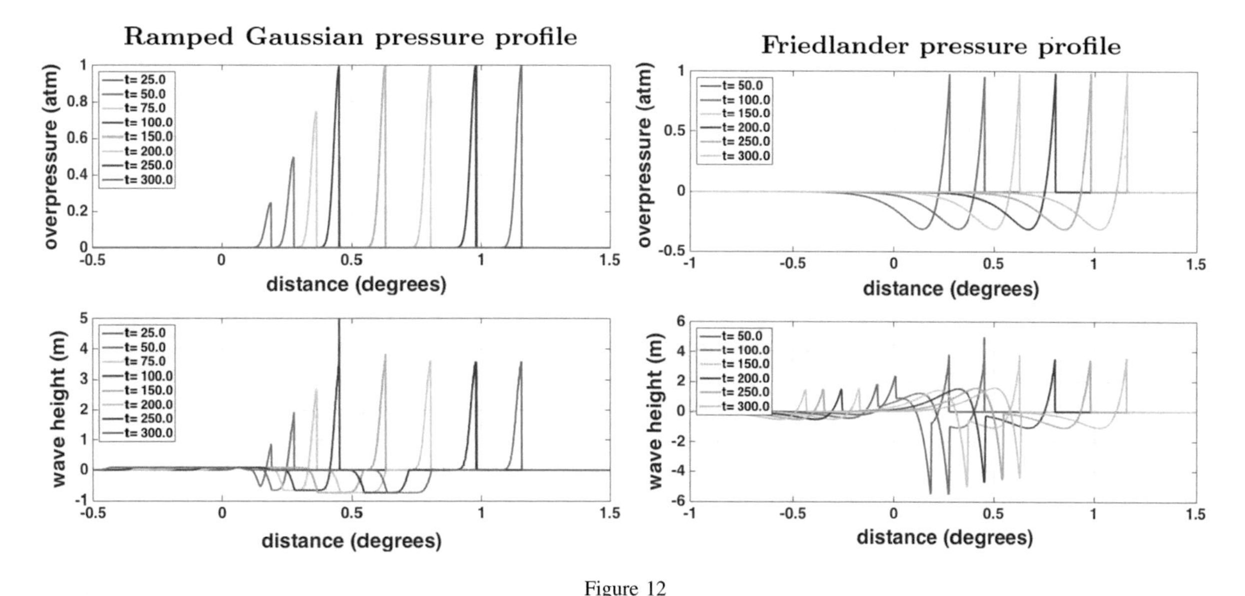

Figure 12
Left figure uses same high-speed Gaussian pressure pulse as in Fig. 11 but with the amplitude linearly ramped up over 100 s. Pressure profile shown on the top; bottom shows water wave heights. Right figure uses a Friedlander blast wave profile instead of a Gaussian. Both figures show the same positive forced water wave (since $s > c_w$) and the expected gravity waves from (8) but they take very different forms

response (5), and when the solution h exists. Dividing (5) by the shallow water speed $c_w^2 = gh_0$, and defining a function $f(H)$ equal to the left hand side, we get

$$f(H) = \frac{s^2}{2c_w^2}\left(1 - \frac{h_0^2}{h(m)^2}\right) + \left(1 - \frac{h(m)}{h_0}\right)$$
$$= \frac{s^2}{2c_w^2}\left(1 - \frac{1}{H^2}\right) + (1 - H)) = \frac{p_e(m)}{\rho_w c_w^2}, \tag{10}$$

where we define $H = h(m)/h_0$.

Figure 13 (left) shows a plot of $f(H)$, for overpressure traveling wave speeds s ranging from 0 to 400 m/s. This includes pressure waves both slower and faster than the water wave speed $c_w \approx 200$ m/s in a 4 km ocean. Note that there is a range where there is no solution to the equations. For an overpressure of 1 atmosphere, the right hand side of (10) is .0025. For the curve with $s = 350$ m/s, the maximum of $f(h) \approx 0.37$, so the equation will have a solution for a much stronger overpressure and/or a much shallower ocean basin. Between 0.8 and 1.2 (which corresponds to a wave response of 0.8 m), there is no solution for a right hand side of (10) outside of the range roughly between -1 and 1.

Figure 13 (right) shows a particular solution $h_r(m)$ for increasing overpressure amplitudes for a Gaussian traveling with speed $s = 350$ m/s in a 4 km deep ocean. As the amplitude of the forcing increases, the solution develops a cusp, and it appears that the derivative approaches infinity. This lack of a smooth solution could precede a bore.

4. Linearized Euler Model

In this section, we analyze a more complete model of the ocean's response to an airburst, to uncover possible shortcomings of the shallow water model of Sect. 3. We model the water using the Euler equations of a compressible fluid, which will bring in the effects of compressibility and dispersion. Another possibility would be to use one of the forms of the Boussinesq equations, but that also assumes incompressible flow, and would be more difficult to analyze. [See, however, a nice comparison of SWE and Serre–Green–Naghdi Boussinesq results in Popinet (2015).] We continue to neglect Coriolis forces, viscosity, friction, the Earth's curvature, etc. We linearize the Euler equations and the boundary conditions, since Fig. 10 suggests that linear approximations are reasonably accurate for these parameters.

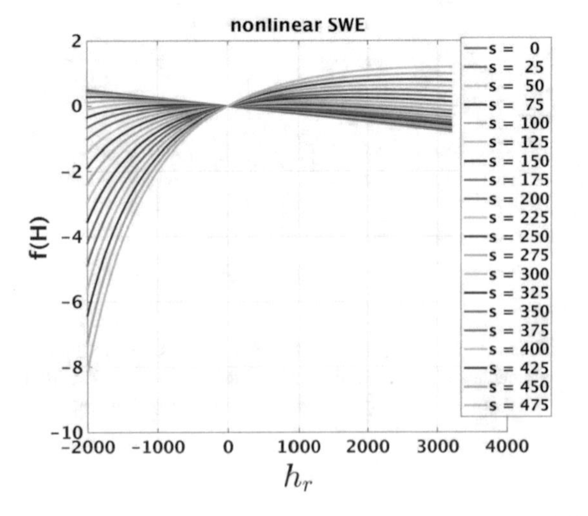

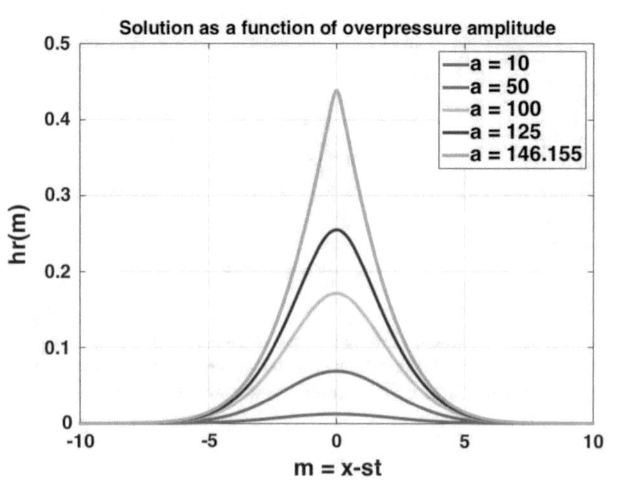

Figure 13
Left shows the region where a solution exists, for a given overpressure amplitude. Right shows the change in solution as the amplitude increases, at the fixed speed $s = 350$ m/s. When the cusp develops, there is no longer a smooth solution, corresponding to the region with no solution on the left plot

4.1. Derivation and Analysis

Our starting point for this section is the linearized Euler equations with linearized boundary conditions. A derivation is given in the Appendix. An explicit solution is not possible, and the results will depend instead on wave number. We will use wave number $k = \frac{2\pi}{L}$, where the length scale L for the atmospheric pressure wave is on the order of 10–20 km. This is very short relative to earthquake-generated tsunamis, which can have length scales on the order of 100 km or more. As before, those not interested in the analysis can skip to the end of the section for a summary of the main points.

The linearized Euler equations and boundary conditions that we use for analysis are

$$\begin{aligned} \widetilde{\rho}_t + \rho_w \widetilde{u}_x + \rho_w \widetilde{w}_z &= 0, \\ \rho_w \widetilde{u}_t + c_a^2 \widetilde{\rho}_x &= 0, \\ \rho_w \widetilde{w}_t + c_a^2 \widetilde{\rho}_z &= -\widetilde{\rho} g , \end{aligned} \tag{11}$$

where c_a is the speed of sound in water. (We use c_a for acoustic to distinguish it from the gravity wave speed $c_w = \sqrt{gh}$). Here, $\widetilde{\rho}$ is a small perturbation of ρ (and the same for the other variables), except for

$$h(x,t) = h_0 + h_r(x,t) \ .$$

where h_r is again the water's disturbance height for consistency with the previous section. The boundary conditions are

$$\text{bottom}: \widetilde{w}(x, z = 0, t) = 0, \tag{12}$$

$$\text{top}: \frac{\partial h_r(x,t)}{\partial t} = \widetilde{w}(x, h_0, t) \tag{13}$$

$$\text{pressure bc:} \quad \begin{aligned} & c_a^2 \widetilde{\rho}(x, h_0, t) - \rho_w g\, h_r(x,t) \\ & = p_e(x,t). \end{aligned} \tag{14}$$

As in Sect. 3, we will assume the atmospheric pressure forcing has the form $p_e(x - st)$, and look for solutions of the same form, functions of $m = x - st$ and z. The system (11) becomes

$$-s\widetilde{\rho}_m + \rho_w \widetilde{u}_m + \rho_w \widetilde{w}_z = 0, \tag{15a}$$

$$-s\rho_w \widetilde{u}_m + c_a^2 \widetilde{\rho}_m = 0, \tag{15b}$$

$$-s\rho_w \widetilde{w}_m + c_a^2 \widetilde{\rho}_z = -\widetilde{\rho} g \ . \tag{15c}$$

The boundary conditions become

$$\widetilde{w}(m, 0) = 0, \tag{16a}$$

$$\widetilde{w}(m, h_0) = -s\, h_{r,m}(m), \tag{16b}$$

$$c_a^2 \widetilde{\rho}(m, h_0) = \rho_w g h_r(m) + p_e(m) \ . \tag{16c}$$

This system now includes the effects of dispersion and water compressibility.

These equations cannot be solved in closed form for general p_e. Therefore, we study the response using Fourier analysis. We will take a Fourier mode of the overpressure

$$p_e(m) = A_k e^{ikm} , \tag{17}$$

with amplitude A_k and compute the response as a function of m. The responses will have the form

$$h_r(m) = \widehat{h}_r\, e^{ikm}, \tag{18a}$$

$$\widetilde{\rho}(m,z) = \widehat{\rho}(z) e^{ikm}, \tag{18b}$$

$$\widetilde{u}(m,z) = \widehat{u}(z) e^{ikm}, \tag{18c}$$

$$\widetilde{w}(m,z) = \widehat{w}(z) e^{ikm} \ . \tag{18d}$$

The hat variables are the Fourier multipliers. The partial differential equations (15a–15c) become ordinary differential equations with wave number k as a parameter.

Note that (15b) depends only on derivatives with respect to m. Integrating it gives

$$-s\rho_w \widetilde{u} + c_a^2 \widetilde{\rho} = 0.$$

The constant of integration is zero for each z since as $m \to \infty$ we know $\widetilde{u} = 0$ and $\widetilde{\rho} = 0$. This gives an expression for $\widetilde{\rho}$ in terms of $\widetilde{u}$,

$$\widetilde{\rho} = \frac{s\rho_w \widetilde{u}}{c_a^2}, \tag{19}$$

which we can use in (15a) and (15c). After substituting for $\widetilde{\rho}$ and dividing by ρ_w, the remaining system of two equations is

$$\begin{aligned} \widetilde{u}_m \left(1 - \frac{s^2}{c_a^2}\right) + \widetilde{w}_z &= 0, \\ -\widetilde{w}_m \quad + \widetilde{u}_z &= -\frac{g}{c_a^2} \widetilde{u}. \end{aligned} \tag{20}$$

Substituting the Fourier modes (18c–18d) into (20)

and differentiating $\widetilde{u}$ and $\widetilde{w}$ with respect to m give an ordinary differential equation in z for the velocities,

$$\begin{pmatrix} \widehat{u} \\ \widehat{w} \end{pmatrix}_z = \begin{pmatrix} -g/c_a^2\,\widehat{u} + i\,k\,\widehat{w} \\ -i\,k\,\widehat{u}\,(1 - s^2/c_a^2) \end{pmatrix} = \begin{bmatrix} -g/c_a^2 & i\,k \\ -i\,k\,(1 - s^2/c_a^2) & 0 \end{bmatrix} \begin{pmatrix} \widehat{u} \\ \widehat{w} \end{pmatrix}. \tag{21}$$

The general solution to this 2-by-2 system is the linear combination

$$\begin{pmatrix} \widehat{u} \\ \widehat{w} \end{pmatrix} = a_+ \mathbf{v}_+ e^{\mu_+ z} + a_- \mathbf{v}_- e^{\mu_- z}, \tag{22}$$

where $\mu_\pm$ and $\mathbf{v}_\pm$ are the eigenvalues and eigenvectors of the matrix in (21), and the scalar coefficients $a_\pm$ are chosen to satisfy the boundary conditions. The eigenvalues are

$$\mu_\pm = \frac{\frac{-g}{c_a^2} \pm \sqrt{\frac{g^2}{c_a^4} + 4k^2(1 - s^2/c_a^2)}}{2}. \tag{23}$$

The eigenvectors (chosen to make the algebra easier so they are not normalized) are

$$\mathbf{v}_+ = \begin{pmatrix} \frac{2\mu_+}{-ik} \\ 2(1 - s^2/c_a^2) \end{pmatrix}, \quad \mathbf{v}_- = \begin{pmatrix} \frac{2\mu_-}{-ik} \\ 2(1 - s^2/c_a^2) \end{pmatrix}. \tag{24}$$

The boundary condition at $z = 0$ is (16a). To apply it, note that $\widehat{w}$ corresponds to the second component of the eigenvectors $\mathbf{v}_\pm$. We find that $a_+ = -a_-$. Henceforth, we call this coefficient simply a.

Next, we substitute the Fourier modes (18c–18d) into the remaining boundary conditions (16b) and (16c). We use the pressure forcing equation (17) in the form $\widehat{p}_e = A_k$. The result is

$$\widehat{w}(h_0) = -iks\,\widehat{h}_r, \tag{25a}$$

$$c_a^2\widehat{\rho}(h_0) - \rho_w g \widehat{h}_r = A_k. \tag{25b}$$

Using equation (19) to substitute $\widehat{\rho} = \frac{s\rho_w}{c_a^2}\widehat{u}$ in (25b) gives an expression for $\widehat{h}_r$

$$\widehat{h}_r = \frac{s}{g}\widehat{u} - \frac{A_k}{\rho_w g}. \tag{26}$$

This can be used to replace $\widehat{h}_r$ in (25a) to get

$$\widehat{w}(h_0) = \frac{-iks^2}{g}\widehat{u}(h_0) + \frac{iksA_k}{\rho_w g}. \tag{27}$$

The final steps are using the form of the solution (22) in (27) to solve for the coefficient a. With this, everything is known, and $\widehat{u}$, $\widehat{w}$ and the response height $\widehat{h}_r$ can be evaluated.

Putting it all together, we get

$$2\,a(1 - s^2/c_a^2)\left(e^{\mu_+ h_0} - e^{\mu_- h_0}\right) = \frac{iks^2}{g}\frac{2\,a}{-ik}\left(\mu_+ e^{\mu_+ h_0} - \mu_- e^{\mu_- h_0}\right) + \frac{iksA_k}{\rho_w g}. \tag{28}$$

Grouping terms, the final expression to solve for a (using the definition (23) for $\mu_\pm$) is given by

$$2a\left[(1 - s^2/c_a^2)\left(e^{\mu_+ h_0} - e^{\mu_- h_0}\right) + \frac{s^2}{g}\left(\mu_+ e^{\mu_+ h_0} - \mu_- e^{\mu_- h_0}\right)\right] = \frac{iksA_k}{\rho_w g} \tag{29}$$

To summarize, given an overpressure amplitude A_k with wavelength k, equation (29) gives the scalar coefficient a in the velocity equations, then we solve for $\widehat{u}$ and $\widehat{w}$ using (22), and use (26) to get the Fourier multiplier for the wave height response.

4.2. Linearized Euler Model Computations

We evaluate these results using the following parameters: an ocean with depth $h_0 = 4$ km , ocean sound speed $c_a = 1500$ m/s, $\rho_w = 1025$ kg/m ^{3}, and atmospheric overpressure of $A_k = 1$ atm with pressure wave speed $s = 350$ m/s, faster than the gravity wave speed of about 200 m/s. The responses are linear in the overpressure amplitude A_k, so we do not evaluate these curves for any other overpressures.

Figure 14 (left) shows the surface wave height $\widehat{h}(k)$ as a function of length scale L, and (right) the amplitude of the surface velocities $\widehat{u}(h_0, k)$ and $\widehat{w}(h_0, k)$ are shown. There are two curves in each plot: one uses the physical acoustic water wave speed of $c_a = 1500$ m/s, and the other uses a very large non-physical acoustic speed in the water of $c_a \times 10^8$. The latter corresponds to the intermediate model of finite depth but incompressible water. This should, and does, asymptote in the long wave ($k \to 0$) limit to the result of the shallow water equations. The difference

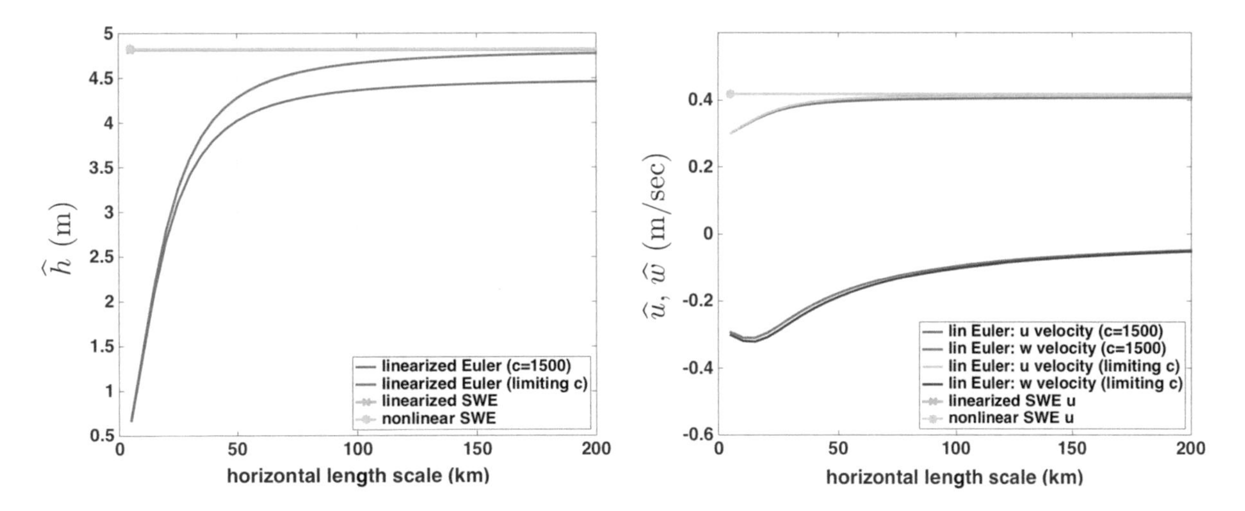

Figure 14
Left figure shows wave height $\widehat{h}(k)$ as a function of wavelength for the linearized Euler equations using an atmospheric overpressure of 1 atmosphere. Also shown is the shallow water solution from Sect. 3. Right figure shows the u and w velocities. Both plots show curves using the physical sound speed of $c_a = 1500$ m/s, and the limiting infinite speed solution. Both figures use the parameters $h_0 = 4$ km, and 1 atmosphere overpressure

between the blue and green curves shows approximately a 20% reduction in the amplitude of the longer length scales due to compressibility (but that this is not the amplitude of the total wave response yet). Note also that the u velocity asymptotes to the shallow water limit, and the w velocity approaches zero. The velocity curves show less of an effect due to compressibility.

For atmospheric forcing from asteroids with airbursts, the length scales of interest are closer to the short end, perhaps 10 or 20 km. In this regime, the compressibility effects are around 10% or less. But at these wavelengths, dispersive effects reduce the response predicted by shallow water theory by nearly half!

This becomes more clear by comparing the forced wave response to a Gaussian pressure pulse instead of using just a single frequency. We use the pressure pulse form (7) as before but with length scale 5 (not $\sqrt{5}$), take the Fourier transform, multiply by the Fourier multipliers shown in Fig. 14, and transform back. Figure 15 shows the results for two different water depths h_0: 4 km and 1 km. The blue curve uses the water wave speed c_a=1500 m/s, and the red curve uses the limiting c_a. Compressibility changes the height by less than 10% in Fig. 15 for both depths. However, in the deeper water, the shallow water response is almost 70% larger than either Euler solution, and has a narrower width. One can think of this in two ways. The broader response from linearized Euler is a result of dispersion, which makes things spread. Alternatively, it is the result of filtering out the shorter wavelengths needed to make a narrow Gaussian. In the right figure, the water is shallower, and the linearized Euler results are closer to the shallow water results.

In Fig. 16, we fix the horizontal length scale at 15 km and instead vary the speed of the pressure wave s. This figure again uses $h_0 = 4000$ m, and $c_a = 1500$ m/s. Three curves are shown: the linearized Euler, and the nonlinear and linearized shallow water responses. There is much more difference in this set of curves, particularly around the regions where resonance occurs. Here too we see that the wave height response to the linearized Euler forcing is negative for pressure forcing speeds $s \lesssim 150$ m/s and again unintuitively, positive for larger s. There is also a section of the red curve that is missing, corresponding to the regions where there is no smooth solution. Note also that the overpressure speed where the resonance occurs is significantly slower for the linearized Euler than for the SWE.

5. Conclusions

We have presented several numerical simulations using the shallow water equations over real

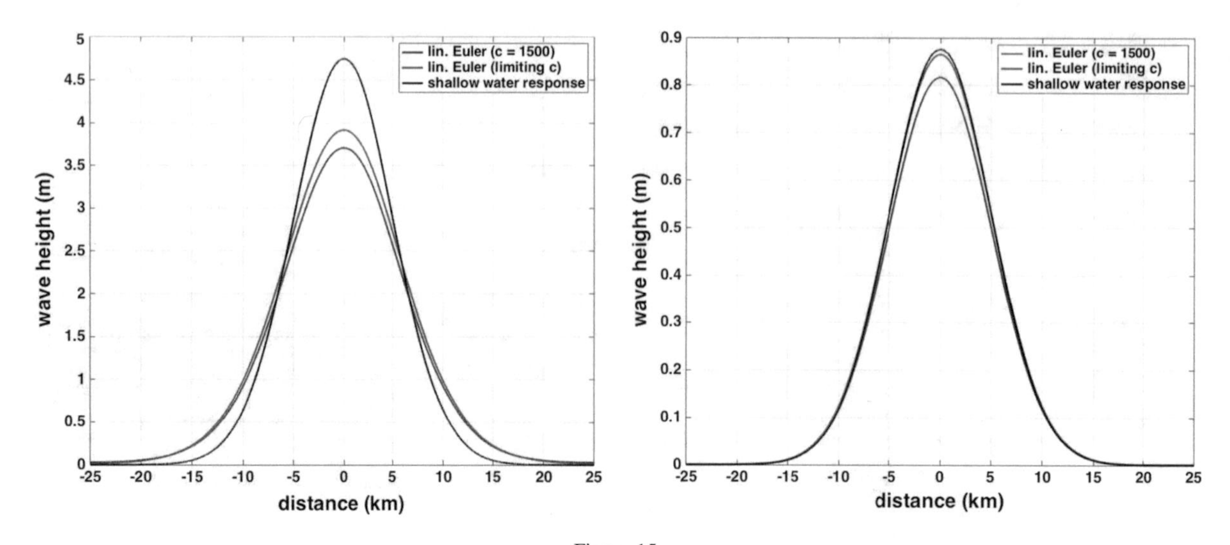

Figure 15
Response to a Gaussian pressure pulse for the linearized Euler equations, using the actual sound speed $c_a = 1500$ m/s, and a limiting sound speed that mimics the incompressible case. The shallow water response is also shown. Left uses depth h_0 = 4 km; right uses h_0 = 1 km, so it is closer to a shallow water wave. Both use a maximum overpressure of 1 atm

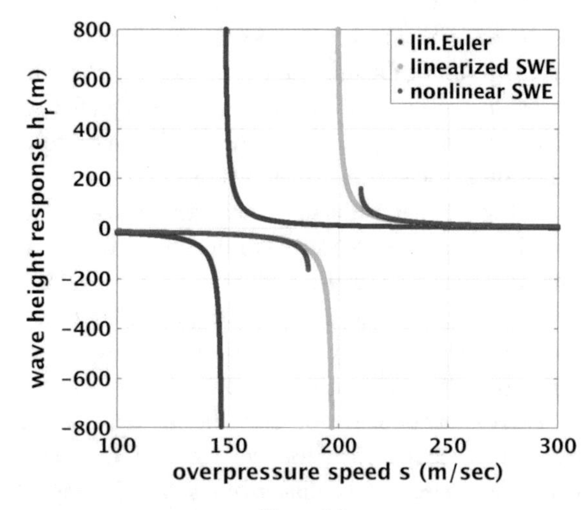

Figure 16
Wave height response as a function of s, the speed of the overpressure front. The depth h_0 is a constant 4 km, and the length scale is held fixed at 15 km. There is a large variation between the models, especially in the location where resonances occur

bathymetry that demonstrate the ocean's response to a 250MT airburst. There is no significant wave response from the ocean, in either the forced wave or the gravity waves after a short distance. Our calculations show that the amplitude of the pressure wave response decreases much more rapidly than the gravity waves do. The blast had to be very close to shore to get a sizeable response. Thus the more serious danger from an airburst is not from the tsunami, but from the local effects of the blast wave itself.

Several unexpected features found in the simulations were explained using a one-dimensional model problem with a traveling wave solution for the SWE. One of the main results, that the wave response height is proportional to the depth of the water, explains why putting the blast on a continental shelf close to shore did not generate more inundation than putting it further away in deeper water.

We also looked at the water's response to an airburst using the linearized Euler equations. In this case, the traveling wave model problem shows that the amplitudes of the important wave numbers in the ocean's response are greatly decreased. We do not yet know what this means for the gravity wave response. In addition, we expect the character of the water's response to be different, since dispersive waves will generate a wave train characterized by multiple peaks and troughs. The effect of this on land, and whether it causes inundation when the SWE response does not, is something we plan to investigate in the future.

Acknowledgements

We are particularly grateful to Mike Aftosmis, Oliver Bühler, and Randy LeVeque for more in-depth discussions. It is a pleasure to thank our colleagues at the Courant Institute for several lively discussions. We thank Michael Aftosmis for providing the blast wave ground footprint model. This effort was partially supported through a subcontract with Science and Technology Corporation (STC) under NASA Contract NNA16BD60C.

Appendix

In this appendix we start with the nonlinear Euler equations for a compressible inviscid fluid with nonlinear boundary conditions at the interface between ocean and air. The static unforced solution to these equations is determined by hydrostatic balance. The hydrostatic pressure is p_0, and the hydrostatic density is ρ_0. Since the static density variation is small (under 2%), we will end up neglecting it and proceed to linearize the equations, deriving Eq. (11)–(14) in Sect. 4.

This time there are two spatial coordinates, a horizontal coordinate x, and a vertical coordinate z. The (flat) bottom is $z = 0$. The moving top surface is $z = h(x, t)$. The horizontal and vertical velocity components are u and w respectively, and the water density is denoted by ρ. The Euler equations are

$$\begin{aligned} \rho_t + (\rho u)_x + (\rho w)_z &= 0, \\ (\rho u)_t + (\rho u^2 + p)_x + (\rho u w)_z &= 0, \\ (\rho w)_t + (\rho u w)_x + (\rho w^2 + p)_z &= -\rho g. \end{aligned} \tag{30}$$

There is a "no flow" boundary condition at the bottom boundary,

$$w(x, z = 0, t) = 0. \tag{31}$$

The kinematic condition at the top boundary Whitham (1974) states that a particle that moves with the surface velocity stays on the surface,

$$h_t + u h_x = w(x, h(x,t), t)\,. \tag{32}$$

The dynamic boundary condition at the top is continuity of pressure,

$$p(x, h(x,t), t) = p_{\text{atm}} + p_e(x,t)\,. \tag{33}$$

The left side of (33) is pressure in the water evaluated at the top boundary. The right side is the atmosphere's ambient pressure, which is the sum of the static background atmospheric pressure p_{atm} and the dynamic blast wave overpressure $p_e(x, t)$.

For static solutions ($p_e = 0$, $u = w = 0$), the nonlinear Eq. (30) reduce to the hydrostatic balance condition

$$\frac{dp_0}{dz} = -g\rho_0(z)\,. \tag{34}$$

Let the water density ρ_w be the density at the water surface. If the density differences are small (as they turn out to be), we may use a linear approximation to the equation of state,

$$p(\rho) = p(\rho_w) + c_a^2(\rho - \rho_w)\,,$$

where c_a is the acoustic sound speed in water at density ρ_w, $c_a^2 = \frac{dp}{d\rho}(\rho_w)$. The behavior of $\rho_0(z)$ is found by substituting this into (34): $\frac{dp_0}{dz} = c_a^2 \frac{d\rho_0}{dz} = -g\rho_0$. Therefore, for any two heights z_1 and z_2, we have

$$\rho_0(z_2) = \rho_0(z_1) e^{-\frac{g}{c_a^2}(z_2 - z_1)}\,.$$

If $z_2 - z_1 = 4$ km , and $c_a = 1500$ m/s, then $\frac{g}{c_a^2}(z_2 - z_1) < .02$. Therefore, the density varies by less than about 2% between the water surface and bottom.

We denote small disturbance quantities with a tilde, except for the wave height response h_r, which we use for continuity with the previous sections. For example, the water density is $\rho_0(z) + \widetilde{\rho}(x, z, t)$. These disturbances are driven by the atmospheric overpressure $p_e(x, t)$. We substitute the expressions $\rho = \rho_0 + \widetilde{\rho}$, $u = \widetilde{u}$, $w = \widetilde{w}$ (since the velocities are linearized around zero), and $p = p_0 + c_a^2 \widetilde{\rho}$ into the Euler Eq. (30) and calculate up to linear terms in the disturbance variables. Using the hydrostatic balance condition (34), this gives

$$\begin{aligned}\widetilde{\rho}_t + \rho_0\widetilde{u}_x + \rho_0\widetilde{w}_z &= 0,\\ \rho_0\widetilde{u}_t + c_a^2\widetilde{\rho}_x &= 0,\\ \rho_0\widetilde{w}_t + c_a^2\widetilde{\rho}_z &= -\,\widetilde{\rho}g\ .\end{aligned}$$

Finally, we replace the (slightly) variable $\rho_0(z)$ with the constant ρ_w. The resulting equations, which we use for analysis, are

$$\begin{aligned}\widetilde{\rho}_t + \rho_w\widetilde{u}_x + \rho_w\widetilde{w}_z &= 0,\\ \rho_w\widetilde{u}_t + c_a^2\widetilde{\rho}_x &= 0,\\ \rho_w\widetilde{w}_t + c_a^2\widetilde{\rho}_z &= -\,\widetilde{\rho}g\end{aligned}\quad . \tag{35}$$

The bottom boundary condition (31) is already linear. For the top boundary conditions, we express the water height as the sum of the background height h_0 and the disturbance height h_r:

$$h(x,t) = h_0 + h_r(x,t)\,.$$

To leading order in h_r, $\widetilde{u}$ and $\widetilde{w}$, the linear approximation to the kinematic boundary condition (32) is

$$\frac{\partial h_r}{\partial t}(x,t) = \widetilde{w}(x,h_0,t)\ . \tag{36}$$

For the dynamic boundary condition (33), which was $p(x,h(x,t),t) = p_{\text{atm}} + p_e$, we use the Taylor expansion and the perturbation approximation

$$\begin{aligned}p(x,h(x,t),t) &\approx p(h_0) + p_{0,z}(h_0)\,h_r(x,t)\\ &\approx p_0(h_0) + \widetilde{p}(x,h_0,t) + p_{0,z}(h_0)\,h_r(x,t)\\ &\approx p_0(h_0) + c_a^2\widetilde{\rho}(x,h_0,t) + p_{0,z}(h_0)\,h_r(x,t)\end{aligned}\quad .$$

For the undisturbed quantities, the pressure at the top is $p(h_0) = p_{\text{atm}}$. The hydrostatic balance relation (34) in the water (applied at the top) is $p_{0,z}(h_0) = -g\rho_w$. Making these substitutions gives

$$p_0(h_0) + c_a^2\widetilde{\rho} + p_{0,z}h_r = p_{\text{atm}} + p_e\,, \tag{37}$$

giving the result

$$c_a^2\widetilde{\rho}(x,h_0,t) - \rho_w g\,h_r(x,t) = p_e(x,t)\,. \tag{38}$$

Summarizing, the linearized Euler equations are (35), with linearized boundary conditions (31), (36) and (38).

References

Aftosmis, M., Mathias, D., Nemec, M., Berger, M.: Numerical simulation of bolide entry with ground footprint prediction. AIAA-2016-0998 (2016)

GeoClaw Web Site. http://www.clawpack.org/geoclaw. http://www.geoclaw.org/ (2017)

George, D. (2008). Augmented Riemann solvers for the shallow water equations over variable topography with steady states and inundation. *J. Comp. Phys.*, *227*(6), 3089–3113.

Gica, E., Arcas, D., & Titov, V. (2014). *Tsunami inundation modeling of Ocean Shores and Long Beach.* NOAA Center for Tsunami Research, Pacific Marine Environmental Laboratory: Washington due to a Cascadia subduction zone earthquake. Tech. rep.

Gisler, G. (2008). Tsunami simulations. *Annu. Rev. Fluid Mech.*, *40*, 71–90.

Gisler, G., Weaver, R., & Gittings, M. (2010). Calculations of asteroid impacts into deep and shallow water. *Pure Appl. Geophys.*, *168*, 1187–1198.

González, F.I., LeVeque, R.J., Chamberlain, P., Hirai, B., Varkovitzky, J., George, D.L.: Geoclaw model. In: Proceedings and Results of the 2011 NTHMP Model Benchmarking Workshop, pp. 135–211. National Tsunami Hazard Mitigation Program, NOAA (2012). http://nthmp.tsunami.gov/documents/nthmpWorkshopProcMerged.pdf

Harkrider, D., & Press, F. (1967). The Krakatoa air-sea waves: an example of pulse propagation in coupled systems. *Geophys. J. R. Astr. Soc.*, *13*, 149–159.

Kranzer, H., & Keller, J. (1959). Water waves produced by explosions. *J. Appl. Phys.*, *30*(3), 398–407.

LeVeque, R. (1997). Wave propagation algorithms for multidimensional hyperbolic systems. *J. Comput. Phys.*, *131*, 327–353.

LeVeque, R., George, D., Berger, M.: Tsunami modelling with adaptively refined finite volume methods. Acta. Numerica. pp. 211–289 (2011)

Liu, P. (2009). Tsunami modeling: propagation. In E. Bernard & A. Robinson (Eds.), *The sea: Tsunamis* (pp. 295–320). Cambridge: Harvard University Press.

Mandli, K.: Finite volume methods for the multilayer shallow water equations with applications to storm surges. Ph.D. thesis, University of Washington (2011)

Melosh, H.: Impact-generated tsunamis: an over-rated hazard. In: 34th Lunar and Planetary Sciences Conference Abstract (2003)

Monserrat, S., Vilibic, I., & Rabinovich, A. (2006). Meteotsunamis: atmospherically induced destructive ocean waves in the tsunami frequency band. *Nat. Hazards Earth Syst. Sci.*, *6*, 1035–1051.

Morrison, D., Venkatapathy, E.: Asteroid generated tsunami: summary of NASA/NOAA workshop. Tech. Rep. NASA/TM-2194363, NASA Ames Research Center (2017)

Petersen, M., Cramer, C., & Frankel, A. (2002). Simulations of seismic hazard for the Pacific northwest of the United States from earthquakes associated with the Cascadia subduction zone. *Pure Appl. Geophys.*, *159*, 2147–2168.

Popinet, S. (2015). A quadtree-adaptive multigrid solver for the Serre-Green-Naghdi equations. *J. Comp. Phys.*, *302*, 336–358.

Popova, O., Jenniskens, P., Emelyanenko, V., Kartashova, A., Biryukov, E., et al. (2013). Chelyabinsk airburst, damage assessment, meteorite recovery, and characterization. *Science*, *342*(6162), 1069–1073. https://doi.org/10.1126/science.1242642.

Proudman, J. (1929). The effects on the sea of changes in atmospheric pressure. *Geophys. Suppl. Mon. Not. R. Astron. Soc.*, *2*(4), 197–209.

Uslu, B., Eble, M., Titov, V., Bernard, E.: Distance tsunami threats to the ports of Los Angeles and Long Beach, California. Tech. rep., NOAA Center for Tsunami Research, Pacific Marine Environmental Laboratory (2010). NOAA OAR Special Report, Tsunami Hazard Assessment Special Series (2)

Vreugdenhil, C. (1994). *Numerical methods for shallow-water flow*. Dordrecht: Kluwer Academic Publishers.

Weiss, R., Wünnemann, K., & Bahlburg, H. (2006). Numerical modelling of generation, propagation and run-up of tsunamis caused by oceanic impacts: model strategy and technical solutions. *Geophys. J. Intl.*, *167*, 77–88.

Whitham, G. B. (1974). *Linear and nonlinear waves*. Hoboken: John Wiley & Sons.

(Received April 12, 2017, revised October 28, 2017, accepted December 1, 2017)

Pure Appl. Geophys.

https://doi.org/10.1007/s00024-017-1729-1

Pure and Applied Geophysics

Odessa Tsunami of 27 June 2014: Observations and Numerical Modelling

JADRANKA ŠEPIĆ,[1] ALEXANDER B. RABINOVICH,[2,3] and VICTOR N. SYTOV[4]

Abstract—On 27 June, a 1–2-m high wave struck the beaches of Odessa, the third largest Ukrainian city, and the neighbouring port-town Illichevsk (northwestern Black Sea). Throughout the day, prominent seiche oscillations were observed in several other ports of the Black Sea. Tsunamigenic synoptic conditions were found over the Black Sea, stretching from Romania in the west to the Crimean Peninsula in the east. Intense air pressure disturbances and convective thunderstorm clouds were associated with these conditions; right at the time of the event, a 1.5-hPa air pressure jump was recorded at Odessa and a few hours earlier in Romania. We have utilized a barotropic ocean numerical model to test two hypotheses: (1) a tsunami-like wave was generated by an air pressure disturbance propagating directly over Odessa ("Experiment 1"); (2) a tsunami-like wave was generated by an air pressure disturbance propagating offshore, approximately 200 km to the south of Odessa, and along the shelf break ("Experiment 2"). Both experiments decisively confirm the meteorological origin of the tsunami-like waves on the coast of Odessa and imply that intensified long ocean waves in this region were generated via the Proudman resonance mechanism while propagating over the northwestern Black Sea shelf. The "Odessa tsunami" of 27 June 2014 was identified as a "beach meteotsunami", similar to events regularly observed on the beaches of Florida, USA, but different from the "harbour meteotsunamis", which occurred 1–3 days earlier in Ciutadella (Baleares, Spain), Mazara del Vallo (Sicily, Italy) and Vela Luka (Croatia) in the Mediterranean Sea, despite that they were associated with the same atmospheric system moving over the Mediterranean/Black Sea region on 23–27 June 2014.

Key words: Meteotsunami, Odessa, Black Sea, sea level oscillations, tide gauge records, numerical ocean modelling, meteorological data.

[1] Institute of Oceanography and Fisheries, Šetalište I. Meštrovića 63, 21000 Split, Croatia. E-mail: sepic@izor.hr

[2] Department of Fisheries and Ocean, Institute of Ocean Sciences, 9860 W. Saanich Rd, Sidney, BC, Canada.

[3] P.P. Shirshov Institute of Oceanology, Russian Academy of Sciences, 36 Nakhimovsky Pr, Moscow, Russia.

[4] Hydrometeorological Center on the Black and Azov Seas, 89 Franzusky Blvd, Odessa, Ukraine.

1. *Introduction*

At the midday of a hot and moist summer day on 27 June 2014, a tsunami-like wave suddenly hit beaches of Odessa and the neighbouring port-town Illichevsk in the northwestern part of the Black Sea. Unsuspected beach goers were swept into the sea by a 1–2-m high wave; six people, including four children, were injured and taken to the central hospital. Beach structures were tossed about and damaged (Fig. 1).

The abrupt occurrence of the wave that arrived from the calm sea, its considerable height, destructiveness, runup and significant inundation on the beach—all these features are typical for tsunami waves. Tsunamis are known in the Black Sea; the strongest were in 1927 (2 events), 1939 and 1966 (Nikonov 1997; Yalçiner et al. 2004; Papadopoulos et al. 2011); however, during June 2014 the region was seismically quiet. Besides, seismically generated tsunamis are widespread events: all known tsunamis in the Black Sea (in particular, 1927, 1939 and 1966) were observed throughout the entire sea (Dotsenko 1995), while the 2014 event in Odessa was local.

Tsunami waves in the Black Sea can also be produced by subaerial or submarine landslides (cf. Ranguelov et al. 2008; Nikonov and Fleyfel 2015), but Odessa is located too far away ($\sim$ 300 km) from the shelf break (the potential area for significant submarine slides) to be severely affected by a hypothetical slide-generated tsunami; besides, a wave generated on the shelf break would be largest on the coasts of the Crimean Peninsula and Turkey, while the Odessa coast would be sheltered by the shelf. Likewise, there were no marks on the Odessa coast indicating possible local subaerial slides. Slides moving downslope produce a leading *positive* (crest) wave propagating *offshore* and a *negative* (trough) wave propagating *onshore* (Fine et al. 2003). Thus, if

Figure 1
Photographs taken by eyewitnesses at Odessa beaches during and after the 27 June tsunami-like event

such a wave was generated by a landslide, then a *set-down* (receding wave) would be first observed on the coast and not a *runup* (inundating wave) as was actually observed on the beaches of Odessa and Illichevsk!

Due to the lack of scientific explanation, rumours and numerous maladaptive hypotheses appeared. For example, the first thought that, unsurprisingly, came to many was: "*This 'tsunami' was generated by an underwater explosion!*" Unlikely as it is, the idea attracted much attention by the over-stressed Ukrainian population, and it spread rapidly through Ukrainian, Russian and international media. Some specialists tried to explain that to create a wave of such power (the entire length of the affected beaches was about 10–12 km), this should be an explosion of a 50-megaton thermonuclear bomb, but rumours are not subject to logic! Several other theories circulated through the media, like "*this wave was generated by a large ship (submarine) passing near the coast*" (in fact, such a ship would need a length of several kilometres!), and others, like a local whirlwind effect, extreme interacting currents, abrupt temperature changes and even "*the great cross of planets*"! Finally, the media declared that "*unfortunately, the scientists are not able to provide any reasonable explanation of this event*".

However, as demonstrated by Šepić et al. (2015b), the Odessa event was not isolated, but rather the last link in a *chain of destructive events* that occurred during 23–27 June 2014 in the Mediterranean and Black Sea regions (Fig. 2), which affected Ciutadella Harbour on the Balearic Islands (Spain), Vela Luka (Croatia) and a number of other bays and harbours in the central and southern parts of the Adriatic Sea (Šepić et al. 2016) and Mazara del Vallo, southwestern coast of Sicily, where its intensity was so high that a 1.5-m high hydraulic jump ("bore"), damaging a number of moored vessels within the harbour (Šepić et al. 2015b, 2017), appeared. Large seiche oscillations were also observed on 25–26 June at other coastal regions of the Central and Eastern Mediterranean, including the coasts of Italy, Greece and Turkey (Šepić et al. 2015b). Finally, as the last episode of a chain of the seemingly unrelated destructive events, the "Odessa tsunami" ocurred on 27 June 2014.

Despite their ferocity, and as opposed to the Odessa event, there was no "surprise effect" related to the Mediterranean events. Media were writing about "*rissagas*" (the Balearic Islands), "*šćigas*" (the Adriatic Sea) and "*marrobbios*" (Sicily) affecting the corresponding coasts: these are local names of destructive tsunami-like waves which have been known in these regions for centuries. They regularly hit precisely the same bays and harbours. It has been shown that these events are of meteorological origin, and they have come to be known as "meteorological tsunamis" or "meteotsunamis" (Rabinovich and Monserrat 1996, 1998; Monserrat et al. 2006;

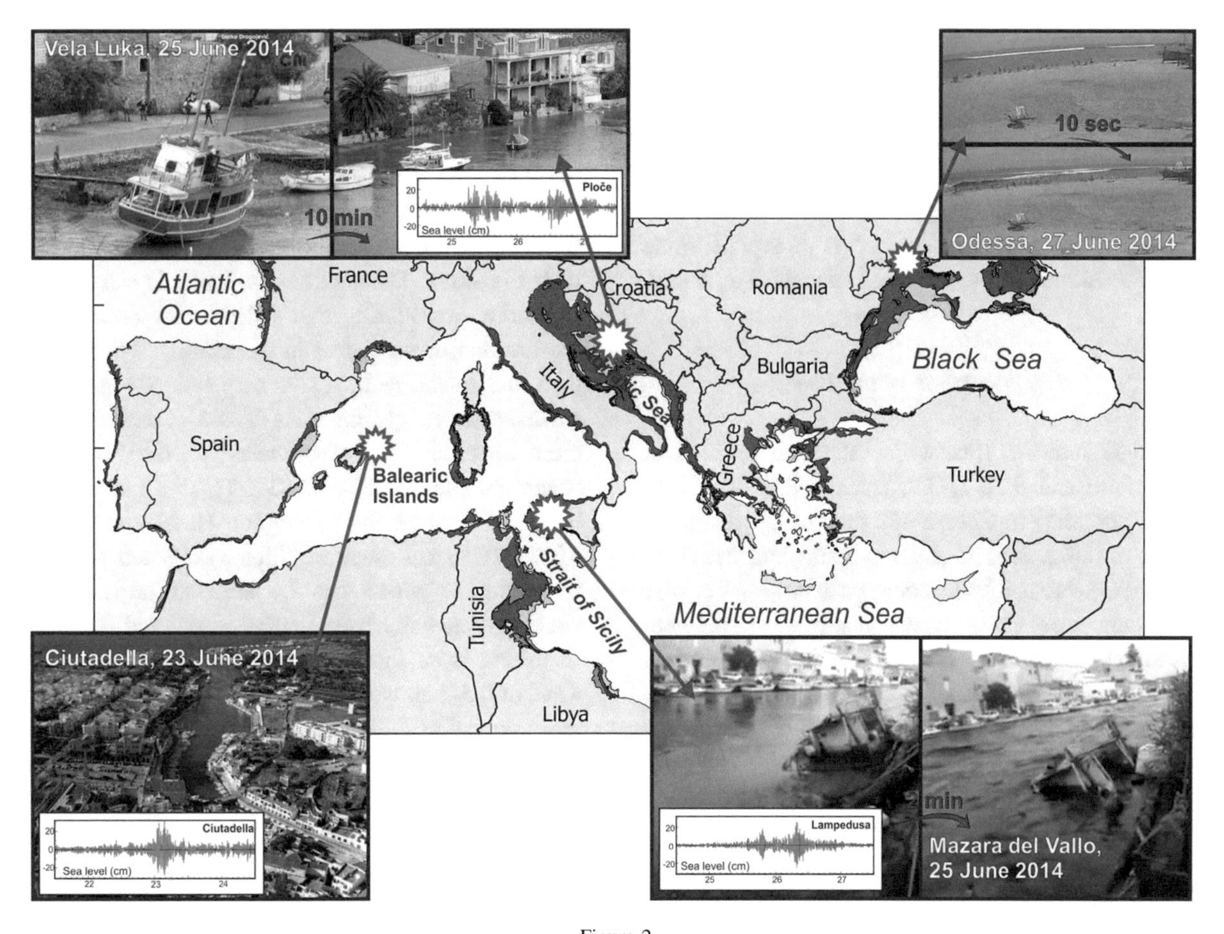

Figure 2
Map showing the Mediterranean and Black Sea locations hit by tsunami-like waves during 23–27 June 2014. Insets show photographs from affected locations, and, where available, sea level time series (high-pass filtered using a 4-h Kaiser Bessel filter) measured at the closest available tide gauge. Areas shaded in red represent parts of the sea for which the Froude number is close to unity ($0.9 < Fr < 1.1$) Modified from Šepić et al. (2015b)

Rabinovich et al. 2009; Vilibić et al. 2014, 2016; Pattiaratchi and Wijeratne 2015). Surprisingly, meteotsunamis in the Mediterranean Sea and some other regions of the world oceans commonly occur not during the extreme weather conditions (hurricanes, typhoons, thunderstorms, frontal passages or tornados), but during warm, quiet and moist summer days (Monserrat et al. 2006), and are, in fact, mostly generated by intense short-lasting, small-scale atmospheric gravity waves (characterized by a sudden air pressure change of $\sim$ 2 hPa/5 min) unfelt by humans but clearly recorded by high-precision microbarographs (Thomson et al. 2009).

Meteotsunamis in the Black Sea are not as well known as in the Adriatic Sea, on the coasts of the Balearic Islands or on the southwestern coast of Sicily (that is why, just after the event this mechanism had not been even considered by local authorities); however, every now and then they do occur in this region. The Black Sea tsunami catalogues of Yalçiner et al. (2004) and Papadopoulos et al. (2011) describe many strong events of "unknown origin", which are probably meteotsunamis. Likewise, following the 2014 Odessa tsunami, eyewitnesses recalled that similar waves occasionally hit Odessa, with one of the eyewitnesses, Andrey Plahonin, stating that a similar event occurred in September 1987. The best documented Black Sea meteorological tsunami is the one of 7 May 2007 in the western Black Sea that destructively affected the coast of Bulgaria (Vilibić et al. 2010). Thus, the event of 27 June 2014 in Odessa is rare, but not unique. At the same time, it still remains a puzzle. The main purpose of the present paper is to examine this event in detail, to provide numerical experiments that can reveal the physical mechanism responsible for formation of this particular event and to explain some specific features of the observed tsunami-like waves.

2. *Description of the Event*

On 27 June 2014, between 12:30 and 12:50 local time (9:30 and 9:50 UTC), tsunami-like waves hit several beaches in Odessa and the nearby port-town Illichevsk (Fig. 3). The day was sunny and windless, and there were a lot of people on beaches. Luckily, the water was cold, so not many of them were swimming when the waves arrived. Still, 15 persons, out of which at least 4 children, were drawn into the sea and slightly injured. A 4-year boy was hospitalized with bruises and contusions. Waves were strongest at the beach complex "Sunrise" where they were more than 1 m high and inundated 50 m of beach (Nikonov and Fleyfel 2015) (Fig. 3). According to some eyewitness reports, the waves were even higher: "Water rose to a height of 2.5 m. It covered the entire beach. There were victims, mostly children. There were broken legs. A lot of things were left scattered on the beach, and a lot of them floated away with the wave. Damage to the beach is very large, especially considering that Odessa city and nearby sea are one of the safest in the world". No wonder then the event came as a surprise! Tsunami-like waves were caught on surveillance cameras at two more locations to the southwest of the "Sunrise" beach complex: first, at 9:33 UTC at Chernomorka beach in front of the Sauvignon Hotel and then at 9:37 UTC 5 km away at Illichevsk beach (Fig. 3). Chernomorka video can be seen at https://www.youtube.com/watch?v=cQjAKz1-amI, and selected snapshots from this video are shown in Fig. 4. A wave of 0.5–1 m height suddenly came from the open sea, inundating the beach and causing panic among sunbathers. From the time the wave appeared on the horizon till the time it started retreating, about 50 s

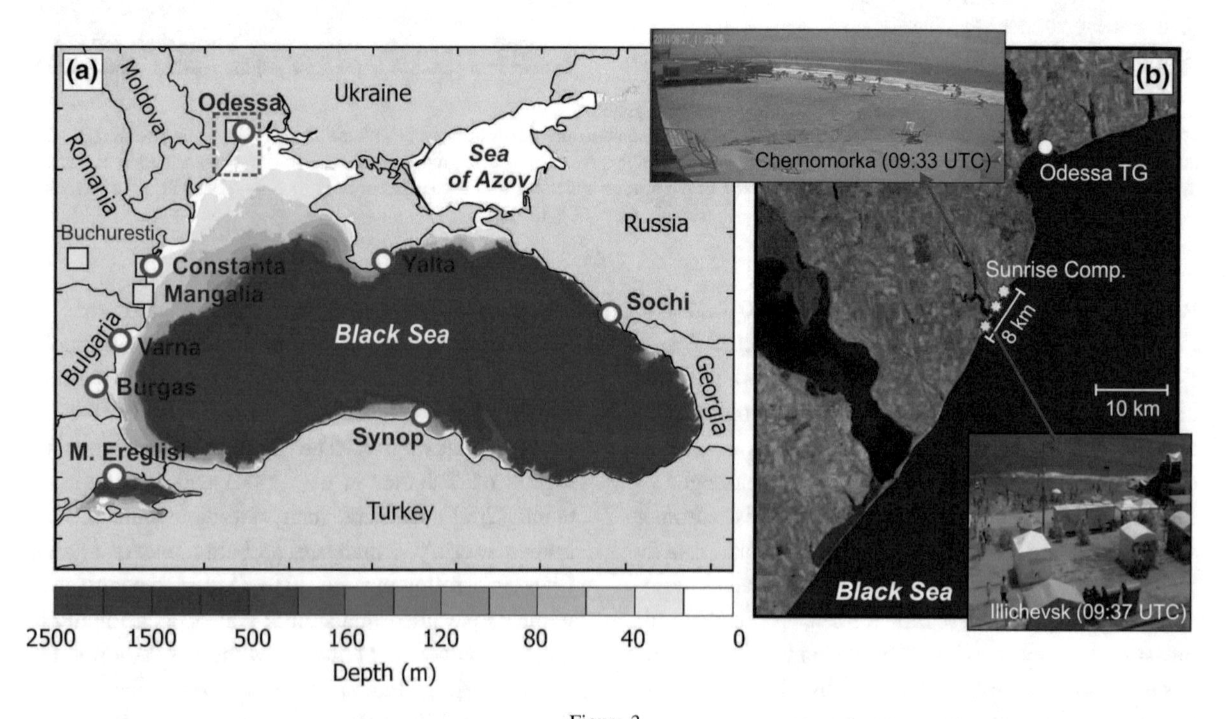

Figure 3

a Map showing bathymetry of the Black Sea and locations of meteorological (square) and tide gauge (circle) stations. **b** The Odessa coast hit by the 2014 tsunami waves, including positions of three relevant sites: "Sunrise" beach complex, Chernomorka beach and Illichevsk beach

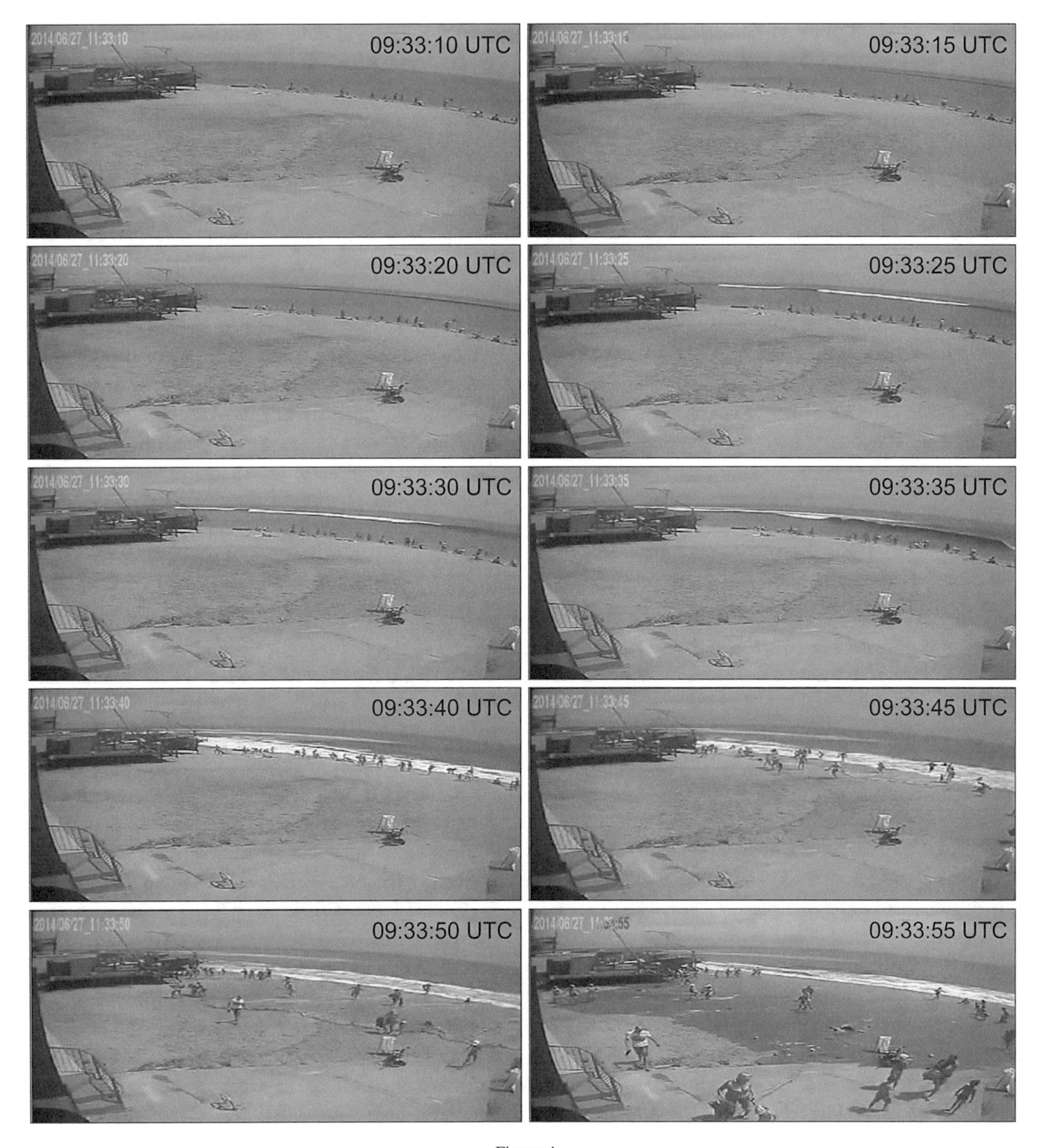

Figure 4
Snapshots from the Chernomorka surveillance camera showing tsunami-like wave approaching and hitting the beach

had passed. The first wave was followed by another one which luckily broke before hitting the beach, thus causing no further flooding and damage. Taking into account the beach orientation (northeast-southwest) and video, we can conclude the wave came from the southeast. Four minutes later, similar waves affected Illichevsk beach (https://www.youtube.com/watch?v=ub1BRmJvWus), which is located 5 km to the southwest of the Chernomorka beach (Fig. 3). Another eyewitness, who was in Illichevsk beach when the wave hit, recalls: "*I was walking through the sea for a while, when suddenly bottom 'failed' under my feet, and sea started to pull me down. At the same time water struck me in the back and I was instantly thrown to the shore, and then dragged back to the depth. I saw sea hitting the breakwater and people jumping completely wet, saving things, and running to higher ground*". The wave at Illichevsk came from the southeast as well. This wave was of a similar period but of smaller height (< 0.5 m) than the wave at Chernomorka. Nikonov and Fleyfel (2015) inspected an additional 14 min of Illichevsk surveillance camera video noting that after the first wave, water continued to oscillate alongshore (following the coastline) for 12 min with a period of ~ 2.5 min.

3. Atmospheric Conditions

Šepić et al. (2015b) established that the chain of events of 23–27 June 2014 was a result of anomalous atmospheric conditions over the Mediterranean/Black Sea region characterized by: (1) inflow of warm and dry air from Africa at heights of ~ 1500 m; (2) a strong southwesterly jet stream with wind speeds > 20 m/s at heights of ~ 5000 m; and (3) the presence of unstable atmospheric layers at heights of 4000–6000 m characterized by a small Richardson stability number, $Ri < 0.25$. These conditions support the generation of numerous intense, small-scale atmospheric disturbances ("tumultuous atmosphere"), which subsequently trigger the meteotsunamis. The corresponding atmospheric pattern was tracked propagating eastward from 23 to 27 June; major meteotsunami events were observed exactly at the time of the most intense atmospheric instability and a well-developed jet stream at a height of ~ 5000 m over each respective area. It arrived at the northwestern Black Sea region on 27 June, coincident with the time of the "Odessa tsunami" (at ~ 12:30 UTC).

This moving synoptic pattern was responsible for short-lived atmospheric pressure perturbations which drifted with the jet stream-like bubbles and produced tsunami-like waves, which became destructive in specific areas. As demonstrated by Šepić et al. (2015b), the governing parameter determining the sea level response to atmospheric disturbances is the Froude number, which can be defined as $\mathrm{Fr} = U/c$, i.e. the ratio of the atmospheric gravity wave speed, U, to the phase speed of long ocean waves, $c = \sqrt{gh}$, where g is the gravity acceleration and h is the water depth. The resonance conditions, known as the "Proudman resonance" (Proudman 1929), occur when $U \approx c$, and $\mathrm{Fr} \approx 1.0$. In this case, ocean waves begin to actively absorb atmospheric energy during their propagation and, as a result, are strongly intensified (Hibiya and Kajiura 1982; Vilibić 2008). The map by Šepić et al. (2015b), showing values of Fr (Fig. 2), indicates that extreme events of 23–27 June 2014 occurred in the regions with the most favourable conditions for meteotsunami generation ($0.9 < \mathrm{Fr} < 1.1$). One of these regions is the northwestern part of the Black Sea ("Odessa Gulf"), i.e. the region where the "Odessa tsunami" took place.

3.1. General Synoptic Situation

During the period of 23–26 June 2014, a tsunamigenic synoptic setting slowly advanced from the western Mediterranean towards the eastern Mediterranean, generating multiple small-scale atmospheric disturbances ("bubbles") and associated meteotsunamis on its way (Fig. 2 and Šepić et al. 2015b). According to the upper air sounding data, the system arrived at the area of Bucharest (Romania) at 00:00 UTC on 27 June 2014, some 200 km to the west of the Black Sea (Fig. 5). The lower tropospheric air mass (up to 3500 m) was dry and stable and had two embedded temperature inversions, at heights of approximately 1000 and 3000 m. This air mass was overtopped by unstable moist air spanning heights from about 4000 to 5500 m. Wind direction was westerly throughout the troposphere, strengthening from the surface (~ 5 m/s) towards the unstable layer (~ 25 m/s).

The described situation favours generation of atmospheric gravity waves at an interface of stable and unstable air mass when they coincide with areas of pronounced wind gradients (Plougonven and Zhang 2014). Furthermore, the described vertical atmospheric profile closely matches an idealized profile favourable for ducting of atmospheric gravity

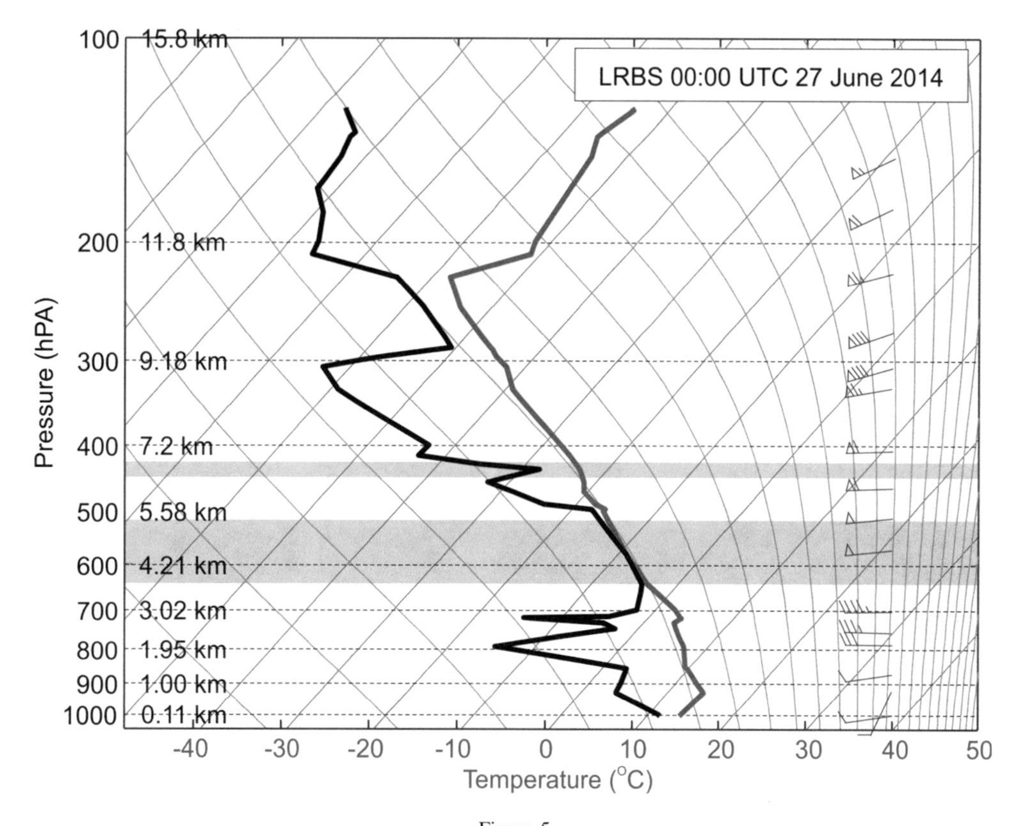

Figure 5
Sounding profile from Bucharest (Romania). Red curve stands for temperature, black curve for dew point temperature and shaded areas for unstable air pressure levels at which $Ri < 0.25$

waves: when a stable atmospheric layer is overtopped by an unstable layer, and when the propagation speed of atmospheric gravity waves in the stable layer equals the wind speed in the unstable layer, waves can become *ducted* and propagate over great distances without noticeable dissipation of their energy (Monserrat and Thorpe 1996). Atmospheric gravity waves are one of the main known sources of meteotsunamis, and the longer they propagate over the sea, the more energy they can pump into the ocean waves and the higher the probability of intense meteotsunami generation (Vilibić 2008).

Throughout the morning hours of 27 June 2014, this pattern continued to propagate eastward, first towards and then across the Black Sea. Unfortunately, no upper air sounding data were available for the northwestern Black Sea and the region of Odessa for the date in question. However, the propagation of a characteristic pattern can also be tracked from the ERA-Interim reanalysis data. Evolution of the temperature field at a pressure level of 850 hPa (height of $\sim$ 1500 m), wind field at 500 hPa ($\sim$ 5000 m) and the Richardson number field at pressure levels between 700 and 400 hPa ($\sim$ 3000–6000 m), all between 00:00 and 12:00 UTC on 27 June, is shown in Fig. 6. Once more, we notice: (1) a lower troposphere warm air mass advected from the west-southwest; (2) unstable mid-troposphere air mass; and (3) mid-troposphere westerly jet stream (> 20 m/s). We can thus conclude that the tsunamigenic synoptic conditions were present over the Black Sea between 06:00 and 12:00 UTC, i.e. precisely during the time period spanning the Odessa event. It is salient that the most favourable tsunamigenic conditions (strongest horizontal temperature gradient, winds with highest speeds and air mass of lowest stability) were positioned not exactly over Odessa, but rather right over the shelf break which stretches from the Romanian coast towards the southern end of the

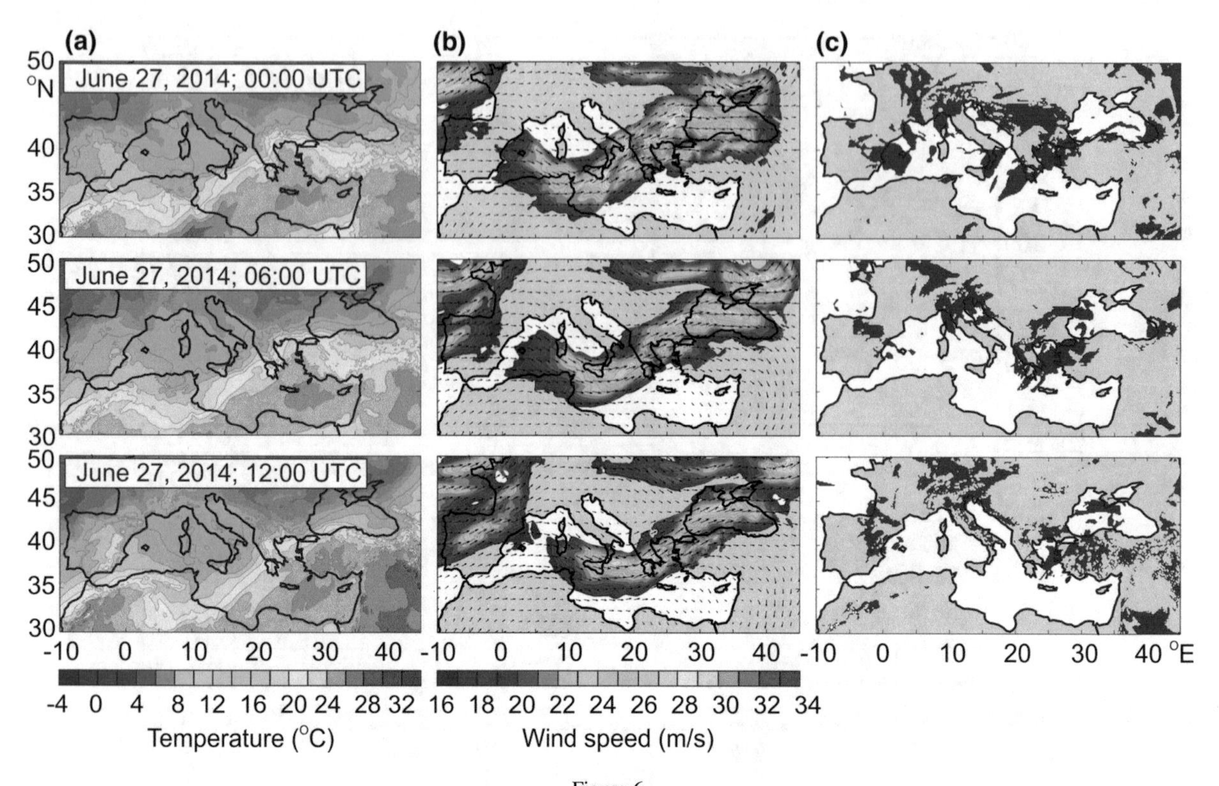

Figure 6
Synoptic situation over the Black Sea on 27 June 2014: **a** temperature at 850 hPa, **b** wind at 500 hPa, and **c** the Richardson number smaller than 0.25 (shaded) Similar figure but for 23–27 June 2017 is shown by Šepić et al. (2015b)

Crimean Peninsula. Following the Odessa event, the tsunamigenic synoptic pattern moved to the east over the Crimean Peninsula and to the northeastern Black Sea, where it was between 12:00 and 18:00 UTC.

It is worthwhile to compare the herein described synoptic situation with the one observed over the Black Sea during the Bulgarian meteotsunami of 7 May 2007 (Vilibić et al. 2010). The closest upper air sounding data, which again comes from Bucharest, reveal a remarkably similar vertical profile: dry and warm air from the surface up to $\sim$ 3500 m, where it was overtopped by an unstable air mass, within which the southwesterly winds reached speeds up to 21 m/s. Furthermore, the ERA-Interim reanalysis fields again disclose the presence of warm air of African origin at pressure levels of $\sim$ 850 hPa, with a warm air tongue stretching from the southwest towards the northeast and a temperature front located right over the western Black Sea. This warm air mass was overtopped by a wide unstable layer (at $\sim$ 500 hPa) in which the southwesterly winds (reaching speeds of 15–25 m/s) were embedded. What was important for the 2007 event is that the atmospheric pressure disturbances propagated with the speed and direction closely matching the speed and direction of the 500-hPa wind. In contrast to the 2014 event, during the 2007 event, the major atmospheric tsunamigenic pattern was centred exactly over the main affected region.

3.2. Satellite Imaginary

Air pressure disturbances, responsible for generating meteotsunamis, are often associated with convective clouds that appear to be an indicator of atmospheric gravity waves (Belušić et al. 2007; Šepić et al. 2009). This is especially true for the Mediterranean region where meteotsunamis at the Balearic Islands (Jansà et al. 2007), in the Adriatic Sea (Belušić et al. 2007; Šepić et al. 2009) and in the Black Sea (Vilibić et al. 2010), have been previously found to correlate with prominent convective clouds.

To check whether something similar occurred during the Odessa tsunami of 27 June 2014, we have inspected relevant satellite imaginary originating from the EUMETSAT Meteosat-9 satellite. Convective clouds are best detected in the images composed from the RGB channels 05–06, 04–09, 03–01, with yellowish colours indicating severe convection. A series of composite satellite images taken with the Meteosat-9 satellite between 03:57 and 10:57 UTC is shown in Fig. 7. We can clearly track the trajectory, growth, propagation and dissipation of pronounced convective clouds. The convection was strongest between 05:57 and 07:57 UTC; during this period of time, convective clouds propagated from the west to the east coast of the northwestern Black Sea. Much like other tsunamigenic conditions, these clouds were not positioned directly over Odessa, but rather over the shelf break, some 200 km to the south of Odessa. According to the hourly METAR reports from the Constanta airport and the thunderstorm data, the passage of convective clouds over Romania was accompanied by rains and here and there by severe thunderstorms (Fig. 8).

4. Data Analysis

The air pressure disturbances, described in the previous section, are clearly seen in 6-min digital atmospheric pressure records from two Romanian stations: Mangalia and Constanta (Fig. 9; the locations are shown in Fig. 3): a train of high-frequency oscillations propagated over eastern Romania between 04:00 and 06:00 UTC. The oscillations

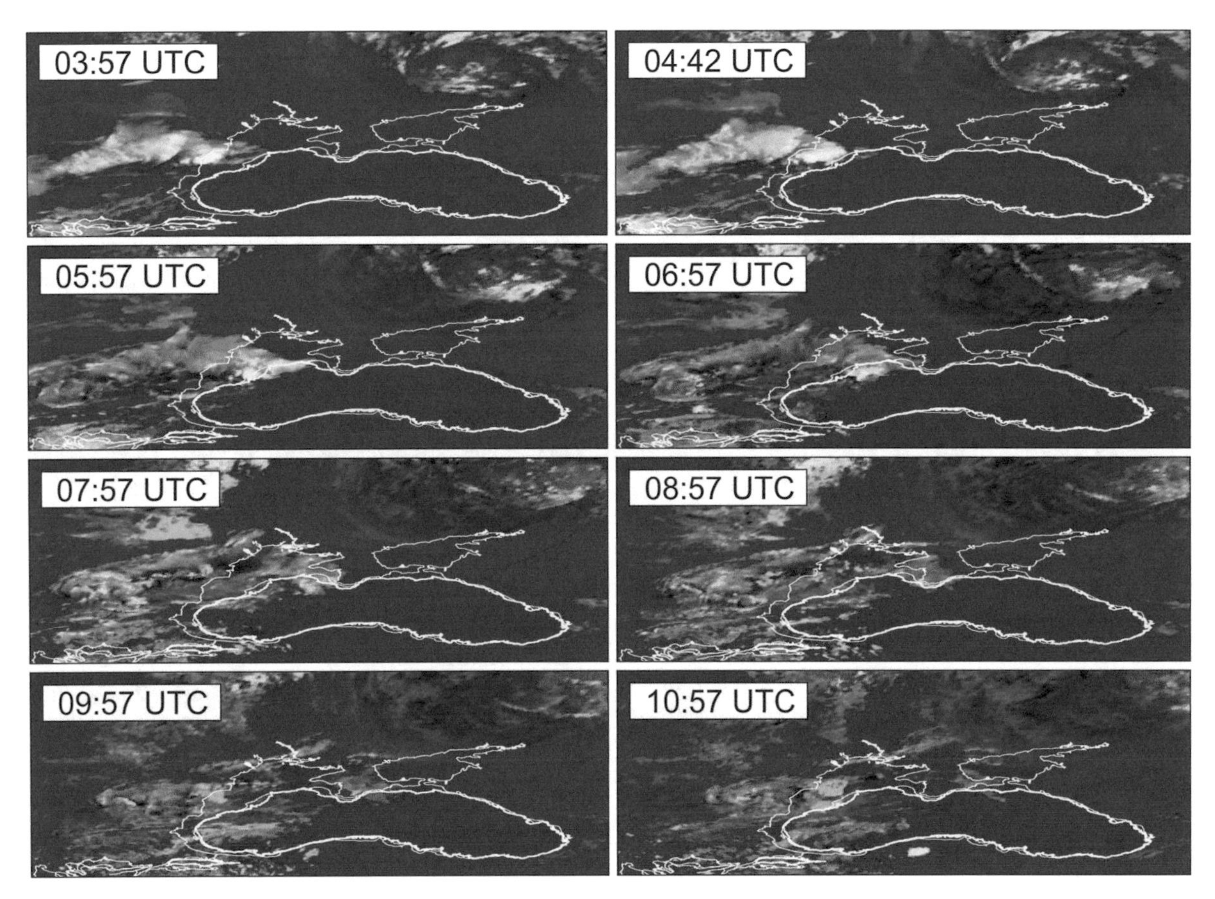

Figure 7
Convective clouds propagating over the Black Sea during the morning hours of 27 June. The yellowish colours indicate areas of severe convection. Coastline and depth contours of 100 and 200 m are shown

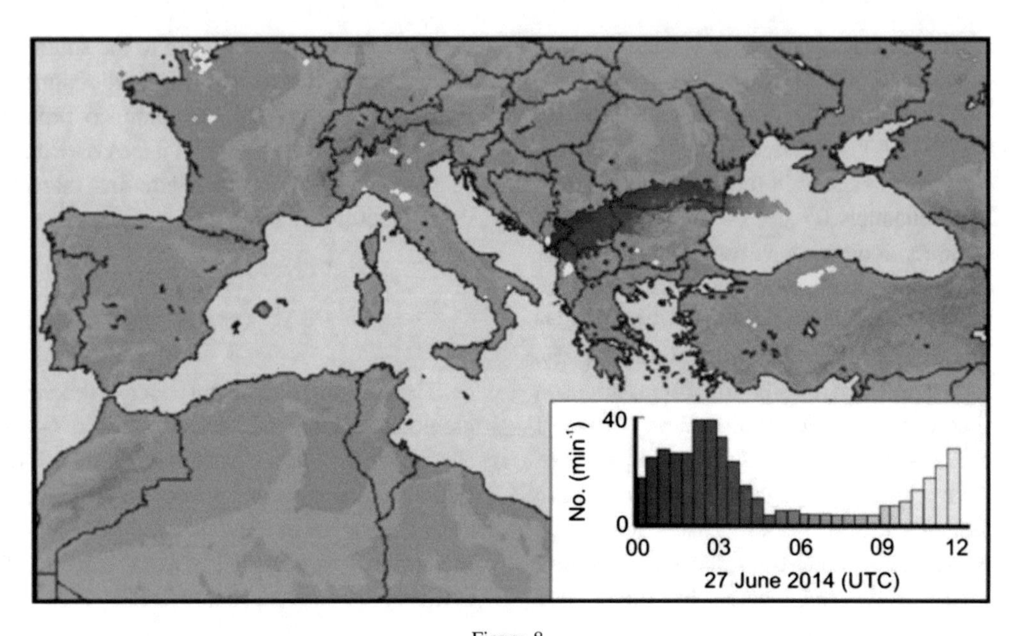

Figure 8
Locations of lightning strikes from 00:00 to 12:00 UTC on 27 June 2014. Colours point to time of lightning strikes, while height of column indicates the number of lightning strikes per minute in given period

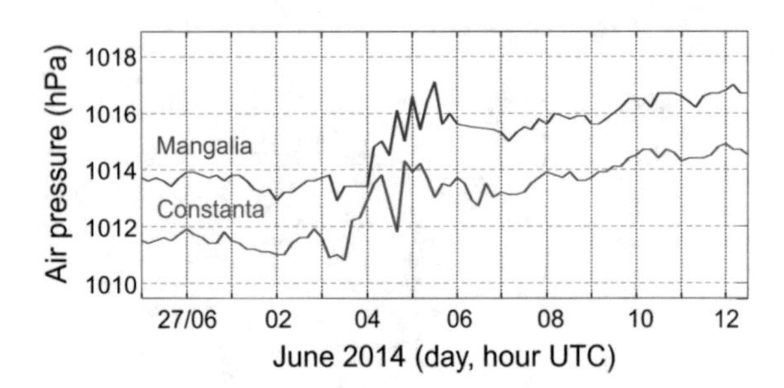

Figure 9
Air pressure records at Mangalia and Constanta (Romania) on 27 June 2014

coincide in time with the tsunamigenic synoptic conditions over Romania (Fig. 6), including rain events, thunderstorms (Fig. 8) and convective clouds (Fig. 7). Air pressure oscillations were up to 3.0 hPa over 6 min at Constanta and up to 1.0 hPa over 6 min at Mangalia. In general, the oscillations started a little earlier at Constanta, indicating the eastward propagation of the atmospheric system. The destructive meteotsunami of 7 May 2007 on the Bulgarian coast of the Black Sea was produced by similar oscillations (Vilibić et al. 2010).

The air pressure oscillations observed in Romania on 27 June 2014 are very similar to those recorded at the Balearic Islands on 23 June and in Croatia and Italy on 25–26 June (Šepić et al. 2015b, 2016, 2017). Certainly, this does not mean that the same disturbances came to the Black Sea from the Mediterranean Sea; the typical life cycle of such atmospheric disturbances is only a few hours (Monserrat and Thorpe 1996; Thomson et al. 2009). Thus, not the specific disturbances, but the entire anomalous atmospheric system ("tumultuous atmosphere") moved eastward and supported the generation of numerous intense, small-scale atmospheric disturbances (Šepić et al. 2015b). In the morning hours of 27 June 2014, the system arrived at the northwestern Black Sea. The strongest atmospheric disturbances appear to have propagated along the shelf break. As was established earlier, such atmospheric disturbances were responsible for the generation of meteotsunamis in Spain (23 June 2014) and in Croatia/Italy (25–26 June 2014) (Šepić et al. 2015b, 2016, 2017). It is logical to expect that, likewise, on the coast of the Black Sea these disturbances would also produce strong sea level oscillations.

The available data for the particular region of Odessa were very limited. After much effort, we could find only two analogue paper records: an air pressure barogram and a sea level record from the

Odessa tide gauge. Locations of the instruments are indicated in Fig. 3. The corresponding time series, digitized with a 2-min time step, are presented in Fig. 10. The air pressure time series (Fig. 10a) shows an abrupt 1.5 hPa pressure jump that occurred over a period of a few minutes during an otherwise calm day, characterized by a slowly increasing air pressure. The time of this jump, 09:30 UTC, precisely coincided with the time of the observed tsunami-like wave on beaches of Odessa and Illichevsk. The simultaneous occurrence of the air pressure jump and the devastating sea wave supports an assumption that this jump was the main source of the observed "Odessa tsunami". Similar examples of hazardous meteotsunamis produced by abrupt jumps of the atmospheric pressure had been observed and numerically simulated in various world regions, including the Balearic Islands (Jansà et al. 2007; Vilibić et al. 2008) and the Adriatic Sea (Orlić et al. 2010).

Further evidence of the atmospheric origin of the event was obtained from the Odessa tide gauge record (Fig. 10b). Continuous 70-min oscillations with wave height of about 15 cm are the typical feature of this record. The oscillations are most likely associated with the fundamental eigenmode of Odessa Gulf (Maramzin 1985). The regular character of these oscillations was suddenly interrupted by an intense train of short-period waves that began shortly before 10:00 UTC on 27 June 2014, i.e. about 15–25 min after the pressure jump and the observed tsunami-like waves on beaches of Odessa and Illichevsk. This train of waves started with a sudden 40 cm drop and then equal rise of sea level (Fig. 10b), followed by a few shorter period oscillations. The most likely reason for these waves is the arrival of an external impulse-like

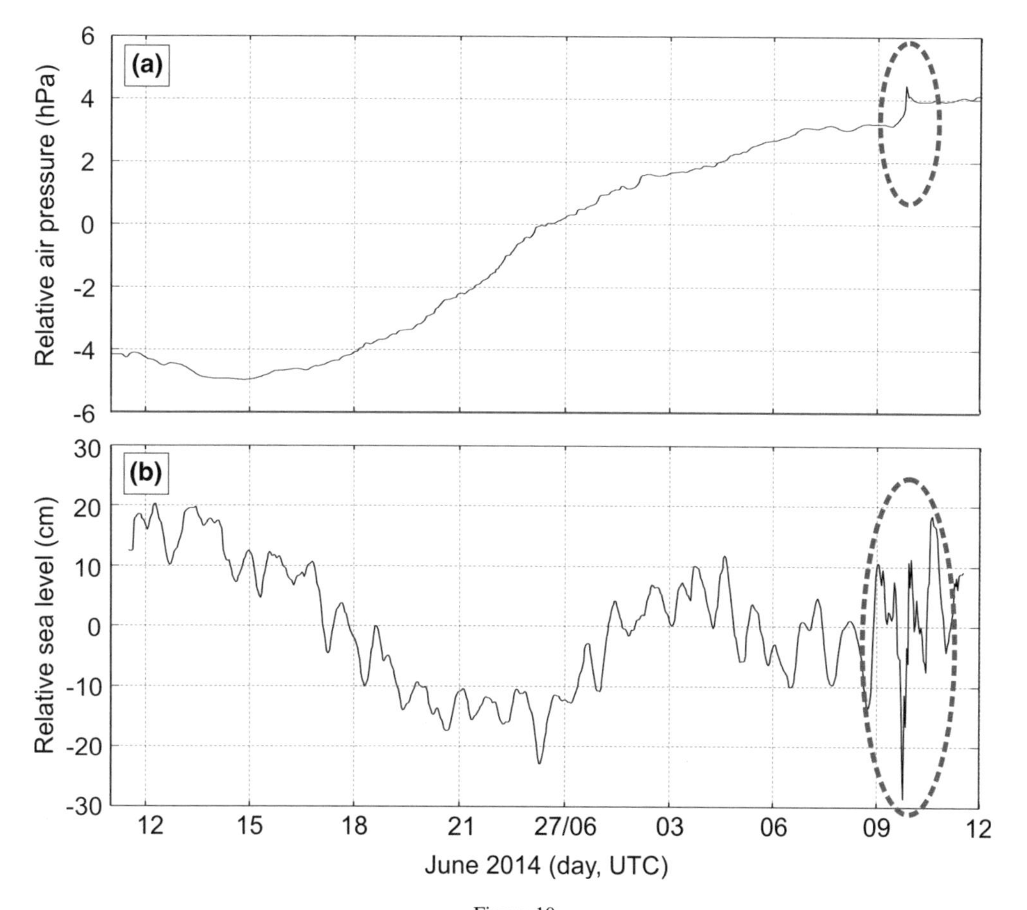

Figure 10
a Air pressure and **b** sea level time series measured at Odessa on 26 and 27 June 2014

wave, which induced harbour oscillations (cf. Rabinovich 2009) in the port of Odessa, where the tide gauge is located. High correlation between the atmospheric pressure disturbance and intense sea level oscillations in the port of Odessa is evident in the frequency–time (*f-t*) diagrams (Fig. 11) computed by the multiple-filter method, which is similar to the wavelet analysis (cf. Thomson and Emery 2014; Šepić and Rabinovich 2014).

In addition, we collected several tide gauge records from various sites of the Black Sea coast: two digital 1-min records from Bulgaria and three analogue (pen-and-paper) tide gauge records from Romania, the Crimean Peninsula and the Caucasian region of Russia. We digitized the analogue series with 1-min time step and thoroughly examined the digital records. The typical characteristic of almost all records is significant amplification of seiche oscillations on 27 June with most intensified waves observed between 12:00 and 18:00 UTC (Fig. 12). At Varna (Bulgaria) the maximum trough-to-crest wave height was 33.9 cm, at Burgas (Bulgaria) 31.6 cm, at Constanta (Romania) 24.5 cm, at Yalta (Crimea) and Sochi (Caucasian coast of Russia) about 10–15 cm (see Fig. 3 for tide gauge locations). A spectacular feature of certain records was an abrupt beginning of

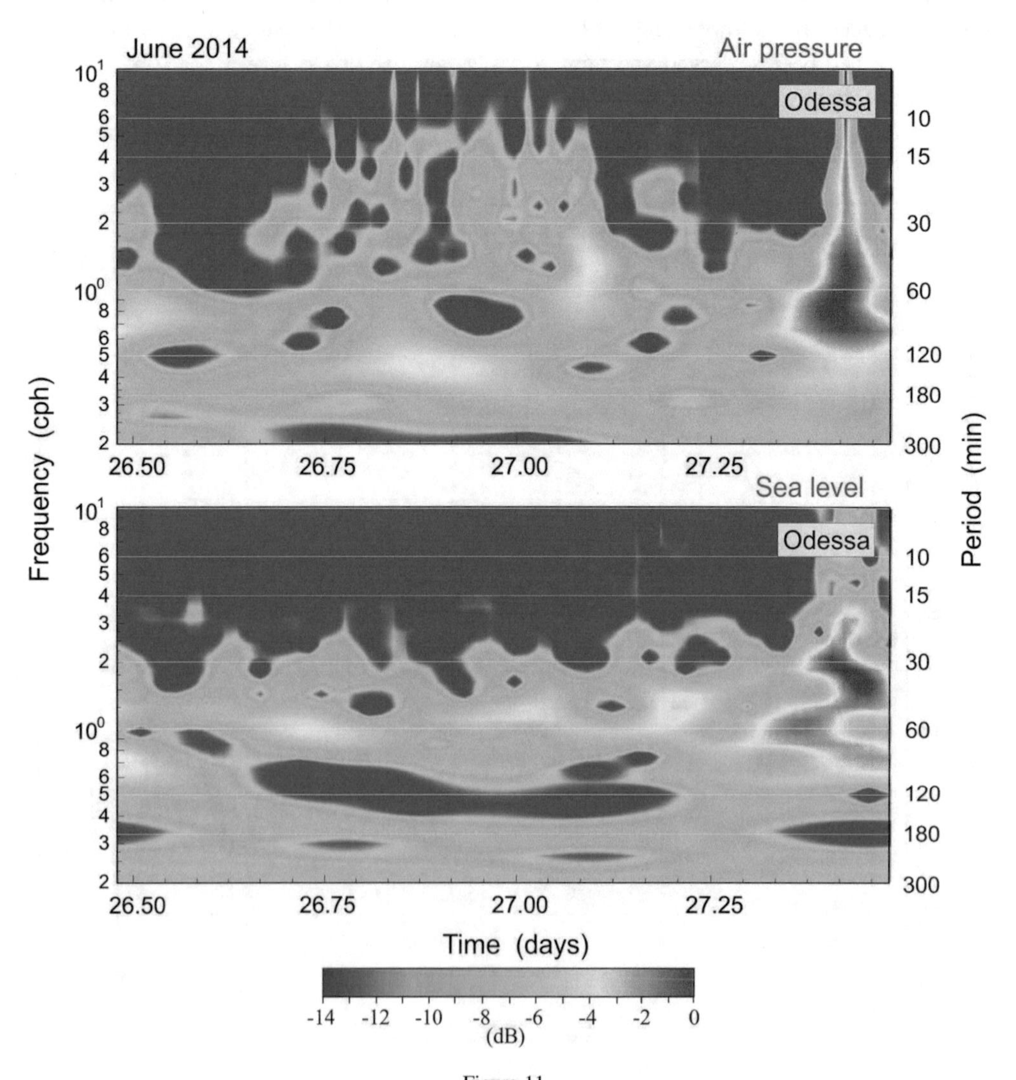

Figure 11
Frequency–time plots (*f-t* diagrams) for the 27 June 2014 Odessa air pressure and sea level digitized records

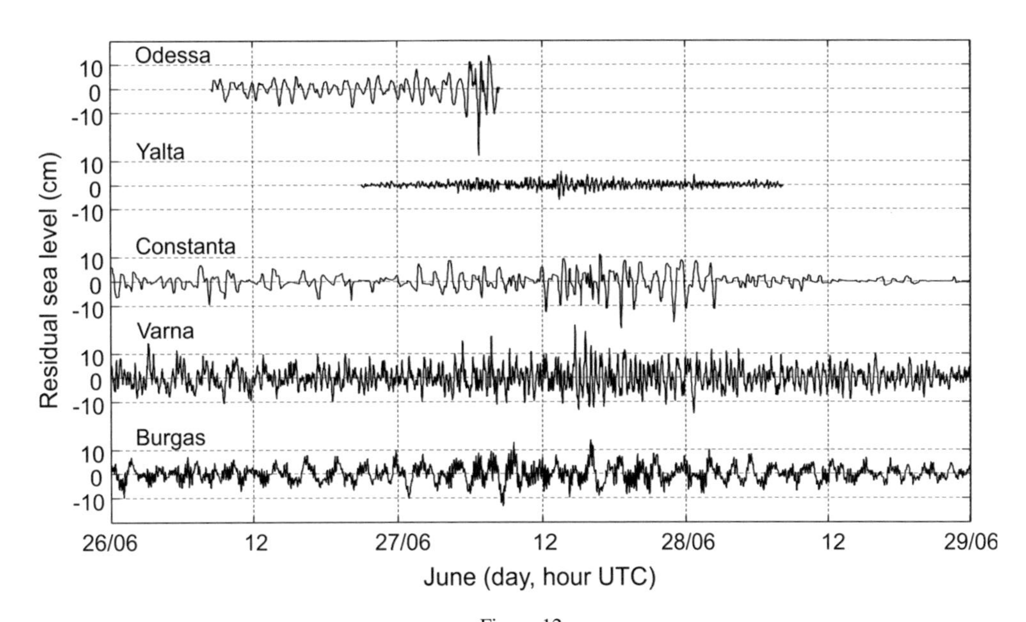

Figure 12
Measured and digitized sea level records from the Black Sea tide gauges for 26–29 June 2014. The tide record at Sochi was not digitized because it was too noisy

pronounced oscillations. At Varna, the first train started at ~ 9:00 UTC and then the second, most prominent train arrived at ~ 15:00 UTC. At Constanta, strong oscillations started at ~ 12:00 UTC, and at Yalta at ~ 14:00 UTC. Such a sharp ("tsunami-like") beginning of a seiche is typically associated with a pressure jump or frontal passage (Monserrat et al. 2006). The intensification of oscillations at Yalta was concurrent with the arrival of the tsunamigenic synoptic pattern to the area of the Crimean Peninsula.

Significant seiche activity at various tide gauge sites is clearly seen in the *f-t* plots of the sea level records (Fig. 13). The *f-t* plots reveal the obvious train structure of the observed oscillations, which is probably determined by small-scale atmospheric disturbances responsible for forcing these oscillations. The intensive oscillations occurred in the middle of the day of 27 June 2014. The wave energy at each site is mainly concentrated at specific frequency bands, which appear to be related to the resonant frequencies of the respective bays and harbours. In particular, at Yalta the dominant oscillations had a period of 20 min, at Constanta 35 and 65 min, at Varna 25 min and at Burgas 12, 45, 100 and 180 min.

Thus, the extreme waves observed on beaches of Odessa and Illichevsk on 27 June 2017 ("Odessa tsunami") were not a unique local event. At approximately the same time, significant seiches were recorded at some other sites in the Black Sea. Comparison of these records with the available atmospheric data, as well as with the results of analysis of similar events of 23–26 June 2014 in the Mediterranean region (Šepić et al. 2015b, 2016, 2017), clearly indicates the meteorological origin of the observed waves. At the same time, Odessa was the only known place in the Black Sea where these oscillations (waves) had a destructive character. The natural questions are: Why specifically here? What is the physical mechanism responsible for strong amplification of atmospherically generated waves? To reply to these questions and to reproduce the actual event, we made several numerical experiments.

5. *Numerical Modelling of the Odessa Tsunami*

Numerical models have already been successfully used to reproduce specific meteotsunamis in various regions of the world oceans and to examine their

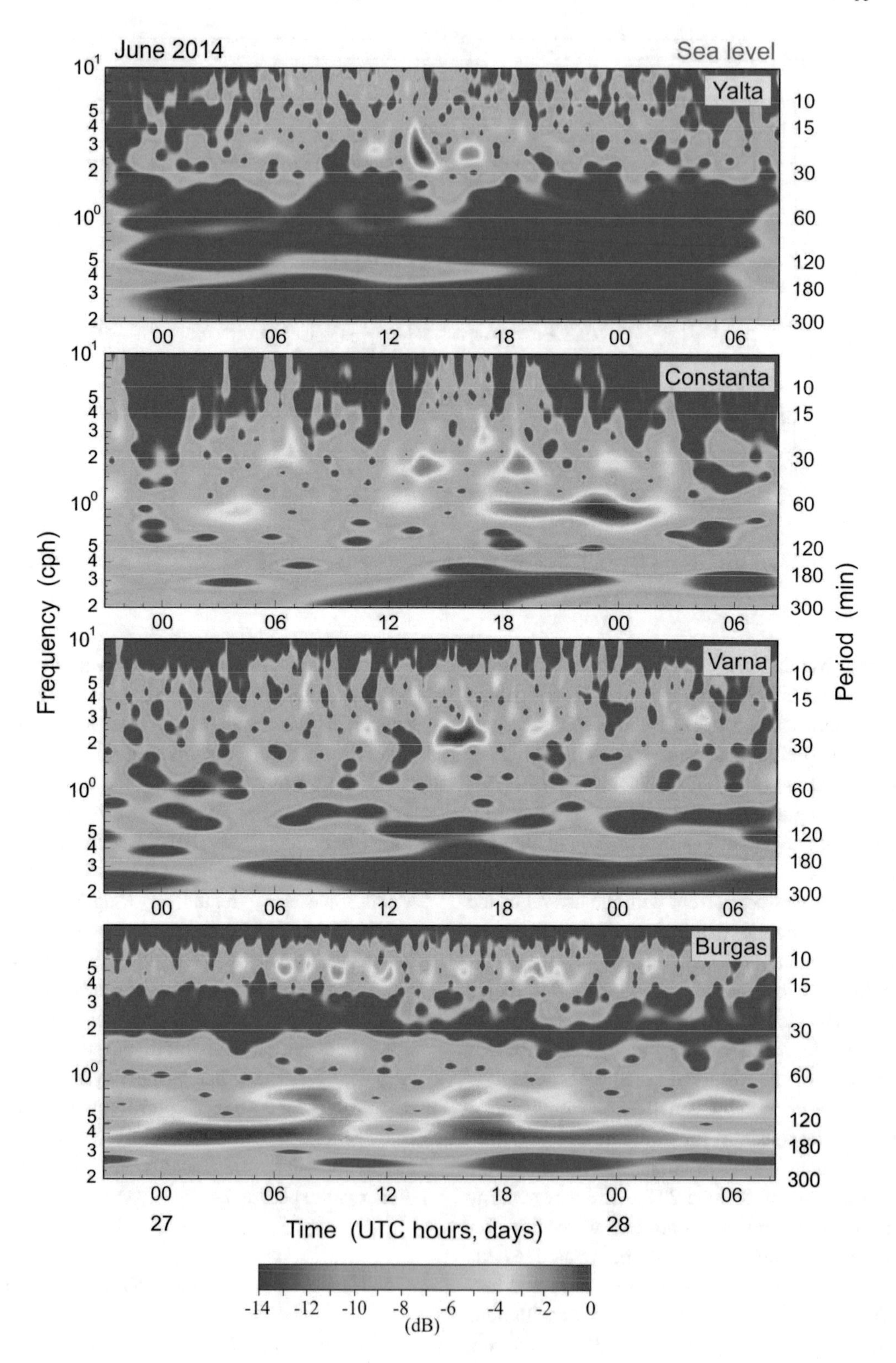
June 2014
Sea level
Yalta
Constanta
Varna
Burgas
Frequency (cph)
Period (min)
Time (UTC hours, days)
27
28
(dB)

◀Figure 13
Frequency–time plots (f-t diagrams) for the 27–28 June 2014 sea level oscillations from four Black Sea tide gauges

generation mechanism. In particular, numerical experiments were applied to inspect several extreme events in the Adriatic Sea, including the hazardous floods of 21 June 1978, 27 June 2003 and 25 June 2014 on the Dalmatian coast (Vilibić et al. 2004; Orlić et al. 2010; Šepić et al. 2016), the destructive event of 15 June 2006 in Ciutadella Harbour, the Balearic Islands, Spain (Vilibić et al. 2008) and the two catastrophic meteotsunamis of 26 June and 6 July 1954 in Lake Michigan, USA (Bechle and Wu 2014). In the present study, we considered two hypotheses, which are based on results of the data analysis: (1) sea level oscillations at Odessa were generated by a single air pressure disturbance propagating towards and over Odessa; and (2) these oscillations were generated by a train of air pressure disturbances propagating along the northwestern shelf break. A number of numerical experiments were done to test both hypotheses.

5.1. Numerical Model

To numerically simulate the Odessa meteotsunami, we used an ocean model based on depth-integrated shallow water equations with the Coriolis term and all nonlinear terms except bottom friction omitted:

$$\frac{\partial U}{\partial t} = -\frac{c}{h^2} V\sqrt{U^2 + V^2} - gh\frac{\partial \eta}{\partial x}; \qquad (1)$$

$$\frac{\partial V}{\partial t} = -\frac{c}{h^2} V\sqrt{U^2 + V^2} - gh\frac{\partial \eta}{\partial y}; \qquad (2)$$

$$\frac{\partial \eta}{\partial t} = -\frac{\partial U}{\partial x} - \frac{\partial V}{\partial y} + \frac{\partial \zeta_A}{\partial t}, \qquad (3)$$

where t is time, (x, y) are the Cartesian coordinates, (U, V) are the vertically integrated water flows, h is the depth, g is the gravity acceleration, η is the sea level reduced for the inverse barometric (IB) effect, ζ_A is the sea level equivalent to the air pressure inverted barometer (IB) effect, and C is the nonlinear bottom friction coefficient (set to 0.0025 in this study). For wave heights up to 1–2 m, the difference between tsunami-like waves modelled with linear vs. nonlinear models is negligible, whereas the Coriolis force for processes on a meteotsunami/tsunami time scale is of no importance (Kowalik 2012). The effect of air pressure on sea level has been embedded as a time derivate into the continuity Eq. (3). It is more common to add this effect as a spatial gradient into the momentum Eqs. (1) and (2). The herein used approach, however, allows the conservation of the bell shape of the air pressure disturbance and is less sensitive to the resolution of the modelling grid. This is particularly important when dealing with an air pressure disturbance characterized by a high temporal change. In our numerical code, the time derivate of air pressure is strictly an input parameter, independent of other model outputs and parameters. This model code had been developed by Isaac Fine (Institute of Ocean Sciences, Sidney, BC, Canada) and already successfully used in various meteotsunami modelling experiments (cf. Monserrat et al. 2014; Šepić et al. 2015a).

At the coastal boundary (G), we assumed a vertical wall with zero normal flow component (no-slip condition):

$$U_n = 0 \quad \text{on } G, \qquad (4)$$

and on the open boundary (Γ) we used the radiation condition:

$$\frac{\partial \eta}{\partial n} - \frac{1}{c}\frac{\partial \eta}{\partial t} = \frac{\partial \zeta_A}{\partial n} - \frac{1}{c}\frac{\partial \zeta_A}{\partial t} \quad \text{on } \Gamma, \qquad (5)$$

where $c = \sqrt{gh}$ is the long wave speed and n is directed normally to the boundary Γ.

The explicit finite-difference method was used to solve Eqs. (1)–(3) with boundary conditions (4) and (5). We applied the Arakawa C-grid approximation and used the General Bathymetric Chart of the Oceans (GEBCO) bathymetry (http://www.gebco.net/) with the horizontal resolution of 30'' ($\sim$ 650 m in longitude and 920 m in latitude). The model domain size was 650 $\times$ 451 grid points (Fig. 14). According to the Courant–Friedrichs–Lewy (CFL) criterion, the time step (Δt) was chosen to be 4.6 s.

The model was forced by a bell-shaped travelling air pressure disturbance given by the following equation:

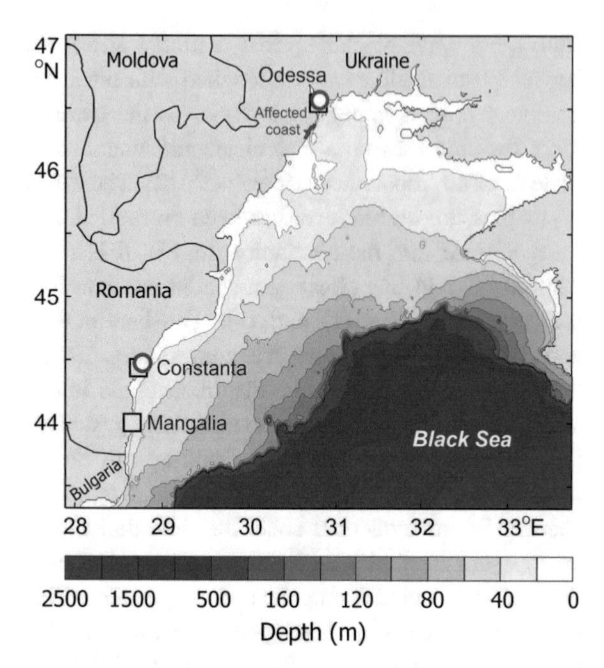

Figure 14
a Computational domain of the ocean numerical model. The red-bordered circles indicate positions of tide gauges and yellow squares of air pressure stations

$$\zeta_A = \frac{A}{1 + (t/a)^{2\gamma}}, \quad (6)$$

where $A = 3$ hPa is the disturbance amplitude, $a = 5$ min is the half width of the disturbance, and $\gamma = a/b$ is the slope parameter, where b is the half width of the rising (decreasing) part of the pressure impulse ($b = 100$ s). This function fits well to the air pressure and wind changes of about $\sim$ 3 hPa/10 min and $\sim$ 10 m/s, respectively, typically observed during meteotsunami events (Vilibić et al. 2004; Jansà et al. 2007).

The bell-shaped air pressure disturbance (6) was imposed to propagate over the domain with predefined speed and direction in two sets of experiments. In the first set, the air pressure disturbance was assumed to propagate directly over Odessa (Experiment 1); in the second set, the disturbance propagated over the shelf break (Experiment 2). The disturbance propagation parameters in both sets of experiments were chosen to vary in the following ranges:

- Speed: from 5 to 35 m/s with an increment of 1 m/s;
- Direction: from 180° to 290° with an increment of 5° (i.e. the disturbance was supposed to have an eastward propagation component).

The speed and direction ranges were chosen based on the wind speeds and directions at the geopotential height of 500 hPa (about 5500 m above the sea surface) between 06:00 and 12:00 UTC on 27 June 2014, according to the ERA-Interim reanalysis data. This height was selected because tsunamigenic pressure disturbances generally propagate in agreement with the wind velocity at this specific height. The air pressure disturbance was set to exponentially decay in the cross-propagation direction with a half "bell" width of 20 km. The length of each simulation series was 14 h. The air pressure disturbance was set to start propagating 60 min after the model initialization. In addition, for Experiment 2, the sensitivity of sea level response to the rate of air pressure change and to cross-propagation width of the air pressure disturbance was tested. The main purposes of these modelling experiments were:

1. To identify specific parameters of the propagating air pressure disturbance producing the strongest sea level response to the atmospheric forcing;
2. To compare the numerical results (computed tsunami wave heights) with those known from the witness reports and existing videos/photographs of the 27 June 2014 event; to estimate how well the constructed model may reproduce the observed tsunami wave heights.

A similar numerical approach and modelling experiments were used by Orlić et al. (2010) to examine the catastrophic flood of 21 June 1978 in Vela Luka Bay, on Korčula Island in the Adriatic Sea.

5.2. *Direct Forcing Experiment*

Maximum predicted wave heights as a function of the speed and direction of the atmospheric disturbance for the *direct forcing experiment* (Experiment 1) are shown in Fig. 15. When a disturbance propagates directly towards Odessa (Fig. 15a), the modelled waves reach the maximum height of 52.7 cm (Fig. 15b). This wave height corresponds

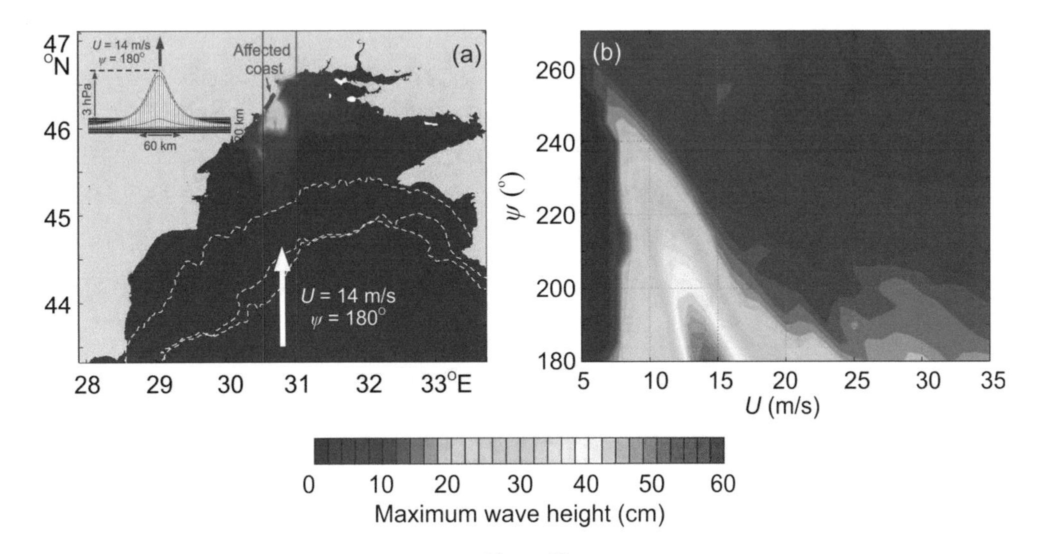

Figure 15
a Maximum simulated wave height across the Odessa Gulf for the air pressure disturbance speed of 14 m/s and direction of 180°. Maximum heights greater than 25 cm are constricted to the coastal points and are not easily distinguishable in the entire domain plot. **b** Maximum simulated wave heights at Odessa as function of speed, U, and direction, ψ, of the propagating air pressure disturbance. Red lines in (**a**) border the propagation path of the air pressure disturbance; white dashed lines denote depth contours of 50, 100 and 150 m; red line indicates stretch of coast hit by tsunami waves. The schematic of the air pressure disturbance is given in (**a**)

to the air disturbance speed $U = 14$ m/s and direction of 180°. Figure 15a shows the maximum wave heights for the entire domain specifically for this set of parameters. Simulated meteotsunami waves amplify northward along the propagation path and become larger than 20 cm north of the latitude 45.7°N. Sea depths in this area range from 20 to 40 m, corresponding to the long wave phase speed $c = 14.0 - 19.8$ m/s. Thus, for this particular experiment, the Froude number, $\mathrm{Fr} = U/c$, i.e. ratio of the disturbance speed to the long wave speed, ranges from 1.0 to 1.4, pointing out the favourable, but not perfect, conditions for the Proudman resonance.

The propagation of the simulated waves is more clearly seen in Fig. 16. Long ocean waves travel as forced waves towards Odessa. Although the air pressure disturbance is set to move over the domain from the very southern border, significant sea level oscillations begin to be noticeably generated only after the air pressure disturbance crosses the shelf break, i.e. once the air disturbance speed becomes comparable with the speed of long ocean waves. Final amplification of the waves occurs in a narrow coastal strip and is due to the shoaling effect. Maximum wave heights of 52.7 cm are reached precisely at that part of the Odessa coast, where the strongest tsunami-like waves had been observed. Simulated meteotsunami waveforms for the locations of the most severe destruction (Sunrise complex) and for the Odessa tide gauge are shown in Fig. 17. Sudden onset of tsunami-like oscillations is well reproduced, as well as the 30-min time lag between the oscillations recorded at the Odessa tide gauge and those observed at Sunrise complex (SC). However, depending on location, the model underestimates the observed/recorded wave heights by 1.3–3 times. At the Odessa tide gauge (TG), the maximum modelled oscillations reach a wave height of 30.3 cm, whereas the maximum recorded oscillations had wave height of 39.4 cm. At Sunrise complex, the maximum simulated oscillations had a height of 52.7 cm, whereas, according to the available videos, the tsunami wave height here was $\sim$ 1 m, and according to eyewitnesses even 2 m. Further on, the 26-min oscillation recorded at the tide gauge location was clearly reproduced by the model (25-min period), whereas a shorter period (1–2 min) wave that was responsible for the main damage on beaches of Odessa and Illichevsk was not reproduced.

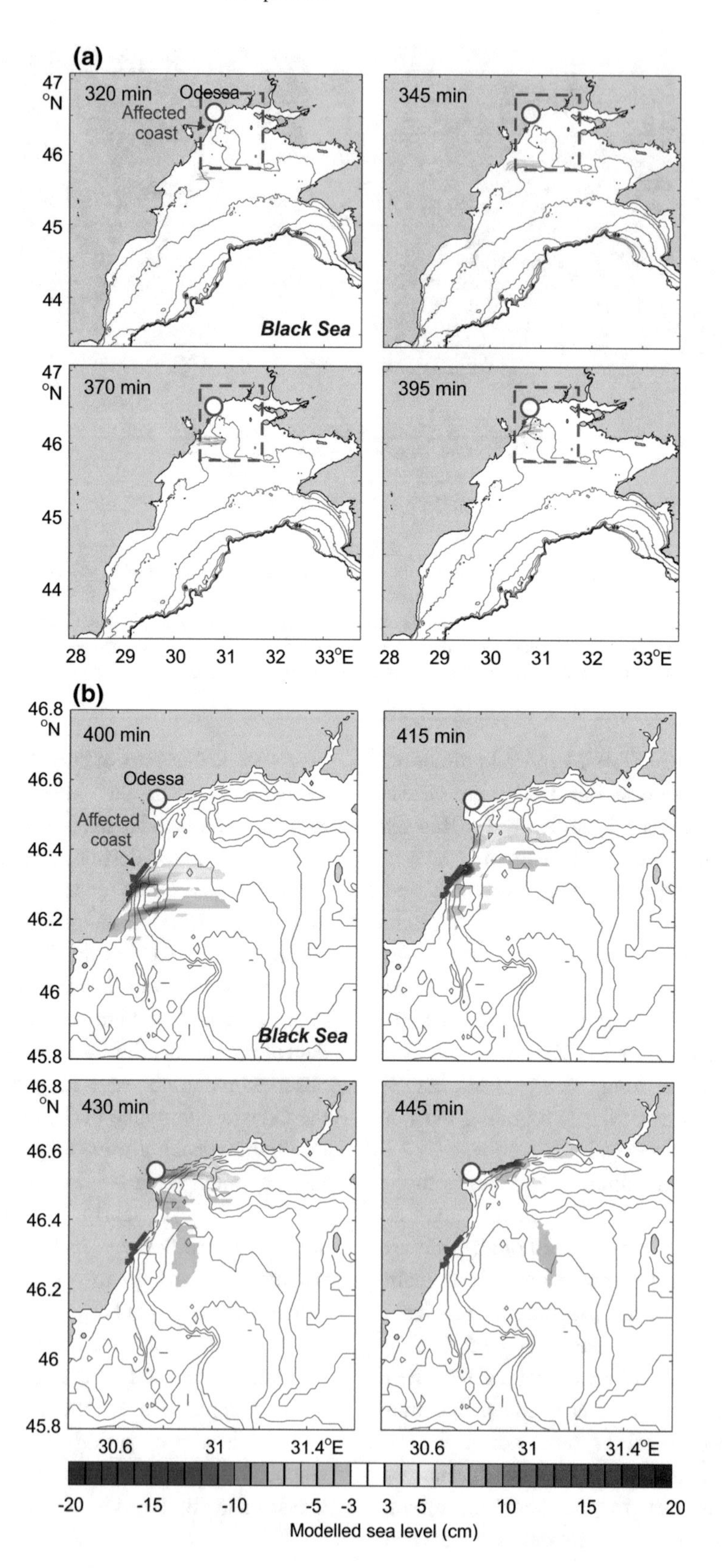
(a)
320 min
Odessa
Affected coast
345 min
370 min
395 min
Black Sea
47 °N
46
45
44
28 29 30 31 32 33°E
(b)
400 min
415 min
430 min
445 min
Odessa
Affected coast
Black Sea
46.8 °N
46.6
46.4
46.2
46
45.8
30.6 31 31.4°E
-20 -15 -10 -5 -3 3 5 10 15 20
Modelled sea level (cm)

◀Figure 16
a Simulated meteotsunami wave in the northwestern part of the Black Sea (Odessa Gulf) at 320; 345; 370; and 395 min since the model start; 0–200 m depth contours are shown, with a step of 20 m. **b** Zoomed snapshots of the simulated meteotsunamis for the Odessa region at 400; 415; 430; and 445 min since the model start; 0–50 m depth contours are shown, with a step of 5 m; red line indicates stretch of coast hit by tsunami waves

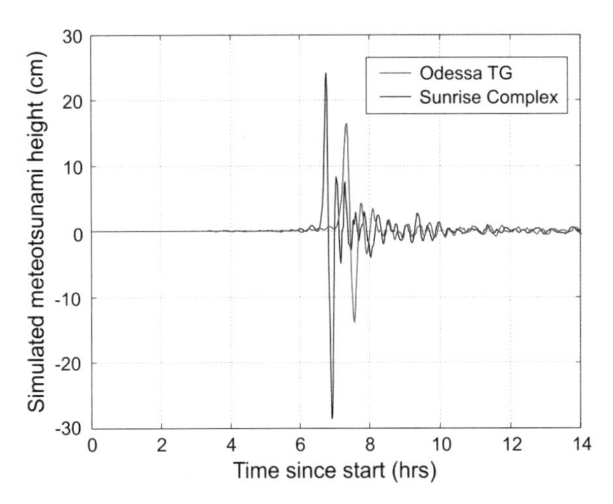

Figure 17
Simulated meteotsunami heights at two sites—Odessa tide gauge and Sunrise complex—for the numerical experiment with the air pressure disturbance propagating directly towards Odessa with speed $U = 14$ m/s and direction $\psi = 180°$

5.3. Shelf Break Experiment

The authors also examined an alternative mechanism of the “Odessa tsunami” generation in the northwestern part of the Black Sea. Maximum modelled wave heights at the Odessa coast most affected by the tsunami-like waves for *the shelf break forcing experiment* (Experiment 2) are shown in Fig. 18 as a function of the direction and speed of the atmospheric disturbance. There is a distinct point of speeds and directions for which the sea level oscillations are most intense: $U = 27$ m/s and $\psi = 230°$. For these particular disturbance parameters, simulated waves at Odessa coast have heights of 28.3 cm, while for the entire domain modelled waves along the shelf break reach maximum heights of 50.0 cm (Fig. 18a).

The transformation of long ocean waves is better perceived in Fig. 19. The waves significantly intensify while travelling along the shelf break, in particular in the region with depths of 70–80 m. Precisely over these depths, the Proudman resonance occurs: the long wave speed is $c = 26.2$–28.0 m/s, and thus for the air pressure disturbance propagating with speed $U = 27$ m/s, Froude number is $\mathrm{Fr} = 0.97$–1.04, pointing out the almost perfect

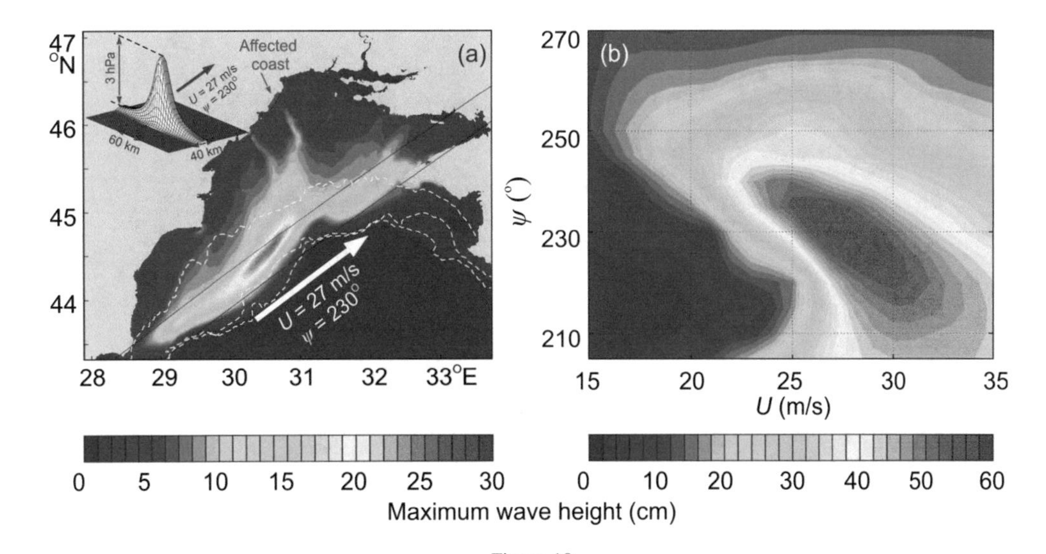

Figure 18
a Maximum modelled wave height across model domain for experiment in which an air pressure disturbance propagates with a speed of 27 m/s and a direction of 230°. Maximum heights greater than 25 cm are constricted to the coastal points and are not easily distinguishable for the entire domain plot. **b** Maximum simulated wave heights at Odessa as function of speed, U, and direction, ψ, of the propagating air pressure disturbance. Red lines in (**a**) mark propagation paths of air pressure disturbance; white dashed lines indicate depth contours of 50, 100 and 150 m; red line indicates stretch of coast hit by tsunami waves. Schematic of air pressure disturbance is given in (**a**)

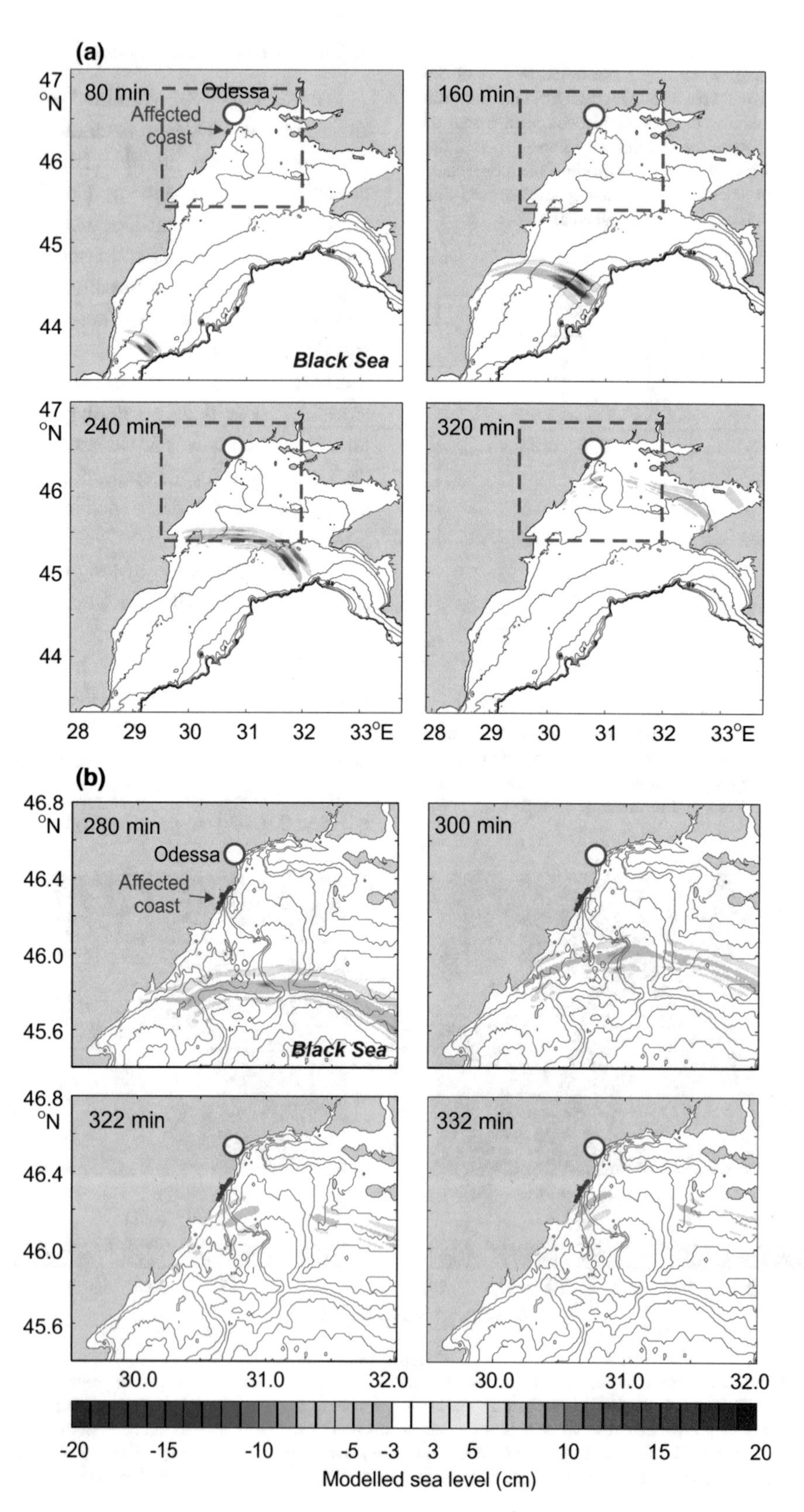
(a)
80 min
Odessa
Affected coast
Black Sea
160 min
240 min
320 min
47 °N
46
45
44
28
29
30
31
32
33°E
(b)
280 min
300 min
322 min
332 min
46.8 °N
46.4
46.0
45.6
30.0
31.0
32.0
-20
-15
-10
-5
-3
3
5
10
15
20
Modelled sea level (cm)

◀Figure 19
a Simulated meteotsunami wave in the northwestern part of the Black Sea (Odessa Gulf) at 80, 165, 240 and 320 min since the model start; 0–200 m depth contours are shown, with a step of 20 m. **b** Zoomed snapshots of the simulated meteotsunami for the Odessa region at 280, 300, 322 and 332 min since the model start; 0–50 m depth contours are shown, with a step of 5 m; red line indicates stretch of coast hit by tsunami waves

resonant conditions. Since the air pressure disturbance propagates over these depths for a considerable distance ($\sim$ 100 km), long ocean waves have enough time to accumulate the forcing energy and significantly intensify. Once the air pressure disturbance crosses over the shelf break into the deeper sea region, part of the propagating long sea waves reflects from the break, heading precisely towards the 200-km distanced Odessa coast along a western edge of a submarine canyon (Fig. 3). Approaching the coast, waves further amplify due to the shoaling effect.

Similar to the first experiment, we simulated meteotsunami waveforms for the SC and TG locations for the shelf break experiment (Fig. 20). The corresponding model also reproduces well the sudden onset of the oscillations, 30-min time lag between oscillations at SC and TG, and the lower period oscillations visible in the TG record (23 min modelled vs. 26 min recorded). However, once again, the model underestimates the tsunami wave height, both for the TG location (10.4 cm modelled vs. 39.3 cm observed) and for the SC location (28.3 cm modelled vs. 1–3 m observed). The remarkable fact is that those parts of the Odessa coast which were hit by the meteotsunami waves are located exactly at the convergence point of long ocean waves reflected from the shelf.

To estimate the influence of the air pressure parameters on generated ocean waves, we ran an additional set of experiments, in which cross-propagation half width of the air pressure disturbance (propagating with $U = 27$ m/s and $\psi = 230°$) was set to change from 5 to 40 km with a step of 5 km, and from 40 to 80 km, with a step of 10 km. Maximum simulated meteotsunami heights at two sites (Odessa tide gauge and Sunrise complex) are shown in Fig. 21. At the shelf break, wave height first steeply grows with increases in the cross-propagation width (from 5 to 40 km), reaching its maximum of 62.8 cm at 50 km, and then it slowly decays. Similarly, for the Odessa coast, the maximum simulated wave height abruptly grows with increases in the cross-propagation width (from 5 to 20 km), reaching its maximum of 28.7 cm at 25 km, after which it slowly decays. Wave heights decrease as the cross-propagation disturbance width increases, likely due to the smaller spatial pressure gradients and less effective interactions of forced waves travelling over the shelf in the northeastward direction (following the direction of the air pressure disturbance) and free waves which are reflected from the shelf break and propagate to the northwest.

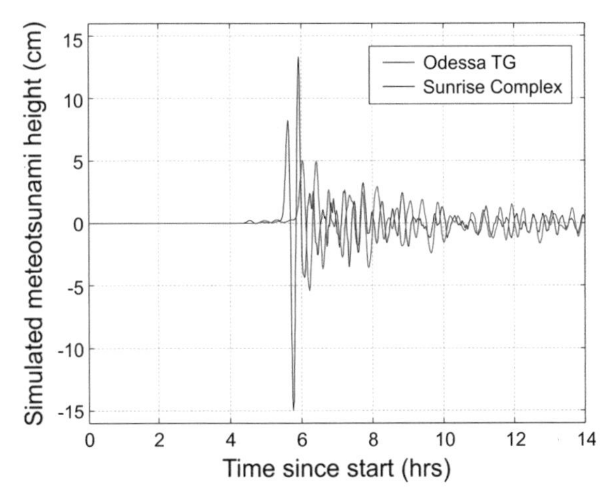

Figure 20
Simulated meteotsunami heights at two sites—Odessa tide gauge and Sunrise complex—for the numerical experiment in which the air pressure disturbance propagates over the shelf break with the speed $U = 27$ m/s and direction $\psi = 230°$

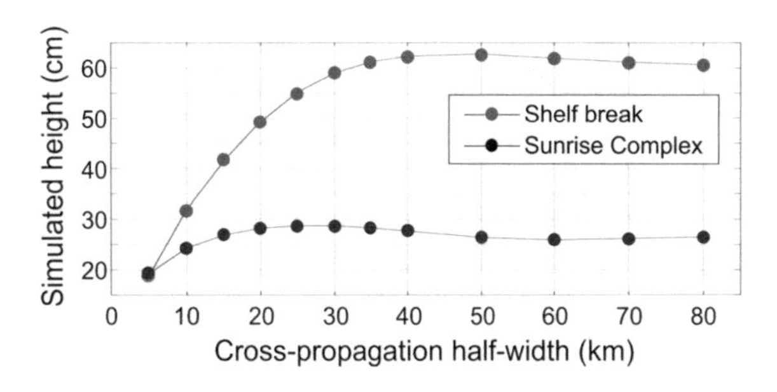

Figure 21
Simulated meteotsunami height at two sites—shelf break and Sunrise complex—for a series of numerical experiments in which the air pressure disturbance propagates over the shelf break with the speed $U = 27$ m/s and direction $\psi = 230°$, as a function of the cross-propagation width of air pressure disturbance

6. Discussion

All information and data that we could collect for the area of Odessa and other regions of the Black Sea, as well as the comparison with the chain of extreme events that occurred in the Mediterranean Sea during four preceding days (23–26 June 2014), indicate the meteorological origin of the "Odessa tsunami" of 27 June 2014. We can see that prominent seiche oscillations were also observed during this day in several other ports of the Black Sea. However, the exact generation mechanism responsible for such strong waves specifically on the Odessa beaches remains unclear. We considered two hypotheses to explain the tsunami-like waves which hit the Odessa coast: (1) the waves were generated by an atmospheric disturbance which propagated directly over Odessa; (2) the waves were generated by an atmospheric disturbance which propagated along and over a shelf break 200 km away (south) and then came to the area of Odessa being amplified by shoaling effects.

We ran two sets of numerical experiments ("Experiment 1" and "Experiment 2" in the further text) to examine these hypotheses. By both modelling experiments, we were able to reproduce: (a) the sudden onset of tsunami-like oscillations; (b) a half-hour time delay between the "tsunami waves" which affected the Odessa beaches and the Odessa tide gauge; and (c) the 26-min period oscillations recorded in Odessa harbour. However, neither of these modelling experiments was able to explain: (1) the actual height of the observed waves and (2) the 1–2-min period of the waves which did most of the damage on beaches. The 1–2 m maximum wave heights are better reproduced in Experiment 1 ($\sim$ 53 cm) than in Experiment 2 (28.3 cm); however, the latter experiment is in better agreement with some other observed features of the event.

Before concluding which experiment is more likely, we need to consider some properties of atmospheric disturbances that are typically responsible for the generation of meteotsunamis. Such pressure disturbances are commonly advected by mid-tropospheric winds at the 500-hPa geopotential height (Thomson et al. 2009; Vilibić and Šepić 2009). According to the ERA-Interim data, between 06:00 and 12:00 UTC on 27 June 2014, i.e. during the event, the wind at 500 hPa along the northwestern Black Sea coast and over Odessa had speeds of 17–21 m/s and directions of 225°–270° (Fig. 6). This means that the atmospheric disturbance was supposed to propagate northeastward or eastward with a similar speed. For this particular atmospheric situation, the maximum modelled wave heights for Experiment 1 ("direct forcing") at the Sunrise complex reached only 4.5–10.5 cm (Fig. 15). Thus, for this specific experiment, we have a clear disagreement between the actual parameters of the propagating atmospheric disturbances and those parameters that produce the maximum sea level response. For Experiment 2 ("shelf"), the situation is different. The actual 500 hPa winds over and along the shelf break were slightly stronger (24–27 m/s) but kept almost the same directions (225°–260°) as the 500-hPa winds along the northwestern coast (Fig. 6). In this case, the speed and direction of the atmospheric disturbance providing maximum modelled wave heights of 10.5–28.3 cm (Fig. 18) coincide with the actual speed and direction of propagating atmospheric waves. Thus, from a qualitative point of view Experiment 2 gives results that are in good agreement with the observed atmospheric features but underestimate the observed tsunami heights. Furthermore, convective (Fig. 7) and thunderstorm (Fig. 8) activities, often related to tsunamigenic air pressure disturbances, as well as a tsunamigenic synoptic pattern (Fig. 6), were centred precisely over the shelf break area.

Both shelf break reflection and amplification along a canyon edge are known effects stimulating meteotsunami generation. Meteotsunamis which hit the US East Coast are occasionally produced by atmospheric pressure disturbances propagating eastward (offshore): upon reaching the shelf break, the "forced" long ocean waves (locked to the atmospheric disturbance) partly reflect from the shelf and propagate backward towards the US East Coast as free waves (Pasquet and Vilibić 2013; Šepić and Rabinovich 2014; Wertman et al. 2014). Likewise, it has been shown by numerical modelling that seismically generated tsunami waves diminish along the centre of submarine canyons but increase along their edges (Aranguiz and Shibayama 2013; Iglesias et al. 2014).

Destructive meteorological tsunamis are commonly generated through a series of successive steps (Hibiya and Kajiura 1982; Monserrat et al. 2006; Rabinovich 2009): (1) Proudman resonance—resonant amplification of long waves propagating in the open ocean with a speed of the atmospheric pressure disturbance (Fr $\sim$ 1.0); (2) shelf amplification/resonance—increase in the meteotsunami wave height due to the depth decrease (Green's law) and resonant effects over the shelf; (3) shoaling effect—additional increase in long ocean waves approaching the coast, frequently with formation of a bore; (4) harbour resonance—amplification in inlets, bays and harbours with a high Q-factor, i.e. with strong resonant amplification and slow energy decay. Additional factors leading to marked intensification of meteotsunamis are offshore depth structures that promote energy convergence, and narrowing V-shaped bays focusing the incoming wave energy. As illustrated by Hibiya and Kajiura (1982) and Monserrat et al. (2006), Nagasaki Bay is a classic example of a V-shaped basin that leads to substantial wave amplification. In this case, a 3-cm wave propagating from the open part of the East China Sea on 31 March 1979 was shown to undergo successive amplifications that eventually resulted in a 5-m wave at the head of the bay that killed three people. Another example is in V-shaped Vela Luka Bay, Korčula Island, Croatia, where a 6-m meteotsunami occurred in June 1978 in the adjacent harbour.

The particular sites with specific features favourable for the meteotsunami generation may be identified as "hot spots" (cf. Šepić and Rabinovich 2014). Most of them are V-shaped bays, elongated narrow inlets or bays with tight entrances. Three Mediterranean locations stricken by the June 2014 event, Ciutadella (Spain), Vela Luka (Croatia) and Mazara del Vallo (Sicily, Italy), are examples of such basins (Šepić et al. 2015b). In all three harbours, destructive meteotsunamis had been regularly observed in the past (Rabinovich and Monserrat 1996, 1998; Orlić 2015; Candela et al. 1999) and probably will occur in the future.

Other locations, like the Odessa beaches, are quite different. First, the meteotsunami took place not in a bay or harbour with high Q-factor and prominent resonant properties, but along the open coast. This means that an important amplification factor, crucial for formation of extreme events at Ciutadella, Vela Luka and Mazara del Vallo on 23–26 June 2014, and in Nagasaki Bay in March 1979, was absent for the Odessa event. Second, as was indicated above, meteotsunamis in the four listed regions are common, while in Odessa this was a totally unexpected event. However, similar extreme tsunami-like waves are known to occur now and then in various regions of the World Ocean. In particular, "*On a clear calm evening during July 1992, an anomalously large wave, reportedly 6 m high, struck Daytona Beach, Florida…*" (Sallenger et al. 1995). The wave injured 75 people and severely damaged properties along Daytona Beach, Atlantic coast of Florida. Following investigations demonstrated that this wave was caused by a squall line associated with a pressure jump travelling with a speed of long ocean waves (Sallenger et al. 1995; Churchill et al. 1995). Three years later, on 25 March 1995 a solitary wave of $\sim$ 3 m height hits the coast of Tampa Bay, the Gulf Coast of Florida; the wave was found to be generated and amplified by a large-amplitude atmospheric gravity wave transiting southeastward over the eastern Gulf of Mexico (Paxton and Sobien 1998). Some other similar examples are solitary tsunami-like waves that attacked the beach of Praia do Cassino in Brazil on 9 February 2014 and the beach of Panama City (Florida, USA) on 28 March 2014 (Vilibić et al. 2016). Recently, on 17 January 2016, one more meteotsunami event occurred on the Gulf Coast of Florida (http://www.orlandosentinel.com/news/breaking-news/os-tornadoes-florida-killed-meteotsunami-20160118-story.html). A solitary wave of about 1.7 m high hits the beach of Naples, FL. It was associated with a squall line that resulted from a cold front; the maximum gust wind on this line was 84 mph ($\sim$ 37.5 m/s).

All five of these "beach" events were identified as "meteorological tsunamis", and they have a strong similarity with the "Odessa tsunami":

1. All of them had a character of a solitary wave.
2. They were observed at the open beaches.
3. There are wide shallow water shelves adjacent to the respective beaches.

4. Each event was "new", i.e. occurred in a place where it had never been observed before.
5. Their generation was associated with a squall line and/or an evident jump of atmospheric pressure.

There are no doubts that the "Odessa tsunami" is from the same family of "beach events". It is, thus, different from the "harbour events" of 23–26 June 2014 in the Mediterranean (Ciutadella, Vela Luka and Mazara del Vallo), as well as from other events at specific places of the World Ocean (e.g. Nagasaki Bay, Japan, and Longkou Harbour, China) (cf. Monserrat et al. 2006; Rabinovich 2009). The main difference is that "beach events", in contrast to "harbour events", are not connected to a particular site and are not amplified by harbour resonance or the fuelling effect of V-shaped external basins. Despite the absence of these amplifying factors, "beach" meteotsunamis can be quite strong, even devastating (as shown by the listed examples). These meteotsunamis must be created over the shelf and the only explanation of their intensity is that they are associated with a *sharp resonance*, strongly magnifying onshore waves arriving from the open sea. Such resonance occurs only in very rare situations of close agreement of the long wave velocities determined by depth characteristics in a specific shelf area and parameters of a moving atmospheric disturbance. Certain regions, for example the shelf of Florida, are favourable for such situations; however, exclusive sharp resonances strongly depend on the particular features of air pressure disturbances and are not tied to an exact site: depending on the speed and direction of propagating atmospheric waves, such resonance can be achieved in quite different places. In other words, the reason why the "Odessa tsunami" affected specific beaches of Chernomorka, Illichevsk and the Sunrise complex is not because these beaches were exceptional, but because a moving atmospheric disturbance went into sharp resonance in a specific shelf region adjacent to these beaches.

On a qualitative level, our numerical Experiment 2 reproduced well the main physical features of the Odessa event and was in good agreement with the propagating parameters of the atmospheric disturbance, but underestimated the observed height. The most probable reason is insufficient information about the disturbance itself. Hibiya and Kajiura (1982) suggested a simple formula that can be used to estimate the *resonant generation* of meteotsunami by a moving atmospheric disturbance. If $\Delta\bar{\zeta}$ is the static open ocean inverted barometer response to the pressure jump ΔP_a, i.e. if $\Delta\bar{\zeta} = \Delta P_a/\rho g$, where ρ is the water density, and g is the gravity acceleration, then the resonantly amplified open ocean long wave may be presented as

$$\Delta\zeta = \frac{\Delta\bar{\zeta}}{L_1}\frac{x_f}{2}, \tag{7}$$

where L_1 is the linear scale of the pressure change (squall width), $x_f = Ut$ is the distance travelled by the pressure jump ΔP_a during time t with speed U. In fact, the first fraction characterizes the gradient of atmospheric pressure: the higher this gradient, the longer is the travel distance, the larger is the generated wave. In particular, if we assume $L_1 = 3$ km (very sharp pressure jump) and take a peak value of $\Delta P_a = 3$ hPa ($\Delta\bar{\zeta} \approx -3$ cm) then for $x_f = 200$ km we get $\Delta\zeta \approx 1$ m, in good agreement with what was observed on the beaches of Odessa and Illichevsk.

This means that to create the strong tsunami-like wave which hits Odessa, the respective atmospheric pressure disturbance has to be extraordinary sharp and strong. If we take the propagation speed $U = 25$ m/s, then for the considered example the rate of air pressure change must be very high, ~ 3 hPa over 2 min. This disturbance must further propagate over a shelf of slowly varying depth with the required resonant speed and direction for a sufficient time interval (~ 2.2 h). This can probably be realized very rarely and only for a few short coastal segments. However, we should not consider such a situation as impossible! Actually, the short period of the observed wave indicates the sharpness of the pressure jump that generated this wave.

To test the above estimate, we repeated the experiment in which air pressure disturbance was set to propagate over the shelf break with $U = 27$ m/s and $\psi = 230°$, but with a more abrupt rate of air pressure change of 3 hPa over 2 min (as opposed to originally used 3 hPa over 5 min). The maximum simulated wave height of 101.2 cm at the shelf break was more than double of 50.0 cm obtained in the original experiment, while at the Odessa coast the computed height was 22.3 cm, i.e. smaller than 28.3 cm originally obtained in Experiment 2. This experiment shows that over relatively deep waters, where the resonant conditions are fulfilled ($U \approx c$) and the induced ocean wave propagates as a forced wave, there is an excellent agreement between analytical (7) and numerical results. However, when the wave starts moving onshore, it propagates and transforms as a free wave: it slows down, reduces its length and amplifies as depth decreases according to the Green's law. It appears that the main reason for the disagreement between numerically simulated and observed wave heights is a model grid that is too coarse ($\sim$ 650 $\times$ 920 m), which does not enable us to reproduce various shallow water/coastal effects when the wave length becomes comparable with the grid size. The accuracy of near-shore and coastal bathymetry data is the most important factor for accurate prediction of the coastal impact of tsunami (cf. Rabinovich et al. 2017); this is especially important for meteorological tsunamis, which are commonly a small-scale phenomenon. Finally, near the shore additional effects due to the nonlinearity become important. We assume our coarse-grid near-coast bathymetry used for this study produced uncertainty in the model results and did not reproduce such effects as the runup formation and wave inundation, which are probably the key factors responsible for the observed "Odessa tsunami". The high-resolution modelling would allow more realistic estimates of coastal meteotsunami dynamics and local topographic amplification.

7. *Conclusions*

During the period 23–27 June 2014, a distinctive synoptic system propagated from the Western to the Eastern Mediterranean and further over the Black Sea. This system was characterized by an intrusion of warm air from Africa in the lower troposphere (up to 1500 m) and exceptionally strong southwesterly to westerly upper air winds embedded into unstable atmospheric layers (at heights of $\sim$ 5000 m) (Šepić et al. 2015b). Surface weather was warm and moist, with occasional thunderstorms. A particular feature of this system ("tumultuous atmosphere") was the continuous production of small-scale but intense atmospheric pressure disturbances (Šepić et al. 2015b, 2016); in their turn these disturbances were found to generate high-frequency sea level oscillations in the Mediterranean and Black seas. At some locations, precisely at the time when this synoptic system traversed over them, hazardous events occurred: (i) 23 June in Ciutadella Inlet (Balearic Islands, Spain); (ii) 25 June at Mazara del Vallo (Sicily); (iii) 25–26 June on the Croatian coast of the Adriatic Sea; and (iv) 27 June in the area of Odessa, northwestern part of the Black Sea. For the first three regions, these hazardous events were thoroughly examined in a cycle of papers by Šepić et al. (2015b, 2016, 2017) and it was shown that they all had characteristics of "*harbour meteotsunamis*", i.e. some local events strongly amplified by harbour resonance.

The "Odessa tsunami" is the main subject of the present study; our investigations showed that this event was not isolated, but that significant seiches on the same day were detected at a number of other sites in the Black Sea. The intensification of these seiches was highly correlated with atmospheric disturbances propagating through this region; and the tsunami-like waves observed on the beaches of Odessa exactly coincided in time with a recorded air pressure jump. All these findings support the assumption of Šepić et al. (2015b) that the "Odessa tsunami" had an atmospheric origin. However, this tsunami was observed not in a bay or harbour, as were the meteotsunamis of 23–26 June 2014 in Spain, Italy and Croatia, but on open beaches of Odessa and Illichevsk, in close similarity to several meteotsunamis that occurred in 1992, 1995, 2014 and 2016 on beaches in Florida and in February 2014 on the Praia do Cassino beach in Brazil. Thus, we can identify the "Odessa tsunami" as a "beach meteotsunami".

To examine the specific generation mechanism of the "Odessa tsunami" of 27 June 2014, we ran two sets of numerical experiments: Experiment 1 ("direct forcing") based on an assumption that the tsunamigenic atmospheric disturbance propagates onshore; and Experiment 2 ("shelf") in which the atmospheric pressure disturbance moves along the shelf break, from Romania in the west to the Crimean Peninsula in the east. The "direct forcing" experiment gives a higher induced wave that better matches the observations, but the "shelf" experiment is in much better agreement with the actual atmospheric situation. For both modelling experiments, we were able to reproduce: (a) sudden onset of the tsunami-like oscillations; (b) a 30-min time delay between the "tsunami waves" observed on the Odessa beaches and beginning of strong oscillations at the Odessa tide gauge; and (c) the 26-min period oscillations recorded in Odessa harbour. However, neither of them could reproduce the 1–2 m heights of the waves that attacked the Odessa beaches; the modelled waves were only up to 53 cm.

We assume that there are two main reasons responsible for this disagreement between the observed and modelled wave heights. (1) The "too coarse" spatial grid of 650 $\times$ 920 m that was used in our computations; it does not allow resolution of the bathymetric and topographic details that play the crucial role in local formation of extreme waves. A fine grid and good bathymetric resolution could probably increase the computed heights by 2–3 times. (2) Insufficient knowledge of the exact source, i.e. the "tsunamigenic" atmospheric disturbance responsible for the formation of destructive waves. Our simple analytical estimates confirmed with an additional numerical exercise show that a "sharp" pressure disturbance (of about 3 hPa per 3 km) moving with a resonant speed ($U \sim \sqrt{gh}$) can produce 1-m sea level response.

In general, we can say that on 27 June 2014 there were conditions for occurrence of sharp resonance over the northwestern Black Sea shelf: a tsunamigenic synoptic pattern and intensive atmospheric pressure disturbances moving in this region with a speed of long ocean waves. A pronounced air pressure disturbance was recorded at the Odessa meteorological station right at the time of the event. This disturbance was characterized by a step-like air pressure jump of > 1.5 hPa occurring over a few minutes. It is likely that in the open sea (in the area of the shelf break) there were stronger and sharper disturbances than measured on the coast. Numerical modelling results confirm that air pressure disturbance propagating in the northwestern Black Sea can produce significant long ocean waves. Our numerical experiments were done for coarse model bathymetry, as well as based on a condition of a minimum depth of $\sim$ 4 m; it is expected that for a finer grid and for a model accounting for runup the results could be several times larger.

It was our intention to put an end to the rumours about the "origin" of the Odessa event and to demonstrate that the "Odessa tsunami" of 27 June 2014 was actually a meteotsunami closely linked with the Mediterranean events of 23–26 June 2014. Synoptic conditions, air pressure and sea level records, as well as our numerical modelling results support this conclusion. We have tested two distinct hypotheses of generation of meteorological tsunamis, each of which has its "pros" and "cons". In fact, the only possible final answer about the exact generation mechanism of extreme wave heights could come from more detailed air pressure and sea level measurements, as well as from the more detailed bathymetry and fine grid resolution that can be used for the further modelling.

Acknowledgements

The authors would like to thank Fred Stephenson of the Canadian Hydrographic Service, Institute of Ocean Sciences (Sidney, British Columbia) for his support and helpful comments. We would like to thank all organisations that kindly provided us the data used in this study: European Centre for Middle-range Weather Forecast, Reading (www.ecmwf.int); European Organization for the Exploitation of Meteorological Satellites (www.eumetsat.int); Romanian National Meteorological Administration, Bucharest; Bulgarian National Oceanographic Data Centre (BNODC), Varna; Hydrometeorological Centre of the Black and Azov Seas, Odessa; Sevastopol Hydrometeorological Observatory and Specialized Center for Hydrometeorology and Monitoring of Environment of Black and Azov Seas (SC HME BAS), Sochi. The work of JS has been supported by the Croatian Science Foundation under the project MESSI (UKF Grant No. 25/15) and for AR by the Russian Science Foundation (Grant 14-50-00095).

References

Aranguiz, R., & Shibayama, T. (2013). Effect of submarine canyons on tsunami propagation: a case study of the Biobio canyon, Chile. *Coastal Engineering Journal, 55*(4). http://dx.doi.org/10.1142/S0578563413500162.

Bechle, A. J., & Wu, C. H. (2014). The Lake Michigan meteotsunamis of 1954 revisited. *Natural Hazards, 74*(1), 155–177. https://doi.org/10.1007/978-3-319-12712-5_9.

Belušić, D., Grisogono, B., & Klaić, Z. B. (2007). Atmospheric origin of the devastating coupled air-sea event in the east Adriatic. *Journal of Geophysical Research, 112,* D17111. https://doi.org/10.1029/2006JD008204.

Candela, J., Mazzola, S., Sammari, C., Limeburner, R., Lozano, C. J., Patti, B., et al. (1999). The "Mad Sea" phenomenon in the Strait of Sicily. *Journal of Physical Oceanography, 29,* 2210–2231.

Churchill, D. D., Houston, S. H., & Bond, N. A. (1995). The Daytona Beach wave of 3–4 July 1992: a shallow water gravity wave forced by a propagating squall line. *Bulletin of the American Meteorological Society, 76,* 21–32.

Dotsenko, S. F. (1995). The Black Sea tsunamis. *Izvestiya Atmospheric and Oceanic Physics, 30,* 483–489.

Fine, I. V., Rabinovich, A. B., Thomson, R. E., & Kulikov, E. A. (2003). Numerical modeling of tsunami generation by submarine and subaerial landslides. In A. C. Yalçiner, E. N. Pelinovsky, C. E. Synolakis, & E. Okal (Eds.), *Submarine Landslides and Tsunamis, NATO Adv. Series* (pp. 69–88). Dordrecht: Kluwer Academic Publishers.

Hibiya, T., & Kajiura, K. (1982). Origin of the 'Abiki' phenomenon (a kind of seiche) in Nagasaki Bay. *Journal of Oceanographical Society of Japan, 38,* 172–182.

Iglesias, O., Lastras, G., Souto, C., Costa, S., & Canals, M. (2014). Effects of coastal submarine canyons on tsunami propagation and impact. *Marine Geology, 350,* 39–51.

Jansà, A., Monserrat, S., & Gomis, D. (2007). The rissaga of 15 June 2006 in Ciutadella (Menorca), a meteorological tsunami. *Advances in Geosciences, 12,* 1–4.

Kowalik, Z. (2012). Introduction to Numerical Modeling of Tsunami Waves, Institute of Marine Science, University of Alaska, Fairbanks. 195 pp. https://www.sfos.uaf.edu/directory/faculty/kowalik/Tsunami_Book/book_sum.pdf.

Maramzin, V. Y. (1985). Computation of seiche oscillations by the finite element method in basins of arbitrary shape. In V. M. Kaistrenko & A. B. Rabinovich (Eds.), *Theoretical and Experimental Investigations of Long Wave Processes* (pp. 104–114). Vladivostok: Far Eastern Scientific Center, USSR Academy of Sciences.

Monserrat, S., Fine, I., Amores, A., & Marcos, M. (2014). Tidal influence on high frequency harbor oscillations in a narrow entrance bay. *Natural Hazards, 74,* 143–153. https://doi.org/10.1007/s11069-014-1284-3.

Monserrat, S., & Thorpe, A. P. (1996). Use of ducting theory in an observed case of gravity waves. *Journal of Atmospheric Science, 53,* 1724–1736.

Monserrat, S., Vilibić, I., & Rabinovich, A. B. (2006). Meteotsunamis: Atmospherically induced destructive ocean waves in the tsunami frequency band. *Natural Hazards and Earth System Sciences, 6,* 1035–1051.

Nikonov, A. A. (1997). Tsunami occurrence on the coasts of the Black Sea and the Sea of Azov. *Izvestiya, Physics of Solid Earth, 33,* 77–87.

Nikonov, A. A., & Fleyfel, L. D. (2015). Tsunami in Odessa: Natural or manmade phenomenon? *Priroda, 4,* 36–43. **(in Russian)**.

Orlić, M. (2015). The first attempt at cataloguing tsunami-like waves of meteorological origin in Croatian coastal waters. *Acta Adriatica, 56,* 83–96.

Orlić, M., Belušić, D., Janeković, I., & Pasarić, M. (2010). Fresh evidence relating the great Adriatic surge of 21 June 1978 to mesoscale atmospheric forcing. *Journal of Geophysical Research, 115,* C06011. https://doi.org/10.1029/2009JC005777.

Papadopoulos, G. A., Diakogianni, G., Fokaefs, A., & Ranguelov, B. (2011). Tsunami hazard in the Black Sea and the Azov Sea: a new tsunami catalogue. *Natural Hazards and Earth System Sciences, 11,* 945–963.

Pasquet, S., & Vilibić, I. (2013). A survey of strong high-frequency sea level oscillations along the US East Coast between 2006 and 2011. *Natural Hazards and Earth System Sciences, 13,* 473–482.

Pattiaratchi, C. B., & Wijeratne, E. M. S. (2015). Are meteotsunamis an underrated hazard? *Philosophical Transactions of the Royal Society, A, 373,* 20140377.

Paxton, C. H., & Sobien, D. A. (1998). Resonant interaction between an atmospheric gravity wave and shallow water wave along Florida's west coast. *Bulletin of the American Meteorological Society, 79,* 2727–2732.

Plougonven, R., & Zhang, F. Q. (2014). Internal gravity waves from atmospheric jets and fronts. *Reviews of Geophysics, 52,* 33–76. https://doi.org/10.1002/2012RG000419.

Proudman, J. (1929). The effects on the sea of changes in atmospheric pressure. *Geophysical Supplement, Monthly Notices of the Royal Astronomical Society, 2,* 197–209.

Rabinovich, A. B. (2009). Seiches and harbor oscillations. In Y. C. Kim (Ed.), *Handbook of Coastal and Ocean Engineering, Chapter 9* (pp. 193–236). Singapore: World Scientific Publishing.

Rabinovich, A. B., & Monserrat, S. (1996). Meteorological tsunamis near the Balearic and Kuril Islands: Descriptive and statistical analysis. *Natural Hazards, 13,* 55–90.

Rabinovich, A. B., & Monserrat, S. (1998). Generation of meteorological tsunamis (large amplitude seiches) near the Balearic and Kuril Islands. *Natural Hazards, 18,* 27–55.

Rabinovich, A. B., Titov, V. V., Moore, C. W., & Eblé, M. C. (2017). The 2004 Sumatra tsunami in the Southeastern Pacific Ocean: New global insight from observations and modeling. *Journal of Geophysical Research - Oceans,* 122. https://doi.org/10.1002/2017JC013078.

Rabinovich, A. B., Vilibić, I., & Tinti, S. (2009). Meteorological tsunamis: Atmospherically induced destructive ocean waves in the tsunami frequency band. *Physics and Chemistry of the Earth, 34*(17/18), 891–893.

Ranguelov, B., Tinti, S., Pagnoni, G., Tonini, R., Zaniboni, F., & Armigliato, A. (2008). The nonseismic tsunami observed in the Bulgarian Black Sea on 7 May 2007: Was it due to a submarine landslide? *Geophysical Research Letters, 35,* L18613.

Sallenger, A. H., List, J. H., Gelfenbaum, G., Stumpf, R. P., & Hansen, M. (1995). Large wave at Daytona Beach, Florida, explained as a squall-line surge. *Journal of Coastal Research, 11,* 1383–1388.

Šepić, J., Međugorac, I., Janeković, I., Dunić, N., & Vilibić, I. (2016). Multi-meteotsunami event in the Adriatic Sea generated by atmospheric disturbances of 25–26 June 2014. *Pure and Applied Geophysics, 173*(12), 4117–4138. https://doi.org/10.1007/s00024-016-1249-4.

Šepić, J., & Rabinovich, A. B. (2014). Meteotsunami in the Great Lakes and on the Atlantic coast of the United States generated by the ''derecho'' of June 29–30, 2012. *Natural Hazards, 74,* 75–107. https://doi.org/10.1007/s11069-014-1310-5.

Šepić, J., Vilibić, I., & Belušić, D. (2009). Source of the 2007 Ist meteotsunami (Adriatic Sea). *Journal of Geophysical Research, 114.* https://doi.org/10.1029/2008JC005092.

Šepić, J., Vilibić, I., & Fine, I. V. (2015a). Northern Adriatic meteorological tsunamis: assessment of their potential through ocean modeling experiments. *Journal of Geophysical Research, 120,* 2993–3010. https://doi.org/10.1002/2015JC010795.

Šepić, J., Vilibić, I., Rabinovich, A. B., & Monserrat, S. (2015b). Widespread tsunami-like waves of 23–27 June in the Mediterranean and Black Seas generated by high-altitude atmospheric forcing. *Scientific Reports, 5,* 11682. https://doi.org/10.1038/srep11682.

Šepić, J., Vilibić, I., Rabinovich, A. B. & Tinti, S. (2017). Meteotsunami (“marrobbio”) of 25 June 2014 on the southwestern coast of Sicily. *Pure and Applied Geophysics* (in Preparation).

Thomson, R. E., & Emery, W. J. (2014). *Data Analysis Methods in Physical Oceanography* (3rd ed., p. 716). New York: Elsevier Science.

Thomson, R. E., Rabinovich, A. B., Fine, I. V., Sinnott, D. C., McCarthy, A., Sutherland, N. A. S., et al. (2009). Meteorological tsunamis on the coasts of British Columbia and Washington. *Physics and Chemistry of the Earth, 34,* 971–988.

Vilibić, I. (2008). Numerical simulations of the Proudman resonance. *Continental Shelf Research, 28,* 574–581.

Vilibić, I., Domijan, N., Orlić, M., Leder, N., & Pasarić, M. (2004). Resonant coupling of a traveling air pressure disturbance with the east Adriatic coastal waters. *Journal of Geophysical Research, 109,* C10001. https://doi.org/10.1029/2004JC002279.

Vilibić, I., Monserrat, S., & Rabinovich, A. B. (2014). Meteorological tsunamis on the US East Coast and in other regions of the World Ocean. *Natural Hazards, 74,* 1–9. https://doi.org/10.1007/s11069-014-1350-x.

Vilibić, I., Monserrat, S., Rabinovich, A. B., & Mihanović, H. (2008). Numerical modelling of the destructive meteotsunami of 15 June 2006 on the coast of the Balearic Islands. *Pure and Applied Geophysics, 165*(11/12), 2169–2195.

Vilibić, I., & Šepić, J. (2009). Destructive meteotsunamis along the eastern Adriatic coast: Overview. *Physics and Chemistry of the Earth, 34,* 904–917.

Vilibić, I., Šepić, J., Rabinovich, A. B., & Monserrat, S. (2016). Modern approaches in meteotsunami research and early warning. *Frontiers in Marine Science, 3*(57), 1–7. https://doi.org/10.3389/fmars.2016.00057.

Vilibić, I., Šepić, J., Ranguelov, B., Strelec Mahović, N., & Tinti, S. (2010). Possible atmospheric origin of the 7 May 2007 western Black Sea shelf tsunami event. *Journal of Geophysical Research, 115,* C07006. https://doi.org/10.1029/2009JC005904.

Wertman, C. A., Yablonsky, R. M., Shen, Y., Merrill, J., Kincaid, C. R., & Pockalny, R. A. (2014). Mesoscale convective system surface pressure anomalies responsible for meteotsunamis along the U.S. East Coast on June 13th, 2013. *Scientific Reports 4.* https://doi.org/10.1038/srep07143.

Yalçiner, A., Pelinovsky, E., Talipova, T., Kurkin, A., Kozelkov, A., & Zaitsev, A. (2004). Tsunamis in the Black Sea: Comparison of the historical, instrumental, and numerical data. *Journal of Geophysical Research, 109,* C12023. https://doi.org/10.1029/2003JC002113.

(Received June 4, 2017, revised November 15, 2017, accepted November 17, 2017)

Pure Appl. Geophys.

https://doi.org/10.1007/s00024-018-1827-8

Pure and Applied Geophysics

Meteotsunami ("Marrobbio") of 25–26 June 2014 on the Southwestern Coast of Sicily, Italy

Jadranka Šepić,[1] Ivica Vilibić,[1] Alexander Rabinovich,[2,3] and Stefano Tinti[4]

Abstract—A major tsunami-like event, locally known as 'marrobbio', impacted the southwestern coast of Sicily on 25–26 June 2014. The event was part of a chain of hazardous episodes in the Mediterranean and Black seas during the last week of June 2014 resulting from an anomalous atmospheric system ("tumultuous atmosphere") propagating eastward over the region. The synoptic patterns and vertical structure of the atmosphere over Sicily at the time of the event indicate that atmospheric wave ducting was responsible for the generation of tsunamigenic air pressure disturbances that produced especially high sea level responses ("meteotsunamis") at certain sites along the Sicilian coast. The strongest sea level oscillations were observed at Mazara del Vallo, where a 1-m meteotsunami bore, propagating upstream in the Mazaro River, was generated. The combined effects of external resonance (Proudman resonance on the western Sicilian shelf) and internal resonant conditions (bathymetric and topographic characteristics of specific sites) were found to be the key factors that caused the meteotsunami (marrobbio phenomenon) on the coast of Sicily and the meteobore at Mazara del Vallo.

Key words: Meteotsunami, marrobbio, Sicily, atmospheric oscillations, tidal bore, meteobore, seiches, tide gauge records.

1. Introduction

As never before documented for the Mediterranean and Black seas, the last week of June 2014 has been characterized by multiple occurrences of tsunami-like waves of atmospheric origin affecting the entire region (Fig. 1a). A prominent "rissaga" phenomenon (local name for a meteotsunami, cf. Rabinovich and Monserrat 1996, 1998; Monserrat et al. 2006) was observed on 23 June on the coasts of the Balearic Islands (Spain), with 1-m oscillations recorded at Ciutadella Harbour, Menorca Island. That was not unusual, as destructive rissagas occur now and then in this region during the late spring and summer periods (Gomis et al. 1993; Jansà et al. 2007). However, a couple of days later, on 25 June, severe tsunami-like waves hit a number of bays in the eastern Adriatic Sea (Šepić et al. 2016). Almost simultaneously with the Adriatic events, strong sea level oscillations were reported on the southwestern coast of Sicily, and later in Greece and Turkey. Finally, on 27 June, a destructive "tsunami" wave inundated beaches of Odessa (northwestern Black Sea), causing injuries to people and damage to coastal infrastructure (Šepić et al. 2018).

Šepić et al. (2015a) demonstrated that all these events are linked and were initiated by an atmospheric system propagating eastward over the Mediterranean region from 23 to 27 June. The system supported the generation of numerous intense, small-scale atmospheric disturbances ("tumultuous atmosphere"), which subsequently induced meteotsunamis. The governing parameter determining the sea-level response to atmospheric disturbances is the *Froude number*, which can be defined as

$$Fr = U/c; \tag{1}$$

i.e. the ratio of the atmospheric gravity wave speed, U, to the phase speed of long ocean waves, $c = \sqrt{gh}$, where g is the gravity acceleration and h is the water depth. The so-called "*Proudman resonance*" (Proudman 1929) occurs when $U \approx c$ and $Fr \approx 1.0$. In this case, ocean waves begin to actively absorb energy from the atmospheric gravity waves, and are strongly intensified (Hibiya and Kajiura 1982; Vilibić

[1] Institute of Oceanography and Fisheries, Šetalište I. Meštrovića 63, 21000 Split, Croatia.
[2] Fisheries and Oceans Canada, Institute of Ocean Sciences, 9860 W. Saanich Rd., Sidney, BC, Canada. E-mail: A.B.Rabinovich@gmail.com
[3] P.P. Shirshov Institute of Oceanology, Russian Academy of Sciences, 36 Nakhimovsky Pr., Moscow, Russia.
[4] Dipartimento di Fisica e Astronomia, Università di Bologna, Viale Carlo Berti Pichat 8, Bologna, Italy.

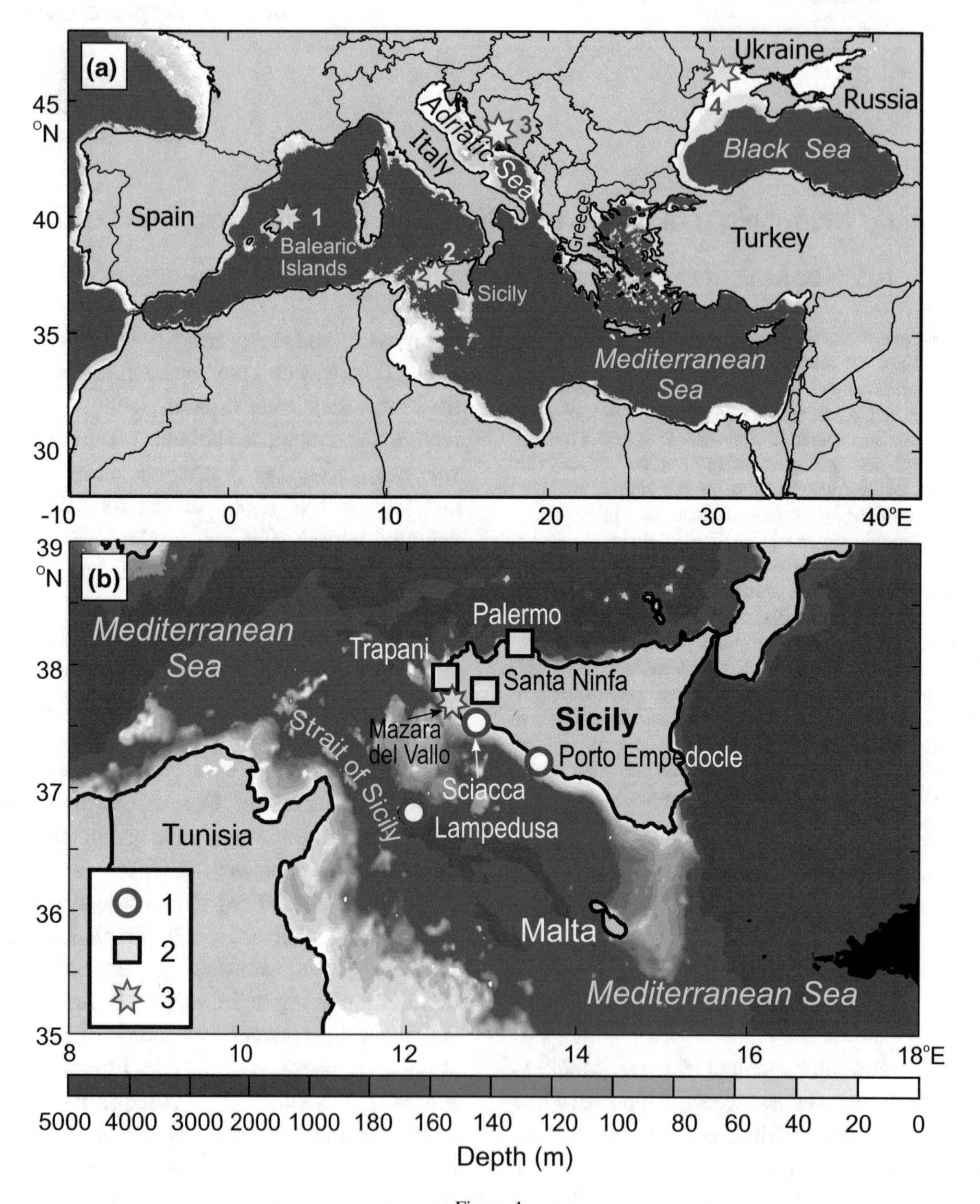

Figure 1

a The Mediterranean Sea and locations where the strongest meteotsunamis were recorded during the 23–27 June 2004 multi-meteotsunami event: (1) Ciutadella, the Balearic Islands; (2) Mazara del Vallo, Sicily; (3) Vela Luka, the Adriatic Sea; (4) Odessa, the northwestern Black Sea. **b** Location of tide gauges (1), atmospheric stations (2), Mazara del Vallo site affected by a meteobore (3), and the bathymetry of the Sicilian region

2008). Thereafter, the generated open-sea waves approaching the coast are additionally topographically amplified over the shelf and further amplified in harbours, inlets or bays through harbour resonance (Wilson 1972; Rabinovich 2009; Vennell 2007, 2010). The extreme events of 23–27 June 2014 were found to occur in regions with the most favourable open-sea conditions for meteotsunami generation ($0.9 < Fr < 1.1$) and mainly at "hot sites", i.e. in harbours with prominent resonant properties and strong amplification factors (Šepić et al. 2015a, 2016).

Šepić et al. (2015a) established that there is a *teleconnection* between meteotsunami events that occurred on 23–27 June 2014 in various parts of the Mediterranean/Black Sea region that spanned almost 3000 km: all these events had been generated by singular uncorrelated short-life[1] atmospheric disturbances, but these disturbances have the same genesis and certain mutual features. At the same time, the distinct character of individual disturbances and unique topographic structures of specific regions determine substantial differences in major physical properties of the detected meteotsunamis: the "Odessa tsunami" (Šepić et al. 2018) was quite different from those in the Adriatic Sea (Šepić et al. 2016) and both of them were strongly dissimilar from the extreme event on the southwestern coast of Sicily. The similarities and differences between prominent meteotsunami events observed in various parts of the region were the main stimulus of the present study.

Among all meteotsunamis that occurred on 23–27 June 2014 in the Mediterranean and Black seas, the extreme event on the coast of Sicily (Fig. 1b) is one of the most important and interesting. The meteorological tsunamis, known locally as "marrobbio" or "mad sea", regularly occur on this coast, occasionally reaching destructive heights (Colucci and Michelato 1976). The key site repeatedly suffering from marrobbio is Mazara del Vallo, an ancient coastal town (~ 3000 years old) in southwestern Sicily that lies on the estuary banks of the Mazaro River (Fig. 2), and considered as one of the most important fishing centres of Italy. This particular town was strongly affected by the event of 25–26 June 2014.

This phenomenon has been known at Mazara del Vallo for centuries. As was indicated on the local blog "*Sicilia Terre d'Occidente*": ...*The Marrobbio or Marrubbio is a strictly physical phenomenon that affects the sea and consists of a sudden and highly sensitive variation in the sea level, observed typically in spring and autumn...The Marrobbio ...is similar to tsunami; for this reason the Arabs, involuntary witnesses of the Marrobbio at the time of their historic landing in Mazara del Vallo, called the river Mazaro, i.e. "possessed"* (http://siciliaterredoccidente.blogspot.ca/2016/06/leggendari-fenomeni-mazara-del-vallo.html).

[1] A typical life cycle of these disturbances is from a few tens of minutes to a few hours (Thomson et al. 2009).

Figure 2
Satellite images of **a** the Mazara del Vallo area and **b** the estuary/harbour of the Mazaro River. Numbers 1 to 4 in **b** denote locations where videos of the meteotsunami bore were taken. Images taken during the 25 June 2014 marrobbio event are shown in the corresponding insets

Candela et al. (1999) assumed that the marrobbio waves are coastal amplification of normal modes of the Strait of Sicily, generated by air pressure disturbances travelling from the southwest with resonant speeds of 24–30 m/s. Normal topographic modes, in particular, edge waves and shelf seiches, in specific areas of the world oceans can induce significant sea level oscillations (Liu et al. 2002; Monserrat et al. 2006). At the same time, recent modelling studies (cf. Vilibić 2008; Šepić et al. 2015b) indicate that the meteotsunami generation processes require considerable time and that substantial waves can be produced only when the resonant pathway of propagating atmospheric waves is several times longer than the wavelength of the individual disturbance. Extensive numerical experiments are required to identify the exact formation mechanism of marrobbio in the region of Sicily. However, the purpose of the present study is not to numerically simulate the meteotsunami of 25–26 June 2014 for the southwestern coast of Sicily, or to discuss the general physical mechanism of marrobbio waves (this can be the subject for a future study), but to document measurements, video materials and synoptic conditions related to the observed event. We tried to collect and examine all available air pressure and sea level data in this region and to use atmospheric soundings, satellite and reanalysis data in the central Mediterranean to map atmospheric patterns associated with the event.

One of the spectacular features of the 25–26 June 2014 event in Mazara del Vallo was the formation of a bore propagating upstream along the harbour and the Mazaro River. We provide documentation and an explanation of this bore. In general, the structure of the paper is the following: Sect. 2 provides an overview of media reports and videos describing and filming the event; Sect. 3 examines the synoptic patterns and conditions related to the event; Sect. 4 presents the results of analysis of available sea level and air pressure records in the region under study; Sect. 5 uses various video materials to track the propagating bore and suggests a simple model explaining its principal features; finally Sect. 6 discusses the nature of the event and summarizes the main findings.

2. *Description of the Event*

In the evening of 25 June 2014, the sea at the mouth of the Mazaro River (southwestern Sicily,

Figure 3
Time slices of a meteotsunami bore propagating upstream in the Mazaro River on 25 June 2014 (taken from the video by Vaccaro 2014)

Fig. 2) suddenly started to “dance” (Vaccaro 2014). The “dancing” was accompanied by a bore moving fast upstream in the river, resembling the tidal bores observed in a number of rivers worldwide (Chanson 2012; Pelinovsky et al. 2015; Bonneton et al. 2015). But tides in the Strait of Sicily are weak (up to 10 cm), as in most of the Mediterranean Sea (Tsimplis et al. 1995), and cannot produce propagating bores.

There are several videos of the event. Probably the most instructive was by Vaccaro (2014) that captured the bore arrival around 19:00 on 25 June 2014. The bore moved upstream initiating a sea level rise of about 1.5 m in less than 2 min (Fig. 3). The bore produced some damage on boats and flooded the promenade in Mazara del Vallo on the river embankment (TeleIBS—L’Opinione 2014). However, the destructive effects of this event were not as severe as those caused by the meteotsunami (“*rissaga*”) of 15 June 2006 in Ciutadella Harbour, the Balearic Islands (Jansà et al. 2007), or a catastrophic flood in the harbour of Vela Luka, Korčula Island, Croatia, on 21 June 1978 (Vučetić et al. 2009).

Repeated sea level oscillations continued inside Mazara del Vallo harbour and the Mazaro River during the entire next day (26 June), creating no noticeable damage. Prominent oscillations had also been observed in neighbouring harbours and on beaches along the southwestern coast of Sicily. The media reported that the sea retreated on Tonnarella Beach for more than 30 m and then returned suddenly in a few minutes, inundating the shore and reaching the coastal road. The wave was strong enough to wash tons of *Posidonia* seagrass onto the beach.

Local media tracked the event, capturing all essential aspects of the phenomenon. An example is the Mazara online internet magazine that reported the following (http://www.mazaraonline.it/?p=67390):

> “Yesterday in the afternoon the phenomenon known as “marrobbio” occurred at Mazara del Vallo: “marrobbio” stands for a sudden,

00'20"
00'34"
00'40"
00'47"
01'00"
01'13"
01'27"
01'33"
02'00"
02'26''

unforeseeable and remarkable change of the sea level and, more specifically, for the Mazaro River, a rise of the water level which floods the docks. As opposed to other marrobbio events of recent years, yesterday afternoon's event, in addition to the river, affected also vast areas of the beaches of Tonnarella and San Vito. Sea level fluxes were such that the sea, for several hours, advanced beyond, and receded from, the normal shoreline by distances of over 30 m. An occurrence that very few people remember!
Luckily, especially if one considers the power of the event, damage was limited to anchored small pleasure craft—their mooring lines were not strong enough. People from Mazara have experienced this type of event since ancient times, but a peculiarity of this event was that it occurred in summer, whereas usually it takes place in spring or fall."

3. Synoptic Setting

Based on results of Šepić et al. (2015a), we assume that the critical factor responsible for the formation of tsunami-like waves, which affected Mazara del Vallo on 25–26 June 2014, was a particular atmospheric system that travelled over the region in an eastward direction. To specify the exact synoptic conditions in the region of western Sicily, we examined high-altitude atmospheric parameters of the system. Figure 4 shows the evolution of the

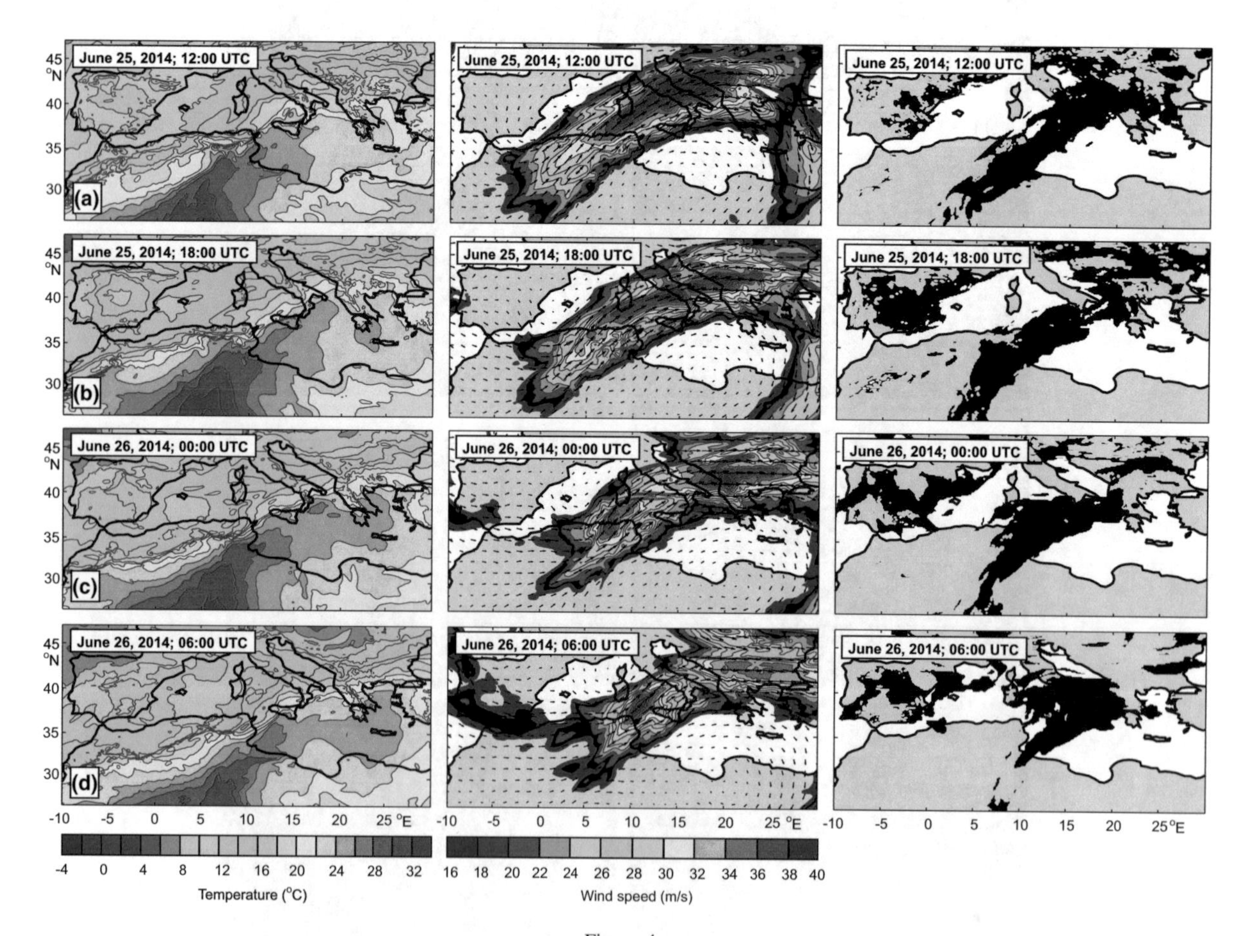

Figure 4
Temperature at 850 hPa, winds at 500 hPa and instability of the mid-troposphere (blue areas denote a Richardson number smaller than 0.25 in the layer 400–700 hPa) over the central Mediterranean, taken from ECMWF ERA-Interim reanalysis at **a** 25 June 2014 12:00 UTC; **b** 25 June 2014 18:00 UTC; **c** 26 June 2014 00:00 UTC, and **d** 26 June 2014 06:00 UTC

850 hPa temperature fields (roughly at a height of 1.5 km), high-altitude 500 hPa winds (~ 5500 m height) and instability of the mid-troposphere for the period between 25 June 12:00 UTC and 26 June 06:00 UTC. In our analysis the mid-troposphere has been considered unstable if the Richardson number was lower than 0.25 in the layer between 400 and 700 hPa. The Richardson number was estimated as:

$$Ri = \frac{N^2}{(du/dz)^2}, \quad (2)$$

where N is the Brunt–Väisälä frequency (in s^{-1}), u is the wind speed (in m/s) and z is the height (in m). The Brunt–Väisälä frequency was calculated as the "moist frequency" (Durran and Klemp 1982) for levels with relative humidity above 70% and as the "dry frequency" otherwise. The data derived from the European Centre for Middle-range Weather Forecast (ECMWF) interim reanalysis dataset (http://www.ecmwf.int) have been used for this analysis.

The temperature fields show a low-troposphere thermal front moving from WSW to ENE with the strongest gradients exactly over the western part of the Strait of Sicily. From the afternoon hours of 25 June to the morning hours of 26 June the front was quasi-stationary, indicating only very slow propagation eastward. The front was overtopped by an intense mid-troposphere jet-stream peaking over central Italy and the Adriatic Sea between 12:00 and 18:00 UTC on 25 June with wind speeds stronger than 40 m/s. The jet-stream began to weaken in the next 12 h and divided into two separate maxima, one of which was located west of Sicily. Throughout 25 and 26 June, the frontal (eastern) segment of the jet was unstable, while the rear (western) segment was stable. Similar synoptic patterns have been observed during most previous Balearic and almost all Adriatic meteotsunami events (Ramis and Jansà 1983; Jansà et al. 2007; Vilibić and Šepić 2009; Šepić et al. 2009b). It has been shown that these conditions favour ducting of atmospheric gravity waves (Monserrat and Thorpe 1996), which are a known source of meteotsunamis (cf. Šepić et al. 2015a). Atmospheric gravity waves can effectively generate a meteotsunami, but only when they propagate with a resonant speed over an open sea along a path which is several times longer than their wavelength (Vilibić, 2008). To effect this prolonged propagation, a certain mechanism of gravity wave energy maintenance and/or renewal must be active. Wave ducting is precisely such a mechanism. When a stable atmospheric layer is capped by an unstable layer (called "duct"), atmospheric gravity waves can travel within the stable layer, for a long distance with little energy dissipation, provided that their phase speed is either equal or close to the wind speed within the duct (Lindzen and Tung 1976; Monserrat and Thorpe 1996). As evident from Fig. 4, the favourable conditions for ducting of atmospheric gravity waves over the Strait of Sicily existed from 12:00 UTC on 25 June to 00:00 UTC on 26 June: a stable (dry and warm) air mass of northern African origin was overtopped by an unstable atmospheric layer within which wind speeds were from 22 to 26 m/s.

Radio-sounding profiles at Trapani (Fig. 5),[2] reveal a presence of vertical atmospheric structure supportive of a wave duct: (1) the presence of a very dry air mass between 850 and 700 hPa overtopped by (2) a near-saturated air mass centred at 500 hPa, within which (3) SSW winds reach speeds of 25 m/s. In addition, SSW winds had a very low surface speed and a temperature inversion existed between 900 and 850 hPa; both these features are commonly observed during other Mediterranean meteotsunami events (Jansà et al. 2007; Vilibić and Šepić 2009). This structure had not markedly changed between 25 June 12:00 UTC and 26 June 00:00 UTC, but had deteriorated by 26 June 12:00 UTC, when a colder air mass from the northwest advected into the lower troposphere.

Satellite images of top-clouds taken between 19:30 and 20:30 UTC on 25 June (not shown because of insufficient resolution of the images), ~ 1.5 h after the start of the major meteotsunami event, show a train of clouds which is stretching along the mid-troposphere jet but is quite narrow across the jet: < 20–30 km. It is obvious that these clouds are related to atmospheric gravity waves propagating from northern Africa towards western Sicily along a 300-km-long track with estimated wavelengths $\lambda \sim$ 30–40 km and speeds $U \approx$ 22–26 m/s (which are resonant for the western Sicilian shelf,

[2] The data was downloaded from the University of Wyoming website, http://weather.edu/upperair/sounding.html.

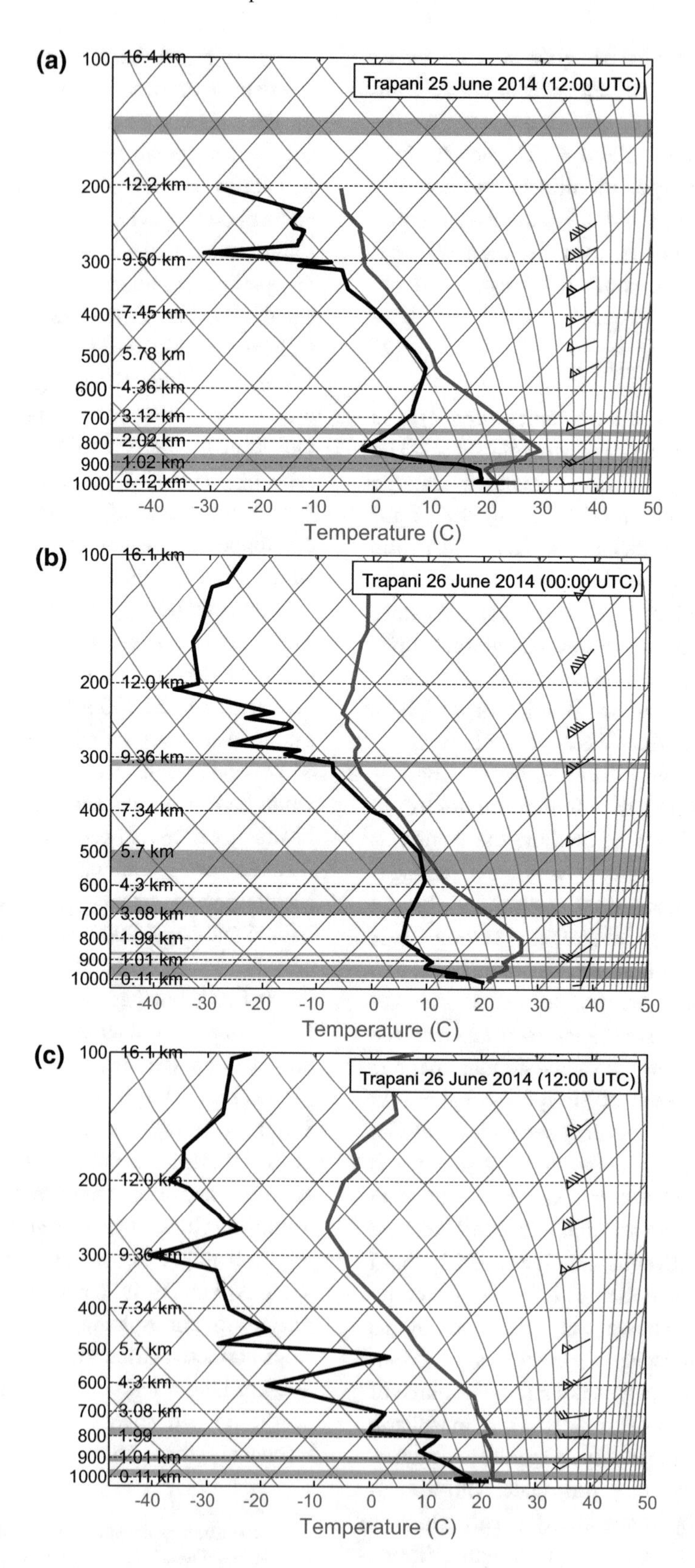
(a)
Trapani 25 June 2014 (12:00 UTC)
16.4 km
12.2 km
9.50 km
7.45 km
5.78 km
4.36 km
3.12 km
2.02 km
1.02 km
0.12 km
100
200
300
400
500
600
700
800
900
1000
-40
-30
-20
-10
0
10
20
30
40
50
Temperature (C)
(b)
Trapani 26 June 2014 (00:00 UTC)
16.1 km
12.0 km
9.36 km
7.34 km
5.7 km
4.3 km
3.08 km
1.99 km
1.01 km
0.11 km
Temperature (C)
(c)
Trapani 26 June 2014 (12:00 UTC)
16.1 km
12.0 km
9.36 km
7.34 km
5.7 km
4.3 km
3.08 km
1.99
1.01 km
0.11 km
Temperature (C)

◀ Figure 5
Radio-sounding profiles of air temperature (red line), dew point temperature (black line) and the Richardson number $Ri < 0.25$ (horizontal grey shaded bands) taken from Trapani station on **a** 25 June 2014 at 12:00 UTC; **b** 26 June 2014 at 00:00 UTC, and **c** 26 June 2014 at 12:00 UTC. Background thin lines represent isotherms (red, sloping) and saturation adiabats (black, slightly curved). Wind speeds and directions are indicated by standard symbols

$U \sim c = \sqrt{gh}$). Such particular wavelengths and speeds are typical for atmospheric disturbances commonly found responsible for producing meteotsunamis in various regions of the world oceans (cf. Monserrat and Thorpe 1992, 1996; Thomson et al. 2009). Although this specific wave train was observed *after* the major meteotsunami event, it does point out the fact that the atmosphere above the Strait of Sicily was susceptible for persistent propagation of atmospheric gravity waves. According to the satellite images, these gravity waves were most likely generated over the Atlas Mountains in North Africa.

We can thus conclude that the atmospheric field over the Strait of Sicily was conductive for ducting and supporting of atmospheric gravity waves propagating during the afternoon hours of 25 June and night and morning hours of 26 June. We also can assume that the source area of these waves, the Atlas Mountains, plays the key role in their formation, in a similar way that the Olympic Mountains (northwestern Washington State, USA) appear to be crucial in inducing tsunamigenic atmospheric disturbances for the regions of Juan de Fuca Strait and the Strait of Georgia (Thomson et al. 2009).

4. *In Situ Measurements*

Altogether, we found three tide gauge (TG) and three atmospheric pressure (AP) records for the region of study. The digital TG data were collected through the Intergovernmental Oceanographic Commission (IOC) Sea Level Monitoring Station portal at www.ioc-sealevelmonitoring.org; the sampling interval of all TG records (Δt) was 1 min. Two tide gauges—Porto Empedocle and Sciacca—are located on the southwestern coast of Sicily; the third gauge—Lampedusa—is an island station situated in the middle of the Strait of Sicily (Fig. 1b). The AP data were downloaded from http://www.wunderground.com; all AP stations are located in the western part of Sicily (Fig. 1b). The nominal accuracy of the AP instruments was $\pm$ 0.15 hPa, the sampling interval was different for different stations: Δt = 5 min for Palermo, 6 min for Santa Ninfa and 10 min for Trapani. Unfortunately, there was no tide gauge at Mazara del Vallo—the site of our primary interest. Also, there were no sites where sea level (SL) and atmospheric pressure were recorded together; therefore, we could not provide direct comparison of SL and AP measured at the same site and had to use neighbouring stations. Another problem was that the sampling intervals of the AP records (5–10 min) were not fine enough to evaluate the true parameters of the tsunamigenic atmospheric disturbances; the real oscillations were probably 30–50% greater (see Candella et al. 2008 for discussion of the influence of sampling intervals on tsunami measurements).

The SL stations operational in the western Strait of Sicily recorded strong high-frequency oscillations that peaked in the evening hours of 25 June 2014 and again in the morning of the next day (Fig. 6). Two wave trains were evident at Lampedusa, with maximum trough-to-crest wave heights of about 30 cm around 19:00 UTC on 25 June and 50 cm at 8:00 UTC on 26 June 2014. Station Porto Empedocle has generally more intense high-frequency oscillations, since it is located in a harbour with a strong amplification factor; the recorded oscillations were up to 75 cm in the afternoon of 25 June and 65 cm in the

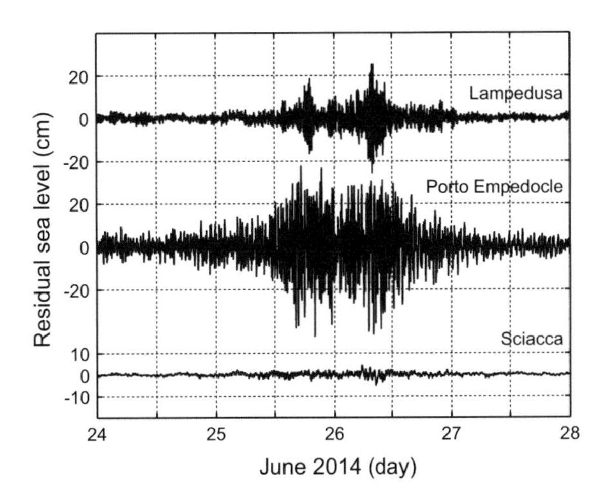

Figure 6
High-frequency sea level records (cutoff period at 4 h) from three tide gauges on or near the coast of southwestern Sicily

morning of 26 June. The observed oscillations east from Mazara del Vallo harbour were much weaker; in particular, at Sciacca, the station located on the open coast ~ 50 km southeast of Mazara del Vallo, the maximum heights were below 10 cm. In general, the spatial variations of maximum wave heights along the coast were considerable, being determined either by alongshore changes in bathymetry/coastal topography, or by local variations of atmospheric forcing, or by both factors.

Intense air pressure oscillations were recorded by all three automatic weather stations (Fig. 7). A train of air pressure high-frequency waves propagated over the region between 18:30 and 20:00 UTC, initiating a rapid air pressure change with a rate of up to 4 hPa per 20 min. The 4–5 major waves at Santa Ninfa and Trapani were alike; except that they arrived at Santa Ninfa 10–12 min earlier. Similar, but 1.5–2 times smaller waves were also recorded at Palermo. These AP waves occurred concurrently with SL oscillations observed in the region. It appears that this train of AP waves, which probably was even stronger than recorded by the 5–10 min AP sensors, was the main reason for the *marrobbio* and the observed bore in the Mazara del Vallo harbour and river estuary. Using the isochronal method (Vilibić et al. 2008), we estimated the speed of this train to be 22 m/s and the direction northward. These values are within the range of speeds (22–26 m/s) and directions (north-eastward) of measured 500 hPa winds blowing over the Strait of

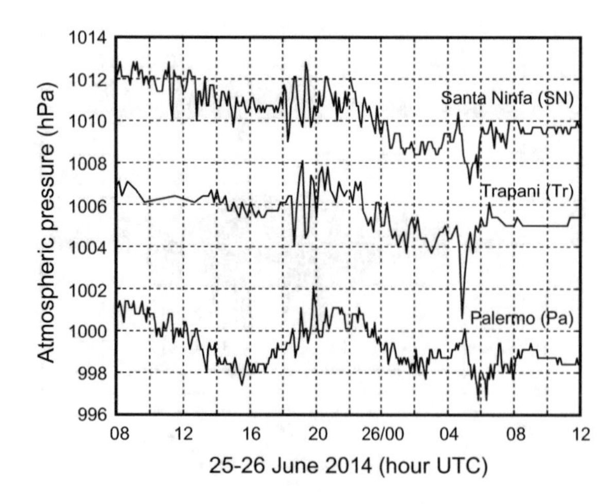

Figure 7
Atmospheric pressure oscillations recorded at three stations in western Sicily

Sicily between 12:00 UTC on 25 June and 06:00 UTC on 26 June (Figs. 4, 5), thus indicating that the recorded disturbances were likely *ducted atmospheric gravity waves*. Moreover, the given speed range corresponds to the speed of long ocean waves at 50–70 m depths, which are precisely the depths of the shelf located in front of Mazara del Vallo (Fig. 1), implying that $Fr \sim 1.0$ in this region and the long ocean waves appear to be intensified via the *Proudman resonance mechanism*, explaining the probable reason of the marrobbio event.

The next day (26 June) at about 04:30 UTC an abrupt AP drop of > 5 hPa was recorded at Trapani (Fig. 7) and weaker drops were also observed at Palermo and Santa Ninfa. Due to the noticeable differences between disturbances at these three sites, we could not estimate their speed and direction by the isochronal method. Furthermore, these AP drops did not coincide exactly with the time of the most intense sea level oscillations at TG Lampedusa and Porto Empedocle that began approximately 2.5 h later (Fig. 6), probably due to the remote distance of these stations from the AP observation sites and the fast transformation of AP small-scale disturbances.

As was indicated above, resonant properties of coastal topography and shelf bathymetry play a principal role in the formation of meteotsunamis. To examine the frequency characteristics of the generated SL oscillations, their evolution in time and possible correlation with the AP disturbances, we used wavelet analysis and constructed frequency–time (*f–t*) diagrams. For comparative analysis of SL and AP records, the latter were interpolated to 1 min. The results are shown in Figs. 8 and 9.

The *f–t* plot for Lampedusa clearly demonstrates that sea level oscillations at this station have monochromatic character with a steady period of about 10 min, related to the fundamental (Helmholtz) mode of the harbour. The intensification of sea level oscillations with this period began at about noon on 25 June and continued for about 1.5 days. There are two distinctive energy maxima: the stronger in the morning hours of 26 June and the weaker in the evening hours of 25 June. Both maxima are precisely associated with the meteotsunami events we are discussing. It is important that the meteotsunamis do not have particularly different frequencies from those of

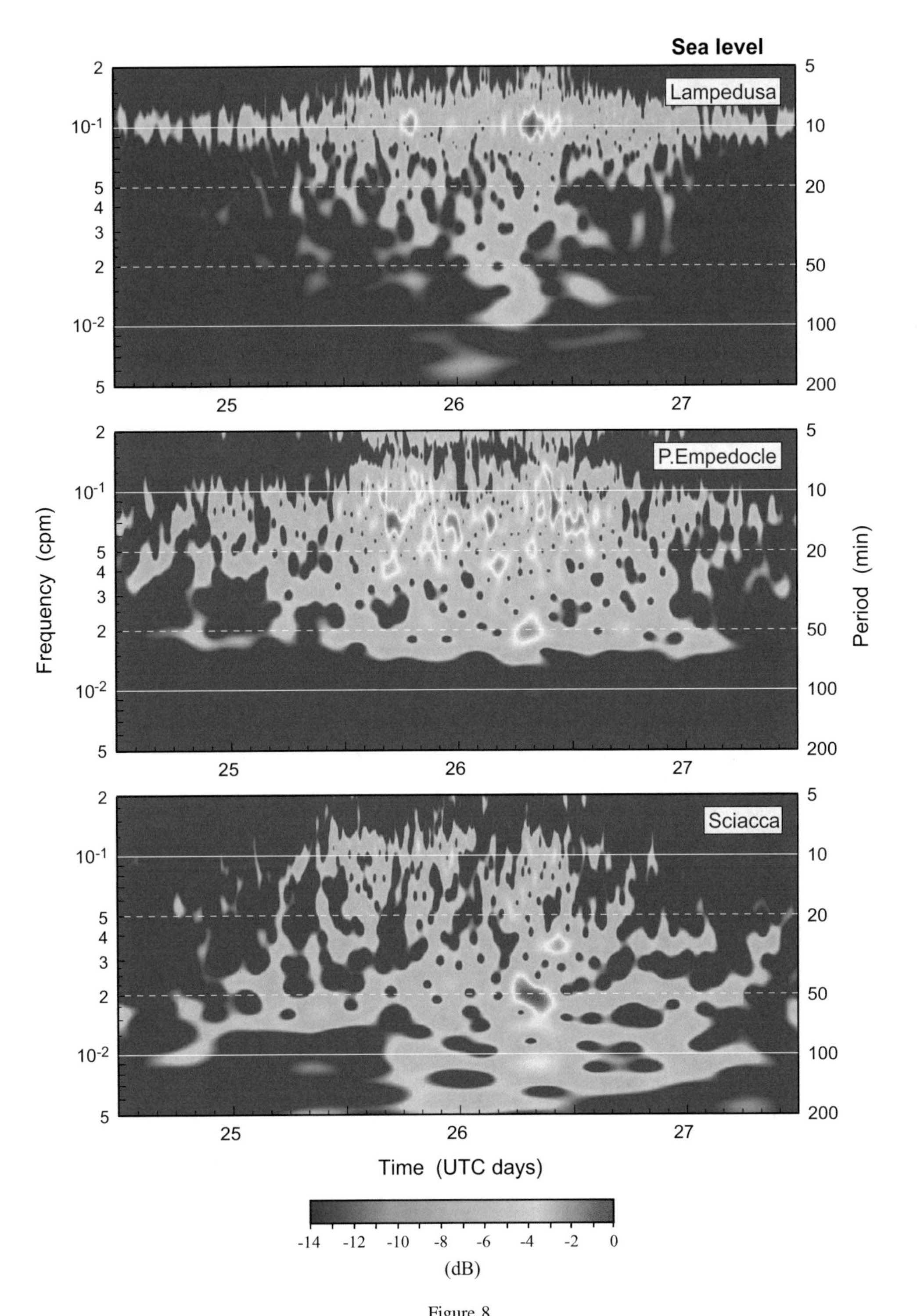

Figure 8
Frequency–time plots (*f–t* diagrams) for the 24–27 June 2014 sea level oscillations recorded at three tide gauges on the coast of southwestern Sicily

Reprinted from the journal

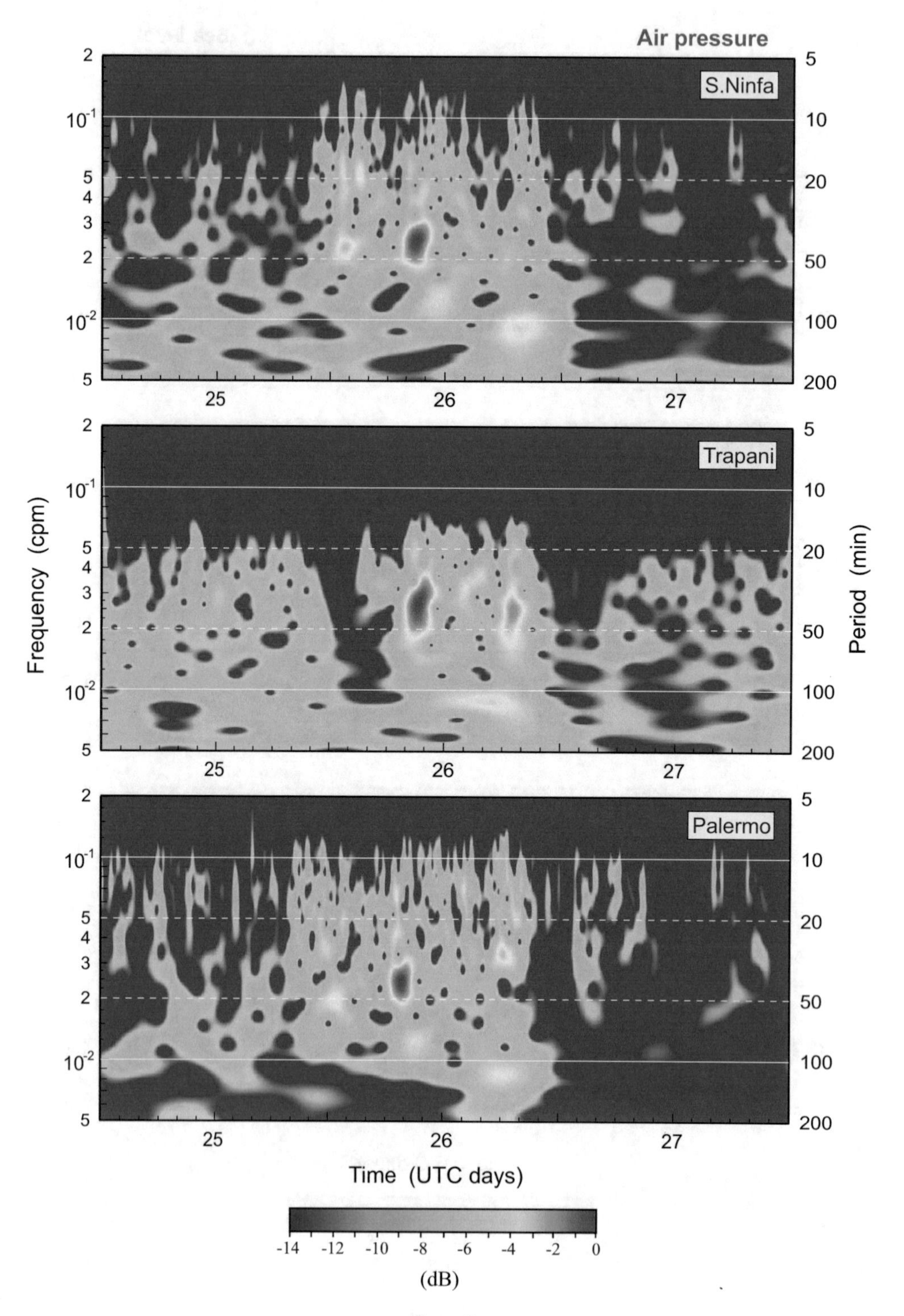

Figure 9
Frequency–time plots (*f–t* diagrams) for the 24–27 June 2014 atmospheric pressure oscillations recorded at three stations in western Sicily

background oscillations associated with harbour eigen oscillations, but are strongly amplified.

In contrast, the Porto Empedocle sea level oscillations have polychromatic character and much more complex frequency structure. There is no prominent narrow frequency band, as at Lampedusa; the sea level energy occupies a broad frequency range of 0.015–0.15 cpm[3] (periods of 7–70 min). However, certain frequency bands stand out, in particular: 0.042 cpm (24 min) and 0.075 cpm (13 min). Also, in contrast to Lampedusa, there were no distinctive energy maxima; the entire period from 12:00 UTC on 25 June to 00:00 UTC on 27 June was characterized by high energy and long sea level ringing. Moreover, the peak values corresponded to approximately the same periods of time as at Lampedusa: evening hours of the 25th and morning hours of the 26th. During the first event (25th), the energy maxima had periods of 13 and 24 min; for the second event (morning hours of the 26th) the strongest maximum was at a period of 50 min, while the maxima with periods of 13 and 24 min were apparent, but much weaker.

As is evident from Fig. 6, sea level oscillations at Sciacca were much weaker than at Lampedusa and Porto Empedocle; however, their frequency structure and time evolution were similar to those at the two other TG stations. Likewise, as at Porto Empedocle and Lampedusa, the increase of sea level intensity at Sciacca was observed during approximately 1.5 days (12:00 25/06 to 00:00 27/06) with the peak value in the morning hours of the 26th. The main and very pronounced maximum had the same period as Porto Empedocle: 50 min; the secondary peak was at 30 min (Fig. 8).

As indicated above, there were no available SL and AP records at the same sites. Nevertheless, it is interesting to compare time and frequency properties of the recorded AP waves with those of SL oscillations. The AP *f–t* diagrams shown in Fig. 9 demonstrate obvious similarities: intensive AP waves occupy approximately the same frequency band of 0.008–0.08 cpm (periods from 2 h to 13 min) and the same time interval from 12:00 UTC on the 25th to 12:00 on the 26th (except at Trapani where significant AP oscillations began about 6 h later). Also, the air pressure energy distribution was remarkably coherent in the evening hours of 25 June, when a pronounced energy maxima occurred at all three stations at periods of ~ 40–50 min. The peak AP energy is concurrent with the meteobore at Mazara del Vallo and with a minor energy peak at Lampedusa, but not with the character of SL oscillations at two other stations (Sciacca and Porto Empedocle). It appears that the tsunamigenic air pressure disturbance, responsible for the marrobbio of 25 June, was concise and propagated over the westernmost part of Sicily, affecting the Mazara del Vallo area. High consistency and slow dissipation of the AP disturbance, prolonged stability of its shape and strength when travelling over the western shelf of Sicily were probably the key factors responsible for the generation of the 25 June marrobbio in the harbour of Mazara del Vallo.

The situation with the second event that occurred in the morning hours of 26 June is slightly different. In contrast with the first event, the disturbances of 26 June were less coherent and travelled more to the east, thus causing no destructive waves in the area of Mazara del Vallo. The AP *f–t* diagrams for this event (Fig. 9) show several simultaneous maxima with different periods: ~ 100, 50, 35 and 20 min. Polychromatic structure and considerable differences between AP oscillations at the three AP stations (Figs. 7, 9) indicate that they were not as steady and consistent as for the first event. For that reason, presumably no "*milghuba*" phenomenon—Maltese name for a meteotsunami (Drago 2009)—was recorded on the coast of Malta located relatively close to the affected coast of Sicily (Fig. 1).

5. Meteotsunami Bore

As was emphasized above, one of the most spectacular features of the Mazara del Vallo event, was a *meteobore* propagating upstream in the Mazaro River. This bore was very similar to a *tidal bore*, a phenomenon which regularly occurs in specific locations worldwide (Chanson 2011, 2012), and also to a *tsunami bore*, which is observed in estuaries of certain rivers during major tsunami events, in particular, the 2004 Sumatra and 2011 Tohoku tsunamis

[3] cpm = cycles per minute.

(cf. Abe 1986; Tsuji et al. 1991; Grue et al. 2008; Tolkova and Tanaka 2016). Wave bores (shock waves) associated with meteotsunamis are, in fact, quite poorly known. Colucci and Michelato (1976) mentioned a number of historical atmospherically induced bores exactly at Mazara del Vallo, and Tappin et al. (2013), in their study of the 27 June 2014 English Channel meteotsunami, documented a weak undular bore advancing through the Yealm River (Devon England). However, these events have never been investigated in detail. Several videos taken during the Mazaro River marrobbio of 25 June 2014 enabled us to examine the bore and to evaluate its characteristics, including propagation speed, amplitude and devastating effects.

A number of natural questions arise:

1. Why did the bore occur specifically at Mazara del Vallo? What is the physical mechanism responsible for the formation of this bore? Why did not similar bores occur at some other nearby sites and at other affected coasts?

There are many bays and inlets along the coast of Sicily that were probably affected by the June 2014 meteotsunami, but the bore was reported only at Mazara del Vallo. The same is true for all past Sicilian marrobbio events: bores have never been observed at any other sites except the Mazaro River. Also, the meteotsunami of 23–27 June 2014 influenced a great number of sites along the coasts of Spain, Croatia, Italy, Greece, Ukraine and other countries (Šepić et al. 2015a). However, a bore had nowhere else been registered.

2. What triggered the bore of 25 June 2014 at Mazara del Vallo? Why do some other atmospheric disturbances, frequently propagating over western Sicily, not produce bores?

Various types of atmospheric disturbances are common in this region. Nevertheless, significant marrobbio events accompanied by bores occur only occasionally (once in several years). That is why, the phenomenon of marrobbio was described as "completely unpredictable" (http://siciliaterredoccidente.blogspot.ca/2016/06/leggendari-fenomeni-mazara-del-vallo.html).

These inquiries are in fact closely related. We have little statistics on bores at Mazara del Vallo and the explicit atmospheric processes responsible for their generation. However, we may assume that the generation mechanism of marrobbio meteobores is similar to that of tidal bores and we can use the existing knowledge on the latter to try to explain the nature of the former.

1. Tidal bores occur only in specific locations, which are *always the same* (except river training, dredging and damming that can cause bore disappearance or weakening). Usually these are the areas with a large tidal range (typically more than 6 m) and where incoming tides are funnelled into a shallow, narrowing river mouth or estuary via a broad-entrance V-shaped bay (Chanson 2011, 2012).
2. A tidal bore develops as a strong tidal surge pushing up the river, against the current. The river must be fairly shallow and have a narrow outlet to the sea. However, the estuary (river mouth) must be wide and flat.

These conditions specify the exact places where tidal bores are observed (Chanson 2012; Bonneton et al. 2015; Pelinovsky et al. 2015), in other words reply to the question: *Where?* It is obvious that this particular site—Mazara del Vallo—has a unique set of conditions, favourable for the meteobore generation: a shallow and narrowing waterway with relatively wide V-shaped mouth and a river with a strong current directed southwestward to the sea. It appears that such conditions are absent at other sites on the Sicilian coast, or in the bays and harbours of the Adriatic Sea and the Balearic Islands, and this is the reason why the bore occurs solely at Mazara del Vallo.

In addition to the inlet shape, the river inflow is another crucial factor conducive to the bore formation. A bore (hydraulic jump) is produced in very shallow water as a result of faster propagation of the wave crest, moving with the speed C_b, than the wave trough, causing sharpening of the frontal wave edge (Fig. 10). The river counterflow (directed against the incoming wave) V_1, increases this difference and the wave steepness. The key parameter, describing the character of tsunami wave propagation through a

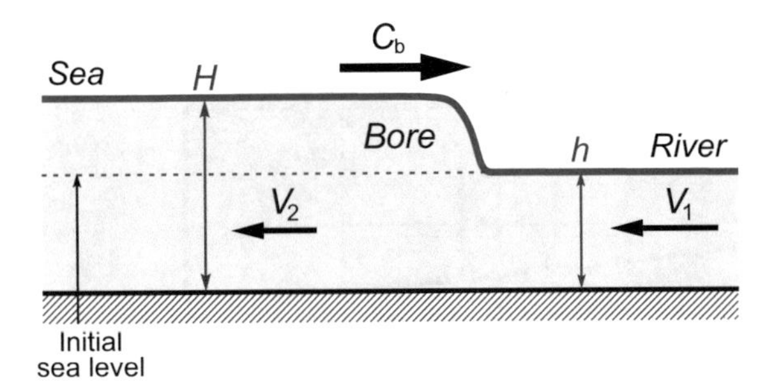

Figure 10
A sketch of a wave bore propagating upstream in a river

river estuary and the bore generation, is the *Froude number* (Chanson 2012):

$$Fr_1 = \frac{C_b - V_1}{\sqrt{gh}}, \tag{3}$$

where g is the gravity acceleration and h is the mean river water depth before the bore. In contrast to the "external" Froude number, determined by (1) (Sect. 1), which characterizes the effectiveness of longwave generation on the shelf, the number Fr_1 defined by expression (3), is the "internal" Froude number that describes the character of the bore within the river/estuary. Whenever there is a bore, Fr_1 is always greater than unity (C_b and V_1 have opposite signs, while $C_b \geq \sqrt{gh}$) and is a *measure of the strength of the bore*; if $Fr_1 < 1.0$, the incoming wave cannot become a bore (Chanson 2012).

We thus conclude that specific *internal* conditions in the estuary of the Mazaro River promote the occasional generation of meteobores in this region. The question, however, is: *When*?, i.e. what are the *external* conditions that cause such an event? Once again we can use the analogy of tidal bores:

3. Chanson (2011, 2012) indicates that typically "A tidal bore... forms during *spring tide conditions...*", i.e. during fortnightly maximum tides. Moreover, as emphasized by Pugh and Woodworth (2014), the largest bores occur around the times of high *equinoctial spring tides* (yearly maximum tides). Thus, the higher the tide, the stronger is the tidal bore and, also, for each particular site there is a certain critical tidal height value: the bore is formed only if the high-water tidal level exceeds this value.

We may assume that a similar situation is observed in the area of Mazara del Vallo for meteobores: they occur only if the incoming long-wave surge is higher than a specific critical level. It appears that such a situation took place on 25 June 2014, as well as during past bore episodes described by Colucci and Michelato (1976). As was indicated by Šepić et al. (2015a), the western Sicilian shelf is favourable for the resonant generation of atmospherically induced long waves when the "external" Froude number is close to unity, $Fr \sim 1.0$, as it was in June 2014, and when the atmosphere supports generation of intense small-scale air pressure disturbances. However, such situations are infrequent; consequently, marrobbio and meteobores are destructive, but luckily relatively rare events.

We will next apply a simple model for tsunami bores to reproduce the dynamics of the meteotsunami bore in the Mazaro River (Fig. 10). A step-like shock bore propagating along a one-dimensional flat-bottom river may be described by the system of the following equations valid for stationary bores (Tolkova and Tanaka 2016):

$$hv_1 = Hv_2, \tag{4}$$

$$v_1 v_2 = g(h + H)/2, \tag{5}$$

with (4) derived from the equation of continuity, and (5) derived from the conservation of momentum. Here

$$v_1 = C_b - V_1 \quad \text{and} \quad v_2 = C_b - V_2,$$

where V_1 and V_2 are the depth-averaged flow velocities at the front and rear segments of the bore and h and H are the respective water depths. These equations describe the 'shock conditions' (Stoker 1957; Henderson 1966). Equations (3), (4) and (5) can be combined to estimate the bore height relative to the mean sea level, $\Delta h = H - h$:

$$\frac{H}{h} = \frac{1}{2}\left(\sqrt{1 + 8Fr_1^2} - 1\right). \tag{6}$$

We can approximate the Mazaro River mouth and estuary by a simple model with exponentially decaying width and depth (Fig. 11a, b). The bore celerity, C_b, and the river flow speed, V_1, can be estimated from video footage and then we can use

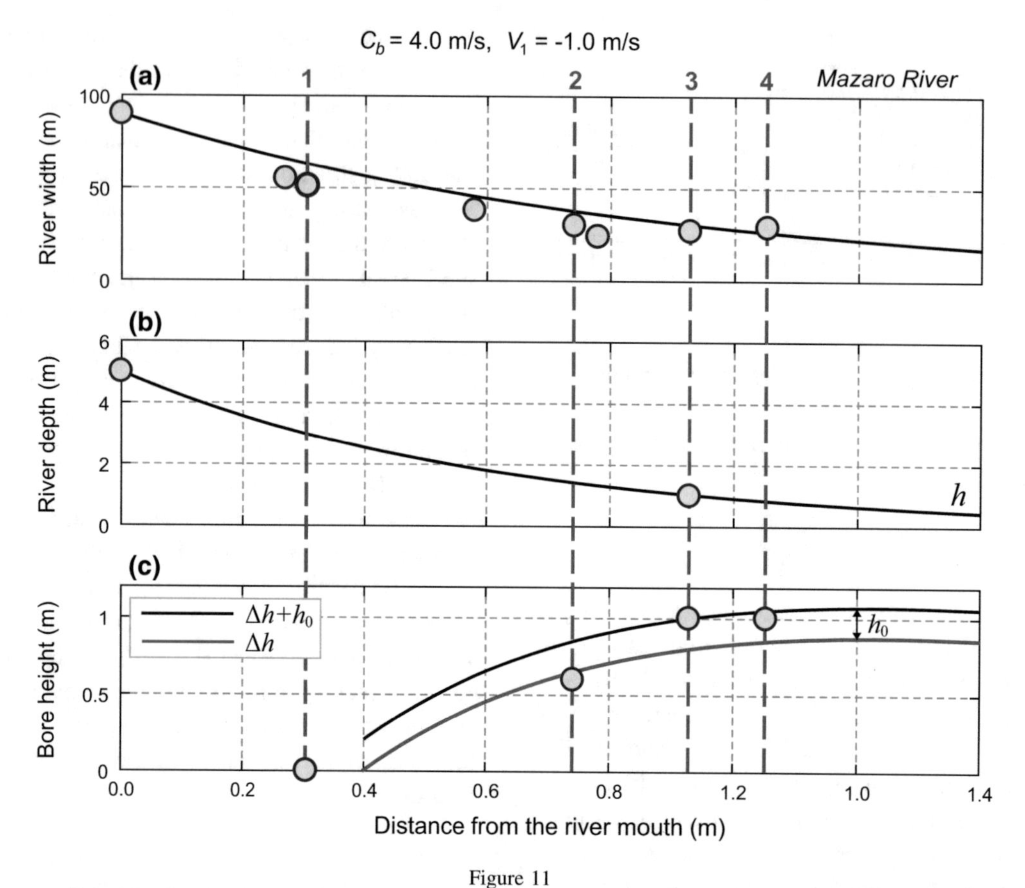

Figure 11
Approximate profiles of the Mazaro River **a** width and **b** depth; **c** estimated meteobore heights. Light blue circles denote data available on the river bathymetry and from video footages (the latter are indicated by vertical dashed red lines). Black and red solid lines indicate meteobore heights estimated by the model with and without incoming open-sea wave height (h_0), respectively

Eqs. (4)–(6) to theoretically evaluate the meteobore height. Parameters of exponential functions describing upstream changes of the river width and depth were obtained by fitting the known values of width (from Google Maps) and depth (from videos and bathymetry maps). The specific locations of the videos are indicated in Fig. 2. Location 1 is close to the river mouth, where depth and width of the river are the largest. The meteobore was not captured here (https://www.youtube.com/watch?v=nZ4mZ0x9saw). The breaking bore was first filmed at Location 2 (https://www.youtube.com/watch?v=7jSY1yyIPlI), where the Mazaro River is much shallower and narrower than at Location 1. The meteobore height appears to be ~ 0.6 m at this location. An increase of the bore height to ~ 1.0 m was captured further upstream at Location 3 (https://www.youtube.com/watch?v=uGHtSTxIrCY&t=8s) and Location 4 (https://www.youtube.com/watch?v=LTjOdN067Zo&t=3s). The distances between objects seen in the videos allowed us to determine the meteobore speed, as well as the speed of the river flow. The meteobore is best defined at Locations 3 and 4, where the estimated speed is: $C_b \approx 4.0$ m/s. The downstream river speed values are estimated as: $V_1 \approx -1.0$ m/s and $V_2 \approx 1.0$ m/s. The theoretical meteobore height is now calculated based on (6). To compare the estimated value with measurements, the height of the incoming ocean wave must be added to the theoretical bore height. According to eyewitness reports and video footage from the nearby Tonnarella beach, the open-sea wave was ~ 20 cm. Taking this value into account, we can reconstruct the transformation of the meteotsunami wave within the Mazara del Vallo harbour. The agreement between theoretical and observed bore parameters is surprisingly good, despite all simplifications inherent to the utilized tsunami bore model (Fig. 11c). The model correctly reproduces the first location of the bore's appearance, as well as the bore growth rate and the upstream height increase.

The open-channel flow can be of two types: *supercritical* and *subcritical*. The type depends on the relationship between the flow depth and the critical depth, h_c, which is defined as:

$$h_c = \left[\frac{Q^2}{b^2 g}\right]^{\frac{1}{3}}, \quad (7)$$

where Q (in m^3/s) is the river flow rate and b is the river width. If $h < h_c$ then the flow is supercritical; if $h > h_c$, then the flow is subcritical. At the location where $Fr_1 = 1.0$, the supercritical upstream flow meets the subcritical downstream flow, $h = h_c$, and the bore starts to develop (at ~ 400 m from the river mouth in Fig. 11c). Depending on the Froude number, two types of bores can be formed in the shallow-water natural conditions: *undular* and *breaking* (Chanson 2012; Pelinovsky et al. 2015). For the first type, the bore front is followed by a train of quasi-periodic waves ("undulations"). For the second type, as a result of non-linearity and river counterflow, the leading edge of the incoming bore becomes steeper and steeper, until it breaks. According to Chanson (2012), the bore breaks when

$$Fr_1 > 1.5 - 1.8. \quad (8a)$$

Another criterion, proposed by Stoker (1957) and supported by analysis of many actual tidal bores (Pelinovsky et al. 2015), is

$$H/h > 1.5. \quad (8b)$$

The multiple videos made in Mazara del Vallo on 25 June 2014 persuasively show the propagation of a *breaking bore*. This agrees with our estimates: under the assumption of the idealized river bathymetry, both conditions (8a) and (8b) are satisfied at ~ 700 m from the river mouth, i.e. close to Location 2 (Fig. 2b). Specifically, near this location the bore became *breaking*.

Conclusively, it appears that during the meteotsunami event of 25 June 2014, both *external* ($Fr \sim 1.0$) and *internal* ($Fr_1 > 1.0$) conditions were favourable to generate a meteobore in the Mazaro River. *External* conditions are related to the atmospheric situation over the western shelf of Sicily, and are, therefore, considerably variable in time, whereas *internal* conditions are related to bathymetric and topographic characteristics of particular regions, and therefore are strongly variable in space. Thus, assuming the favourable *external* conditions (atmospheric situation), meteotsunami waves will be generated over the entire affected area, but their final strength and destructiveness will be determined by the internal conditions.

6. Discussion and Conclusions

The main focus of our study is the meteotsunami events (*marrobbio*) along the southwestern coast of Sicily that peaked in the late hours of 25 June 2014 (Event No. 1) and again in the morning hours of 26 June 2014 (Event No. 2). The analyses of these events supplement the examination of other meteotsunami events that occurred in the Mediterranean and Black Sea regions between 23 and 27 June 2014 (Šepić et al. 2015a, 2016, 2018). Two distinctive events observed on the coast of Sicily were separated by a 'calm' period that lasted for several hours, similar to what was observed on the same dates in the central and southern parts of the Adriatic Sea (Šepić et al. 2016). Our analysis was concentrated on in situ sea level and air pressure records collected during these events and supported by the analysis of synoptic patterns and remote sensing atmospheric data. For further understanding of the generation mechanism of extreme long-wave oscillations in the area of Mazara del Vallo and for quantifying the resonant influence of bathymetry/local topography, a detailed numerical modelling is crucial, similar to what was done for the Adriatic Sea (Šepić et al. 2016). The corresponding numerical experiments have to particularly quantify the effects of (1) a steep and narrow along-coast canyon in the middle part of the Strait of Sicily (400–500 m deep); (2) a wide (~ 70 km) shelf off the western and southwestern coasts of Sicily (depths between 50 and 100 m), and (3) a narrow (~ 10 km) shelf with steep break towards 300–700 m depths off the southern coast of Sicily. The latter factor probably prevents the propagation of the generated meteotsunami waves from the western shelf of Sicily towards the Islands of Malta in front of which there is a shelf with similar bathymetric properties: width of $\sim$ 90 km and depth of 70–130 m.

There are several important conclusions following from our study:

1. Almost simultaneous occurrence of intense air pressure and sea level oscillations supports the preliminary assumption that the recorded marrobbio phenomenon at Mazara del Vallo was a meteotsunami.
2. Synoptic patterns and vertical structure of the atmosphere in the region of western Sicily are similar to those that were found to be responsible for the Balearic and Adriatic meteotsunamis, thus indicating that the atmospheric wave ducting mechanism (Lindzen and Tung 1976; Monserrat and Thorpe 1992, 1996) might have played an important role in conservation and propagation of atmospheric tsunamigenic disturbances.
3. The intensity of the recorded sea level oscillations alongshore of southwestern Sicily was strongly variable: these oscillations were destructively strong at Mazaro del Vallo (marrobbio hot spot), but quite weak at Sciacca, located 50 km to the southeast. Two possible reasons for that are: particular effects of bathymetry/coastal topography and spatial narrowness of meteotsunami-generating atmospheric disturbances, as visible on satellite imagery.
4. A strong meteotsunami bore associated with the marrobbio of 25 June 2014 was found propagating for several kilometres inland, being dependent on the river depth and bathymetry, similar to what is commonly observed for tsunami and tidal bores (Yasuda 2010; Tolkova and Tanaka 2016; Chanson, 2011, 2012).

Abrupt air pressure changes and small spatial extent of a single tsunamigenic air pressure disturbance have often been observed during other destructive meteotsunamis, occurring in various regions of the world oceans, like, for example, during the 2007 Ist Island (Croatia) meteotsunami, when the disturbance width was less than 50 km (Šepić et al. 2009a). For that reason, it may happen that meteotsunami waves are destructive at one "hot spot" and insignificant in another nearby "hot spot", despite pronounced resonant properties and large amplification factor of that bay or harbour. These are the reasons, in addition to specific topographic properties of the Mazaro River estuary and a break between the western and eastern parts of the Sicilian shelf, why a marrobbio phenomenon does not necessarily match in time with a similar phenomenon, called *milghuba*, observed on the Maltese Islands (Airy 1878; Drago 2009).

This paper also documents the generation and propagation of a bore (upstream) in the Mazaro River induced by an incoming meteotsunami wave. Upstream bores in rivers are normally driven by tides (cf. Chanson 2011, 2012; Bonneton et al. 2015; Pelinovsky et al. 2015) and tsunamis (Abe 1986; Tsuji et al. 1991; Grue et al. 2008; Yasuda, 2010; Tolkova and Tanaka 2016). Yet, significant similarity was found between the meteobore observed at Mazara del Vallo on 25 June 2014 and tidal/tsunami bores. This finding further endorses the deep physical resemblance between tsunamis and meteotsunamis (cf. Monserrat et al. 2006). Obviously, such strong bores, as observed during the June 2014 event, can affect coastal structures (Wei et al. 2015) and the morphology of the river bed (Chanson and Lubin 2013).

The formation of marrobbio on the southwestern coast of Sicily is similar to meteotsunami generation processes at other worldwide locations. The strongest meteotsunamis usually occur at, so-called, "meteotsunami hot spots". In the Mediterranean, in addition to Mazara del Vallo, these are:

1. Ciutadella Harbour, Menorca Island (Balearic Islands, Spain);
2. Vela Luka, Korčula Island (Croatia);
3. Stari Grad, Hvar Island (Croatia).

A number of devastating meteotsunamis with wave heights exceeding several metres occurred at these sites (cf. Rabinovich and Monserrat 1996; Monserrat et al. 2006; Jansà et al. 2007; Vilibić and Šepić 2009; Vučetić et al. 2009), including those associated with the June 2014 chain of events (Šepić et al. 2015a, 2016). However, meteobores have never been reported at these sites, probably because there are no rivers flowing into these harbours or bays.[4] Strong amplification of open-sea long waves at these harbours is primarily due to the distinct resonant properties of the corresponding basins (i.e. high *quality*-factors; cf. Rabinovich 2009) and to the matching of the frequencies of incoming waves and natural harbour modes.

It appears that, during the June 2014 multi-meteotsunami events, several types of meteotsunamis were observed at different sites: (1) "classical" meteotsunamis in Ciutadella and in bays of the Adriatic islands (with all the bays characterized by large *Q*-factors and robust resonant properties) (Šepić et al. 2016); (2) meteobore at Mazara del Vallo, and (3) open-beach meteotsunami at Odessa, the north-western Black Sea (Šepić et al. 2018).

Meteotsunamis (marrobbio) occurred in this region in the past and will definitely occur in the future. It is evident that the southwestern coast of Sicily has favourable conditions for this phenomenon. Despite the *Sicilia Terre d'Occidente* statement that the phenomenon of marrobbio is "completely unpredictable" (http://siciliaterredoccidente.blogspot.ca/2016/06/leggendari-fenomeni-mazara-del-vallo.html), these destructive waves can and should be forecast! Meteotsunami Warning Systems already exist in two Mediterranean countries strongly affected by such hazardous waves: Spain and Croatia (Vilibić et al. 2016). In particular, synoptic forecast of rissagas (Balearic meteotsunamis) has been issued operationally by the Meteorological Centre of the Instituto Nacional de Meteorologia (Palma de Mallorca, Baleares, Spain) since 1984 (Jansà et al. 2007; Renault et al. 2011). The creation of a similar system in Croatia began in late 2015 within the MESSI (Meteotsunamis, destructive long ocean waves in the tsunami frequency band: from observations and simulations towards a warning system) project (http://www.izor.hr/messi). These systems are based on the existence of specific tsunamigenic synoptic patterns—recognition of these patterns allows the timely forecast of corresponding atmospheric situations for up to a week in advance, a time frame for which synoptic forecasts usually work well (Simmons and Hollingsworth 2002). Mostly, these synoptic patterns propagate from west to east and, therefore, some crucial information from Spain and France may be used to facilitate meteotsunami forecasts in Sicily. As was shown by Šepić et al. (2009b), there are teleconnections between meteotsunami events occurring in various parts of the Mediterranean Sea. Actually, the June 2014 events clearly demonstrated this; the *rissaga* impacted the Balearic Islands (Spain) about 2.5 days earlier than the marrobbio affected the west coast of Sicily and strong meteotsunamis (locally known as *šćiga*) were observed in the middle Adriatic Sea (Šepić et al. 2015a).

For a reliable and timely forecast of marrobbio, a statistically robust relationship between this phenomenon and synoptic conditions has to be first established. From this point of view, a catalogue of marrobbio events, similar to the one recently created for the Adriatic Sea (Orlić 2015), would surely be helpful. An ultimate goal is that the Meteotsunami Warning System for the west coast of Sicily, as well as the Balearic and Adriatic systems, will become a part of the more general North Eastern Atlantic, the Mediterranean and Connected Seas (NEAMTWS) Tsunami Warning System (Tinti et al. 2012).

[4] Actually, some photos made during the catastrophic meteotsunamis (rissagas) of 1984 and 2006 in Ciutadella Harbour showed bore-like wave disturbances.

Acknowledgements

The authors would like to thank Fred Stephenson, Richard Thomson and Isaac Fine, all from the Institute of Ocean Sciences (Sidney, British Columbia) and Alex Sheremet from the University of Florida, USA (Gainesville, FL) for their helpful comments and suggestions. We would also like to thank Francesco Vacarro from Mazara del Vallo for filming and making public videos of marrobbio. We thank all organisations that kindly provided us the data used in this study: European Centre for Middle-range Weather Forecast, Reading, UK (www.ecmwf.int); European Organization for the Exploitation of Meteorological Satellites (www.eumetsat.int); the Intergovernmental Oceanographic Commission (IOC) Sea Level Monitoring Station portal (www.ioc-sealevelmonitoring.org); Weather Underground (http://www.wunderground.com). The work of JS and IV has been supported by the Croatian Science Foundation under the project MESSI (UKF Grant No. 25/15), and for AR by the Russian Science Foundation (Grant 14-50-00095).

References

Abe, K. (1986). Tsunami propagation in rivers of the Japanese Islands. *Continental Shelf Research, 5*(6), 655–677.

Airy, G. B. (1878). On the tides at Malta. *Philosophical Transactions of the Royal Society of London, 169,* 123–138.

Bonneton, P., Bonneton, N., Parisot, J.-P., & Castelle, B. (2015). Tidal bore dynamics in funnel-shaped estuaries. *Journal of Geophysical Research Oceans, 120,* 923–941.

Candela, J., Mazzola, S., Sammari, C., Limeburner, R., Lozano, C. J., Patti, B., & Bonnano, A. (1999). The "Mad Sea" phenomenon in the Strait of Sicily. *Journal of Physical Oceanography, 29,* 2210–2231.

Candella, R. N., Rabinovich, A. B., & Thomson, R. E. (2008). The 2004 Sumatra tsunami as recorded on the Atlantic coast of South America. *Advances in Geosciences, 14*(1), 117–128.

Chanson, H. (2011). Current knowledge in tidal bores and their environmental, ecological and cultural impacts. *Environmental Fluid Dynamics, 11,* 77–98.

Chanson, H. (2012). *Tidal Bores, Aegir, Eagre, Mascaret, Pororoca: theory and observations.* London: World Scientific.

Chanson, H., & Lubin, P. (2013). Mixing and sediment processes induced by tsunamis propagating upriver. In T. Cai (Ed.), *Tsunamis: Economic Impact, Disaster Management and Future Challenges* (pp. 65–102). Hauppauge: Nova Science Publishers.

Colucci, P., & Michelato, A. (1976). An approach to the study of the "Marrubbio" phenomenon. *Bollettino di Geofisica Teorica ed Applicata, 19,* 3–10.

Drago, A. F. (2009). Sea level variability and the 'Milghuba' seiche oscillations in the northern coast of Malta, Central Mediterranean. *Physics and Chemistry of the Earth, 34,* 948–970.

Durran, D. R., & Klemp, J. B. (1982). On the effects of moisture on Brunt-Väisälä frequency. *Journal of the Atmospheric Sciences, 39,* 2152–2158.

Gomis, D., Monserrat, S., & Tintoré, J. (1993). Pressure-forced seiches of large amplitude in inlets of the Balearic Islands. *Journal of Geophysical Research, 98,* 14437–14445.

Grue, J., Pelinovsky, E. N., Fructus, D., Talipova, T., & Kharif, C. (2008). Formation of undular bores and solitary waves in the Strait of Malacca caused by the 26 December 2004 Indian Ocean tsunami. *Journal of Geophysical Research Oceans, 113,* C05008. https://doi.org/10.1029/2007jc004343.

Henderson, F. M. (1966). *Open channel flow.* New York: Macmillan Publishing Co., Inc.

Hibiya, T., & Kajiura, K. (1982). Origin of 'Abiki' phenomenon (a kind of seiche) in Nagasaki Bay. *Journal of Oceanographical Society of Japan, 38,* 172–182.

Jansà, A., Monserrat, S., & Gomis, D. (2007). The rissaga of 15 June 2006 in Ciutadella (Menorca), a meteorological tsunami. *Advances in Geosciences, 12,* 1–4.

Lindzen, R. S., & Tung, K.-K. (1976). Banded convective activity and ducted gravity waves. *Monthly Weather Review, 104,* 1602–1617.

Liu, P. L. F., Monserrat, S., & Marcos, M. (2002). Analytical simulation of edge waves observed around the Balearic Islands. *Geophysical Research Letters, 29,* 1847. https://doi.org/10.1029/2002gl015555.

Monserrat, S., & Thorpe, A. J. (1992). Gravity-wave observations using an array of microbarographs in the Balearic Islands. *Quarterly Journal of the Royal Meteorological Society, 118,* 259–282.

Monserrat, S., & Thorpe, A. J. (1996). Use of ducting theory in an observed case of gravity waves. *Journal of the Atmospheric Sciences, 53,* 1724–1736.

Monserrat, S., Vilibić, I., & Rabinovich, A. B. (2006). Meteotsunamis: Atmospherically induced destructive ocean waves in the tsunami frequency band. *Natural Hazards and Earth System Sciences, 6,* 1035–1051.

Orlić, M. (2015). The first attempt at cataloguing tsunami-like waves of meteorological origin in Croatian coastal waters. *Acta Adriatica, 56,* 83–96.

Pelinovsky, E. N., Shurgalina, E. G., & Rodin, A. A. (2015). Criteria for the transition from a breaking bore to an undular bore. *Izvestiya, Atmospheric and Oceanic Physics, 51*(5), 530–533.

Proudman, J. (1929). The effects on the sea of changes in atmospheric pressure. *Geophysical Supplement to Monthly Notices of the Royal Astronomical Society, 2*(4), 197–209.

Pugh, D., & Woodworth, P. (2014). *Sea-level science: understanding tides, surges, tsunamis and mean sea-level changes* (p. 395). Cambridge: Cambridge University Press.

Rabinovich, A. B. (2009). Seiches and harbour oscillations. In Y. C. Kim (Ed.), *Handbook of coastal and ocean engineering* (pp. 193–236). Singapore: World Scientific.

Rabinovich, A. B., & Monserrat, S. (1996). Meteorological tsunamis near the Balearic and Kuril Islands: descriptive and statistical analysis. *Natural Hazards, 13,* 55–90.

Rabinovich, A. B., & Monserrat, S. (1998). Generation of meteorological tsunamis (large amplitude seiches) near the Balearic and Kuril Islands. *Natural Hazards, 18,* 27–55.

Ramis, C., & Jansà, A. (1983). Condiciones meteorológicas simultáneas a la aparición de oscilaciones del nivel del mar de amplitud extraordinaria en el Mediterráneo occidental. *Revista de Geofísica, 39,* 35–42. **(in Spanish)**.

Renault, L., Vizoso, G., Jansá, A., Wilkin, J., & Tintoré, J. (2011). Toward the predictability of meteotsunamis in the Balearic Sea using regional nested atmosphere and ocean models. *Geophysical Research Letters, 38,* L10601. https://doi.org/10.1029/2011gl047361.

Šepić, J., Međugorac, I., Janeković, I., Dunić, N., & Vilibić, I. (2016). Multi-meteotsunami event in the Adriatic Sea generated by atmospheric disturbances of 25–26 June 2014. *Pure and Applied Geophysics, 173,* 4117–4138.

Šepić, J., Rabinovich, A. B., & Sytov, V. N. (2018). Odessa tsunami of 27 June 2014: observations and numerical modelling. *Pure and Applied Geophysics.* https://doi.org/10.1007/s00024-017-1729-1.

Šepić, J., Vilibić, I., & Belušić, D. (2009a). The source of the 2007 Ist meteotsunami (Adriatic Sea). *Journal of Geophysical Research, 114,* C03016. https://doi.org/10.1029/2008JC005092.

Šepić, J., Vilibić, I., & Monserrat, S. (2009b). Teleconnection between the Adriatic and the Balearic meteotsunamis. *Physics and Chemistry of the Earth, 34*(928-937), C03016. https://doi.org/10.1016/j.pce.2009.08.007.

Šepić, J., Vilibić, I., Rabinovich, A. B., & Monserrat, S. (2015a). Widespread tsunami-like waves of 23-27 June in the Mediterranean and Black Seas generated by high-altitude atmospheric forcing. *Scientific Reports, 5,* 11682. https://doi.org/10.1038/srep11682.

Šepić, J., Vilibić, I., & Fine, I. (2015b). Northern Adriatic meteorological tsunamis: assessment of their potential through ocean modeling experiments. *Journal of Geophysical Research Oceans, 120,* 2993–3010.

Simmons, A. J., & Hollingsworth, A. (2002). Some aspects of the improvement in skill of numerical weather prediction. *Quarterly Journal of the Royal Meteorological Society, 128,* 647–677.

Stoker, J. J. (1957). *Water waves, the mathematical theory with applications*. New York: Interscience Pub Inc.

Tappin, D. R., Sibley, A., Horsburgh, K., Daubord, C., Cox, D., & Long, D. (2013). The English Channel 'tsunami' of 27 June 2011—a probable meteorological source. *Weather, 68*(6), 144–152.

TeleIBS—L'Opinione. (2014). 2014-06-25 Il Marrobbio anche sulla spiaggia di Tonnarella fa ritirare le acque di circa 30 metri. https://www.youtube.com/watch?v=ue3XgxxccUc. Accessed on 31 March 2017.

Thomson, R. E., Rabinovich, A. B., Fine, I. V., Sinnott, D. C., McCarthy, A., Sutherland, N. A. S., & Neil, L. K. (2009). Meteorological tsunamis on the coasts of British Columbia and Washington. *Physics and Chemistry of the Earth, 34,* 971–988.

Tinti, S., Graziani, L., Brizuela, B., Maramai, A., & Gallazzi, S. (2012). Applicability of the decision matrix of North Eastern Atlantic, Mediterranean and Connected Seas Tsunami Warning System to the Italian tsunamis. *Natural Hazards and Earth System Sciences, 12,* 843–857.

Tolkova, E., & Tanaka, H. (2016). Tsunami bores in Kitakami River. *Pure and Applied Geophysics, 173,* 4039–4054.

Tsimplis, M. N., Proctor, R., & Flather, R. A. (1995). A two-dimensional tidal model for the Mediterranean Sea. *Journal of Geophysical Research, 100,* 16223–16239.

Tsuji, Y., Yanuma, T., Murata, I., & Fujiwara, C. (1991). Tsunami ascending in rivers as an undular bore. *Natural Hazards, 4,* 257–266.

Vaccaro, F. (2014). Marrobbio a Mazara del Vallo del 25 giugno 2014 IV video. https://www.youtube.com/watch?v=7jSY1yyIPlI. Accessed on 31 March 2017.

Vennell, R. (2007). Long barotropic waves generated by a storm crossing topography. *Journal of Physical Oceanography, 37,* 2809–2823.

Vennell, R. (2010). Resonance and trapping of topographic transient ocean waves generated by a moving atmospheric disturbance. *Journal of Fluid Mechanics, 650,* 427–442.

Vilibić, I. (2008). Numerical simulations of the Proudman resonance. *Continental Shelf Research, 28,* 574–581.

Vilibić, I., Monserrat, S., Rabinovich, A. B., & Mihanović, H. (2008). Numerical modelling of the destructive meteotsunami of 15 June 2006 on the coast of the Balearic Islands. *Pure and Applied Geophysics, 165,* 2169–2195.

Vilibić, I., & Šepić, J. (2009). Destructive meteotsunamis along the eastern Adriatic coast: Overview. *Physics and Chemistry of the Earth, 34,* 904–917.

Vilibić, I., Šepić, J., Rabinovich, A., & Monserrat, S. (2016). Modern approaches in meteotsunami research and early warning. *Frontiers in Marine Science, 3*(57), 1–7. https://doi.org/10.3389/fmars.2016.00057.

Vučetić, T., Vilibić, I., Tinti, S., & Maramai, A. (2009). The Great Adriatic flood of 21 June 1978 revisited: An overview of the reports. *Physics and Chemistry of the Earth, 34,* 894–903.

Wei, Z., Dalrymple, R. A., Hérault, A., Bilotta, G., Rustico, E., & Yeh, H. (2015). SPH modeling of dynamic impact of tsunami bore on bridge piers. *Coastal Engineering, 104,* 26–42.

Wilson, B. (1972). Seiches. *Advances in Hydrosciences, 8,* 1–94.

Yasuda, H. (2010). One-dimensional study on propagation of tsunami wave in river channels. *Journal of Hydraulic Engineering, 136*(2), 93–105.

(Received February 5, 2018, revised February 27, 2018, accepted March 2, 2018)

Printed in the United States
By Bookmasters